$21.00 F-8

D0023162

Textbook of Work Physiology

**McGRAW-HILL SERIES
IN HEALTH EDUCATION,
PHYSICAL EDUCATION, AND RECREATION**

DEOBOLD B. VAN DALEN, *Consulting Editor*

ÅSTRAND AND RODAHL
Textbook of Work Physiology: Physiological Bases of Exercise

JENSEN AND SCHULTZ
Applied Kinesiology

JERNIGAN AND VENDIEN
Playtime: A World Recreation Handbook

METHENY
Movement and Meaning

SHIVERS AND CALDER
Recreational Crafts: Programming and Instructional Techniques

SINGER
Coaching, Athletics, and Psychology

Textbook of Work Physiology

PHYSIOLOGICAL BASES OF EXERCISE
Second Edition

PER-OLOF ÅSTRAND, M.D.
Professor, Department of Physiology
Swedish College of Physical Education
Stockholm, Sweden

KAARE RODAHL, M.D.
Director, Institute of Work Physiology
Professor, Norwegian College of Physical Education
Oslo, Norway

McGraw-Hill Book Company
New York　St. Louis　San Francisco　Auckland
Bogotá　Düsseldorf　Johannesburg
London　Madrid　Mexico　Montreal
New Delhi　Panama　Paris　São Paulo
Singapore　Sydney　Tokyo　Toronto

TEXTBOOK OF WORK PHYSIOLOGY
Physiological Bases of Exercise

Copyright © 1977, 1970 by McGraw-Hill, Inc.
All rights reserved.
Printed in the United States of America.
No part of this publication may be reproduced,
stored in a retrieval system, or transmitted,
in any form or by any means,
electronic, mechanical, photocopying,
recording, or otherwise,
without the prior written permission of the publisher.

234567890DODO78321098

This book was set in Times Roman by Black Dot, Inc.
The editors were Nelson W. Black and James R. Belser;
the cover was designed by Nicholas Krenitsky;
the production supervisor was Robert C. Pedersen.
New drawings were done by J & R Services, Inc.
R. R. Donnelly & Sons Company was printer and binder.

Library of Congress Cataloging in Publication Data

Åstrand, Per Olof.
    Textbook of work physiology.

    (McGraw-Hill series in health education, physical
education, and recreation)
    Includes bibliographies and index.
    1. Work—Physiological aspects. 2. Exercise—
Physiological effect. I. Rodahl, Kåre, date
joint author. II. Title.
QP301.A23   1977   612'.042   76-55390
ISBN 0-07-002406-5

To

PROFESSOR ERIK HOHWÜ CHRISTENSEN,
who first introduced us to the field of work physiology.
It is to a large measure due to his encouragement
and continuous and active interest
that the writing of this book was undertaken.

Contents

Preface

The purpose of this new, revised edition of the *Textbook of Work Physiology* is the same as that of the original text: to bring together into one volume the various factors affecting human physical performance in a manner that is comprehensible to the physiologist, the physical educator, and the clinician. Contrary to most of the conventional textbooks of physiology, in which the emphasis is on the regulation of the various functions of the body at rest, the regulatory mechanisms studied during physical activity have been especially emphasized in this book. It is assumed that the reader has some knowledge of elementary physics and chemistry, as well as human anatomy and physiology. However, to facilitate the understanding of some of the physiological and biochemical events encountered during work stress and physical exercise, a certain amount of basic physiology and biochemistry has been included.

In the selection of the material, an attempt has been made to meet the modern needs of the student of physical education at both the undergraduate and the postgraduate levels. More references have been included than is customary in most textbooks. Inevitably, since the submission of our revised manuscript, new developments have taken place which we were not able to include in this edition.

We are aware of the fact that the curriculum in many physical education programs does not permit such a comprehensive study of physiology as this book may entail. For this reason, each chapter has been written as a fairly complete entity, relatively independent of the rest of the book. With this arrangement, the book may also be useful for those students who wish to penetrate more deeply into a particular field or a limited area of study.

It is our hope that this text may be useful not only in the teaching of physical education but also in the teaching of clinical and applied physiology and that it may serve to stimulate the appreciation of the role of physical education for young and old, in health and disease.

Much of the unpublished material included in this book has been gathered in collaboration with our colleagues at the College of Physical Education in Stockholm, and at the Institute of Work Physiology and the College of Physical Education in Oslo. Their kind cooperation is gratefully acknowledged. We have also benefited greatly from personal association and frequent discussions with our many colleagues in these institutions. We are especially indebted to O. Grönneröd and O. Vaage for their valuable contribution in the revision of Chapter 2 and to O. Vaage and N. Secher in preparing the section on rowing in Chapter 16.

We are also very grateful for the technical assistance given us by Karin Marina and Joan Rodahl in the preparation of the manuscript.

Per-Olof Åstrand
Kaare Rodahl

Textbook of Work Physiology

Physical
Activity

Chapter 1

Physical Activity

In the simplest forms of animal life, such as the amoeba, all essential functions (metabolism, response to stimuli, movement, and reproduction) are developed in a single cell. Because of the cell's minute size, foodstuffs, waste products, electrolytes, and dissolved gases can be distributed within the cell and between it and its surroundings mainly by diffusion and osmosis. And when each individual cell has not become any larger than it originally was, the reason is that there is a limit to how large a cell can be without choking itself. Krogh (1941) has estimated that diffusion can provide sufficient oxygen only to organisms with a diameter of less than 1 mm when the metabolism of the organism is fairly high. It is therefore evident that each cell has to remain small.

In the evolution of higher organisms, the overall size of the organism increased. And since each cell could not become much larger, more cells had to be used. These cells had to collaborate; they had to be organized and coordinated. Different cells had to specialize in the performance of particular functions and had to build up specialized tissues and organs. This process eventually caused each individual cell (Fig. 1-1) to lose its intimate contact with the external environment of the organism. Furthermore, this environment, in the course of evolution, changed from water to air. Thus, evolution created the need for the organism to develop a system which could transport matter to and

3

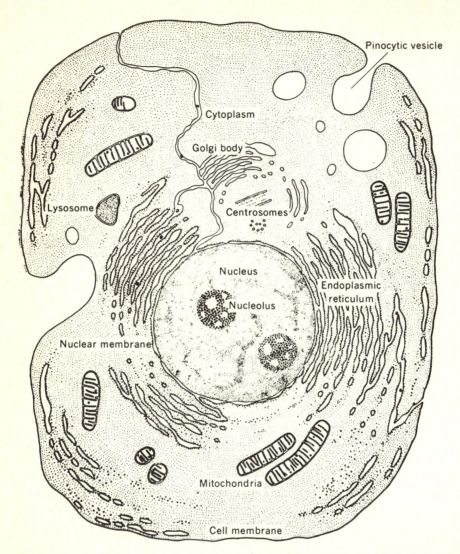

Figure 1-1 A typical cell. (*From Jean Bracket,* Scientific American, **205**:3, *1961.*) The cell membrane plays a crucial role in almost all cellular activity. It can actively regulate the internal environment of the cell and transport substances in and out of it. The structural framework of the membrane is a double layer of lipid molecules. The individual lipid molecule has a head and two tails. Hydrophilic heads (soluble in water) form the outer and inner membrane surface, and hydrophobic tails (with poor affinity to water) meet in the membrane interior. This structure is then the anchorage for other components of the membrane, like proteins and glycoproteins. These molecules can also provide additional support, i.e., they can act as enzymes or function as "pumps" and carriers of material across the cell membrane. Many of the lipids in mammalian cell membranes are unsaturated and are liquid at body temperature. Therefore, the membrane has the consistency of a light oil within the sheetlike structure, and both lipid and protein molecules are relatively free to move about within the membrane. Protein molecules may lie close to either membrane surface; they may penetrate the membrane surfaces and eventually bridge them completely. The inner membranes of mitochondria have a similar structure and many of the enzyme proteins combine to

from the external environment and within the body itself, and a system which would enable the organism to orient itself and to communicate.

During the transition from sea to land, the organism brought the sea water with it, so to speak, in a bag made of skin. As a result, the human body consists of 50 to 70 percent water. In this body fluid, the different ions are found in about the same relative proportions as in the ancient ocean. Here the individual cell, like the amoeba, can bathe in fluid. The composition of this extracellular or interstitial fluid is of the utmost importance to the function of the cell. Its content of organic compounds, such as fatty acids, glucose, hormones, and enzymes, and of inorganic substances, exerts a profound influence upon the cell in one way or another.

The main objective of most organ functions is to maintain the internal equilibrium of the single cell in spite of primary changes or disturbances in the animal's internal or external environment, in accordance with the "one for all and all for one" concept. A continuous exchange of materials between interstitial fluid and blood plasma is necessary for the normal function of the cell. This exchange across the capillary membrane creates an enormous traffic problem around the cell when the exchange suddenly increases many fold, as it does when a resting individual suddenly starts vigorous physical exercise and the energy metabolism instantaneously increases by a factor of 30 or more.

Higher animals are basically designed for mobility. This also applies to humans. Consequently, our locomotive apparatus and service organs constitute the main part of our total body mass. The shape and dimensions of the human skeleton and musculature are such that the human body cannot compete with a gazelle in speed or an elephant in sturdiness, but in diversity human beings are indeed outstanding.

The basic instrument of mobility is the muscle. It is unique in that it can vary its metabolic rate to a greater degree than any other tissue. In fact, the working skeletal muscles may increase their oxidative processes to more than 50 times the resting level (Asmussen et al., 1939). Such an enormous variation in metabolic rate must necessarily create serious problems for the working muscle cell, for while the consumption of fuel and oxygen increases 50-fold,

---

form supramolecular aggregates organized in an orderly array throughout the membrane (Capaldi, 1974). Encapsulated within the boundaries of the cell membrane is the cytoplasm, which contains a number of formed and dissolved elements, including enzymes which support the anaerobic metabolic processes of the cell. The mitochondria, which take up oxygen, are rod-shaped bodies, surrounded by a double-walled membrane and represent the "powerhouse" of the cell. Here fuel and oxygen enter into energy-yielding processes resulting in the formation of ATP. In the endoplasmic reticulum, a network of canaliculi formed by a system of membranes may extend all the way from the outer surface of the cell to the membrane surrounding the nucleus. Through these canaliculi, substances may move from the outer membrane of the cell to the membrane of the nucleus. The dots that line the endoplasmic reticulum are ribosomes. These are the sites of protein synthesis. The nucleus contains the chromosomes, which contain genes and deoxyribonucleic acid and are the carriers of the hereditary factors. In cell division, the pair of chromosomes, shown in longitudinal section (rods) and in cross section (circles), parts to form two poles of an apparatus that separates two duplicate sets of chromosomes.

the rate of removal of heat, carbon dioxide, water, and waste products must be increased similarly. To maintain the chemical and physical equilibrium of the cells, there must be a tremendous increase in the exchange of molecules between intra- and extracellular fluid; "fresh" fluid must continuously flush the exercising cell. When muscles are thrown into vigorous activity, the ability to maintain the internal equilibria necessary to continue the work is entirely dependent on those organs which service the muscles. This dependence is especially true of the circulatory and respiratory organ functions which strive to keep, so to speak, the muscle cell in indirect instantaneous contact with the surrounding air at all times.

Since the heat production may increase from about 4 kJ $\cdot$ min^{-1} at rest to perhaps 200 kJ $\cdot$ min^{-1} during maximal work, or from about 80 to several thousand watts, the temperature-controlling mechanisms must come into play to arrange for the excess heat to be transported from the muscles to the skin. Profuse sweating may cause water and salt losses which secondarily may affect the circulation and the renal function. To restore the energy content of the body working at maximal capacity, up to 4 times more food must be digested daily than when the individual is at rest. During exercise, many of the hormone-producing glands are involved in the regulation of metabolic and circulatory functions. Parts of the central nervous system are specialized in receiving sensory information from muscles and joints and sending impulses to the muscles. In the final analysis, all the external evidence of the activity of the brain is eventually manifested by muscular movement.

One fascinating aspect of the physiology of the human being at work is that it provides basic information about the nature and range of the functional capacity of different organ systems. Physiological and clinical studies on human beings cannot be restricted to a resting or basal condition, because the functional capacity of an organ can be evaluated only when the organ is subject to functional loads. A theory on the regulation of a function must consider and explain the adaptation to various physiological conditions, including muscular activity.

Manual labor, sometimes under adverse environmental conditions, still exists in all countries, and will probably always remain an essential part of society. Furthermore, individuals continue to find satisfaction and enjoyment in their leisure time through sports or other types of muscular activity. Important objectives of physiological research are to study the effects of various activities and environmental factors on different organ functions; to investigate the capacity of individuals to meet the demands imposed upon them; and finally, to determine how this capacity can be influenced by factors such as training and acclimatization (Table 1-1).

In a very broad sense, physical performance or fitness is determined by the individual's capacity for energy output (aerobic and anaerobic processes and oxygen transport), neuromuscular function (muscle strength, coordination and technique), joint mobility, and psychological factors (e.g., motivation and tactics). These factors play a more or less dominating role, depending upon the

Table 1-1 Physical Performance

	Function	Structural basis	Biochemical processes involved	Modifying factors
Locomotor organs	1. Muscular strength	Motor unit: (a) muscle cell, contractile elements, type of fiber	Contraction-tension (static, dynamic)	Genetic endowment, sex, age, training, use-disuse
			Energy metabolism: chemical energy, mechanical work aerobic-anaerobic processes, enzymatic reaction fuels: carbohydrate, fat, nutritional intake, storage	Training
				Diet, training
		(b) motoneuron, synapses, endplate	Excitation, impulse propagation, membrane depolarization.	Psychic factors, CNS, training
	2. Joint mobility	Skeleton: (a) muscular attachment (b) skeletal levers (c) joints and ligaments, articul. cartilage, synovia, bursae		Training of joint mobility
	3. Coordination	Neuromuscular apparatus: afferent-efferent pathways, senses	Propagation of nerve impulses, facilitation, inhibition	Training, practice, psychic factors, drugs
Service organs	4. Endurance	Oxygen-transporting organs: (a) pulmonary ventilation (b) O_2 binding capacity of the blood (Hb, blood vol., etc.) (c) cardiac output: stroke volume heart rate cardiac pump: myocardium valves pacemaker (d) a-$\bar{v}O_2$ diff.: local milieu in muscle cell (shift of the dissociation curve, etc.) venous return (muscle pump) negative pressure in the thorax redistribution of the blood volume to the muscle (e) fluid balance		Training, altitude, air pollution, smoking, iron intake
				Genetic endowment, state of health, training
				Environment (heat)
				Fluid intake and loss, heat
Central nervous system	5. Will to win	Reticular formation, etc.		Psychic factors, attitude

```
Physical Performance

    Energy output

        Aerobic processes

        Anaerobic processes

    Neuromuscular function

        Strength

        Technique

    Psychological factors

        Motivation

        Tactics
```

Figure 1-2   Factors constituting physical performance.

nature of the performance. In golf, there is no need for a high energy output, but a good technique is essential. In a high jump, muscular strength and technique are of primary importance. In long-distance running and cross-country skiing, the capacity of the aerobic energy-yielding processes essentially determines the speed. The technique is more important in skiing than in running; psychological stamina and motivation are essential to fight the feeling of fatigue. Various activities, sports, and exercises can thus be analyzed by a variety of methods in order to determine the specific requirements for optimal performance.

REFERENCES

Asmussen, E., E. H. Christensen, and M. Nielsen: Die O_2-Aufnahme der Ruhenden und der Arbeitenden Skelettmuskein, *Skand. Arch. Physiol.*, **82**:212, 1939.
Capaldi, R. A.: A Dynamic Model of Cell Membranes, *Sc. Am.*, **230**:26, 1974.
Krogh, A.: "The Comparative Physiology of Respiratory Mechanisms," University of Pennsylvania Press, Philadelphia, 1941.

Energy Liberation and Transfer

2

CONTENTS

Chapter 2

Energy Liberation
and Transfer

The physiology of muscular work and exercise is basically a matter of transforming bound energy into mechanical energy.

Many similarities exist between the "human engine" and the combustion engine constructed by human beings. In the combustion engine, gasoline and air are introduced into the cylinder. The spark from the spark plug initiates the explosive combustion of the gas mixture. Chemical energy is transformed into kinetic energy and heat. The expansion of the gas forces the piston to move, and a system of mechanical devices can transfer this motion to the wheels. The motor is cooled by fluid or air to prevent overheating. The waste products are expelled with the exhaust. As this motor can work only in the presence of oxygen, its function is aerobic. When the gasoline tank is empty, the engine can no longer continue to run, since the operation of the combustion engine is dependent upon a continuous supply of fuel. In an automobile, the self-starter provides the energy for the first movements of the pistons. This energy comes from the electrical accumulator (battery); the starter can thus work in the absence of oxygen, or anaerobically. The stored energy of the battery is quite limited, however, so that the battery must be frequently recharged.

"Living organisms, like machines, conform to the law of the conservation of energy, and must pay for all their activities in the currency of metabolism" (Baldwin, 1967). In the human machine, the muscle fibers are the pistons. When fuel is available and a spark is introduced to start the breaking down of the fuel, part of the energy which is thus liberated can cause movement of the pistons. Heat and various waste products are produced.

In the following paragraphs we shall summarize the chemical processes involved in the human machine, omitting, for the sake of simplicity, the more complicated steps of the reactions and placing less emphasis on the actual chemistry involved than on the account of where and how the energy is released. For a more complete study, the reader is directed to more detailed textbooks and reviews (Conn and Stumpf, 1972; Lehninger, 1973, 1975; and Needham, 1971).

ENERGY TRANSFORMATION

Energy is required for the various kinds of biological work done by living organisms. There is a demand for energy, for we have: (1) a synthesis of new cell materials from simpler precursors so that, in the mature organism, the rate of formation of new molecules balances the rate of degradation of the "old" ones; (2) a transport of materials against gradients of concentration, and this active movement of molecules and ions is often called osmotic work; (3) a mechanical work particularly evident in contracting muscles; and (4) production of heat to maintain a body temperature of about 37°C. This heat is in many situations just a by-product of the above mentioned processes.

Originally, the energy from the sun is trapped by the green plants and conserved in the form of foodstuffs. In the course of this process, carbon dioxide and water are consumed by the plant and oxygen is liberated into the atmosphere. When the plant is consumed by an animal, the energy it contains may be utilized by the animal cells where the foodstuffs, through combination with oxygen, are once again brought back to carbon dioxide and water, while part of the energy is utilized for the types of biological work described in (1) through (3), and the rest is liberated as heat (4).

Chemical reactions involving transfer of energy may be divided into two kinds:

   1   Reactions that liberate energy in some form (most often as heat). These reactions may take place spontaneously and are called *exergonic.*
   2   Reactions that cannot take place unless energy in some form is added to the system; these are *endergonic* reactions.

In biological chemistry we are dealing with more or less complex organic molecules where the atoms are bound together with covalent bonds. Each bond has a certain energy level. As a rule, bond breakage is exergonic, whereas bond formation (synthesis) is an endergonic process. Different molecules will yield different amounts of released energy as they are broken down to simpler molecules; they have a different *chemical potential.**

---

*Thermodynamically, the chemical potential has a more precise definition, based upon the G-function.

The chemical potential may be thought of as an equivalent of the more familiar potential energy of a waterfall. The water in a lake located 1,600 m above sea level has a potential energy of 16 kJ (1600 kpm) per liter, which may be utilized as electrical power as the water flows down to the sea. Analogously, glucose has a chemical potential of roughly 16 kJ (some 4 kcal) per gram that may be used for work as it is broken down into carbon dioxide and water. A resynthesis needs added external energy, just as energy has to be added in order to get the water back from sea level into the lake.

This waterfall model may be useful to clarify other points concerning chemical energy and transformation. Let us assume that water from a top reservoir A flows down to a bottom reservoir B, thereby losing potential energy; but some of this energy will be used to drive a turbine. This A → B process is exergonic, very much like a chemical reaction where reactant A gives product B.

Without a stopper of some kind, the water from the lake would keep falling at maximal speed and at a maximal rate of energy liberation until the lake would be completely empty. However, in the operation of an electric power plant, as in living organisms, there are times, such as during the night, when comparatively little energy is needed, and when, therefore, a full-speed operation of the power plant would represent an enormous waste of potential energy. Both systems (power plant and living organisms) have therefore built in a valve device that will automatically regulate the flow of water or molecules according to the energy need at any given moment. Water from the bottom reservoir may be lifted to another reservoir with higher potential energy, i.e., from B to C. This B → C process is endergonic, and obviously does not happen by itself. Now the energy released in the exergonic A → B process may be used to drive the endergonic B → C process. The same can be accomplished in biochemical reactions, but in both cases some type of coupling device is needed. In our waterfall model, we can couple the turbine directly to the pump shafts by some mechanical device. Alternatively, we may let the turbine drive a generator; the electric power from this generator can then directly drive an electric pump motor or charge accumulators which later may be used for driving the pump motor.

Regardless of what type of coupling we select, it is impossible to pump the water to an energy level of A or higher with the energy obtained from the waterfall alone. If the system is to work, the water level C has to be somewhat lower than A, unless we could construct a *perpetuum mobile*, which is impossible as it would violate the second law of thermodynamics. There is always a loss of energy during transformation, a frictional loss mostly coming out as heat. This is the case with waterfalls as it is in biochemical reactions. An exergonic reaction may drive an endergonic reaction through some type of coupling, but *all* the energy in the exergonic reaction cannot be regained as potential energy. There will always be a heat loss.

In biological chemistry one will find several directly coupled reactions. In our exergonic A → B and endergonic B → C processes, the following combination or coupling may well work:

$$A + B \rightarrow B + C \ (+ \text{heat})$$

The coupling between two processes or reactions may be a common intermediate (I) (some type of carrier, like the molecular complex RH_2), as in the following scheme:

$$A + B \rightarrow I \rightarrow B + C \ (+ \text{heat})$$

This would be analogous to a gearbox placed between the turbine and the pump in our waterfall model.

However, nature has chosen more flexible solutions to the problem of how to get an endergonic reaction going. Instead of direct coupling, we find systems analogous to our power plants: the energy is transformed and stored in another and much more readily available form. It is not very practical to have our vacuum cleaner directly coupled to a water turbine, but it works nicely via electric power!

A classical equation for an exergonic oxidative metabolic process is the oxidation of glucose:

$$C_6H_{12}O_6 + 6O_2 \quad \underset{\text{photosynthesis}}{\overset{\text{oxidation}}{\rightleftharpoons}} \quad 6CO_2 + 6H_2O + \text{energy}$$

The glucose has a much higher energy level than CO_2 and H_2O, and this energy is liberated by the oxidation and made available for processes that require energy. (One mole of glucose can yield a maximum of 2870 kJ, or 686 kcal, of chemical energy.) The reaction is reversible. With chlorophyll as a catalyst, CO_2 and H_2O can combine again, the necessary energy for this endergonic reaction being provided by light. Thus the circle is completed: Animals dissimilate or break down carbohydrates (catabolism), and plants can also assimilate or synthesize carbohydrates (anabolism).

The animals require the energy-rich and complex products of photosynthesis to pay the costs of the biological work mentioned. They can do this by degrading the structures of such molecules as glucose. Secondly, they also need complex carbon compounds, such as glucose, as building blocks for the synthesis of their own cellular materials since they cannot, like plants, use CO_2 for this purpose.

Oxidation is defined chemically as the loss of electrons from an atom or molecule; *reduction* is defined as a gain of electrons. In many reactions the electrons are carried in the form of hydrogen atoms, and the oxidized compound is thereby dehydrogenated. In animals, organic fuels, such as glucose, lipids, and proteins, constitute the major electron donors, and sooner or later oxygen is the final electron acceptor or oxidant of the fuel; that is, it is an *aerobic* oxidation and is also called *respiration*. As an alternative pathway, the glucose and glycogen molecules can be broken down into two or more fragments, and one of these fragments then becomes oxidized by another. This energy yield is *anaerobic*, and the processes are named glycolysis and glycogenolysis, respectively.

In biological oxidation-reduction reactions, an intermediary carrier of electrons usually acts together with catalysts (enzymes and coenzymes). In the *mitochondria* the aerobic energy-transducing system is bound into its structures. The mitochondria has two membranes, of which the inner one has inward folds, or invaginations, called cristae, which greatly increase the surface area. (It has been estimated that the mitochondria of blowfly flight muscle has 400 m^2 of inner membrane surface per gram of mitochondria protein.) This inner membrane surface contains enzymes (e.g., the cytochromes) that act in a chain to carry electrons from nutrient molecules to oxygen. Each one of the enzymes specializes in carrying atoms or electrons in just one range of the molecular chain in the substrate. In other words, there is a lock-and-key fit of the substrate molecule to a small patch on the surface of the enzyme molecule.

An enzyme can split a larger molecule into smaller pieces, or it can construct larger molecules from smaller fragments. A mitochondria with a weight of 10^{-8} can contain $3 \cdot 10^{14}$ molecules of water and 10 billion (10^9) enzyme molecules!

In one minute a single enzyme molecule may carry out as many as several million catalytic cycles (Lehninger, 1973). The inner compartment of the mitochondria, the matrix, contains about 50 percent protein.

High-Energy Phosphates

Electric power may be stored in accumulators or batteries. Our living cells have solved the problem of storing energy in a similar way. The most abundant "battery packs" used in the cell are the compound called adenosinetriphosphate (ATP). This compound belongs to a class of molecules called high-energy phosphates. This term is somewhat imprecise, but the crucial point is that in such molecules a great deal of the total energy content is concentrated in one (or two) phosphate bonds. (We have already mentioned that the energy in a molecule lies in the bonds between the atoms.) There exist other energetically equivalent compounds of this type (GTP, UTP, etc.), but ATP is the one most often used when energy is called for in biological systems. (See Table 2-1 for these and other abbreviations and their meanings.) Furthermore, such phosphate bonds donate their energy by very simple reactions, i.e., hydrolysis or phosphate transfer, thereby providing energy for driving endergonic processes.

Almost every energy-demanding process in a cell is associated with the use of the energy from ATP stored in a rapidly usable form such as ATP.

A battery contains a limited amount of energy; it needs continuous recharging to be constantly operative. The same holds for ATP. Here the charging consists of combining ADP and phosphate in an endergonic reaction to ATP, and the thus-stored energy is liberated by the typical exergonic reaction:

$$ATP + H_2O \rightarrow ADP + P_i + energy$$

Table 2-1   Abbreviations

ATP,	:	adenosine triphosphate
GTP		guanosine triphosphate
UTP		uridine triphosphate
ADP,	:	adenosine diphosphate
AMP		adenosine monophosphate
P_i	:	orthophosphate (PO_4^{3-})
PP_i	:	pyrophosphate
G-l-P	:	glucose-l-phosphate
G-6-P	:	glucose-6-phosphate
F-6-P	:	fructose-6-phosphate
F-1, 6-P_2 :		fructose-1, 6-diphosphate
PEP	:	phosphoenol pyruvate
α-KG	:	α-ketoglutarate
SuccCoA:		succinylcoenzyme A
OAA	:	oxaloacetate
NAD	:	nicotinamide-adenine-dinucleotide
FAD	:	flavin-adenine-dinucleotide
GS	:	glycogen synthetase
GP	:	glycogen phosphorylase
HK	:	hexokinase
TCA	:	tricarboxylic acid
PFK	:	phosphofructokinase
FDPase	:	fructosediphosphatase
LDH	:	lactate dehydrogenase
PDH	:	pyruvate dehydrogenase

ATP contains, in fact, one more energy-rich bond, and in a separate chemical reaction, the transformation may proceed further to adenosine monophosphate (AMP):

$$ATP + H_2O \rightarrow AMP + PP_i + energy$$

We can conclude that ATP is one universal intracellular vehicle of chemical energy; it saves or conserves some of the energy yielded in the degradation of fuel molecules and can transfer its energy, by donation of its terminal high-energy phosphate group, to processes requiring energy within the cell: biosynthetic processes, active transport of material against gradients, and muscle contraction. Without the ATP-ADP system, no cell could function or survive!

ADP is like a discharged battery. In order for it to be used again, it has to be recharged. ATP, then, operates as the immediately usable "energy pack" for driving endergonic processes, but it is certainly no energy *store* in the usual sense of the term. The muscle cell contains a type of "accumulator" with a somewhat greater capacity than the ATP "energy pack," namely, phosphocreatine (CP). This is also a high-energy phosphate compound, and is in equilibrium with ATP through the following reactions:

$$ATP + C \rightleftharpoons ADP + CP$$

CP operates as an immediate store for ATP regeneration, but also this energy store is fairly rapidly depleted (within a matter of seconds or minutes in strenuous muscular work). The resynthesis of ATP (reversal of the reaction $ATP + H_2O \rightarrow ADP + P_i + energy$) therefore has to go on continuously. The energy source for this process is the breakdown of more complex molecules (foodstuffs) to simpler ones, and ultimately to CO_2 and H_2O. Organic compounds containing carbon and hydrogen may be totally combusted to CO_2 and H_2O by oxygen (Fig. 2-1). This can be done either in a bomb calorimeter or by catalytic oxidation with the aid of enzymes in a living cell. It should be noted

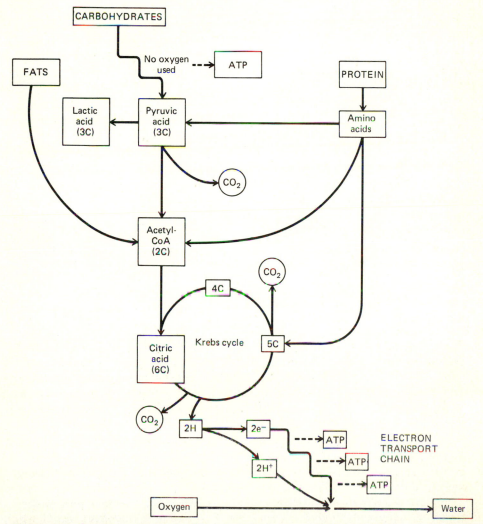

Figure 2-1 Organic compounds (fats, carbohydrates, proteins) containing carbon and hydrogen may be totally combusted to CO_2 and H_2O by oxygen, that is, by the process of a catalytic oxidation with the aid of enzymes in the living cell. For some of the steps the number of carbon atoms involved are given.

that nitrogen is never oxidized in animal cells (although much energy lies in such a process!). It is mostly excreted as urea ($NH_2 CO NH_2$), in itself an energy-demanding process.

A total combustion of a compound in a bomb calorimeter gives off all the energy in one burst. In our waterfall model, for example, all the water in the lake would suddenly fall down into the sea. Obviously, such a burst would then make it technically very difficult to catch all the available energy by a turbine. On the other hand, it is possible to build several power plants in a river, taking out energy in parts as the water flows downward and gradually loses its initial potential energy. In fact, the cell uses a similar technique. Combustible material (foodstuffs, fuel) is broken down in discrete steps, thereby giving the system an opportunity to catch much of the released energy step by step.

A combustion is an oxidation, but an oxidation does not necessarily involve the direct participation of oxygen. Most of the single-step oxidations in biological pathways are, in fact, dehydrogenations. The hydrogen atoms are later carried through different reactions and ultimately combined with oxygen to give water. This is of particular importance for our working muscle cell, for several energy-yielding breakdown reactions of our fuel do take place anaerobically, that is, without the presence of oxygen.

The main stored energy-yielding fuels are carbohydrate in the form of glycogen, which is a glucose polymer, and fatty acids in the form of acylglycerols (triglycerides).

The third class of foodstuffs, proteins, are metabolized in a rather complex manner. The carbon skeletons of the different amino acids will, after deamination and transport of nitrogenous compounds to the urea cycle, enter the glycolysis–Krebs cycle scheme at different points. Some amino acids give pyruvate, some acetyl-CoA, some different Krebs cycle intermediates, etc. Glycerol from fats will enter glycolysis as a triosephosphate. Normally, the contribution from these compounds to the total energy expenditure in skeletal muscles during work is very small and may be ignored.

THE OXIDATION OF FUEL

We may now take a closer look at how the main stored fuels metabolize and give rise to ATP resynthesis.

In the total carbohydrate combustion, we have an initial anaerobic sequence, glycolysis, followed by an aerobic phase. The glycolytic enzymes are located in the cytoplasm of the cells, whereas the aerobic oxidation of carbohydrates and also fatty acids takes place inside the mitochondria. The last step of the entire mitochondrial oxidation is oxygen-dependent, while the cytoplasmic glycolysis is not. Therefore, whereas the breakdown of fatty acids is strictly aerobic, the breakdown of glucose may proceed either aerobically or anaerobically.

Figure 2-2 shows a simplified flow chart that presents the main reactions operating and controlling the breakdown of glycogen and fatty acids in a muscle cell. It is not stoichiometrically correct, and only part of the coenzymes, entering at different steps, is included. Besides ADP and ATP, which partici-

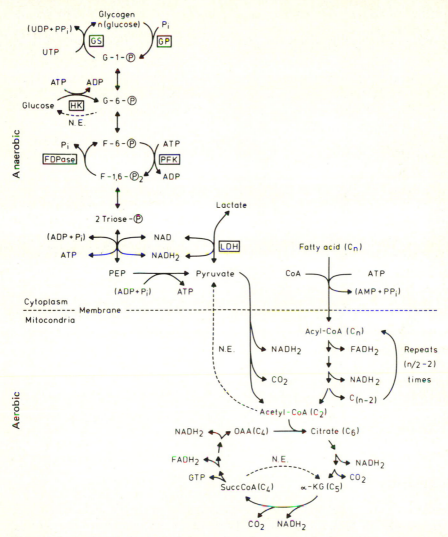

Figure 2-2 A simplified scheme for glycolysis, fatty acid oxidation, and the Krebs citric acid (tricarboxylic acid) cycle in a muscle cell. (For abbreviations, see Table 2-1). N.E = nonexistent.

pate directly in some reactions, the *coenzymes* (hydrogen carriers) NAD and FAD play a most important role in trapping energy. In this context we neglect the fact that almost every intermediary metabolite exists in the cell in an ionized state (that is, all acids are anions). The oxidized form of NAD exists as the cation NAD^+ and the reduced, hydrogenated form as $NADH + H^+$ (one of the two hydrogens is transported as a proton), but here we designate it $NADH_2$ for the sake of simplicity. A generalized dehydrogenation reaction will be:

$$RH_2 + NAD \rightarrow R + NADH_2$$

(In some reactions the coenzyme FAD is used instead of NAD.) The steady-state concentrations of these coenzymes are rather low. To keep such dehydrogenations going, the reduced (hydrogenated) form must be reoxidized by means of some other hydrogen acceptor. This happens in the mitochondria via the electron-transport chain in which oxygen serves as the final hydrogen acceptor, giving water. As we shall see, $NADH_2$ may also be reoxidized (dehydrogenated) in the cytoplasm, giving lactate as the final product.

Anaerobic Energy Yield

Let us first consider in more detail the anaerobic condition, i.e., when oxygen is not at the disposal of the cell. The breakdown of glycogen and glucose to pyruvic acid and lactic acid can occur in the absence of oxygen. We shall analyze only a few of the 11 steps and 11 enzymes involved in this anaerobic energy yield.

Starting with glycogen stored in the muscle cell (Fig. 2-2), a phosphorolytic cleavage gives glucose-1-(P), which isomerizes to glucose-6-(P). Glucose from outside the cell will be phosphorylated by ATP to glucose-6-(P) and follows the same pathway as that derived from glycogen. The glucose-6-(P) isomerizes to fructose-6-(P), which then enters the irreversible phosphofructokinase (PFK) step. This enzymatic step represents the main control of the carbohydrate metabolism in the muscle cell. The catalytic activity of an enzyme can be sensitive to the concentration of some crucial metabolites, which can act as modulators. A *negative modulator* will inhibit its catalytic activity, but a *positive modulator* will stimulate the reaction.

The rate of the enzymatic reaction forming fructose-1,6-diphosphate is greatly accelerated if ADP is present in a high concentration, but it is inhibited if the ATP concentration is high. ADP therefore serves as a positive modulator in a situation when energy is needed for the regeneration of ATP from ADP. ATP is a negative modulator and can inhibit the system that produces it!

The product fructose-1,6-P_2 splits into two triosephosphates which, through a sequence of reactions, donate hydrogens to NAD and also are coupled to a direct resynthesis of ATP, giving PEP (phosphoenol pyruvate), which in itself is "high-energetic" and resynthesizes one more ATP as it is converted to pyruvate.

The fate of pyruvate now depends on whether or not we are dealing with an anaerobic metabolism. In anaerobic metabolism, part of the $NADH_2$ which has been generated in the cytoplasm donates the hydrogens to pyruvate, NAD becomes free to reenter the dehydrogenation of triosephosphate, and lactate will be the final product by the lactate dehydrogenase (LDH) reaction. These events allow the cell to obtain energy from glycogen and glucose when oxygen is lacking, and the energy-balance scheme for the anaerobic part will be:

$$(n + 1) \text{ glucose} + \text{ATP} \rightarrow (n) \text{ glucose} + 4 \text{ ATP} + 2 \text{ lactate}$$

that is, three moles ATP for every glucose monomer in glycogen. If the reaction starts with the entrance of glucose into the muscle cell from the outside, this glucose has to be activated by ATP initially, and only two ATPs are given off as the net result.

This anaerobic glycolysis solves the problem of how to reoxidize the $NADH_2$ when the oxygen supply is insufficient, but the price that has to be paid is a steady accumulation of the end product—lactate—in the cell. This accumulation sooner or later will slow down and eventually stop the process when the concentration of lactic acid becomes too high. The muscle cell will be "poisoned" by its own metabolic product, so to speak, and will cease working, somewhat like the yeast cell that is poisoned by the accumulation of its own end product, methanol.

The enzyme catalyzing this reaction is *lactate dehydrogenase* (LDH). There are two distinct subunits of this enzyme in the tissues of most animals (Kaplan and Goodfriend, 1964.) One form (H) predominates in the heart, and the other (M) in many skeletal muscles. They differ in the rate at which they reduce pyruvate to lactate so that the H type is inhibited at low concentrations of the substrate pyruvate. In organs with a domination of the H form of LDH, a complete aerobic oxidation of glucose would be favored, but the potential for a rapid lactate production from pyruvate is very limited. The M form of LDH, on the other hand, is inhibited at a much higher substrate concentration and it can therefore reduce pyruvate to lactate at a high rate. Actually, five different molecular forms of LDH, so-called isozymes, with different proportions of the M and H forms, have been identified. They are symbolized as M_4, M_3H, M_2H_2, MH_3, and H_4. In the muscle cells characterized by a well-developed aerobic potential, like the heart muscle and slow twitch (type I, or red) fibers in the skeletal muscles, the H isozyme dominates. In fast twitch fibers (type II, or white muscle fibers) with a more pronounced potential for an anaerobic work, the M isozyme is at a higher concentration than the H form (Sjödin, 1976).

Summary   In the reactions by which carbohydrates are broken down to lactates, ATP serves as a phosphate carrier and NAD as an electron carrier. All the intermediates of the glycolysis between glucose and pyruvate are phosphorylated compounds. The phosphorylation makes these intermediates unable to penetrate through the cell membrane; they are trapped inside the cell and cannot diffuse out. Ultimately the phosphates become the terminal phosphate group of ATP as the glycolysis and glycogenolysis proceed. ATP can then transform some of the yielded energy to various cell activities. Figure 2-3 gives, in a schematic way, the ATP cost and the ATP yield for an anaerobic breakdown of one 6-carbon unit into pyruvate. Starting with glycogen, the net energy yield covers the resynthesis of three ATP, but with glucose as the substrate, the net ATP formation is limited to two molecules per molecule of glucose.

Aerobic Energy Yield

The *aerobic oxidation of glucose or glycogen* is identical down to pyruvate. If the supply of oxygen is sufficient, only unimportant quantities of pyruvate will be reduced to lactate, if anything at all (see the lower section of Fig. 2-2). Instead, pyruvate will enter the mitochondria through a very complex enzyme system (pyruvate dehydrogenase, PDH). This system catalyzes a reaction

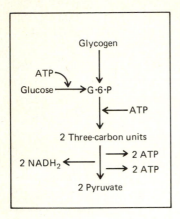

Figure 2-3 A summary of the demand and yield respectively of ATP molecules in the anaerobic breakdown of carbohydrates to pyruvate in the cell cytoplasm. Note the formation of two $NADH_2$.

known as oxidative decarboxylation, using NAD as H-acceptor and, after CO_2 release, attaching the rest of the pyruvate molecule to coenzyme A, whereby the compound acetyl-coenzyme A is formed. This reaction is essentially irreversible, and no other backward reaction involving the same reactants and products is known. Acetyl-CoA is also the product of the initial fatty acid β–oxidation (Fig. 2-1), and combines with oxaloacetate (OAA) to form citrate, thereby entering the *tricarboxylic acid cycle* (also named the *Krebs citric acid cycle*). In this cycle a cyclic sequence of reactions catalyzed by a multienzyme system accepts the acetyl-CoA as a fuel and dismembers it to yield CO_2 and hydrogen atoms.

In this cycle we find another irreversible oxidative decarboxylation: from α-ketoglutarate (KG) to succinyl-CoA, i.e., a one-way traffic around the cycle. The α-KG formation is also associated with a decarboxylation: The two carbons that enter the cycle as acetyl will leave it as two CO_2's. These facts have rather important implications: Fatty acids, which inevitably are broken down to acetyl-CoA, can never give rise to a net synthesis of carbohydrate. On the other hand, since acetyl-CoA is also the precursor for fatty acid synthesis, carbohydrates can be broken down to this compound and act as precursors for fat.

For every acetyl unit (whether it stems from carbohydrate or fat) consumed in the cycle, there are two NAD-linked dehydrogenases and one FAD-linked dehydrogenase. In addition, there is one direct GTP (guanosinetriphosphate)-synthesizing step (GTP and ATP contain equivalent amounts of energy).

The reduced coenzymes $NADH_2$ and $FADH_2$ from the different dehydrogenating steps will enter the so-called respiratory chain or electron-transport chain, which is schematically presented in Fig. 2-4. This chain is a complex device consisting of lipoproteins with different cytochromes, metals, and other cofactors. The flow of two electrons through the chain will release energy for the phosphorylation of ADP to ATP at three different sites, as indicated, and at the end of the chain, each pair of electrons is combined with two protons (H^+)

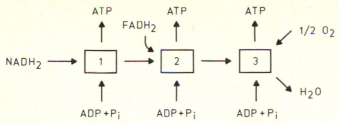

Figure 2-4   Schematic presentation of the respiratory or electron-transport chain.

and oxygen, forming water. $NADH_2$ enters the first "box," giving rise to NAD and three ATP, while $FADH_2$ will enter the second "box" and give FAD and two ATP. The coenzymes thus released are then able to participate in dehydrogenations once again.

Fat: We have mentioned the β-oxidation of fatty acids with resulting acetyl-CoA. Before fatty acids are able to enter this pathway, they have to be activated, i.e., transformed to CoA esters. This activation is energy-dependent and uses one ATP with hydrolysis to AMP; this is equivalent to the energy from two ATP when hydrolyzed to two ADP. Once activated, the fatty acids are ready for complete combustion. The β-oxidation is again performed as a dehydrogenation. There is one FAD-linked and one NAD-linked enzyme operating, which results in one $FADH_2$ and one $NADH_2$ for every acetyl-CoA cut off (Fig. 2-2). The splitting is associated with a reaction with a new CoA molecule, resulting in the formation of a CoA ester of the remaining fatty acid molecule, and the process repeats itself. The final result is a division of the entire fatty acid molecule in C_2 fragments (acetyl-CoA) ready for further oxidation in the Krebs cycle, and formation of one $NADH_2$ and one $FADH_2$ for every division.

In summary, we can identify three main steps in these reactions: (1) the formation of acetyl-CoA from fatty acids or pyruvate; (2) the degradation of the acetyl residue of acetyl-CoA by the tricarboxylic acid cycle to yield CO_2 and hydrogen atoms; and (3) the transport of the electrons from the corresponding hydrogen atoms to molecular oxygen in the respiratory chain. Some of the yielded energy is caught by the formation of ATP.

It should be noted that the 4-carbon unit oxaloacetate also enters the tricarboxylic acid cycle "race," and at the finish of one revolution, oxaloacetate arrives in complete isolation, ready for a new lap of bringing more fuel into the cycle. One molecule of oxaloacetate can therefore assist in the oxidation of an infinite number of acetic acid molecules to CO_2 and H_2O simply because it is regenerated by the cycle at the end of each turn.

The enzyme complex guiding the entrance to the cycle, that is, catalyzing the formation of acetyl-CoA, is inhibited by a high concentration of ATP in the mitochondria. However, whenever the ADP concentration is high and ample pyruvate is available, the enzyme complex is turned on and more acetyl-CoA, which is the fuel for the cycle, is formed. This reaction is also enhanced by Ca^{2+}. The formation of acetyl-CoA from pyruvate is an irreversible step; in other words; pyruvate cannot be synthesized from acetyl-CoA.

ATP Formation

We are now in a position to calculate the energy trapped as resynthesized ATP when glucose or fatty acids are totally converted to CO_2 and H_2O. Figure 2-5 presents a condensed scheme of the actual events in the combustion of one glucose unit from glycogen. To the left we find the ATPs associated with glycogenolysis, and in the middle the reduced equivalents that are reoxidized, with concomitant ATP production to the right.

In summary, we get 3 ATP from the glycogenolysis (whether pyruvate is reduced to lactate or oxidized in the mitochondria), while we get 36 ATP from the aerobic mitochondrial oxidation, resulting in a total of 39 ATP. There is at least one uncertain point in this calculation: the $NADH_2$ produced glycolytical-ly in the cytoplasm cannot freely enter the mitochondria. Some transport mechanism must be used, and some of these "shuttle mechanisms" will consume energy, with the consequence that this $NADH_2$ enters analogs to $FADH_2$ at box 2 in Fig. 2-4, and gives rise to only two ATP per $NADH_2$ and not

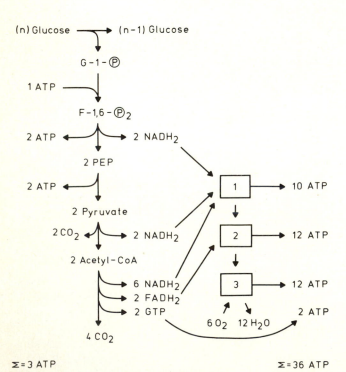

Figure 2-5 A condensed scheme for the events involved in the combustion of one glucose unit from glycogen, and the ATP produced.

three ATP. Thus, we may obtain a total of only 37 ATP per glucose monomer (McGilvery, 1975).

A similar calculation for a saturated fatty acid with n-carbon atoms is presented in Fig. 2-6. Here we consider only fatty acids with an even number of carbon atoms (odd-numbered fatty acids will give rise to a propionyl-CoA as final product, with a more complicated combustion scheme). As the acid splits into C_2 fragments, we obtain $n/2$ acetyl-CoA, and as we need one less division than the total number of pieces, we obtain $(n/2-1)$ $NADH_2$ and $FADH_2$ in the β-oxidation. Adding up and subtracting the initial activating ATP consumption, we will have a total of $8.5n - 7$ ATP for this combustion. For palmitate, the most abundant fatty acid in the body, we will obtain $8.5 \cdot 16 - 7 = 129$ ATP when this C_{16} acid is metabolized to CO_2 and H_2O.

$$C_{15}H_{31}COOH + 129\ P_i + 129\ ADP + 23\ O_2 \rightarrow$$
$$16\ CO_2 + 16\ H_2O + 129\ ATP$$

It is estimated that the average yield is 138 moles of ATP per mole of a mixture of fatty acids resembling the composition of human adipose tissue.

If we now take a closer look at the stoichiometry in the bomb calorimeter-combustion of glucose:

$$C_6H_{12}O_6 + 6\ O_2 \rightarrow 6\ CO_2 + 6\ H_2O$$

we will find that this fits in with the scheme in Fig. 2-5, except for water. We put one glucose unit into the reaction, get out 6 CO_2, use 6 O_2, but obtain 12 H_2O. This means that somewhere

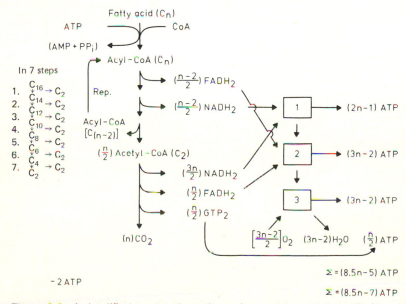

Figure 2-6 A simplified combustion scheme for a saturated fatty acid with n-carbon atoms.

in the scheme, six H_2O molecules are added. This is so because all the oxygen consumed is concentrated in one single reaction, namely, the cytochrome oxidase step that terminates the electron transport chain (Fig. 2-4). Oxygen, as such, participates in no other reaction in these sequences. Furthermore, all this molecular oxygen is used to burn hydrogen to water. Evidently we are dealing with a carbon combustion to CO_2, but we do not find any reaction similar to: $C + O_2 \rightarrow CO_2$. Instead, the cell utilizes systems that in their consequences are like a "water gas" reaction: $C + 2 H_2O \rightarrow CO_2 + 2 H_2$. This implies that except for the acid group, all the oxygen atoms which we find in CO_2, after oxidation of a fatty acid, stem from water, and when glucose is oxidized, one half comes from the oxygen already present in the glucose molecule and the other half from water. The hydrogen liberated in the water-splitting reaction is transported on NAD or FAD and finally reacts with molecular oxygen back to H_2O. The stoichiometry, since all the consumed oxygen is used for water production, would be:

$$C_6H_{12}O_6 + 6 O_2 + 6 H_2O \rightarrow 6 CO_2 + 12 H_2O$$

The same arguments are valid for fatty acid oxidation. Here we have the equation:

$$C_nH_{2n}O_2 + \left(\frac{3n}{2} - 1 \right) O_2 \rightarrow {}_nCO_2 + n H_2O$$

to be converted to:

$$C_nH_{2n}O_2 + \left(\frac{3n}{2} - 1 \right) O_2 + (2n-2)H_2O \rightarrow nCO_2 + (3n-2)H_2O$$

In fact, the total water balance in these reactions is even more complicated. In several of the reactions, water will participate either as reactant or as product. For every ATP resynthesized, a water molecule is released, and so forth.

ENERGY YIELD

It has been emphasized that in all chemical transformation there is a loss of energy, a "friction" loss. In the degradation of fuel molecules, the energy content in the remainder is reduced. The symbol ΔG denotes the energy which is available in a chemical reaction to do work as the system reaches equilibrium at a given temperature, pressure, and volume. The $\Delta G^{0'}$ symbol for a chemical reaction denotes the maximal work it can theoretically perform.

In terms of thermodynamics in biological systems, the most useful expression of energy is the Gibbs free energy, G, which gives an expression of energy changes at constant pressure. Potential energy has no absolute magnitude, but always describes the difference between two energy levels, and ΔG gives the difference in chemical potential between two states:

$$\Delta G = \Delta G^0 + RT \, 1n \, \frac{[P]}{[R]}$$

where R is the gas constant, T is the temperature, [P] is the product concentration, [R] is the reactant concentration, and ΔG^0 refers to a standard state.

At equilibrium there is no free energy change and therefore $\Delta G = O$, and since $[P_{eq}] / [R_{eq}] = K_{eq}$, we get:

$$\Delta G^0 = -RT \, 1n \, K_{eq}$$

From the equations above, it follows that an exergonic reaction, with an equilibrium far to the right, is associated with a high K_{eq} and consequently with a highly negative ΔG^0. Whereas ΔG is difficult to measure, ΔG^0 is of a definite magnitude for each given enzymatic reaction. Therefore, ΔG^0 is usually listed in the various tables and handbooks and may be taken as an expression of the free energy change of a reaction.

The concept of ΔG^0 may tell us a great deal about the energy changes in the cell, but it has to be viewed with some caution. Not only is it difficult to measure; it is equally difficult to interpret. It is pH-dependent in all biochemical reactions where protons are liberated or captured. As an example, we may take a closer look at the hydrolysis energy of ATP. The reaction:

$$ATP^{4-} + H_2O \rightarrow ADP^{3-} + P_i^{2-} + H^+$$

is highly exergonic (equilibrium far to the right) with a highly negative ΔG^0, and consequently difficult to measure directly. However, the free energy of hydrolysis may be estimated from a set of reactions whose sum is equivalent to the particular reaction expressed above. By such an estimate, we arrive at a $\Delta G^{0\prime}$ of about -30 kJ ($-7, 3$ kcal) mole^{-1} at 25°C. The sign ' indicates that the ΔG^0 value is standardized to pH 7.0. This reference is necessary because the [H$^+$] participates in the reaction. Furthermore, the ionization constants (pK-values) for ATP, ADP, and P$_i$ have to be taken into account. As a consequence, the ΔG^0 for the ATP hydrolysis will be highly pH-dependent, rising almost linearly to some -80 kJ (-19 kcal) mole^{-1} at pH 12.

The concentrations of all the reactants and products in the above equations inside the cell are not precisely known; they may even fluctuate and may perhaps be compartmentalized; in any case, they are certainly far from the standard concentrations of 1 M. The temperature in the muscle cell is not 25°C but probably around 37°C and may exceed 40°C during intense muscular work. The pH inside the cell and within its different compartments can only be estimated; it may be around 7.0 at rest but may perhaps drop to below 6.5 during strenuous muscular work. Consequently, the ΔG for ATP hydrolysis inside the cell is not exactly known, but it is usually estimated to be somewhere between -40 and -50 kJ (-10 and -12 kcal) mole^{-1}. Thus, the calculation of the efficiency of the cell's combustion of its fuels is hampered by severe uncertainty. On the other hand, it is not very difficult to obtain ΔG values for the fuels (glycogen or fatty acids) and the final combustion products (lactate or CO$_2$ and water), but in this case we must also consider some difficulties concerning ionization and [H$^+$]. Lactic acid is almost completely dissociated into H$^+$ and lactate (the anion), but the hydrogen ions will be captured to a certain extent by the total buffer system of the cell and the pH will therefore drop less than expected. CO$_2$ will be present not only as dissolved gas, but also bound as bicarbonate (HCO$_3^-$). Furthermore, the cell utilizes water electrolysis, i.e., $H_2O \rightarrow H^+ + OH^-$, in energy transformations.

With due respect to all these reservations, and assuming a $\Delta G^{0\prime}$ value of 40 kJ (9.6 kcal) mole^{-1} for the ATP hydrolysis, the calculations presented in the following may be valid.

The complete oxidation of glucose to CO$_2$ and water has a $\Delta G^{0\prime}$ of 2.87 MJ (686 kcal) mole^{-1}. During aerobic conditions, 38 ATP is regenerated and the efficiency will be [(38 × 40)/2870] × 100 = 53 percent. During anaerobic conditions, $\Delta G^{0\prime}$ for the transformation of glycogen to lactate is about 219 kJ (or 52.4 kcal) mole^{-1}, which yields energy for the regeneration of three mole ATP. The calculated efficiency is [(3 × 40)/219] × 100 = 55 percent, or very similar to the efficiency of the aerobic oxidation of glycogen.

At a given energy expenditure, the glycogen stores will be depleted at a much faster rate when the muscles must work anaerobically than when they can work aerobically. With a gross energy requirement of 110 kJ (26 kcal), 0.5 mole of the 6-carbon unit of glycogen has to be used for the formation of one mole of lactate anaerobically. If, on the other hand, enough oxygen were available for an aerobic oxidation, only 0.038 mole of glycogen would have to be oxidized, with a consumption of about 5.2 liters of oxygen yielding the required 26 kcal. Glycogen can actually generate 39/3 = 13 times more ATP per mole of glucose unit aerobically than anaerobically.

$\Delta G^{0\prime}$ for the oxidation of palmitate is about 9.79 MJ (2340 kcal) per mole. With 129 mole ATP produced, the efficiency of this fatty acid metabolism will be close to 53 percent if we use 40 kJ (9.6 kcal) as $\Delta G^{0\prime}$ for ATP.

CONTROL SYSTEMS

Biochemical reactions may be controlled by several means:

   1 By controlling the effective concentrations of the reactants and/or products in a particular reaction.
   2 By allosteric feedback control of the enzyme catalyzing the reaction. That is a mechanism by which some particular compound will bind to the enzyme protein at site other than the substrate binding site, thereby changing the catalytic power. Such modulators may be positive (activators) or negative (inhibitors).
   3 By differences in the amount of the particular enzyme involved in the reaction.

   In order to appreciate the magnitude of the energy yield involved in the different metabolic reactions, it may be useful to take a closer look at the first two of these control mechanisms:
   A reversible reaction

$$A + C \rightleftharpoons B + D$$

will come to an equilibrium when the reaction rate in the forward direction equals that of the backward direction, whether we start with A + C or with B + D. From the law of mass action, we can formulate an equilibrium constant

$$K_{eq} = \frac{[D]\,[B]}{[A]\,[C]}$$

If we now either add more reactant (A or C) or remove some of the products (B or D), the reaction in either case will proceed forward. Even if the equilibrium lies far to the left (having small concentrations of B and D), the reaction may proceed until all the reactants A and C have been converted, provided that the products are continuously removed. This is indeed what happens in several biological reaction sequences. Products in one reaction are entering as reactants in another new reaction, and we get a flow through several steps (i.e., glycolysis), even if some of the steps have an equilibrium which favors a backward direction of the reaction.

In our scheme (Fig. 2-2), we have marked the enzymes catalyzing irreversible steps in boxes. The enzymes included are those that are most closely controlled, both externally (by hormonal control, often via cyclic AMP and protein kinases) and internally (by allosteric feedback). As all these reactions are designed primarily to meet the energy demand of the cell, it is not surprising that the key energy-exchanging nucleotides ATP, ADP, and AMP are directly involved in the feedback regulation. It is beyond the scope of this section to present all these complicated controls in detail here. We shall therefore limit our discussion to certain points.

As already mentioned, the electron-transport/oxidative-phosphorylation device is a very tightly coupled one. However, several compounds may act as uncouplers; that is, they cause a reoxidation of the entering $NADH_2$ or $FADH_2$ without ATP resynthesis. This, incidentally, is what happens physiologically in the brown fat in hibernators and other animals. Here the reoxidation energy is directly utilized as heat, and not used for ATP resynthesis. Normally, however, the ADP concentration will directly control the O_2 consumption; there will be no reoxidation as long as the ADP concentration is low. Low ADP normally means high ATP, and the control depends not so much on the concentration of one of these nucleotides as on the ratio between them. It is the ATP/ADP ratio (or, to be more precise, some expression of $[ATP]/([ADP] + [AMP])$ that primarily determines how the flow through glycolysis and mitochondrial oxidation will take place. This ratio is also of the utmost importance in controlling the glycolysis in the phosphofructokinase (PFKase)-fructose-diphosphatase (FDPase) step. These two reactions represent most of the control of the glycolysis and the gluconeogenesis (the reverse of glycolysis, i.e., resynthesis of carbohydrate). The PFK step is irreversible, and one ATP is used to synthetize fructose-1,6-diphosphate from fructose-1-phosphate. The backward reaction catalyzed by FDPase is an irreversible hydrolysis of fructose-1,6-diphosphate with the liberation of inorganic phosphate (Fig. 2-2). The two reactions work as a "futile cycle" with only ATP hydrolysis and liberation of heat as a result. The possibility exists that this is a necessary design for maintaining our body temperature, for even at rest this cycle will spin somewhat faster than does the flow through glycolysis in general. Another possibility is that such a design is necessary in order to obtain the restricted control by the ATP/(ADP + AMP) ratio. In fact, it is known that the PFK protein is furnished with several allosteric sites for ATP, ADP, AMP, P_i and also citrate, and the same is true for FDPase. For example, ATP is an inhibitor and ADP an activator for PFK, while the opposite is true for FDPase. The complete story is quite complex, but the main effect is that a high ATP concentration over the ADP concentration will slow down the glycolysis by inhibiting PFK, activate FDPase, and thereby favor glyconeogenetic activity. Heavy muscular work will favor glycolysis as the ATP/ADP ratio is markedly displaced toward a low ATP concentration and a high ADP concentration. PFK becomes much more active, while FDPase will be more inhibited, resulting in some 100-fold rise in the flux from glycogen to pyruvate.

INTERPLAY BETWEEN ANAEROBIC
AND AEROBIC ENERGY YIELD

It should be emphasized that we have at our disposal a very accurate method for measuring the body's total aerobic metabolic rate in the determination of the oxygen uptake. On the other hand, good methods for a quantification of the anaerobic energy yield are presently not available.

It is the availability of oxygen in the cell that determines the extent to which the metabolic processes can proceed aerobically and anaerobically. In other words, the procedure depends on whether the cytoplasmic $NADH_2$ can be reoxidized in the mitochondria at the same pace as it is formed. At rest and during moderate exercise, the oxygen supply is sufficient and the energy metabolism is essentially aerobic. The ATP concentration is high and the ADP concentration is low. With increasing severity of exercise, ADP accumulates, the breakdown of glycogen speeds up, and the reduction of NAD is correspondingly faster. At some critical intensity, the oxygen-transporting system cannot provide enough oxygen to the cells, and part of the pyruvic acid which is formed must act, in addition to oxygen, as a hydrogen acceptor. Under these conditions, some of the cytoplasmic coenzyme $NADH_2$ is reoxidized anaerobically by pyruvate, which is transformed to lactic acid, and the rest of the cytoplasmatically formed $NADH_2$ is reoxidized in the mitochondria in the aerobic process. When the work is intensified further, an increasingly greater proportion of the $NADH_2$ (formed in the cytoplasm) will have to be reoxidized by pyruvate, and consequently a larger part of the work is done anaerobically. When the intensity of the work exceeds 100 percent of the maximal oxygen uptake, all further increase in intensity will have to be done anaerobically. Since anaerobic glycolysis furnishes one-thirteenth as much ATP per mole glucose-unit from glycogen as does aerobic oxidation, the glycolysis will have to be speeded up 13 times more for each percentage point of increase in work intensity at work loads above 100 percent of maximal oxygen uptake than at work loads below about 60 percent of maximal oxygen uptake.

At the very start of the exercise, even if the exercise intensity is below 60 to 70 percent of the maximal oxygen uptake, some anaerobic energy metabolism does take place while the oxygen supply to the working muscle cells is being adjusted to meet the actual demand, that is, during the first 45 to 90 seconds when the cardiac output is changing from a resting condition to the level required by the work load. After the cessation of exercise, the metabolic "generators" continue to run for a while to replenish the energy in an immediately accessible form and at the same time to get rid of the troublesome lactate that has accumulated during the work period.

It should be noted that one molecule of extra oxygen is consumed in the formation of about 6.5 ATP when glycogen is completely oxidized (6 O_2 will produce 39 ATP, or 6.5 ATP per O_2), but only 5.6 ATP when a fatty acid is oxidized (23 O_2 will produce 129 ATP, or only 5.6 ATP per O_2). When the oxygen supply to a muscle becomes limited during heavy exercise, glycogen contributes relatively more to the energy yield than does fat. This obviously gives a better utilization of the oxygen transported to the muscle.

Although it is the availability of oxygen in the cell that determines the extent to which the metabolic processes can proceed aerobically or anaerobically, the exact regulatory mechanism is unknown. However, the intracellular enzymes engaged in the aerobic metabolism can still operate maximally at oxygen tensions in the order of about 1 mm Hg. If this is a critical value, the capillary oxygen pressure must be higher than 1 mm Hg to secure an effective diffusion gradient in order to supply the cell with the oxygen needed for the aerobic processes to proceed. An increase in concentration of the phosphate ADP, and AMP, which may also be formed, will enhance the oxidation. The positive effects in various enzyme systems may not be linearly related to the ADP concentration. McGilvery (1975) has suggested that the rate of conversion of triosephosphates to pyruvate may vary as the square, and the rate of oxidative phosphorylation as the cube of the ADP concentrations in its low ranges. It means that high concentrations of ADP are needed for a strong stimulation of the glycogenolysis. It should be emphasized that the variations in ADP concentrations are modest in absolute terms, but with a reduction in the concentration of phosphocreatine to one-tenth the initial level (the ATP concentration may be reduced to 0.7 of its resting level), there is a 20-fold change in the concentration of ADP and a nearly 700-fold increase in the concentration of AMP at a constant pH (McGilvery, 1975). The effects of such dramatic changes are difficult to predict.

When oxygen is not available in sufficient amount, which is particularly true at the beginning of heavy exercise, the glycogenolysis is, so to speak, the only alternative to provide energy for the regeneration of ATP. It probably takes seconds before the oxidative system in the mitochondria can be thrown into full swing; it takes time for the ADP that is accumulating in the cytoplasm to diffuse into the matrix of the mitochondria.

RELATIVE IMPORTANCE OF THE DIFFERENT ENERGY STORES

A given weight of an organic compound contains a fixed amount of potential energy locked up in the bonds between the atoms of its molecules. Knowing the amount of such available compounds within the body and their energy content gives us the energy stores of the human machine. Table 2-2 summarizes data which, however, are approximate and subjected to large individual fluctuations, particularly the content of fat. In spite of these individual differences, some general statement can be made about the relative importance of the different compounds. Their specific importance depends heavily on both the intensity and the duration of the work. During maximal exercise the energy demand may exceed 200 kJ (50 kcal) min^{-1}. The supply from a breakdown of all the ATP available would cover only a maximal effort of about one second, and the generation of ATP from the breakdown of all the phosphocreatine would cover another few seconds of maximal effort (Bergström et al., 1971). It is well established that maximal speed can be maintained for less than 10 seconds, that is, for less time than it takes to run the 100-m dash, and the explanation may be that "rapid energy" is no longer available because of an exhaustion of phosphocreatine and eventually also of ATP. In addition to the

Table 2-2 The figures are very approximate. The glycogen concentration can be anything from almost zero up to 250 mmol kg^{-1}, and certainly the fat content is subject to large variations. It is assumed that only part of the muscle mass is activated.

	Energy mole^{-1}		Concentration	Total energy in humans	
	kJ	kcal	mmol kg^{-1} wet muscle	(body weight 75 kg, muscle weight 20 kg) kJ	kcal
ATP	42	10	5	4	1
Phospho-creatine	44	10.5	17	15	3.6
Glycogen	2900	700	80	4600	1100
Fat	10,000	2400	-	300,000	75,000

energy coming directly from the ATP and phosphocreatine stores, some of the energy in a 100-m dash will come from the glycogenolysis which is rapidly speeding up during the effort, giving rise to increasing amounts of lactate.

Lactate is not normally considered a form of stored energy. However, when a certain amount of anaerobic work has been done, the concomitant production of lactate is by no means wasted. If the intensity of the work is reduced to aerobic conditions, lactate is readily converted back to pyruvate in the working muscles and can be oxidized in the mitochondria, replacing glycogen as fuel. If, on the other hand, the anaerobic work is followed by rest, the lactate-via-pyruvate is converted back to glycogen in the liver (the Cori cycle), and probably also in the muscles themselves (Vaage et al., personal communication, Fig. 2-7).

The total amount of energy that can be derived from the ATP, phosphocreatine, and lactate stores is of limited importance when the work period exceeds 15 to 30 min. In this situation the energy demand may be in the order of 20 to 40 kJ · min^{-1}, and it is observed that both the ATP and the phosphocreatine concentrations are only moderately lowered. The levels of lactate are also modest compared with what is found in maximal work. It is therefore evident that during prolonged work, the rapid and continuous energy production from the oxidation of glycogen and fatty acids is extremely important. The rather complicated regulation of the ratio between the utilization of these two main fuels will be discussed at length in Chap. 14.

Whereas maximal work of short duration in essence depends only on the ATP and phosphocreatine stores, and whereas prolonged exercise depends only on the oxidation of glycogen and fat (free fatty acids), exercise with duration from 1 to 10 min is much more complex from the viewpoint of fuel utilization. When exercise is performed to exhaustion within this time interval, probably all fuel stores are utilized at the same time, but the relative amount of each fuel changes from second to second. At the start of vigorous exercise, utilization of ATP and phosphocreatine is predominant; then the anaerobic

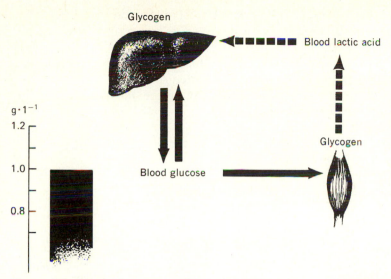

Figure 2-7 Blood glucose can form glycogen in the liver and the muscles (enzyme hexokinase at the first step) but can be released only from the liver into the blood because of a lack of the proper enzyme for this formation in the muscles (glucose-6-phosphatase). The formation of glycogen is induced by insulin and by a rise in blood sugar. The formation of glucose is induced by epinephrine and glucagon, and a formation of glucose from glycogen is stimulated by a fall in blood sugar level.

conversion of glycogen to lactate takes over more and more, and toward the end of the exercise, the oxidation of glycogen and eventually of fat will predominate.

SUMMARY

High-energy phosphate compounds represent the common currency for the transfer of energy within the living organism. The ATP-ADP system is the primary carrier of chemical energy in each and every cellular reaction. As quickly as ADP is formed, it is rephosphorylated, and the metabolic generation of ATP provides a general mechanism for the coupling of energy-yielding and energy-requiring processes. If the process is anaerobic, ADP is rephosphorylated by phosphocreatine, or, at the expense of glycogenolysis or glycolysis, lactate is formed. In aerobic conditions, ADP is rephosphorylated during oxidative phosphorylation by the mitochondria using glycogen, glucose, or free fatty acids as fuel.

In the absence of oxygen the skeletal muscles can work only for short periods of time, and the total energy available is very limited compared with the aerobic work situation. Theoretically, ATP and phosphocreatine could cover the energy demand alone, but only for a few seconds of heavy exercise.

Glucose has a slightly lower energy content than glycogen; in addition, ATP must provide energy to "lift" it into the glycogen-pyruvic acid system.

The fatty acids contribute significantly in the aerobic oxidation by entering the tricarboxylic acid cycle. The amino acids, after deamination, can enter this cycle via pyruvate or acetyl-CoA and can be completely oxidized or synthesized to glycogen or fat. As a fuel for muscular contractions, however, the oxidation of proteins normally plays a very limited role.

The main steps in the energy exchange in the muscle cell can be summarized in the following way:

$$
\begin{array}{lll}
& (1) & \text{ATP} \rightleftharpoons \text{ADP} + P_i + \text{free energy} \\
\text{Anaerobic} & (2) & \text{Phosphocreatine} + \text{ADP} \rightleftharpoons \text{creatine} + \text{ATP} \\
& (3) & \text{Glycogen or glucose} + P_i + \text{ADP} \rightleftharpoons \text{lactate} + \text{ATP} \\
\text{Aerobic} & (4) & \text{Glycogen and free fatty acids} + P_i + \text{ADP} + O_2 \rightarrow \\
& & \qquad\qquad\qquad\qquad\qquad CO_2 + H_2O + \text{ATP}
\end{array}
$$

In this schematic presentation, all quantitative aspects and the efficiency of the processes are disregarded.

REFERENCES

Baldwin, E.: "Dynamic Aspects of Biochemistry," 5th ed., Cambridge University Press, New York, 1967.

Bergström, J., R. C. Harris, E. Hultman, and L. O. Nordensjö: Energy Rich Phosphagens in Dynamic and Static Work, in B. Pernow and B. Saltin (eds.), "Muscle Metabolism during Exercise," p. 341, Plenum Press, Plenum Publishing Corporation, New York, 1971.

Conn, E. E., and P. K. Stumpf: "Outlines of Biochemistry," 3d ed., Wiley International Edition, New York, 1972.

Kaplan, N. O., and T. L. Goodfriend: Role of the Two Types of Lactatic Dehydrogenase, *Adv. in Enz. Reg.*, **2:**203, 1964.

Lehninger, A. L.: "Bioenergetics," W. A. Benjamin, Menlo Park, Calif., 1973.

Lehninger, A. L.: "Biochemistry," 2d ed., Worth Publisher, New York, 1975.

McGilvery, R. W.: The Use of Fuels for Muscular Work, in H. Howald and J. R. Poortmans (eds.), "Metabolic Adaptation to Prolonged Physical Exercise," p. 12, Birkhäuser Verlag, Basel, 1975.

Needham, D. M.: "Machina Carnis. The Biochemistry of Muscular Contraction in Its Historical Development," Cambridge University Press, Cambridge, 1971.

Sjödin, B.: Lactate Dehydrogenase in Human Skeletal Muscle, *Acta Physiol. Scand.*, Suppl. 436, 1976.

Muscle Contraction

CONTENTS

Chapter 3

Muscle Contraction

ARCHITECTURE

The previous chapter dealt with the principles of the energy-yielding reactions. Now we shall turn to the question of how chemical energy is transformed into mechanical work and describe the mechanical design of the muscle.

Standard textbooks of anatomy and histology (e.g. Bloom and Fawcett, 1975) or more specialized texts should be consulted for detailed descriptions of the muscle (Bourne, 1973; Carlson and Wilkie, 1974; Fuchs, 1974; Needham, 1971). In this review, we shall simply present a summary of the architecture of the muscle tissue, with some details of the working unit within the muscle. Of the three types of mammalian muscles—smooth muscles, heart muscle (striated involuntary muscle), and skeletal muscles (striated voluntary muscles)—only the skeletal muscle tissue will be discussed here.

The *muscle groups*, known by their Latin names, such as the brachialis, are groups of muscle bundles that join into a tendon at each end. On the outside the muscle is covered by a fascia of fibrous connective tissue known as the *epimysium*. Each bundle is separately wrapped in a sheath of connective tissue called *perimysium*. The bundle is made up of thousands of muscle fibers, each embedded in a fine layer of connective tissue (*endomysium*) (Fig. 3-1). The various sheaths of connective tissue blend with the tendon in a way that is determined by function and space.

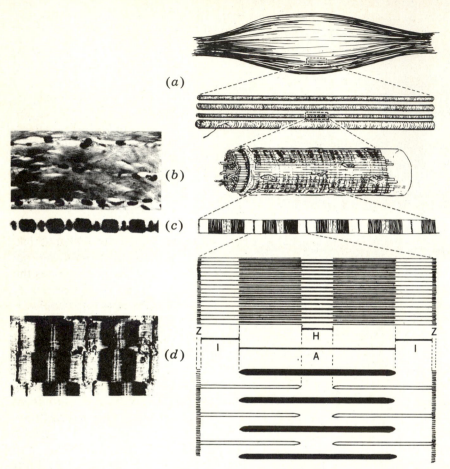

(a)

(b)

(c)

(d)

Figure 3-1 Schematic drawing of striated muscle (right) with corresponding photomicrographs (left). The striated muscle (a) is made up of muscle fibers (b), which appear striated in the light microscope. The small branching structures at the surface of the fibers are the "endplates" of motor nerves, which signal the fibers to contract. Each single muscle fiber (c) is made up of myofibrils, beside which lie cell nuclei and mitochondria. In a single myofibril (d), the striations are resolved into a repeating pattern of light and dark bands. A single unit of this pattern consists of a Z line, than an I band, than an A band which is interrupted by an H zone, then the next I band, and finally, the next Z line. This repeating band pattern is due to the overlapping of thick and thin filaments (bottom part of diagram). (Redrawn from H. E. Huxley, 1958.)

The functional unit within a muscle is the group of muscle fibers innervated by a single motor nerve fiber, the *motor unit.* The individual muscle fiber is, however, anatomically separated from the neighboring fibers by the endomysium.

The amount of connective tissue (collagen fibers, elastic fibers, and other cells) varies in different muscles and in different species of animals. In humans, the number of muscle fibers in a muscle group probably is finally established after the embryo has reached the age of four to five months (MacCallum, 1898). However, the thickness of a fiber can vary. At birth the fiber is about twice as thick as in the fourth fetal month, but has only one-fifth the adult thickness (Lockhart, 1973). (See also Chap. 12.)

The skeletal muscle fiber is a cylindrical, more or less elongated cell. Its thickness varies in different muscles or even in the same muscle, and may be from 10 to 100 μm. In many muscles, the length of the individual cell extends all the way from the tendon of origin to the tendon of insertion. Whether this is so in the longest muscles, such as the sartorius, is not clear. However, cells more than 30 cm long have been traced in this muscle. The muscle cell is multinucleated, with sometimes as many as several hundred nuclei in a single fiber.

Other elements in the cell are the sarcolemma, myofibrils, and sarcoplasm. The *sarcolemma* is a thin, elastic noncellular membrane, less than 10 nm thick, enveloping the striated muscle fiber. Its structure is very similar to the internal membranes of other cells (nerve cells, Schwann cell, etc.). In some places the sarcolemma has tunnels, cavelike invaginations, or open vesicles. These structures are morphological manifestations of active transport mechanisms. The sarcolemma also has remarkable electrical properties.

The contractile element of the cell consists of the *myofibrils.* The structure and function of these fibrils have been extensively studied and described by A. F. Huxley and H. E. Huxley and coworkers (reference papers in Cold Spring Harbor Symposia on Quantitative Biology, 1972; A. F. Huxley, 1974). Methods which have been used in the study of these structures have ranged from electron microscopy and small-angle x-ray diffraction to light microscopy. Furthermore, biochemical studies of the protein components of skeletal muscles have contributed to the present understanding of muscular function.

In each muscle fiber (or cell) there are many myofibrils, each 1 to 3 μm thick, arranged parallel to one another. The individual myofibrils are aligned within the sarcolemma so that points with the same density lie at the same level, giving the appearance of disks crossing the whole thickness of the muscle fiber. This arrangement is illustrated in Figs. 3-1 and 3-2. Each repeat is called a *sarcomere* and is bordered by a narrow membrane called the Z line, which, like a disk, crosscuts the myofibril in units. In the middle region of the sarcomere, there is a dark band, the A band (detected by using the deep-focusing position on the microscope). A stands for anisotropic, which refers to the optical property of the tissue. The alternate light bands are isotropic, or I bands. The Z line is located in the middle of the I band. The region in the center of the A band is called the H zone, and is less refractile (lighter) than the rest of the A band. In the center of the H zone there is a darker structure, called the M line.

This terminology may be summarized as follows (Fig. 3-2): Two bands of protein rods or filaments are distributed in the myofibrils in parallel order. Filaments of *myosin* (about 1.6 μm in length and 15 nm thick), arranged lengthwise, fill the A band. The other band of filaments consists of the protein

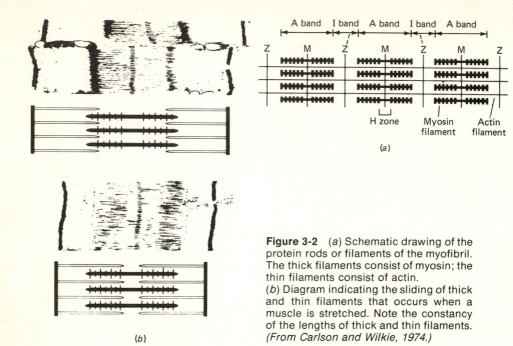

Figure 3-2 (*a*) Schematic drawing of the protein rods or filaments of the myofibril. The thick filaments consist of myosin; the thin filaments consist of actin.
(*b*) Diagram indicating the sliding of thick and thin filaments that occurs when a muscle is stretched. Note the constancy of the lengths of thick and thin filaments. *(From Carlson and Wilkie, 1974.)*

actin. They are thinner (about 6 nm), and run from the area of the *Z* line into the *A* band as far as the beginning of the *H* zone. The length of the actin filaments is about 1.0 μm. on each side of the *Z* line. Evidently the *I* band is occupied only by thin filaments, the *H* band only by the thicker filaments, and the outer parts of the *A* band by both. A cross section of a muscle fiber in the overlapping zone reveals a regular pattern. The thick filaments lie about 45 nm apart with thin filaments in between so that each of the thin ones is "shared" by three thick filaments (Figs. 3-2, 3-3).

The Sarcoplasm   The filaments just described are embedded in a fluid in which there are soluble proteins (such as myoglobin), glycogen droplets, fat droplets, phosphate compounds, other small molecules, and ions. This aqueous phase is called the *sarcoplasmic matrix.*

A fraction of the sarcoplasm is called the *sarcoplasmic reticulum.* It consists of elaborately anastomosing tiny tubular sacs, vesicles, and channels of varying caliber extending in the spaces between the myofibrils, resembling a lace cuff or sleeve (Fig. 3-4). The membranes bounding the channels and vesicles are similar in structure to the sarcolemma. Adjacent to the *Z* lines are transversely oriented compartments or cisternae (*terminal cisternae*) running across the myofibrils in close contact with a tubular system (the *T system*). The excitation of a muscle moving along the sarcolemma of the fibril can penetrate into the fibril via this tubular system. Actually, the *T* system starts as inward extensions of the sarcolemma at each *Z* line. They can therefore conduct

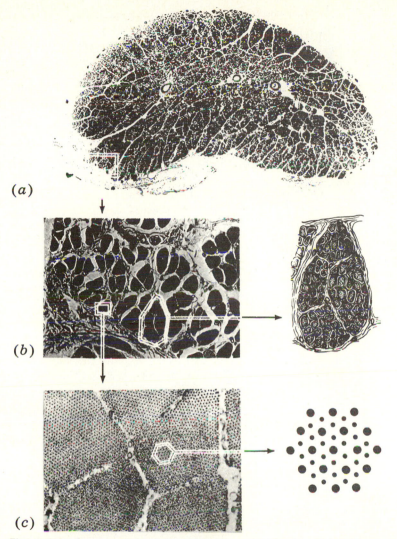

Figure 3-3 Cross section of (a) striated muscle. The larger magnification shows (b) the individual fibers and (c) the fibrils with the very regular pattern of thick and thin filaments.

electrical messages and also serve as inlets through which fluid outside the cell can flow into the cell. The sarcoplasmic reticulum can also transport lactic acid to the surface of the cell for further transport by the bloodstream to the liver and other tissues. In other words, the sarcoplasmic reticulum is the plumbing and fueling system of the muscle, with structures which support, control, regulate, and excite the contractile material itself, the actomyosin complex. As we shall see, Ca^{2+} ions play a key role as trigger of the actin-myosin interaction. We shall now present a more detailed description of the working units.

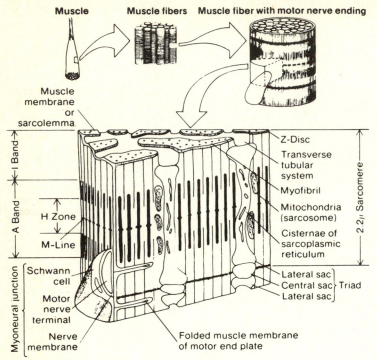

Muscle   Muscle fibers   Muscle fiber with motor nerve ending

Figure 3-4 Enlarged section of a single striated muscle fiber showing the myofibrillar system, the transverse tubular system, the sarcoplasmic reticulum, and the myoneural junction. *(From Carlson and Wilkie, 1974.)*

MYOFIBRILLAR FINE STRUCTURE

The basic contractile components of the muscle fiber are assembled by four proteins into the two multimolecular aggregates, the mentioned thick myosin and thinner actin filaments. The four proteins are *myosin, actin, tropomyosin,* and *troponin.* Neither protein, by itself, is contractile. In vitro, however, myosin and actin can, under certain conditions, form a complex protein, actomyosin, which can contract. The individual *myosin* molecule (about 50 percent of the muscle protein) resembles a golf club with a "head" (of about one-sixth the total length), and a long shaft or "tail." The site responsible for its enzymatic activity (see below) and its affinity for actin is located in its globular head (of heavy meromyosin), and the sites responsible for its affinity to other adjacent myosin molecules are in its tail (of light meromyosin). Several hundred such molecules are packed in a sheaf with their heads pointed in one direction along half the filament, and in the opposite direction along the other half, leaving a projection-free region midway along their length (Fig. 3-5). Thus, from this cigar-shaped structure the heads project out toward the thin filaments, and these cross-bridges are the only structural and mechanical linkages between the thick and thin filaments. The projections, resembling barbs, are arranged in pairs, each being rotated by about 120° from the preceding pair. In this "screw" there will be six projections in a period of about 43 nm, representing one turn (Fig. 3-5).]

According to H. E. Huxley (1969), the connecting region between the two meromyosin segments may form a flexible "hinge" which allows the head or cross-bridge to rotate outward from the myosin filament backbone. A second hinge region may be located at the junction between the globular terminal of the head myosin molecule (subfragment S-1) and the rest of the molecule. A rotation in the first hinge can bring S-1 into close proximity to the thin filament, and a rotation at the second hinge, that of the S-1, can generate the force or power stroke of the cross-bridge cycle (Fig. 3-6a). Probably the head portion is divided into two S-1 fractions, and so-called light peptide chains are essential components for the ATPase activity and the ability to bind actin. These light chains vary among species, between adult and fetus, and between the two fiber types in skeletal muscles (Tonomura, 1973).

The myosin molecule also contains two heavy chains which are wound around each other, and which are folded at one end into the globular structures that are part of the head. The *M* line, mentioned above, may represent protein which serves to hold the bundles of myosin molecules together.

The other three major proteins involved in the muscular contraction are all incorporated in the thin filament. *Actin* constitutes 20 to 25 percent of the myofibril protein and is the main component of the thin filament. It has the form of a double helix, consisting of two chains of roughly globular subunits (monomers) twisted around each other like two strings of beads (Fig. 3-5). Apparently the actin monomers have the potential to interact in identical fashion with a given myosin cross-bridge. The actin filaments (like the myosin filaments) show a structural polarization—their molecules are assembled into the filaments in a front-to-back manner. A reversal of polarity occurs on either side of the *Z* line (there are interconnections of thin filaments at the *Z* line). With this arrangement of actin and myosin molecules in the two halves of an *A* band, we can expect the actin filaments on either side of the sarcomere to move in opposite directions, that is, toward one another in the middle of the sarcomere.

The two other proteins serve a regulatory function in the making and breaking of the contacts between the thick and thin filaments during contraction. *Tropomyosins* are long polypeptide molecules that attach end to end, forming two very thin continuous strands running along the length of the actin filament. These rod-shaped molecules, about 40 nm long, lie along each of the two grooves in the actin double helix, each tropomyosin molecule being in contact with seven actin monomers (Fig. 3-5).

Finally, we have the protein *troponin*. It has a more or less globular shape and sits astride the tropomyosin molecule close to one of its ends. In the presence of one molecule of tropomyosin, the troponin can regulate the activity of about seven actin monomers. As we shall see, it can switch the actin filament on or off.

The troponin molecule contains three subunits: one is a calcium-binding protein, TN-C or TpC; one is an inhibitory protein, TN-I or TpI, for it can in one way or the other inhibit the interaction of actin with the myosin cross-bridges; the third subunit, TN-T or TpT, is a protein which binds very strongly to tropomyosin.

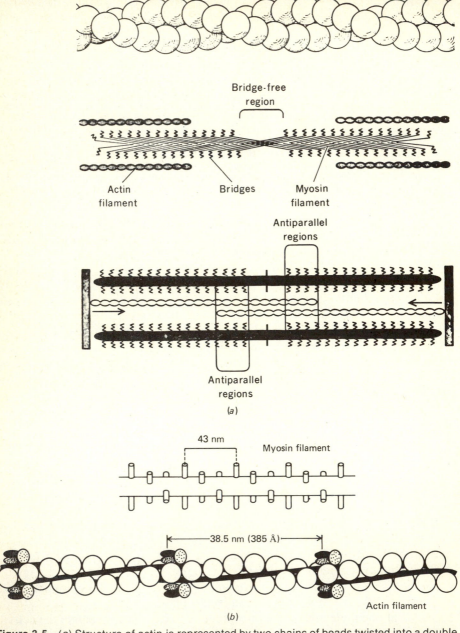

(a)

(b)

Figure 3-5 (*a*) Structure of actin is represented by two chains of beads twisted into a double helix (top). The contact of actin and myosin might be made in the manner schematically illustrated in the middle part of the figure. The thin actin filaments at top and bottom are so shaped that certain sites are closest to the thick myosin filament in the middle. The heads of individual myosin molecules (zigzag lines) extend as cross-bridges to the actin filament at these close sites. At the bottom of the figure is shown double overlap of thin filaments from each side of the sarcomere, which would result if the sliding-filament hypothesis is correct.

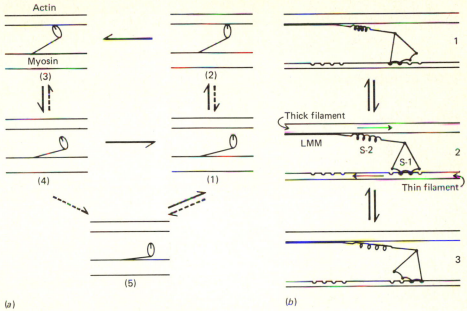

Figure 3-6 (a) A schematic diagram of the physical events thought to take place during the contraction of a striated muscle. In step 1 the myosin cross-bridges are separated from the actin; step 2 is symbolic of attachment at a specific orientation, which then changes to that indicated by step 3, generating tension, and for a submaximal load, shortening. Step 4 represents the cross-bridge redissociated from actin, presumably by the binding of ATP. Step 5 represents the extremely well-ordered cross-bridge array that gives rise to the live, resting myosin x-ray pattern. This last may be only a more ordered form of step 1. *(From Lymn and Huxley, 1972.)*

(b) Development from (a) proposed by A. F. Huxley and Simmons to incorporate the elastic and stepwise-shortening elements for which they gave evidence derived from tension transients. The strength of binding of the attached sites is higher in position 2 than position 1, and in position 3 than position 2. During isometric contraction the myosin head oscillates rapidly between its three stable positions. The myosin head can be detached from position 3 with the utilization of a molecule of ATP; this is the predominant process during shortening. During stretch, the myosin head can dissociate from position 1 without utilization of ATP. *(From A. F. Huxley, 1974.)*

Altogether, the thin filament is built of some 300 to 400 thin molecules and about 40 to 60 molecules of tropomyosin and troponin. (Apparently the thick filaments are of similar length in different species but, in contrast, the thin filaments may have varied length; Close, 1972.)

The tension would fall when thin filaments cross the center and interact with improperly oriented cross-bridges. *(From H. E. Huxley, 1965.)*

(b) Diagrammatic representation of the two filaments of the myofibril. On the myosin filament the heads, with the potential to form cross-bridges with the actin filaments, are arranged in pairs at regular intervals, each being rotated by about $120°$ from the preceding pair. The troponin complex is assumed to consist of equimolar amounts of the calcium-binding protein (black), the inhibitory protein (cross-hatched) and troponin-T (stippled). The tropomyosin (dark line) is represented as lying in each groove of the actin filament. *(From Ebashi et al., 1969.)*

Figure 3-5 summarizes in a schematic way the structure of the myofibrils with the thick myosin filaments and the thin actin filaments with the "regulatory proteins" troponin and tropomyosin attached in a regular pattern.

INTERACTION OF ACTIN AND MYOSIN

It should be emphasized that many details in the function of the muscles are still unrevealed and that there are various hypotheses about some of the mechanisms. So far, it is generally agreed that changes in the length of striated muscle take place predominantly by relative sliding movements of the two sets of filaments, the myosin and actin filaments (Ebashi et al., 1969; Needham, 1971; Weber and Murray, 1973; A. F. Huxley, 1974). During the sliding movements of the arrays of thin filaments inward into the arrays of thick filaments, when the muscle shortens, the length of the filaments remains constant. It is therefore evident that the length of the *A* band also remains constant (Fig. 3-2), but the *I* band becomes narrower, and eventually it will disappear. With the actin filaments attached to the *Z* lines, they will be drawn together during the myosin-actin interaction, and the sarcomere will shorten. In an isometric muscular contraction (which causes no change in length), the length of the *A* and *I* bands remains constant, but when lengthening the muscle, the *I* band will become broader.

At Rest

None of the cross-bridges is attached to the actin filaments, and the "bards" are probably located very near the myosin "backbone," tilted in one of their hinges. It seems as if, in the resting muscle, the tropomyosin rods lie toward the edge of the thin filament groove, and in this position they can block the actin sites which otherwise would react with the cross-bridges. With this organization of the contractile proteins into separate actin and myosin filaments, the resistance to passive extensibility is very modest.

Excitation

(1) The motor nerve stimulates the muscle, and the propagated action potential depolarizes the muscle cell membrane (see Chap. 4). There is then an inward spread of action potential along the T system.

(2) This event will make the membrane of the terminal cisternae of the sarcoplasmic reticulum permeable for Ca^{2+} (see Ebashi, 1976). At rest, the sarcoplasm surrounding the myofibrils is almost free from Ca^{2+} (concentration $< 10^{-7}M$). Ca^{2+} ions are quickly released from stores in the sarcoplasmic reticulum, and they become bound to the troponin on the thin filament (actually, two Ca^{2+} ions attach themselves to one troponin molecule).

Contraction

(3) This binding of calcium to troponin causes a change in the troponin-tropomyosin-actin complex which will remove the inhibition for an interaction

between the myosin head and actin. There is evidence (H. E. Huxley, 1972) that the tropomyosin rods are now moved from their blocking positions. In other words, with the binding of Ca^{2+} to troponin, the tropomyosin strands are drawn toward the center of the groove of the actin filaments, thus allowing actin to react with myosin; the actin monomers are now released from the preexisting inhibitory influence of the troponin-tropomyosin complex.

(4) The heads of myosin molecules move out perpendicularly from the thick filament core toward the actin filament and attach (the S-1 subunits) to the actin molecules within reach. The cross-bridges are "energized" by Mg-ATP (or its intermediate), which is bound to the myosin head.

(5) The heads undergo an energy-yielding conformational change, so that the cross-bridges change their angular relationship to the axis of the myosin core: the actin filaments are moved or pulled along the myosin filaments. This is the power stroke of the myosin heads as they flip into a different conformational state (Fig. 3-6a).

(6) The ATP (or its intermediate) becomes hydrolyzed, and the ADP and free phosphate will leave the binding sites on the myosin head.

Relaxation

(7) Fresh ATP is taken up by the myosin head, which will promptly dissociate actin from myosin. Ca^{2+} is released from the troponin and transported across the membrane back into the cisternae of the sarcoplasmic reticulum [eventually also into mitochondria (Carafoli, 1974)]. This transport is energy-consuming, requiring hydrolysis of ATP (one ATP for transport of two Ca^{2+}). The tropomyosin again changes its position relative to the actin subunits and inhibits actin from interaction with the cross-bridges. The cross-bridges switch back to their original conformation.

In Summary Ca^{2+} is the link between excitation and contraction. When a nerve signal arrives at the muscle cell, it causes a release of calcium into the fluids surrounding the filaments from special storage vesicles in the sarcoplasmic reticulum. By combining with troponin, it initiates the removal of a hindrance for an interaction between actin and myosin filaments. Tropomyosin, which in the resting muscle has turned off the active site of actin, will now switch it on. This allows myosin heads to attach to sites on the actin filament. An attached bridge, during its action, exerts a longitudinal force for a certain distance in which probably one molecule of ATP is split, and that pulls the actin filament along toward the center of the A band. The system is switched off when Ca^{2+} is rebound by the sarcoplasmic reticulum. In the absence of calcium, the troponin-tropomyosin complex again prevents the interaction between actin and myosin filaments.

These events are repeated as long as the muscle is stimulated, and the cross-bridges attach, swivel, and detach cyclically, thus propelling the thin filaments past the thick ones shortening the muscle. The active actin site may successively react with several linearly arranged myosin groups, and the actin filament travels alongside the myosin filament. If the muscle is working isometrically, the same molecular groups may react with one another repeatedly, but eventually the actin filament will slip back if enough tension cannot be generated. It is easy to understand why troponin and tropomyosin are called regulatory proteins.

Murray and Weber (1974) present an analogy:

> The process of liberating the energy of ATP hydrolysis can be likened to the firing of a gun. The gun must first be loaded by placing an appropriate cartridge (ATP) in a specific chamber (the myosin head). This combination (the myosin-ATP) is converted to a special metastable form by cocking the gun (the second step in the hydrolysis). If the cocked gun (or charged form) is left alone, it is more or less stable. If the trigger is squeezed (or if an actin molecule is available), however, the stored energy is rapidly released and work is done on a bullet (or a cross-bridge). The process is completed by ejecting the spent cartridge (the products of hydrolysis, ADP and phosphate) and reloading.

In the closing remark at the symposium on "The Mechanism of Muscle Contraction" at Cold Spring Harbor in 1972, A. F. Huxley stated: "However, there is a very large gap in our present knowledge, unfortunately right at the heart of the whole problem. It concerns the nature of the structural changes in the myosin head accompanying the different stages of ATP-splitting which are thought to produce the changes in the effective angle of attachment to actin during the operation of the cross-bridge."

According to Huxley (1974), the filaments themselves are stiff, but within each cross-bridge there is an elastic element in series with a nonlinear element. The attachment of a cross-bridge to actin is followed by a movement in a small number of steps (Fig. 3-6). The total range of movement may be 10 to 12 nm, but part of this movement (some 5 nm) will stretch the elastic element even in an isometric contraction. By this mechanism, some of the yielded energy is stored in a springlike fashion, and then the extended spring may shorten and give out its stored energy.

It is assumed that the tension generated by an attached cross-bridge can vary in a series of steps depending on the affective angle of attachment of the S-1 subunit to actin and the degree of extension of the elastic S-2 linkage (connecting S-1 to the light-meromyosin backbone of the thick filament). The attachment of bridges may be the rate-limiting step that controls the speed of shortening at moderate loads. A consequence is that the number of attached cross-bridges at any particular moment decreases as the velocity of shortening increases. Even during maximal activation only part of the cross-bridges are effectively linked to sites on actin (less than 50 percent?), while the remaining move randomly about in the interfilament space.

Apparently the equilibrium position of a bridge is sensitive to changes in its environment, and there may be "messages" passed between neighbouring bridges affecting the position and eventual movements, even if the heads are not in a position to interact with actin (H. E. Huxley, 1972).

It is not known which structure is the elastic element in the cross-bridge, what structure undergoes the stepwise change, what kind of bonds hold the myosin head to the actin filament or how the binding of ATP causes myosin to dissociate from actin (A. F. Huxley, 1974). It was mentioned that one ATP is probably yielding energy for one working cycle of a cross-bridge (but it may cost two ATP, Tonomura, 1973). The energy demand of the transport of Ca^{2+} is an additional expense requiring ATP as an energy source. (For a discussion of the ATP hydrolysis and cross-bridge movement, see Fuchs, 1974.) In the resting muscle cell, ATP undergoes a very slow hydrolysis when bound to myosin. It may be partly reacting with myosin (M), forming a "charged" $M \cdot ADP \cdot P$ complex. It can then react with actin (A), which will markedly accelerate the ATP hydrolysis by combining with the $M \cdot ADP \cdot P$ intermediate. (Tropomyosin also exerts an activating effect on actomyosin ATPase.) The resting rate of ATP hydrolysis is probably less than 1 percent of the maximal rate during contraction. The sequence is:

$$A + M \cdot ADP \cdot P \rightarrow A\text{-}M \cdot ADP \cdot P \rightarrow A\text{-}M + ADP + P_i$$

Binding of ATP (plus Mg^{2+}) to the actomyosin complex is very rapid and the dissociation of the A-M complex follows also very rapidly:

$$A\text{-}M + ATP \rightarrow A + M \cdot ATP$$

and again we obtain a M $\cdot$ ADP $\cdot$ P intermediate.

Normally, there is plenty of ATP available to keep the system going, but if the concentration of Mg $\cdot$ ATP is very low, the cross-bridges will remain attached to actin, and the muscle is brought into sustained contraction (rigor). In this situation the cross-bridge is apparently locked at the end of the working stroke, the point at which the next ATP molecule normally comes along to dissociate the bridge.

When the muscle shortens, the ends of the thin filaments will slide toward one another in the center of the *A* band, and they may even overlap (see the bottom illustration in Fig. 3-5). (A dense zone appears in the center of the *A* band, and in a transverse section of this zone, an electron micrograph shows twice as many thin filaments as in a relaxed muscle, proving that actin filaments may overlap during contraction.) This may explain the observed decrease in maximal tension generated by a muscle as it shortens. The projections are absent at the center of the thick filaments, and as the thin filaments from one *Z* line continue into the "wrong" part of the *A* band, the orientation of the molecules becomes abnormal from a functional viewpoint. In this region, the filaments would not be expected to contribute to the development of tension by the muscle, and they may even interfere in a negative way with the interaction of the correctly oriented actin and myosin molecules.

It remains to be explained why a muscle can develop its maximal tension as it is stretched beyond its resting length. This stretch does diminish the distance of overlap between the thick and the thin filaments, and therefore the more they slide apart, the fewer cross-bridges between the two types of filaments are possible. During a variation in the sarcomere length, there is also a variation in the actin-to-myosin distance. In toad muscle, it has been estimated that the surface-to-surface distance between actin and myosin was about 5 nm at sarcomere length 3.5 μm but 13 nm at sarcomere length 1.8 μm (Elliott, 1967). It could well be that there are mechanical disadvances in generating force with the increasing distance between the actin and myosin filaments as the muscle shortens. The flexible cross-bridges can extend from the myosin backbone, thanks to their hinges, and reach the actin filaments despite an increase in the interfilament distance, but the leverage for the interaction must be different. When a muscle fiber is stretched so that the length of the sarcomere exceeds 3.5 μm, no tension can develop when the muscle is stimulated. The explanation is that the two sets of filaments have ceased to overlap; the reacting groups do not "reach" each other.

DIFFERENT FIBER TYPES

From a functional point of view, the muscle cells do not constitute a homogeneous tissue. Most muscles are built up of muscle fibers with different mechanical properties. There is a correlation between these properties and the histochemical, morphological characteristics of the fibers (Burke and Edgerton, 1975). There is presently a debate about classification and nomenclature (Close, 1972; Engel, 1974). Strong evidence shows that the quality of ATPase bound to myosin is rate limiting in the shortening process, and by exposing a muscle sample to buffers with different pH's and histochemical staining, one can identify two main groups of muscle fibers (Table 3-1).

The time course of an isometric twitch and the kinetic properties of the two fiber types differ markedly; hence, they have been called *slow twitch fibers* and *fast twitch fibers.* A more "neutral" nomenclature names them *Type I* and *Type II*, respectively. A low activity of myosin ATPase in the muscle fiber is usually associated with a relatively slow contraction time and the glycolytic enzyme system is less well developed, but there is a high mitochondrial content and thereby a high potential for oxidative enzyme activities. The cell is well adapted for prolonged work. With a high activity of myosin ATPase, the picture is reversed: the contraction time is relatively short, the fiber has a well-developed glycolytic enzyme system, the mitochondrial content and oxidative activity are lower, and the fiber fatigues rapidly. However, type II muscle fibers with high myosin ATPase activity may be divided into several subgroups (Type IIa, IIb, and IIc) on the bases of different susceptibility to treatment with different buffers prior to incubation (Brooke and Kaiser, 1970). Type IIa has a high oxidative potential and an intermediate glycolytic power. It is relatively resistant to fatigue. Type IIb is the "typical" fast twitch fiber with a low aerobic potential. Type IIc is most commonly considered as a relatively little differentiated fiber. In contrast to myosin ATPase, the metabolic enzymes

Table 3-1 Classification of human skeletal muscle fibers based on histochemical staining and some functional properties (subtypes not included). (Training can modify some of the properties, and the difference between "high" and "low" is not always large; see Chap. 12.)

Property	Type I Slow twitch fiber	Type II Fast twitch fiber
Myofibrillar ATPase activity (at pH 9.4)	Low	High
Mitochondrial enzyme activity	High	Low
Glycogenolytic enzyme activity	Low	High
Glycogen content	No difference	
Myoglobin content	High	Low
Capillary density	High	Low
Contraction speed	Low	High
Endurance	High	Low

seem to be influenced by the level of activity and may thus change in response to endurance training (see Chap. 12). It is quite possible that the proportions between the fibers of different subgroups within the type II family may vary at different times for a single individual. On the other hand, the proportions between type I and type II fibers seem to be a genetic matter. (This is a generalization, but it is not possible at present, on the basis of conventional histochemical methods, to describe accurately the kinetic properties of muscle fibers. It is also evident that one cannot directly apply data obtained on rat and cat muscles to the function of human muscles.)

Not only is there a difference in the myosin ATPase activity of the two fiber types, but the head of the myosin, like that of the tropomyosin, may also have a different structure. Apparently there are more differences in the myosin properties of fast and slow twitch fibers than in the actin molecules. There appears to be twice as much sarcoplasmic reticulum in the fast muscle (Type II) compared with the slow fibers (Type I). The consequence is supposed to be that the rate of uptake of Ca^{2+} by the sarcoplasmic reticulum and also by troponin is faster in the fast muscle fiber, which may promote a quick reactivation of the fiber.

Differences may exist in the structures of the Z line and M line between the Type I and Type II fibers. The ratio myosin/actin is apparently not different in the two types of muscle fibers, and the intrinsic strength of the contractile material, i.e., the maximal tension per unit cross-sectional area of the muscle, is about the same for the two muscles. The functional difference is evident mainly in the kinetic properties.

The differentiation in different fiber types seems to start after about 20 weeks in the human fetus (with maturation occurring at different times in different muscles in the same animal), but it is not established until after birth (Saltin, 1977). There is strong evidence showing that neural influences determine the fundamental dynamic properties of the contractile material; in other words, the nerve has a trophic influence on the contractile properties of the muscle. It has been shown that an operative cross-union of motor nerves to the two fiber types of muscles in juvenile or adult animals will lead to reciprocal changes in the twitch time course of a contraction. Muscle fibers formerly fast became slow, and the speed of contraction of muscle formerly slow was increased. These events were mirrored in changes in the myosin of the reinnervated muscle. The changed speed of the muscle contraction can be explained by a change in the kinetic properties of the myosin ATPase (Bullen et al., 1969; Close, 1972; Luff, 1975; and Chap. 4).

It should be mentioned that the increase in muscle length during growth is brought about mainly by an increase in the number of sarcomeres without marked change in the individual sarcomere length. The increase in the cross-sectional area of a muscle fiber during growth is the result of longitudinal splitting of myofibrils, and thereby of an increase in the number of myofibrils when they reach a critical size (Goldspink, 1970).

Additional aspects of the muscle function will be discussed in Chap. 4.

REFERENCES

Bloom, W., and D. W. Fawcett: "Textbook of Histology," 10th ed., Saunders, Philadelphia, 1975.

Bourne, G. H. (ed.): "The Structure and Function of Muscle," vols. 1–2, Academic Press, Inc., New York, 1973.

Brooke, M. H., and K. K. Kaiser: Muscle Fiber Types: How Many and What Kind?, *Arch. Neurol.*, **23**:369, 1970.

Bullen, A. J., W. F. H. M. Mommaerts, and K. Seraydarian: Enzymic Properties of Myosin in Fast and Slow Twitch Muscles of the Cat Following Cross-innervation, *J. Physiol.*, **205**:581, 1969.

Burke, R. E., and V. R. Edgerton: Motor Unit Properties and Selective Involvement in Movement, *Exercise and Sport Sci. Rev.*, **3**:31, 1975.

Carafoli, E.: Mitochondrial Uptake of Calcium Ions and the Regulation of Cell Function. *Biochem. Soc. Symp.*, **39**:89, 1974.

Carlson, F. D., and D. R. Wilkie: "Muscle Physiology," Prentice-Hall, Inc., Englewood Cliffs, N.J., 1974.

Close, R. I.: Dynamic Properties of Mammalian Skeletal Muscles, *Physiol. Rev.*, **52**:129, 1972.

"Cold Spring Harbor Symposia on Quantitative Biology," vol. 37, Cold Spring Harbor, N.Y., 1972.

Ebashi, S.: Excitation-Contraction Coupling, *Ann. Rev. Physiol.*, **38**:293, 1976.

Ebashi, S., M. Endo, and I. Ohtsuki: Control of Muscle Contraction, *Quart. Rev. Biophys.*, **2**:351, 1969.

Elliott, G. F.: Variations of the Contractile Apparatus in Smooth and Striated Muscles. X-ray Diffraction Studies at Rest and in Contraction, *J. Gen. Physiol.*, **50**:171, 1967.

Engel, W. K.: Fiber-Type Nomenclature of Human Skeletal Muscle for Histochemical Purposes, *Neurology*, **25**:344, 1974.

Fuchs, F.: Striated Muscle, *Ann. Rev. Physiol.*, **36**:461, 1974.

Goldsprink, G.: The Proliferation of Myofibrils during Muscle Fiber Growth, *J. Cell. Sci.*, **6**:593, 1970.

Huxley, A. F.: Muscular Contraction, *J. Physiol.*, **243**:1, 1974.

Huxley, H. E.: The Contraction of Muscle, *Sci. Am.*, **19**:3, 1958.

Huxley, H. E.: The Mechanism of Muscular Contraction, *Sci. Am.*, **213**:(6):18, 1965.

Huxley, H. E.: The Mechanism of Muscle Contraction, *Science*, **164**:1356, 1969.

Huxley, H. E.: Structural Changes in the Actin- and Myosin-containing Filaments during Contraction, *Cold Spring Harbor Symp. Quart. Biol.*, **37**:361, 1972.

Lockhart, R. D.: Anatomy of Muscles and Their Relation to Movement and Posture, in G. H. Bourne (ed.), "The Structure and Function of Muscle," vol. 1, p. 1, Academic Press, Inc., New York, 1973.

Luff, A. R.: Dynamic Properties of Fast and Slow Skeletal Muscles in the Cat and Rat Following Cross-reinnervation, *J. Physiol.*, **248**:83, 1975.

Lymn, R. W., and H. E. Huxley: X-ray Diagrams from Skeletal Muscle in the Presence of ATP Analogs, *Cold Spring Harbor Symp. Quart. Biol.*, **37**:449, 1972.

MacCallum, J. B.: On the Histogenesis of the Striated Muscle Fiber, and the Growth of the Human Sartorius Muscle, *Bull. Johns Hopkins Hosp.*, **9**:208, 1898.

Murray, J. M., and A. Weber: The Cooperative Action of Muscle Proteins, *Sci. Am.*, **230**(2):59, 1974.

Needham, D. M.: "Machina Carnis: The Biochemistry of Muscular Contraction in Its Historical Development," Cambridge University Press, Cambridge, 1971.

Saltin, A. S. (1977), see: Tomanekh, R. J., and A. S. Colling-Saltin: Cytological Differentiation of Human Fetal Skeletal Muscle, *Am. J. Anat.* (In press, 1977.)

Tonomura, Y.: "Muscle Proteins, Muscle Contraction and Cation Transport," University Park Press, Baltimore, 1973.

Weber, A., and J. M. Murray: Molecular Control Mechanisms in Muscle Contraction, *Physiol. Rev.*, **53**:612, 1973.

Neuro-muscular Function

4

CONTENTS

Chapter 4

Neuromuscular Function

The central nervous system (CNS) receives information concerning the outside world via *exteroceptors* as they react to light, sound, touch, temperature, or chemical agents, and *interoceptors* that are stimulated by changes within the body. The latter include *proprioceptors* (such as muscle spindles, tendon-end organs of Golgi, joint receptors, and vestibular receptors), *chemoreceptors*, and *visceroceptors* (Fig. 4-1). The CNS is equipped to receive, interpret, and handle information, and then eventually to transform the result into movements. Even the process of feeding information to the CNS may involve some muscular activity; for example, the eye muscles are almost continuously active, especially while the individual is awake. Reaction to various stimuli includes gestures, speech, writing, and sometimes very vigorous muscular contractions, as in the event of an approaching danger. There is scarcely any stimulus that does not, through reflexes, affect smooth muscle, heart muscle, or skeletal muscle.

A presentation of the physiology of work and exercise would justify a very detailed discussion of the function of the CNS. In fact, it is difficult to decide which aspects are more essential and which are less essential for the understanding of muscular movements, athletic techniques, and training. However, we shall limit our discussion mainly to the units more directly involved in the activation of the skeletal muscle. For a more comprehensive description of the CNS, the reader should consult basic physiological textbooks and reviews (Eccles, 1973b; Granit, 1970).

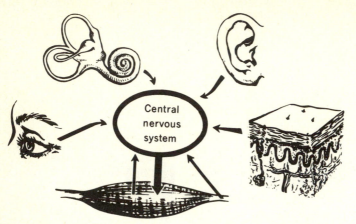

Figure 4-1 The central nervous system receives information via afferent nerves from the various receptors. Impulses are transmitted through the "final common path," which is the only available route to the skeletal muscle.

THE MOTONEURON AND TRANSMISSION OF NERVE IMPULSES

Anatomy

The task of the single nerve cell is to receive a message and then to pass it on to other cells; the message may go to a nerve cell, muscle cell, or cells of other types. Some nerve cells are specialized in "holding the message back" and preventing other cells from reacting to impulses (Fig. 4-2). It is a happy coincidence that the nerve cells that finally control the skeletal muscles (the *motoneurons*) are the most studied ones. Consequently, these events are fairly well understood.

Principally, the motoneuron in the spinal cord is similar to other nerve cells in its anatomy and physiology. The cell body, or *soma*, is approximately 70 μm across. A number of branching processes (*dendrites*) radiate from the cell, and these dendrites soon split up into terminal branches (Fig. 4-3). One long branch, the *axon*, or motor nerve, leaves the spinal cord in the ventral root and runs to the muscle, a distance that may be 90 cm in humans. A membrane (5 nm thick) covers the soma and the processes. Mitochondria in the cell indicate a high level of metabolic activity. In fact, the oxygen uptake of nerve cells is probably of the same order as that of maximally contracting skeletal muscles. In the nerve branches, fine threads, known as neurofibrils, run parallel to the axis and are bathed in the axoplasm. The membrane, or *axolemma*, constitutes a barrier between the intra- and extracellular fluid. In the peripheral part, the axon of the motoneuron is covered by a *myelin sheath.* This myelin sheath is interrupted at intervals by the nodes of Ranvier. (Numerous nerves in the CNS, the thinner nerve fibers in the somatic nervous system, and postganglionic fibers of the autonomic nervous system are unmyelinated.) For the nutrition of the dendrite and axon, an uninterrupted connection with the cell body is essential, and

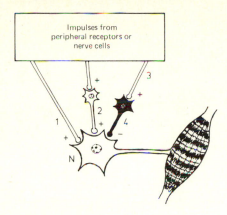

Impulses from
peripheral receptors or
nerve cells

Figure 4-2 The motoneuron, N, in the ventral horn of the gray matter of the spinal cord can be excited (+) directly (1) or via an interneuron (2). Thereby, an impulse is propagated in the nerve fiber, and the muscle is stimulated, causing muscular activity. However, other nerve terminals can prevent the motoneuron from responding to the exciting impulse. Schematically, this is illustrated as follows: Nerve end (3) stimulates an interneuron nerve cell (4). But this cell inhibits (−) the motoneuron's reaction to the stimulation by nerve 1. The mechanisms involved are discussed in the text.

axoplasm flows peripherally from the cell body. If the nerve fiber is cut, the peripheral part will degenerate.

Axonal terminals (*synaptic knobs*) of other nerves end at the surface of the soma and at the basal regions of the dendrites (Fig. 4-3). Actually, the surface of the soma, the dendritic stumps, and even the axonal origin may be almost completely covered by synaptic knobs from thousands of other nerve cells. The space between the membranes of the synaptic knobs and the contacted nerve cell (postsynaptic membrane), known as the *synaptic cleft*, is about 20 nm.

Resting Membrane Potential

In the nerve cells, as in other cells, the chemical composition of the fluid on each side of the semipermeable cell membranes is different. In its composition the external fluid medium is an ultrafiltrate of blood; i.e., it has a concentration of various particles very similar to a protein-free filtrate of blood plasma. However, the internal fluid contains a lower concentration of sodium and chloride ions and a higher concentration of potassium. These differences in concentration of ions would cause a diffusion across the membrane if perfusibility and electrochemical gradients could permit this. The membrane, however, does not permit a free passage of ions. The cell contains a high concentration of protein anions which cannot escape through the membrane into the fluid outside the cell where the concentration of protein is low. Electric charges of opposite sign attract each other, and the excess protein anions inside the cell therefore attract cations. The membrane is much less permeable to Na^+ than to K^+; the tendency for K^+ to diffuse out of the cell according to a difference in concentration gradient is counteracted by the attraction from the protein anions inside the cell. As long as the Cl^+ is in diffusional equilibrium, the outward flux of K^+ is slightly larger than the inward flux, and the diffusional inward flux of Na^+ would be many times greater than the outward flux. To maintain the difference in concentration of ions across the resting cell membrane, there must be a forced movement of Na^+ outward and K^+ inward,

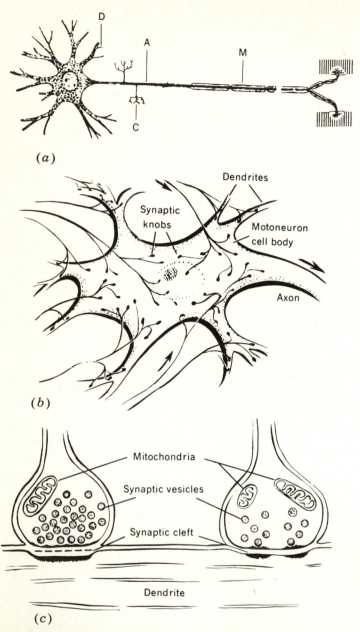

Figure 4-3 (a) Motoneuron: A, axon; C, collateral; D, dendrite; M, myelin. *(K. E. Schreiner and A. Schreiner, 1964.)*
(b) A motoneuron cell body and its dendrites covered with synaptic knobs of which only a few are included in the figure. These knobs are the terminals of the impulse-carrying nerve fibers (axons) from other nerve cells.
(c) When stimulated, the synaptic knobs deliver a chemical transmitter substance into the synaptic cleft, where it can act on the surface of the opposite nerve cell membrane. The chemical transmitter substance is stored in numerous vesicles. [(b) *and* (c) *from Eccles, 1965.*]

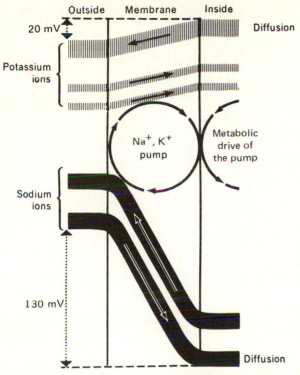

Figure 4-4 Schematic diagram showing the K^+ and Na^+ fluxes through the surface membrane of the nerve cell in the resting state. The slopes in the flux channels across the membrane represent the respective electrochemical gradients. The voltage drop across the surface membrane is about 70 mV, with the inside being negative. Under these conditions, chloride ions diffuse inward and outward at equal rates. But the concentration of sodium and potassium has to be maintained by some sort of a metabolic pump. The negative potential inside the nerve cell membrane is 20 mV short of the equilibrium potential for potassium ions. Therefore, the pump has to actively prevent an outward diffusion of potassium ions. For sodium ions, the potential across the membrane is 130 mV in the opposite direction. To keep the sodium out, a very active pump is needed. *(Modified from Eccles, 1965.)*

virtually equaling the diffusion in the opposite directions (Fig. 4-4). Actually, aerobic metabolic processes supply the energy to keep a sodium and a potassium pump going, whereby ions that diffuse or "leak" in or out of the cell are forced back.

This pump is driven by energy derived from a hydrolysis of ATP. The essential enzyme ATPase is in the cell membrane and it is highly activated by an increased Na^+ concentration *inside* the cell and an increased concentration of K^+ *outside* the cell; the ATPase is Na^+-K^+ dependent. Supposedly there are two carriers coupled together with binding sites for Na^+ and K^+, respectively. On the inner side of the membrane, Na^+ is picked up, and a rotation of the carrier may occur whereby K^+ is delivered to the interior of the cell. Then the carrier moves to the outer side of the membrane for the delivery of Na^+ and uptake of K^+. It is suggested that an internal transfer system may exist so that the binding sites do not have to move all the way across the membrane (Glynn and Karlish, 1975). There is a continuous leakage of Na^+ into the cell along its concentration gradient, and it is speculated that a carrier, traveling with Na^+, also has binding sites for other substances, e.g., specifically amino acid or other nutrients, which thereby get a "free ride" into the cell. The cost of the ticket required for the transport of Na$^+$ out of the cell is paid for by ATP. Similar Na$^+$-coupled carriers may exist with specific binding sites to other molecules which are transported across the cell membrane.

The net effect of this different distribution of ions on the two sides of the cell membrane is a *higher concentration of anions in the interior of the soma and axon.* With a microelectrode inserted in the cell, the potential difference has been measured and found to be about 70 mV over the membrane, the interior being negative with respect to the exterior. The cell contains nondiffusible protein; a disturbance of the resting potential difference would essentially occur if the sodium and potassium ions were allowed to move according to the diffusion gradients (Fig. 4-4). Such a flux of ions across the surface membrane would disturb the resting potential of 70 mV. Hence, an inward flow of positive ions, for example, sodium, would decrease the potential and cause a *hypopolarization*; an outward flow of positive ions, say potassium, would give rise to a *hyperpolarization.* Actually, such fluxes of ions occur normally when the nerves terminating at the motoneuron are stimulated.

Facilitation-Excitation

Let us first consider the events taking place when the axon (Fig. 4-2) is stimulated. The electron microscope has revealed the construction of the synaptic junction: the synaptic knob of about 2 μm diameter, the synaptic cleft, and the subsynaptic membrane covering the motoneuron. Within the synaptic knob there are numerous vesicles approximately 30 nm in diameter (Fig. 4-5). They contain the chemical substance that is responsible for the transmission of the impulse across the synaptic junction. The arrival of an impulse at the synaptic knob causes a synchronous ejection of many vesicles. A corresponding number of uniform packages, quanta, of transmitter substance is liberated. A trigger for this almost instantaneous chemical "injection" into the synaptic cleft is the entry of Ca^{2+} into the presynaptic terminal. In one way or another, the nerve impulse (causing a depolarization of the cell membrane) will open gates in the membrane, allowing Ca^{2+} to diffuse rapidly down a very large

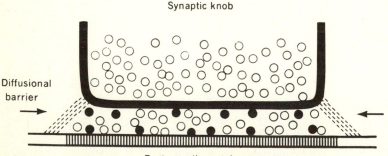

Figure 4-5 Synaptic vesicles. Unfilled dots denote transmitter substance (e.g., acetylcholine), filled dots symbolize an inactivator (e.g., cholinesterase). *(Redrawn from Eccles, 1957.)*

electrochemical gradient. In a presently unknown way, Ca^{2+} will allow a large number of vesicles to move to the cell membrane and fuse with it, releasing the transmitter into the junctional gap within 1 ms (Axelrod, 1974). The transmitter substance diffuses readily to receptors on the subsynaptic membrane which are specifically affected by this transmitter substance. With a microelectrode inserted in the cell, the effect of this event can be recorded: a hypopolarization can be assigned to an inflow of sodium ions. Eventually the active sodium pump will restore the resting membrane potential of -70 mV by extracting the Na^+ ions. However, if the membrane potential is reduced to a critical value, about -60 mV, there is suddenly a free passage through the membrane of sodium ions, and they move in the direction of their concentration gradient, causing a depolarization of the membrane. The equilibrium potential for Na^+ ions will actually be about $+60$ mV (it varies in different cells), with the interior of the motoneuron 60 mV positive to the extracellular fluid. However, the potential charge or height of the recorded "spike" (*action potential*) will not be $70 + 60 = 130$ mV, but only about 100 mV, for, immediately after the beginning of the influx of sodium, there is an outward flow of potassium ions along their electrochemical gradient. Consequently, the number of positive ions within the cell will again decrease. The efflux of potassium ions actually begins early during the "rising" limb of the spike and reduces the spike potential. Within 1 ms the resting membrane potential of -70 mV is almost restored by trading potassium ions for sodium ions across the cell membrane. The movement of ions back to where they belong is now affected by a "slow" process requiring metabolic energy in running the Na^+-K^+ pump. Following the action potential, there is a prolonged phase (15 to 100 ms) when the permeability of the neural membrane for potassium is still increased, and during this phase there is an *after-hyperpolarization* adding 5 mV to the resting membrane potential of -70 mV (Fig. 4-6).

The events just described have far-reaching effects: The electrochemical charges initiated by the transmitter mechanism, if exceeding the given threshold of about -60 mV (it varies in different cells), will cause similar sequential

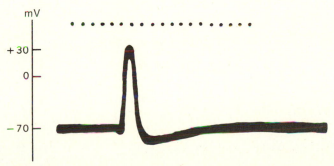

Figure 4-6 Changes in membrane potential during activity along an axon. The interior voltage changes from -70 mV to about $+30$ mV with respect to the exterior. Note the hyperpolarization after the "spike." Time base, 2,000 cps. *(Modified from Kuffler et al., 1951.)*

changes in the ionic permeability of the membrane of the whole nerve axon. This traffic is associated with the passage of a change in membrane potential (an action potential) during activity along the axon cell interior, changing from -70 mV to about $+30$ mV with respect to the exterior (Fig. 4-6). A propagated nerve impulse, traveling at a speed of up to 100 m $\cdot$ s^{-1}, will reach the muscle and stimulate it to contract.

The hypopolarization, after a stimulus of the different excitatory nerves, may not be strong enough to depolarize the postsynaptic membrane. It is then *subliminal* and gives rise only to a local potential change under the activated synaptic knob. This condition is referred to as an *excitatory postsynaptic potential* (EPSP); it is also called *facilitation.* After a few milliseconds the sodium pump has restored the normal membrane potential of -70 mV, and a new impulse of the same strength at an interval of 5 or more ms will merely repeat the local hypopolarization, say to -63 mV, the interior still being negative with respect to the exterior. If, however, the next stimulus arrives at the synaptic knob close to the preceding volley and initiates an additional ejection of transmitter substance while the subsynaptic membrane is still hypopolarized, say to -66 mV, the new sodium influx will be superimposed. This will decrease the potential to about -59 mV. This voltage charge can evoke a depolarization and generate a propagating spike potential in the axon. If the single stimulus is below the minimal strength, successive stimuli can elicit a depolarization by a summation of the receptor potential. In this case it is called a *temporal summation.* With several synaptic knobs terminating close together on a nerve cell, a simultaneous excitation may deliver the quanta of transmitter substance sufficient to produce the sieve for sodium and potassium on the subsynaptic membrane. This is an example of *spatial summation.*

If the stimulus is adequate to elicit a propagated spike, the magnitude of the spike potential of a single nerve fiber is independent of the strength of the stimulus; the nerve responds to the utmost of its ability. The answer to a stimulus is an all-or-none response. There is no alternative. During the spike potential, neither excitability nor conductivity is present in the nerve; it is in an *absolute refractory period.* Apparently, for a new Na$^+$ flux to occur as a consequence of a nerve stimulation, the potential must become more negative than about 50 mV. For the next few milliseconds a partial refractoriness exists, during which time a stronger stimulus is necessary to evoke an action potential. The normal excitability is gradually restored in the course of about 100 ms. Nerves with large diameters recover 90 percent of the normal state within 1 ms, with the consequence that they can generate impulses at a frequency of 1,000 s^{-1}. The intermittent impulse traffic in nerves is due to the refractoriness, and the recovery of the nerve cell is certainly enhanced by this mechanism. The maximal frequency by which a nerve can send its messages depends on the time of the absolute refractory period. If it takes 2 ms, this maximum will be 500 impulses per second.

Summary   When an action-potential spike reaches the terminal of a synaptic knob, the postsynaptic membrane is converted into a sieve by a chemical transmitter substance that is released from the knob. This sieve permits an inward flux of sodium ions rather than an outward flux of potassium. This "short circuit" depolarizes the postsynaptic membrane and generates an impulse, carried by similar ion movements, that propagates along the axon. The transmitter is inactivated, and the aerobic metabolism provides energy to extract the sodium ions from the intracellular fluid and to bring back the potassium ions. Thus, the normal voltage difference of 70 mV across the cell membrane is gradually restored. If the stimulus of a single impulse is not strong enough to evoke an action potential, it can provide a local hypopolarization (EPSP). This can add to the hypopolarization produced by preceding (temporal) or concurrent (spatial) impulses, and hence the threshold level for initiating an impulse in the affected nerve may be reached, this threshold being about 10 to 18 mV above the resting potential in the spinal motoneurons (Fig. 4-7).

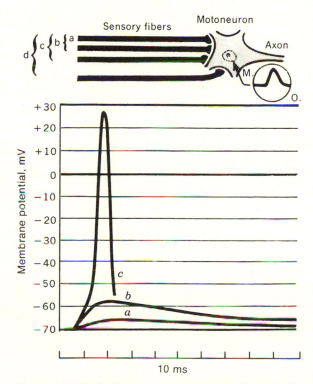

Figure 4-7 Excitation of a motoneuron can be studied by stimulating the sensory fibers that send impulses to it. A microelectrode (M.) implanted in the motoneuron measures the changes in the cell's internal electric potential. These changes, the excitatory postsynaptic potentials (EPSPs), appear on oscilloscope (O.). It is assumed here that one to four sensory fibers can be activated. When only one fiber is activated (a), the potential inside the motoneuron shifts only slightly. When two fibers are activated (b), the shift is somewhat greater. When three fibers are activated (c), the potential reaches the threshold at which depolarization proceeds swiftly, and a spike appears on the oscilloscope. The spike signifies that the motoneuron has generated a nerve impulse of its own. When four or more fibers are activated (d), the motoneuron reaches the threshold more quickly (spike not shown). *(Modified from Eccles, 1965.)*

Inhibition

From a functional viewpoint, there is another type of neuron with short axons ending in synaptic connection with motoneurons of other nerve cells, forming interneurons. When stimulated, their synaptic knobs liberate a chemical transmitter which can open ionic gates, permitting a diffusion of K^+ out of the cell and Cl^- inward. The membrane retains its high degree of impermeability to Na^+ ions, however. The result is an outward current and an increase in the voltage difference between the interior of the cell membrane and its exterior, that is, a *hyperpolarization.* It rises to a summit in 1.5 to 2 ms after its onset and fades exponentially. During this phase, a stronger stimulus is needed to discharge an impulse in the motoneuron; i.e., an *inhibitory postsynaptic potential* (IPSP) has developed. The IPSP can bring the voltage to -80 mV (which is the equilibrium potential of the inhibitory ionic mechanism). The further the membrane is removed from the equilibrium potential of -70 mV, the larger is the IPSP. Recovery will occur as the intracellular potassium ion concentration is restored by the operation of a potassium pump.

In Fig. 4-2, nerve 3 represents such an inhibitory interneuron. Within a motoneuron there is a spatial as well as a temporal summation of inhibition, as is true for excitatory stimuli. The IPSP is virtually a mirror image of the EPSP, differing in its longer latency and shorter time constant of decay. The total duration for the inhibition by the IPSP is 2 to 3 ms at the most.

Summary   A specific transmitter substance causes an inhibitory synaptic response on areas of the motoneuron by changing the ionic flux of potassium and chloride ions through the subsynaptic membrane. During the hyperpolarization caused by this ion flux, a stronger excitatory action is necessary for initiation of a depolarization, evoking the discharge of an impulse. The terminology is summarized in Fig. 4-8.

Neurotransmission

Billions of nerves communicate with one another by means of *transmitter substances* manufactured in the nerve terminals. Each nerve has its own biochemical competence to make a specific product. When the transmitter diffuses across the synaptic cleft, its effect depends on a receptor on the postsynaptic membrane to recognize and quickly combine with the transmitter. Like a key it must fit into a lock, and the interaction between transmitter and receptor has the effect that two kinds of ionic gates and associated channels momentarily open, allowing the fluxes of Na^+ and K^+. Other transmitters have the ability to trigger the opening of gates for K^+ and Cl^- exclusively, and their effect will be inhibition of the contacted cell.

In some excitatory synapses, acetylcholine (ACh) is identified as the transmitter substance. Some sort of barrier keeps it fairly effectively within the synaptic cleft, but some of the molecules will disappear by diffusion.

It is clear that when the transmitter has fulfilled its job on the postsynaptic membrane by opening the ionic gates, its action must be rapidly terminated; otherwise, its effect would last too long and a precise control would be lost. It is assumed that the Glia cells surrounding the nerve cells can function as chemical insulators, preventing a diffusion of transmitters away from sites of release. They may metabolize them and so keep the extracellular spaces of the CNS free from disturbing transmitters floating around. Most important for the inactivation of ACh is a breakdown to much less active choline and acetic acid by an enzyme, cholinesterase, present in the postsynaptic structure (Fig. 4-5).

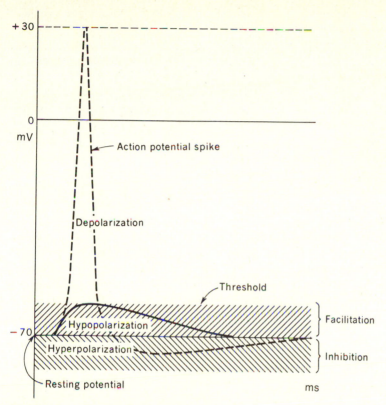

Figure 4-8 Summary of the terminology used when discussing the electrical activity in a nerve cell under various conditions.

After 1 to 10 ms (for some cells, up to 50 ms), the ejected ACh is inactivated and the resting membrane potential is gradually restored by the ion pumps. It should be noted that both the strength of the stimulus of the nerve and the rate of application of the stimulus are of critical importance for the membrane potential. This is readily explained, as the delivered quanta of ACh are continuously destroyed, and its concentration on the subsynaptic membrane may not become high enough for an effective hypopolarization. (At rest, there is a release of a few quanta of ACh in a random manner, causing a minute local membrane potential change. The passage of a nerve impulse produces a synchronous ejection of a great number of such quanta, however, and finally the subsynaptic nerve cell does not have the capacity to "defend" itself against its depolarizing effect.)

Much of the choline formed from the enzymatic breakdown of ACh is recaptured from the synaptic cleft and brought back into the nerve terminal, and so it becomes available for synthesis of ACh, which will be packed into vesicles and available for reuse. Actually, a release of ACh will accelerate its synthesis so that, under optimal conditions, the level of the ACh depot is relatively constant (Hebb, 1972). There is no evidence that under normal conditions, fatigue is the result of a failure of the transmitter release mechanism in the CNS; but it may possibly result from failure at the neuromuscular junction (see below).

The motor nerve is an example of a nerve which communicates with other nerves and with the skeletal muscle by means of ACh. Such nerves are called cholinergic. Another example is the parasympathetic nerve terminal.

There is accumulating evidence (Axelrod, 1974; Krujevic, 1974) that most neurotransmitters are not inactivated by an enzymatic degradation but that they are simply recaptured back into the presynaptic nerve terminals by a very fast and efficient transport process.

In the central nervous system, few transmitters are identified according to specified criteria (Axelrod, 1974). Evidently the amino acid glycine and gamma amino butyric acid (GABA) can exert inhibitory effects. Dopamin is supposed to exert an excitatory effect. Norepinephrine (alternatively named noradrenaline, with the nerve releasing it called nor-adrenergic) may have receptors mediating excitation on some postsynaptic membranes, but inhibition on others.

In primitive animals a direct electrical transmission exists between nerves. In higher vertebrates, however, the rule is that the gap between nerve cells and nerves and muscles is bridged by a chemical transmission. This provides greater safety and more exactness in the communication, compared with an electrical transmission.

Within the CNS, the quantic number of transmitter substances released at an impulse to an individual nerve end is quite low, compared with the several hundred quantas liberated at the synapse at the neuromuscular junction.

Inhibition and Excitation

Figure 4-9 gives a quantitative evaluation of the effect of an excitatory impulse on the membrane potential of a motoneuron simultaneously subjected to a stimulus of an inhibitory nerve fiber.

Motoneurons are subjected to a continuous bombardment of impulses

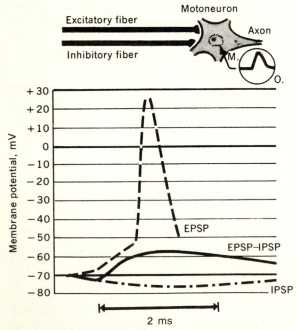

Figure 4-9 With a microelectrode (M.) implanted in a motoneuron, the changes in the cell's internal electric potential can be displayed on an oscilloscope (O.). When the excitatory nerve fiber is stimulated, an EPSP is evoked, and the threshold for producing a spike is eventually reached (dotted line at top). Stimulation of the inhibitory nerve fiber gives rise to an IPSP (dotted line at bottom). Inhibition of a spike discharge is an electrical subtraction process. The IPSP widens the gap between the cell's internal potential and the firing threshold (here about −55 mV). Thus, if the motoneuron is simultaneously subjected to both excitatory and inhibitory stimulation, the IPSP is subtracted from the EPSP, and no spike occurs (full line). *(Modified from Eccles, 1965.)*

which expose their membranes to both inhibitory and excitatory transmitter substances. They discharge impulses only when the excitatory synaptic activity is momentarily so dominating that the hypopolarization exceeds the threshold value, causing a short circuit of the membrane. The synapses are distributed fairly uniformly over the soma and dendrites of the motoneuron, but at some areas of the motoneuron, the threshold for a depolarization is lower than in other areas.

The excitatory synapses tend to be located out on the dendrites, while the inhibitory synapses are likely to be concentrated on the soma and at the axonal origin. It is a good strategic design, for the inhibitory synapses located on the soma thus have the final control of whether impulses shall be allowed to propagate down the axon (Eccles, 1973b).

As a result of the longer latency for inhibitory impulses (about 1 ms), they are most effective when they precede the excitatory volley.

Functionally, there are only two types of nerve cells, excitatory and inhibitory. Any one class of nerve cells operates by the same chemical transmitter substance at all its synapses, and the substance usually has the same synaptic action on all connected nerve cells: It causes either excitation or inhibition. All inhibitory nerve cells are short axon neurons lying in the gray matter, whereas all transmission pathways, including the peripheral afferent and efferent pathways, are formed by the axons of excitatory neurons. Many short axon interneurons also belong to the class of excitatory neurons.

If one inhibitory interneuron, preventing a contacted nerve cell from responding to stimuli, becomes subjected to the influence of other inhibitory interneurons, it may be forced to remain inactive. This means that its inhibitory effect has been neutralized and the nerve cell which it contacts synaptically, thus released from inhibition, can now respond. This mechanism is called *disinhibition.*

Nerve Conduction Velocity

The conduction velocity in a nerve fiber would be relatively slow if it were not wrapped in a myelin sheath around its "internode" sections. At the nodes this wrapping is interrupted (Fig. 4-3), and apparently it is only at these zones that the bare nerve membrane has channels for Na^+ and K^+ ions that are controlled by gates. In the propagation of an impulse in such a myelinated fiber, the current flow through the membrane is therefore restricted to the nodes. The impulse jumps from node to node (a saltatory transmission) with the long internodal sections remaining passive, which greatly enhances the velocity of the transmission. Furthermore, by ionic fluxes restricted to less than 1 percent of the surface area of the nerve fiber, there is an enormous metabolic advantage of the myelination (for details, see Eccles, 1973b).

The normal range of velocity in myelinated fibers is 12 to 120 m $\cdot$ s^{-1}. The fastest fibers are concerned in the control of movement. Unmyelated fibers have, as mentioned, much slower conduction velocities, or 0.2 to 2 m $\cdot$ s^{-1}. Such nerves are typical pathways carrying less urgent visceral information.

TRANSMISSION OF IMPULSES FROM NERVE AXON
TO SKELETAL MUSCLE: THE MOTOR UNIT

As the axon of the motoneuron approaches the muscle fiber, it loses its myelin sheath. The axis cylinder branches and the terminals make close contact with the sarcoplasm of the muscle, forming *motor endplates* (Fig. 4-10). Each motoneuron supplies from about five (eye muscles) up to several thousand muscle fibers (limb muscles). The motoneuron and the muscle fibers it supplies function as a *motor unit*, since an impulse in the axon will activate all the fibers almost simultaneously. The gain in muscular tension with the activation of one motor unit will depend on the number of muscle fibers included in the unit. The fewer there are, the finer can be the grading of the contraction. The fibers in a motor unit may be scattered and intermingled with fibers of other units, and they can be spread over an approximately circular region with an average diameter of 5 mm (Buchtal et al., 1957). Single muscle fibers may be innervated by more than one motoneuron, but the majority of the muscle fibers have only one endplate, which is supplied by a single nerve fiber. The terminal branches of the axon end near the middle of the muscle fiber. (The motoneurons in the human spinal cord and their dependent motor units number about 200,000 altogether.)

Figure 4-11 presents a schematic drawing of an endplate. The surface membranes of the nerve fiber and muscle fiber are comparable with the pre- and postsynaptic membranes, respectively, of a central synapse. In the synaptic ending there are a great number of minute vesicles similar to those found in the synaptic knobs in the CNS. They contain acetylcholine, and this chemical transmitter substance is responsible for the depolarization of the postjunctional membrane.

At rest, a miniature endplate potential can be recorded and is probably produced by ejections of a few quanta of ACh from the prepacked vesicles that spontaneously burst on the surface of the membrane. The arrival of an impulse from the motoneuron to the nerve terminal causes an almost synchronous ejection of many quanta of ACh. The released ACh diffuses rapidly across the narrow junctional gap and attaches to the specific receptors on the muscle fiber membrane, producing an endplate potential. When it reaches a certain critical magnitude, it depolarizes the surface membrane of the muscle fiber and evokes a propagated muscle action potential, which travels simultaneously to both ends of the muscle fiber. The electrochemical events are actually very similar to the ones described for the excitatory processes in the CNS. The ACh effect causes an opening of the gates which control the membrane channels and thereby drastically increases the permeability of the postjunctional membrane to sodium ions and later on to potassium ions. This effect is very short-lived, however, for in the junctional gap there is also the ACh-destroying enzyme cholinesterase. Thus the membrane permeability can again be reduced, and the membrane potential is gradually restored to its resting level. The membrane of the muscle fiber is now sensitive to a subsequent release of transmitter substance. Choline formed at the breakdown of ACh is partly taken up by the nerve terminal and quickly reacetylated for further use. (It is estimated that 100 to 200 quanta, each one containing up to 2×10^4 ACh molecules, are released through the membrane into the synaptic gap. Both receptors and enzyme molecules are present in vast excess, i.e., there may be as many as 2×10^7 both of receptor molecules and of active centers of ACh-esterase.)

A key result of the depolarization of the nerve terminal is an entry of Ca^{2+} into the

Figure 4-10 Motor endplate. In the motor endplate, the knoblike terminations of the telodendron of a motor axon fit into depressions of a multinucleated mass of sarcoplasm. *(After Willy Schwartz, from Elias and Pauly, 1961.)*

terminal, for this ion movement will powerfully but briefly accelerate the ongoing ACh release. As discussed in Chap. 3, Ca^{2+} also plays an important role as a trigger for the actin-myosin interactions during muscular contraction. (For details concerning the neuromuscular transmission, see Hubbard, 1973.)

It should be recalled that within the CNS, the algebraic sum of the electrochemical effects of impulses arriving at excitatory and inhibitory synaptic knobs in contact with a neuron determines whether or not it will become excited. In the neuromuscular function there is, however, *only one transmitter*, and *its effect is excitatory.* The quanta of ACh released is usually more than adequate to excite the muscle fiber. It should be emphasized that the only way to promote relaxation of the muscle fibers is to decrease or stop the discharge of their motoneurons.

The motor unit also obeys the all-or-none law: An effective stimulus of the motoneuron causes a maximal muscle action potential, and the muscle fibers react "to the best of their ability." However, the strength developed will depend on the length of the muscle fiber, on temperature and oxygen supply, and on the frequency of stimuli. These factors will be discussed later. The action potentials can be recorded with needle electrodes or surface electrodes attached to the skin over the active muscle, and the obtained *electromyogram* (EMG) reveals the sum of the motor unit activity (for references, see Milner-Brown and Stein, 1975).

Summary   An impulse started in a motoneuron propagates in the axon and is transmitted to the motor endplate where ACh is released. The ACh affects the muscle membrane, giving rise to an ion flux reversing the resting membrane potential. Similar to the impulse traffic between nerve cells, we have in the neuromuscular transmission a transducing mechanism, where electrical signals (nerve impulses) are transduced to chemical signals (ACh), and then back to electrical signals (muscle action potentials). The muscle action potential propagates over the membrane and initiates the mechanical-chemical mechanisms which cause the myosin and actin in the sarcolemma of the muscle to react (Chap. 3). The developed muscle tension will depend on (1) the number of motor units activated and (2) the frequency with which they are stimulated.

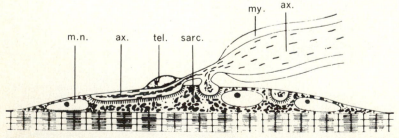

Figure 4-11 Schematic drawing of endplate: ax., axoplasm with its mitochondria; my., myelin sheath; tel., teloglia; sarc., sarcoplasm with its mitochondria; m.n., muscle nuclei. *(From R. Couteaux, 1960.)*

REFLEX ACTIVITY AND SOME OF THE BASIC FUNCTIONS OF PROPRIOCEPTORS

It has been pointed out that various nerves, either directly or through interneuronal relays, expose the motoneurons to a continuous bombardment of transmitter substances.

There is a minimum of two neurons in a reflex chain: the *afferent* (receptor) neuron and the *efferent* (effector) neuron. The nutrient cell of the afferent neuron is in the dorsal root ganglion (or cranial equivalent), and it conveys information from cutaneous, muscular, or special senses. The cell body of the efferent nerve of the motoneuron is located in the ventral horn (or motor cranial nucleus). The afferent fibers entering the dorsal root of the spinal cord may thus end monosynaptically around motoneurons, but generally several or many connecting interneurons intervene between the afferent and efferent neurons (Fig. 4-12).

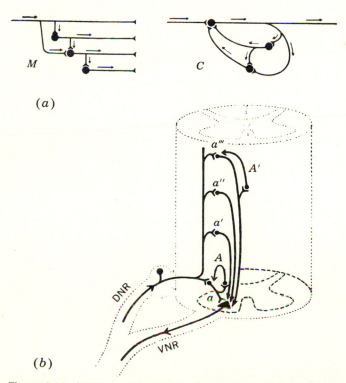

Figure 4-12 (a) Plans of the two fundamental types of neuron circuits. *M*, multiple chain; *C*, closed chain. (After Lorente de No, 1938.) (b) Role of internuncial neurons in reflex action. Diagrammatic section of spinal cord. *DNR* = dorsal (posterior) nerve root; *VNR* = ventral (anterior) nerve root; *a, a', a'', a''', A, A'* = internuncial neurons.

The shortest reflex arc is via DNR, *a*, VNR. Longer reflex arcs involving delay paths are shown via *a', a''*, or *a'''*. *A, A'* are "reverberators." Note how an impulse passing from DNR to VNR along *a* may branch to excite *A*, which in turn reexcites *a*; similarly, an impulse along *a'''* may branch to excite *A'*, which in turn reexcites *a'''*. *(From Keele and Neil, 1971.)*

Most neural control of skeletal muscles is reflex in nature. The excitability of the motoneurons is increased or decreased, depending on the algebraic sum of the excitatory and inhibitory activity in the synaptic knobs. If the muscles are the slaves of the motoneurons, the motoneurons are slaves of spinal and supraspinal mechanisms. It may be pointed out that most of the pyramidal tract fibers from the cerebral cortex area evidently terminate on interneurons. Two systems of interneurons are involved, one at the cortical level and one at the spinal stage of pyramidal tract innervation. "It can be stated that by establishing synaptic connections with interneurons rather than with motoneurons, the pyramidal tract and the other descending tracts (from the cerebellum, the red nucleus, the reticular formation) are able to operate through the coordinative mechanisms at the segmental levels of the spinal cord" (Eccles, 1957). Otherwise, the only alternative of an impulse traffic in the tracts would be an excitation of the motoneuron. As it is, the intervening interneuron can give rise to excitation *or* inhibition. The descending impulses from the higher levels of the brain, including those voluntarily evoked, will necessarily be modified on the basis of coincident information from all kinds of receptors. It also seems important that, via interneuronal circuits, information based on past experiences will permit modification of responses. Thus, the *impulse traffic in the final common path, the axons of the motoneurons, will reflect an integration of synaptic activity based on both past and present experience.*

The synapses may be created in the same segment of the spinal cord as the afferent fiber enters, or collaterals may pass up or down the spinal cord, eventually reaching the brain (Fig. 4-12). Theoretically, there are innumerable pathways which an impulse can travel via nerve fibers, but some tracts are preferred, partially on an anatomic basis. One nerve may branch and terminate with several synaptic knobs on the same nerve cell and its dendrites.

In the synapse there is a delay of about 0.5 ms. "It used to be thought that one of the distinguishing properties of synapses was fatigue, but this is belied by the performance of synaptic mechanisms in the normal functioning of the brain" (Eccles, 1973b). Actually, synapses and neurons of the central nervous system are well adapted metabolically to respond continuously at frequencies as high as 100 s^{-1} with peaks of 500 s^{-1} or, for some species of neurons, as high as 1000 s^{-1}.

The most important features of the synaptic functions may be summarized as follows: If fibers from two afferent nerves ending on the same nerve cell are stimulated separately, neither causes a reflex response if the stimulus is subliminal (below threshold). However, when excited, each afferent nerve liberates sufficient chemical transmitter to cause an EPSP. When both nerves are stimulated simultaneously, the liberated quanta of transmitter substance may be enough to discharge a spike potential, and a reflex response is evoked (*spatial summation*). The other type of summation is the *temporal summation*: Individual stimuli may be ineffective, but if repeated at a high frequency, a depolarization of the membrane of the connected nerve cell occurs.

Afterdischarge is another example of spatial summation. If the axon of a motoneuron is stimulated with an electrical shock of sufficient strength, the muscle supplied responds with a single twitch. If the stimulus is applied to the afferent nerve, the tension may remain for a much longer time because the muscle responds with repetitive twitches. Figure 4-12 explains how the motoneuron can be persistently stimulated, since volleys continuously arrive after having passed through the shorter or longer delay paths formed by the intricate maze of interneurons.

Renshaw Cells and Other Inhibitory Interneurons

Motoneurons give off collateral branches as they traverse the spinal cord to emerge in a ventral root. They form excitatory synaptic contacts with interneurons located in the ventromedial region of the ventral horn. These *Renshaw cells* send axons which affect inhibitory synaptic connections with the same and other motoneurons of that segmental level in an overlapping and diffuse fashion (Fig. 4-13). The Renshaw cells may also be excited by afferent impulses from muscles and skin. The Renshaw cells provide a feedback, and a single volley in the axon of the motoneuron may evoke a repetitive discharge of the Renshaw cell with the consequent tendency to dampen the motoneural

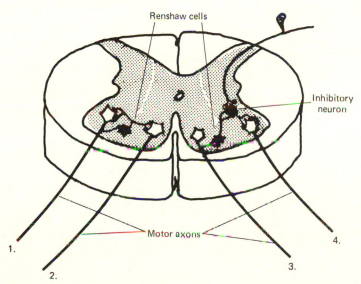

Figure 4-13 Some motoneurons give off branches, called recurrent collaterals, before they leave the gray matter of the spinal cord. These collaterals synapse with inhibitory interneurons named Renshaw cells. They synapse in turn with the same or other motoneurons. To the left in the figure, the activated motoneuron (1) stimulates the Renshaw cell, which then inhibits both motoneurons (1) and (2). To the right, we have an example of inhibition of inhibition: an impulse in the afferent nerve stimulates the inhibitory neuron, and therefore, the motoneuron (4) will become inhibited; if the motoneuron (3) is now stimulated, it excites via its recurrent collaterals the Renshaw cell, which in turn inhibits the inhibitory neuron, and as a consequence, the motoneuron (4) will be released from the inhibition and may more easily respond to excitatory impulses. *(Inhibitory neurons are shown in black.)*

activity. The effect is thus a reduction in the motoneural discharge frequency. This may protect against convulsive activity and overloading of the muscles.

In other words, the Renshaw cells and other inhibitory interneurons will keep down the level of excitation by suppressing discharges from all weakly excited neurons. Only the strongly excited neurons will survive this inhibitory barrier. In addition, as pointed out by Eccles (1973b), they participate very effectively in a neuronal integration.

Figure 4-13 illustrates how an inhibition itself may be inhibited. Many motoneurons are normally subjected to a steady background inhibition by the activity of inhibitory nerve cells, eventually driven from higher levels of the CNS. Renshaw cells may form synapses with such inhibitory neurons. When these Renshaw cells are stimulated, they will inhibit or depress the activity of the inhibitory neurons and free the motoneurons from inhibition. The effect will eventually be a "recurrent facilitation" or *disinhibition* of the motoneurons.

These are examples of postsynaptic inhibition and how it can be suppressed. There is evidence that a *presynaptic inhibition* may also occur. Some inhibitory cells form synapses with the terminals of axons of other nerve cells, and their transmitter substance reduces the amplitude of nerve impulses traveling along the contacted fiber. The result is a liberation of less than the normal amount of transmitter at the synapse and a reduced effect on the postsynaptic membrane. This type of inhibition is very widespread and powerful at lower levels of the brain. It is often found to be acting on terminals of large afferent fibers from the skin and from muscles, regardless of function. It creates a general suppression of weakly supported discharges which is not selective (Eccles, 1973b).

Renshaw cells are thus subjected to both excitatory and inhibitory inputs. The main spinal excitatory input comes through the recurrent collaterals, whereas impulses mediated through afferent nerves are predominantly inhibitory. Descending impulses from supraspinal structures (e.g., the cerebellum, the brain stem) can inhibit, facilitate, or excite the Renshaw cells. In many instances, these pathways converge on excitatory or inhibitory interneurons forming synapses with the Renshaw cells. (Granit, 1975.)

The Gamma Motor System

The axons of the motoneurons so far discussed are of the so-called *alpha* (α) *type* (12- to 20-μm diameter), and they supply the skeletal muscles (*extrafusal* fibers). In the ventral horn of the spinal gray matter, there are also other types of motoneurons giving rise to axons which are thinner, and they are of the *gamma* (γ) type. These gamma fibers supply the *intrafusal* muscle fibers of the *muscle spindles.* This proprioceptor sense organ is shown diagrammatically in Fig. 4-14a. It is an elongated fusiform capsule lying *parallel* to the extrafusal muscle fibers. Its central part consists of a noncontractile elastic "nuclear bag" above or elastic "nuclear chain" (below, in Fig. 4-14a). On either side, some intrafusal muscle fibers are attached (consequently named nuclear bag fibers and nuclear chain fibers, respectively). The other ends of these fibers terminate either at a tendon or in the endomysium of extrafusal muscle fibers.

Activation of the gamma fibers causes the intrafusal muscle fibers to contract, and the equatorial parts between the polar regions become stretched. Receptors are located at these central parts, and if the stretch is strong enough, they will react with action potentials which, via the afferent nerve fibers, are propagating toward the spinal cord. Some impulses are monosynaptically transmitted to alpha neurons of the same muscle (and its synergists), providing an excitation. When the muscle contracts, the ends of the muscle spindle come

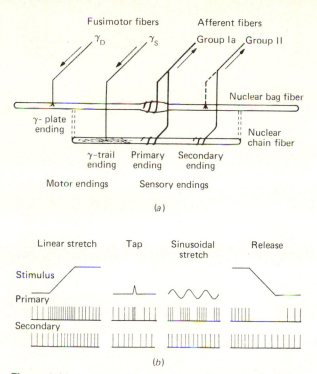

Figure 4-14 (a) Diagram of the mammalian muscle spindle mechanoreceptors showing the main functional component involved in the transduction of muscular displacement into a receptor potential that is considered to be the direct governor of the afferent discharge pattern (γ_D = dynamic gamma axon; γ_S = static gamma axon). *(From Rudfjord, 1972.) (b)* Diagrammatic comparison of the responses of "typical" primary and secondary endings to various stimuli. The responses are drawn as if the muscle were under moderate initial stretch and as if there were no fusimotor activity. (From Matthews, 1964.)

closer together, and the stretch on the nuclear bag and chain is reduced. The stimulus of the receptors will be less or will cease completely (Fig. 4-15). The afferent impulses, via an interneuron, can also cause an inhibitory effect on the alpha neuron of the antagonists.

As will be discussed later, the α and γ motoneurons will often be activated together (coactivation) and the γ fibers can adjust the length of the muscle spindle so that it can maintain an impulse traffic through its afferent nerve fiber during shortening of the extrafusal musculature. A passive stretch of the muscle will also cause the muscle spindle to fire, and by reflex the same muscle will respond by contracting, thus counteracting the stretching. Tapping the quadriceps tendon, as is routinely done in medical examinations, stretches the muscle and its contained muscle spindle, causing a jerk of the lower leg in the "knee reflex." The explanation for this is the one presented above.

The reflex can be evoked by (1) stretch of the muscle and (2) increased activity in the gamma fiber.

Regarding the latter, it should be emphasized that the activation of the

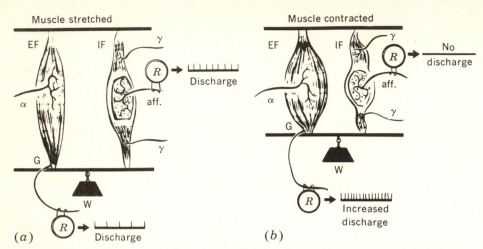

Figure 4-15 Schematic illustration of the activity in afferent fibers from muscle spindle (aff.) and Golgi tendon organ (G, lower record) when the muscle is (*a*) stretched and (*b*) contracted. EF = extrafusal muscle fibers, IF = intrafusal fibers innervated by gamma fibers. *R* = recording instrument with its electroneurogram. The muscle spindle is stimulated by stretch, but there is eventually a pause in its discharge during an extrafusal muscle contraction, during which activity there is an accelerated rate of discharge from the tendon organ.

gamma fibers per se does not cause any increase in the muscular tension. An intact afferent nerve from the muscle spindle is necessary to induce a contraction as a response to gamma activity or stretch. Adaptation in the muscle spindle takes place very slowly; that is, the discharge in the afferent nerve fiber continues for as long as the muscle is stretched, though the frequency of the discharge gradually declines. A stimulus which increases slowly in intensity produces a lower frequency of impulses than a stimulus which rises very rapidly to the same level. Thus, a pull on a muscle of a cat with a given force attained within 1 s produces an afferent impulse frequency of more than $100 \cdot s^{-1}$. But a slower increase in stretch, continuing until the same force is applied, within 6 s will give a peak volley of about 40 impulses $\cdot s^{-1}$ (Granit, 1962).

It may thus be concluded that the muscle has an excellent instrument to measure its length (or more precisely, the difference in length between the parallel extrafusal and intrafusal muscle fibers), the extent of a mechanical stimulation, and the rate with which the stretch is applied. Furthermore, the spindle control enormously increases the number of functions that a slightly activated alpha pool of motoneurons can perform (Granit, 1970, 1975; Matthews, 1972).

The structure and function of the muscle spindles are much more complicated than revealed by the discussion presented. There are actually *two types of gamma efferents* and it is not settled exactly how the nuclear bag and nuclear chain are innervated (see Stein, 1974). Most likely the so-called *dynamic gamma axons* are innervating nuclear bag fibers while axons of *static gamma axons* can innervate both nuclear bag and chain fibers. The intrafusal muscle fibers of the nuclear bag are slow fibers, they build up their state of contraction slowly, and they possess damping. The muscle fibers of the nuclear chain are fast twitch fibers and they respond by twitches but with little viscous damping (see Granit, 1970). It is only the static gamma fibers which adapt the length of the spindle muscle fibers during a contraction of the extrafusal muscle fibers thus preventing it from slackening. The *afferent* nerves from the spindles are also of two different kinds: the *primary*, fast conduction *nerve fibers* (group Ia) which are equatorically located in both bag and chain fibers; and the *secondary*, slower conducting *nerve fibers* (group II). The large majority of the secondaries are found on the chain fibers, with a minority on the bag fibers.

Fig. 4-14b illustrates the different properties of the responses of primary and secondary nerve endings to various stimuli. (1) The primaries are particularly sensitive to a stretch. They cease to fire when the stretch is released. The secondaries on the other hand are better adapted for signaling the actual length of the muscle and they do not show a period of silence after release of pull. (2) The primaries respond strongly to a "tap" on the tendon of the muscle which is also in contrast to the secondary nerve endings. (3) If the muscle or its tendon is subjected to sinusoidal stretches or high-frequency vibrations, it is only the primaries which, to a marked degree, vary their action potential frequency. (When recording from single afferent nerve fibers, this characteristic of the primaries can be used for identification.) It should be pointed out that γ activity during the release of pull can adapt the length of the muscle spindles, and the pause noticed in firing from the primary nerve endings in Fig. 4-14b may be filled out. As will be discussed, both static and dynamic γ fibers are activated or inhibited in voluntary muscle activities.

It is the primary nerve fibers that monosynaptically can activate the motoneurons supplying their own and synergistic muscles. In fact a primary afferent fiber makes contact with virtually all homonymous motoneurons. They inhibit via interneurons those of the antagonistic muscles. This is the "classical" function of the muscle spindles. The effect of impulses from secondary endings is more complicated and partly unknown. They may polysynaptically excite flexor motoneurons irrespective of whether the endings themselves lie in flexor or in extensor muscles. The effect on extensors may be an inhibition. We can conclude by stating that in general muscle spindle primaries and secondaries have been found to behave similarly, but the primary nerve endings are more sensitive to the velocity at which a muscle is being stretched (dynamic conditions), and the secondary nerve endings more precisely report the muscle length (static conditions). On a subconscious level, they can so provide the CNS with essential information about the state of the muscle. The final effects on the α motoneurons from the ingoing impulse traffic of the different nerve endings may evidently be essentially similar or antagonistic.

The Golgi Tendon Organs

These spindle-shaped corpuscles, a few millimeters in size, are connected in *series* with extrafusal muscle fibers and inserted between the muscles and their tendons. The relatively few muscle fibers attached to a tendon organ belong to different motor units. A stretch of the Golgi tendon organ will give rise to action potentials which will be conducted centrally in relatively thick nerve fibers. For mechanical reasons the receptor will be particularly sensitive to "active" tension caused by a contraction of the extrafusal motor units which are attached in series to the portion of the tendon in which the Golgi tendon organ is located. A "passive" tension created by pulling the muscle is a much less effective stimulus on the tendon organ because the force is exerted on a much larger tissue area.

When stimulated, the afferent nerve fibers cause a reflex inhibition of their corresponding muscle that is elicited via interneurons. Excitatory interneurons are also stimulated, but they form synapses on the motoneurons to the antagonistic muscles. It should be emphasized that the reflex arc of the gamma loop is monosynaptic, but the Golgi tendon organ stimulates via interneurons, and this delays the impulse. The tendon receptors inhibit not only the alpha motoneuron of the same muscle, but also the gamma motoneuron. This is a wise arrangement if the elicited reflex is effectively to prevent the development of a dangerous tension in muscle and tendon. The Golgi tendon reflex also causes the smooth retardation of a movement.

Reciprocal Inhibition (Innervation)

If a muscle is stretched, impulses are evoked in the afferent fibers from the muscle spindles, as already mentioned. These fibers, ending directly at the α motoneurons of the muscle containing the excited muscle spindles, exert an excitatory effect. Their collaterals, releasing the same transmitter to excite adjacent interneuron cells, form a direct inhibitory pathway by sending their axons to the motoneurons of the antagonistic muscles. The effect of this stretch stimulus will be a stretch reflex, since the extensors contract and the flexors relax. This reciprocal inhibition is an "inborn" mechanism in the CNS.

Joint Receptors

The ligaments and the capsules of the joints contain different kinds of receptors (Ruffini end organs and receptors of the Pacinian and Golgi type). Some of these proprioceptors are specialized to respond to movement of the joint; others show an impulse discharge that varies with the exact position of the joint but are less sensitive to movement. The afferent nerve fiber's synapse in the spinal cord and other neurons can transmit impulses to the cerebellum, thalamus, and sensory cortex respectively. These impulses are extremely important for the coordination of movement and for our information concerning the exact position of the joint. The ability, with the eyes shut, to touch one's nose with a finger tip or the left knee with the right heel is based mainly on information originating in the joint receptors.

The Resting and Activated Muscle

At rest, the skeletal muscles are relaxed and normally no EMG can be recorded. Vallbo (1973) reports that, when he was recording from afferent nerves of spindle primaries in finger muscles, only a small proportion of the nerves studied were continuously discharging in his conscious human subjects sitting in comfortable resting positions.

In animals, it has been noted that muscle spindles may be activated by a steady or slowly fluctuating tonic firing of gamma neurons which create a general state of arousal or readiness to move. The activation seems to be independent of the alpha activity and it is not related in time to specific movement (Granit, 1975). Vallbo (1973) concludes that in the human, the *onset* of voluntary movement is controlled through the alpha motoneurons alone, without the support of an afferent impulse traffic from the muscle spindles. However, there is a simultaneous *coactivation* of α and γ motoneurons to the engaged muscles. The impulses from the spindles appear after the onset of the contraction of the extrafusal muscles and they stop firing at the cessation of contraction.

Lateral Inhibition

If the skin is touched by an object, touch and eventually pain receptors are stimulated. The most strongly stimulated receptors may, through collaterals from the afferent fibers, stimulate inhibition interneurons. These inhibitory interneurons, in turn, may abolish the weaker afferent nerve impulses by making the nerve cells involved sufficiently hyperpolarized. Functionally, this is an advantage when there are several synaptic relays for afferent impulses from the various receptors. At every synaptic transmission, there is a possibility of an inhibitory action that sharpens the neuronal signals by eliminating the weaker excitatory actions which are elicited, for example, when an object touches the skin. In this way the touch stimulus may be more precisely located and evaluated when the more sharply defined signal reaches the cortex. Also, the efferent pathways down from the cerebral cortex, via interneurons, can exert inhibitory blockage at the relays in the spinocortical pathways. Thereby, the cortex can "protect itself from being bothered by stimuli that can be neglected" (Eccles, 1973b). When intensely engaged in an activity, even quite strong stimuli may be kept below the level of consciousness. Thus, in the heat of combat, severe injuries may be ignored. Noise, which otherwise may be quite disturbing, may under such circumstances be unnoticed. Acupuncture is no doubt an example of the stimulation of cutaneous pathways which then inhibit pain impulses via some relays. The effects of hypnosis and yoga are other examples of lateral inhibition.

Supraspinal Control of Motoneurons

In the biological evolution, primarily a development of the CNS and, to a lesser extent, a reconstruction of existing parts have taken place. The cerebellum has been developed, so to speak, parallel with other parts of the brain. In particular, the evolutionary building of a linked system between the cerebrum and the cerebellum gave humans their immense superiority for survival. The more recently developed parts of the brain have assumed a functional dominance, compared with the older parts. This is especially true in the case of the cerebral cortex.

In the introduction to his book *The Understanding of the Brain*, Eccles (1973b) claims that "this human brain that doesn't look too distinguished on the outside, weighing about 1.5 kilograms, is without any qualification the most highly organized and most complexly organized matter in the universe." He believes that attempts to understand the brain will occupy hundreds of years of our future.

Generally speaking, it appears that the nerve cells in the brain have a tendency to fire all the time; there is an incessant, irregular discharge from billions of nerve cells, giving a massive "background" noise. The number of synapses on a nerve cell may vary from a few hundred to more than 200,000. In other words, it is a matter of a pronounced convergence. Probably none of the neurons in the brain is exposed only to excitation, and certainly no nerve cells are affected solely by inhibitory reception. It is therefore often a question of a modification of a basic frequency more or less, when a nerve cell is exposed to a synaptic effect. Eccles (1973b) has pointed out that the "voice" of a single neuron in the brain would be lost in the "chorus" of all the neurons humming together. It appears that the same kinds of neurons are organized together, that they receive the same kinds of coded message, and that they will eventually forward the message to another cluster of neurons. "The neurons have to shout together, as it were, to get the message across and so make a reliable signal despite all the background noise" (Eccles, 1973b). It may be compared with the situation at a fully packed stadium during an exciting athletic event. The single shouts are drowned in the roar of the masses but many voices joining in a claque of yellers can be identified!

Cerebral Motor Cortex

The region just anterior to the central fissure has a key position in the voluntary control of muscular movements. It is not the prime initiator of movement, but it is the final relay station receiving instructions from widely dispersed areas in the brain and putting them into effect via the motoneurons. Most of the pyramidal cells are arranged in narrow columns (about 0.5 to 2.0 nm in diameter) which are vertically oriented to the cortical surface. One column (with a few hundred pyramidal cells and some 10,000 other types of cells) projects on a specific muscle group or to synergistic muscle groups. Thus, stimulation of a column can produce one single movement and other stimuli may give rise to several movements, depending on which part of the column is stimulated. The afferent input to such cell columns from skin, joint, and muscle spindle receptors is clearly related to the direction of the movement. Thus, all the various parts of the body are represented in the striplike map established in the motor cortex. The muscles controlling the tongue, lips, larynx, and thumb have a particularly large representation in such columns.

Afferent fibers reaching the motor cortex can exert inhibitory or excitatory actions on the pyramidal cells (which are themselves excitatory). Interneurons can secure a fast, vertical spread of excitation through the whole depth of a column. They can, by a positive feedback, build up a strong, amplified excitation in the column. With multiple synapses, one afferent fiber can excite many cells. Inhibitory neurons in a column, which can probably be activated by recurrent collaterals of the pyramidal cells, often have their axons distributed to the immediately surrounding parts of the cortex. By this design, activation of a column may suppress the nerve activities in adjacent columns, often with antagonistic functions, in a reciprocal fashion. Thus, the activated column may cause a flexion of a finger and the inhibited surrounding columns may be projecting on extensor muscles which will now relax. It should be pointed out that the pyramidal cells are not tied together in rigid relationships. They can plastically operate reciprocally and simultaneously and can combine in different patterns of activity.

It is estimated that the pyramidal tract in the human contains about 1.2 million fibers (in a cat, less than 200,000). In the higher primates, many of the pyramidal endings in the spinal cord do form monosynaptic contacts with both α and γ motoneurons. Many such fast-conducting fibers connect with the muscles of the hands. With the α motoneurons excited to discharge impulses, the innervated muscle fibers will contract. At approximately the same moment, the γ motoneurons will also discharge, thus exciting the muscle spindles in the same muscle, and the γ loop is put in operation. There is a *coactivation of the* α *and* γ *motoneurons* with the exception of very fast movement, which probably is carried out by pure α activity. The two pools of motoneurons are also *coinhibited*. The sequence is: (1) The skeletal muscle is activated by the α motoneuron discharge; (2) the thinner γ fibers conduct impulses more slowly than the α fibers, and therefore the muscle spindle contraction will begin somewhat later than the extrafusal contraction; and (3) the afferent discharge from the muscle spindles follows, with the triple role of directly supporting the α activity to the same muscle and inhibiting the antagonists as well as reporting to the brain about the progress of the initiated movement.

One important feature of the pathways from the cerebral motor cortex is their profound collateral branching to brain stem nuclei (particularly to relays in the pontine gray matter but also to those in the lateral reticular nucleus and the olive) (Fig. 4-16). Many of the relay nuclei project to the cerebellum. The result is that the cerebral cortex cannot keep any secrets with regard to its signal traffic to the motoneurons!

Pathways from the motor cortex can also, via an interneuron, powerfully *inhibit both* α *and* γ *motoneurons*. Clough et al. (1971) point out the importance of a control of γ activity to muscle spindles of "unwanted" synergists and of withdrawal of support from the spindles of antagonistic muscles that might otherwise undergo autogenic stiffening when lengthened passively by the action of the prime movers.

In summary, we have discussed an inhibition of muscles not needed for a special movement on two levels: by a lateral inhibition in the motor cortex, and by interneurons at the spinal level.

The motor area, i.e., the precentral cortex of the brain, is not only a source of motor output but also a target for sensory input. The afferent input is strongest to a particular subdivision of the motor cortex from the part of the body controlled by that subdivision (e.g., from receptors in one hand to the area of the motor cortex that can induce movements in that hand). Monkeys were trained to reposition a movable handle into a correct zone and gained reward when the task was performed within a given time (Evarts, 1973). Recordings were obtained from neurons in the postcentral (sensory) cortex as well as the precentral (motor) cortex. It was noted that the minimal latency from perturbation of handle to discharge of postcentral neurons was 10 ms. Precentral neurons not classified as pyramidal tract neurons could respond after 14 ms. Changes in pyramidal tract neurons began still later, i.e., there was a decreased activity at 20 ms and an increased activity with a minimal latency period of 24 ms. In Evarts' experiments EMG records indicated muscular activity at a latency of 12 ms in muscles whose length was changed by the abrupt movement of the handle. This early response was probably mediated by the muscle spindle afferents making monosynaptic connections with motoneurons. After a period of inhibition, there came a second muscle response at a total latency of 30 to 40 ms followed by a third phase at about 80 ms. There are experimental evidences backing up the conclusion that the cerebral cortex and its sensorimotor neurons are involved in the muscle activity occurring at a latency as short as 30 to 40 ms.

In experiments on man reported by Hammond (1956), the subject flexed his forearm against a steady force. A sudden pull of constant rate was then applied to the wrist to extend the forearm. Instructions, to "resist" or "let go" when pulled, were given several seconds before the pulls. At a latency of 18 ms there occurred an EMG response in the stretched biceps regardless of the instruction followed by a second phase of muscular activity which could begin at about 50 ms (mean value 70 ms) in subjects who had been instructed to "resist." This response was usually absent in subjects who were told to "let go." Marsden et

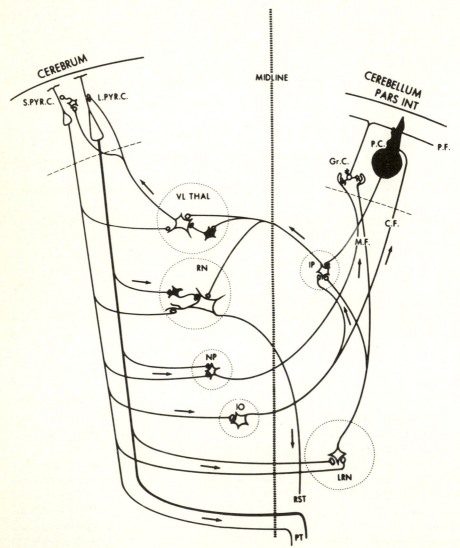

Figure 4-16 Pathways linking the sensorimotor areas of the cerebrum with the cerebellum. [C.F. = climbing fiber; Gr.C. = granule cell; I0 = interior olive, IP = interpositus nucleus (intracerebellar nucleus); L.PYR.C. = large pyramidal cell; LRN = lateral reticular nucleus; M.F. = mossy fiber; NP = nuclei pontis; P.C. = Purkinje cell; P.F. = parallel fiber; PT = pyramidal tract; RN = red nucleus; RST = rubrospinal tract; S.PYR.C. = small pyramidal cell; VL THAL = ventrolateral thalamus.] *(From Eccles, 1973b.)*

al. (1967) studied flexion movements of the top joint of the thumb (humans), and the movement was unexpectedly interfered with in various ways. They noted that 50 ms after a perturbation the muscle's activity (EMG) altered in such a sense as to tend to compensate for the perturbation. They interpreted the responses as manifestations of automatic servoaction based on the stretch reflex. The sense organs could discriminate a misalignment of only a few hundreds of a millimeter. In light of Evarts' results one can assume that there also was a transcortical pathway in operation in the human subjects. The difference in reaction time noted for monkey and man corresponds well with the differences in body size. *In summary*, these data give an idea of the fast flow of information from hand and arm to the brain and back again to the arm muscles. *In less than 0.1 s a message delivered from peripheral sense organs can reach the brain, be interpreted, and provoke a muscle action.* (For further discussion see Porter, 1976.)

Cerebellum

The cerebellum has a key function in the smooth and efficient control of movements. It integrates and organizes information flowing into it along the various pathways, from peripheral receptors (in skin, joints, muscles, tendons, ears, and eyes), and from various nerve centers in the brain. (For further studies, see Eccles, 1973b; Allen and Tsukahara, 1974.)

There are two distinct *afferent tracts* feeding the cerebellum with information from the periphery: (1) the direct pathways via the spinocerebellar tracts, (2) and an indirect tract which first projects to the cerebrum (via relays in the thalamus). After subjection to modifications, the information goes through the mentioned relay stations in the brain stem to the cerebellum. In addition, the cerebellum receives "original" inputs from many association areas as well as the sensorimotor cortex of the cerebrum. Most of the end points of the afferent fibers are located in the cerebellar cortex.

The *efferent tracts* originate from the numerous "deep" intracerebellar nuclei located in the interior of the cerebellum. They project to relay stations in the thalamus, the red nucleus, and the vestibular and reticular nuclei of the brain stem. The cerebello-cerebral pathways (via the thalamus) are traveled in a few milliseconds, and using these routes, the cerebellum is capable of activating pyramidal neurons and modifying their impulse discharges. The cerebellum also has the more direct route to the spinal motoneurons via the rubrospinal and vestibularspinal tracts which can monosynaptically or polysynaptically activate both α and γ motoneurons (Grillner et al., 1969).

We may conclude that there are several available loops which can tie the cerebellar nuclei to the locomotor organs. One of them is more peripherally oriented, and the other includes the cerebral motor cortex. Both of them have the potential to operate the motoneurons and participate in the control of movements. It should be noted that the pathways between the cerebellum and cerebrum permit a two-way traffic. The same is true for the connections between the cerebellum and the spinal level.

The function of the cerebellar nerve cells is relatively well known. Their details will therefore be discussed in order to illustrate an architectural principle in nerve connections.

Figure 4-17 presents, in a schematic way, the organization of the elements within the cerebellar cortex. There are two afferent fiber types, *climbing fibers* and *mossy fibers*. The

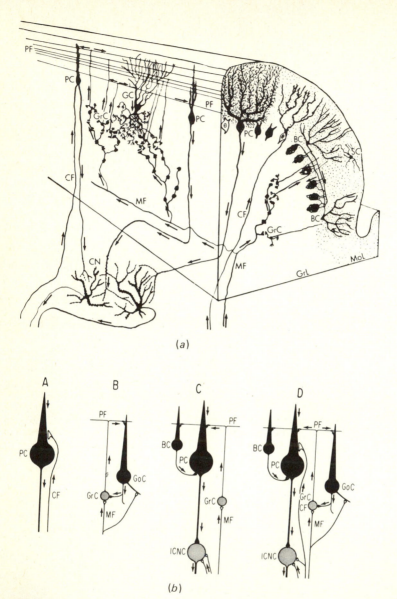

(a)

(b)

Figure 4-17 *Top:* Schematic drawing of a segment of a cerebellar folium. *(From C. A. Fox, 1962.)*
Below: The most significant cells and their synaptic connections in the cerebellar cortex. The component circuits of A, B, and C are assembled in D. Arrows show lines of operation. Inhibitory cells are shown in black. *(From Eccles, 1973b.)*
(BC = basket cell; CF = climbing fiber; CN or ICNC = intracerebellar nuclear cell; GoC or GC = Golgi cell; GrC = granule cell; GrL = granular layer; MF = mossy fiber; MoL = molecular layer; PC = Purkinje cell; PF = parallel fiber; SC = stellate cell.)

only pathway out of the cerebellar cortex is via the fibers of the *Purkinje cell* axons that contact the intracerebellar nuclear cells.

The afferent fibers exert their effects on the Purkinje cell in different manners. One climbing fiber innervates one Purkinje cell by forming hundreds of synapses on its dendritic tree. The climbing fibers exert a very powerful excitatory action when discharging. Originating in the olive, they relay messages from muscles and skin as well as from the precentral and postcentral gyri of the cerebral cortex. The mossy fibers branch enormously and synapse on the *granule cells* (one mossy fiber innervates some 450 granule cells). Their axons bifurcate to form the densely packed *parallel fibers* that extend along the folium for about 1 mm in each direction from the bifurcating point. These fibers form excitatory synapses to Purkinje cells, and it is estimated that one Purkinje cell can receive as many as 80,000 parallel fibers, or, eventually, up to 200,000 fibers. (Some 400,000 parallel fibers may pass through the dendritic tree of one Purkinje cell. To give an idea of the richness of nerve cells: There are about 30,000 million granule cells and 30 million Purkinje cells in the human cerebellum; Eccles, 1973a). In addition, the parallel fibers can *activate* inhibitory cells: the *basket cells*, the *stellate cells*, and the *Golgi cells* (which can also be stimulated by mossy fibers.) The Golgi cells have the potential to inhibit the granulate cells in a negative feedback fashion; they have a profuse branching, so that one Golgi cell can inhibit some 10,000 granule cells. *Functionally*, the Purkinje cells are arranged in columns, and, in some areas, there is a strict topographic representation of distinct movements similar to the arrangement of the pyramidal cells in the cerebral cortex. In other words, specific Purkinje cells can integrate both peripheral and cortical inputs that represent a limb.

The Purkinje cells are "spontaneously" discharging, driven by a constant bombardment via the numerous mossy fibers. With a more or less synchronous activation of mossy fibers, generated from (1) peripheral receptors (and initiated by movements) and/or (2) other parts of the brain, a compact granule cell channel will excite a cluster of Purkinje cells via the parallel fibers. They will cover a rectangular field extending about 2 mm. In addition, the climbing fibers transmit an excitation conveying much the same information as the mossy fibers. However, the *granule cell-basket cell channel will inhibit the Purkinje cells on each side of the field*, and an inhibitory state will develop on either side of the beam of excited Purkinje cells. Depending on their location, Purkinje cells will increase or decrease their discharge rate. They can respond to an input with all gradations of excitation and inhibition. A single climbing fiber has a strong influence. The mossy fibers work by their large number, each with a very weak, almost insignificant, influence on the Purkinje cell. An individual Purkinje cell performs an independent computation on a specific subset of inputs available to it. It can integrate inputs from several areas of the cerebral cortex and from peripheral receptors. There appear to be many centers for many kinds of integration, with much overlap and replication in an irregular manner. However, "No part of the cerebellum knows in general what other parts are doing" (Eccles, 1973b).

It was mentioned that the axons of the Purkinje cells form the sole connection between the cerebellar cortex and the intracerebellar nuclear cells. (As an additional example of convergence in the neuronal connections, up to 200 Purkinje cells may contact one nuclear neuron.) The effect on these nuclei is *inhibitory* and the neurons therefore can be further inhibited or released from inhibition (by disinhibition) because of the modified Purkinje cell discharge. After about 10 ms, the Purkinje cells in the "on-beam" region become inhibited, probably by the negative feedback from basket cells and Golgi cells.

Because all the neurons of the cerebellar cortex except the granule cells are inhibitory, the inhibition has dominance. Eccles (1973b) points out that "there can be no prolonged chattering in chains of excitatory neurons. . . . within 0.1 s after some computation, that area of the cerebellar cortex is 'clean', ready for the next computation. This automatic 'cleansing' is very important in giving reliable performance during quick movements." Excitation actions have to fight their way against the opposite inhibitory action. The lateral inhibitions by the basket cells, stellate cells, and the Golgi cells can sharply focus the Purkinje cell response to limited cortical areas, and in a reciprocal fashion keep "unwanted" nerve cells in adjacent foci silent. With Purkinje cells inhibited, the synaptically contacted nuclear cells are relieved from a depression and the net effect is an excitation. The Purkinje cells that are activated will consequently reduce the discharge rate of projected nuclear cells or eventually silence them. During rapid movements, many different patterns of excitatory and inhibitory interactions will be completed within a second.

The continuous irregular activity, even under resting conditions, in all neurons of the cerebellar cortex and nuclei makes them primed to respond to changes in the afferent input by transitory variants in frequency that constitute signals for computation. A large number of parallel lines carrying similar information converge on the same target neuron. This onslaught gives an automatic averaging which will ensure an improved reliability and an elimination of "noise." The role of the climbing fibers is not clear (see the subsection "Learning"). The signal traffic mediated by these fibers is delayed, and the Purkinje cell response comes at least 10 ms later than the response elicited by the mossy fibers.

Slow-conducting mossy fibers give off collaterals to the intracerebellar cells (Fig. 4-17). These excitatory fibers can provide a strong background, a nonspecific drive onto the nuclear cells, which the Purkinje cells can modulate downward by inhibition. A specific stimulus of the nuclear cells by this route is delayed and arrives at approximately the same time as the Purkinje cell inhibitory control. Eccles points out that this simultaneity is important for an effective computational performance by the nuclear cells; there is a closely timed clash of the opposed influences of excitation and inhibition.

To give an idea of the time relations, it may be mentioned that a tap on a cat's forepaw pads can, via mossy fibers, stimulate Purkinje cells after a latency of about 12 ms, and after 16 to 18 ms, the response comes from nuclear cells (Eccles, 1973a).

In summary, the cerebellum receives nerve impulses from peripheral receptors and other parts of the brain via mossy fibers and climbing fibers. In the cortex there are five neuronal species, four of them being inhibitory in their effects. The only output from the cortex is via the axons of the Purkinje cells. These cells can be fired by either cortical or peripheral inputs alone, but in general, the two inputs are cooperative. The Purkinje cells "have the role of reading out the computation by the neuronal machinery of the cerebellar cortex and transmitting their coded message to the nuclear cells" (Eccles, 1973a). These intracerebellar nuclei then feed back "messages" to the cerebral cortex and to the periphery.

Various Nuclei in the Brain Involved in Movement

The pyramidal tract originating in the cerebral motor cortex has been mentioned. The *extrapyramidal tracts* constitute other pathways contacting the motoneurons mono- or polysynaptically. They run from the basal ganglia, red nucleus, mesencephalic and pontine reticular formation, vestibular nucleus, and cerebellum. The *basal ganglia* (consisting of striatum and pallidum) form important relay stations for pathways between cerebral cortex, thalamus, and brain stem nuclei. The *red nucleus* is specially concerned in transmitting the cerebellar output down the spinal cord in the rubrospinal tract.

The *reticular formation* is a poorly defined part of the brain stem (mainly midbrain, pons, and the legmentum of the medulla). It is formed by many nuclei scattered throughout the central part of the brain stem. It is also characterized by an interlacing network of nerve fibers giving a reticular appearance, hence the name: reticular formation. It sends and relays excitatory and facilitatory extrapyramidal reticulospinal fibers to the spinal neurons, both to α and γ neurons. A stimulus of the facilitatory reticular neurons increases the rate of discharge of the gamma efferents. Conversely, the activated inhibitory reticular neurons reduce or annul the gamma activity. Similar effects can be elicited from the hypothalamus. Movements induced either by stimuli in the motor cortex or by reflex mechanisms can be abolished by impulses from the inhibitory reticular nerves. The net effect of the descending impulse traffic normally is a facilitation of spinal neurons, especially those innervating the extensor (antigravity) muscles.

The *thalamus* is a large, ovoid mass of gray matter. It is an important sensory relay station and projects efferent fibers to the primary cortical sensory areas. Part of the thalamus is concerned in the transmission between the cerebellum and cerebrum. It also receives important impulses from the reticular formation.

Integration of the Neuronal Activity in Movement

What is the beginning of the neural events that lead to a muscular movement? It has been noted that there is a slowly rising negative electrical potential over a wide area of the cerebral cortex as early as about 0.8 s before the onset of a voluntary movement (a "readiness potential"). About 30 to 150 ms prior to the onset of muscular contraction, there appear sharper potentials, the so-called motor potentials, located more specifically over an area of the cerebral motor cortex concerned in the movement. An inhibition of the antagonist may precede the excitatory response of the agonists by some 15 to 20 ms. In other words, nerve cells in the motor cortex are important components in a circuit that initiates muscular contractions.

It has been pointed out that the cerebrum cannot begin instituting any movement without the cerebellum's immediately knowing about it. In fact, the cerebellum, many association areas in the cerebral cortex, and the basal ganglia are engaged in planning the movement and in translating the idea to move into a patterned activation of certain motor cortical columns. It should be recalled that the entire cerebral cortex sends fibers to both the cerebellum and the basal ganglia, and that these two structures in turn have massive connections back to the motor cortex via the thalamus. By recording the electrical activity in single nerve cells, it has been shown that both the cerebellum and the basal ganglia become active in advance of muscular contraction (Evarts, 1973).

There are apparently "memory stores" in the cerebellar nuclei related to specific innate as well as learned movements. The cerebellum continuously receives messages from peripheral receptors about joint positions, muscle length and tension, movements, environment, and so on. When a call for movement is reported to the cerebellum, it has all the necessary requirements for execution and control of the movement. At the time the motor command descends to the motoneurons, the cerebellum updates the intended movement on the basis of the somatosensory description of body position and velocity on which the movement is to be superimposed. Within a fraction of a second, the return circuit from the cerebellum can take over the output from the cerebrum and modify the discharges down the pyramidal tract. In addition, the cerebellum can affect the motoneurons via extrapyramidal tracts.

The muscle spindles and other proprioreceptors are part of a mechanism for checking the execution of movement in relation to command (Granit, 1970, 1972). The cerebellum, in particular, has the potential to compute and integrate the total sum of information about a movement and to execute a follow-up correction. It has been emphasized that there is usually a coactivation of both α and γ motoneurons. In fact, one descending nerve axon can activate both α and γ motoneurons, at least those innervating the intercostal muscles (Granit,

1975). If the muscle spindle's own contraction is set for shortening that is suddenly prevented by an unexpected increase in the load, an accelerated afferent impulse traffic from the spindles driving the motoneuron pool automatically compensates for the extra load. This drive stops when the intrafusal and extrafusal lengths are functionally equal. The muscle spindles can serve as quick error detectors. In taking compensatory measures, the cerebellum has a key function.

We can see how a movement planned within the cortex in close cooperation with the cerebellum and basal ganglia is finally executed by the "upper motoneurons" in the cerebral motor cortex via the "lower motoneurons," most of them located in the spinal cord. There is a plan-ahead activity based on previous experience (memories) or memory fragments of fixed movement patterns, modified by a continuous updating of the motor command at its beginning and throughout its duration. There is always some kind of preprogram available. This is particularly important in very rapid movements, for they cannot be adequately updated once they begin. Therefore, in learning an exercise, one must first execute it slowly; the cerebral cortex continually intervenes, but the cerebellum does participate and it is capable of "learning." With practice the movement can become preprogrammed and eventually can be executed more and more rapidly. This means that the feedback from the periphery becomes less essential for a successful execution. For learned movements, the cerebellum provides an internal substitute for the external world (for details, see Eccles, 1973a, b; Allen and Tsukahara, 1974). The cerebellum's way of functioning has been compared with the servomechanisms commonly used in modern technology (the automatic pilot, industrial control systems, antirobot weapons, and so on).

At various levels of the nervous system, there are "generators" that can initiate and coordinate movements. The reciprocal innervation is a basic feature of such movements and, as we have seen, it is established all the way down to the spinal level. A brief description of some reflexes will clarify this point. The stretch reflex has been discussed in connection with the muscle spindle presentation. If pain receptors in the skin are stimulated (by a needle prick in the sole of the foot, for example), the limb is withdrawn from the source of the pain even before any sensation of pain has been experienced. The response is executed by means of excitatory connections in the spinal cord between the afferent fibers whose nerve endings were stimulated and the motoneurons of the flexor muscles of the extremity. Several interneurons are included in the reflex arc. Some of the intercalated interneurons will have an inhibitory influence on the motoneurons of the extensors of the extremity. This is an example of an ipsolateral flexor reflex.

If the nocuous stimulation is strong enough, the extensor muscle of the opposite limb of an animal is contracted and its flexor muscles are reciprocally inhibited almost simultaneously with the development of flexion on the ipsolateral limb. This crossed-extensor reaction (or contralateral extensor reflex) will further assist in removing the foot from the irritating needle. A stimulus of the hind limb may spread to the forelimbs via long nerve collaterals within the spinal cord. Examples of muscular activities that are normally executed and controlled by generators in the brain stem are breathing, swallowing, blinking, and coughing. At any moment one can voluntarily interfere with most such normally subconscious muscle actions. Motoneurons innervating the respiratory muscles can be activated by two parallel pathways, an "automatic" one serving the metabolic function of breathing, and a voluntary behavioral one originating in the cerebral cortex.

Grillner (1975) reports that cats with chronic spinal transections can generate the essential and basic features of the step cycle as well as coordinate the limbs and shift from one type of gait to another. This also holds true for the deafferented animal (with the dorsal roots cut). It means that locomotor patterns are generated in the spinal cord. There are also several structures in the brain stem from which locomotion can be initiated, and these patterns are exerted via pathways descending from the brain stem. The afferent input is important when, for one reason or another, the locomotor movements are disturbed. Grillner (1975) noticed that a light touch on the dorsum of a spinal kitten's paw during the flexion phase in walking gave an additional flexion. However, the same stimulus during the extension phase produced no response in the flexor muscles but enhanced the extensor activity. In other words, there was a reversal of the reflex. In one position, the afferent pathway to the flexor motoneurons was wide open; in a different position, it was shut off. Anesthesia of the skin abolished these reflexes. Also, deafferented higher primates like monkeys can develop motor skill; they can walk and run, and the timing of the muscles can be normal so the afferent signals from a limb are not necessary for movement. The eyes can take over part of the information input to the CNS. However, in patients with nonfunctioning dorsal roots, one notices a retardation and prolongation of the execution of voluntary movements, with particularly the slower movements being affected. More rapid movements are least influenced. This can be explained by the basic importance of a central programming, especially of the fast movements. It should be emphasized that the engagement of specific muscle groups and the demand for force in the performance of a given movement in a joint (at a given speed) are dependent on whether gravity works for or against the movement, the heaviness of extra external loads, the degree of muscle fatigue, etc. Numerous sense organs of different kinds provide a feedback mechanism that can report from the periphery, and they are therefore essential for very efficient and skilled movements. Allen and Tsukahara (1974) point out that the cerebellum with its "computer" provides the dominant input to the pyramidal neurons, whereas the direct somatosensory and cortical association areas exert subsidiary influences on these neurons.

The traditional point of view has been that the cerebral cortex reigns at the highest level in the hierarchical organization for the brain's motor function. However, Evarts (1973) emphasizes that the cerebral motor cortex is at a rather low level of the motor control system, close to the muscular apparatus itself.

The cerebellum and basal ganglia are at a higher functional level in the neural chain of command that initiates and controls movement. The primary function of the cerebral motor cortex may not be volition, but rather, the refined control of motor activity. Naturally, the evolution has created a specialization. The cerebellum is particularly involved in rapid ballistic movements, whereas the basal ganglia are preferentially active in slow movements. The motor cortex is engaged in both types of activities. In addition, the basal ganglia are important (1) to inhibit muscular tone and (2) to provide a "background activity" which is necessary for all movements. For example, if someone in a standing position performs a precision movement with the fingers, these movements are primarily guided by the motor cortex, but the leg, trunk, and arm muscles serve to stabilize the hand; their engagement takes place subconsciously, probably largely through the basal ganglia. A malfunction of these ganglia causes postural disturbances, muscular tremor at rest, increased tonus, muscular rigidity, and difficulty in the initiation of movement. Such symptoms are typical in Parkinson's disease, which is a neurological disorder resulting from damage to the basal ganglia. With damage to the cerebellum, a muscular tremor is also one symptom, but it is most severe during voluntary movement. The patient's movements are clumsy, carried out in a more or less disorderly fashion, and slow. The person also suffers from poor equilibrium. (It should be recalled that the cerebral cortex in the left hemisphere is "connected" with the right part of the cerebellum and skeletal muscles and with peripheral sense organs also on the right side of the body, and vice versa.) Higher primates with the cerebral motor cortex destroyed can still perform many activities, but there is a loss of fine motor control, particularly of the fingers, and decreased voluntary movement. In most individuals the left cerebral hemisphere is dominant, in liaison with the conscious self, and analytic and sequential; the right hemisphere has no liaison with consciousness, and it is nonverbal, synthetic, and musical, to list a few characteristics. Apparently the two hemispheres are complementary (Eccles, 1973b). Thanks to the massive impulse traffic through the corpus callosum, the minor hemisphere achieves consciousness by its communication with the dominant hemisphere.

To summarize this section, one may quote Eccles (1973b):

Let us now try to visualize what would be happening in the cerebro-cerebellar circuits during some skilled action, for example, a golf stroke. In the first place we will assume a loop time of the cerebro-cerebellar circuit of a fiftieth of a second, so that a motor command to start the stroke will result in a "wise" comment from the cerebellum to modify the pyramidal tract (PT) discharge in accord with its learnt performance. The modified (PT) discharge is reported to the cerebellum and this in turn evokes a further corrective comment from the cerebellum. Thus there is this continuous on-going cerebellar modification of PT discharge. Furthermore, it has to be remembered that, before the initial PT discharge that started off the golf stroke, there was already the combination background discharge of the PT cells. Moreover, continuous background discharges would also be occurring from Purkinje cells and nuclear cells (in the cerebellum), and in fact for all the cells in the circuit. Thus, when the movement is about to start, the whole circuit is in dynamic operation. Every action that you make is already superimposed on background discharges in the various neuronal circuits. They are "ticking over" all the time, and the cell discharges at all stages of the loop are modulated up or down in the process of the continuous flow of information. I think this concept of dynamic loop operation gives the essence of the motor control by the cerebellum. It is very important to recognize that all pyramidal tract discharges are provisional and are unceasingly subject to revision by the feedback operation of this cerebellar loop. It should be mentioned that there is also a feedback loop from the periphery that is concerned in motor control, but the loop time is much longer, about a tenth of a second. Presumably the cerebellum carries an immense store of information coded in its specific neuronal connectivities so that, in response to any pattern of pyramidal tract input, computation by the integrational machinery of the cerebellum leads to an output to the cerebrum that appropriately corrects its PT discharge. These hypotheses of the manner of cerebellar action in the control of movement provide great challenges for future research.

Learning

When unfamiliar and complicated movements are performed, they are performed clumsily and with difficulty. With proper practice they become smooth and easy. However, we are far from understanding the reasons for this in physiological terms, how new connections are formed during learning, etc. The movements of the newborn are characterized by uncoordinated movements of many parts of the body, but gradually, coordinated reflexes develop (postural reflexes, tonic neck reflexes, writing reflexes, walking, and so on). The extrapyramidal and pyramidal motor centers and tracts develop, so that the movements become more complex. Nevertheless, when beginning school, the child's extrapyramidal centers dominate the movements, and specific practical tasks directed toward external objects constitute the activity in play, imitations, and other activities. The child usually fails in the performance of more complicated "artificial" movements. It is during puberty that the pyramidal system first attains full functional maturity and the child becomes capable of developing the fine coordinated movements necessary for writing and other precise actions, based on the integration of nervous activity from various levels of the CNS and the impact from all peripheral receptors. The changing body dimensions during adolescence necessitate a continuous modification of the interpretation of impulses exchanged between muscles and CNS to secure a correct innervation pattern for given tasks. Apparently there are definite anatomic and physiological limitations to the complex movements that can be performed during early adolescence.

The CNS, like other tissues, is characterized by a plasticity in its structural, biochemical, and functional properties (Eccles, 1973b; Gilbert, 1975; Buser, 1976). The gross neuronal connectivities are established in their final form before birth. After the initial formation, no change occurs in the neuronal pathways or growth in the brain of mammals. A genetic instruction secures the development, and there is some kind of mechanical guidance and, at the end, a chemical sensing and recognition. However, in adults as well as children a structural plasticity is evident at the microlevel, that is, some synapses may develop while others regress. Frequent impulse traffic through synapses seems primarily to result in a maintenance of the dendritic spine synapses, but it can also modify the synapses by a branching of the spines to form secondary synapses. There may be a hypertrophy of the synaptic knobs. In other words, the process of learning seems to stimulate a protein synthesis of activated synapses which gives a selective synaptic hypertrophy. In one way or another, repetition of a movement can be coded and memory engrams can be stored and played back with great accuracy and precision. Available results suggest that the cerebellum is directly involved in the learning of motor actions (Gilbert, 1975).

In view of the very sharply defined projection of impulses within and between different areas of the brain, it is difficult to understand how a "general movement pattern" may function. The organization does not appear to favor a "transfer" effect, i.e., the learning by practice of a certain movement pattern does not in itself enhance the performance of another movement pattern, not even one that is relatively similar. However, the technique of learning new tasks can be improved. One can learn and memorize specific activities which can then be utilized and woven together in different combinations. The pianist, having practiced many hours at the keyboard, has the potential to learn new pieces of music, quickly.

Learned movements can easily be performed subconsciously, but *the motor cortex is still engaged in the control of the movement regardless of whether the movement is innate or learned*. The same nerve cells seem to control the pattern of muscle contraction regardless of the circumstances or context of the movement (Evarts, 1973).

There is an interesting hypothesis that special nerves at their synapses "instruct" other activated synapses on the same dendrite to grow (Eccles, 1973b; Gilbert, 1975). In the cerebellum the climbing fibers synapsing on Purkinje cell dendrites, when activated, may say "Now learn" to the spine synapses that are simultaneously activated by the parallel fibers (see the subsection "Cerebellum"). This means that a trophic effect is exerted by the climbing fibers on the synaptic spines for parallel fibers changing the strength of the synapses. This provides a *short-term memory* of the motor output. From the periphery, reports will arrive telling whether or not the movement was satisfactory. If the report is favorable, a *third* event will take place: Axons from an area (the nucleus locos coeruleus) will signal to the unit of Purkinje cells engaged in the computation and indicate that the outcome of the movement was satisfactory and that the movement was suitable for storage. This signal will therefore produce a *long-term memory* of the motor output (Fig. 4-18). The axons of the locus coeruleus cell produce norepinephrine which, in one way or another, should initiate the transformation of short-term synaptic changes into long-term changes. These cells also seem to be connected with a general reward, a positive reinforcement system of the brain, and their terminals are also distributed to regions other than the cerebellum. If the movement was unsatisfactory, the locus coeruleus cells will remain silent, the short-term changes in the synaptic strength will vanish, and there will be no memory of the particular motor output left behind (Gilbert, 1975). Considering the unceasing nerve activity going on continuously in the CNS, it is an attractive thought that several systems must be engaged simultaneously to achieve a structural modification giving a long-term memory for the event that triggered this coordinated impulse traffic in the systems.

There is evidence that the information required in the learning process is encoded in the molecular structure of ribonucleic acid (RNA) of a special type in macromolecules derived from the nucleus of nerve cells (Hydén and Egyhazi, 1962). Rats were trained 45 min per day for 3 to 4 days to walk a tight wire to obtain food. In the trained group, a special type of RNA was formed in the neurons which were assumed to have been engaged in the wire balance walk.

There is further evidence reported in support of the phenomenon that the injection of brain-derived fractions prepared from trained donors results in significant behavioral changes in the injected animals, shortening the time required to learn a given task. This means that it is possible to transfer information among animals via brain extracts (Byrne and Samuel, 1966).

Little by little a "sense" is developed, based on both endowed and acquired capabilities. Granit (1972) reports old experiments showing that a person, when lifting two equally heavy spheres, one 4 cm and the other 10 cm in diameter, feels the larger sphere to be lighter. Subjects who lifted equally heavy objects (112 g) with volumes ranging from 10 ml to 2,000 ml felt that the largest ones were also lighter; they were then asked to add weights to the apparently lighter object (2,000 ml in volume) until it felt as heavy as the 10 ml one. It turned out that the amount required averaged about 112 g!

Granit and Burke (1973) quoted Hagbarth's observation: "If, in a completely darkened room, a normal subject swings his arm in an arc around the elbow joint and the arm is stroboscopically illuminated only when it is at 90°, the subject has a strong sensation that he is not really moving his arm at all. On closing his eyes, this illusion immediately disappears." No doubt the visual input has a relatively dominant influence for the determination of position. It can also as mentioned partially compensate for defects in the afferent signal input.

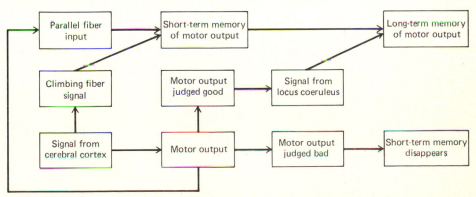

Figure 4-18 The proposed scheme for the storage in the cerebellum of information relating to movements is shown. Initially a movement is carried out through signals from the cerebral cortex. These signals cause climbing fiber signals in the memorizing units of the cerebellum projecting to the muscles involved in the movement. The simultaneous firing of the parallel and climbing fibers produce changes in the parallel fiber synapses on the Purkinje cells, giving a short-term memory of the movement. Only if the movement is judged to be satisfactory will a signal be sent from the locus coeruleus to convert the memory into a long-term form. *(From Gilbert, 1975.)*

PROPERTIES OF THE MUSCLE
AND MUSCULAR CONTRACTION

The Motor Unit

The motor unit consists of the alpha motoneuron and the muscle fibers which it innervates. As discussed earlier, the number of muscle fibers in a unit varies from about 5 to about 2,000. A single muscle fiber rarely has a polyneuronal innervation. When a nerve impulse reaches the motor endplate, ACh is liberated and the membrane of the muscle fibers is locally depolarized. This evokes an action potential propagating along the muscle fiber at a speed of about 5 m · s^{-1}. The whole length of each muscle fiber and all the fibers of the motor unit are therefore brought into action almost instantly, even if the unit is acted upon by a nerve fiber at only one point in its length. The main feature of the muscle action potential is very similar to the action potential of the nerve. It is generated and propagated by basically the same mechanisms. A single, adequately strong, stimulus of the motor nerve gives rise to a twitch of the innervated muscle. After a short period the tension developed by the muscle increases and thereafter it decreases again. The muscle action potential is completed during the early phase of the contraction (Fig. 4-19a). If the motor nerve of the muscle is stimulated repeatedly and the second impulse reaches the muscle before it has relaxed after the first stimulus, it contracts again. Since the second twitch starts from a higher tension level, the tension resulting from the two stimuli will be considerably greater than that from a single stimulus of the same strength (summation). At high rates of stimulation, the muscle does not relax before the next contraction, and the muscle fibers are in full tetanus; that is, there is a complete mechanical fusion of the contractions. The tension developed by a muscle in tetanus may be 4 to 5 times greater than that exerted during a single twitch. At lower rates of stimulation, the mechanical fusion is incomplete and the developed tension is not maximal (Fig. 4-19b). Even in a tetanus the muscle action potential develops in response to each stimulus, revealing the rate at which the nerve is being stimulated (Fig. 4-19c). It was mentioned in Chap. 3 (and shown in Table 3-1) that the skeletal muscle is composed of two main types of muscle fibers with different mechanical properties: Type I or *slow twitch fiber*, and Type II or *fast twitch fiber*. In the cat gastrocnemius, the contraction time course for a single isometric twitch averages about 75 ms in the slow fibers and 35 to 40 ms in the fast fibers (see Burke and Edgerton, 1975). The speed of contraction is proportional to the activity of the myofibrillar (myosin) ATPase activity of the fibers. In a preparation, staining of the ATPase activity (after incubation at pH 9.4) can be applied for an identification of the fibers. An example is given in Fig. 4-20. The fast twitch fibers are innervated by larger motoneurons than the slow fibers and *each motor unit is composed of only one kind of fiber*. Because of a relatively long after-hyperpolarization for the slow fibers, they will fuse their contractions into tetanus at a lower frequency (around 20 Hz) than do the fast fibers (tetanus at about 50 Hz). Apparently, during a single twitch, the active state of

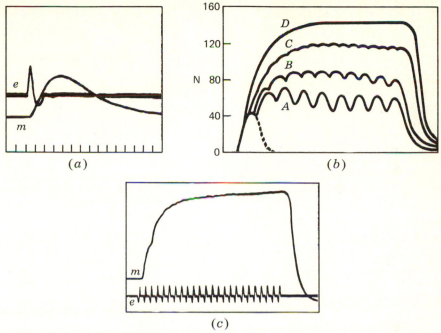

Figure 4-19 Action potentials (*e*) and mechanical changes (*m*) in skeletal muscle in response to stimulation of its motor nerve (from mammalian nerve-muscle preparation).
(*a*) A single stimulus gives rise to a single twitch. The diphasic action potential is completed in the early part of the contraction phase. Each mark on the abscissa denotes 0.05 s.
(*b*) Time (abscissa) is not marked; the ordinate gives tension in Newton(N). The lower curve, partly dotted, shows one response of the muscle to a single maximal stimulation to its nerve. Curves *A* to *D* show the responses to repetitive, maximal stimuli: *A* at 19, *B* at 24, *C* at 35, and *D* at 115 Hz. Note that the higher the frequency of stimulation, the greater is the tension which is developed, and it is maintained more steadily. From a state of partial or incomplete tetanus (curves *A* to *C*), the muscle goes into full tetanus (curve *D*).
(*c*) At a high frequency of stimulation (in this case 67 Hz), the muscle goes into tetanus, but the action potentials appear at the same frequency as the stimulation rate. [(a) and (c) *after Sherrington et al., 1932*, (b) *after Cooper and Eccles, 1930*.]

the muscle fibers is not long enough to permit the mechanical rearrangement essential for maximal tension to develop.

In the cat gastrocnemius, the number of muscle fibers in a motor unit is not very different but the fiber area of the fast twitch fibers is much larger than that of slow twitch fibers. Therefore the maximal stimulus of a motor unit of a large α motoneuron gives an increase in the tension of about 60 g (average) in contrast with about 5 g for a small α motoneuron (Burke and Edgerton, 1975).

The slow twitch fibers are much more resistant to fatigue than the fast twitch fibers. This can be explained by a higher potential for aerobic metabolism in the slow fibers. They are rich in mitochondria and myoglobin, and the density of the capillary network is more developed around the slow fibers. The fast twitch fibers are richer in glycolytic enzymes. (The fast twitch fibers are also referred to as phasic, or white fibers, while the slow ones are tonic or red fibers.)

Histochemically and functionally, identification can be made of fast twitch (Type II a or intermediate) fibers which are better equipped for aerobic energy yield and are less fatigable than the "classical" fast twitch fibers (Type II b). It should be pointed out that these two fiber types are similar in their kinetic properties. The modification of the functional potential can be an effect of training. It should also be noted that different animal species differ in their muscle "composition." In human muscles there are normally few of the Type II a fibers. (For references, see Burke and Edgerton, 1975; Close, 1972; Essen et al., 1975; Saltin et al., 1977).

In human muscles, the distribution of slow and fast twitch fibers is relatively homogeneous throughout the muscle. However, the proportion of the

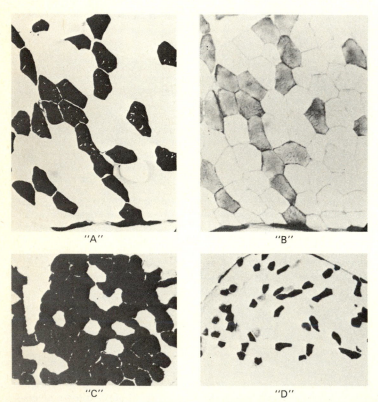

Figure 4-20 *Top:* Serial cross sections of human skeletal muscle samples. "A" (stained for myofibrillar ATPase) showing fast twitch fibers, Type II (dark), and slow twitch fibers, Type I (light); "B" (stained for glycogen with PAS reaction staining glycogen, dark). By comparison with "A," the fibers can be identified in "B." Note the selective depletion of glycogen in the slow twitch fibers.
Bottom: Sections of skeletal muscle (stained for myofibrillar ATPase as in "A"). "C" was obtained from a male sprinter who ran 100 m in 10 s; "D" is from an elite long-distance runner. Note the dominance of fast twitch fibers in the sprinter's muscle and the reversed picture for the endurance athlete. *(By courtesy of Gollnick and Saltin).*

two fiber types within a given muscle varies greatly from individual to individual. Thus, the proportion of slow fibers can be anywhere from about 10 up to 95 percent. On the average, the percentage is around 50 percent in the vastus lateralis quadriceps and deltoidus but as high as about 85 percent in the soleus (Saltin, et al. 1977). Top athletes in endurance events have a high percentage of slow twitch fibers, but this is not as a consequence of the prolonged training. So far, all studies have indicated that the *fiber composition is genetically determined*, and the differentiation into fast and slow fibers takes place before or soon after birth. Identical twins, for example, have a very similar proportion of fiber types in given muscles, in contrast with nonidentical twins (Komi et al., as reported by Sjödin, 1976). As mentioned, the motoneurons differ for the two fiber types and apparently the nerve has a trophic (chemically mediated) influence on the contractile properties of the muscle cells it innervates. The pattern and frequence of the impulse traffic may also play a role in the characteristics of the muscle fibers.

An α motoneuron, via its axon, will establish functional endplates on denervated muscle fibers and excite them effectively. Buller et al. (1960) describe experiments on kittens and cats in which the nerves to a slow muscle (such as the soleus) and a fast muscle (such as the flexor digitorum longus) were divided and cross-sutured. When a nerve from the "fast" motoneurons had innervated the slow muscle, the muscle was gradually transformed to a fast muscle, even in the adult cat. Likewise, "slow" motoneurons converted fast muscles to slow ones. Later experiments have shown that a changed speed of the muscle contraction after reinnervation is due to changes in the kinetic properties of myofibrillar ATPase and other enzymes. There are also changes in the myosin, troponin, and sarcoplasmic reticulum, for example, the contractile material itself and the regulatory protein system after cross-innervation (for references, see Amphlett et al., 1975; Close, 1972). These effects are probably mediated by the molecular transport along the motor axons and across to the muscle fibers via the neuromuscular synapses (see Gutmann, 1976).

Because of this trophic function of the nerve on a muscle's metabolism and protein synthesis, it is easy to understand that a muscle is much more severely affected by a denervation than by a long-lasting block of nervous impulses or disuse for other reasons (Guth, 1968). A stimulation of denervated muscle, even a very intensive one, will not lead to a hypertrophy, as noticed in a normal animal subjected to training (Gutmann et al., 1961).

Figure 4-21 shows how a single short interval in an otherwise relatively low-frequency stimulus train can produce marked and long-lasting tension enhancement. Similar "double" discharging can apparently occur in the human nerve-muscle system (Burke and Edgerton, 1975).

Summary The human skeletal muscle is composed of two main types of muscle fibers with different contractile properties, the slow twitch fibers (Type I) and the fast twitch fibers (Type II). They differ in some of the morphological and histochemical properties, and of particular significance is the higher activity of the myofibrillar ATPase in the fast fibers. One α motoneuron innervates a homogenous group of muscle fibers (fast *or* slow twitch fibers) and forms a motor unit. The development of muscle fibers is guided by a trophic influence from the motoneuron. Most motor units composed of fast twitch fibers can produce a higher maximal tension than the slow units, but they lack

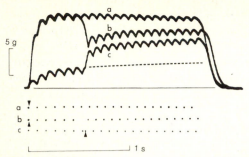

Figure 4-21 The "catch property" in a single muscle unit. Mechanical responses of a type I unit in cat MG (superimposed records) to three stimulation sequences, each with mean frequency of about 12.8 Hz (basic interstimulus interval of 78 ms). Sequence *a* included one extra stimulus following the first with an interval of 5 ms (arrow), which resulted in marked tension enhancement over what would have resulted from a constant frequency train without the extra impulse (dashed line). Sequence *b* also had the same short interval to start (arrow) plus a later, longer interval which caused tension to drop to a different level, at which it was "caught" by resumption of the 12.8 Hz stimulus rate. Sequence *c* included an interstimulus interval of about 20 ms (arrow), which resulted in less tension enhancement than the shorter intervals of sequences *a* and *b*. *(From Burke and Edgerton, 1975.)*

endurance. Repeated discharges in the motor nerve produce a higher tension than a single twitch (Figs. 4-19, 4-21). The contractions will fuse in a tetanus at a frequency of about 20 Hz for the slow twitch units and at about 50 Hz for the fast twitch units. The higher the stimulation frequency and the larger the number of active motor units, the greater will be the tension by the muscle.

Figure 4-19 is based on results obtained from nerve-muscle preparations. Normally the activated motoneurons discharge asynchronously, and the muscle fibers of the different motor units are in different phases of activity. The net effect is a smooth muscle contraction, even if the individual motor units were activated at a subtetanus frequency and therefore showed a tremulous response.

Mechanical Work

In the activated muscle the contractile components, i.e., the myofibrils, shorten and stretch the elastic components (connective tissue, tendon). When both ends of the muscle are fixed and no movement occurs in the joint involved, the contraction is called *isometric*. If the muscle varies its length when activated, the contraction is *isotonic* (*dynamic*). In the latter case, external work can be done, and the amount of work calculated from the product of weight lifted times the distance. Since the distance is zero in an isometric contraction, no mechanical work is done according to physical laws. However, isometric activity demands energy and can be very fatiguing. From a physiological viewpoint, the "work" performed is definitely related more to developed force times contraction time than to force times the displacement. In the dynamic exercise, the muscle can shorten, in which case the work is labeled *positive* or *concentric*, or the muscle can lengthen, in which case the work is *negative* or *eccentric*.

An activation of muscle fibers is always associated with an increased heat production (Hill, 1958). During isometric contraction, all the extra energy output is converted to heat. In dynamic exercises, various fractions of the chemically available energy is transformed to heat. The chemical reactions providing the energy for contraction of the muscles are in some way controlled by the change in the length of the muscle and by the tension placed on the muscle.

The mechanical efficiency of work (ME), expressed as a percentage, is the ratio of external work performed, W, expressed in energy units, to the extra energy production:

$$(ME) = \frac{W \times 100}{E - e}$$

where E represents the gross energy output and e is the resting metabolic rate expressed in energy units.

When a person exercises on a bicycle ergometer, climbs stairs, or engages in similar performances, the mechanical efficiency rises to 20 to 25 percent; i.e., 75 to 80 percent of the energy is dissipated as heat. In fast running and jumping, the efficiency can be still higher. In isometric work and many other types of activity, the mechanical efficiency is 0 percent. It should be emphasized that for a work that is partially anaerobic, the energy output or oxygen uptake must be measured not only during the work period but also during the recovery period following such work (oxygen debt).

The produced heat increases the temperature of the muscle. Within limits the elevated temperature improves the performance of the muscle. This can be explained on both chemical and physical bases. The increased tension produced by the muscle fibers upon repeated stimulation (Fig. 4-19b) can be explained partially by the beneficial effect of increased tissue temperature. To prevent overheating during prolonged activity of the muscle fibers, an increased local blood flow is essential as well as an increased heat conductance of the skin (Chaps. 6 and 15).

Posture

The upright position is maintained by muscular activity against the force of gravity. In the erect posture the line of gravity runs in the midline through (1) the mastoid processes (in front of the atlanto-occipital articulation); (2) a point just in front of the shoulder joints; (3) the hip joints (or just behind); (4) a point just in front of the center of the knee joints; and (5) a point (3 to 7 cm) in front of the ankle joints (Basmajian, 1967). None of the joints engaged in the erect position is moved to the extreme of its mobility, and the body therefore does not "hang" in the ligaments or capsules of the joints. However, when loads are supported so that traction is exerted across a joint, ligaments, and not muscles, normally maintain the integrity of the joint. For example, the muscles that are able to support the arches in the foot are generally inactive in standing at rest.

With a subject seated upright with the arms hanging in the relaxed neutral position and with heavy downward pull applied to an arm, the muscles that cross the shoulder joint and the elbow joint are not active in preventing dislocation of these joints (Basmajian, 1967). The center of gravity of the head and trunk is very close to the supporting column of bones, so that the antigravity muscles are only very slightly loaded. Basmajian points out that among mammals, the human being has the most economical antigravity mechanisms once the upright posture is attained. (A quadruped, when standing with its joints partially fixed, has a much more wasteful antigravity machinery, and a direct comparison of the function of antigravity muscles of humans and animals may therefore be misleading.) The so-called antigravity muscles of the human have the very important function of producing the powerful movements necessary for the changes from a lying to a sitting or standing position and of providing a firm foundation for the variety of muscular activities of everyday life.

The stretch reflex is the basic reflex in postural control. The muscles antagonizing the pull of gravity are stretched, and thereby the muscle spindles located in the muscles are also stretched. Afferent impulses are evoked, and the muscle contracts so that the pull of gravity is counterbalanced. Since the intrafusal muscle fibers of the muscle spindle can be activated from higher centers via the gamma fibers, its receptor may be more or less prone to respond to a stretch. The hypothalamus is probably one important relay center. A feeling of happiness, alertness, or attention may increase the gamma activity, whereas unhappiness, drowsiness, or lack of attention may decrease the activity in gamma fibers. In this way the very noticeable relationship between an individual's mood and posture may be explained.

Many of the antigravity muscles are of the slow twitch fiber type (tonic muscles), and they are more affected by the gamma loop than the fast twitch fibers. The afferent discharge from the muscle spindles increases in direct proportion to the degree of extension of the muscle, but the efferent rate of discharge from the motoneurons is fairly constant. A stronger afferent impulse traffic recruits more motoneurons rather than increasing the rate of discharge. In addition to the gamma system, many other reflexes contribute to the integrated activity regulating the normal posture. On the ventral horn, nerve fibers converge from dorsal nerve roots and from all levels of the brain and spinal cord. Impulses from the eye, vestibular apparatus, and sole of the foot are especially important for the modification of the activity of the alpha motoneurons. The cerebellum and reticular formation constitute important relay stations and links between the alpha and gamma systems, and the cerebral cortex exhibits an overall control (partly inhibitory).

The free normal posture is characterized by a "postural sway," so that the center of gravity varies with respect to its projection on the ground with a frequency of 5 to 6 "cycles"/min. With the eyes closed (or in the dark), the swaying is more pronounced. The muscle spindles are actually pulled upon irregularly, and their rate of discharge is therefore highly irregular. Electromyographic studies reveal that the antigravity muscles may be activated with a frequency of 5 to 20 $\cdot$ s^{-1}. The alternating activity-inactivity in the motor units involved and the postural sway will prevent fatigue and facilitate the blood flow through the muscles; it also assists the venous return. It should be emphasized that fatigue in the well-balanced standing individual is usually not due to muscular fatigue, but is more likely caused by an inappropriate distribution of the blood. The energy output in the erect position is only slightly elevated, say from 0.25 liter $\cdot$ min^{-1} in the supine to 0.30 to 0.35 liter $\cdot$ min^{-1}, but the cardiac output and stroke volume are decreased and the heart rate is elevated. In free positions where the center of gravity of limbs and/or trunk is shifted from a balanced position, activity in counteracting muscle groups must compensate; this increases the load on the muscles. The slow muscles respond with tetanic contractions even if the discharge frequency in their motoneurons is low; hence the blood flow may be obstructed, causing local fatigue. It is an interesting finding that during activity with forward flexion of the spinal column, there is a marked muscular activity until flexion is extreme, at which time the ligamentary structures assume the load and discharge from the trunk muscle ceases (Floyd and Silver, 1955; Basmajian, 1967). (For a general discussion see Carlsöö, 1972.)

Muscle Length and Speed of Contraction

The tension exerted by a stimulated muscle, either during a single twitch or during incomplete or full tetanus, depends on the initial muscle length. An unattached, unstimulated muscle is at its equilibrium length, and the tension is zero. If stretched, the passive elastic tension increases as an exponential function of length, actually over a range up to 200 percent of the equilibrium length (Fig. 4-22, lower curve). Normally, when attached by its tendons to the skeleton, the muscle is under slight tension, since it is moderately stretched (at resting length). Measurements of the tension developed by the stimulated muscle show that isometric tension is maximal when the initial length of the muscle at the time of stimulation is about 20 percent above the equilibrium length (relative length 1.2:1). The active tension falls roughly linearly at lengths below this optimum and is zero when the muscle is maximally shortened. When stretched beyond the relative length 1.2:1, the active tension produced by the stimulated muscle becomes progressively smaller and is zero when the muscle is elongated about twice its resting length. This failure to yield tension when overstretched can be explained on the sliding hypothesis of muscular contraction. When the actin and myosin filaments cease to overlap, the cross-linkages between them cannot be formed upon stimulation, no sliding movement can occur, and therefore no active tension develops. Figure 4-22 shows experimental data from studies in which measurements were made of (1) the tension produced by passive stretching of the triceps muscle in the human (lowest curve) and (2) the tension developed during maximal voluntary effort (top curve). The explanation of why less tension is exerted when the muscle shortens is not known. The tension also decreases as the speed of shortening

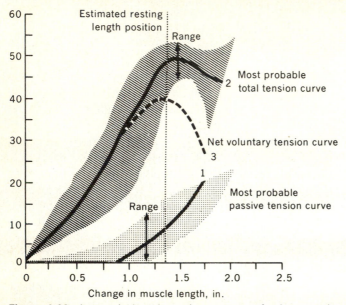

Figure 4-22 Isometric length-tension summary for human triceps muscle. (The ordinate gives the maximal tension in pounds.) To obtain the "net voluntary tension curve," the values from the "passive tension curve" were subtracted from the "total tension curve." For further explanation, see text. (After University of California, "Fundamental Studies of Human Locomotion and Other Information Relating to Design of Artificial Limbs," **2**, 1947.) *(See Ruch and Fulton, 1960.)*

increases. The velocity in a muscle contraction is maximal with zero load. With a load which the muscle just fails to lift, the velocity is zero, and the maximal isometric tension develops.

Striated muscles can, within a second, shorten at a rate up to 10 times their length. In the intact body, anatomic limitations of the joints usually restrict the lengthening and shortening of the muscles, and the permitted range is normally within 0.7 to 1.2:1 (for some muscles up to 1.4:1) of the equilibrium length. The maximal tension can therefore usually be exerted when the muscle is maximally stretched, and as it shortens, the tension produced decreases. Since the skeletal muscles exert their effect on external resistance via levers, the geometric arrangement of the bony levers must be included in an analysis of the optimal work positions and the most effective utilization of the forces of muscle contractions. Figure 4-23 illustrates a position in which the flexor of the arm holding a weight must produce a force that is about 10 times greater than the force of gravity acting on the load. The optimal lever arm for the biceps muscle is obtained with the arm flexed at an angle of about 90° in the elbow joint. An extension or a further flexion of the arm from the right angle causes a decrease in the lever arm of the biceps. If a constant force acts on the forearm against the pull of the biceps, this muscle must increase its exerted tension to balance the resistance as the lever arm decreases in length. If the tendon of

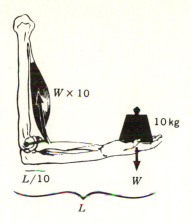

$W \times 10$

10 kg

$L/10$ W

L

Figure 4-23 In order to hold a weight of 10 kg in the hand in the position shown in the figure, the arm flexor must exert a force 10 times greater (or about 100 kg or 1000 N), since the lever of the weight (L) is 10 times the length of the lever of the arm flexors.

biceps were closer to the hand, its ability to flex the forearm against resistance would certainly be favored—but at the expense of speed of movement.

Summary In any analysis of the most favorable position, if maximal force is to be exerted on an external resistance, consideration must be given to the following facts: (1) The maximal tension which any muscle fiber can develop depends on the relative length of the muscle fiber at the time of contraction. The tension has a maximum at a relative length of about 1.2:1 and decreases at lower and higher lengths. (2) The lever arms in the body, through which the muscle tensions are transformed into pulls, pushes, etc., alter with changing positions of the movable joints.

With regard to terminology, there can be no confusion about the term *isometric muscle contraction*. However, since the lever arm usually alters during a movement in a joint, very seldom is a muscular contraction purely isotonic (having constant tension). Even if the external load is kept constant, the force developed by the muscle varies as the lever arms become shorter or longer. Therefore it is more correct to use the term *dynamic exercise* than *isotonic exercise* when there is a movement in joints involved.

In functional muscle activities, several anatomically different muscles collaborate. The parts of the muscle group that act in synergism may change with position of the limb. Consequently, it is very difficult to predict, from theoretical considerations, the most efficient work position which will produce the greatest strength.

Figure 4-24 illustrates the variation in maximal force developed by muscles during contraction with different speeds, showing the well-known S-shape relationship, i.e., the classical *force-velocity curve*. In agreement with previous discussions, we note that the highest force can be developed in a fast eccentric contraction and that the maximal force then declines to a minimum when the muscle is activated in a concentric contraction at high speed. When measuring maximal dynamic strength, it is therefore extremely important to control the speed of contraction carefully. For example, in a strength test performed

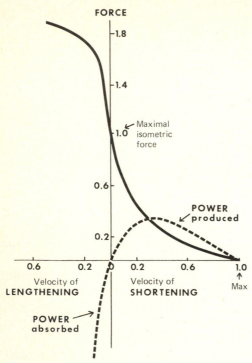

Figure 4-24 The classical force-velocity curve (full line) obtained on an isolated muscle, showing the maximal force that can be developed when a muscle is contracting at various speeds. Note that the maximal force in a concentric activity is less than in an isometric contraction. Highest force can be attained in a rapid eccentric contraction. The dotted line gives the maximal power, i.e., the force times the velocity of contraction. Curves differ according to the muscles studied. In vivo experiments, the analysis is complicated because in movements, the muscles are rarely working isotonically; the leverage changes during a dynamic contraction and therefore the force demand on a muscle is not constant even when the external load is kept constant.

before and after a training program, a slight reduction in velocity of a concentric contraction in the second test can erronously simulate an increase in muscle strength. There has been developed an apparatus for measurement of force potentials at given velocities, i.e., an isokinetic (same speed) loading dynamometer (for ref. see Thorstensson, 1976). Figure 4-24 also shows the power, i.e., the force times distance per unit of time, that is developed or absorbed in maximal contraction at various speeds. The highest power is attained when velocity of contraction is 25 to 30 percent of the maximal value; the force is then about 30 percent the maximal isometric strength.

Electromyographic (EMG) studies (Bigland and Lippold, 1954) show the following: (1) When a voluntary isometric contraction of a muscle is made, the electrical activity, measured by integrating the action potentials from surface electrodes, bears a linear relation to the tension that is being exerted. (2) At constant velocity of shortening or lengthening, the electrical activity in the muscle is directly proportional to the tension. However, the slope of this correlation falls off during lengthening (Fig. 4-25). This means that the degree of muscle excitation required to produce a given force of contraction is smaller when the active muscle is forcibly stretched than it is when the muscle shortens at the same velocity. (3) At constant tension, the electrical activity increases linearly with velocity of shortening, but it decreases when the muscle is being lengthened. (See also Milner-Brown and Stein, 1975.) The integrated EMG activity is the same in maximal contraction of different speeds, indicating the same degree of motor-unit recruitment in maximal efforts; it does not matter whether the contraction is concentric, isometric, or eccentric (Komi, 1973; Rodgers and Berger, 1974). Roughly, the oxygen consumption of the active muscles should reflect the number of active motor units and their frequency of excitation, that is, the electrical activity displayed. As a matter of fact, the oxygen uptake increases linearly with the work load in many types of activities. A comparison of the oxygen uptake at a given rate of work, in exercise where the muscle groups work concentrically ("positive work") and eccentrically ("negative work"), respectively, reveals that the negative work demands much less oxygen (Fig. 4-26) (Asmussen, 1953; Bigland and Lippold, 1954). These results fit with the electromyographic studies and the force-velocity characteristic of a muscle.

"Regulation of Strength"

Muscle strength is a very complicated function, and the number of muscle fibers available is only part of the story. It is not surprising that even in well-standardized measurements of muscle strength, the standard deviation of

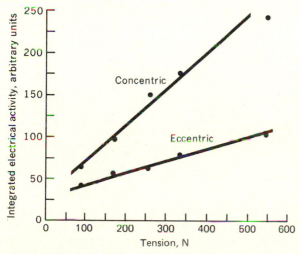

Figure 4-25 The relation between integrated electrical activity and tension in the human calf muscles. Recording from surface electrodes. Shortening at constant velocity (above) and lengthening at the same constant velocity (below). Each point is the mean of the first 10 observations on one subject. Tension represents weight lifted and is approximately one-tenth of the tension calculated in the tendon. *(Bigland and Lippold, 1954.)*

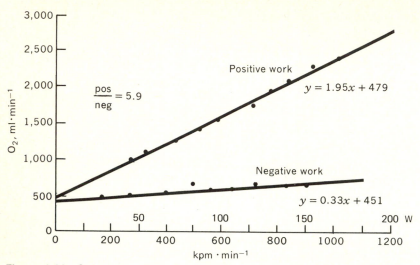

Figure 4-26 Oxygen uptake in positive (upper curve) and negative (lower curve) work consisting of riding a bicycle on a motor-driven treadmill, uphill in positive and downhill in negative work (with the movements of the pedals reversed). The work load was measured as the product of weight of subject plus bicycle times the vertical distance that this weight was lifted or lowered. Oxygen uptake at zero load was measured during free-wheeling. Rate of pedaling = 45 rpm. Note that the cost of positive work was on an average 5.9 times higher than that of negative work. *(From Asmussen, 1953.)*

the results obtained in repeated tests on the same subject is of the order of ± 10 percent or even higher (Table 4-1) (Tornvall, 1963; Simonson, 1971, p. 244).

The main objection against the method for the testing of muscle strength with the aid of the muscle dynamometer is its dependence upon the motivation of the subject to make the maximal effort. In fact, the practical application of this method may ultimately depend upon objective criteria for the evaluation of the state of motivation of the subject to be tested. In an effort to search for such criteria, use was made of the observation made by athletic performers that there is a relationship between the rise in heart rate and the level of endeavor or exertion immediately before the start of an athletic competition. Heart rates were recorded with the aid of a recording electrocardiograph for a 10-s period immediately before and again immediately after the muscle dynamometer test, in five normal subjects, with a total of 137 observations, and correlated with the measured muscle strength. A statistically significant correlation ($r = 0.738$, $P < 0.001$) was observed between the subject's heart rate before the test and his measured muscle strength; and also between his heart rate immediately after the test and his muscle strength ($r = 0.798$, $P < 0.001$). This finding suggests the possibility of using the heart rate as an indication of motivation in the selection of motivated subjects for experiments involving testing of muscle strength.

From studies on individuals of different body size, age, and sex, Asmussen et al. (1965) and Lambert (1965) concluded that the correlation r between symmetrical muscle groups (right and left) is quite high ($r = 0.8$). Between flexors and extensors of the same extremity, it is fairly high, but between muscles from different parts of the body, the correlation is rather low ($r =$ about 0.4 or less). Therefore, they conclude, the general muscle strength should not be evaluated from measurements in one single muscle group, such as the

Table 4-1 Testing of Muscle Strength in Four Subjects at Half-Hourly Intervals for One Day*

Subject	Right Arm	Left Arm
F.A.	33.1 ± 1.0 (13) 3.6	33.8 ± 2.0 (13) 6.1
G.B.	31.1 ± 1.1 (13) 3.9	28.9 ± 1.2 (13) 4.3
E.D.	37.5 ± 1.5 (10) 4.6	38.6 ± 1.1 (10) 3.5
J.S.	30.6 ± 1.0 (13) 3.5	29.9 ± 0.8 (13) 2.9

*Figures denote mean, standard error of the mean, number of observations, and standard deviation.

finger flexors in a handgrip, but from application of a battery of selected, well-standardized muscle tests. Since the correlation between dynamic and isometric strength was high (r = about 0.8), the maximal dynamic strength can be roughly predicted from the simpler measurements of the maximal isometric strength of the same muscles, provided that the subject is not especially well trained in a particular type of exercise.

With this in mind, we may proceed to discuss the "regulation of strength." In a completely relaxed muscle, no motor units are active and the muscle is electrically silent. As a result of the elasticity of the myofibrils and fibrous tissues, there is, however, a certain tension (tonus) even in the relaxed muscle. Data have been accumulated illustrating the recruitment pattern of different motor units in human muscles activated in voluntary contraction (studied with EMG and electroneurogram). Hannerz (1974) and Grimby and Hannerz (1976) have reported experiments in which they mapped out the discharge pattern and recruitment order of single motor units in voluntary contraction of human anterior tibial muscle. Low-threshold units were recruited in mild, sustained contractions starting at a frequency of 5 to 10 Hz. During such activity, when the exerted tension reached a certain value, a particular motor unit started to work and continued working until the tension again dropped below the threshold level. The same unit usually started its activity at about the same tension level and at a regular rate. With increasing contraction strength, the units increased their discharge rate up to a relatively low maximum (around 25 Hz). New motor units were recruited but they started at a higher frequency (in some cases exceeding 30 Hz), and they attained a higher maximum (up to 65 Hz). It was noted that new motor units were recruited at all levels of tension up to maximal effort. Units with high threshold, coming in later, tended to exhibit a discontinuous discharge, particularly when the tension exceeded 80 percent of the maximum. As an example, a motor unit recruited at a sustained contraction with 15 percent of maximal isometric strength had a minimal frequency of about 10 Hz and at maximal voluntary tension it reached 30 Hz. A unit starting at 60 percent load at about 20 Hz increased its rate up to 45 Hz, and the "90 percent unit" began at 35 Hz and finished at about 60 Hz. In twitch contractions and rapid movements, the recruitment order of the motor units differed considerably from the pattern noticed in sustained contractions, and high-threshold units could be activated first with rapid firing in bursts. One can assume that the low-threshold units are slow twitch fibers innervated by small α motoneurons, and that the high-threshold motor units are composed of fast twitch fibers with nerve axons from large α motoneurons. From experiments in which they manipulated with the proprioceptive afferent input, the authors conclude that the low frequency motor units have more support by the afferent feedback than the high-threshold units. The recruitment order can be modified or reversed by various procedures changing this feedback. Different muscles may exhibit different orders of motoneuron recruitment, but the pattern just described seems to be quite basic.

In conclusion, slowly contracting motor units are recruited first in activities with low demands on tension (both in reflex and voluntary contractions). Their frequency is relatively low. With increased tension the "old" motor units increase their discharge rate and new motor units are recruited. The fast-contracting motor units gradually start working and then at a relatively high frequency. The *gradation of a muscle contraction is brought about by varying the number of active motor units (recruitment) and their frequency of excitation (rate coding).* The recruitment of slow twitch fibers in sustained contractions of low tension is a "wise" arrangement, for the fibers have the potential to work aerobically and are fatigue resistant in their activity at a relatively low frequency. By recruiting new motor units when larger tension is demanded, the metabolic load on the individual muscle fiber can be kept low. Stein (1974) points out that rate coding would be less useful than recruitment for maintaining a steady force, because there is only a limited range of rates at which motor units can produce a reasonable steady force without fatigue, i.e., at low force levels. Rate coding becomes more prominent when higher tension is demanded, however. (See also Milner-Brown et al., 1974, who give an example of subjects who were able to increase the force of their hand muscles from 15 to a maximum of 20 N without recruiting new motor units.)

The relatively stereotyped recruitment order of given motor units in specific tasks must be a consequence of the movement programs established in the cerebral and cerebellar centers (innate or learned). These centers are capable of selecting in advance the appropriate recruitment order for the task intended. There is a continuous interaction of central commands with sensory feedback which can modify the number of motor units recruited and their frequency of contraction.

As previously discussed, motoneurons are exposed to both excitatory and inhibitory impulses from various levels in the CNS and from the muscle spindles of the innervated muscle, as well as from the Golgi tendon organs of the muscle tendons and the Renshaw cells.

In below-elbow amputees who had the distal tendon of the biceps brachii separated from its lateral attachment, the force that could be developed in maximal voluntary contractions at various lengths of the muscle was invariably greater with the elbow flexed 90° than with the elbow extended (Blaschke et al., 1952). Apparently, impulses from structures in or around the elbow joint could inhibit the motoneurons innervating biceps brachii if the elbow was extended.

Ralston (1957) points out that an already shortened muscle cannot be activated as fully as an artificially stimulated muscle because the alpha motoneuron excitability is reduced as a result of the lack of facilitation via the spindles. Neither is the stretched muscle able to produce as much tension as an artificially stimulated one because of the inhibition of some of the alpha motoneurons via Golgi tendon organs and tendon afferents.

It has been mentioned that a contracted muscle has elastic properties. If an activated muscle is stretched, part of the added energy is degenerated into heat but another part can be absorbed by the "series elastic elements" of the muscle. This stored energy can be made available during a concentric contraction of the muscle immediately following the stretch. This release of stored energy can support the chemically yielded energy during the contraction. In fact, under certain conditions the developed tension in concentric work can exceed the maximal isometric strength. Cavagna et al. (1971) measured the work during sprint running by means of a platform (force-platform) sensitive to the force impressed by the foot. They noticed that up to a speed of 6 to 7 m · s^{-1} the contractile component was alone responsible for the power output of the muscles. In other words, the muscle behavior followed the power-velocity curve (Fig. 4-24). It should be recalled that the maximal power is "normally" attained at about one-third the maximal speed of the shortening of a muscle. At higher velocities, the maximal force and power should decrease and this reduction must soon limit the running speed. However, at speeds greater than 6 to 7 m · s^{-1}, the good sprinter could provide an "abnormal" increase in power (Fig. 4-27). Cavagna and his coworkers (1971) explained this finding thus: (1) The extensor muscles of the leg were activated before the foot reached the ground. (2) There was a phase of deceleration and kinetic energy stretched these muscles. (3) Part of this energy was stored in the elastic elements during stretching of the contracted muscles, doing negative work. (4) It was then utilized in the next phase of the step when the muscles did positive work. Thereby, the muscle can produce more tension than when it merely contracts.

The calculated average power developed during a push at each step reached 2,200 to 3,000 W at the highest speed. A technique factor is evidently involved, for not every subject could make good use of the energy absorbed by the muscles at the higher speeds. Apparently the properties of the activated muscles have some qualities of a springboard, which also can be utilized in a more or less efficient way by the diver.

Asmussen and Bonde-Petersen (1974a, b) have provided additional data on storage of elastic energy in muscles. Maximal vertical jumps were performed with and without countermovement (bouncing) and after jumps down to the force-platform from various heights. When the subjects bounced, an average of 23 percent of the negative work could be recovered in the positive work. When they jumped down from 0.40 m before the upward jump, the stored energy was 10 percent of the total negative work. The actual gain in jumping height at the most averaged 10 percent, attained after the jump down from 0.40 m, in comparison with the jump from a semisquatting position. In another series of experiments, the authors (1974b) measured the energy cost of lowering and lifting the body by flexing and extending the legs from a standing or sitting position, that is, with or without rebound in the deepest position. Figure 4-28 shows the marked difference in efficiency in the two types of exercises. At a work power of 105 watt (25 kcal min⁻¹), the energy output was 52 kJ (12.5 kcal min⁻¹) in knee bending without rebound but only 40 kJ (9.4 kcal min⁻¹) with rebound. The mechanical efficiency of one subject was around 25 percent and 40 percent respectively. The authors conclude that the differences in efficiency must depend on the possibility of using the energy, absorbed and stored in the muscles as elastic energy during a phase of negative exercise, when the positive phase follows immediately after the negative phase.

It should again be emphasized that the energy cost of eccentric, negative work is relatively modest (Fig. 4-26). Secondly, when the muscle is stretched it can produce more tension, that is, it moves up the length-tension curve (Fig. 4-22). Therefore, at least two advantages exist in the rebound technique so common in various activities. Cavagna et al. (1975) also believe that there may

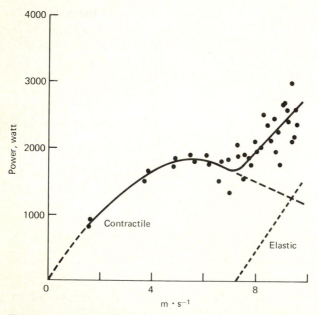

Figure 4-27 The average power developed by the muscles during the push in sprinting, as a function of the speed of the run (abscissa). The continuous line is traced by hand through the experimental points. In the tracing the "contractile" curve (interrupted line partially overlapping the continuous line) indicates the mechanical power developed by the contractile component of the muscles. The curve "elastic" was obtained by subtracting the "contractile" curve from the continuous line. It indicates the part of the total power output due to the mechanical energy stored in the series elastic elements during the stretching of the contracted muscles. *(Cavagna et al., 1971.)*

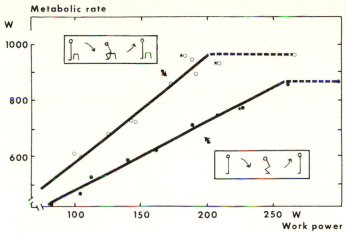

Figure 4-28 Metabolic rate against work power in rhythmic, deep leg flexions and extensions at varying frequencies, with (closed circles) and without (open circles) rebound. Points marked with a star signify experiments in which the subject carried an extra load weighing 10 kg on the shoulders. Dotted lines indicate that a maximum value for oxygen uptake has been reached. *(Modified from Asmussen and Bonde-Pedersen, 1974b.)*

be a greater or more efficient mobilization of chemical energy by the contractile component induced by previous stretching of the muscle. The activation of extensor muscles *before* the foot reaches the ground in walking and running is an effect of a central pattern generator discussed above. When the body weight then stretches the muscles, there is eventually no need for an additional recruitment of motor units to take the load (Grillner, 1975). There is a similar predetermined pattern of neuromuscular activity in a repetitive, rhythmic hopping movement, i.e., gastrocnemius muscle is activated before contact with the ground (Melwill Jones and Watt, 1971a). A stretch reflex would occur too late to avoid a "catastrophic landing." The preferred frequency of hopping is apparently chosen so that the takeoff phase of muscular activity is timed to make maximal use of the stretch reflex, that is, of support from the muscle spindles.

It has been noted (Carlsöö and Johansson, 1962) that when one breaks a fall to the ground by landing on an outstretched hand, there is a strong activity in all muscles which surround the elbow joint even before the hand touches the ground. This reflex can efficiently protect the involved joint. In subjects dropped from an electromagnetic suspension at unexpected moments, the muscular response began about 75 ms after the fall, independent of the height (Melwill Jones and Watt, 1971b). These reflexes most likely originate in the otolith apparatus. A stretch reflex would come too late to activate the muscles.

This initial peak of muscle activity is found in muscles throughout the body. If a longer time (more than 200 ms) elapses before landing, there is a second activation in the muscles of the lower limbs with a timing related to the timing of landing (Greenwood and Hopkins, 1975).

Ikai and Steinhaus (1961) conducted experiments in which the subjects made a maximal arm flexion every minute during a 30-min period. They found that the firing of a 22-caliber gun 2 to 10 s before a pull could significantly modify the exerted maximal strength. The same result was achieved by shouting, giving various drugs, or exposing the subject to hypnosis. Thus, the performance was distinctly higher after the gunshot than before. Shouting, hypnosis, epinephrine, and amphetamine also tended to improve performance as compared with the controls. The positive effect on strength was noticeable on untrained subjects, but slight or absent on well-trained athletes, such as weight lifters. Ikai and Steinhaus cite Pavlov's statement that "any unusual sensory experience or excitement may inhibit inhibitions." They emphasize that their findings "support the thesis that in every voluntarily executed, all-out maximal effort, psychological rather than physiological factors determine the limits of performance. Because such psychological factors are readily modified, the implications of this position gravely challenge all estimates of fitness and training effects based on testing programs that involve measures of all-out or maximal performance."

It is a well-known phenomenon that individuals in a stress situation can perform better than otherwise; they become exceptionally powerful. In controlled experiments, it is established that epinephrine and norepinephrine increase both excitability and contractility above normal values, but the mechanism for such effects is not clear. During stress the production of these hormones is increased.

Ikai et al. (1967) report experiments with the same muscle group involved, showing that the maximal strength caused by an electrical stimulation of the peripheral nerve to the muscle gave a tension that was about 30 percent greater than the strength developed during maximal voluntary effort (isometric contraction, electrical stimulation of the ulnar nerve in the elbow region with 50 to 60 volts and 50 Hz). The situation is very artificial, however, with all motor units working synchronously in tetanus.

In summary, muscle strength is a very complicated function. It depends on the number of motor units activated and their frequency of contraction. With increasing load, recruitment of more motor units is most important until the load becomes heavy; then an increase of the firing rate becomes the most prominent mechanism for the development of more force. The maximal tension can be produced when the muscle is lengthened, and it declines as the muscle shortens. The maximal strength in rapid eccentric movements exceeds the maximum in isometric work. It is less in concentric movements, and particularly in rapid movements.

A given tension can be produced at lower energy yield in eccentric work than in concentric activities. When stretching an activated muscle, part of the mechanical energy can be stored because of the elastic properties of the muscle. In the immediately following contraction, this energy can be released and can support the contractile energy. In our opinion, there is overwhelming evidence to show that a voluntary maximal muscle effort, in most situations and with unconditioned subjects, does not engage all the motor units of the active muscle at tetanus frequency. Effective inhibitions of varying degree exist on some motoneurons, depending on supraspinal and proprioceptor activity. In a specific situation, say an emergency, and perhaps as an effect of training, inhibition decreases (or facilitation increases), and the muscle mass can become more completely utilized in the contraction. The reader is reminded of the rather massive inhibitory interaction between interneuronal paths in the spinal cord and the many inhibitory or excitatory impulses that may descend from higher levels of the CNS. Many athletes shout in the critical stage of an effort, and in the light of the experiments by Ikai and Steinhaus (1961), this action may increase the motoneuron activity. Training in technique may empirically teach the athlete how to take advantage of facilitating stimuli and reflex mechanisms and how to avoid inhibitions which reduce the performance. The bouncing maneuver is typical in many activities. A proper understanding of the underlying mechanism is very important and useful, not only for sport activities, but also in physical therapy.

The thesis that central factors are of decisive significance for the development of strength is also supported by the observation that the strength can increase without a proportional hypertrophy of the muscles. When striving for muscle strength for a particular activity, the best training is that activity. The gain in strength when one is engaged in "unfamiliar" procedures is comparatively modest even when they activate the muscles which are being trained.

The explanation of these results may be that a gain in strength after a training program is due not only to changes in the muscle tissue but also to a modification of the impulse traffic reaching the motoneurons. In a different procedure, receptors, nerves, and synapses which are not "trained" are engaged. Furthermore, a slight change in a movement may reduce the load on one muscle and increase it on another.

Basmajian (1967), in his studies of the elbow flexors (the biceps brachii, the brachialis, and the brachioradialis), found that these muscles show a wide range of individual patterns of activity during flexion and extension of the forearm. These muscles also differ in their flexor activity in the three positions of the forearm: prone, semiprone, and supine (Fig. 4-29). The speed of a movement

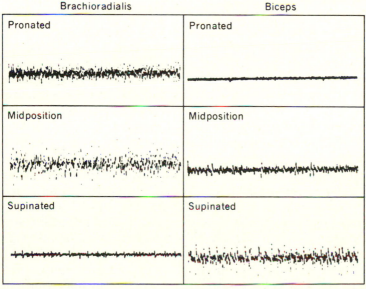

Figure 4-29 Electromyograms recorded from brachioradialis and biceps brachii during maximal isometric contraction (flexion) in three different positions of the forearm. Note the high electrical activity in brachioradialis and the "silence" in biceps when the forearm is pronated. In the supinated position, the biceps is very active but the brachioradialis contributes much less than in the pronated position. A training of the flexor muscles of the arm with the forearm pronated cannot be so effective if the aim is to improve the ability to perform with the arm supinated, since this position partly engages different muscles. *(By courtesy of S. Carlsöö.)*

also influences the activity pattern. Therefore, a training of the elbow flexors with the forearm in a prone position may be quite ineffective for a performance done with a supine forearm.

Asmussen and Heebøll-Nielsen (1955) pointed out that functions of importance for physical performance that depends on maximal muscular exertion increase with height at a much greater rate than predicted for boys seven to seventeen years of age. Their results pointed to the fact that an increasing ability to mobilize and coordinate the muscles of the body is the reason for the rapidly increasing physical capability of schoolboys.

All athletes have experienced variability in their performance from day to day; it may be a matter of a more or less successful inhibition of the inhibition on a subconscious level.

Coordination

In fast (ballistic) movements, at least a spurt of activity in the agonist produces momentum and kinetic energy in the segment, and then it relaxes as the limb proceeds by its own momentum. By reciprocal inhibition the antagonist relaxes completely except perhaps at the end of a movement or when the movement is stopped by the limits of the joint or an external force. Also, in slow movements in some activities, discrete bursts of neural activity are observed in agonists and antagonists to produce muscular impulses, which alternately act independently to accelerate and decelerate the segment (Hubbard, 1960). An integration of central programs and the feedback loops from the proprioceptors are probably responsible for a periodicity and modulation of the motoneural activity. In fast movements, proprioceptor and visual stimuli are probably relayed too late to correct a misdirected movement but in time to make adjustments in succeeding movements. In slow movements, a continuous close control is possible.

Hubbard (1960) also points out that skilled pitching, shot-putting, and discus throwing are excellent examples of developing moments serially to stretch agonists successively. In terms of efficient production, the important factor is to develop tension under conditions as like the isometric (or even eccentric) as possible and to maintain this condition as long as possible. This can be accomplished by moving the proximal segment ahead of the distal segment so that the agonist develops tension while lengthening or remaining at the same length as long as possible. The difference between a good discus thrower and a poor one is that the poor one uncoils while spinning across the circle, and the good one stays coiled until set to throw.

During a strong effort in a particular muscle, there is a high incidence of activity (in a predictable pattern) in far-removed muscles of the same limb and trunk musculature, an especially frequent occurrence in children (Basmajian, 1967).

Degeneration and Regeneration of Nerves

After damage to peripheral nerve fibers, the axis cylinder and myelin sheath distal to the site of injury degenerate. A regeneration can take place, and it is facilitated by the presence of uninterrupted endoneural tubes. The regenerating axons grow into the peripheral endoneural tubes at a rate of several millimeters per day. How the regenerating nerve finds its way is not exactly known. It is believed, however, that chemical factors released from the degenerated nerve fibers stimulate and guide the collateral growth of the axons. The remyelination closely follows, and the anatomic recovery is completed within one year. However, the functional recovery takes longer. Many of the sprouting central axons may get lost before reaching the peripheral sheath; other axons will establish faulty connections. A touch fiber originally supplying one receptor may connect with a touch receptor in a different region or may even connect with a temperature receptor. The effect will be reduced sensitivity, false localization, or complete misinterpretation of a stimulus of the receptor, such as touch giving a sensation of heat. Prolonged practice may be necessary for a return of the finer and discriminative sensibility. A functional regeneration of damaged nerves within the CNS will not take place.

The reinnervation of the muscle by axonal sprouts can reestablish the old pattern or innervation, but if the collateral sprouts happen to be derived from inappropriate parent fibers, transient or permanent functional disorder may be the consequence. To some extent, eventual inappropriate neuromuscular connections can be mastered by a reeducation and, also to some extent, by a reorganization in the CNS. Nerve fibers in the CNS can be stimulated to grow and achieve new functional connections.

Missiuro and Kozlowski (1963) have shown that transplantation of a knee flexor of the hind leg of a rabbit to the distal tendon of a severed extensor modifies the coordinating relations reflected in electromyograms. The transplanted flexor begins to act as a knee extensor. The first electromyographic manifestations that the transplanted muscle is taking over the new function sometimes become evident within a few days after the transplantation and gradually become stronger and more firmly established. Over a period of 4 months after the operation, the transplanted muscle periodically displays a tendency to revert to its original function. On retransplantation the muscle immediately resumes the original function. The authors' explanation for this is that the nervous system is capable of functional adaptation of the transplanted muscle and corresponding nerve centers to a new situation.

Muscle Fatigue

Fatigue is a very complex conception, especially since heavy exercise does load respiration and circulation, as well as the neuromuscular function. Feelings of fatigue without preceding exercise are not uncommon. Simonson (1971) has published an extensive discussion of various explanations for muscular fatigue.

Fatigue in static work produces a sensation of discomfort and sometimes even pain. The disposition to subdue the feeling of fatigue is very different among individuals. Cooperative and well-motivated subjects can maintain a muscular contraction to the point of muscular fatigue, whereas others terminate the activation before reaching that point. It should be emphasized that experiments on fatigue factors must be very carefully controlled with respect to initial length of the involved muscles and the position of the joints.

Figure 4-30 summarizes data on maximal holding time in sustained isometric contractions at various levels of tensions related to the maximal force

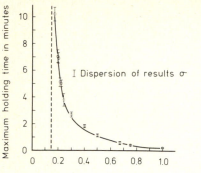

Maximum holding time in minutes

I Dispersion of results σ

Force developed in fractions of maximum force

Figure 4-30 Maximal work time plotted versus force in isometric muscular contraction against a load expressed in percentage of maximal isometric strength in the same type of work. Average of results obtained in studies on different muscle groups. Note that a 50 percent load can be maintained for just 1 min, but as long as the muscular force is less than 15 percent of the maximal force (dashed line), the contraction may be maintained almost "indefinitely." The subjects experienced discomfort and aches in the working muscle some time before they were compelled to terminate the effort because of muscular exhaustion. *(From Rohmert, 1968.)*

of contraction (MFC). The maximal force can be maintained for only a few seconds, 50 percent of MFC for about 1 min, but at the 15 percent level and below. Rohmert and others found that the isometric contraction could be held for more than 10 min and even up to hours (Simonson and Lind, 1971). Some studies indicated, however, that the upper limit for isometric contraction maintained for "indefinite" time is below 10 percent of MFC (Björksten and Jonsson, 1976). One must expect a variation depending on the muscle groups studied, fiber types, and individual variations in maximal strength.

For many years, physiologists have been debating whether muscular fatigue is central or peripheral in origin. Presently, there are no data indicating that the various synaptic connections should be subjected to fatigue. (The effects of inhibition is not a fatigue factor, for, by disinhibition, it can be removed!) From his review, Simonson (1971) concludes that transmission fatigue at the neuromuscular junction can be excluded. However, Stephens and Taylor (1972) have presented data on humans which they interpret as giving evidence that in a maximal voluntary contraction, neuromuscular junction fatigue is most important at first (during the first minute), but later, contractile element fatigue increases. In view of the quantities of transmitter substance released at each transmission of an impulse (in contrast to the situation in synapses within the CNS), these authors' hypothesis seems reasonable. They also noted that "neuromuscular junction fatigue is believed to be most marked in high-threshold motor units, while contractile element fatigue more especially affects low-threshold units." As stated earlier, the high-threshold motor units are activated at a high contraction rate, and consequently there must be a rapid turnover of acetylcholine. During the first minute of a maximal voluntary contraction, the force falls to about 50 percent. EMG recordings indicate that during this phase there is a loss of active motor units with a high-frequency rate of firing and a recruitment of units with a lower rate (Lloyd, 1971).

There are characteristic changes of the EMG in muscle fatigue, indicating a change in both the impulse traffic in the motor nerve and the muscle reaction to the discharge. The amplitude increases and the rhythm slows down. A grouping and synchronization of the discharges appear which, at least partly, can be attributed to a decrease of the proprioceptive afferent impulses from muscle spindles, as shown by Kogi and Hakamada (1962). These authors found that the quotient of the electrically integrated amplitude of slower components divided by that of the faster components increased gradually and steadily in fatigue experiments of isometric-isotonic contractions of various strengths. The appearance of a high "slow wave" ratio was significantly related to the onset of a local fatigue sensation, to the feeling of pain, and to the subject's incapability of maintaining the intended tension.

In a maximal contraction for *more than one minute*, we have apparently to look for fatigue factors in the contractile element.

Blood Supply In its contraction, the muscle consumes energy. Metabolites are formed, and oxygen, if available, is used, with a concomitant production of carbon dioxide, water, and heat. A restoration of the internal equilibrium necessitates an adequate blood supply. During contraction the active muscle swells and becomes hard. In maximal static contractions of the quadriceps femoris in humans, the pressure within the muscle can be several hundred millimeters of mercury. Since the peak arterial blood pressure at rest is some 120 mm Hg and during exercise below 200 mm Hg in most cases, the blood flow through the active muscle will be partially or completely blocked. According to Edwards et al. (1972), the intramuscular pressure in the quadriceps muscles exceeds the systolic arterial pressure already at a tension of 25 percent of MVC. Lind and his coworkers have shown that at 5 and 10 percent of MVC, the blood flow (forearm) increased to a steady state and dropped immediately after exercise (Simonson and Lind, 1971). As mentioned, contractions at that level can be held for a very long time and the energy yield is most likely aerobic. At tensions of 20 to 30 percent of MVC, the blood flow increased steadily during the activity and increased further immediately after the end of contraction. Apparently, there was a "blood flow debt" and the muscle fibers had to pay part of the energy cost by anaerobic processes. At tensions exceeding 30 percent of MVC, there was a *decrease* in the blood flow and it was completely arrested at about 70 percent of MVC. As noted earlier, Edwards has reported that in the quadriceps, the circulation could be occluded at contraction forces greater than 20 percent of MVC.

With the blood flow occluded by a cuff just before and during a vigorous contraction, there was no difference in the initial tension developed as compared with the controls. This was the case with contraction tensions from 60 to 70 percent and higher in Lind's studies—the blood flow was then occluded anyway. At lower tensions, an occlusion of the local blood flow reduced the maximal isometric contraction time. It is not surprising that the reduction of endurance by external occlusion of blood flow is much more pronounced at lower tensions than at tensions closer to MVC. The normal occlusion of blood flow can be both a result of "nipping" of the arteries between moving and nonmoving tissues (Simonson and Lind, 1971) and an effect on the capillary flow due to the increased intramuscular pressure.

There are indications that stronger subjects are less able to maintain isometric contractions at the same percentage of their maximal strength than are weaker subjects (Kroll, 1968; Mundale, 1970; Thorstensson, 1976). First, from a mechanical point of view, an impediment of blood flow should be more dependent on the absolute than on the relative muscle tension; that is, the strong subject will be handicapped by an impaired circulation at a tension that is relatively low in percent of the maximum. Furthermore, one can speculate that the weaker subject has a larger proportion of slow twitch muscle fibers (Type I) and that they are activated at a relatively low discharge rate, which means a lower energy demand. Third, their richer content of myoglobin could make more oxygen available for an aerobic energy yield.

With a blood flow below the level required for an adequate supply of oxygen and removal of carbon dioxide, metabolites, and heat, there must inevitably be a lack of oxygen and an accumulation of metabolites and heat. According to Ahlborg et al. (1972), the lactate accumulation at exhaustion is maximal between 30 to 60 percent of MVC. The accumulation rate is linear with respect to contraction force above 20 percent of MVC.

Tesch & Karlsson (1977) analyzed the lactate concentration in single fibers and also reported peak content at a sustained maximal contraction at about 50 percent of MVC. The concentration was higher than at the 25 and 75 percent strength levels. No significant difference was observed in lactate content of slow and fast twitch fibers. Apparently both types of fibers were engaged in the contractions and both types worked at least partially anaerobically.

Dynamic contractions also periodically hinder the passage of blood, partly or totally. The work load in relation to the duration of the contraction periods, and the intervals between the periods of contraction, determine the length of time the work can be endured. In exercises including frequent dynamic concentric contractions, the energy output for a given tension is relatively high. According to Asmussen (1973), this type of exercise can probably be performed for long periods of time only if the developed strength does not exceed 10 to 20 percent of the maximal isometric strength.

Figure 4-31 shows results from experiments in which the subjects performed rhythmic, maximal isometric contractions on a dynamometer in pace with a metronome (Molbech, 1963). The tensions gradually decreased because of fatigue, but they finally leveled off at a value that could be maintained for a long time. With 10 contractions/min, about 80 percent of the maximal isometric strength could be applied without impairment. With 30 contractions/ min the maximal load was reduced to 60 percent. The values seemed to be independent of the size of the activated muscle group. There appears to be an optimum capacity to work when the ratio between period of work and period of rest is 1:2.

In summary, the ability of the muscle fibers to maintain a high tension and the individual's subjective feeling of fatigue are highly dependent on the blood flow through the muscle. At the beginning of exercise, there is a time lag between blood demand and blood supply. In very short spells of isometric contraction, ATP and phosphocreatine can yield energy and the oxygen present in the muscle (bound to myoglobin) also makes possible an energy delivery from aerobic processes. A maximal contraction can, however, be sustained for

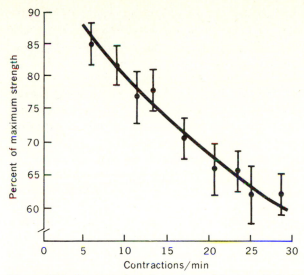

Figure 4-31   Percent of maximum isometric strength that can be maintained in a steady state during rhythmic contractions. Points are averages for finger muscles, hand muscles, arm muscles, and leg muscles, combined. Vertical lines denote ± standard error. *(Molbech, 1963.)*

only a few seconds. In isometric contractions with less than 15 percent of maximal tension or at higher tensions with appropriately spaced pauses, the blood flow can secure the supply of oxygen and energy-rich compounds and remove the formed metabolites, and the work can proceed aerobically for long periods of time. At heavier loads, there is an impaired blood flow due to more or less occlusion of the blood vessels. Therefore the oxygen need will exceed the oxygen supply, and the anaerobic processes must contribute markedly to the energy yield. The impaired blood flow limits not only the oxygen supply but also the removal of metabolites and heat. The exact factors that limit the performance are not known. One could be a "fatigue" at the neuromuscular junction in maximal efforts during less than a minute. At a tension of roughly 50 percent of MVC (which can be sustained for about 1 min), accumulation of lactic acid and/or H^+ may negatively interfere with the contractile elements. In isometric contractions demanding more than 15 but less than 50 percent of MVC, there is no knowledge available as to which factors may limit the performance.

It should be emphasized that different muscle groups behave differently because of mechanical design in structure and blood vessel arrangement as well as composition with respect to fiber types.

The relation of developed strength to maximal voluntary contraction of the whole muscle may introduce erroneous conclusions about the relative tension of the few muscle fibers studied by EMG or thermistor probe, or sampled by a biopsy needle. They may be submitted to a different strain; the local blood flow may vary in different parts of the engaged muscle group.

Finally, it should be kept in mind that prolonged, purely isometric muscle contraction is primarily an artificial laboratory exercise which does not commonly occur in everyday life. For the understanding of the neuromuscular function, such artificial experiments are essential, however.

Effect of Prolonged Exercise   In heavy, prolonged exercise maintained for hours, the work output during maximal effort gradually decreases. After 1 hour's rest, a work load that normally could be tolerated for 6 min had to be terminated after about 4 min because of exhaustion. The peak lactate level in the blood was correspondingly decreased. It is believed that the limiting factor in this case must be sought at the cellular level in the exercising skeletal muscles, and that it may be anything from a change in the properties of the membrane of muscle fibers, a disturbed ATP-ADP "machine," etc., to a depletion of the glycogen stores or a reduced capacity to neutralize the metabolites produced.

In prolonged work requiring up to 90 percent of the individual's maximal oxygen uptake, it has been noted that the slow twitch fibers in engaged muscle groups are the first to be glycogen-depleted. As work continues, fast twitch fibers start to become recruited, and at least they too will be glycogen-depleted (see Burke and Edgerton, 1975; Piehl, 1974). A drop in tension as the slow fibers drop out may temporarily be prevented by an increased drive on high-threshold motoneurons induced by an afferent muscle spindle discharge. Their recruitment could compensate for the failing slow twitch fibers.

Nöcker (1964) points out that exercise prolonged to exhaustion decreases the potassium concentration within the active muscle cells, for example, from 635 to 460 mg/100 ml in rats. An increase in the hydrogen ion concentration increases the permeability of the cell membrane. The coupled Na^+-K^+ pump may be less efficient in prolonged activity of the muscles. Since the potassium-sodium balance is of the utmost importance for the excitability and the recovery of the muscle fibers, it is reasonable to assume that the muscle's decreased ability to contract can be linked to a disturbed ion balance and eventually to a hyperpolarization of the cell membrane. There is also a possibility of modifying the afferent impulses from a muscle subjected to prolonged severe exercise, with an increased inhibition of the motoneurons as a consequence. In emergency situations, this inhibition can, however, eventually be inhibited. A direct stimulus of the fatigued muscle (prolonged work) has increased the force of contraction in some experiments (Ikai et al., 1967).

"Fatigue" in Joints   Basmajian (1967) subjected a sitting person's upper limbs to a strong downward pull. The muscles crossing the shoulder joints and the elbow joints were electrically quiescent, but the subject felt local fatigue. Basmajian thinks that such fatigue originates from the painful feeling of tension in the articular capsule and ligaments.

Motivation   With the tools now available, the conductor of physiological exercise experiments need not rely exclusively on the subject's subjective statements for an evaluation of the relative load placed on the muscles, the circulation, or other functions. The production of lactic acid, the pH of the blood, EMG, heart rate, blood sugar level, tissue temperature, and other findings may be used as an indication of how the subject "should" feel. In healthy subjects, young as well as old, untrained and well-trained, the problem during maximal effort is, as a rule, that the muscles eventually stiffen and refuse to obey the subject's will.

Recovery Muscle fatigue is a reversible phenomenon, provided that proper periods of rest or mild exercise are inserted. Actually, light exercise with other muscle groups than those already fatigued can promote the recovery.

Sore Muscles

If an untrained individual performs vigorous exercise, the engaged muscles may become very painful and hard. The symptoms usually appear after about 12 hr, become more severe during the next day, and gradually fade away, so that the muscles are symptom-free after 4 to 6 days. The pain is probably caused by injuries, especially of connective tissues within the muscle and its attachment to the tendon. Secondarily, histamine and other substances are produced which may cause edema and pain. The repair of the damaged tissue results in a stronger muscle much less susceptible to further injuries, even if the subsequent exercise is much more severe. A sore muscle is more likely to develop in the beginner if the muscle works eccentrically, even if the metabolic rate is much lower than during concentric contractions; the exerted tension may, however, reach much higher values in eccentric work. Sore muscles are particularly noticeable after prolonged eccentric exercise.

The localized, sustained, and painful cramp that occasionally can throw a muscle into vigorous involuntary contraction, e.g., after prolonged monotonous exercise or even during sleep, cannot be explained at present. A stretch of the muscle usually relieves it from the cramp, probably due to inhibitory impulses from the stimulated Golgi tendon organs (Sherrington's lengthening reaction).

MUSCLE STRENGTH, SEX, AND AGE

Muscle strength depends on many factors, some of which have been discussed in the preceding pages. In tests of muscle strength, the day-to-day variation is usually in the order of ± 10 to 20 percent. The correlation between the strength of different muscle groups in the same individual is, as mentioned, low, moderate, or fairly high, depending on which muscle groups are compared. In isolated muscles, the maximal strength is related to the cross-sectional area of the muscle, but many factors make the deviations relatively large. This is also true for muscles *in situ*. Applying ultrasonic measurements, Ikai and Fukunaga (1968) noticed almost the same maximal isometric strength of the elbow flexors in male and female subjects regardless of their age (twelve to twenty years; about 60 N cm^{-2}). Costill et al. (1976) present data on fiber size in untrained women and men and, on average, the cross area for slow twitch fibers in women was about 70 percent of the men's size and for fast twitch fibers 85

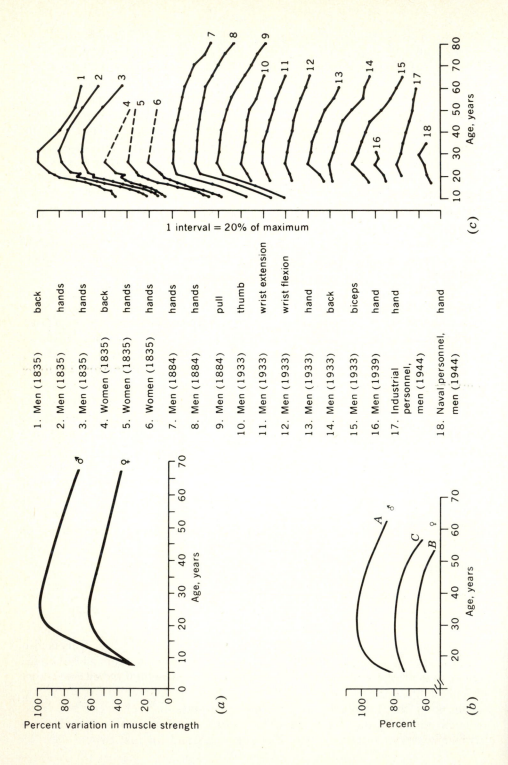

1 interval = 20% of maximum

(c)

1. Men (1835)   back
2. Men (1835)   hands
3. Men (1835)   hands
4. Women (1835)   back
5. Women (1835)   hands
6. Women (1835)   hands
7. Men (1884)   hands
8. Men (1884)   hands
9. Men (1884)   pull
10. Men (1933)   thumb
11. Men (1933)   wrist extension
12. Men (1933)   wrist flexion
13. Men (1933)   hand
14. Men (1933)   back
15. Men (1933)   biceps
16. Men (1939)   hand
17. Industrial personnel, men (1944)   hand
18. Naval personnel, men (1944)   hand

(a) Percent variation in muscle strength

Age, years

(b) Percent

Age, years

percent. The scatter was, however, very large, e.g., from 3,400 to 8,700 μm^2 for one fiber type. It is not possible to judge the degree of shortening of the biopsy specimen and therefore the "true" relationship under standardized conditions cannot be evaluated. Figure 4-32 presents data from several studies on average muscle strength for different muscle groups in female and male subjects of different ages. The maximal strength is reached between the ages of twenty and thirty, after which it decreases gradually, so that the strength of the sixty-five-year-old is approximately 80 percent of that attained between the ages of twenty and thirty (Fisher and Birren, 1947; Hettinger, 1961). In both sexes the rate of decline with age in the strength of the leg and trunk muscles is greater than in the strength of arm muscles. It should be emphasized that training of muscle strength, and therefore the degree of engagement of muscle synergists in the daily routine, will highly influence the results.

In adult women, the strength of any muscle group is lower than in men of the same age. On the average, the muscle strength of women is about two-thirds that of men.

A difference in body dimensions must necessarily be considered when evaluating the variation in strength with sex and age. With strength proportional to the transverse sectional area of the muscle, the isometric strength should vary with the individual's height raised to the second power, provided there is a geometrical similarity between the individuals of different body size (Chap. 11). If the height of a child increases by a factor of 1.5 (for example, from 120 to 180 cm), the developed strength during a maximal pull or push should increase by a factor of $1.5^2 = 2.25$. In measurements of strength as torque, the distance over which the muscle can shorten, being proportional to the height, must also be considered. Therefore, in the given example, the torque should increase by a factor of $1.5^3 = 3.375$.

In Fig. 4-32, the strength in women has been corrected for body size, and thereby the difference in strength between women and men has been reduced, so that the average strength of adult women is about 80 percent of that of men.

From studies on children seven to seventeen years of age, Asmussen (1973) reports that when muscle strength is plotted in relation to body height on a double logarithmic scale, the data usually fit straight lines. However, for some muscle groups, especially the arm muscles, the slope of this line increases suddenly in boys at a height of about 155 cm, that is, at about the age of thirteen. A similar tendency was not observed in girls. (See also Clarke, 1976.) This sudden increase in strength in boys may be of a hormonal origin.

Figure 4-32 Variation in maximal isometric strength for various muscle groups with age.
(a) Changes in strength with age in men and women.
(b) Curve A is for men and curve B for women. The data in curve C are derived from curve B but corrected for the sex difference in body height with the assumption that muscular strength is related to H— where H = body height. The average strength for the twenty-two-year-old man = 100 percent. (By courtesy of E. Asmussen.)
(c) [Data compiled by Fisher and Birren (1947). Figures in parentheses indicate year in which study was made.]

In performances of more functional character, such as sprinting and jumping, muscle strength is of importance. In such events, girls below about twelve years of age perform on a par, or slightly under par, when compared with boys. Between the ages of twelve and eighteen, the boys show an improvement in results that is much greater than predicted from the increase in their body size. In girls, a comparable improvement does not take place (Asmussen, 1973).

Asmussen concludes that in children, age affects muscle strength by (1) increased size of the anatomic dimensions; (2) the results of aging itself (one extra year of age increases the strength by 5 to 10 percent of the average strength of the same height group, a gain that may be attributed to the maturation of the CNS); and (3) development of the child's sexual maturity (with the male sexual hormones probably of special importance for this effect). As a matter of fact, about one-third of the increase in body height occurs between the ages of six and twenty, but during the same period, four-fifths of the development of strength takes place.

Even with correction for body size, age, and sex, large individual differences in muscle strength are observed, and a standard deviation from a mean value of $\pm$ 15 to 20 percent must be accepted as a normal finding.

REFERENCES

Ahlborg, B., J. Bergström, L.-G. Ekelund, G. Guarnieri, R. C. Harris, E. Hultman, and L.-G. Nordesjö: Muscle Metabolism during Isometric Exercise Performed at Constant Force, *J. Appl. Physiol.*, **33:**224, 1972.

Allen, G. I., and N. Tsukahara: Cerebrocerebellar Communication Systems, *Physiol. Rev.*, **54:**957, 1974.

Amphlett, G. W., S. V. Perry, H. Syska, M. D. Brown, and G. Vrbova: Cross Innervation and the Regulatory Protein System of Rabbit Soleus Muscle, *Nature*, **257:**602, 1975.

Asmussen, E.: Positive and Negative Work, *Acta Physiol. Scand.*, **28:**364, 1953.

Asmussen, E.: Growth in Muscular Strength and Power, in G. L. Rarick (ed.), "Physical Activity Human Growth and Development," p. 60, Academic Press, Inc., New York, 1973.

Asmussen, E., and K. Heebøll-Nielsen: A Dimensional Analysis of Physical Performance and Growth in Boys, *J. Appl. Physiol.*, **7:**593, 1955.

Asmussen, E., O. Hansen, and O. Lammert: The Relation between Isometric and Dynamic Muscle Strength in Man, *Communications from the Testing and Observations Institute of the Danish National Association for Infantile Paralysis*, no. 20, 1965.

Asmussen, E., and F. Bonde-Petersen: Storage of Elastic Energy in Skeletal Muscles in Man, *Acta Physiol. Scand.*, **91:**385–392, 1974a.

Asmussen, E., and F. Bonde-Petersen: Apparent Efficiency and Storage of Elastic Energy in Human Muscles during Exercise, *Acta Physiol. Scand.*, **92:**537, 1974b.

Axelrod, J.: Neurotransmitters, *Sci. Am.*, **230**(6):58, 1974.

Basmajian, J. V.: "Muscles Alive," The Williams & Wilkins Company, Baltimore, 1967.

Bigland, B., and O. C. J. Lippold: The Relation between Force, Velocity and Integrated Electrical Activity in Human Muscles, *J. Physiol.*, **123:**214, 1954.

Björksten, M., and B. Jonsson: Endurance Limit of Force in Long Term Intermittent Static Contractions, *Scand. J. Work Environ. & Health*, 1977. (In press.)

Blaschke, A. C., H. Jampol, and C. L. Taylor: Biomechanical Considerations in Cineplasty, *J. Appl. Physiol.*, **5**:195, 1952.

Buchtal, F., C. Guld, and P. Rosenfalck: Multielectrode Study of the Territory of a Motor Unit, *Acta Physiol. Scand.*, **39**:83, 1957.

Buller, A. J., J. C. Eccles, and R. M. Eccles: Interactions between Motoneurons and Muscles in Respect of the Characteristic Speeds of Their Responses, *J. Physiol.*, **150**:417, 1960.

Burke, R. E., and V. R. Edgerton: Motor Unit Properties and Selective Involvement in Movement, *Exercise and Sport Sci. Rev.*, **3**:31, 1975.

Buser, P.: Higher Functions of the Neurons System, *Ann. Rev. Physiol.*, **38**:217, 1976.

Byrne, W. L., and D. Samuel: Behavioral Modification of Injection of Brain Extract Prepared from Trained Donor, *Science*, **154**:418, 1966.

Carlsöö, S.: "How Man Moves," Heinemann Educational Books, Ltd., London, 1972.

Carlsöö, S., and O. Johansson: Stabilization of Load on the Elbow Joint in Some Protective Movements, *Acta Anat.*, **48**:224, 1962.

Cavagna, G. A., L. Komarek, and S. Mazzoleni: The Mechanics of Sprint Running, *J. Physiol.*, **217**:709, 1971.

Cavagna, G. A., G. Citterio, and P. Jacini: The Additional Mechanical Energy Delivered by the Contractile Component of the Previously Stretched Muscle, *J. Physiol.*, **251**:65P, 1975.

Clarke, H. H.: "Application of Measurement to Health and Physical Education," 5th ed., Prentice-Hall Inc., Englewood Cliffs, N. J., chap. 7, 1976.

Close, R. I.: Dynamic Properties of Mammalian Skeletal Muscles, *Physiol. Rev.*, **52**:129, 1972.

Clough, J. F. M., C. G. Phillips, and J. D. Sheridan: The Short-Latency Projection from the Baboon's Motor Cortex to Fusimotor Neurones of the Forearm and Hand, *J. Physiol.*, **216**:257, 1971.

Cooper, S. and J. C. Eccles: The Isometric Responses of Mammalian Muscles, *J. Physiol.*, **69**:377, 1930.

Costill, D. L., J. Daniels, W. Evans, W. Fink, G. Krahenbuhl, and B. Saltin: Skeletal Muscle Enzymes and Fiber Composition in Male and Female Track Athletes, *J. Appl. Physiol.*, **40**:149, 1976.

Couteaux, R.: Motor End Plate Structure, in G. H. Bourne (ed.), "The Structure and Function of Muscle," vol. 1, Academic Press, Inc., New York, 1960.

Eccles, J. C.: "The Physiology of Nerve Cells," The Johns Hopkins Press, Baltimore, 1957.

Eccles, J. C.: The Synaps, *Sci. Am.*, **212**(1):56, 1965.

Eccles, J. C.: The Cerebellum as a Computer: Patterns in Space and Time, *J. Physiol.*, **229**:1, 1973a.

Eccles, J. C.: "The Understanding of the Brain," McGraw-Hill Book Company, New York, 1973b.

Edwards, R. H. T., D. K. Hill, and M. J. McDonnell: Myothermal and Intramuscular Pressure Measurements in Man, *J. Physiol.*, **224**:58P, 1972.

Elias, M. H. and J. E. Pauly: "Human Microanatomy," 2d ed., Da Vinci Publishing Co., Chicago, 1961.

Essén, B., E. Jansson, J. Henriksson, A. W. Taylor, and B. Saltin: Metabolic Characteristics of Fibre Types in Human Skeletal Muscle, *Acta Physiol Scand.*, **95**:153, 1975.

Evarts, E. V.: Motor Cortex Reflexes Associated with Learned Movement, *Science*, **179**:501, 1973.

Fisher, M. B., and J. E. Birren: Age and Strength, *J. Appl. Physiol.*, **31**:490, 1947.

Floyd, W. F., and P. H. S. Silver: The Function of the Erectores Spinae Muscles in Certain Movements and Postures in Man, *J. Physiol.*, **129**:184, 1955.

Fox, C. A.: "Correlative Anatomy of the Nervous System," The Macmillan Co., New York, 1962.

Gilbert, P.: How the Cerebellum Could Memorize Movements, *Nature*, **254**:688, 1975.

Glynn, I. M., and S. J. D. Karlish: The Sodium Pump, *Ann. Rev. Physiol.*, **37**:13, 1975.

Granit, R.: Muscle Tone and Postural Regulation, in K. Rodahl and S. M. Horvath (eds.), "Muscle as a Tissue," p. 190, McGraw-Hill Book Company, New York, 1962.

Granit, R.: "The Basis of Motor Control," Academic Press, London, 1970.

Granit, R.: Constant Errors in the Execution and Appreciation of Movement, *Brain*, **95**:649, 1972.

Granit, R., and R. E. Burke: The Control of Movement and Posture, *Brain Research*, **53**:1, 1973.

Granit, R.: The Functional Role of the Muscle Spindles—Facts and Hypotheses, *Brain*, **98** (part IV): 531, 1975.

Greenwood, R., and A. Hopkins: Muscle Response during Sudden Falls in Man, *J. Physiol.*, **254**:507, 1975.

Grillner, S.: Locomotion in Vertebrates: Central Mechanisms and Reflex Interaction, *Physiol. Rev.*, **55**:247, 1975.

Grillner, S., T. Hongo, and S. Lund: Descending Monosynaptic and Reflex Control of γ-Motoneurones, *Acta Physiol. Scand.*, **75**:592, 1969.

Grimby, L., and J. Hannerz: Disturbances in Voluntary Recruitment Order of Low and High Frequency Motor Units on Blockades of Proprioceptive Afferent Activity, *Acta Physiol. Scand.*, **96**:207, 1976.

Guth, L.: "Trophic" Influences of Nerve on Muscle, *Physiol. Rev.*, **48**:1968.

Gutmann, E.: Neurotrophic Relations, *Ann. Rev. Physiol.*, **38**:217, 1976.

Gutmann, E., R. Béránek, P. Hnik, and J. Zelená: Physiology of Neurotrophic Relations, *Proc. 5th Nat. Congr. Czech. Physiol. Soc.*, 1961.

Hagbarth, K.-E.: Excitatory and Inhibitory Skin Areas for Flexor and Extensor Motoneurons, *Acta Physiol. Scand.*, **26** (Suppl. 94): 1952.

Hammond, P. H.: The Influence of Prior Instruction to the Subject on an Apparently Involuntary Neuro-Muscular Response, *J. Physiol.*, **132**:17P, 1956.

Hannerz, J.: Discharge Properties of Motor Units in Relation to Recruitment Order in Voluntary Contraction, *Acta Physiol. Scand.*, **91**:374, 1974.

Hebb, C.: Biosynthesis of Acetylcholine in Nervous Tissue, *Physiol. Rev.*, **52**:918, 1972.

Hettinger, T.: "Physiology of Strength," Charles C Thomas, Publisher, Springfield, Ill. 1961.

Hill, A. V.: The Priority of the Heat Production in a Muscle Twitch, *Proc. Roy. Soc. (Biol.)*, **148**:397, 1958.

Hubbard, A. W.: Homokinetics: Muscular Function in Human Movement, in W. R. Johnson (ed.), "Science and Medicine of Exercise and Sports," Harper & Row, Publishers, Incorporated, New York, 1960.

Hubbard, J. I.: Microphysiology of Vertebrate Neuromuscular Transmission, *Physiol. Rev.*, **53**:674, 1973.

Hydén, H., and E. Egyhazi: Nuclear RNA Changes of Nerve Cells during Learning Experiment in Rats, *Proc. Nat. Acad. Sci., U.S.*, **48**:1366, 1962.

Ikai, M., and A. H. Steinhaus: Some Factors Modifying the Expression of Human Strength, *J. Appl. Physiol.*, **16**:157, 1961.

Ikai, M., K. Yabe, and K. Ischii: Muskelkraft und Muskuläre Ermüdung bei Willkürlicher Anspannung und Electrischer Reizung des Muskels, *Sportarzt und Sportmedizin*, **5**:197, 1967.

Ikai, M., and T. Fukunaga: Calculation of Muscle Strength per Unit Cross-sectional Area of Human Muscle by Means of Ultrasonic Measurement, *Int. Z. angew. Physiol. einschl. Arbeitsphysiol.*, **26**:26, 1968.

Keele, C. A., and E. Neil: "Samson Wright's Applied Physiology," 12th ed., Oxford University Press, Fair Lawn, N.J., 1971.

Kogi, K., and T. Hakamada: Slowing of Surface Electromyogram and Muscle Strength in Muscle Fatigue, *Rep. Inst. Sci. Labour* (Tokyo), **60**:27, 1962.

Komi. P. V.: Relationship between Muscle Tension, EMG and Velocity of Contraction under Concentric and Eccentric Work, in J. E. Desmedt (ed.), "New Developments in Electromyography and Clinical Neurophysiology," vol. 1, p. 596, Karger, Basel, 1973.

Kroll, W.: Isometric Fatigue Curves under Varied Intertrial Recuperation Periods, *Res. Quart. Amer. Assoc. Health Phys. Educ.*, **39**:106, 1968.

Krujevic, K.: Chemical Nature of Synaptic Transmission in Vertebrates, *Physiol. Rev.*, **54**:418, 1974.

Kuffler, S. W., C. C. Hunt, and J. P. Quilliam: Function of Medullated Small-nerve Fibres in Mammalian Ventral Roots: Efferent Muscle Spindle Innervation, *J. Neurophysiol.*, **14**:29, 1951.

Lambert, O.: The Relationship between Maximum Isometric Strength and Maximum Concentric Strength at Different Speeds, *Intern. Fed. Phys. Educ. Bull.*, **35**:13, 1965.

Lloyd, A. J.: Surface Electromyography during Sustained Isometric Contractions, *J. Appl. Physiol.*, **30**:713, 1971.

Lorente de No, R.: Analysis of Activity of Chains of Internuncial Neurons, *J. Neurophysiol.*, **1**:207, 1938.

Marsden, C. D., P. A. Merton, and H. B. Morton: Servo Action in the Human Thumb, *J. Physiol.*, **257**:1, 1976.

Matthews, P. B. C.: Muscle Spindles and Their Motor Control, *Physiol. Rev.*, **44**:219, 1964.

Matthews, P. B. C.: "Mammalian Muscle Receptors and Their Central Actions," The Williams & Wilkins Company, Baltimore, 1972.

Melwill Jones, G., and D. G. D. Watt: Observations on the Control of Stepping and Hopping Movements in Man, *J. Physiol.*, **219**:709, 1971a.

Melwill Jones, G., and D. G. D. Watt: Muscular Control of Landing from Unexpected Falls in Man, *J. Physiol.*, **219**:729, 1971b.

Milner-Brown, H. S., R. B. Stein, and R. Yemm: Changes in Firing Rate of Human Motor Units during Voluntary Isometric Contractions, *J. Physiol.*, **230**:371, 1974.

Milner-Brown, H. S., and R. B. Stein: The Relation between the Surface Electromyogram and Muscular Force, *J. Physiol.*, **246**:549, 1975.

Missiuro, W., and S. Kozlowski: Investigations on Adaptive Changes in Reciprocal Innervation of Muscles, *Arch. Phys. Rehabil.*, **44**:1963.

Molbech, S.: Average Percentage Force at Repeated Maximal Isometric Muscle Contractions at Different Frequencies, *Communications from the Testing and Observations Institute of the Danish National Association for Infantile Paralysis*, no. 16, 1963.

Mundale, M. O.: The Relationship of Intermittent Isometric Exercise to Fatigue of Hand Grip, *Arch. Phys. Med. Rehabil.*, **51**:532, 1970.

Nöcker, J.: "Physiologie der Leibesübungen," Ferdinand Enke Verlag, Stuttgart, 1964.

Piehl, K.: Glycogen Storage and Depletion in Human Skeletal Muscle Fibers, *Acta Physiol. Scand.* (Suppl. 402), 1974.

Porter, R.: Influences of Movement Detectors on Pyramidal Tract Neurons in Primates, *Ann. Rev. Physiol.*, **38**:121, 1976.

Ralston, H. J.: Recent Advances in Neuromuscular Physiology, *Am. J. Phys. Med.*, **36**:94, 1957.

Rodgers, K. L., and R. A. Berger: Motor-unit Involvement and Tension during Maximum, Voluntary Concentric, Eccentric, and Isometric Contractions of the Elbow Flexors, *Med. and Sci. in Sports*, **6**:253, 1974.

Rohmert, W.: Die Beziehung zwischen Kraft und Ausdauer bei Statischer Muskelarbeit, Schriftenreihe Arbeitsmedizin, Sozialmedizin, Arbeitshygiene, Band 22, p.118. A. W. Gentner Verlag, Stuttgart, 1968.

Ruch, T. C., and J. F. Fulton: "Medical Physiology and Biophysics," 18th ed., W. B. Saunders Company, Philadelphia 1960.

Rudfjord, T.: Model Study of Muscle Spindles Subjected to Static Fusimotor Activation, *Kybernetik*, **10**:189, 1972.

Saltin, E., J. Henriksson, E. Nygaard, P. Andersen, and E. Jansson: Fiber Types and Metabolic Potentials of Skeletal Muscles in Sedentary Man and Endurance Runners, *Annals N.Y. Acad. Sci.*, 1977. (In press.)

Schreiner, K. E., and A. Schreiner: "Menneskeorganismen," 6th ed., J. Jansen (ed.), Universitetets Anatomiske Inst., Oslo, 1964.

Sherrington, C., R. S. Creed, D. Denny-Brown, J. C. Eccles, and E. G. T. Liddell: "Reflex Activity of Spinal Cord," Clarendon Press, Oxford, 1932.

Simonson, E. (ed.): "Physiology of Work Capacity and Fatigue," Charles C Thomas, Springfield, Ill., 1971.

Simonsen, E., and A. R. Lind: Fatigue in Static Work, in E. Simonsen (ed.), "Physiology of Work Capacity and Fatigue," p. 241, Charles C Thomas, Springfield, Ill., 1971.

Sjödin, B.: Lactate Dehydrogenase in Human Skeletal Muscle, *Acta Physiol. Scand.* (Suppl. 436), 1976.

Stein, R. B.: Peripheral Control of Movement, *Physiol. Rev.*, **54**:215, 1974.

Stephens, J. A., and A. Taylor: Fatigue of Maintained Voluntary Muscle Contraction in Man, *J. Physiol.*, **220**:1, 1972.

Stréter, F. A., J. Gergely, S. Salmons, and F. Romanul: Synthesis by Fast Muscle of Myosin Light Chains Characteristic of Slow Muscle in Response to Long-term Stimulation, *Nature New Biol.*, **241**:17, 1973.

Tesch, P., and J. Karlsson: Lactate in Fast and Slow Twitch Skeletal Muscle Fibers of Man during Isometric Contraction. *Acta Physiol. Scand.*, **99**:230, 1977.

Thorstensson, A.: Muscle Strength, Fibre Types and Enzyme Activities in Man, *Acta Physiol. Scand.*, Suppl. 443, 1976.

Tornvall, G.: Assessment of Physical Capabilities, *Acta Physiol. Scand.*, **58** (Suppl. 201): 1963.

Vallbo, Å. B.: Muscle Spindle Afferent Discharge from Resting and Contracting Muscles in Normal Human Subjects, in J. E. Desmedt (ed.), "New Development in Electromyography and Clinical Neurophysiology," vol. 3, p. 251, Karger, Basel, 1973.

Blood and Body Fluids

CONTENTS

Chapter 5

Blood and Body Fluids

The function of the individual cells within the body is dependent on the constancy of their internal and surrounding environment. Claude Bernard recognized that an evolution of higher forms of organisms could not have taken place without the establishment of a stable *milieu interne*, its composition being guarded by regulatory mechanisms.

The muscle cell is unique in regard to its ability to increase its metabolic rate. The maintenance of a constant *milieu interne* of the cell during the transition from rest to vigorous exercise necessarily represents, at times, a tremendous challenge to the circulation. The result may be that the muscle cell must cease working or that it is forced to slow down the intensity of work as the changed composition and property of the fluid within the cell or surrounding it interfere with the various processes which are necessary for the cell to function and to perform work.

BODY FLUIDS

Generally speaking, we are dealing with a large water pool of about 60 percent of the body weight in men, 50 percent in women. As adipose tissue contains very little water, the percentage of total water in the obese individual is lower than in the nonobese (Fig. 5-1). The relation between total water and fat-free body weight (lean body mass) is fairly constant; in an adult the total water is about 72 percent of the lean body mass (Hernandez-Peon, 1961).

We can divide the water space into three compartments. The intracellular fluid is separated by the cell membrane from the interstitial fluid, while the third water compartment, the intravascular fluid, is circulating within the blood vessels (Fig. 5-2). In the vascular system, only the walls of the capillaries and, to some degree, of the postcapillary venules permit an exchange of materials.

The fluid outside the cells, i.e., the intracellular fluid, constitutes about 30 percent of the total body water. Extracellular fluid has an ionic composition similar to that of sea water. The ions are found in approximately the same relative proportions, although the total ionic concentration of sea water is several times that of extracellular fluid. It has been postulated that this resemblance suggests that extracellular fluid was originally derived from the ancient oceans, which were more dilute than those of today (Hernandez-Peon, 1961).

The property of the capillary wall allows free passage of all substances in the blood except the plasma proteins and blood corpuscles. On the other hand, there is a marked difference in concentration of various electrolytes between the extra- and intracellular fluid. The membrane of the resting cell acts as a "barrier," especially for positively charged ions (cations), but processes within the cells must continuously work to maintain this barrier property by throwing out some intruding ions (mainly Na^+) and by holding back others (mainly K^+). It is somewhat of a paradox that the prerequisite for homeostasis in the body is based on a "heterostasis" at the cellular level.

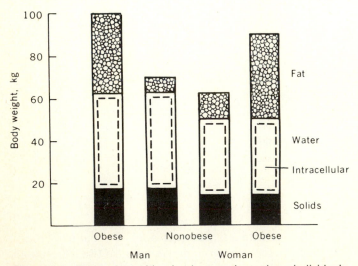

Figure 5-1 Body composition in obese and nonobese individuals.

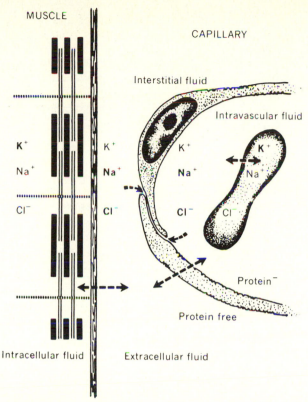

MUSCLE

CAPILLARY

Interstitial fluid

Intravascular fluid

K⁺   K⁺   K⁺   K⁺

Na⁺   Na⁺   Na⁺   Na⁺

Cl⁻   Cl⁻   Cl⁻   Cl⁻

Protein⁻

Protein free

Intracellular fluid   Extracellular fluid

Figure 5-2   Schematic drawing of capillary wall and the exchange of material through it.

BLOOD

The blood and the lymph take care of the transportation of material between the different cells or tissues. The blood brings food materials from the digestive tract to the cells for catabolism or synthesis of molecules in tissue structures or for depots which are later mobilized and redistributed. Heat and the chemical products of catabolism are removed, carbon dioxide is expelled through the lungs, heat is dissipated through the skin, and metabolites are transported to the kidneys and liver for further processing. The blood circulation has a key position in the maintenance of a proper water balance and fluid distribution. As a carrier of hormones produced by endocrine glands and other active chemical agents, the blood can, in various ways, modify the function of cells and tissues.

Volume

The most common procedure to determine the blood volume in the human body is to inject into a vein a measured amount of a substance that does not easily escape from the blood vessels. After some time, when complete mixing has taken place, a blood sample is secured for determination of the concentration of the "tracer substance," and the subject's volume of blood can be calculated.

The individual variations in blood volume are large. Therefore, the figure presented in textbooks is nothing more than the mean value from determinations on a certain number of individuals under certain experimental conditions. The blood volume varies with the degree of training. A volume of 5 to 6 liters for men and 4 to 4.5 liters for women can be considered as normal (about 75 ml · kg^{-1} body weight for men, 65 ml · kg^{-1} body weight for women, and 60 ml · kg^{-1} body weight for children).

Cells

The solid elements of the blood which are visible under the ordinary microscope are red corpuscles, or *erythrocytes*; white cells, or *leukocytes*; and platelets, or *thrombocytes*. When a blood sample in a tube containing an anticoagulant is centrifuged, these elements are separated from the liquid portion of the blood, known as the *plasma*. The relative amount of plasma and corpuscles in blood is known as the *hematocrit*. For men the hematocrit is usually about 47 percent, and for women and children it is 42 percent.

The erythrocyte is a biconcave circular disk without a nucleus. It has an average diameter of 7.3 μm, and a thickness of 1 μm in the center and 2.4 μm near the edge (Fig. 5-2). The average cell count in the adult male is $5.5 \cdot 10^{12} \cdot$ $liter^{-1}$ (5.5 million · mm^{-3}); in females and children it is about $4.8 \cdot 10^{12} \cdot liter^{-1}$ (4.8 million · mm^{-3}). The life span of a red cell is about 4 months. Red cells are formed in the bone marrow at a speed that normally matches the rate of destruction. It can be estimated that the formation of erythrocytes proceeds at a rate of 2 million to 3 million per second. The color of the red cell (and the blood) is due to its content of hemoglobin, which is a protein (globin) united with a pigment (hematin). This pigment contains iron; each Fe^{2+} atom can combine with one molecule of O_2. This combination is not a chemical oxidation but loose and reversible. In an oxygen-free medium the *hemoglobin* (Hb) is reduced, and with O_2 available, it forms *oxyhemoglobin* (HbO_2). Another property of Hb is its affinity for CO. The Fe^{2+} atom reacts with CO, and when given a choice it prefers CO (its affinity is about 250 times greater for CO than for O_2), resulting in a proportionally decreased capacity to take up O_2. This phenomenon explains the high toxicity of CO.

In adult men the average Hb content is 158 g · $liter^{-1}$ of blood; in women it is 139 g · $liter^{-1}$. From a statistical point of view, a normal value could be within 140 to 180 and 115 to 160 g · $liter^{-1}$ for men and women respectively, with a range of 95 percent. Each gram of Hb can maximally combine with $1.34 \cdot 10^{-3}$ liter of oxygen. With 150 g Hb per liter of blood, the fully saturated blood can carry 0.201 liter oxygen per liter plus some 0.003 liter oxygen dissolved in the plasma (oxygen pressure 100 mm Hg).

The concentration of Hb in blood can be determined according to two principles: (1) The blood sample is saturated with O_2, and the content of O_2 in a measured volume of blood is determined. The volume of dissolved O_2 can be calculated, and as the O_2-combining power of

1 g Hb is fixed to $1.34 \cdot 10^{-3}$ liter, the concentration of Hb in the sample can easily be calculated. (2) The second method is based on the typical and specific light absorption spectra of Hb and its derivatives. A rough estimation of the Hb content of blood can be obtained from the hematocrit value, or from blood counts, with the presumption that each red corpuscle has a normal Hb content.

Plasma

The plasma occupies about 55 percent of the total blood volume. It contains about 9 percent solids and 91 percent water. The concentration of protein in the plasma is 60 to 80 g · liter^{-1}. Three types of proteins are present: albumin (4.8 percent), globulin (various fractions, 2.3 percent), and fibrinogen (0.3 percent). If blood is permitted to clot, the fibrinogen, in the presence of thrombin, forms fibrin. The fluid "squeezed" from the clot is called *serum*.

The proteins of the blood have several important functions: fibrinogen is necessary for blood coagulation, and the globulin (especially gamma globulin) is necessary for the formation of antibodies. All fractions, but especially the albumin, have the important transport functions of carrying ionic and nonionic substances to the sites of need or elimination. The blood proteins are active in the buffer action and constitute in a way a mobile reserve store of amino acids. Finally, they play an important role in the plasma volume and tissue fluid balance.

All plasma proteins can pass through the capillary wall in small amounts, the albumin most readily. However, as most of the protein molecules are kept within the capillaries, they will exert an osmotic pressure of about 25 mm Hg. This protein or colloidosmotic pressure is an important factor in the exchange of fluid between the intravascular and interstitial spaces. (See Chap. 6.)

The total amount of electrolytes in the plasma is 9 g · liter^{-1}. The main ions are Na$^+$ and Cl$^-$. The plasma normally contains about 5 mmol · liter^{-1} (100 mg · 100 ml^{-1}) of glucose. It also contains free fatty acids (FFA), amino acids, hormones, various enzymes, and about 25 different electrolytes in varying amounts.

Specific Heat

The specific heat of whole blood is 3.85 J (0.92 cal) per gram; that is, the change in temperature of 1 g blood by 1°C will require or deliver 3.85 J (0.92 cal), or about 3.8 kJ (0.9 kcal) per liter of blood.

Buffer Action—Blood pH—CO$_2$ Transport

Apart from its many other functions, the blood also acts as a buffer (for a detailed discussion, see Davenport, 1958). A buffer solution contains an acid or base that is only slightly ionized (weak) and a highly ionized salt of the same acid or base. If we take a weak acid (HA), we have an equilibrium

$$HA \rightleftharpoons H^+ + A^-$$

If a strong acid is added to the solution, there is an increase in the hydrogen concentration, pushing the reaction to the left. As long as the buffer salt can provide A$^-$ ions, thereby forming undissociated HA, a change in the pH of the solution is prevented, or "buffered."

The pH of arterial blood is 7.40, and that of mixed venous blood collected at rest is about 7.37. In the catabolism of the cells, CO_2 is formed, and in the case of anaerobic oxidation, lactic acid is formed. The oxidation of P and S in protein leads to the formation of phosphoric and sulfuric acid. The predominantly acid nature of the metabolites could easily explain the shift in pH as blood passes the capillary bed.

The isoelectric point of the proteins in the blood is on the acid side of the blood pH. Thus, suspended in an alkaline solution, the proteins ionize as acids and form negatively charged anions (NH_2—R—COOH $\rightleftharpoons$ NH_2—R—COO$^-$ + H$^+$ where R symbolizes the protein radical; for simplicity we may write protein$^-$ for the anion). The proteins, therefore, act as hydrogen acceptors. With proteins ionized as weak acids, we have a buffer system: H protein $\rightleftharpoons$ H$^+$ + protein$^-$.

The number of ionizable groups in the protein is large, and as whole blood contains about 190 g protein per liter (with about 70 g in the plasma), its capacity to accept hydrogen without pH changes is considerable. In this respect, the potency of the plasma proteins is only one-sixth that of Hb.

There is another reaction of considerable physiological importance. H—HbO$_2$ is a stronger acid than reduced H—Hb; i.e., it dissociates more completely than does H—Hb. Hence, when blood is giving off O_2, the hydrogen-ion concentration of the blood falls, and pH rises.

Before we discuss the hydrogen exchange in the capillaries, we must draw attention to the second important buffer system of the blood. The CO_2 formed in the cells is dissolved and diffuses freely into the erythrocytes. Catalyzed by carbonic anhydrase, present only in the red cells, CO_2 forms, with water, H_2CO_3. This weak acid is dissociated as follows:

$$CO_2 + H_2O \rightleftharpoons H_2CO_3 \rightleftharpoons H^+ + HCO_3^- \tag{1}$$

The equilibrium in the reactions in Eq. (1) is determined by the concentration of the various molecules and ions. If free H$^+$ can be removed from the system, the reaction will be pushed to the right. Potentially, this reaction gives place for more CO_2 in the solution without a change in the pH. Though more H_2CO_3 will be formed, this is only an intermediate step in the formation of H$^+$ and HCO$_3^-$. In this sense, the supply of a hydrogen acceptor will actually determine the final equilibrium. Protein$^-$ anions may serve as such a hydrogen acceptor. The CO_2 is produced in the tissue and diffuses into the red cell. At the same time, O_2 is diffusing out to the tissue, where the O_2 concentration is lower than in the capillaries (HbO$_2$ $\rightarrow$ Hb + O$_2$). Reduced Hb is a weaker acid than HbO$_2$, and when reduced, it "binds" some of the dissociated H$^+$ ions (H_2CO_3 + protein$^-$ $\rightleftharpoons$ HCO$_3^-$ + H protein).

We may now summarize the main functions involved in *the CO_2 transport from the muscle tissue to the lungs* (Fig. 5-3):

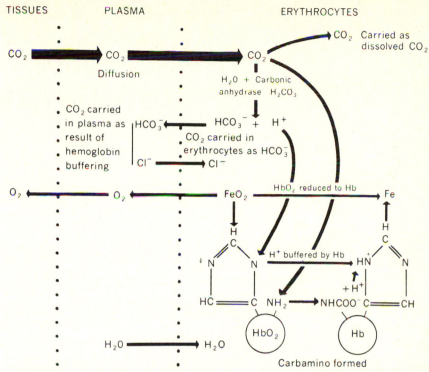

Figure 5-3 Schematic presentation of the processes occuring when carbon dioxide passes from the tissues into the erythrocytes. At the bottom the effect of oxygenation and reduction upon buffering action of the imidazole group of hemoglobin is illustrated. An increase in the acidity of the blood drives the reaction to the right, and oxygen is given off. (A decrease in the acidity of the solution would drive the reaction to the left, and oxygen would be taken up by reduced hemoglobin.) In other words, the reduction of oxyhemoglobin causes the hemoglobin to become a weaker acid and to take up hydrogen ions from the solution. *(From Davenport, 1969.)*

1 The most important reactions in the O_2 and CO_2 exchange between blood and tissue are (*a*) the formation of a weak acid, reduced Hb, from the stronger one, HbO_2; (*b*) the formation of $H^+ + HCO_3^-$ from $CO_2 + H_2O$ with the carbonic anhydrase serving as an enzyme in the intermediate formation of H_2CO_3. The interplay between (*a*) and (*b*) can in a quantitative way be illustrated by the following example: For each millimole of oxyhemoglobin reduced, about 0.7 mmole of H^+ can be taken up, and consequently 0.7 mmole CO_2 can enter the blood without causing any change in pH. If the CO_2 were derived only from fat combustion, 0.7 mole of CO_2 would be produced per mole of O_2 used [respiratory quotient (RQ) = 0.7]. This means that the formation of reduced Hb from the oxygenated HbO_2 would completely buffer the CO_2 uptake. However, at rest, 0.82 to 0.85 mole of CO_2 is formed per mole O_2 used. During heavy exercise, this figure is close to 1.00.

2 Some of the remaining CO_2 can combine directly with Hb forming carbaminohemoglobin ($CO_2 + HbNH_2 \rightleftharpoons HbNHCOOH$), and the simultaneous reduction of HbO_2 greatly favors this formation.

3 There is an increase in the volume of dissolved CO_2, so that the venous blood at rest contains about 0.003 l $CO_2 \cdot l^{-1}$ more than the arterial blood.

Together, these three factors explain how the CO_2 can be transported with the very small change in pH of 0.03 to the acid side.

The net result of the reaction under item 1 is an increase in bicarbonate ions (HCO_3^-) in the red cells. HCO_3^- can freely diffuse across the cell membrane and enter the plasma. Actually, there is simultaneously a decrease in the concentration of the anions as undissociated HHb is formed. The consequent excess of cations within the cell and lack of positive ions in the plasma cannot be compensated by a simple diffusion of a cation (in this case K^+) out to the plasma. The cellular activity "prohibits" an exchange of metallic cations. To restore the electrochemical equilibrium (Donnan equilibrium), the Cl^- ions diffuse into the cell, thereby replacing its loss of anions. This chloride shift allows about 70 percent of the formed HCO_3^- to be transported in the plasma.

If these reactions are understood, the events that take place in the lung capillaries should be easily comprehensible. The reactions run "backward," and in an ingenious way the O_2 uptake and formation of the relatively strong acid HbO_2 facilitate the exclusion of CO_2. In brief, the increase in free H^+ with the dissociation of HbO_2 forces the reaction in Eq. (1) to the left. As the concentration of bicarbonate within the cell decreases, the plasma is ready to send in more HCO_3^- ions. In exchange, the plasma gets back the Cl^- ions. At the same time, carbaminohemoglobin gives up some of its CO_2. The degree of these reactions is dependent on the O_2 and CO_2 tension of the alveolar air.

At rest, about 0.2 liter of CO_2 is produced per min in the tissues, and the cardiac output to transport this volume is about 5 l $\cdot$ min^{-1}. Thus one liter of blood transports 0.040 l of this CO_2. The CO_2 content of one liter of arterial blood is still about 0.5 l after delivery of CO_2 to the lungs. About 0.025 l is dissolved in the plasma, which is in equilibrium with a CO_2 tension of 40 mm Hg. Most of the remaining CO_2 is in the plasma in the form of bicarbonate.

The buffer system composed of CO_2 and HCO_3^- is, in itself, of relatively low capacity. However, as CO_2 can be expired and actually stimulates the respiration, the physiological importance is significant. For simplicity, the buffer action may be illustrated with the following example. In heavy muscular exercise, lactic acid (HLa) is formed. After diffusion into the blood,

$$Na^+ + HCO_3^- + H^+ + La^- \rightleftharpoons Na^+ + La^- + H_2CO_3(\rightleftharpoons H_2O + CO_2)$$

the reaction goes to the right, as H_2CO_3 is a weaker acid (a stronger acceptor of H^+) than lactic acid. The increase in free CO_2 and the decrease in pH of the blood stimulate to a hyperventilation, thereby decreasing the CO_2 content of the blood and body. This causes the buffer base to decrease. As the lactic acid is later oxidized or transformed into glycogen, the blood becomes more alkaline, but this tendency is opposed by a withholding of CO_2 (depression of breathing).

The powerful buffer capacity of the blood has been emphasized, and the various proteins play an important role in this function. As the protein content of many tissues is high (averaging about 15 percent), we should expect that such tissues could contribute to the acid-base equilibrium. In fact, it appears that about 5 times as much acid is neutralized by other tissues as by the blood. The body of an average man might neutralize one equivalent (in other words, 1 liter of a 1.0 M solution) of a strong acid, such as HCl, before the blood pH would fall below 7.0, close to the lowest pH value compatible with life (Hitchcock, 1960). Without buffers the pH would be about 1.6!

The buffer capacity of blood (and tissues) must be considered as a "first aid" service not capable of maintaining a constant acid-base equilibrium for long periods. The *kidney function* stands as a final guard over the body pH. This is accomplished by several mechanisms:

1 In the tubuli and the collecting ducts, a certain amount of H^+ is secreted into the lumen, causing an increase in the acidity of the urine. This increase may cause the pH of the urine to drop to as low as 4.5. If the CO_2 content of the body fluid increases, more H_2CO_3 is formed in the kidneys, and it is in turn dissociated into $H^+ + HCO_3^-$. The hydrogen ions diffuse through the tubuli wall into the tubular fluid; HCO_3^- diffuses more or less back into the blood.

2 In the tubuli there is also a secretion of NH_3, which, together with H^+, may form NH_4, thereby trapping hydrogen ions. If this proton were to be eliminated as, for instance, $H^+ + Cl^-$, the acidity would be intolerable. But as $NH_3^+ + H^+ + Cl^-$, the elimination takes place in the form of a neutral salt NH_4Cl.

3 In the blood there are $Na^+ + Na^+ + HPO_4^{2-}$ which, during the passing through the tubuli, pick up hydrogen ions, with the formation of $Na^+ + H_2PO_4^-$ as the result. Na_2HPO_4 is more alkaline than NaH_2PO_4, and in the tubuli the ratio Na_2HPO_4/NaH_2PO_4 may be varied according to the need for the elimination of H^+.

The formation and excretion of acidic phosphate, as well as the other two processes, assist the conservation of sodium by the body. On the other hand, excess alkali may be eliminated with the urine as bicarbonate or basic phosphate.

Viscosity

The viscosity, or resistance to flow, of blood depends mainly on the plasma proteins and the cell content. The higher the hematocrit value, the higher the viscosity. Blood with a normal hematocrit of 45 percent is 2.1 times as viscous as water. With an increase in hematocrit to 55 percent, the viscosity is 2.6 times higher than that of water. A polycythemia that occurs after acclimatization to high altitude will evidently influence the work of the heart, as the resistance of the blood to flow will vary with its viscosity. At 0°C, the viscosity of blood is 2.5 times as great as at 37°C, which is an important factor in reducing the circulation in tissues exposed to cold, as in frostbite (Burton, 1965).

Blood has an "anomalous viscosity." The red cells tend to accumulate in the axis of the blood vessels, which leaves a zone near the wall relatively free from cells. This axial accumulation of the erythrocytes may result in small side branches of a blood vessel containing a volume of red cells that is considerably less than that for mixed blood (effect of "plasma skimming"). Another effect is a lower viscosity of the blood than expected. The axial accumulation of cells is more pronounced with an increase in velocity of the blood. However, in the physiological range of flow, the axial accumulation is complete, and therefore, the effective viscosity is constant (i.e., blood behaves as if it were a Newtonian fluid). There is a third factor that influences the effective viscosity of blood. In the very narrow vessels (arterioles and capillaries), the blood behaves as if the viscosity were reduced (Fåhraeus-Lindquist effect), which also contributes to reducing the load on the heart. This effect is of great importance during heavy muscular activity (Burton, 1965).

REFERENCES

Burton, A. C.: Hemodynamics and the Physics of the Circulation, in T. C. Ruch and H. D. Patton, (eds.), "Physiology and Biophysics," pp. 523–542, W. B. Saunders Company, Philadelphia, 1965.

Davenport, H. W.: "The ABC of Acid-Base Chemistry," 5th ed., The University of Chicago Press, Chicago, 1969.

Hernandez-Peon, R.: Physiology in Body Fluids, in D. S. Dittmer (ed.), "Blood and Other Body Fluids," Federation of American Societies for Experimental Biology, Washington, D.C., 1961.

Hitchcock, D. I.: Physical Chemistry of Blood, in T. C. Ruch and J. F. Fulton (eds.), "Medical Physiology and Biophysics," pp. 529–551, W. B. Saunders Company, Philadelphia, 1960.

Circulation

CONTENTS

HEART
Cardiac muscle   Blood flow   Pressures during a cardiac cycle   Innervation of the heart

HEMODYNAMICS
Mechanical work and pressure   Hydrostatic pressure   Tension   Flow and resistance

BLOOD VESSELS
Arteries   Arterioles   Capillaries   Capillary structure and transport mechanisms   Filtration
and osmosis   Diffusion—filtration   Veins   Vascularization of skeletal muscles

REGULATION OF CIRCULATION AT REST
Arterial blood pressure and vasomotor tone   The heart and the effect of nerve
impulses   Control and effects exerted by the central nervous system   Mechanoreceptors
in systemic arteries   Posture   Other receptors

REGULATION OF CIRCULATION DURING EXERCISE

CARDIAC OUTPUT AND THE TRANSPORTATION OF OXYGEN
Efficiency of the heart   Venous blood return   Cardiac output and oxygen uptake   Oxygen
content of arterial and mixed venous blood   Stroke volume   Heart rate   Blood
pressure   Type of exercise   Heart volume   Age   Training and cardiac output

SUMMARY

Chapter 6

Circulation

HEART

It is assumed that the reader is familiar with the anatomy and basic physiology of the heart.

Cardiac Muscle

Under a microscope the heart muscle shows transverse striation essentially similar to that of the skeletal muscle, but the nuclei are placed centrally within the cell; the fibers give off branches which anastomose with adjacent fibers, giving an impression of a syncytial continuity of the muscle fibers. However, electron microscope studies give a different and clearer picture of the structure. Apparently the heart muscle is subdivided into numerous cell territories by cell boundaries fanning transversely across the branches of the heart muscle tissue (Sjöstrand and Andersson-Cedergren, 1960). These boundaries are associated with the intercalated disks of the heart, which form wavy "membranes" about 10 nm apart. Because of this "cutting" of the muscle fibers, the heart muscle does not represent a true syncytium.

The conductive system of the heart and the effect of the refractory period after an excitation give the heart its rhythmical activity, unceasing from early embryonic life until death.

Blood Flow

Normally the blood vessels in the heart do not form anastomoses if the diameter is larger than 40 μm (the size of small arterioles). Therefore, in experiments on animals, a sudden occlusion of an artery will be followed by an infarction of the tissue supplied by that artery. On the other hand, a gradual narrowing and final occlusion of a coronary artery will promote the development of a very rich network of anastomotic vessels. Experiments on dogs with coronary obstruction (Eckstein, 1957) have shown that effective collateral vessels develop more profusely if the dogs are exercised. Eckstein found that the greater the coronary narrowing, the greater the development of collateral circulation. This is in agreement with the clinical experience that development of collateral circulation does occur in cardiac patients (see Gregg, 1974). The capillary network in the heart is extraordinarily rich, with 2500 to 3000 capillaries per mm².

The contraction of the heart muscle interferes mechanically with the blood flow through the heart. In the left ventricle of the dog at rest, there is a complete stop during the isometric contraction, and a backflow is actually established, but a peak flow is reached when the systolic pressure is at maximum, during the early ejection phase. A second peak flow is reached during early diastole, after which it drops gradually to about 70 percent of maximum just at the end of diastole (Gregg and Green, 1940) (Fig. 6-1). In the right heart, where pressure is lower during systole, the fluctuations are not so pronounced as in the left ventricle. On the average, the volume flow of blood through the coronary vessels during diastole is about 2.5 times larger than during systole; during exercise there is an increased coronary inflow per heartbeat. The oxygen uptake rate is reported to be higher during systole than diastole (Gregg and Coffman, 1962).

Oxygen deficiency has a strong dilating effect on the arterioles of the heart, but the mechanism is unknown. Adenosine is a potent coronary vasodilator. Excesses of CO_2, H^+, and lactic acid have a milder vasodilating effect. Epinephrine has some dilating influence on the vessels, but this is at least partly a secondary effect, caused by metabolites liberated by the increased metabolism induced by epinephrine. Norepinephrine has a similar, but weaker, effect. No evidence is reported for a reflex effect on coronary blood flow. (See Lundgren and Jodal, 1975.)

Pressures during a Cardiac Cycle

Figure 6-1 describes the pressure variations in the left heart chambers as well as in the aorta. The systole of the left ventricle starts with an isometric contraction, for the mitral valves close rapidly as the pressure in the ventricle exceeds that of the atrium (phase within the vertical lines). Within about 0.05 s, the ventricular pressure is brought up to and above the level in the aorta, and the aortic valves open. The isotonic contraction increases the pressure further. The peripheral resistance does not permit the same volume of blood to escape

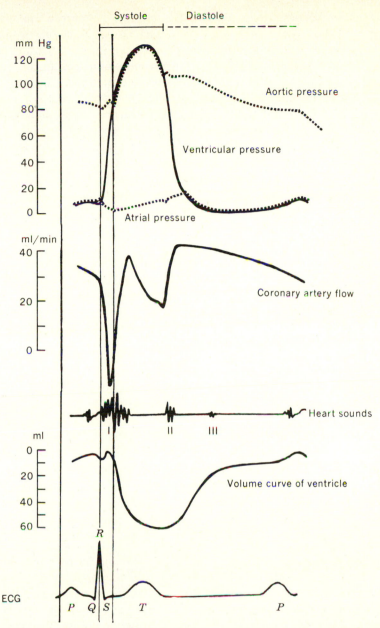

Figure 6-1 Pressure variations in the left heart chambers and aorta during the cardiac cycle. See text for details. [The heart sounds can be objectively analyzed under various pathological conditions by phonocardiography. Closure of the atrioventricular valves contributes largely to the occurrence of the first sound (I), but so do vibrations in the tissues and turbulence of the blood flow; closure of the aortic valves is of prime importance for the occurrence of the second heart sound (II); vibrations of the chamber walls caused by movement of blood into the relaxed ventricle produce the third sound (III).] *(Modified in part from Wiggers, 1952.)*

from the aorta as is ejected into it. Part of this volume is "stored" in the distended aorta and its large branches ("windkessel vessel"). Then, as the pressure falls in the ventricle during the relaxation of the muscle, the aortic valves close, and the elastic property of the aortic wall can propel the stored blood out into the arterial tree. The intermittent energy outbursts of the heart would give an intermittent flow if the vessels were rigid tubes. Part of the potential energy is, however, taken up by the elastic arterial wall and then released during diastole of the heart, keeping the hydraulic energy level close to the heart continuously high. During the systole, blood has returned to the large veins close to the heart and the atrium. There is, perhaps, a passive lengthening of the atrium as the ventricle contracts, which may facilitate the filling of the atrium. During diastole there is a period of rapid filling of the ventricle after the opening of the mitral valves. The period of diastasis follows, during which the filling is much less rapid. The next cycle begins with the atrial contraction, which more or less empties the atrium.

> With a heart rate of 75, the time for diastole is about 0.48 s and for systole 0.32 s (40 percent of the heart cycle); with a heart rate of 150, the periods are 0.19 and 0.21 s (52 percent), respectively. It should be noted that an increase in heart rate occurs mainly at the expense of the length of the diastole and its period of diastasis. As the ventricle may eventually offer resistance to filling at the end of diastole, the pressure gradient between atrium and ventricle drops. An increase in heart rate can, in such a case, improve the filling of the heart and increase the output.

The physical events in the right heart are essentially similar to those just described, but the ventricular and pulmonary artery pressures during systole are about one-fifth of those in the left heart.

In the *aorta*, the pressure at rest varies between 120 mm Hg during systole and 80 mm Hg at the end of diastole. In the *pulmonary artery* the values are 25 and 7, respectively. It should be emphasized that the sphygmomanometer cuff technique does not always give the same value as direct measurement or measurement via a catheter in the artery. This holds true particularly during exercise.

Innervation of the Heart

Many *parasympathetic* nerve fibers from the vagus terminate in the region of the pacemaker, or sinoatrial node. When stimulated, they deliver acetylcholine, causing a slowing of the heart rate (inhibition). *Sympathetic* nerves are the efferent fibers from the upper part of the paravertebral ganglia which join the cardiac sympathetic nerves. They end in a dense network within the heart muscle. The effect upon stimulation is (1) an increase (acceleration) of the heart rate and (2) an increase in contractile force of the muscle fibers (a positive, so-called "inotropic effect"). Both epinephrine and norepinephrine cause a more rapid increase in the systolic pressure and a more complete emptying of the heart, i.e., the diastolic volume of the heart becomes smaller. The vagus does not seem to influence the contractility of the heart muscle, but it may slightly depress the atrial contractility. Via this automatic innervation of the heart, the inherent rate of beating can be highly modified; the healthy heart in a young, fully grown individual can cover a range from about 40 at rest to 200 beats/min during heavy exercise.

HEMODYNAMICS

We shall present in this summary a few definitions and also call attention to some of the physical laws governing the behavior of fluids, and especially of blood in motion. For a more complete review, the reader should consult, for example, Folkow and Neil, 1971; Ruch and Patton, 1974. (Taylor, 1973, recently gave a summary of techniques to study hemodynamics of the heart and the arterial system including measurements of blood flow.)

Definitions:

Heart rate (HR) is the number of ventricular beats per minute as counted from records of the electrocardiogram or blood pressure curves. The heart rate can also easily be determined by auscultation with a stethoscope or by palpation over the heart, both during rest and exercise.

Pulse rate is the frequency of pressure waves (waves per minute) propagated along the peripheral arteries, such as the carotid or radial arteries. In normal, healthy individuals, pulse rate and heart rate are identical, but this is not necessarily so in patients with arrhythmias. In such cases, the output of blood by some beats may be too small to give rise to a detectable pulse wave.

Cardiac output is the volume of blood ejected into the main artery by *each* ventricle, usually expressed as liters per minute ($\dot{Q}$). With small fluctuations, the cardiac outputs of the right and left ventricles are identical. The cardiac output divided by the estimated surface area gives the "cardiac index," which relates the cardiac output to the body size.

Stroke volume (SV) is the volume of blood ejected into the main artery by each ventricular beat. The stroke volume is usually calculated by dividing the cardiac output by the heart rate ($\dot{Q}/HR$).

Oxygen uptake is the volume of oxygen (at 0°C, 760 mm Hg, dry = STPD*) extracted from the inspired air, usually expressed as liters per minute ($\dot{V}_{O_2}$). If the oxygen content of the body remains constant during the period of determination, the oxygen uptake equals the volume of oxygen utilized in the metabolic oxidation of foodstuffs. One liter of oxygen, then, corresponds to 19.7 to 21.1 kJ (4.7 to 5.05 kcal) of energy liberation.

Arteriovenous oxygen difference is usually expressing the difference in oxygen content between the blood entering and that leaving the pulmonary capillaries ($a-\bar{v}O_2$ diff.). Usually the oxygen content of the mixed venous blood ($C\bar{v}_{O_2}$) is determined in blood withdrawn from a long thin tube (catheter) introduced into a cubital vein and then passed through the right atrium and ventricle into the pulmonary artery. The arterial oxygen content ($C_{a_{O_2}}$) is analyzed in blood

---

*STPD = standard temperature and pressure, dry.

samples taken from a systemic artery, usually the femoral, brachial, or radial artery.

The relationship between the functions discussed so far can be summarized as follows:

Cardiac output ($\dot{Q}$)   = stroke volume (SV) × heart rate (HR)
Oxygen uptake ($\dot{V}_{O_2}$) = cardiac output ($\dot{Q}$)
                                                      × arteriovenous O_2 difference ($a\text{-}\bar{v}O_2$ diff.)

$$\text{or } \dot{V}_{O_2} = SV \times HR \times (C_{a_{O_2}} - C_{\bar{v}_{O_2}})$$

Mechanical Work and Pressure

Chemical energy is transformed into mechanical (external work) energy plus heat by the contraction of the heart muscle. The mechanical efficiency of the heart, i.e., the external work divided by the total energy exchange, is rather low, or only some 10 percent at rest.

The external work is calculated from data on force ttimes distance moved, or for a fluid, pressure times volume moved. The pressure can create kinetic energy and a flow of the fluid. The higher the velocity of the fluid, the higher the kinetic energy (related to velocity squared). On the other hand, it may be concluded that in the part of a vessel where the velocity is highest, the fluid pressure against the wall is smallest, since total hydraulic energy = pressure energy + kinetic energy (Bernoulli's principle). If the pressure is measured in a vessel with a catheter or tube connected to a manometer with the opening against the flow, the kinetic energy of flow is reconverted to pressure (*end pressure*) and the total energy is measured. With only a side hole in the catheter the *side* or *lateral pressure* is measured, and it will be less than the end pressure by an amount equivalent to the kinetic energy of flow, according to the formula given above. It can be calculated that, at rest, the kinetic factor in the aorta flow is only about 3 percent of the total work of the heart, but during exercise with a cardiac output 5 times the resting level, it may be an important part of the total work of the heart, possibly as much as 30 percent. This means that a simultaneous measurement of end pressure and side pressure in the aorta may show a difference of some 75 mm Hg, the end pressure being highest (Burton, 1965) (see discussion below). In the other arteries and in the smaller vessels, the kinetic energy factor is normally negligible, at least at rest. It should be pointed out that the sphygmomanometer cuff method for the measurement of blood pressure includes the end pressure.

Hydrostatic Pressure

The pressure in a vessel is created by the continuous bombardment by the molecules of the fluid against the inner surface of the vessel. The pressure is equal at all points lying in the same horizontal plane in a static liquid. The hydrostatic pressure in a fluid at rest, under the influence of gravity, increases uniformly with depth under the free surface (Pascal's law). It is evident that in the supine position the hydrostatic pressure is of similar magnitude in all parts of the body, but in the standing person it is higher (perhaps 100 mm Hg) in the arteries of the feet than in the head. The hydrostatic factor in a standing individual is modified by the action of the valves in the veins of the extremities. Arterial pressure (e.g., for diagnostic purposes) should be measured at the horizontal level of the heart, or the value should be corrected to correspond to heart level. For practical purposes, measurement of the blood pressure with a sphygmomanometer cuff around the arm of a seated subject gives satisfactory values.

When a person is tilted from a supine position to an erect position with the feet down, the veins, and to some extent the other vessels below heart level, will dilate passively as a result of the hydrostatic force. Temporarily the blood is pooled, but when the vessels are filled, the flow continues unhindered by the uphill flow.

Tension

The pressure within the heart and vessels will more or less distend the walls (Fig. 6-2). The resistance to this force, producing *tension*, depends on the thickness of the wall and its content of elastic and collagen fibers and of active smooth-muscle fibers. The elastic tissue can balance the pressure without any energy output, but the contraction of the muscles requires a continuous expenditure of energy. The tension T is roughly proportional to the pressure difference P on the inside and outside of the wall (transmural pressure) and the radius r of the vessel, or $T = aPr$, where a is a constant (Laplace's law). Therefore, a given pressure applied in a small vessel does not produce the same tension it does in a vessel with a larger radius (Fig. 6-2). The very thin membrane of the capillary, essential for the exchange of materials, can withstand the capillary blood pressure because its radius is so small.

The same law may, with some reservations, be applied to the heart. With a given tension developed by the heart muscle, a lower pressure is produced if the heart is dilated (increased in size, i.e., radii of curvature). If the diameter of the heart is doubled, the tension per unit length of ventricular wall must be

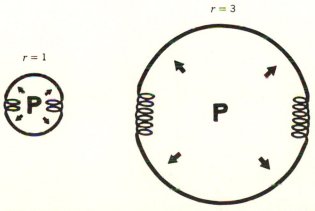

Figure 6-2 The vascular tension at a given transmural pressure P is roughly proportional to the radius of the vessel, r. (This is a simplification neglecting the thickness and properties of the wall.) Thus, if the radius is tripled, 3 times more tension will be required to maintain the same pressure.

about twice as great to produce the same pressure. The energy cost, i.e., the oxygen uptake, of the heart is related to the tension that must be developed and the time this tension is maintained. Therefore, an increase in the size of the heart increases the load on the heart.

As pointed out by Burton (1965), an increase in heart rate also increases tension work, since the time for rest (no tension developed) is relatively shorter. He also emphasizes that an evaluation of the load on the heart must be based on the tension-time integral of the heart muscle, and that the external work, pressure times volume flow, reflects only a small fraction of this load (see the subsection "Mechanical Work and Pressure"). If the external work is calculated, the integral of pressure with respect to flow should be used, not the mean pressure times flow. When using the aortic pressure instead of the ventricular, one must add the kinetic energy factor (end pressure); this is especially important during exercise.

It should be borne in mind that the heart muscle follows the same law as the skeletal muscle: Its ability to produce tension increases with length. By applying Laplace's law, one finds that the heart has to pay more to maintain a given pressure if a reduction of muscular strength is compensated for by a dilatation (increase in length of the fibers).

The maximal active tension developed by the cardiac muscle is attained at a sarcomere length of about 2.2 μm but it decreases rapidly both at sarcomere lengths below and when stretched beyond this optimum length. In skeletal muscle fibers there is more of a tension plateau with changes in sarcomere length of approximately between 2.0 and 2.2 μm. The cardiac muscle is also relatively stiff and tension rises extremely rapidly as the muscle is stretched over a sarcomere length range for which resting tension is minimal in skeletal muscle (Sonnenblick and Skelton, 1974).

Flow and Resistance

Since the difference in the hydraulic energy of a fluid in two parts of a system cannot be resisted by the fluid, it flows with a rate proportional to the energy difference or pressure head. The resistance to flow of blood results from the inner friction or viscosity of the blood. There is a cohesive force between the blood and the wall of the vessel which "retards" the flow of the molecule layers close to the wall. The nearer the center of the vessel, the higher the speed of each lamina of fluid, and this "friction" phenomenon results in a maximal speed in the very center of the vessel. The hydraulic energy provided the blood by the contracting heart muscle is gradually spent and transformed to heat. Fluid flow can be *laminar* (streamlined); that is, each lamina of liquid slips over adjacent laminae without mixture or interchange of fluid. At a critical velocity (expressed by Reynolds number) the laminae of fluid move irregularly and start to mix with one another; the flow becomes *turbulent*. During turbulence the energy loss is larger, and for a given pressure gradient the flow is lower (Fig. 6-3). The velocity of flow in the ventricles of the heart and aorta is normally turbulent during the early phase of contraction. This turbulence produces vibrations, causing high-pitched sounds which can easily be heard by listening over the heart (Fig. 6-1). If the velocity is abnormal, as in mitral stenosis, the heart sounds are abnormal. Similarly, a shunt between aorta and the pulmonary

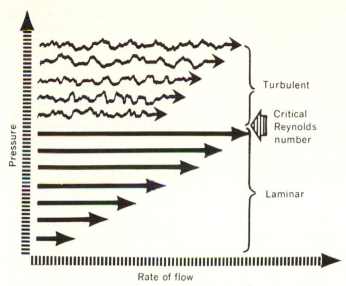

Figure 6-3 In turbulent flow, the energy loss is greater and the rate of flow slower than in laminar flow at a given pressure gradient. Length of arrows indicates rate of flow. *(From Haynes and Rodbard, 1962.)*

artery (open ductus Botalli) can give rise to a turbulent flow in that shunt vessel. In measurement of the blood pressure by applying a measured pressure over the brachial artery (with the sphygomomanometer cuff), the peak blood pressure suddenly overcomes the resistance of the gradually reduced compression. The blood flows with a high velocity through the narrowed artery, and a turbulence gives rise to sounds detectable by the stethoscope.

The radius of a tube is a deciding factor in the flow through it; the resistance to flow is a reverse function of its radius to the fourth power. A decrease to half the radius, other things being equal, will actually decrease the flow to one-sixteenth of the original value. A dilatation of 10 percent increases the blood flow about 50 percent compared with the flow through the same vessel constricted. The variation in activity of the smooth muscles of the vessels gives a very sensitive instrument for the control of blood flow.

Flow in the blood vessels is normally laminar except close to the heart. The laminar blood flow (F) is actually proportional to the driving force (ΔP), inverse to the viscosity (η), proportional to the radius of the vessel raised to the 4th power (r^4), and inverse to the length of the vessel (ℓ):

$$F = \Delta P \times \frac{\pi}{8} \times \frac{1}{\eta} \times \frac{r^4}{\ell} \quad \text{(Poiseuille-Hagen formula)}$$

The resistance to laminar flow (R) is proportional to pressure gradient per rate of flow. This simple formula can help to evaluate the resistance offered to the blood flow in the vascular bed. If the blood pressure in the right atrium is zero, an increase in mean aortic blood pressure (BP_{mean}) at constant cardiac output ($\dot{Q}$) gives evidence of an increased peripheral resistance to flow (as $R \times Q = BP_{mean}$).

The complex composition of blood and the distensible blood vessels complicate the situation in some ways. If a given pressure gradient ("driving pressure") is maintained between artery and vein, the flow is not, as expected, constant at varying pressure levels. At low pressures the flow becomes zero, even though there is still a considerable driving pressure. The pressure at which the vessels close completely is called the critical closing pressure. The decisive factor here is the transmural pressure. When the pressure outside the inner layer of the wall (tissue pressure and active muscular contraction in the vessel wall) exceeds the intravascular pressure, a collapsible vessel collapses (this happens "artificially" in the regular blood pressure measurement commented on earlier). At pressures well above the critical closing pressure, the flow is essentially proportional to the driving pressure. (For detailed discussion see Folkow and Neil, 1971.)

Summary The pressure-flow curve obtained in studies of the hemodynamics of circulation is not linear. The effect of driving pressure can be modified by the transmural pressure, especially in the arterioles and capillaries, and at low intravascular pressure. The resistance to flow can be profoundly altered by active and passive variations of the radius of the blood vessels, the flow being directly proportional to the fourth power of the radius (r^4).

BLOOD VESSELS

Anatomically, and to some extent functionally, the blood vessels may be classified into arteries, arterioles, capillaries, and veins. The exchange of water, electrolytes, gases, etc., can occur only across the semipermeable membrane of the capillaries and, to some degree, in the postcapillary venules; the other vessels are only transport channels. The amount of elastic fibers, collagen fibers, and smooth muscles varies in the different vessels; in the big arteries the walls are thickest, but in veins of the same size the walls are thin, with few elastic and muscle fibers. Endothelial cells cover the inner surface of the vessels; they are formed as flattened plates where pressure is high, and as cuboidal or rounded plates where pressure is low. The vascular system adapts to the forces acting on the walls, modifying their mechanical property.

It has been emphasized that the local blood flow is mainly determined by the pressure head and the diameter of the actual vessels. The capacity of the different vessels to influence vascular resistance, blood flow, and blood distribution is, in a functional way, described by the following subdivision: windkessel vessels (main arteries), resistance vessels (arterioles are the most important, but capillary and postcapillary sections are also of significance), precapillary sphincters (determining the functioning capillary surface area), shunt vessels (e.g., arteriovenous anastomoses), capacitance vessels (veins), and exchange vessels (capillaries) (Folkow, 1960). Most of this terminology is illustrated in Fig. 6-4.

Arteries

The arteries serve as windkessels, or pressure tanks, during the ejection of blood from the heart, and the elastic tissue tends to recoil when stretched. This

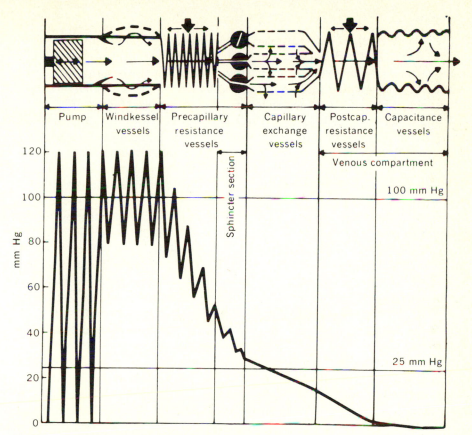

Figure 6-4 A schematic illustration of the functionally different, consecutive sections of the vascular bed, related to the blood pressure fall along the circuit (ordinate blood pressure in mm Hg). Note the marked pressure drop and absorption of the pulse amplitudes in the precapillary resistance vessels. The line at 25 mm Hg represents the plasma protein osmotic pressure (colloid-osmotic pressure). A similar illustration can be made of the sections of the vascular bed in the "low pressure system" (pulmonary circulation), but the pressure in the right pump at rest is alternating between about 25 and 0 mm Hg, and in the pulmonary artery (windkessel vessels) between 25 and 10 mm Hg. *(Modified from Folkow and Neil, 1971.)*

property enables it to store and release energy generated by the heart and convert an intermittent flow to a continuous flow. The resistance to flow in the large arteries (and veins) is very small. The diameter of the vessels is large and the velocity of flow is high.

The expansion of the arterial wall during the ejection of blood causes a pressure wave that travels along the peripheral blood vessels at a speed of 5 to 9 m · s^{-1} (the velocity of the blood in the aorta at rest is 0.5 m · s^{-1}). The more elastic the arterial wall, the lower the speed of the pulse wave. In practical application, the frequency of this wave is counted as the pulse rate, since it can easily be felt over the radial or carotid arteries.

Arterioles

In arterioles the peripheral resistance is high, causing a marked pressure drop. The velocity is still high because the total cross section of the arterioles is not so large. The individual vessels are narrow, from 0.1 mm to 60 μm. In the wall are transversely oriented smooth-muscle fibers that can be activated by stimuli of their nerves, transmitter substance, or by other chemical factors. Figure 6-5, *A*, *B*, and *C*, shows schematically various degrees of constriction of an arteriole caused by contraction of the smooth-muscle fibers. It should be stressed that the vessels do not have muscles that can actively dilate them.

Figure 6-6 illustrates how the arterioles (and capillaries) are arranged in parallel-coupled circuits between the arteries and the veins. They can effectively alter the total resistance against the outflow of blood from the arteries and thereby the arterial blood pressure and work of the heart; they have a decisive influence on the distribution of blood flow to the various organs. An increase in caliber of the arterioles in a muscle will decrease the resistance in that area and hence increase the flow. If this local vasodilatation is not compensated for by a vasoconstriction in another area or an increase in cardiac output, the arterial blood pressure will inevitably fall. The arterioles are arranged in a series of parallel channels joining the capillary bed (and veins) with the arterial side, effectively "regulating" the blood flow through the organs and tissues. The pressure is still high when the blood enters the arterioles, but then drops to 30 to 40 mm Hg (the pressure is somewhat higher in the kidneys).

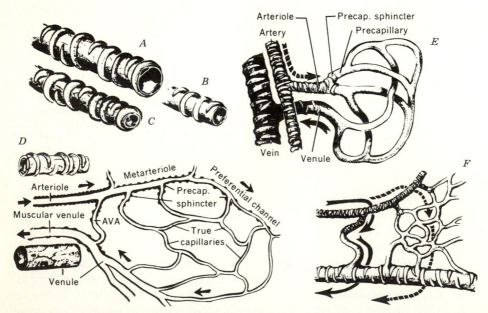

Figure 6-5 A schematic representation of the structural pattern of the capillary bed. The distribution of smooth muscle is indicated in the vessel wall; see text for discussion. AVA = arteriovenous anastomose. *(Modified from Elias and Pauly, 1960.)*

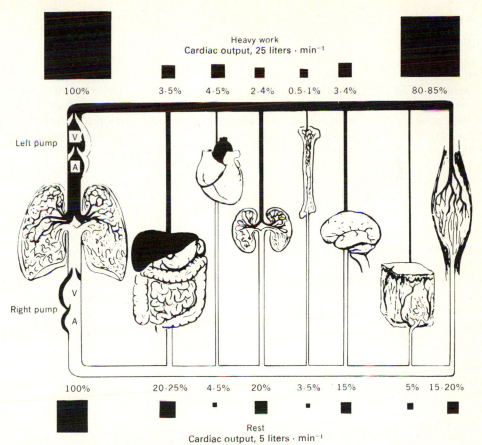

Figure 6-6 Schematic drawing showing how the arterioles and capillaries are arranged in parallel-coupled circuits between the arteries (top) and the veins. The cardiac output may be increased fivefold when changing from real to strenuous exercise. The figures indicate the relative distribution of the blood to the various organs at rest (lower scale) and during exercise (upper scale). During exercise the circulating blood is primarily diverted to the muscles. The area of the black squares is roughly proportional to the minute volume of blood flow. Not included is an estimated blood flow of 5 to 10 percent to fatty tissues at rest, about 1 percent during heavy work.

The systolic and diastolic pressure variations will usually disappear before the blood reaches the capillaries (Fig. 6-4).

Capillaries

The microarchitecture of the capillary network is different in different organs (Fig. 6-5). The precise pattern of the capillary arrangement is not completely known. It may be assumed to be as follows (see Zweifach, 1973):

1 The arterioles branch off into terminal or final arterioles. There the smooth-muscle fibers are arranged in circular or spiral alignment and constitute

a single layer. The terminal arteriole ends in a capillary-like channel or "thoroughfare channel" (preferential channel), from which the capillaries branch (Fig. 6-5, *D*). In the proximal wall portions of the thoroughfare channel is a discontinuous coat of thin muscle cells. There is frequently a group of muscle fibers forming precapillary sphincters, especially at the origin of the capillary. The preferential channel finally loses its muscle coat completely, but can be differentiated from adjoining capillaries by the presence of a somewhat thicker supporting connective-tissue coat. The true capillaries are thin-walled endothelial tubes with no muscle or fibrous fibers. They are kept in place by fine connective-tissue fibers. In skeletal muscles the true capillaries are at least 8 to 10 times as numerous as the preferential channels. In other tissues there are relatively fewer capillaries. The capillaries and the preferential channels drain into the venules, which in turn drain into the larger veins.

2 Another pattern of the microcirculation may be a net of arterioles which divide into anastomosing true capillaries (thin endothelial tubes without muscle cells) (Fig. 6-5, *E*). Precapillary sphincters can be seen. The capillaries converge into venules. No vessels resembling a thoroughfare channel are found.

3 Arteriovenous anastomoses or shunts are typical for the vascular bed in the skin but are also present in other tissues. Anatomically, this may be a vessel with a relatively thick muscle coat running from an artery or arteriole to a vein or venule (Fig. 6-5, *F*). The diameter can, in the latter case, be as small as 15 to 20 μm. Similar shunts may be located proximally to the terminal arterioles or form a direct continuation from one of the branches of an arteriole to the venous circulation. In both cases, this shunt provides a good possibility of maintaining blood flow and low resistance, but the capillary vascular bed is bypassed (dark arrows in Fig. 6-5, *F*).

Summary The circulation of blood from arteries to veins can take various routes via thoroughfare channels, true capillaries, and arteriovenous or arteriolovenous shunts. The pressure gradient and the activity of the smooth-muscle cells in the walls or the precapillary sphincters decide which route the blood flow will take.

Capillary Structure and Transport Mechanisms

Let us now take a closer microscopic and ultramicroscopic look at the architecture of the capillary. Its length is less than 1 mm (down to 0.4 mm) and its inner diameter ranges from 3 to 20 μm; therefore, there is sometimes hardly room for an erythrocyte (diameter 7.2 μm), which may temporarily stop the flow and be squeezed out of shape as it is forced through. The capillary wall is made of very thin endothelial cell plates of protein and lipid, with a rather homogeneous cytoplasm and long granular nucleus. The wall has no muscle cells or other contractable elements but has elastic properties.

The most important processes for the exchange of substances across the capillary membrane are filtration and diffusion. Isotope studies have shown that gases, water, and molecules in water and lipid can rapidly diffuse back and forth in the direction of the concentration gradients through the capillary wall.

The thin-walled capillaries and venules behave as porous membranes, and there are morphological evidences for pores, in some 4 to 5 nm width, in the intercellular substance. They are numerous in skeletal muscles. In other areas, e.g., in the renal glomeruli, in glands, and the intestinal mucosa, there are thinned-out portions of the endothelium, the so-called fenestrae. These structures are most likely involved in the blood-tissue exchange of various substances. There are also numerous vesicles (20 to 30 nm in size) in the cytoplasm of the endothelial cells which may function as carriers for macromolecules between blood and tissue compartments (see Zweifach, 1973). On the other side of the endothelium is a "rigid jacket" consisting of a fine mesh of fibrillary elements embedded in an amorphous, mucopolysaccharide matrix ("basement membrane"). Lipid-soluble substances such as O_2 and CO_2, can pass through the endothelial cells, but water-soluble substances can only traverse the walls through the pores, if the pore radius is large enough.

Filtration and Osmosis

Figure 6-7 shows the driving forces in a process so well outlined by Starling in 1896. Within the capillary the hydrostatic pressure (h.p.) is normally well above the pressure in the interstitial fluid. As the plasma proteins cannot pass through the capillary wall (which is a statement more didactic than true; within 1 to 2 days the entire quantity of plasma protein may be exchanged), they will exert a colloid-osmotic pressure much higher than that existing outside the capillary. For simplicity, the pressures outside the capillary are disregarded (or corrected for) in Fig. 6-7. The gradual change in colloid-osmotic pressure as water escapes or returns to the vessel is also omitted.

When entering the capillary, the net hydrostatic pressure may be 40 mm Hg, and this pressure drops gradually to 10 mm Hg on the venous side of the capillary. The net colloid-osmotic pressure is about 25 mm Hg (see Fig. 6-4).

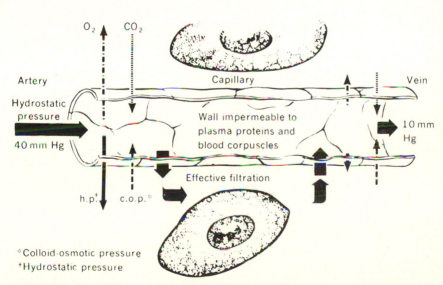

Figure 6-7 Driving force in capillaries. The hydrostatic pressure depends on the degree of vasoconstriction in precapillary vessels. In a resting skeletal muscle it may be only 15 mm Hg at the beginning of the capillary. For details, see text.

The net filtration pressure is then $40 - 25$ mm Hg "proximally," giving an outflow of fluid, but it is $10 - 25$ "distally," giving a pressure differential of 15 mm Hg. This means that fluid is sucked back with a force of 15 mm Hg. It is easy to see how small variations in pressures can profoundly change the exchange of fluid. Three examples are given: (1) If the arterial pressure falls, for instance during blood loss, there is also a reduced capillary pressure. Thereby the balance between the hydrostatic pressure and protein osmotic pressure becomes disturbed, with an increased absorption of fluid from the interstitial fluid as a consequence. (2) A decrease in concentration of plasma protein (prolonged starvation or loss due to kidney disease causing albuminuria) lowers the plasma protein osmotic pressure and the return of fluid from the tissue. An increase in the extravascular fluid follows (edema). (3) A rise of the venous pressure increases the mean capillary pressure and hence the outward passage of fluid. Furthermore, the reabsorption of fluid at the venous side of the capillary is decreased (it may be $15 - 25 = -10$ mm Hg). It is easy to understand how a failure of the left ventricle of the heart can cause pulmonary edema. When the right ventricle eventually becomes uncompensated by the increased load, the edema "moves" to the systemic circulation area. The liver and dependent parts of the body, which are most affected by the hydrostatic pressure, are sensitive to an elevation of pressure in the systemic veins.

It was mentioned that many capillaries are "too small" for the red cells. Consequently the red cell becomes temporarily squeezed in the bulging capillary, with no plasma separating the red cell from the protein-poor interstitial fluid. The red cell sucks up water, and the loss of oxygen and uptake of carbon dioxide are also enhanced. The deformation of the red cell, eventually down to a 3μm diameter, also promotes the exchange of various substances through its membrane. When returning to the vein or lung capillary, the red cell will release the excess water to the surrounding plasma (Hansen, 1961).

When metabolites are formed in active cells, the osmotic pressure outside the capillaries might temporarily increase. The pore-bound passage of the solute particles is less rapid than for water, and there is temporarily an egress of water from the blood.

Diffusion–Filtration

The importance of a net filtration of fluid across the capillary membrane, as illustrated in Fig. 6-7, has probably been overemphasized in the past. It is now believed that diffusion accounts for the major exchange of fluid. It should be noted that diffusion can take any direction regardless of how filtration moves the fluid. It is important to keep in mind that these two processes are separate.

The transport of CO_2 from the tissue to the capillary starts immediately as the blood reaches the thin membrane tube and continues as long as there is a pressure gradient for CO_2. In the same way, oxygen diffuses from the capillary throughout its length, irrespective of the direction of the net water flow across the membrane.

Veins

The collecting venules have a supporting coat of connective tissue and irregularly spaced smooth-muscle cells (Fig. 6-5, *E* and *F*). In the distal part of the venules a well-defined muscle layer is developed. Also, the larger veins have muscles in their walls which can constrict the lumen of the vessel. Valves are frequent in the veins of the extremities of the body.

The veins within a skeletal muscle will be compressed mechanically during the muscular contraction. Blood is squeezed from the veins, and because of the higher resistance on the capillary side and the design of the valves, the blood return to the heart is facilitated by this "muscle pump." During muscular relaxation, the venous pressure drops, and therefore blood may flow from the superficial veins, anastomosing the deep veins. The blood flow to a working muscle is higher during the period of relaxation than during contraction. The venous pressure has its minimum at the time of relaxation, which increases the arteriovenous pressure difference and thus the perfusion.

It has been calculated (since direct measurements are presently impossible) that the venous systems normally contain some 65 to 70 percent of the total blood volume. The veins are therefore often referred to as capacitance vessels (Fig. 6-4). The pulmonary veins may account for about 15 percent of the total blood volume. (At rest, the blood in the capillaries is estimated to be only 5 percent of the total; the spleen plays only a minor role in the human as a blood or red-cell depot.)

The resistance to flow in the larger veins is very small, with a pressure gradient between the foot and the groin of only 6 mm Hg in the horizontal position (Ochsner et al., 1951). Haddy et al. (1962) have analyzed the resistance to flow in the vessels of the limb. The fraction caused by veins larger than 0.5 mm was about 10 percent of the total under normal conditions.

Vascularization of Skeletal Muscles

Nerves and vessels enter the muscle at a neuromuscular hilus often located at half-length of the muscle. The artery enters the muscle substance and branches freely in its course along the perimysium (Fig. 6-8). By anastomoses, a primary arterial network is established. Finer arteries arise and create a secondary network infiltrating the muscle tissue. The smallest arteries and terminal arterioles branch off, usually transversely to the long axis of the muscle fibers and at fairly regular intervals of 1 mm. The arterioles then supply the capillary network oriented parallel to the individual muscle fibers, but also form frequent transverse linkages over or under the intervening fibers, thereby forming a delicate oblong mesh. The veins have valves (from a caliber of 40 μm or larger) directing the blood flow toward the heart, and they follow the course of the arterioles and arteries. The capillary network at the motor endplate is especially well developed.

The exact capillary density in skeletal muscles in the human body is not known. Brodal (in his unpublished results) counted, with the aid of the electron microscope, the number of capillaries per mm² in cross sections of needle

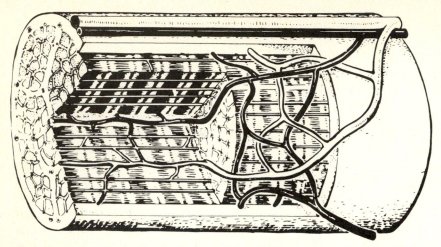

Figure 6-8 Vascularization of skeletal muscle. Capillaries run parallel to the fibrils which make up the fiber.

biopsies taken from the vastus lateralis of the quadriceps muscle from 12 untrained and 11 endurance-trained individuals. The mean number of capillaries per mm² was 585 ± 40 in the untrained and 820 ± 28 in the well-trained individuals. Correcting for shrinkage and artificial spaces between the fibers, he arrived at a mean number of capillaries per mm² of 325 for the untrained and 460 for the trained group (Andersen, 1975). Thus up to 500 capillaries/mm² may be open during heavy muscular activity. The total surface area of the capillary bed would then be about 250 m² in an individual with 35 kg of muscles. With an average capillary diameter of 8 μm, the area of the capillaries would be 2.5 percent of the total tissue area; the distance from a capillary to the most remote cell would be about 10 μm if the capillaries are evenly distributed. The precapillary sphincters can close the inlet to the capillary, and at rest the number of open capillaries is markedly reduced. The opening or closing is supposed to be operated by local chemical factors of a hypoxic or metabolic nature, and only to a smaller degree by nerve activity (see below) (Folkow and Neil, 1971). In a given degree of metabolic activity in the tissue, the number of working capillaries may be fairly constant, but the individual capillary is intermittently open or closed. In well-trained athletes a muscle fiber may be surrounded by as many as 5 to 6 capillaries (see Saltin et al., 1977). Fast- and slow-twitch fibers may share common capillaries, but—in average—the capillary density around the slow-twitch fibers is higher than around the fast-twitch fibers.

It is still uncertain which type of capillary system is predominant in the muscle (Fig. 6.5): whether it is the thoroughfare channel (preferential channel) system or a system in which capillaries form a parallel-coupled circuit between terminal arterioles and venules. There are probably no real arteriovenous or arteriolovenous shunts.

The abundance of capillaries in the muscles provides good facilities for the supplying of oxygen and nutritive materials to the cells "bathing" in the interstitial fluid, and for the removal of products from metabolic activity. Krogh (1929) calculated that with 100 open capillaries/mm², a gradient in oxygen tension of 12 mm Hg from a capillary to the most distant point would be sufficient during resting conditions to provide those distant parts with oxygen by diffusion. With all capillaries open, a lower oxygen gradient would be enough to drive a diffusion; the mitochondria are functioning with an oxygen pressure of 1 mm Hg or even less.

REGULATION OF CIRCULATION AT REST

This could be a very confusing discussion, for the theories on "regulation of circulation" are almost equal to the number of physiologists working in the field. Moreover, there is no unanimous agreement on the terminology of *regulation* and *control*. From a demonstration of an effect on heart rate or vasomotor tone of a stimulation of some part of an animal, it may be tempting to construct a hypothesis concerning regulatory mechanisms. On the contrary, the noticed effect may represent an interference with, or a disturbance of, the normal function and by no means a reflection of a regulatory mechanism.

It should be remembered that by following nerve fibers, one can travel from one cell to almost any other cell within an animal's body. The anatomic basis for a regulation, or interference, in the function of cells and organs is therefore definitely present. Hormones and other chemical substances (including anesthetic agents) can exert a purposeful effect on some cells but, by accident, disturb the function of others.

The evolution of Homo sapiens has passed through many stages. Nerve tracts and organs have probably changed their functions, but remnants of once very essential functions and mechanisms may still remain, normally suppressed or transiently noticeable only in specific situations. Similar obsolete mechanisms may eventually be revived under special conditions. If so, this is most likely to occur when points in the central nervous sytem or nerves are artificially stimulated, which may rouse centra or tracts from "sleep" or inhibition. The effect thus displayed cannot, however, be considered as a mechanism of regulation. In research, artificial situations and tools are essential in studying the cells, organs, and even the intact organism. The most critical and crucial parts of a study are the interpretation and application of results obtained. With these difficulties and pitfalls in mind, we shall now discuss the regulation of circulation at rest and during exercise.

Any change in cellular activity should be met by a corresponding variation in local blood flow through the capillary bed. If an individual cell, in one way or another, could control its environment by varying the blood supply in balance with the actual nutritional demand, that cell would benefit. But other cells or tissues might suffer if some cells selfishly take more than their share, leaving others with little or nothing. Hence, coordinative mechanisms are essential if the distribution of blood is to be balanced properly. An active regulatory mechanism ensures that more active and less active cells, as well as more susceptible and less susceptible organs, are supplied according to their need and to the capacity of the whole circulation.

Arterial Blood Pressure and Vasomotor Tone

The blood pressure in the aorta is maintained by an integration of the following factors: (1) cardiac output, (2) peripheral resistance, (3) elasticity of the main arteries, (4) viscosity of the blood, and (5) blood volume. Evidently a regulation of the arterial blood pressure can use factors 1 and 2, since tools for 3 through 5 are normally not at its disposal for rapid modifications.

The local blood flow is mainly determined by the pressure head and the diameter of the actual vessels. The smooth muscles of the arterioles and veins in many regions continuously receive nerve impulses that keep the lumen of the vessels more or less constricted. This vasomotor tone is provided by the sympathetic vasoconstrictor fibers driven from the vasomotor area in the medulla oblongata. The transmitter substance is norepinephrine. (The membrane receptors of alpha type have a fairly general distribution in the vascular tree, and the effect on the smooth muscles by the transmitter substance norepinephrine is a constriction of the vessels. There are β-receptors in some precapillary resistance sections, such as in skeletal muscle and myocardium where epinephrine can relax the smooth muscles.) The heart and brain receive few vasomotor fibers; the supply to the abdominal organs (by splanchnic nerves) and skin is very rich. The muscles have an intermediate position. The vasomotor tone at rest can be demonstrated by section or blocking of the sympathetic nerve fibers in an animal. The arterioles dilate, and the arterial blood pressure falls. The effect of such an inhibition of vasomotor tone is very marked in the skin, but less pronounced in skeletal muscles. From a basal blood flow of 3 to 5 ml $\cdot$ 100 ml^{-1} $\cdot$ min^{-1} of muscle tissue, a doubling of the flow may occur. During exercise it can, however, amount to some 100 ml $\cdot$ min^{-1} or more. The smooth muscles in precapillary vessels may exhibit spontaneous rhythmic contractions creating a basal vascular tone. The intravascular pressure may be a stimulating factor. Some of the smooth muscle fibers, predominantly localized in the most narrow vessels, may actually serve as stretch receptors and pacemakers and thereby as triggers for the neighboring cells. Thus a wave of depolarization and contraction is propagated in the proximal direction (see Folkow and Neil, 1971). The vasomotor tone is important in keeping arterial blood pressure and cardiac output on an economical level. The splanchnic area could contain the whole blood volume after maximal dilatation of the vessels. Also, the vascular bed of skin and muscles has a similarly large capacity. Fainting (vasovagal syncope) may be the result of a central inhibition of the vasomotor efferent impulses.

The cardiac output can certainly not exceed the venous return. A constriction of the postcapillary vessels (the capacitance vessels) with their large content of blood will increase the blood flow toward the heart, making an increase in cardiac output possible. A decrease (inhibition) in the vasomotor tone, which actually causes a relaxation of the smooth muscles in the vessel wall and therefore a vasodilatation, can be obtained principally in three ways: (1) The smooth muscles can be affected by chemical substances liberated locally from neighboring cells or delivered from the blood. More or less effective as dilating agents are hypoxia, lowered pH, an excess of CO_2 and lactic acid, adenosine compounds, an increase in extracellular potassium, P, or hyperosmolarity (heat can exert a similar effect on the skin vessel, but only to a small degree in the skeletal muscles). It should be noted that smooth muscles in the precapillary vessels in the skeletal muscles are effectively relaxed by metabolites. A constriction of veins induced by sympathetic nerve activity is, on the other hand, well maintained even at extensive metabolite accumulation (Kjellmer, 1965; Folkow and Neil, 1971). Vasodilatation may also be caused (2) by a decreased discharge in the sympathetic vasomotor nerves and (3) by liberation of acetylcholine from the nerve endings of active sympathetic vasodilator fibers (cholinergic effect which is blocked by atropine). The final common path of those nerve fibers starts in the lateral spinal horns, i.e., they have from here on a similar anatomy to the sympathetic nerves. However, these sympathetic cholinergic vasodilator fibers are distributed only to the vessels of the skeletal muscles.

Summary   The peripheral resistance to blood flow is determined by the vasomotor tone. The degree of contraction of the smooth muscles in the arterioles is of special importance for the local blood flow as well as the total resistance. In some tissues (e.g., muscles) this tone is probably partly spontaneous, the smooth muscle contractions (myogenic activity) being triggered by mechanical stretch induced by the intravascular pressure, but a sympathetic vasomotor tone is superimposed. In other regions (such as the skin) this sympathetic vasomotor tone is definitely dominating. A vasodilatation can occur after a local inhibition of the sympathetic effect or the myogenic activity (by metabolites, heat, activity in the sympathetic vasodilator fibers) or by central inhibition of the sympathetic impulse traffic. The spontaneously active mechanoreceptors in the precapillary vessels can induce vasoconstriction. With the capillary flow reduced, tissue metabolite concentrations will increase and exert a vasodilator influence on the sphincter mechanism. Soon the pacemakers will become activated again, etc.

The Heart and the Effect of Nerve Impulses

The heart has its own pacemaker, initiating about 70 impulses/min if left alone. Both sympathetic and parasympathetic nerve impulses can modify the heart rate. The parasympathetic activity from a cardioinhibitory center via the vagus nerve (and acetylcholine) causes a slowing of the heart rate (bradycardia), and the sympathetic cardiac nerves (norepinephrine) can produce an increased heart rate (tachycardia). In a resting subject, a blocking of the vagus nerve will cause the heart to beat faster, indicating a predominating parasympathetic tone.

There are indications that the slowing of the heart rate during a standard work load after a period of physical training is induced by an increase in vagus tone and a reduction of the sympathetic drive (see Ekblom et al., 1973). The sympathetic nerves can increase the contractile force of the heart muscle fibers, but sympathetic nervous control of the vasomotor tone in the heart vessels is probably insignificant. The blood vessels of the heart dilate willingly when affected by hypoxia and metabolites.

(The so-called adrenergic β-receptors dominate in the heart muscle, and it is possible pharmacologically to reduce or block those receptors to respond to sympathetic stimuli, e.g., by propranolol.)

Control and Effects Exerted by the Central Nervous System

As we have now analyzed the tools available for a variation and redistribution of blood flow within the body, the actual control and regulating mechanisms may be discussed. (For a more detailed analysis of the overall cardiovascular regulation, see Öberg, 1976.)

Important and essential nuclei are located in the medulla oblongata. The *medullary vasomotor area* can be divided into a *vasoconstrictor center* and a *vasodepressor area*, the latter operating through inhibition of the sympathetic vasoconstrictor outflow. Efferent nerve fibers extend to sympathetic connector cells in the lateral horns of the spinal cord (thoracic region and first two segments of the lumbar region). The neurogenic vasomotor tone of the blood vessels originates essentially in the vasomotor area, and a continuous, somewhat rhythmical discharge can be detected in the nerve cells. This discharge is probably caused by the influence of the chemical composition of the interstitial fluid that bathes the cells. The spontaneous sympathetic vasomotor activity can then be modified by impulses from the vasodepressor area (inhibitory) or from higher levels of the CNS (inhibitory or facilitating) (Fig. 6-9). The negative feedback operates via inhibitory neurons directly on the tonically active cell bodies of the constrictor center. It also damps the preganglionic cell bodies in the spinal cord directly.

The vasodepressor area is essentially a relay station without spontaneous activity, but it is activated by afferent impulses, especially from the baroreceptors in the aortic and carotid bodies.

On higher levels of the CNS, there are areas, especially in the cerebral cortex and diencephalon, from which cardiovascular reactions can be elicited. Although these higher centers do not contribute to the continuous vasomotor tone, many adjustments are initiated primarily from the brain above the level of the medullary centers (see Korner, 1971). Of special interest are the *sympathetic cholinergic vasodilator fibers* (Folkow and Neil, 1971; Uvnäs, 1960). They can be traced from the motor cortex and followed through the anterior hypothalamus and mesencephalon (relay stations). The nerve fibers bypass the vasomotor center and run to the lateral spinal horns and the sympathetic final common path. When stimulated, they can be activated in synergism with vasoconstrictor fibers. The combined effect is then a vasodilatation of precapillary resistance vessels in the skeletal muscles and vasoconstriction of the

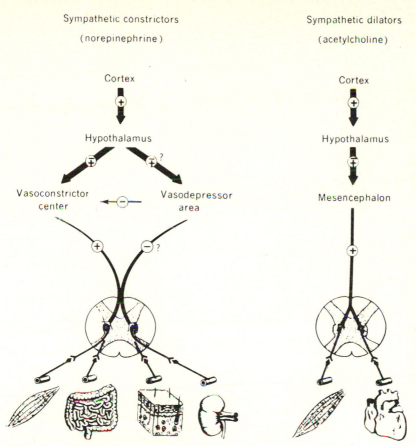

Figure 6-9 Central vasomotor integration. The inhibition on the vasoconstrictor activity exerted from the vasodepressor area may operate directly on the center or eventually at the spinal level. The vessels in skeletal muscles and the gastrointestinal tract are, at rest, easier to involve in a constrictor activity ("low-threshold areas") than are the cutaneous and kidney vessels. For detailed description, see text.

vessels of the abdominal organs and skin. "With very little shift in arterial pressure this automatic activation pattern leads to a remarkable and instantaneous redistribution of cardiac output to favor skeletal muscles" (Folkow, 1960). Simultaneously, accelerator fibers to the heart may also be stimulated, and the medulla of the suprarenals may give rise to a liberation of epinephrine. This hormone dilates the resistance vessels of the skeletal muscles and excites the smooth muscles of the capacitance vessels; norepinephrine strongly contracts both resistance and capacitance vessels. (These various responses are consequences of the different receptors, α and β.)

Similar reaction patterns are characteristic for emergency conditions, such as fear, and can be elicited by electrical stimulation of the hypothalamus (Abrahams et al., 1960; Folkow, 1960; Uvnäs, 1960) and even by cutaneous stimulation (Abrahams et al., 1960).

Figure 6-9 describes the central vasomotor integration in a schematic way.

Mechanoreceptors in Systemic Arteries

The afferent input is far from being completely mapped out. Important afferent fibers come from mechanoreceptors in blood vessels and in the heart. The systemic arterial receptors are located in the tissue of the carotid sinus, aortic arch, right subclavian artery, and common carotid artery (Heymans and Neil, 1958; Neil, 1960). Mechanical deformation of the walls of the vessels is the normal stimulus of the receptors, and they respond to the rate of rise of blood pressure as well as amplitude of pulse pressure.

The receptors are actually stretch receptors, and pressure as such is not the adequate stimulus. (The commonly used name "baroreceptors" is therefore somewhat misleading.) However, an increase in the intravascular pressure does expand the vessel wall and stretch the receptors. They respond with a discharge transmitted to the CNS. If the wall of the vessel, where the stretch receptors are located, becomes less distensible, owing to increased activity of smooth muscles in the wall or a progressive structural change (in a hypertensive patient or aging person), a given pressure would induce less deformation of the receptors and a reduced impulse output (Peterson, 1967). In fact, pathways from suprabulbar areas are probably involved in a central resetting of the mechanoreceptor reflex during increased activity of other receptor groups, e.g., during exercise (see Korner, 1971). This can augment or depress the reflex responses mediated through mechanoreceptors.

Pulsatile pressure about a given mean blood pressure is more effective than a steady mean pressure to set up an impulse traffic in the afferent nerves (sinus branch of the glossopharyngeal nerve and afferent fibers of the vagus). The threshold at which a stimulus is effective varies for the different receptors. A recording of activity in the sinus nerve normally reveals a continuous discharge and a variation in the impulse traffic with each pulse beat.

The baroreceptors can report a fall as well as a rise in blood pressure to the cardiovascular centers, primarily the medullary vasomotor area. At rest, the baroreceptors exert a restraining influence on the cardiovascular system, causing a reflex bradycardia and reflex inhibition of the medullary vasomotor center (Fig. 6-10).

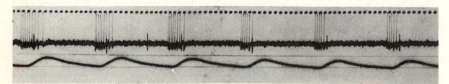

Figure 6-10 Action potentials from mechanoreceptors in the carotid sinus recorded from the sinus nerve in the cat. Dots at the top give the time, 50 Hz; the bottom line shows the blood pressure recorded in the carotid artery with calibration lines at 100 and 150 mm Hg respectively. Note that on the left side, when pressure is high, there are 9 to 10 large "spikes" per heartbeat. When the pressure is lower, there are only 5 spikes. (The spikes occur at the pressure rise, but there was some delay in the pressure recording.)

Posture

The physiological interplay of the various factors involved in the maintenance of an adequate arterial blood pressure can be elucidated by the following experiment:

On a tilting table, a subject is tilted from supine to a head-up position (about a 60° angle to the horizontal). Owing to the force of gravity, blood is pooled in the parts of the body below the heart level. Thus, the venous return to the heart is temporarily reduced. Consequently, the cardiac output decreases and so does the arterial blood pressure. The strain exerted on the baroreceptors is reduced, and fewer impulses are transmitted from them to the CNS. The impulse output from the parasympathetic cardioinhibitory center is diminished (which results in an increase in heart rate); the vasodepressor area becomes inhibited, and from the adrenergic sympathetic centers there is an increased impulse traffic (the effect is a vasoconstriction in resistance vessels and capacitance vessels, especially in the splanchnic area, and an increase in heart rate). The precapillary vessels in skeletal muscles are also important targets for this baroreceptor reflex (see Rowell, 1974). Thus, the peripheral resistance becomes higher, the cardiac output can be restored to an adequate level, and the arterial blood pressure can increase. The variation in heart rate in a subject passively tilted to different body positions is illustrated by Fig. 6-11 (Asmussen et al., 1939). If blood pressure cuffs are placed around the upper parts of the thighs and a pressure of about 200 mm Hg is applied when the subject is in a horizontal or head-down position, the heart rate response to the head-up position is less pronounced (Fig. 6-11). The hydrostatic forces are acting on a shorter "column," as the blood is prevented from circulating to the legs. If the

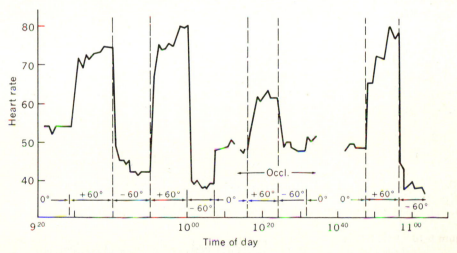

Figure 6-11 Variation in heart rate (ordinate) in a subject passively tilted to different body positions. +60° = tilting to a head-up position; −60° = head-down position; Occl. = experiments in which inflated blood pressure cuffs were placed around the upper parts of the thighs. For further details, see text. *(From Asmussen et al., 1939.)*

pressure within the cuffs is suddenly released, the fall in arterial blood pressure may be very pronounced, and eventually the subject faints as the blood, and thereby the oxygen supply to the brain, becomes inadequate. The capacity of the legs to retain blood has increased, for its arterioles and capillaries dilate as anaerobic metabolites accumulate during the period of circulatory occlusion. Tilting of the subject to a head-down position will quickly restore the circulation and consciousness as the legs are drained. Figure 6-12 shows how the heart rate can be lowered some 10 beats/min if the subject in a head-up position (on a tilting table) contracts the leg muscles. The massaging effect of the repeatedly contracting muscles on the capillaries and veins enhances the venous return to the heart, and the heart rate is lowered. Most likely the nerves from the cardiovascular mechanoreceptors form an important link in the reflex chain.

The beneficial effect of the muscle pump on the venous return should certainly be stressed for people who work in a fixed sitting or standing position. Bandaged legs can partially reduce the hydrostatic shift of fluid to the legs in the upright position, and thereby the circulation is facilitated (Lundgren, 1946; Arenander, 1960). A sudden standstill after prolonged exercise, particularly in a hot environment, can cause fainting, as the blood pools in the dilated vessels in exercised legs and in the skin. The unexpected fall of the tall soldier during a military parade can also be explained by an inappropriate distribution of blood.

There are two important factors counteracting an edema formation in the legs in erect position: (1) a pressure-induced facilitation of the rate of myogenic contractions of the smooth muscles of the arterioles and precapillary sphincters, decreasing the surface area of the capillary bed available for

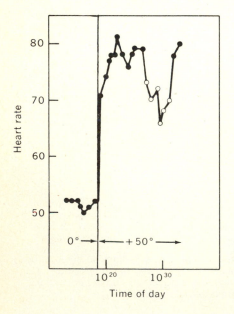

Figure 6-12 Effect of voluntary contraction of leg muscles on heart rate during passive standing. ● = passive standing; ○ = voluntary contraction of leg muscles. *(From Asmussen et al., 1939.)*

filtration; (2) an effective reabsorption of fluid via the red cells squeezed in the narrow capillaries. The close contact between red cells and tissue fluid will promote an uptake of water in the capillaries as well as the exchange of gases, as discussed earlier in this chapter.

Summary   A change in body position will inevitably affect the circulation as long as the individual stays under the influence of the pull of gravity. A head-up position will primarily increase the blood volume in the legs and decrease the central blood volume and cardiac output. Secondary variations in arterial blood pressure are reported from the mechanoreceptors in some arteries to inform the cardiovascular centers in the brain. The activity in sympathetic and parasympathetic nerves varies by reciprocal innervation in such a way that the arterial blood pressure and cardiac output return to a level fairly close to the one typical for the individual in a supine position. The nerves from the mechanoreceptors were appropriately called "buffer nerves" by Samson Wright.

Other Receptors

In the walls of the pulmonary artery there are *mechanoreceptors* with reflex effects on the systemic circulation and heart similar to those caused by the systemic arterial baroreceptors. A third group of mechanoreceptors are located in the walls of the atria and ventricles of the heart. When stimulated, they cause by reflex a vasodilatation, bradycardia, and systemic hypotension, so that the load on the heart diminishes. Exceptions are other "low-pressure" receptors in the left atria which, when stimulated, will reflexly accelerate the heart. Stretch receptors also serve as one sensory mechanism in a reflex regulation of blood volume by control of urine output (see Folkow and Neil, 1971). The filling volume of the cardiac atria, related to the circulating or thoracic blood volume, is likely to be the appropriate stimulus; a variation in the production of antidiuretic hormone from the hypophysis is likely to be the tool.

The *chemoreceptors* in the carotid and aortic bodies, stimulated by low oxygen tension in the circulating blood, not only influence the pulmonary ventilation, but also, indirectly, the circulation. In artificially ventilated dogs, stimulus of the chemoreceptors by hypoxia causes an increase in sympathetic vasoconstrictor discharge, a bradycardia, and decrease in cardiac output (Daly and Scott, 1958; Daly, 1964). In a spontaneously breathing animal, however, a hypoxia causes a tachycardia, probably secondary to the hyperventilation. Consequently, it can be concluded that the tachycardia of systemic hypoxia does not have a chemoreceptor reflex origin.

In this context it is of interest that one of the primary *adjustments which take place during a dive* is circulatory in nature (Scholander et al., 1962; Irving, 1963). Most animals, including man, display a diving bradycardia, which usually develops gradually. Thus during the dive, the frequency may drop to one-half of normal in some species and to one-tenth in others. As a rule, the bradycardia stays with the animal whether it exercises or not during the dive. Blood pressure is maintained at a normal, or even an elevated, level as a result of peripheral vasoconstriction. In the diving animal, the reduction and selective redistribution of the circulation can save oxygen for vital organs such as the brain and heart. In the human, the heart rate increases immediately at the start of a dive and slows down very markedly as the dive progresses, even if the diver exercises vigorously (Scholander et al., 1962; Olsen et al., 1962; Irving, 1963). Prompt return to normal sinus rhythm occurs with the first breath.

This bradycardia during diving is not a result of apnea only, since it is reinforced by water submersion. Skin receptors, wetting of the nose, and asphyxial reflexes may contribute to the development of the bradycardia. The onset of the decline in heart rate is very rapid. However, it may be that the chemoreceptors gradually contribute to the bradycardia by a progressive hypoxic drive, since the drop in alveolar oxygen tension is very marked, at least during exercise combined with breath-holding (P.-O. Åstrand, 1960). In any case, there is a striking similarity in the circulatory response of isolated chemoreceptors in dogs to a stimulus of hypoxia and the effect induced by diving on the circulation in the intact animal.

Phylogenetically, the chemoreceptors may be a very ancient receptor mechanism of importance for diving animals, but they assumed a new function when the animal left the water for good. *In summary* stimulation of arterial stretch receptors inhibits the sympathetic discharge, but chemoreceptor activation enhances the sympathetic discharge. Both of them, however, excite the vagal cardio-inhibitory "center."

Afferent impulses to the cardiovascular centers from receptors in the skin, muscles, and joints have been suggested but have not as yet been shown to exist. Receptors in skeletal muscles may sense local chemical changes due to an increased metabolism and therefore create stimuli proportional to the metabolic rate. The evidences for muscle-heart reflexes are discussed by Hollander and Bouman, 1975.

Any damage to superficial tissues causes a regional vasodilatation response. The afferent fibers of pain receptors branch, and impulses pass along those branches to arterioles, causing a dilatation (antidromic activation or *axon reflex*).

REGULATION OF CIRCULATION DURING EXERCISE

The conclusions of the preceding discussions are as follows: The blood flow in the precapillary vessels of the muscles is "regulated" locally, for the most part, by the level of metabolism in the muscle cells. The actual nutritional demand is the determining factor and will be superimposed on any nervous influence; the vessels then become functionally sympathectomized. The blood flow in the splanchnic area is, on the other hand, under control of the CNS. The flow is varied so that the systemic arterial blood pressure is maintained at an adequate level to supply brain, heart, and some other vital organs with blood. In this case, neural vasoconstrictor activity can be superimposed upon the local dilator control. A stimulation of the vasoconstrictor fibers can increase the resistance in the vessels of the gastrointestinal tract about 20 times (Folkow, 1960). The vessels of the skin also subserve centrally controlled mechanisms. Impulses from the vasomotor area are important, but the final control is probably exerted by the temperature-regulating centers, at least while the subject is at rest or during submaximal exercise.

At rest, the kidneys receive about 25 percent of the cardiac output. There are no tonic impulses from the CNS to renal blood vessels, but electrical stimulation of the renal nerves causes intense renal vessel constriction with associated changes in blood flow (down to 250 ml/min) and in excretion of water and electrolytes (Pappenheimer, 1960). Exercise, postural changes, and circulatory stress in general may cause profound alterations in renal function, mediated through the hemodynamic effects of renal nerves.

This summary gives the main tools available for the regulation of the circulatory system during exercise. The approximate blood distribution to the various organs is illustrated in Fig. 6-6.

The changes in heart function and circulation, from the moment muscular exercise begins (or even before), are initiated from the brain levels above the medullary centers (probably the cerebral cortex and diencephalon). By a reciprocal innervation there is a simultaneous increase in the sympathetic activity and decrease in the parasympathetic impulse traffic. The skeletal muscles may receive an increased share of the cardiac output because of their innervation with sympathetic cholinergic vasodilator fibers. The activity in the sympathetic adrenergic vasoconstrictor fibers reduces the blood flow to skin, kidneys, and splanchnic area (Fig. 6-13). Rowell (1974) calculated that at maximal vasoconstriction of the splanchnic and renal blood vessels, about 2.2 liters · min^{-1} can be redistributed to the working muscles. This could increase their oxygen uptake by about 0.5 liters · min^{-1} without any additional increase in the cardiac output. It should be emphasized that a vasoconstriction of precapillary vessels will cause secondarily a reduction in the transmural (distending) pressure in the postcapillary vessels. The veins collapse easily due

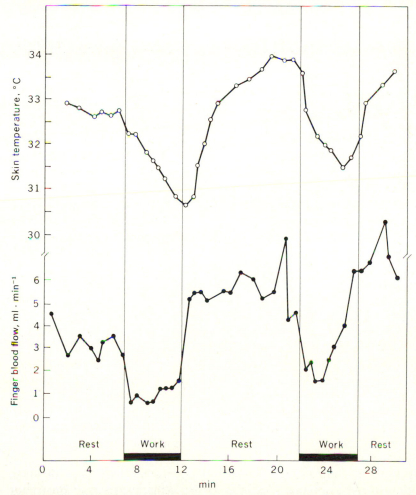

Figure 6-13 Blood flow in index finger (measured by means of a plethysmographic method) and skin temperature on the same finger at rest and during two periods of work on a bicycle ergometer (1080 kpm · min^{-1}, or 180 watts). *(From Christensen and Nielsen, 1942.)*

to the passive recoil of the venous wall. In doing so, their cross-sectional profile alters from one nearly circular to one which is elliptical. The change in geometrical configuration will cause a major fraction of its volume of blood to be passively dispelled towards the heart (see Folkow and Neil, 1971). The reduction in blood flow to the splanchnic tissues and kidneys (and the increase in heart rate) during exercise is more related to the severity of the exercise in relation to the individual's maximal oxygen uptake than to the absolute rate of oxygen uptake (see Clausen, 1976; Rowell, 1974). The balance in the redistribution of blood can be such that there is no (or only a slight) decrease in the systolic blood pressure during the initial period of exercise, despite a marked dilatation of the resistance vessels in the muscles (Holmgren, 1956). Constriction of the veins (capacitance vessels), pumping action of the working muscles, and forced respiratory movements assist the venous return to the heart. Since the heart escapes from the vagal inhibition, and sympathetic impulses can increase the force and frequency of the muscle contractions, the heart gains capacity to take care of the increased inflow of blood, and if necessary, pumps it out against an elevated resistance. The volume of blood in the heart and lungs actually increases, at least during exercise in the supine position (Mitchell et al., 1958a; Braunwald and Kelly, 1960).

In the working muscles, the effect of the increased metabolism is a local change in pH and in composition of the interstitial fluid, causing an opening of capillaries and those arterioles not already dilated by the sympathetic vasodilator activity. For several reasons, this local control of blood flow is the most important factor in securing an efficient blood supply to the working muscles. It appears that the mechanisms controlling distribution of open capillaries in the muscle are more responsive to muscular exercise than to hypoxia (Otis, 1963). This difference in response can be explained, since during exercise the potassium ions released from the intracellular space reach such a high extracellular concentration that they can account for a major part of the vascular dilatation accompanying muscular activity (Kjellmer, 1965).

Furthermore, exercise leads to hyperosmolarity by release of particles (metabolites) from the striated muscle fibers into interstitial fluid space. This change in the environment of the smooth muscles of the blood vessels may inhibit vascular tone (Mellander and Johansson, 1968).

Skinner and Costin (1970) point out that the concentration of oxygen in the tissue can substantially influence the reaction of the vascular bed to potassium. Oxygen-deficient blood strengthens the vasodilating effect of potassium at all levels of potassium concentration in the tissue.

An eventual neurogenic vasodilatation in skeletal muscles affects the resistance vessels almost exclusively, but it is not restricted to working muscles.

The capacitance vessels are rather sensitive to constrictive influence. Actually, at a given low impulse traffic in the sympathetic vasoconstrictor fibers, the constrictor response from the capacitance vessels is relatively much more pronounced than from the resistance vessels (Mellander, 1960; Kjellmer, 1965).

In experiments on humans, Bevegård and Shepherd (1967) have observed an increase in tension of the venous walls in both exercising and nonexercising limbs. This increase in tension persists throughout the exercise and is proportional to the severity of the work. They emphasize that this venoconstriction is not released by metabolites produced in the working muscles.

At least in experiments on animals under various conditions, the activity in vasodilator fibers is of short duration (Bevegård and Shepherd, 1967). It makes physiological sense to interpret the initial vasodilatation in muscles as a state of general preparedness; arterioles and thoroughfare channels are flooded by a rapid bloodstream just at the doors of the capillaries. In the metabolically active areas, the capillary sphincters open and the nutritive vessels can immediately be perfused by blood (part of an emergency reaction). After perhaps some 10 s, the sympathetic vasodilator activity ceases. In the active muscles, the vessels remain open in proportion to the metabolic rate, but in the resting muscles, the arterioles constrict by the now-dominating activity in sympathetic vasoconstrictor fibers. The metabolic control of the blood flow in muscles is illustrated by the finding that the oxygen content of venous blood from a resting extremity falls as low as that from one that is exercising (Donald et al., 1954; Mitchell et al., 1958b; Carlson and Pernow, 1959).

It should be recalled that epinephrine, liberated from the activated suprarenal medulla, excites the smooth muscles of the capacitance vessels, but dilates the resistance vessels of skeletal muscles. Norepinephrine from the sympathetic vasoconstrictor fibers contracts muscles of both capacitance and resistance vessels (the threshold being different).

As the exercise proceeds, the blood vessels of the skin, especially the arteriovenous anastomoses, dilate, so that the produced heat can be transported to the surface of the body. The heavier the exercise and the higher the environmental temperature, the more pronounced is this secondary vasodilatation in the skin. Indirectly, impulses in sympathetic nerve fibers are partly behind this dilatation, and the temperature-regulating center in the hypothalamus is guiding the impulse traffic. These nerve fibers stimulate the sweat glands to sweat production by acetylcholine, but the glands also deliver an enzyme acting on proteins in the tissue fluid, and a substance with vasodilator effect is formed, identical with or related to bradykinin (Barcroft, 1960). The local skin temperature also affects the lumen of the vessels. (Circulatory aspects of a heat stress are discussed in Chap. 15; see also Rowell, 1974.)

During work, the integrated effect of neural and chemical factors (including hormones) gives a cardiac output that may be markedly higher and with quite a different distribution than when the subject is at rest (Fig. 6-6).

Most investigators report an elevated *arterial systolic* and *pulse pressure* during exercise (Fig. 6-20), and the effect of such increased pressure on the mechanoreceptors in the arterial walls has been discussed. There are, however, important aspects to consider when interpreting blood pressure curves and the effects of the pressure on mechanoreceptors.

A simultaneous measurement of intra-arterial blood pressure in a peripheral artery and in the aorta during exercise gives a significantly higher systolic end pressure in the peripheral artery, but the mean and diastolic pressures are about the same as in the aorta (open-ended catheter directed against the flow axis) (Kreeker and Wood, 1955; Holmgren, 1956; Marx et al., 1967). The systolic blood pressure in a peripheral artery is higher in a resting than in a working limb (P.-O. Åstrand et al., 1965). The progressive increase in systolic pressure (and pulse pressure) along an artery is at least in part due to a distortion in the transmission because of summation of the centrifuge wave and the reflected waves from the periphery. The importance of the wave reflection increases when the peripheral resistance is high, as is the case in a resting limb.

Marx et al. (1967) report that the pressure measured in the aorta with a catheter with side holes gives a significantly lower pressure during exercise than does measurement with an open-ended catheter directed upstream. The reader is reminded of the discussion earlier in this chapter. The total energy of the blood is the sum of the kinetic energy and the pressure energy. The side-hole catheter is measuring only the pressure energy (side pressure), but the catheter with the opening directed upstream also includes the kinetic energy. Since the kinetic energy factor of the blood in the aorta is high during exercise with a pronounced increase in cardiac output, the two catheters should give quite different pressure readings. Marx et al. have convincingly shown that this is the case.

It can be calculated (1) that the blood pressure measured in a peripheral artery during exercise does not give a true picture of the *systolic* blood pressure in the aorta; (2) a measurement in the aorta against the flow (end pressure) gives an evaluation of the pressure-load on the heart, but (3) it does not reflect the side pressure and strain on the vessel walls. The mechanoreceptors cannot sense the kinetic energy factor of the passing blood. Lateral distending pressure should therefore be used for an evaluation of the strain on the mechanoreceptors (data on mean pressure are of no value in this connection).

As emphasized above, the mechanoreceptors are lying in the vessel walls and they convert distortion of the wall into nerve impulses. A sympathetic vasoconstrictor activity involving the smooth muscles of the vessel wall makes it stiffer. Under such conditions, a given intravascular pulsatile pressure gives less deformation of the wall than when the smooth muscles are less active; therefore the discharges from the mechanoreceptors are reduced (Peterson, 1967).

At least during submaximal exercise, the mechanoreceptor mechanism continues to oppose the rise in heart rate and blood pressure through a negative feedback. By creating subatmospheric pressures in a Plexiglas box enclosing the neck, researchers have found that blood pressure, heart rate, and cardiac output were significantly reduced at rest as well as during exercise (with cardiac output elevated up to 17 liters min^{-1}) (Bevegård and Shepherd, 1964, 1967).

If the *cardiac output* ($\dot{Q}$) during exercise is 4 times the resting value, giving a 25 percent increase in arterial mean pressure (P_{mean}), it follows that the resistance to flow (R) is reduced to more than one-third of what it was at rest, since $Q \times R = P_{\text{mean}}$. The lowered resistance is, however, due to the dilated vascular bed in the working muscle, and the vessels in the abdomen are still obeying signals in the sympathetic vasoconstrictor fibers (Wade et al., 1956). It should be recalled that if nonmuscular tissues receive 80 percent of the cardiac output at rest but only 20 percent during heavy exercise, their blood flow would in both cases be some 4 liters · min^{-1}. The distribution of this flow of 4 liters · min^{-1} is probably different at rest than during exercise (Fig. 6-6).

The linear increase in heart rate and, within a wide range, cardiac output with the increase in oxygen uptake suggests a "neat" regulatory mechanism with the metabolic activity as an important guide. Whether or not the heart muscle has a capacity to pump out more blood than it actually does during the most severe exercise but is prevented from doing so by an inhibition elicited from mechanoreceptors, for instance, is still an open question.

The importance of impulses from higher levels of the CNS to the first circulatory adjustments to exercise has been emphasized. On the other hand, for the muscular blood flow the neural control of the vessels of the muscles probably has minor practical importance. Barcroft and Swan (1953) have shown that the hyperemia which follows standard exercise is essentially the same in a normal and in a sympathectomized muscle (Donald et al., 1970). This fits with the previous statement that local factors are superimposed on any nervous influence on the vascular bed of the muscles.

The cardiac nerves are not of essential importance for the ability to perform heavy exercise. Donald et al. (1964) found that dogs with chronic cardiac denervation showed an almost unchanged capacity for work as measured by oxygen uptake or time for a race over a given distance.

Summary   We can schematically describe the circulatory response to exercise by considering four stages:

1   At rest the skeletal muscles receive only some 15 percent of the minute blood flow, and their arterioles are constricted by a continuous vasoconstrictor activity and some sort of a spontaneous vascular tone. Few capillaries are open, but the individual capillaries open and close alternatively. The heart rate is kept down by a parasympathetic outflow via the vagal nerve.

2   When, or even before, exercise begins, there are an inhibition of the parasympathetic activity and an increased sympathetic impulse traffic. The heart escapes from its inhibition and beats faster and with increased force. Eventually impulses from higher levels of CNS, transmitted by sympathetic cholinergic vasodilator fibers, dilate arterioles in the muscles, thereby increasing their blood flow. On the other hand, sympathetic adrenergic vasoconstrictor fibers act on the vessels of abdominal organs and skin so that a decreasing share of the cardiac output flows through those tissues. The veins become constricted passively, and by activity in the constrictor fibers. This constriction of veins, together with the pumping action of the working muscles and the forced respiratory movements, facilitates the blood return to the heart, making an increased cardiac output possible.

3   The appropriate adjustment of the circulation occurs. In the working muscles, the increased metabolism and potassium ions cause changes in the environment which locally dilate arterioles and open capillaries. The sympathetic vasodilator fibers are probably inactive or without effect, and in the resting muscles, the arterioles constrict. Hormones contribute in the constriction of vessels in nonactive areas.

4   For temperature balance within the body, the produced heat is transported to the skin, since skin vessels become dilated.

The resistance vessels, particularly the precapillary sphincters in muscles, are dominated by local vasodilative factors, while the capacitance vessels are more sensitive to the constrictive influence. Therefore, the blood flow to the muscle and the distribution of blood within it is determined by the metabolic requirements, and pooling of blood in the active muscle is prevented by nervous activity (Kjellmer, 1965).

The raised capillary pressure leads to a net outward filtration of fluid, the flow of which is facilitated by a simultaneous increase of the capillary area. The mean capillary pressure may increase from 15–20 to 25–35 mm Hg in activated skeletal muscles. In addition, the osmolarity increases in the exercising muscles due to the breakdown of larger molecules into smaller units. This will contribute to an increased fluid volume in those muscles. Conversely, in tissues where the precapillary vessels are constricted the mean capillary pressure becomes lower. This factor, plus an increased arterial osmolarity, favor a mobilization of extravascular fluid, and the plasma volume can be relatively well maintained. Lundvall et al. (1972) calculated that the total fluid loss into the active muscle mass was about 1,100 ml during heavy exercise (bicycling) but that loss was partly compensated for by a fluid absorption of some 500 ml from inactive tissues. The capillary permeability does not change during exercise (Kjellmer, 1965). (For a more detailed discussion see Clement and Shepherd, 1976; Rowell, in Fulton and Patton, 1974, p. 200.)

CARDIAC OUTPUT AND THE TRANSPORTATION OF OXYGEN

Efficiency of the Heart

Studies on cardiac output and the energy output and work efficiency of the heart have shown that (1) a given stroke volume can be ejected with a minimum of myocardial shortening if the contraction starts at a larger volume; (2) energy losses in the form of friction and tension developed within the heart wall are also at a minimum in a dilated heart; (3) the stretched muscle fiber can, within limits, provide a higher tension than the unstretched one; (4) loss of energy is larger when the contraction occurs rapidly, i.e., with heart rate high as compared with a slower contraction time; (5) on the other hand, the greater the volume of the heart, the higher is the tension of the myocardial fibers necessary to sustain a particular intraventricular pressure (as a consequence of Laplace's law, as discussed earlier in this chapter). The energy need for a contraction is closely related to the tension that has to be developed.

There are a few reports on the energy output and mechanical efficiency of the heart muscle when heart rate and stroke volume are altered. In the isolated heart, it has been found that an increase in cardiac output caused by an increased stroke volume markedly improved the efficiency of the heart work (blood pressure and heart rate were kept constant) (Evans, 1918). This means that the increase in oxygen uptake was not proportional to the increased performance.

If the cardiac output was maintained unaltered, variations in heart rate were followed by an almost linear variation in oxygen uptake of the heart; the high heart rate was associated with a low mechanical efficiency.

The *individual with a high capacity for oxygen transport* because of natural endowment and/or training is characterized by a large stroke volume and a slow heart rate. Cardiac output and systolic/diastolic blood pressures at given work loads or at rest are not noticeably different from normal. Of the factors listed above, the first four tend to act in the individual's favor as far as each single heart beat is concerned, while the fifth factor acts against the person. However, as relatively few contractions are performed per minute, the total energy cost to maintain a given work level (flow times pressure) may be relatively low and the efficiency high. At maximal work levels, the person with the large diastolic heart volume and stroke volume may have as high a heart rate as the individual with a small diastolic filling. In that situation, factors 1 through 4 still favor the fit person, but factor 5 tends to decrease the efficiency when performance per unit time is considered.

It should be borne in mind that at a given maximal heart rate, the heart which has the capacity to provide the largest stroke volume can attain the highest cardiac output. One condition is a good diastolic filling, which inevitably leads to the drawback of increased tension required to produce the pressure.

A rich capillary network in the myocardium would provide the capacity to meet the demand of an increased metabolism. For the exercising skeletal muscles, however, it is essential that the increased demand for blood flow through the myocardium does not consume too many of the extra liters of blood that the heart eventually can manage to pump out. No data are available which would permit an analysis of this problem.

From a general viewpoint, it is considered an advantage if a given level of cardiac output can be maintained with a low heart rate, i.e., a large stroke volume. The reason why a training with "overload" apparently is necessary to improve the efficiency of the heart is presently unknown.

For the *cardiac patient*, the situation is different if a large diastolic filling is combined with a small stroke volume and a high heart rate. The dilatation of the heart muscle gives improved ability for the muscle fibers to produce tension as long as they are stretched, as in factor 3. Factors 1 and 2 help in keeping the efficiency high. Factors 4 and especially 5 tend to reduce the efficiency. Undoubtedly the heart has to pay for its compensating of the basically reduced myocardial strength. Sooner or later there will be a critical equation for energy requirement and energy supply of the heart, and the vascular bed in the myocardium is the key in that formula. At rest, the capillary blood flow may cover the need, but during even mild exercise, a discrepancy arises, and the patient is forced to stop because of symptoms such as angina pectoris, which is a pain probably caused by hypoxia in the myocardium. It should be emphasized that with a high heart rate, the tension time, i.e., the time during which the heart muscle is contracted per minute, is prolonged, and the blood flow through the heart muscle is therefore reduced (Fig. 6-1).

According to Kitamura et al. (1972), there is in young, healthy male subjects a high correlation ($r = 0.88$ to 0.90) between coronary blood flow and myocardial oxygen consumption versus the product of the heart rate and systolic blood pressure measured in the aorta. (As pointed out, the systolic blood pressure measured indirectly over a peripheral artery is not identical with the aortic pressure.)

Summary A large dilated heart may provide tension when the muscle fibers are stretched, but if tension decreases markedly as the muscle shortens, the frequency of contractions must be high. A high heart rate lowers the mechanical efficiency of the heart and increases the oxygen uptake for a given cardiac output. An impaired circulation in the myocardium can further complicate the situation. The difference in fitness of the athlete's large heart and that of the cardiac patient is evident.

Venous Blood Return

Certainly, the cardiac output cannot exceed the flow of blood returning to the heart. In the supine position, the hydrostatic pressure on the venous side of most of the capillaries is about 10 mm Hg, but the pressure gradient to the right atrium is increased by the negative intrathoracic pressure. At rest, this pressure is about 5 mm Hg less than the ambient barometric pressure because the elastic tissue of the lungs is expanded to the size of the thoracic cavity and the recoil of the tissue exerts a tension on the thin-walled vessels within the thorax. The inspiration increases this pulling force, and blood is sucked into the thorax. At the same time, the abdominal veins are compressed as the diaphragm contracts. The variations in intrathoracic and intra-abdominal pressures with breathing will significantly enhance the venous return because a backflow in the veins is hindered by the capillary resistance and venous valves. During an expiratory effort with the glottis closed (Valsalva's maneuver), the intrathoracic pressure increases, impairing both the venous return and the cardiac output; this is usually the case in weight lifting. Hyperventilation followed by Valsalva's maneuver may cause fainting (for the explanation, see Chap. 7). There is evidence that the ventricular action does contribute to the subsequent filling of the ventricle by exerting a suction force on the atrium (see Folkow and Neil, 1971).

A third important factor improving the venous return is dynamic activity of the skeletal muscles (Figs. 6-12, 6-14). As pointed out earlier in this chapter, the muscle pump is very effective in propelling the blood toward the heart. When the leg muscles contract rhythmically, there is a decrease in the blood volume of the legs, indicating the emptying of blood; and blood can thus be driven from a segment against a resistance of 90 mm Hg (Barcroft and Swan, 1953). The action of the muscle pump is especially important in the erect position. If exercise is started (walking), the pressure in the veins of a foot can decrease from 100 mm Hg in passive standing to about 20 mm Hg. The increase in venous return to the heart as muscular work starts will promote the possibility of an immediate increase in cardiac output. The bigger the muscle mass involved, the more pronounced is this effect. As pointed out above, when the exercise stops, the blood stays temporarily in the dilated vascular bed, and the decrease in venous return to the heart may cause such a drop in cardiac output and arterial blood pressure that fainting occurs. (Potentially about one liter of blood may be present in the muscles during maximal vasodilatation, Folkow and Neil, 1971.) This transient pressure drop is more likely to occur if the skin vessels are dilated to secure heat elimination and the person remains motionless in a standing position.

At rest, the venous system (capacitance vessels) contains about 65 to 70

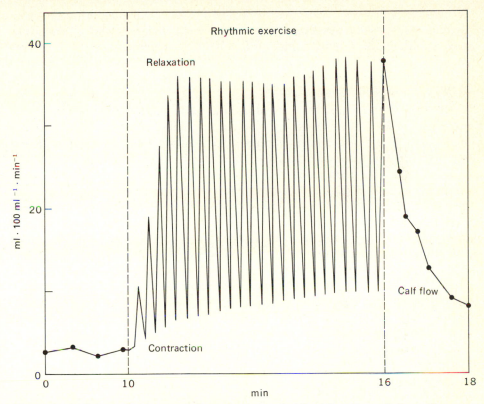

Figure 6-14 Schematic representation of changes in blood flow through the calf muscle during strong rhythmic contractions. During contraction the blood within the muscles is emptied toward the heart but at the same time the inflow is greatly reduced. *(From Barcroft and Swan, 1953.)*

percent of the total blood volume. By constriction of the venules and veins, close to half of that blood volume may be mobilized and emptied toward the heart. Vasomotor activity in the skeletal muscles, the largest interstitial fluid depot in the body, is especially effective as a tool for a reflex control of the blood volume. Variations in blood volume by blood loss, dehydration, or prolonged inactivity may influence the filling of the heart.

To complete the picture, it should be emphasized that the normal heart has the capacity to pump all the blood that is returned to the heart into the arteries. Increases in heart rate and force of contraction can thus keep the atrial pressures low even during heavy exercise. A failure of the left or right ventricle would cause a rise in the central venous pressure and severe disturbances in the capillary fluid exchange (congestive heart failure).

Summary The venous return to the heart is determined by the balance between the filling pressure and the distensibility of the heart, i.e., the intraventricular pressure minus the intrathoracic pressure. The filling is enhanced by (1) the variation in intrathoracic and intra-abdominal pressures during the respiratory cycle, (2) the effect of the muscle pump during muscular movements, and (3) a vasoconstriction in the postcapillary vessels. Changes in body position will, at least temporarily, affect the volume of blood in central veins.

Cardiac Output and Oxygen Uptake

At rest in the supine position, the cardiac output is 4 to 6 liters $\cdot$ min^{-1} with an extraction of 40 to 50 ml O_2/liter of blood and a total oxygen uptake of 0.2 to 0.3 liter $\cdot$ min^{-1}. When a subject strapped to a tilting table is tilted from the horizontal to the feet-down position, the cardiac output may fall from 5 to 4 liters $\cdot$ min^{-1}. This decrease is due to the previously discussed venous pooling. Stroke volume is reduced and heart rate is usually increased. Activation of the muscle pump propels the blood toward the heart, and the heart rate may even decrease as stroke volume increases (Fig. 6-12). In the passive feet-down position, the oxygen uptake is unchanged, and hence the arteriovenous O_2 difference is increased.

   During exercise, the cardiac output increases with the increase in oxygen uptake, but not linearly, if a range from rest value up to maximal is considered (Fig. 6-15). On 11 women and 12 men, well-trained, twenty to thirty years of

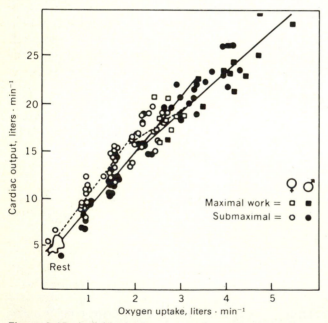

Figure 6-15   Individual values on cardiac output in relation to oxygen uptake at rest, during submaximal, and during maximal exercise on 23 subjects sitting on a bicycle ergometer. Regression lines (broken lines for women) were calculated for experiments where the oxygen uptake was (1) below 70 percent and (2) above 70 percent of the individual's maximum. *(From P.-O. Åstrand et al., 1964.)*

age, the cardiac output, oxygen uptake, heart rate, and oxygen content of arterial blood were determined at rest sitting on a bicycle ergometer and on four to five different work loads up to the maximum that could be maintained for 4 to 6 min. Dye dilution technique with indocyanine green was used for determination of cardiac output. Two to three measurements were done at each load, the first after about 5 min of exercise (except for maximal load). The mean values were used to calculate cardiac output, stroke volume, and $a\text{-}\bar{v}O_2$ difference (P.-O. Åstrand et al., 1964). The physically very fit man can increase his oxygen uptake from 0.25 to 5.00 liters $\cdot$ min^{-1} or more when working on a bicycle ergometer or treadmill. This increase is, let us assume, met by an increase in heart rate from 50 to 200, and in stroke volume from 100 to 150 ml, which means that during maximal exercise the cardiac output has increased from 5.0 to 30.0 liters $\cdot$ min^{-1}, or 6 times the resting level. Since the oxygen uptake increased 20 times, the $a\text{-}\bar{v}O_2$ difference must have changed from 50 to 165 ml $\cdot$ liter^{-1} of blood, or 3.3 times the resting level (and 3.3 times 6 is close to 20). This better utilization of the oxygen transported by the blood is reached principally in two ways: (1) The blood flow is redistributed during exercise so that skeletal muscles, with their pronounced ability to extract oxygen, may receive 80 to 85 percent of the cardiac output, as compared with some 15 percent at rest. (2) The oxygen dissociation curve is shifted so that more oxyhemoglobin is reduced than normally at a given pressure for oxygen, i.e., the percentage of saturation is less. This "Bohr effect" is easier to understand if one studies Fig. 6-16. In the working muscles, the temperature may exceed 40°C and the pH may be lower than 7.0, so there is really not a fixed relation between oxygen tension and oxygen saturation of the hemoglobin.

At normal pH (7.40), CO_2 tension (40 mm Hg), and blood temperature (37°C), the blood keeps about 33 percent HbO_2 at an oxygen tension of 20 mm Hg. At a pH of 7.2 and temperature of 39°C, the percentage of HbO_2 at the same oxygen tension is reduced to about 17, which means that 1 liter of blood with an O_2 capacity of 200 ml can deliver 26 ml more oxygen without changes in the pressure gradient for oxygen between capillary and muscle cell. With 24 liters of blood circulating the working muscles per minute, this extra oxygen delivery amounts to about 0.6 liter $\cdot$ min^{-1}, or 12 percent of the total uptake, in the example given above. With pH 7.0 and muscle temperature 40°C, the extra oxygen uptake would, however, not be further increased because the oxygen content of the arterial blood would be affected negatively by the low pH and high temperature.

With carbon monoxide present, carboxyhemoglobin (HbCO) is formed. Such a conversion effects the oxyhemoglobin dissociation curve with a shift to the left. The effect of carbon monoxide on oxygen transport is therefore twofold: it reduces the amount of hemoglobin available for oxygen transport, and it interferes with the unloading of oxygen in the tissues. For smokers, the effect of CO on the hemoglobin becomes a real handicap during exercise.

The shift in the dissociation curve is a result of the heat production by the working cell and the formation of CO_2 and lactic acid during the heavy exercise. The effect of CO_2 in releasing O_2 from the blood is actually twofold: CO_2 lowers the pH of the blood, and by combining with hemoglobin, reduces its affinity for O_2.

By the two mechanisms, based on (1) a regulation of the circulation and (2)

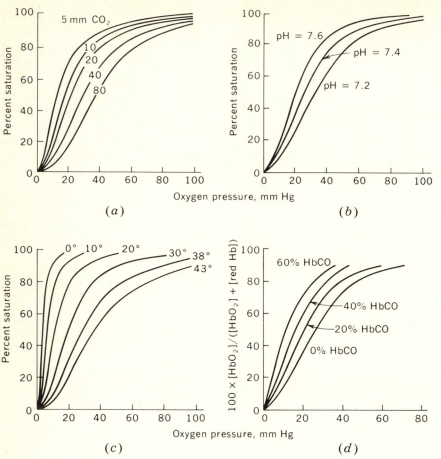

Figure 6-16 Effects of CO_2, pH, temperature (°C), and CO on the oxygen dissociation curve of the blood. Percent of saturation = percent of HbO_2. *(From various sources; see Ruch and Patton, p. 331, 1974.)*

an inherent characteristic of hemoglobin, the oxygen uptake can be elevated 20 times, but the cardiac output has to increase to only 30 liters · min^{-1}, not $20 \times 5 = 100$ liters · min^{-1}.

Oxygen Content of Arterial and Mixed Venous Blood

During exercise, there is a *hemoconcentration* of the blood, which is partly explained by a withdrawal of fluid to the active muscle cells and by the interstitial fluid (receiving the metabolites produced in the cells). Hence the osmotic pressure is highest within and close to the working cells. The raised capillary pressure and surface area also lead to an increased outward filtration. In the experiments on 23 subjects (P.-O. Åstrand et al., 1964), the oxygen capacity of the arterial blood was about 10 percent higher during maximal exercise than at rest. The actual oxygen content of the blood drawn from the brachial artery was 3 percent higher during heavy exercise than at rest. This hemoconcentration makes the blood more viscous, but it also increases the transportation capacity per liter of blood for both oxygen and carbon dioxide.

There is evidently a discrepancy between the increase in oxygen-binding capacity of the blood and the extra oxygen actually taken up during strenuous exercise. In other words, there is a slight reduction in saturation of the arterial blood during maximal exercise, despite a normal or even elevated oxygen tension in the lung alveoli. The arterial pH may, however, be below 7.2 and the blood temperature markedly elevated, and therefore the shift in the oxygen dissociation curve to the right from 97 toward 90 percent saturation (Figs. 6-16, 6-17) is noticeable even at high oxygen tensions. This shift is a disadvantage in the lungs, but the overall effect is, as discussed above, an improved oxygen delivery due to the advantage at the tissue level, both in active and nonactive areas.

The increased extraction of oxygen from the arterial blood as exercise becomes heavier is illustrated in Fig. 6-17. During maximal exercise, the venous

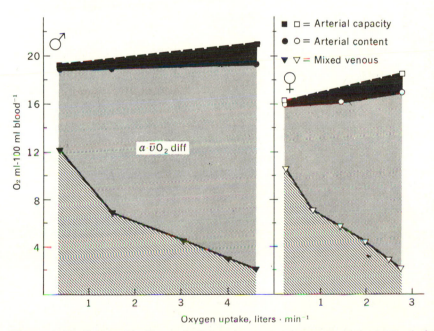

Figure 6-17 Oxygen-binding capacity and measured oxygen content of arterial blood; calculated oxygen content of mixed venous blood at rest and during work up to maximum on bicycle ergometer. Mean values for five female (right part) and five male subjects (left), twenty to thirty years of age, and with high maximal aerobic power. *(From data presented by P.-O. Åstrand et al., 1964.)* During maximal work, the arterial saturation is about 92 percent as compared with 97 to 98 percent at rest, and the venous oxygen content is very low and similar for women and men.

blood leaving the muscles has a very low oxygen content. In this study, the calculated oxygen content in *mixed venous blood* averaged about 2.0 ml · 100 ml^{-1} of blood for both women and men. At rest, the blood flow through most tissues is luxurious as far as the oxygen need is concerned, since other functions determine the flow distribution (e.g., through the vascular bed in kidneys, intestines, and skin; for the oxygen supply to the kidneys, about 50 ml of blood per min would be enough, but at rest the actual flow exceeds 1 liter · min^{-1}). As emphasized in previous discussions, the blood flow, during exercise, is redistributed with the primary object of supplying metabolically active tissues with oxygen and removing the produced carbon dioxide.

Saltin et al. (1968) report the oxygen content in the femoral vein during maximal running on the treadmill to be 1.4 vol percent (average of four subjects). The oxygen tension was 12 mm Hg and the pH 7.09. They noted that the difference in oxygen content between femoral venous and mixed venous blood (containing about 2 vol percent of oxygen) was small, particularly after training.

It is reasonable to assume that some correlation should exist between the oxygen content of the arterial blood and the cardiac output at a given oxygen uptake. For some unknown reason, women have about 10 percent lower concentration of Hb in the blood than men. In the mentioned study (P.-O. Åstrand et al., 1964), the cardiac output required to transport 1.0 liter of oxygen was 9.0 liters for women (O_2 content in arterial blood: 16.7 ml · 100 ml^{-1}), and 8.0 liters for men (O_2 content: 19.2 ml · 100 ml^{-1}) during submaximal work with an oxygen uptake of 1.5 liters · min^{-1}. The cardiac output in males is more effective in its oxygen-transporting function than in women, and this difference can be explained by the Hb content of the blood.

In other experiments, the relationship between oxygen uptake during maximal exercise and the oxygen content of arterial blood has been further analyzed. The subjects were exposed to acute hypoxia by reducing the ambient pressure of the inspired air to simulate an altitude of 4,000 meters ($P_{bar} = 460$ mm Hg). The cardiac output attained during submaximal exercise was higher at high altitude than at sea level, but during maximal effort, no difference in cardiac output was observed. The oxygen uptake during maximal work was, however, reduced in proportion to the decrease in oxygen content of the arterial blood, or to about 70 percent of what it was at sea level (Stenberg et al., 1966; Hartley et al., 1973).

With part of the hemoglobin blocked by carbon monoxide (up to 20 percent), the oxygen transport at a given submaximal rate of work can be maintained. The heart rate is increased and the cardiac output is at control level or somewhat higher. During maximal work the oxygen uptake is reduced more or less in proportion to the varied oxygen content of the arterial blood. However, with 15 percent HbCO the cardiac output averaged 5 percent lower than in the control experiment (see Ekblom et al., 1975).

An increased oxygen tension in the inspired air will increase the maximal oxygen uptake and improve the performance (see Ekblom et al., 1975; Fagraeus, 1974; Nielsen and Hansen, 1937). Recent studies by Ekblom et al. (1975) on eight subjects breathing 50 percent oxygen in nitrogen at sea level showed an average 12 percent increase in maximal aerobic power (in uphill running) (Fig. 6-18a). The cardiac output was in maximal running similar in both hyperoxia and in the control.

With controlled blood loss and reinfusion of red cells, the effect of acute variations in hematocrit can be studied. The effect of blood loss is a deterioration of physical performance,

Figure 6-18   *(a)* Relation between oxygen uptake and transported oxygen. $\dot{Q} \times C_{a_{O_2}}$, during maximal running. Individual values on eight subjects (broken lines) and means (solid lines), at hypoxia induced by about 15 percent HbCO, control experiments, and hyperoxia with the subjects inhaling 50 percent oxygen in nitrogen, respectively. *(from Ekblom et al., 1975).*
(b) Work time at a standard maximal run (maximal work time) and maximal oxygen uptake during control (day "O"), after 800 ml blood loss, and after reinfusion of the packed red cells (day "28") in three subjects. *(From Ekblom et al., 1972a.)*

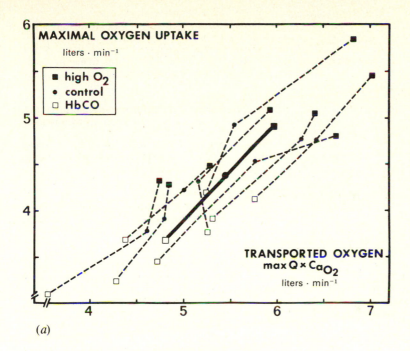

(a)

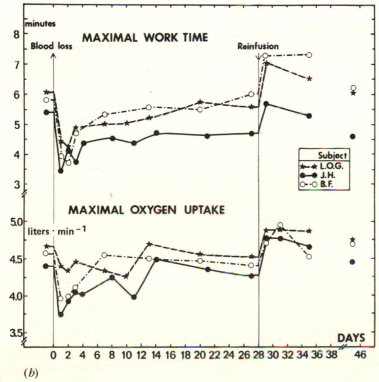

(b)

which is related to the reduced maximal oxygen uptake. A reinfusion of red cells (equivalent to 800 ml of blood) in subjects who had recovered after blood loss could dramatically (overnight) improve the maximal oxygen uptake and the performance to supernormal values (in average an increase in maximal oxygen uptake of 9 percent) (Fig. 6-18b). In five subjects running at maximal speed, which could be maintained for about five minutes, the oxygen content of the arterial blood was in average 16 percent higher after reinfusion of red cells compared with the situation after blood loss. The difference in maximal oxygen uptake was actually about 14 percent (but the individual variations were large). The maximal heart rate and stroke volume respectively were more or less identical in the different experiments (Ekblom et al., 1976). (See Fig. 17-9.)

The purpose of this brief summary is to illustrate that the maximal oxygen uptake (maximal aerobic power) in exercise engaging large muscle groups is apparently not limited by the capacity of the muscle mitochondria to consume oxygen. Slight variations in the volume of oxygen offered to the tissue $\dot{Q} \times C_{a_{O_2}}$ will produce almost proportional changes in the volume of oxygen consumed. Exercise with the arms (in swimming) as well as with one leg (bicycling) does include muscle groups which are also engaged in normal swimming and two-leg work respectively. It is remarkable, however, that the combined exercise does not dramatically increase the maximal oxygen uptake (see Clausen, 1976; Davies and Sargent, 1974; Holmér, 1974).

It should be emphasized that a period of physical conditioning will increase the volume of mitochondria in trained muscles increasing their aerobic energy potential (see Holloszy, 1973; Howald, 1975). Gollnick et al. (1972) have, however, concluded from their studies of enzyme systems in skeletal muscles of untrained and trained men that the metabolic capacity of both the conditioned and unconditioned muscles normally exceeds the actual oxygen uptake of the muscles. The increase in enzymes noticed with conditioning also far exceeds the noticed improvement in maximal oxygen uptake. (See Chap. 12.)

The *myocardial oxygen extraction* is reported to be approximately the same at rest, during exercise (oxygen uptake was increased two- to threefold over the resting value), and during recovery in normal healthy subjects (Messer et al., 1962). The arteriovenous oxygen difference is as high as 16 to 17 vol percent at rest. During exercise, it may increase to 18 to 19 ml · 100 ml^{-1} blood flow (oxygen content of arterial blood 20 vol percent or higher).

Stroke Volume

Factors affecting the stroke volume are (1) the venous return to the heart and (2) the distensibility of the ventricles. The degree of diastolic filling has an anatomic limitation (children-adults, women-men), but within a range various factors, some of which were discussed above, affect the stretching of the muscle fibers. The final factors determining the stroke volume are (3) the force of contraction in relation to (4) the pressure in the artery (aorta or pulmonary artery).

The heart adjusts itself to changing conditions by an inherent self-regulatory mechanism. Starling, using his famous lung-heart preparations, found that the normal heart tended to empty itself almost completely. It was distended to a greater diastolic volume in response to either a greater venous return or an increase of the arterial pressure. In the latter case, there was a transient decrease in stroke volume, but as the force of contraction increased with a greater initial length of the muscle fibers, the stroke volume and cardiac output became normal. By stimulation of the sympathetic cardiac nerves, the contraction force increased from the same initial length, and the arterial resistance could be overcome despite a greater extent of myocardial shortening.

The results from Starling's studies of the isolated heart were also considered applicable in the intact animal. Most earlier textbooks concluded that the diastolic volume of the heart was smaller at rest but increased during exercise, when the venous return increased and the arterial pressure was elevated. The same end-systolic volume could be maintained or was dependent on the strength of the heart muscle. The well-trained person was characterized by a small residual volume of blood in the heart after systole at rest as well as during exercise. The net effect was a substantial increase in stroke volume during exercise, according to these standard texts.

Then a novel approach developed which was in complete disagreement with Starling's heart law. From x-ray studies at rest and during exercise, it was concluded that the end-diastolic volume of the heart was fairly constant during rest and exercise (Reindell, 1943; Kjellberg et al., 1949). The systolic emptying of the ventricles was limited at rest, but during muscular activity the stroke volume increased, resulting in a smaller residual volume. Sympathetic nerve impulses and catecholamines were thought to increase the contraction force essential for the more complete emptying of the ventricle, actually against an elevated arterial blood pressure.

Earlier experimental data on cardiac output and stroke volume were obtained with various indirect methods using nitrous oxide (Krogh and Lindhard, 1912), acetylene (Christensen, 1931), and with other methods. Exercise was performed on a treadmill or in a sitting position on a bicycle ergometer, and the control values were usually secured on the subject in the same position at rest. The typical finding was an increase in stroke volume in transition from rest to work but a tendency for the stroke volume to remain unchanged with further increase in work load. In some studies no increase in stroke volume was found. The data thus obtained did, to some extent, confirm Starling's law, but in essential parts they refuted it.

The present concept on the stroke volume in the human during exercise accepted by most cardiophysiologists can be summarized as follows (Wade and Bishop, 1962; Bevegård, 1962; Bevegård and Shepherd, 1967): When the position is changed from supine to standing or sitting, there is a diminution in end-diastolic size of the heart and a decrease in stroke volume. If muscular work is then performed, the stroke volume increases to approximately the same size as obtained in the recumbent position. The discrepancy between earlier and recent studies is in most cases not due to inconsistent results, but must be blamed on erroneous conclusions and bold extrapolations of the data to conditions not examined. Certainly Starling's law holds true for the isolated heart-lung preparation, but when the heart is functioning in the intact animal, other mechanisms are superimposed.

The importance of the *central blood volume* for the stroke volume was demonstrated in 1939 by Asmussen and Christensen. Subjects in the sitting position were exercising with their arms. In some experiments, the subjects lay down with their legs elevated for about 10 min before the exercise started. The circulation to the legs was then arrested by pressure cuffs around the thighs. When assuming a sitting position following this procedure, there was approximately 600 ml of blood less in the legs compared with sitting without occlusion of the blood flow to the legs. It was noted that the cardiac output was about 30 percent higher when the legs were "blood free," i.e., when the central blood volume was high, as compared with the experiments with blood pooling in the legs. The high cardiac output was due to a high stroke volume, for the heart rate was actually lower than in exercise with reduced central blood volume (and low cardiac output).

In Fig. 6-19, individual data on stroke volume are plotted, expressed in percentage of the maximum reached. The maximal oxygen uptake of the subject is similarly set to 100 percent, and the submaximal loads are defined in percent of this maximum. At rest, the stroke volume is, for most subjects, between 50 and 70 percent (mean 63 percent) of the maximum measured during exercise. Hence, during exercise in the sitting position, there is a definite increase in stroke volume with an "optimum" reached when oxygen uptake exceeds 40 percent of maximal aerobic power. The heart rate at that load is 110 to 120. Methodological errors and biological variations result in a variation in calculated stroke volume of ±4 percent as oxygen uptake further increases. There is no tendency toward a decrease in stroke volume at the peak load. No significant correlation was found between maximal heart rate and eventual decrease in stroke volume during the maximal exercise. In our opinion, this rules out the hypothesis that a high heart rate (about 200) during exercise should interfere with the filling of the heart.

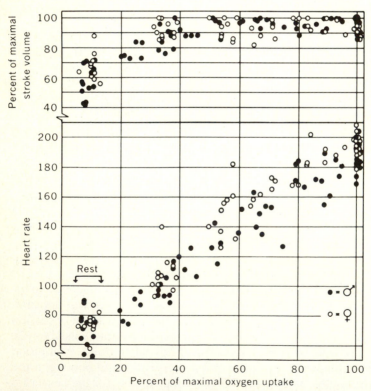

Figure 6-19   Stroke volume in percentage of the individual's maximum, and heart rate at rest and during exercise. The oxygen uptake on the abscissa is expressed in percentage of the subject's maximum. Circled dot at "100 percent" represents 11 of the 23 subjects. Measurements were made with the subjects in the sitting position (same subjects as in Fig. 6-15).

In this discussion of the variation in stroke volume, it should be emphasized that in experiments on animals, reflexes that evoke tachycardia also evoke simultaneously, either directly or indirectly, an increase in contractility of the heart muscle, so that stroke power and stroke work increase (Sarnoff and Mitchell, 1962).

Braunwald et al. (1967) observed in experiments on human subjects that an increase in heart rate increases the velocity of the shortening of the myocardial fibers. They emphasize that the exercise tachycardia as such contributes to an improvement of the contractile state of the myocardium. They conclude that "the normal cardiac response to exercise involves the integrated effects on the myocardium of simple tachycardia, sympathetic stimulation, and the operation of the Frank-Starling mechanism. During submaximal levels of exertion, cardiac output can rise even when one or two of these influences are blocked. However, during maximal levels of exercise, the ventricular myocardium requires all three influences in order to sustain a level of activity sufficient to satisfy the greatly augmented oxygen requirements of the exercising skeletal muscles." (See below.)

Heart Rate

In many types of work, the increase in heart rate is linear with the increase in work load. There are exceptions and those exceptions are perhaps more frequent among untrained subjects. When the subject performs very heavy exercise, the $a \cdot vO_2$ difference may increase so that the oxygen uptake increases relatively more than the cardiac output. The evaluation, from submaximal work loads, of an individual's maximal oxygen uptake or capacity to perform work is, in most test procedures, based on a registration of the heart rate during steady state and then an extrapolation to a fixed heart rate or to an assumed maximal heart rate (see Chap. 10). There are many pitfalls in this method. Here, the following should be noted: (1) The standard deviation for maximal heart rate during exercise is ± 10 beats $\cdot$ min^{-1}. Hence, for twenty-five-year-old individuals, women or men, the maximal heart rate for 5 out of 95 subjects may be below 175 or above 215, since the maximum is about 195 on an average. (2) There is a gradual decline in maximal heart rate with age, so that the ten-year-old attains 210, the sixty-five-year-old only about 165 beats $\cdot$ min^{-1} (Fig. 6-20) (Robinson, 1938; P.-O. Åstrand, 1952; I. Åstrand, 1960; Hollmann, 1963). Furthermore, longitudinal studies have shown a wide individual scatter in the decline in maximal heart rate with age (I. Åstrand et al., 1973).

When 50 percent of the maximal aerobic power is used, the heart rate in the twenty-five-year-old man is about 130, but the same relative work load and feeling of strain are experienced at a heart rate of 110 for the sixty-five-year-old man (Fig. 6-20) (I. Åstrand, 1960). For women, the 50 percent oxygen uptake is attained at a heart rate of about 140 beats $\cdot$ min^{-1} at the age of twenty-five.

Prolonged exercise in a hot environment causes a higher heart rate than exercise at a low room temperature. Emotional factors, nervousness, and apprehension may also affect the heart rate at rest and during work of light and moderate intensity. During repeated maximal exercise, the heart rate is, however, remarkably similar under various conditions, with a standard deviation of ± 3 beats per min (P.-O. Åstrand and Saltin, 1961a).

The heart rate at a given oxygen uptake is higher when the work is performed with the arms than with the legs (Christensen, 1931; P.-O. Åstrand et

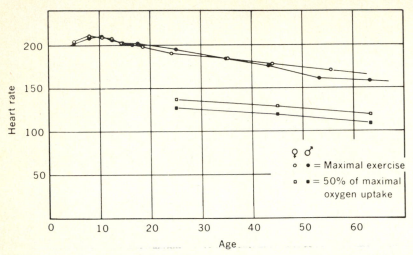

Figure 6-20 The decline in maximal heart rate with age, and heart rate during a submaximal work load. Mean values from studies on 350 subjects. The standard deviation in maximal heart rate is about ±10 beats/min in all age groups. *(From Astrand and Christensen, 1964.)*

al., 1965; Stenberg et al., 1967; Vokac et al., 1975). Static (isometric) exercise also increases the heart rate above the value expected from work load. The mechanism for these differences in heart-rate response to exercise is not understood. However, the elevated heart rate is usually accompanied by a decreased stroke volume. It is known that a variation in heart rate at a given oxygen uptake at rest and during submaximal exercise often produces a change in stroke volume, so that the cardiac output is maintained at an appropriate level. (This information is based on studies in patients with artificial pacemakers or irregular heart rate, and in subjects submitted to various drugs influencing the heart rate.) (I. Åstrand et al., 1963; Bevegård and Shepherd, 1967; Braunwald et al., 1967.)

Fig. 6-21 presents data on subjects submitted to submaximal (bicycle ergometer) and maximal (treadmill) exercise four times: (1) control; (2) after infusion of 10 mg propranolol blocking the adrenergic β-receptors in the heart; (3) after infusion of 2 mg atropine blocking the parasympathetic impulse traffic; (4) after double blockade (propranolol and atropine). At a given oxygen uptake the heart rate varied about 40 beats · min^{-1}, taking the extremes, but the cardiac output was almost similar in the four situations, since the stroke volume compensated for the changes in heart rate. (In the propranolol experiments there was in average 1.5 to 2 liters · min^{-1} reduction in cardiac output.) It should be noted that the subjects reached their normal maximal oxygen uptake despite a reduction in maximal heart rate from 195 to about 160 beats · min^{-1}. The performance time was significantly shorter after β-blockade and the intra-arterial blood pressure was reduced (Ekblom et al., 1972b).

The regulation of the circulation in exercise is probably guided primarily by factors sensitive to an adequate cardiac output. Heart rate and stroke

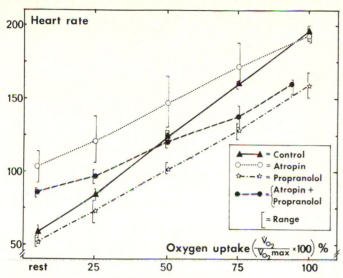

Figure 6-21   Relationship between heart rate (means and ranges) and relative oxygen uptake ($V_{O_2} \cdot V_{O_2}max^{-1} \cdot 100$) in five sets of experiments (four subjects) during normal conditions (control) and after blockade of receptors. Resting heart rate was recorded in the sitting position. *(From Ekblom et al., 1972b.)*

volume are the variables, and the stroke volume is more likely to be directly influenced by such factors as venous return or peripheral vascular resistance.

Blood Pressure

The side pressures (lateral pressures) in the aorta during systole and diastole are reported not to be markedly different during exercise compared with the resting condition (Marx et al., 1967). The aortic (arterial) pressures, also including the kinetic energy factor (end pressure), increase linearly with the increase in oxygen uptake as the exercise becomes heavier.

As a result of the vasodilatation in the vascular bed in the working muscles, the peripheral resistance to blood flow is reduced during exercise, but the elevation in cardiac output causes the blood pressure to rise. The arterial pressure obtained in a peripheral artery at rest, 120 mm Hg in systole and 80 mm Hg in diastole, may exceed 175 and 110 mm Hg, respectively, during exercise (Fig. 6-22).

It should be noted that the arterial blood pressure is significantly higher in arm exercise than in leg work (Fig. 6-22). The high blood pressure at a given cardiac output, when the work is performed by the arms, induces an increased stroke work of the heart. Therefore, for untrained individuals or for cardiac patients, it may be hazardous to work hard with the arms, e.g., to shovel snow, dig in the garden, or carry heavy trunks. The relatively high blood pressure in exercise with small muscle groups is probably due to a vasoconstriction in the inactive muscles. The larger the activated muscle groups, the more pronounced

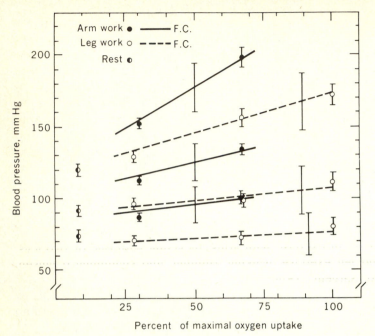

Figure 6-22 Effect of exercise on blood pressure (end pressure). Regression lines of arterial systolic, mean, and diastolic blood pressures, respectively, in relation to oxygen uptake (in percentage of the maximum) during arm and leg exercise in the sitting position for 13 subjects. F.C. = femoral artery catheter. The figure summarizes data from 23 submaximal and 13 maximal work loads with arm work (cranking) and 44 and 13 experiments, respectively, with leg work. The vertical heavy lines represent ±1 SD (standard deviation) around the regression line. The dots and thin lines represent the mean ±1 SEM (standard error of the mean) for three groups of values at different levels of oxygen consumption. (The systolic pressure measured in a peripheral artery is higher than in the aorta, but the mean and diastolic pressures are similar in the two vessels.) *(From P.-O. Åstrand et al., 1965.)*

is the dilatation of the resistance vessels. The lower peripheral resistance is reflected in a lower blood pressure (since $\dot{Q} \times R = P_{mean}$).

We have a similar situation with a considerable ventricular load in isometric work. Donald et al. (1967) describe a powerful cardiovascular reflex causing an unexpectedly high rise in blood pressure in response to sustained contractions above 15 percent of maximal voluntary force. The pressor response appears to be largely independent of the muscle bulk involved in the contraction, provided the relative tension is constant. They point out that relatively moderate and localized isometric work can cause a far higher pressure component than noticed in dynamic work; this "may well be dangerous to a person with a compromised heart or impaired integrity of the arterial wall." (See also McCloskey and Streatfeild, 1975.)

I. Åstrand et al. (1968) studied carpenters using a hammer to nail at different heights. When they were hammering into the ceiling, their heart rate and intra-arterially measured blood pressures were significantly higher than when hammering at bench level. They were also higher than during leg exercise on a bicycle ergometer with a similar level of oxygen uptake (see Chap. 13).

Reindell et al. (1960) report that when comparing the arterial blood pressure response to exercise in subjects of different ages, the older men had consistently higher systolic and diastolic pressures than the younger ones. At rest, the twenty-five-year-olds averaged 125/75 mm Hg, and during exercise at a rate of 100 watts (oxygen uptake about 1.5 liters · min^{-1}), the pressures were 160 and 80 mm Hg in systole and diastole, respectively. For the fifty-five-year-old group, the increase was from 140/85 at rest up to 180/90 mm Hg, the work load being the same. Similar results are reported by Hollmann (1963) and Gerstenblith et al. (1976) in their recent review.

Type of Exercise

The cardiac output at a given submaximal oxygen uptake is in many types of exercise similar, e.g., in exercise with arms, bicycling with one or two legs, combined arm and leg exercise, walking, running, and swimming (Clausen, 1976; Davies and Sargent, 1974; Hermansen et al., 1970; Holmér, 1974; Stenberg et al., 1967). The cardiac output at a given oxygen uptake during submaximal exercise is consistently 1 to 2 liters less in the erect position than when the subject is recumbent; the heart rate is about the same (Reeves et al., 1961; Bevegård, 1962; Wade and Bishop, 1962). The compensation for the lower cardiac output must, by definition, be an increased a-$\bar{v}O_2$ difference when erect. It should be emphasized that the maximal oxygen uptake during cycling in the supine position is about 15 percent lower than during work in the sitting position on the bicycle ergometer (Fig. 6-23) (Åstrand and Saltin, 1961b; Stenberg et al., 1967). Similarly, the cardiac output is somewhat lower during maximal exercise with the legs in the supine position. This difference in response may be explained in the following way. Rhythmic muscular contractions will squeeze out blood from the veins lowering the average venous pressure considerably and hence raising the effective perfusion pressure of flow. This is evident in exercise in the upright position as the arterial pressure at the calf level is then raised by some 70 to 80 mm Hg as compared with the supine position, i.e., in proportion to the distance from the heart because of the increase in hydrostatic pressure. However, the pressure in the calf veins is maintained at a low level by the "milking" action of the muscle pump. Folkow et al. (1971) noted that the calf blood flow in man could be 50 to 60 percent larger when a standard heavy rhythmic exercise was performed in the upright position as compared with the reclining position. Combined arm and leg work in the sitting or supine position reveals almost the same values for maximal oxygen uptake, heart rate, and cardiac output as work in the sitting position with only the leg muscles (Fig. 6-23) (P.-O. Astrand and Saltin, 1961b; Stenberg et al., 1967).

In prolonged exercise in a neutral thermal environment the cardiac output is normally well maintained, but there is a progressive rise in heart rate and fall in stroke volume. Furthermore, there is a gradual reduction in blood pressures

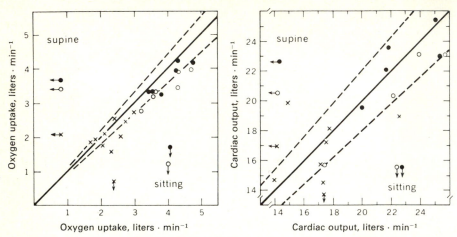

Figure 6-23 Oxygen uptake (to the left) and cardiac output; comparison between the highest individual values attained in the sitting (abscissa) and supine (ordinate) position for arm work (x), leg work (o), and combined arm and leg work (•). Line of identity and lines corresponding to 10 percent deviation are drawn. Symbols with arrows give the mean of the different groups. Note that leg work in the supine position did not bring oxygen uptake and cardiac output to a maximum, but when the arm muscles were also exercised, the oxygen uptake and cardiac output did increase to the same level as in leg work, or in combined arm and leg work in the sitting position. *(From Stenberg et al., 1967.)*

(systemic arteries, pulmonary artery, right ventricular end-diastolic pressures). The mechanism of these cardiovascular shifts are, so far, not established (see Ekelund, 1967; Eklund, 1974; Rowell, 1974).

Heart Volume

The size of the heart can be visualized by means of roentgenograms, and its volume can be computed by application of empirical formulas. A high correlation has been established between heart volume and various parameters, such as blood volume, total amount of hemoglobin, and stroke volume in healthy younger individuals (Kjellberg et al., 1949; Sjöstrand, 1953; Reindell et al., 1960; Mellerowicz, 1962). The difference between the athlete's heart and the dilated heart of the cardiac patient is not only the configuration of the heart, but also the disproportion between heart size and maximal aerobic power, total hemoglobin content, etc. Figure 6-24*a* shows that a group of 30 young girls had a calculated heart volume which in many cases was much larger than expected for their body size. The girls were some of the best Swedish swimmers and were not very likely to suffer from any heart disease. When the heart volume is related to the girls' maximal aerobic power (Fig. 6-24*b*), the findings make functional sense. The girl with the highest oxygen uptake and largest heart was actually also the best swimmer; she was second in the 400-m free-style in the Olympic Games, 1960 (P.-O. Åstrand et al., 1963).

On the average, the heart volume calculated from roentgenograms has

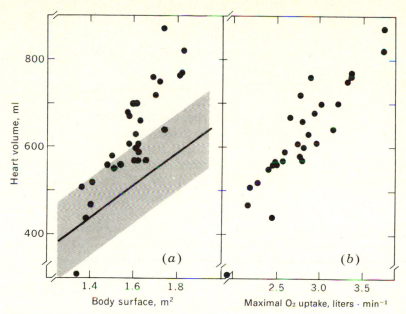

Figure 6-24 Relationship between heart volume and (*a*) calculated body surface area and (*b*) maximal O_2 uptake in 30 young well-trained girl swimmers. Shadowed area gives the 95 percent range for "normal" girls. *(Modified from P.-O. Åstrand et al., 1963.)*

been found to be largest (above 900 ml) for well-trained athletes engaged in events calling for endurance (bicyclists, canoeists, cross-country skiers, long-distance runners). The average for middle-distance runners, swimmers, and soccer and tennis players was between 800 and 900 ml, and for boxers, fencers, gymnasts, jumpers, sprinters, throwers, and untrained controls, below 800 ml (Reindell et al., 1960; Mellerowicz, 1962).

Age

At a given work load or oxygen uptake, the older individual attains, on the average, the same heart rate as the younger one (I. Åstrand, 1960; Hollmann, 1963; Strandell, 1964). Strandell (1964) found that the cardiac output at a given work load was about 2 liters · min⁻¹ lower in the sixty- to eighty-year-old men at any level of oxygen uptake compared with the young ones. Stroke volume was also significantly lower for the older men (about 20 percent). (Age changes in myocardial function were recently summarized by Gerstenblith et al., 1976.) Eriksson (1972) studied 11 to 13 year-old boys and noted that the cardiac output during submaximal exercise was 1 to 2 liters · min⁻¹ less than for adult young men.

As already mentioned, the heart rate reached during maximal exercise decreases with age. The value typical for the ten-year-old girl or boy is 210, for the twenty-five-year-old 195, and for the fifty-year-old 175 beats · min⁻¹ (Fig.

6-20). Therefore the decrease in circulatory capacity in the old individual is more marked than predicted from heart rate, stroke volume, and cardiac output observed during submaximal exercise if the norms are the same as when evaluating young individuals. The old man has a larger heart volume, calculated from roentgenograms taken in supine position, than does the young man; blood volume and total amount of hemoglobin are not different in young and old men (Strandell, 1964; Grimby and Saltin, 1966). These findings should be related to the decrease in maximal stroke volume, cardiac output, and maximal aerobic power in old men (see Fig. 11-6).

Whether the decrease in maximal heart rate with age is a consequence of an arteriosclerosis in the vessels of the heart is not known. The oxygen cost of a cardiac performance involving a high heart rate is great, and therefore the load on the heart is reduced by a lowered ceiling for heart rate. The lower heart rate during maximal exercise in the old individual is probably not a direct response to hypoxia, since breathing pure oxygen instead of room air during the exercise does not further elevate the heart rate (I. Åstrand et al., 1959).

During exercise, the arterial blood pressure (in systemic as well as pulmonary arteries) is higher in the old than in young persons (Reindell et al., 1960; Hollmann, 1963; Strandell, 1964).

Training and Cardiac Output

It is an old observation that the heart rate at rest and during standard exercise, as well as during recovery from such work, is lowered with a training of the oxygen-transporting system. Published reports on the cardiac output during standard exercise repeated during a course of training indicate that cardiac output is maintained at the same level (Musshoff et al., 1959; Rowell, 1974; Wade and Bishop, 1962). The reduction in heart rate should then mirror an increase in stroke volume. (The training effects are further discussed in Chap. 12.)

Top athletes in endurance events are characterized by a very high maximal oxygen uptake (maximal aerobic power). Their maximal values for circulatory parameters must therefore be high compared with those of less athletic individuals. Ekblom and Hermansen (1968) have collected data obtained during maximal work on the treadmill in athletes, using the dye dilution technique. Some data are presented in Table 6-1. Both an intensive training and superb natural endowments contribute to the remarkable circulatory capacity for oxygen transportation in these subjects.

Figure 6-25 summarizes data on 32 subjects with maximal aerobic power ranging from 2.8 to 6.2 liters · min^{-1} of O_2 uptake. This figure shows a clear relationship between maximal cardiac output and oxygen uptake. It also shows that it is the stroke volume which to a large extent determines the maximal cardiac output. The most pronounced difference between the sexes is the smaller stroke volume and higher heart rate during exercise of a given severity for women compared with men. Actually, a similar difference in stroke volume and heart rate is usually observed when comparing individuals of the same age but with a low and high performance capacity, respectively (Fig. 6-26).

Table 6-1

	Maximal values				
Subject	$\dot{V}O_2$, liters $\cdot$ min^{-1}	Cardiac output, liters $\cdot$ min^{-1}	Heart rate	Stroke volume, ml	$a\text{-}\bar{v}O_2$ difference, ml $\cdot$ 100 ml^{-1}
G. P.	6.00	39.8	188	212	15.1
C. R.	5.77	37.8	188	201	15.3
A. H.	5.60	34.4	189	182	16.3
C. S.	5.50	36.2	198	183	15.2
B. T.	5.64	38.0	193	197	14.8
L. R.	6.24	42.3	206	205	14.8
Mean	5.79	38.1	194	197	15.3

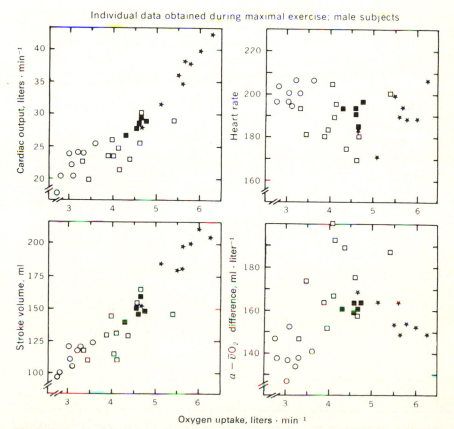

Individual data obtained during maximal exercise; male subjects

Figure 6-25 Cardiac output, heart rate, stroke volume, and arteriovenous oxygen difference during maximal exercise in relation to maximal oxygen uptake in top athletes who were very successful in endurance events (stars), well-trained but less successful athletes (filled squares), and twenty-five-year-old habitually sedentary subjects (unfilled circles). *(From Ekblom, 1969.)* Also included are maximal values on the male subjects presented in Fig. 6-15 (unfilled squares).

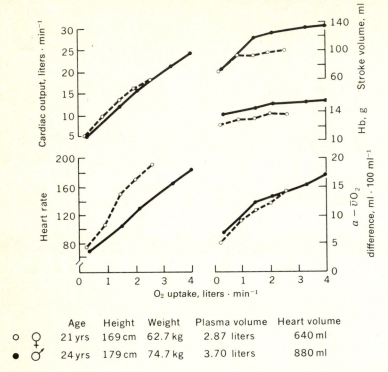

	Age	Height	Weight	Plasma volume	Heart volume
○ ♀	21 yrs	169 cm	62.7 kg	2.87 liters	640 ml
● ♂	24 yrs	179 cm	74.7 kg	3.70 liters	880 ml

Figure 6-26 The figure is based on average values from measurements on 11 women and 12 men, all of them relatively well trained and working on a bicycle ergometer in the sitting position (*P.-O. Åstrand et al., 1964*). The individual data are presented in Figs. 6-15 and 6-19. (Since the abcissa gives the oxygen uptake in absolute values, the calculated mean curves can be misleading. The less fit subjects have both a low maximal oxygen uptake and low stroke volume. Those with a high capacity for oxygen uptake also have a larger stroke volume. A man with a maximal aerobic power of 5 liters · min⁻¹ eventually attains maximal stroke volume first at a work load giving an oxygen uptake of 2 liters · min⁻¹. The one with a maximal oxygen uptake of 3.5 liters · min⁻¹ reaches his plateau for stroke volume when the oxygen uptake exceeds 1.3 liters · min.⁻¹)

A reduced maximal stroke volume is a mechanism limiting the maximal cardiac output, and therefore the maximal aerobic power in many patients with coronary heart disease. During submaximal exercise the cardiac output at a given oxygen uptake is, however, often the same as in healthy subjects of the same age (Bruce et al., 1974; Clausen, 1976; McDonough et al., 1974).

SUMMARY

This discussion of the oxygen-transporting system during muscular activity reveals that a regulation of the circulation at rest and during submaximal exercise is not guided only by the metabolic rate; various other factors may influence the circulatory response to exercise.

Of probable primary importance for the regulatory mechanisms is the relation between volume of oxygen supplied to the metabolically active tissue (cardiac output times the oxygen content of arterial blood) and the oxygen demand of the tissue. Within limits, other demands can be met by compensatory mechanisms; for example, if the stroke volume is reduced, the heart rate may increase so that an adequate cardiac output is still maintained.

During maximal exercise, however, the cardiac output, oxygen uptake, and heart rate are remarkably fixed to values typical for the individual even if the performance is made under adverse conditions. In this situation, apparently all circulatory functions of decisive importance for a maximal oxygen supply to working muscles are actually devoted to this task. Irrespective of environment, external and internal (within limits) maximal vasoconstriction occurs in the blood vessels of the viscera and skin, so that practically the entire cardiac output is diverted to the vigorously working muscles. Maximal exercise involving large muscle groups creates an emergency reaction in the circulatory adjustment which favors the exercising muscles, including the heart, at the expense of all other tissues with exception of the central nervous system.

REFERENCES

Abrahams, V. C., S. M. Hilton, and A. Zbrozyna: Active Muscle Vasodilation Produced by Stimulation of the Brain Stem: Its Significance in the Defense Reaction, *J. Physiol. (London)*, **154**:491, 1960.

Andersen, P.: Capillary Density in Skeletal Muscle of Man, *Acta Physiol. Scand.*, **95**:203, 1975.

Arenander, E.: Hemodynamic Effects of Varicose Veins and Results of Radical Surgery, *Acta Chir. Scand.* (Suppl. 260):1, 1960.

Asmussen, E., and E. H. Christensen: Einfluss der Blutverteilung auf den Kreislauf bei köperlicher Arbeit, *Skand. Arch. Physiol.*, **82**:185, 1939.

Asmussen, E. E. H. Christensen, and M. Nielsen: Pulsfrequenz und Körperstellung, *Skand. Arch. Physiol.*, **81**:190, 1939.

Åstrand, I.: Aerobic Work Capacity in Men and Women with Special Reference to Age, *Acta Physiol. Scand.*, **49**(Suppl. 169):1960.

Åstrand, I., P.-O. Åstrand, and K. Rodahl: Maximal Heart Rate during Work in Older Men, *J. Appl. Physiol.*, **14**:562, 1959.

Åstrand, I., T. E. Cuddy, J. Landegren, R. O. Malmborg, and B. Saltin: Hemodynamic Response to Exercise during Atrial Flutter and Sinus Rhythm, *Acta Med. Scand.*, **173**:121, 1963.

Åstrand, I., A. Guharay, and J. Wahren: Circulatory Response to Arm Exercise with Different Arm Positions, *J. Appl. Physiol.*, **25**:528, 1968.

Åstrand, I., P.-O. Åstrand, I. Hallbäck, and Å. Kilbom: Reduction in Maximal Oxygen Uptake with Age. *J. Appl. Physiol.*, **35**:649, 1973.

Åstrand, P.-O.: "Experimental Studies of Physical Working Capacity in Relation to Sex and Age," Munksgaard, Copenhagen, 1952.

Åstrand, P.-O.: Breath Holding during and after Muscular Exercise, *J. Appl. Physiol.*, **15**:220, 1960.

Åstrand, P.-O., and B. Saltin: Oxygen Uptake during the First Minutes of Heavy Muscular Exercise, *J. Appl. Physiol.*, **16:**971, 1961a.

Åstrand, P.-O., and B. Saltin: Maximal Oxygen Uptake and Heart Rate in Various Types of Muscular Activity, *J. Appl. Physiol.*, **16:**977, 1961b.

Åstrand, P.-O., L. Engström, B. O. Eriksson, P. Karlberg, I. Nylander, B. Saltin, and C. Thorén: Girl Swimmers, *Acta Paediat.* (Suppl. 147), 1963.

Åstrand, P.-O., T. E. Cuddy, B. Saltin, and J. Stenberg: Cardiac Output during Submaximal and Maximal Work, *J. Appl. Physiol.*, **19:**268, 1964.

Åstrand, P.-O., and E. H. Christensen: Aerobic Work Capacity, in F. Dickens, E. Neil, and W. F. Widdas (eds.), p. 295, "Oxygen in the Animal Organism," Pergamon Press, New York, 1964.

Åstrand, P.-O., B. Ekblom, R. Messin, B. Saltin, and J. Stenberg: Intra-arterial Blood Pressure during Exercise with Different Muscle Groups, *J. Appl. Physiol.*, **20:**253, 1965.

Barcroft, H.: Sympathetic Control of Vessels in the Hand and Forearm Skin, *Physiol. Rev.*, **40**(Suppl. 4): 1960.

Barcroft, H., and H. J. C. Swan: Sympathetic Control of Human Blood Vessels, Edward Arnold (Publishers) Ltd., London, 1953.

Bevegård, S.: Studies on the Regulation of the Circulation in Man, *Acta Physiol. Scand.*, **57**(Suppl. 200):1962.

Bevegård, S., and J. T. Shepherd: Circulatory Effects of Stimulating the Carotid Artery Stretch Receptors in Man at Rest and during Exercise, *Clin. Res.*, **12:**335, 1964.

Bevegård, B. S., and J. T. Shepherd: Regulation of the Circulation during Exercise in Man, *Physiol. Rev.*, **47:**178, 1967.

Braunwald, E., and E. R. Kelly: The Effect of Exercise on Central Blood Volume in Man, *J. Clin. Invest.*, **39:**413, 1960.

Braunwald, E., E. H. Sonnenblick, J. Ross, Jr., G. Glick, and S. E. Epstein: An Analysis of the Cardiac Response to Exercise, *Circulation Res.*, **20** and **21:**44, 1967.

Brodal, P., F. Ingjer, and Lars Hermansen: Number and Density of Capillaries in the Quadriceps Muscle of Untrained and Endurance Trained Men. A Quantitative Electron-Microscopical Study, 1976. (Unpublished.)

Bruce, R. A., F. Kusumi, M. Niederberger, and J. L. Petersen: Cardiovascular Mechanisms of Functional Impairment in Patients with Coronary Heart Disease, *Circulation*, **49:**696, 1974.

Burton, A. C.: Hemodynamics and the Physics of the Circulation, in T. C. Ruch and H. D. Patton (eds.), "Physiology and Biophysics," pp. 523–542, W. B. Saunders Company, Philadelphia, 1965.

Carlson, L., and B. Pernow: Oxygen Utilization and Lactic Acid Formation in the Legs at Rest and during Exercise in Normal Subjects and in Patients with Arteriosclerosis Obliterans, *Acta Med. Scand.*, **164:**39, 1959.

Christensen, E. H.: Beiträge zur Physiologie schwerer körperlichter Arbeit. Minutenvolumen und Schlagvolumen des Herzens während schwerer körperlicher Arbeit, *Arbeitsphysiol.*, **4:**453, 470, 1931.

Christensen, E. H., and M. Nielsen: Investigation of the Circulation in the Skin at Beginning of Muscular Work, *Acta Physiol. Scand.*, **4:**162, 1942.

Clausen, J. P.: Circulatory Adjustments to Dynamic Exercise and Effect of Physical Training in Normal Subjects and Patients with Coronary Artery Disease, *Progr. Cardiovas. Diseases*, **18:**459, 1976.

Clement, D. L., and J. T. Shepherd: Regulation of Peripheral Circulation during Muscular Exercise, *Progr. Cardiovasc. Diseases*, **19**:23, 1976.

Daly, M. de B.: Reflex Circulatory and Respiratory Responses to Hypoxia, in F. Dickens and E. Neil (eds.), "Oxygen in the Animal Organism," p. 267, Pergamon Press, New York, 1964.

Daly, M. de B., and M. J. Scott: The Effect of Stimulation of the Carotid Body Chemoreceptors on Heart Rate in the Dog, *J. Physiol. (London)*, **144**:148, 1958.

Davies, C. T. M., and A. J. Sargent: Physiological Responses to One- and Two-leg Exercise Breathing Air and 45% Oxygen, *J. Appl. Physiol.*, **36**:142, 1974

Donald, D. E., S. E. Milburn, and J. T. Shepherd: Effect of Cardiac Denervation on the Maximal Capacity for Exercise in the Racing Greyhound, *J. Appl. Physiol.*, **19**:849, 1964.

Donald, K. W., J. K. Bishop, and O. L. Wade: A Study of Minute to Minute Changes of Arteriovenous Oxygen Content Difference, Oxygen Uptake and Cardiac Output and Rate of Achievement of a Steady State during Exercise in Rheumatic Heart Disease, *J. Clin. Invest.*, **33**:1946, 1954.

Donald, D. E., D. J. Roulands, and D. A. Ferguson: Similarity of Blood Flow in the Normal and the Sympathectomized Dog Hind Limb during Graded Exercise, *Circ. Res.*, **26**:185, 1970.

Donald, K. W., A. R. Lind, G. W. McNicol, P. W. Humphreys, S. H. Taylor, and H. P. Staunton: Cardiovascular Responses to Sustained (Static) Contractions, *Circulation Res.*, **20** and **21**:15, 1967.

Eckstein, R. W.: Effect of Exercise and Coronary Artery Narrowing on Coronary Collateral Circulation, *Circulation Res.*, **5**:230, 1957.

Ekblom, B.: Effect of Physical Training on Oxygen Transport System in Man, *Acta Physiol. Scand.* (Suppl. 328), 1969.

Ekblom, B., A. N. Goldbarg, and B. Gullbring: Response to Exercise after Blood Loss and Reinfusion, *J. Appl. Physiol.*, **33**:175, 1972a.

Ekblom, B., A. N. Goldbarg, Å. Kilbom, and P.-O. Åstrand: Effects of Atropine and Propranolol on the Oxygen Transport System during Exercise in Man, *Scand. J. clin. Lab. Invest.*, **30**:35, 1972b

Ekblom, B., and L. Hermansen: Cardiac Output in Athletes, *J. Appl. Physiol.*, **25**:619, 1968.

Ekblom, B., R. Huot, E. M. Stein, and A. T. Thorstensson: Effect of Changes in Arterial Oxygen Content on Circulation and Physical Performance, *J. Appl. Physiol.*, **39**:71, 1975

Ekblom, B., Å. Kilblom, and J. Soltysiak: Physical Training, Bradycardia and Autonomic Nervous System, *Scand. J. Clin. Lab. Invest.*, **32**:251, 1973.

Ekelund, L. -G.: Circulatory and Respiratory Adaptation to Prolonged Exercise, *Acta Physiol. Scand.*, **70**(Suppl. 292), 1967.

Eklund, B.: Influence of Work Duration on the Regulation of Muscle Blood Flow, *Acta Physiol. Scand.*, (Suppl. 411), 1974.

Elias, H. M., and J. E. Pauly: "Human Microanatomy," Da Vinci, Chicago, 1960.

Eriksson, B. O.: Physical Training, Oxygen Supply and Muscle Metabolism in 11-13-year Old Boys, *Acta Physiol. Scand.*, (Suppl. 384), 1972.

Evans, C. L.: The Velocity Factor in Cardiac Work, *J. Physiol.*, **52**:6, 1918.

Fagraeus, L.: Cardiorespiratory and Metabolic Functions during Exercise in Hyperbaric Environment, *Acta Physiol. Scand.*, (Suppl. 414), 1974.

Folkow, B.: Range of Control of the Cardiovascular System by the Central Nervous System, *Physiol. Rev.*, **40**(Suppl. 4):93, 1960.

Folkow, B., U. Haglund, M. Jodahl, and O. Lundgren: Blood Flow in the Calf Muscle of Man during Heavy Rhythmic Exercise, *Acta Physiol. Scand.*, **81**:157, 1971.

Folkow, B., and E. Neil: "Circulation," Oxford University Press, London, 1971.

Gerstenblith, G., E. G. Lakatta, and M. L. Weisfeldt: Age Changes in Myocardial Function and Exercise Response, *Progr. Cardiovas. Diseases*, **19**:1, 1976.

Gollnick, P. D., R. B. Armstrong, C. W. Saubert IV, K. Piehl, and B. Saltin: Enzyme Activity and Fiber Composition in Skeletal Muscle of Untrained and Trained Men. *J. Appl. Physiol.*, **33**:312, 1972.

Gregg, D. E.: The Natural History of Coronary Collateral Development, *Circ. Res.*, **35**:335, 1975.

Gregg, D. E., and H. D. Green: Registration and Interpretation of Normal Phasic Inflow into Left Coronary Artery by Improved Differential Manometric Method, *Am. J. Physiol.*, **130**:114, 1940.

Gregg, D. E., and J. D. Coffman: Coronary Circulation, in D. I. Abramson (ed.), "Blood Vessels and Lymphatics," chap. 9, p. 269, Academic Press, Inc., New York, 1962.

Grimby, G., and B. Saltin: Physiological Analysis of Physically Well-trained Middle-aged and Old Athletes, *Acta Physiol. Scand.*, **179**:513, 1966.

Haddy, F. J., L. A. Sopirstein, and R. R. Sonnenschein: Arterial and Arteriolar Systems: Biophysical Principles and Physiology, in D. I. Abramson (ed.), "Blood Vessels and Lymphatics," chap. 2, p. 61, Academic Press, Inc., New York, 1962.

Hansen, T.: Osmotic Pressure Effect of the Red Cells: Possible Physiological Significance, *Nature*, **190**:504, 1961.

Hartley, L. H., J. A. Vogel, and M. Landowne: Central, Femoral, and Brachial Circulation during Exercise in Hypoxia, *J. Appl. Physiol.*, **34**:87, 1973.

Haynes, R. H., and S. Rodbard: Arterial and Arteriolar Systems, Biophysical Principles and Physiology, chap. 2, p. 26, in D. I. Abramson (ed.), "Blood Vessels and Lymphatics," Academic Press, Inc., New York, 1962.

Hermansen, L., B. Ekblom, and B. Saltin: Cardiac Output during Submaximal and Maximal Treadmill and Bicycle Exercise, *J. Appl. Physiol.*, **29**:82, 1970.

Heymans, C., and E. Neil: "Reflexogenic Areas of the Cardiovascular System," Churchill, London, 1958.

Hollander, A. P., and L. N. Bouman: Cardiac Acceleration in Man Elicited by a Muscle-heart Reflex, *J. Appl. Physiol.*, **38**:272, 1975.

Hollmann, H.: "Höchst- und Dauerleistungsfähigheit des Sportlers," Johann Ambrosius Barth, Munich, 1963.

Holloszy, J. O.: Biochemical Adaptations to Exercise: Aerobic Metabolism, in J. H. Wilmore (ed.) "Exercise and Sport Sciences Reviews," vol. 1, p. 46, Academic Press, Inc., New York, 1973.

Holmér, I.: Physiology of Swimming Man, *Acta Physiol. Scand.* (Suppl. 407), 1974.

Holmgren, A.: Circulatory Changes during Muscular Work in Man, *Scand. J. Clin. Lab. Invest.* (Suppl. 24), 1956.

Howald, H.: Ultrastructural Adaptation of Skeletal Muscle to Prolonged Physical Exercise, in H. Howald, and J. R. Poortmans (eds.), "Metabolic Adaptation to Prolonged Physical Exercise," p. 372, Birkhäuser Verlag, Basel, 1975.

Irving, L.: Bradycardia in Human Divers, *J. Appl. Physiol.*, **18**:489, 1963.

Kitamura, K., C. R. Jorgensen, F. L. Gobel, H. L. Taylor, and Y. Wang: Hemodynamic

Correlates of Myocardial Oxygen Consumption during Upright Exercise, *J. Appl. Physiol.*, **32**:516, 1972.

Kjellberg, S. R., U. Rudhe, and T. Sjöstrand: The Amount of Hemoglobin (Blood Volume) in Relation to the Pulse Rate and Heart Volume during Work, *Acta Physiol. Scand.*, **19**:152, 1949.

Kjellmer, I.: Studies on Exercise Hyperemia, *Acta Physiol. Scand.*, **64**(Suppl. 244):1965.

Korner, P. I.: Intergrative Neural Cardiovascular Control, *Physiol. Rev.*, **51**:312, 1971.

Kreeker, E. J., and E. H. Wood: Comparison of Simultaneously Recorded Central and Peripheral Arterial Pressure Pulses during Rest, Exercise and Tilted Position in Man, *Circulation Res.*, **3**:623, 1955.

Krogh, A.: "The Anatomy and Physiology of Capillaries," rev. ed., Yale University Press, New Haven, Conn., 1929.

Krogh, A., and J. Lindhard: Measurements of the Blood Flow through the Lungs of Man, *Skand. Arch. Physiol.*, **27**:100, 1912.

Lundgren, N.: The Physiological Effects of Time Schedule Work on Lumber Workers, *Acta Physiol. Scand.*, **13**(Suppl. 41):1946.

Lundgren, O., and M. Jodal: Regional Blood Flow, *Ann. Rev. Physiol.*, **37**:395, 1975.

Lundvall, J., S. Mellander, H. Westling, and T. White: Fluid Transfer between Blood and Tissues during Exercise, *Acta Physiol. Scand.*, **85**:258, 1972.

Lutz, B. R., and G. P. Fulton: Structural Basis of the Microcirculation, in D. I. Abramson (ed.), "Blood Vessels and Lymphatics," chap. 5, p. 137, Academic Press, Inc., New York, 1962.

McCloskey, D. I., and K. A. Streatfeild: Muscular Reflex Stimuli to the Cardiovascular System during Isometric Contractions of Muscle Groups of Different Mass, *J. Physiol.* (London), **250**:431, 1975.

McDonough, J. R., R. A. Danielson, R. E. Wills, and D. L. Vine: Maximal Cardiac Output during Exercise in Patients with Coronary Artery Disease, *Am. J. Cardiol.*: **33**:23, 1974.

Marx, H. J., L. B. Rowell, R. D. Conn, R. A. Bruce, and F. Kusumi: Maintenance of Aortic Pressure and Total Peripheral Resistance during Exercise in Heat, *J. Appl. Physiol.*, **22**:519, 1967.

Mellander, S., Comparative Studies on the Adrenergic Neurohormonal Control of Resistance and Capacitance Blood Vessels in the Cat, *Acta Physiol. Scand.*, **50**(Suppl. 176):1960.

Mellander, S., and B. Johansson: Control of Resistance, Exchange, and Capacitance Functions in the Peripheral Circulation, *Pharm. Rev.*, **20**:117, 1968.

Mellerowicz, H.: "Ergometrie," Urban & Schwarzenberg, Munich, 1962.

Messer, J. V., R. J. Wagman, H. J. Levine, W. A. Neill, N. Krasmow, and R. Gorlin: Patterns of Human Myocardial Oxygen Extraction during Rest and Exercise, *J. Clin. Invest.*, **41**:725, 1962.

Mitchell, J. H., B. J. Sproule, and C. B. Chapman: Factors Influencing Respiration during Heavy Exercise, *J. Clin. Invest.*, **37**:1693, 1958a.

Mitchell, J. H., B. J. Sproule, and C. B. Chapman: The Physiological Meaning of the Maximal Intake Test, *J. Clin. Invest.*, **37**:538, 1958b.

Musshoff, K., H. Reindell, and H. Klepzig: Stroke Volume, Arteriovenous Difference, Cardiac Output and Physical Working Capacity and Their Relationship to Heart Volume, *Acta Cardiol. Brux.*, **14**:427, 1959.

Neil, E.: Afferent Impulse Activity in Cardiovascular Receptor Fibers, *Physiol. Rev.*, **40**(Suppl. 4):201, 1960.

Nielsen, M., and O. Hansen: Maximale Körperliche Arbeit bei Atmung O₂-reicher Luft, *Skand. Arch. Physiol.*, **76**:37, 1937.

Öberg, B.: Overall Cardiovascular Regulation, *Ann. Rev. Phys.*, **38**:537, 1976.

Ochsner, A., Jr., R. Colp, Jr., and G. E. Burch: Normal Blood Pressure in the Superficial Venous System of Man at Rest in the Supine Position, *Circulation*, **3**:674, 1951.

Olsen, C. R., D. D. Fanestil, and P. F. Scholander: Some Effects of Breath Holding and Apneic Underwater Diving on Cardiac Rhythm in Man, *J. Appl. Physiol.*, **17**:461, 1962.

Otis, A. B.: The Control of Respiratory Gas Exchange between Blood and Tissues, in D. J. C. Cunningham and B. B. Lloyd (eds.), "The Regulation of Human Respiration," p. 111, Blackwell Scientific Publications, Ltd., Oxford, 1963.

Pappenheimer, J. R.: Central Control of Renal Circulation, *Physiol. Rev.*, **40**(Suppl. 4):35, 1960.

Peterson, L. H.: Cardiovascular Control and Regulation, in "Les Concepts de Claude Bernard sur le Milieu Intémie, p.191, Masson et Cie, Lebraries de l'Académie de Médecine, Paris, 1967.

Reeves, J. T., R. F. Grover, S. G. Blount, Jr., and G. F. Filley: Cardiac Output Responses to Standing and Treadmill Walking, *J. Appl. Physiol.*, **16**:283, 1961.

Reindell, H.: Uber den Kreislauf der Trainierten. Uber die Restblutmenge des Herzens und über die besondere Bedeutung röntgenologischer (kymographischer) hämodynamische Beobachtungen in Ruhe und nach Belastung, *Arch. Kreislaufforsch.*, **12**:265, 1943.

Reindell, H., H. Klepzig, H. Steim, K. Musshoff, H. Roskamm, and E. Schildge: "Herz Kreislaufkrankheiten und Sport," Johann Ambrosius Barth, Munich, 1960.

Robinson, S.: Experimental Studies of Physical Fitness in Relation to Age, *Arbeitsphysiol.*, **10**:251, 1938.

Rowell, L. B.: Human Cardiovascular Adjustments to Exercise and Thermal Stress, *Physiol. Rev.* **54**:75, 1974.

Ruch, T. C., and H. D. Patton (eds.): "Physiology and Biophysics, Circulation, Respiration and Fluid Balance," vol. II, W. B. Saunders Company, Philadelphia, 1974.

Saltin, B., G. Blomqvist, J. H. Mitchell, R. L. Johnson, Jr., K. Wildenthal, and C. B. Chapman: Response to Submaximal and Maximal Exercise after Bedrest and Training, *Circulation* **38**(Suppl. 7):1968.

Saltin, B., J. Henriksson, E. Nygaard, P. Andersen, and E. Jansson: Fiber Types and Metabolic Potentials of Skeletal Muscles in Sedentary Man and Endurance Runners, *Bull, N.Y. Acad. Med.*, in press, 1977.

Sarnoff, S. J., and J. H. Mitchell: The Control of the Function of the Heart, in W. F. Hamilton and P. Dow (eds.), vol. 1, p. 489, American Physiological Society, Washington, D.C., 1962.

Scheinberg, P., L. I. Blackburn, M. Rich, and M. Saslaw: Effects of Vigorous Physical Exercise on Cerebral Circulation and Metabolism, *Am. J. Med.*, **16**:549, 1954.

Scholander, P. F., H. T. Hammel, H. LeMessurier, E. Hemmingsen, and W. Garey: Circulatory Adjustment in Pearl Divers, *J. Appl. Physiol.*, **17**:184, 1962.

Sjöstrand, F. S., and E. Andersson-Cedergren: Intercolated Discs of Heart Muscle, in G. H. Bourne (ed.), "Structure and Function of Muscle," vol. 1, chap. 12, p. 421, Academic Press, Inc., New York, 1960.

Sjöstrand, T.: Volume and Distribution of Blood and Their Significance in Regulating Circulation, *Physiol. Rev.*, **33**:202, 1953.

Skinner, N. S., Jr., and J. C. Costin: Interactions of Vasoactive Substances in Exercise Hyperemia: O_2, K^+ and Osmolarity, *Amer. J. Physiol.*, **219**:1386, 1970.

Sonnenblick, E. H., and C. L. Skelton: Reconsideration of the Ultrastructural Basis of Cardiac Length-tension Relations, *Circ. Res.*, **35**:519, 1974.

Stenberg, J., B. Ekblom, and R. Messin: Hemodynamic Response to Work at Simulated Altitude, 4,000 m, *J. Appl. Physiol.*, **21**:1589, 1966.

Stenberg, J., P.-O. Åstrand, B. Ekblom, J. Royce, and B. Saltin: Hemodynamic Response to Work with Different Muscle Groups in Sitting and Supine, *J. Appl. Physiol.*, **22**:61, 1967.

Strandell, T.: Circulatory Studies on Healthy Old Men, *Acta Med. Scand.*, **175**(Suppl. 414):1964.

Taylor, M. G.: Hemodynamics, *Ann. Rev. Physiol.*, **35**:87, 1973.

Uvnäs, B.: Sympathetic Vasodilator System and Blood Flow, *Physiol. Rev.*, **40**(Suppl. 4):69, 1960.

Vokac, Z., H. J. Bell, E. Bautz-Holter, and K. Rodahl: Oxygen Uptake/Heart Rate Relationship in Leg and Arm Exercise, Sitting and Standing, *J. Appl. Physiol.*, **39**(1):54, 1975.

Wade, O. L., and J. M. Bishop: "Cardiac Output and Regional Blood Flow," Blackwell Scientific Publications, Ltd., Oxford, 1962.

Wade, O. L., B. Combes, A. W. Childs, H. O. Wheeler, A. Cournand, and S. E. Bradley: The Effect of Exercise on the Splanchnic Blood Flow and Splanchnic Blood Volume in Normal Man, *Clin. Sci.*, **15**:457, 1956.

Wiggers, C. J.: "Circulatory Dynamics," Grune & Stratton, Inc., New York, 1952.

Zweifach, B. W.: Microcirculation, *Ann. Rev. Physiol.*, **35**:117, 1973.

Respiration

CONTENTS

MAIN FUNCTION

ANATOMY AND HISTOLOGY
Airways   Blood vessels   Nerves

"AIR CONDITION"

FILTRATION AND CLEANSING MECHANISMS

MECHANICS OF BREATHING
Pleurae   Respiratory muscles   Total resistance to breathing

VOLUME CHANGES
Terminology and methods for the determination of "static" volumes   Age and
sex   "Dynamic" volumes

COMPLIANCE

AIRWAY RESISTANCE

PULMONARY VENTILATION AT REST AND DURING WORK
Methods   Pulmonary ventilation during exercise   Dead space   Tidal volume-respiratory
frequency   Respiratory work (respiration as a limiting factor in physical work)

DIFFUSION IN LUNG TISSUES, GAS PRESSURES

VENTILATION AND PERFUSION

OXYGEN PRESSURE AND OXYGEN-BINDING CAPACITY OF THE BLOOD

REGULATION OF BREATHING
Rest   Central chemoreceptors   Peripheral chemoreceptors   Simultaneous effect of
hypoxia and variations in P_{CO_2} Exercise   Hypoxic drive during exercise

BREATHLESSNESS (DYSPNEA)

SECOND WIND

HIGH AIR PRESSURES, BREATH HOLDING, DIVING
High air pressures   Breath holding-diving

Chapter 7

Respiration

MAIN FUNCTION

The living cell uses oxygen for its metabolism, and in the process, carbon dioxide is produced. Thus, the concentration of oxygen within the cell is lowered, and oxygen will tend to diffuse toward the place of combustion. Similarly, the carbon dioxide produced will tend to diffuse away. The exchange of O_2 and CO_2 is dependent on the distance the molecules have to travel and the pressure gradient. In the single-cell organism the surface can effectively be utilized for the respiratory exchange. By diffusion, oxygen can easily reach every point within the cell, and CO_2 can be eliminated. The calculations of Krogh (1941) indicated that when metabolism is fairly high, diffusion can provide sufficient oxygen only to organisms with a diameter of 1 mm or less. A spherical cell with a 1-cm radius and an oxygen uptake of $100 \, ml \times kg^{-1} \times hr^{-1}$ would need an external oxygen pressure of 25 atm to secure the oxygen supply to its center by diffusion. In multicelled animals, the problem of gas transports is solved in various ways; e.g., numerous airways, like tracheae in insects, or specialized organs, like lungs or gills, are developed, exposing an enlarged respiratory surface to the external medium to effect exchange of oxygen and carbon dioxide. With the gas exchanger located distant from the sites of

209

metabolism, two transport systems carry O_2 and CO_2 between the respiratory surface and the cells in which the metabolism proceeds: a circulatory system supplies the blood, and a respiratory system supplies the air to the lungs. The spongelike structure of the gas-exchange area provides an enormous contact surface between air and blood. This air-tissue-blood interface of an average adult human lung is estimated to be some 70 to 90 m^2, with a thickness of the tissues varying from 0.2 μm to several μm (average, some 0.7 μm). Fresh air and "new" blood must continuously be supplied to the many million gas-exchange units, since the store of oxygen in the body is very limited and the central nervous system and heart muscle do not tolerate any lag in the supply of oxygen.

In this chapter, we shall discuss the exchange of oxygen and carbon dioxide between ambient air and blood. For a comprehensive discussion of the many aspects of respiration, the reader is referred to the *Handbook of Physiology* (Fenn and Rahn, eds., 1964–1965) and other books (Cherniak et al., 1972; Dejours, 1975; Porter, 1970; West, 1974; Widdicombe, 1974).

The ideal gas exchanger involves four factors. (1) It should provide *a large contact area* between the air and blood with a very thin membrane separating the two media, since diffusion is directly proportional to the area but inversely related to the thickness of the membrane. The membrane should cause a minimal resistance to gas flow. (2) The inspired air must become *saturated with water vapor and heated* to tissue temperature to *protect the delicate membranes from injury*; any particles and agents in the air which may be harmful should be removed during the passage through the airways, and if introduced, should be expelled. (3) Variations in oxygen and carbon dioxide in the blood leaving the lungs should vary only within small limits, and therefore, *the distribution of gas and blood in the many exchange units should be closely matched.* (4) *The gas exchange, and therefore the perfusion of the units, must be proportional to the uptake of oxygen and production of carbon dioxide by the cells.* This demands some sort of regulative mechanism linking the needs of the distant cells with the external respiration. The anatomic location of airways and lungs inevitably causes respiration to be influenced by movements of the trunk. Speaking and singing modify the respiration.

Point (4) of the factors listed is of special interest in this connection, since muscular exercise may increase the oxygen uptake of the body some 20 times the resting level, with a similar rise in CO_2 production. The respiration also plays an important role in the *maintenance of the pH of the blood at normal levels.* Hydrogen ions cannot be exchanged between air and blood, but the acid-base equilibrium of the blood (and tissue) is closely associated with the CO_2 content and pressure in the blood (see Chap. 5). During heavy exercise, lactic acid, an acid stronger than H_2CO_3, is produced in the working muscles and more CO_2 is formed, which stimulates respiration. Elimination of CO_2 during hyperventilation will reduce the effect of lactic acid on the blood pH.

Before we discuss respiration during exercise, a summary of the basic respiratory function will be presented.

ANATOMY AND HISTOLOGY

Airways

Figure 7-1 illustrates the respiratory tree and its subdivision into finer airways until they finally terminate in numerous blind pouches, the alveoli. Figure 7-2 presents schematically the general architecture of the airways. Via nose or mouth, the inspired gases pass through the pharynx, larynx, and trachea into the bronchial tree. The airways branch by asymetric dichotomy and the two daughter branches in turn become parent branches, etc. They terminate in the *alveoli*, polyhedral or cuplike outpouchings of the finer airways. Their diameter is up to 300 μm. The surface of the alveolus is not smooth but is corrugated by the capillaries and by various subcellular structures, like nuclei, bulging into the alveolus (Fig. 7-3). The alveoli share their walls with their neighbors and form the spongelike texture of the lungs (Fig. 7-1).

In the adult man, about 23 such generations of branches canbe traced as the airways subdivide into the periphery of the lung and distribute them among the numerous respiratory units. The first 16 generations roughly constitute the "conductive zone" where practically no gas exchange occurs between blood and air. Then follow the "transitory and respiratory zones": generations 17 to 19 of branching form the respiratory bronchioles with a diameter of about 1 mm, which subdivide to produce about 1 million alveolar ducts. The last of a short series of alveolar ducts terminates in rotundate enclosures: the alveolar ducts with a diameter of about 400 μm. The cylindrical surface of the respiratory bronchioles bears a smaller number of variously spaced alveoli, but the alveolar ducts and sacs are fully alveolated. Therefore, they lack proper walls but open out on all sides into alveoli, some 300 million altogether in the adult (Fig. 7-1). (There are considerable normal variations in those numbers—from 200 million up to 600 million—related to the individual's body height, but the alveoli have similar dimensions in large and small adult human lungs, Angus and Thurlbeck, 1972.) The respiratory and transitory zones, including the alveoli, amount to about 90 percent of the lung volume; about 65 percent of the air in the lungs is in the actual alveoli at three-fourths the maximal inflation of the lungs (Weibel, 1964;1972).

The bronchi and bronchioles that are more than about 1 mm wide have a discontinuous cartilaginous support in the wall. Muscle fibers in circular or crisscrossing bundles are incorporated into a complex connective tissue framework of collagenous reticular and elastic fibers. The inner surface is covered with a ciliated epithelium. Goblet cells occur singly or in groups between the epithelial cells and produce a secretion. In the finest bronchioles, the mucus-secreting elements become sparse and finally absent. They are lacking cartilages, and the ciliated cells also disappear gradually. Muscle fibers as well as elastic, collagenous, and reticular fibers provide the supporting latticework of the interalveolar septum and a framework for the entrance of the alveoli and alveolar sacs and ducts.

In the fetus, the alveoli are atelectatic, but the first influx of air in the newborn child provides a force that stretches the original cuboidal epithelium lining of the alveoli into an extremely thin layer of squamous cells. The lung of the newborn is not fully developed. The airways have subdivided into only some 17 generations of branchings and the number of alveoli is less than

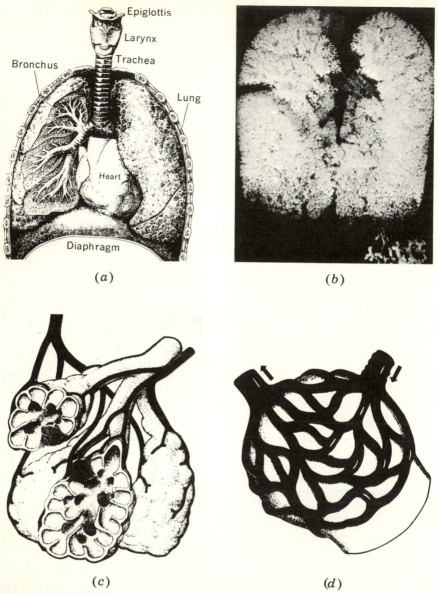

Figure 7-1 (*a*) Principal organs of breathing. In this drawing, the ribs, the large arteries from the heart, and part of one lung have been cut away.
(*b*) A cast of the complete air spaces of the lung, showing the millions of air sacs at the end of the bronchioles. Inset: the terminal portion of a bronchiole magnified. (*From C. M. Fletcher, BBC Publication 3 s., 1963.*)
(*c*) Schematic drawing of a bronchiole and its air sacs and alveoli.
(*d*) An alveolus embraced by the capillary network.

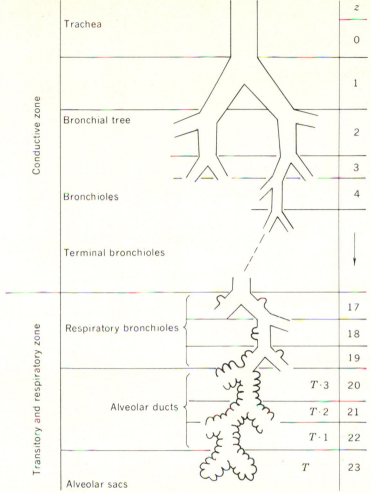

Figure 7-2 General architecture of conductive and transitory airways. The z column designates the order of generations of branching; T, the terminal generation. A more detailed discussion appears in the text. *(Weibel, 1963.)*

one-tenth of that found in the adult. As time goes on, additional branches are added and many more alveoli are formed as new ramifications grow out, so that before ten years of age the adult number is reached. Whether or not strenuous physical efforts may serve to add more branches and alveoli to the mature lung is not known (see below). The relative alveolar volume decreases with age.

Despite the latticework fibers supporting the alveolar walls, the air-tissue-blood interface would provide special problems due to the surface tension which is created. This tension tends to decrease the surface wall to a minimum so that the alveoli may collapse. The alveoli are, however, lined with an insoluble surface film of lipoprotein, about 5nm thick. It is produced in the alveoli and keeps the alveoli open and free from transudate from the blood by

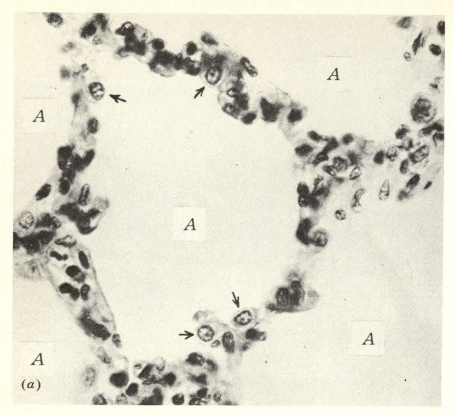

(a)

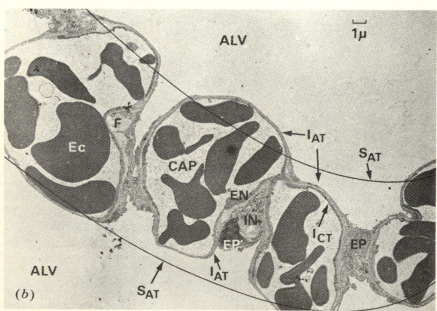

(b)

lowering the surface tension. This function is especially important as the volume decreases, e.g., during forced expiration, which would otherwise empty the small alveoli. During quiet breathing, there may actually be a reduced effect of the surface film and an occasional collapse of alveolar units. Forced inflation of the lungs may cause more material to be provided for the lining film and open the alveoli. A yawn or deep breath may exert such a beneficial function. The surface film thus stabilizes the small alveolar spaces and enables the lung to retain air at low inflation pressure (Pattle, 1965).

Blood Vessels

The pulmonary arteries enter the lungs with the stem bronchi and provide arterial partners to the airways as they subdivide toward the respiratory zones of the lungs. The arterioles follow the bronchioles, alveolar ducts, and sacs, and provide short twigs to capillary networks enveloping the alveoli surrounding the particular airway terminal and to any other alveoli in the immediate vicinity. Each alveolus may be covered by a capillary network consisting of

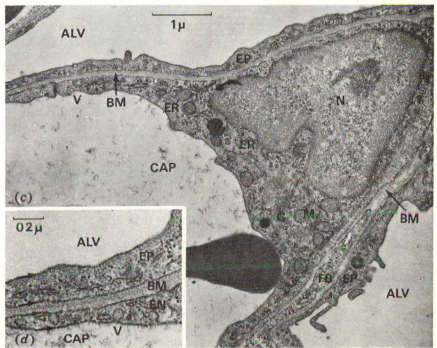

Figure 7-3 (a) Histologic section through one complete alveolar outline (center) and portions of four adjacent alveoli (A); magnification ×650. (*Modified from Krahl, 1964.*) (b, c, d) Electron micrographs of interalveolar septum of rat lung in cross section showing the barrier composed of alveolar epithelial (EP), capillary endothelial cells (EN), and some interstitial elements. (*From Weibel, 1964.*)

(b) Magnification ×5,500. Note the difference between the estimated alveolar surface (S_{AT}) and the real "corrugated" air-tissue interface (I_{AT}). I_{CT} = tissue-blood interface; CAP = capillaries; IN = interstitium; F = fibrous elements; Ec = erythrocytes.

(c) Magnification ×23,500. BM = basal membranes; N = nucleus; EP = endoplasmic reticulum; V = pinocytotic vesicle.

(d) Magnification ×59,000 of the thin portion of air-blood barrier with four membranes.

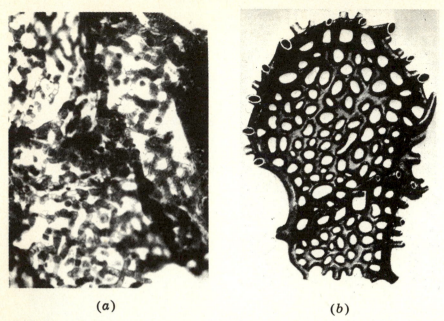

(a) (b)

Figure 7-4 (a) Network of capillaries in alveolar walls. Magnification ×375. *(From Miller, 1947.)*
(b) Blood-filled capillary network in interalveolar septum of human lung. Larger, dark vessel to the right (arteriole) gives off short precapillaries which open at once into pulmonary capillaries. Magnification ×650. *(From Krahl, 1964.)*

almost 2,000 segments; the capillary networks in the lungs are the richest in the body and are more or less continuous throughout large parts of the lungs (Fig. 7-4). The air-blood "barrier" is formed by the continuous alveolar epithelial and capillary endothelial cells with a tenuous interstitium with fibrous elements in between the two cell layers. The thickness of the "barrier" can vary from about 0.2 μm to several μm, and the variations are caused by various structures scattered throughout the continuous cell layers (Fig. 7-3). It is evident that the capillary network should be regarded as a sheet of blood floating along the alveolar surface, the sheet merely connecting the two walls. The capillary surface area is, in fact, of the same order of size as the alveolar surface area.

Nerves

Vagal efferent fibers go to the smooth muscles of the bronchial tree as far as the terminations of the alveolar ducts and sacs, and to the bronchial mucous glands. Nerve impulses stimulate the smooth muscles to contract and activate the glands. *Vagal afferent fibers* carry impulses from special stretch receptors scattered in lungs and pleura. Sympathetic fibers acts as bronchodilators. There are many free nerve endings in the bronchial walls and pleura, but their function is not known.

"AIR CONDITION"

The inspired air may be cold or hot, dry or moist, but owing to the rich blood supply of the mucous membranes of the nose, the mouth, and the pharynx, the air temperature becomes adjusted to body temperature and also moistened. In a person exposed to the cold air in the Arctic, or the hot air in the tropics, the inspired air is about 37°C by the time it reaches the pharynx. As a matter of fact, in experimental animals exposed to -100°C and up to $+500$°C, the air temperature was warmed or cooled during its passage through the upper respiratory tree and attained body temperature in the tracheobronchial tree (Moritz et al., 1945; Moritz and Weisiger, 1945). The mouth and pharynx can perform these air-conditioning functions as effectively as the nose and pharynx. Air saturated with water vapor at 37°C has a $P_{H_2O} = 47$ mm Hg; the content of water is then 44 g $\times$ m^{-3}. At low temperatures the water content in the air is low. Even if saturated, the air at 0°C contains only 5 g H_2O $\times$ m^{-3}. In a normal climate, about 10 percent of the total heat loss of the body at rest or during work takes place through the respiratory tracts by the air conditioning of the inspired air. At -15 to -20°C the percentage would be about 25 (see Chap. 15). The respiratory tract serves as a regenerative system: the heating and humidifying of inspired air cool the mucosa. But during expiration, some of the heat and water are recovered by the mucosa from the passing alveolar air. Body heat and water are conserved (Cole, 1954). On a very cold day, this condensation of water vapor may result in excessive accumulation of water in the nostrils, leading to a runny nose! A cross-country skier breathing 100 liters/min of air at -20°C must, in 1 hr, add about 250 ml of water to this air. Not all of this water volume is expired, however, thanks to the regenerative system.

FILTRATION AND CLEANSING MECHANISMS

If living in a city, we may inhale billions of particles of foreign matter every day. Particles larger than about 10 µm are effectively removed from the inspired air in the nose, where they are trapped by the hair or the moist mucous membranes. Those particles which escape these obstacles usually settle on the walls of the trachea, the bronchi, and the bronchioles. Therefore only a few very small particles are likely to reach the alveoli, and this part of the lung is practically sterile. Alveolar macrophages perform a vital function of maintaining the alveoli clean and sterile. They are of hematogenous origin but thanks to their extraordinary amoeboid mobility they can pass into the alveoli and move freely over the airspace surfaces. They are able to phagocytose large quantities of foreign material, such as dust particles and bacteria. They have a large metabolic activity, also related to the formation of antibodies. The macrophages and the phagocytosed remains can be cleared into the digestive tract via the airways or removed via the blood flow or lymphatics. It should be pointed out that the phagocytic capacity can be diminished by certain influences such as smoking. (See Weibel, 1972.)

As mentioned, the epithelium of the airways within the lungs consists of *ciliated cells*. In the conductive zone each cell carries up to 300 cilia about 6 to 7 μm in length, at the free cell surface. The cilia of many thousands of cells beat in an organized, whiplike fashion in strokes, like oars of a boat, with a rapid upward propulsive stroke followed by a slower recovery downward stroke. This goes on continuously day and night. The cilia are covered by a continuous surface of watery mucus. By the ciliar activity, this fluid carpet with all the entrapped particles moves toward the larynx at a speed of well over 1 cm × min^{-1}. This mucus is either expectorated or swallowed. The ciliary escalator is remarkably resistant to noxious influences. However, cigarette smoke has a deleterious effect on the ciliar function. The cilia slow down or stop their beating when exposed to the smoke.

From time to time we may sneeze or cough, and with our explosive blast (the air moves with a speed approaching the speed of sound), foreign particles may be expelled.

MECHANICS OF BREATHING

Pleurae

The lungs increase and decrease their volume with the reciprocating movement of the bellowslike pump, the thorax. The thoracic cavity is covered by the very thin parietal pleura, and the lungs by the pulmonary (visceral) pleura. These very thin membranes of single layers of flat epithelial cells on fibrous connective sheets continue uninterrupted from one pleural surface to the other across the pulmonary hilus. The two pleurae surfaces are held close together with a thin fluid film in between, providing smooth lubricated surfaces. If the thorax is opened so that the atmosphere pressure prevails in the intrapleural space, the elastic recoil of the lungs causes them to collapse, and the chest expands a little, since a retractive force of the lungs is normally counterbalanced by an outward spring of the chest cage (pneumothorax). Such an injury, of course, makes the lung involved incapable of any respiratory function. Normally, however, the pleurae are in close, but friction-free, contact with each other. Any volume changes in the thorax are completely transmitted to the lungs. The two pleural surfaces may be compared with two flat sheets of glass placed face to face with a thin layer of water between the two opposing surfaces. While the two sheets of glass can easily be slid back and forth, a great force is required to move the two sheets away from one another by forces acting perpendicular to the glass surfaces.

It would appear plausible that gas and fluid from the blood might collect in the intrapleural space because of the discrepancy in the size of the thoracic cavity and the lungs: the opposing forces of the lung and chest wall tend to separate the two pleurae with a pressure of a few centimeters of water lower than atmospheric pressure. Gas is, however, absorbed because the sum of the gas tensions in venous blood (and pleural liquid) is less than arterial (and atmospheric) pressure. Water from the space is effectively absorbed, since the collois osmotic pressure of the plasma proteins in the pulmonary capillaires easily matches the slightly lower hydrostatic pressure in the pulmonary vessels.

Respiratory Muscles

Figure 7-5 shows the contours of the thorax and lungs at the end of an expiration and an inspiration, respectively. During quiet breathing at rest, the diaphragm is the principal muscle driving the inspiratory pump: the abdominal muscles relax, the abdomen protrudes, the thoracic volume increases, and the lungs expand. The contraction of the diaphragm causes its dome to descend some 1.5 cm, and the intra-abdominal pressure increases. This movement is actually of the same order of magnitude both in the so-called costal and in the diaphragmatic type of breathing. During deep breathing, the vertical movements of the diaphragm may exceed 10 cm. The external intercostal muscles assist in the inspiration, especially during exercise. The fibers slope obliquely downward and forward from the caudal margin of one rib to the cranial margin of the rib below. The lower insertion is located more distant from the center of rotation than the upper one. When the fibers contract, the force exerted by the muscle is equal at both insertions, but the longer leverage of the lower rib gives a torque that raises rather than lowers the upper rib. The net effect is a lifting of the ribs when the external intercostal muscles contract. The elevation of the

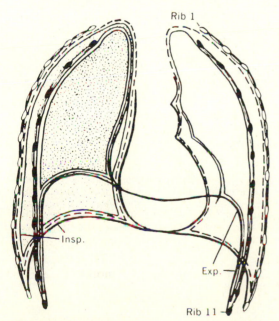

Rib 1

Insp.

Exp.

Rib 11

Figure 7-5 Diagram of frontal section of thorax, based on roentgenograms, showing changes in size of thorax and position of heart and diaphragm with respiration. To the left: inspiration is designated with light stippling, and expiration with heavy stippling. *(From Braus, 1956.)*

ribs, rotating around the axis of their necks, increases the dimensions of the rib cage in both transverse and dorsoventral directions, similar to that of the handle of a bucket when lifted. When the inspiratory muscles relax during quiet breathing, the elastic recoil forces in the lung tissue, thoracic wall, and abdomen restore the chest to the resting position without any help of the expiratory muscles. During exercise or forced breathing at rest, with a ventilation exceeding 2 or 3 times the resting value, these recoil pressures are supplemented by activity of the expiratory muscles. The internal intercostal fibers have a direction opposite to those of the external intercostal muscles. Therefore the function of the inner layer of the intercostal muscles is to facilitate expiration. The muscles of the abdominal wall are essentially expiratory muscles, but do not become engaged forcefully until the pulmonary ventilation reaches high levels.

At a ventilation exceeding 50 liters$\times$min^{-1} and especially at very high ventilation, accessory muscles may assist. The sternocleidomastoids and scaleni are the most important ones during forced inspiration. When the athlete grasps for support after an exhausting spurt, this posture may facilitate the action of the respiratory muscles.

The activity of the inspiratory muscles increases progressively throughout inspiration and they actually continue to contract while being stretched during the early part of expiration. Part of the work done during inspiration is "stored" in the elastic structures of the system and is then available to supply part of the power for the expiration. If an expiration decreases the lung volume below the resting level of the system, the chest wall recoils outwardly, causing a passive inspiration back to the resting volume.

During exercise, the inspiratory and expiratory muscles are activated reciprocally, especially the expiratory ones in the last part of expiration.

Total Resistance to Breathing

The respiratory muscles work mainly against an airway resistance and a pulmonary tissue and chest wall resistance.

The work done against inert forces to accelerate tissues and gases is in this connection negligible. Most of the tissue resistance is offered by elastic forces. But the collagen fibers, providing the supporting framework for the delicate structures of the lungs, also contribute to the resistance when a volume change occurs. Thanks to the soft, yielding tissues of the lungs, the resistance is low. Of the total pulmonary resistance, only about 20 percent is a tissue resistance and 80 percent is airway resistance.

At high flow velocities, as during heavy exercise, the air flow is turbulent in the trachea and the main bronchi, giving a high flow resistance. Owing to the large total cross area of the finest air tubes, the air flow in this region is low and therefore laminar. In fact the greatest part of resistance to airflow within the lungs lies in airways greater than 2 mm internal diameter and particularly in the medium-sized bronchi (the four to ten airway generations, Fig. 7-2). An air flow of 1 liter $\times$ s^{-1} requires a pressure drop along the airways of less than 2 cm H$_2$O.

At rest, the oxygen cost of the breathing is only a small fraction of the total resting energy turnover. It has been estimated to be about 0.5 to 1.0 ml $\times$ liter^{-1} of moved air. With pulmonary ventilation of 6 liters $\times$ min^{-1}, the oxygen uptake of the respiratory muscles would be up to 6 ml, compared to a total resting oxygen uptake of the body as a whole of about 250 to 300 ml. With the high pulmonary ventilation during heavy exercise, the energy cost per liter ventilation becomes progressively greater, and the oxygen cost of breathing may be up to 10 percent of the total oxygen uptake. The air resistance when breathing through the nose is 2 to 3 times greater than that obtained by breathing through the mouth. It is therefore natural for the athlete to breathe through the mouth when performing heavy exercise, since this reduces the airflow resistance.

Summary   When the respiratory muscles are relaxed (resting volume), the chest wall is retracted by the elastic recoil of the lungs. The beginning of an inspiration is assisted by the recoil of the chest wall but the lung tissue is further stretched. During deeper inspiration, there is a retractive force from both the chest wall and the lungs. Part of the energy provided by the inspiratory muscle, mainly the diaphragm and the external intercostal muscles, is "stored" in the elastic structures and is utilized during the expiration. An expiration below the resting volume increases the outward recoil of the chest wall. In any volume position, the lungs and the chest cage behave like opposing springs, and at the resting volume, the forces exerted exactly counterbalance each other. The work done by the respiratory muscles is mainly devoted to doing elastic work and to overcoming the airway resistance.

VOLUME CHANGES

Terminology and Methods for the Determination of "Static" Volumes

Figure 7-6 should be consulted for the terminology. When the respiratory muscles are relaxed, there is air still left in the lungs. This air volume is the *functional residual capacity* (FRC). A forced maximal expiration brings the volume down to the *residual volume* (RV) by expiration of the *expiratory reserve volume*. (Actually, the limit of a maximal expiration is not only the capacity of the expiratory muscles to compress the thoracic cage. Many small airways become occluded during the forced expiration, and the lungs, with trapped gas, are also compressed.) A maximal inspiration from the functional residual capacity adds the *inspiratory capacity*, and the gas volume contained in the lungs is then the *total lung capacity* (TLC). The maximal volume of gas that can be expelled from the lungs following a maximal inspiration is called the *vital capacity* (VC). It follows that vital capacity plus the residual volume constitute the total lung capacity. The volume of gas moved during each respiratory cycle is the *tidal volume* (V_T).

The vital capacity and its subdivisions are commonly measured with the help of a spirometer. With the subject connected to the spirometer via a wide-bore tube, any change in lung volume is reflected in a volume displacement in the spirometer. A calibration factor translates this displacement recorded on a kymograph into liters.

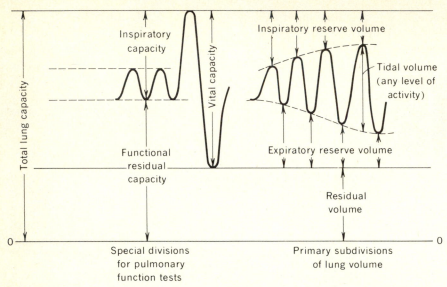

Figure 7-6   Diagram of lung volumes and capacities. *(From Pappenheimer et al., 1950.)*

The functional residual capacity can be measured with the closed-circuit methods (*gas-dilution method*). A closed spirometer contains a small, known amount of helium (or hydrogen). After a normal expiration, the subject is connected to the spirometer and rebreathes from the system. The expired carbon dioxide is absorbed by soda lime. Oxygen is added to the circuit at a rate to keep the volume at the end of expiration at a constant level. This refilling can be adjusted automatically. The concentration of the indicator gas falls in the spirometer and rises in the lungs. The final concentration is a simple function of the added gas volume, i.e., the functional residual capacity. The principle for this method is clarified by Fig. 7-7. The concentration of the indicator is analyzed continuously, for example with a katharometer, and a constant reading for about 2 min indicates a complete mixing. If the subject then performs a maximal expiration, followed by a maximal inspiration to total lung capacity, the recordings permit calculation of residual volume and the subdivisions discussed above (Fig. 7-6). Normally, about 5 min of rebreathing is enough for complete mixing of the indicator gas within spirometer-lungs, but in patients with an impaired lung function, up to 20 min of rebreathing may be necessary. The reason for using helium or hydrogen as indicator gas is that these gases are absorbed by the lung tissues and blood to only a negligible degree.

If He_1 and He_2 are the initial and final concentrations, respectively, of helium and V_s is the volume of gas in the spirometer to the point of the subject's mouth, the functional residual capacity, V_{FRC}, can be calculated from the formula

$$V_{\text{FRC}} = V_s (\text{He}_1 - \text{He}_2)/\text{He}_2$$

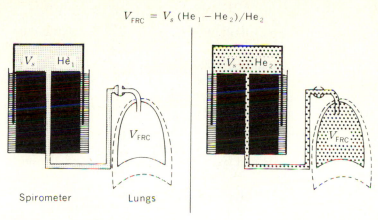

Spirometer   Lungs

Figure 7-7 A spirometer with a measured volume of gas (V_s) contains helium in a small, analyzed concentration (He_1). After a normal expiration (lung volume = functional residual capacity = V_{FRC}), the subject rebreathes from the spirometer until a homogeneous gas mixture is attained with a new and lower helium concentration (He_2) due to its dilution with the air in the lungs. V_{FRC} can now be calculated (see text).

$$V_s \times \text{He}_1 = (V_{\text{FRC}} + V_s)\text{He}_2 \quad \text{or} \quad V_{\text{FRC}} = \frac{V_s(\text{He}_1 - \text{He}_2)}{\text{He}_2}$$

The lung volumes are expressed at BTPS, i.e., gas volume at body temperature and ambient pressure (P_B), saturated with water vapor ($P_{\text{H}_2\text{O}} = 47$ mm Hg), and therefore the gas volumes recorded by the kymograph must be recalculated. For a spirometer temperature of $t°C$ with a water pressure of $P_{\text{H}_2\text{O}}$, we have

$$\text{FRC} = V_{\text{FRC}} \times \frac{310}{273 + t} \times \frac{P_B - P_{\text{H}_2\text{O}}}{P_B - 47} \text{ liters (BTPS)}$$

If the subject is connected with the spirometer after a maximal expiration and then rebreathes deeply three times before being disconnected, the residual volume after a maximal expiration can be directly determined by the application of the same formula.

Besides the gas-dilution method, there are other methods of obtaining fairly accurate measurements of the absolute gas volumes and air spaces in the airways: the gas washout and the body plethysmography methods. The results obtained by these methods are closely comparable and reproducible with a coefficient of variation of roughly ±5 percent. (For methods see West, 1974.)

Age and Sex

Table 7-1 presents data on some of the lung volumes in liters obtained from some fairly well-trained students (physical education), about twenty-five years old (P.-O. Åstrand, 1952). These data were obtained in the standing position. During tilting from standing to supine position, the TLC and VC are reduced 5 to 10 percent because of a shift of blood to the thoracic cavity from the lower part of the body. This illustrates the effect of gravity on the blood distribution within the body (see Chap. 6).

VC, RV, and TLC are related to body size and vary approximately as the cube of a linear dimension, such as body height, up to the age of twenty-five. In other words, these volumes in children are of a size that could be expected from theoretical considerations (see Chap. 11).

The individual dimensions are, however, not exclusively decisive for the size of the lung volumes. The lung volumes are about 10 percent smaller in women than in men of the same age and size. For the average person, the lung volumes are up to 20 percent smaller than the values listed in Table 7-1. Training during adolescence will eventually increase the VC and TLC. After the age of about thirty, the residual volume and functional residual capacity increase and the vital capacity usually decreases. There are observations from a longitudinal study that well-trained individuals attained the same vital capacity at the age of forty to forty-five as twenty years earlier (I. Åstrand et al., 1973). The ratio of RV/TLC $\times$ 100 in the young individual is about 20 percent, but for the fifty- to sixty-year-old individual, this ratio increases to about 40 percent, an increase which can be accounted for almost entirely by changes in lung elasticity with age (Turner et al., 1968).

Athletes have similar or slightly higher values for VC and TLC compared with the data in Table 7-1. The highest recorded value for VC is 9.0 liters for a Danish rower, recorded by Secker and Jackson (personal communication).

The vital capacity has previously been proposed as one method to assess physical work capacity. In a group of about 190 individuals seven to thirty years of age, a significant correlation was found between vital capacity and maximal oxygen uptake (Fig. 7-8). A closer examination of the individual figures reveals, however, that individuals with a vital capacity of approximately 4 liters may have a maximal oxygen uptake from about 2.0 to 3.5 liters $\times$ min^{-1}. From this and similar studies, it is evident that vital capacities of 6.0 liters may be associated with oxygen uptake capacities varying from about 3.5 to 5.5 liters $\times$ min^{-1}. This example shows that one function may appear closely related to another if the data are derived from persons of greatly different size. However, the scattering of the data may still be considerable and sufficiently large to make any prediction of an individual's maximal oxygen uptake from such parameters as vital capacity rather unreliable. The conclusion may be drawn, however, that an oxygen uptake of 4.0 liters $\times$ min^{-1} or more does require a vital capacity of at least 4.5 liters.

Table 7-1

Sex	Number	FRC	VC	RV	TLC
♀	51	2.60	4.25	1.15	5.40
♂	45	3.40	5.70	1.50	7.20

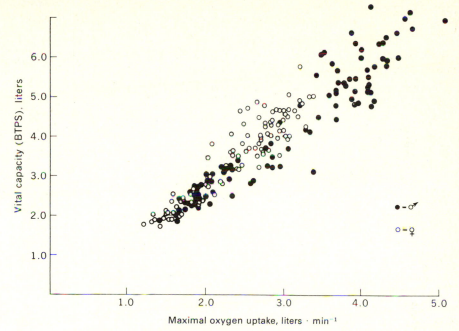

Figure 7-8 Individual data on vital capacity measured in standing position in relation to maximal oxygen uptake during running or cycling in 190 subjects from seven to thirty years of age. *(From P.-O. Åstrand, 1952.)*

The measurement of the vital capacity as part of a larger test battery may yield valuable information, especially concerning the distensibility of the respiratory system. Certain pathological conditions are associated with a reduced vital capacity.

"Dynamic" Volumes

Dynamic spirometry, that is, the determination of ventilatory capacity per unit time, is also used to assess an individual's respiratory function. The subject breathes into a low resistance spirometer, and its displacements are recorded with the aid of a kymograph.

For the determination of *forced expiratory volume* (FEV), the subject first takes a deep breath and inspires maximally. The subject then exhales as forcefully and completely as possible. In this way it is determined how much of the person's vital capacity can be exhaled in the course of 1 s ($FEV_{1.0}$), and this volume is expressed as a percentage of the individual's entire vital capacity. A normal figure for a twenty-five-year-old individual is about 80 percent. The maximal flow is limited by the rate by which the muscles are able to transform chemical energy into mechanical energy and also by a rising flow resistance. Thus, $FEV_{1.0}$ is reduced in persons who have any airway obstructions.

An evaluation of the mechanical properties of the lungs and the chest wall can also be made by determining the *maximal voluntary ventilation* (MVV) (also referred to as maximal breathing capacity). The subject is asked to breathe as rapidly and as deeply as possible during a given time interval, usually 15 s. The individual differences in MVV are large. In the case of healthy twenty-five-year-old men, the mean value is about 140 liters $\times$ min^{-1}, with a range from 100 to 180 liters $\times$ min^{-1}. For women, the normal values range from about 70 to 120 liters. The pulmonary ventilation during maximal work is somewhat lower than that obtained during the determination of MVV.

Since the volume is also affected by the breathing frequency, it may be advisable to have the subject maintain a fixed respiratory rate, such as 40 respirations per minute (MVV$_{40}$), especially in the case of longitudinal studies. The respiratory volume may be recorded with the aid of a spirometer, or the expiratory air may be collected in a Douglas bag. The volume of air is then expressed in liters per minute (BTPS). Since the result depends to a great extent on the complete cooperation of the subject, it is essential that every attempt be made to encourage the subject's exertion to make a maximal effort. MVV depends, among other factors, on the body size of the individual, the forces of the respiratory muscles, the mechancial properties of the thoracic wall and lungs, and on the airway resistance. The measurement of MVV is therefore a measure of the overall capacity of the breathing apparatus to pump air. The maximal air flow during short periods of peak flow during expiration may reach values up to 400 liters $\times$ min^{-1}. One of the limiting factors is the rising air-flow resistance in the tracheobronchial tree, which becomes progressively compressed as the intrathoracic pressure increases during the expiratory effort. Forced expiration tends to collapse the walls of the intrathoracic airways.

COMPLIANCE

The lungs and the thorax are partly made of elastic tissue. During inspiration these tissues are stretched. Because of their elastic nature, they return to their resting position as soon as the inspiratory muscles are relaxed. The more rigid these tissues are, the greater muscular force must be applied in order to achieve a given change in volume. The relation between force and stretch or between pressure and volume can be measured. Thus a measure is obtained of the tissue's elastic resistance to distension, or its so-called compliance. With the aid of a balloon placed in the intrathoracic esophagus, the pressure may be measured at the end of a normal expiration and again after the subject has inhaled a known volume of gas. These measurements may be repeated at different volume changes. The volume changes in liters produced by a unit of pressure change in centimeters H$_2$O gives the lung compliance. If a pressure change of 5 cm H$_2$O produces a change in lung volume of 1 liter, the lung compliance is 1.0 liters/5 cm H$_2$O, or 0.2 liters/cm H$_2$O, which is the normal value at quiet breathing. With a respiratory depth of about 0.5 liter, the pressure variations in overcoming the resistance are, in consequence, a few cm of water. At lung volumes closer to maximal inspiration, or maximal expiration, a greater pressure is required for a given volume change; that is, the compliance is reduced. If, owing to pathological changes, such as interstitial or pleural fibrosis, the lungs are more rigid and less distensible, the compliance is also reduced, and the respiratory work is increased.

In the foregoing, the principle for the measurement of the compliance of the *lungs* has been discussed. It is also of interest to assess the compliance of the thoracic cage. This can be estimated by measuring the compliance of the respiratory system as a whole and then subtracting the compliance of the lungs alone. The chest wall compliance decreases markedly with age.

AIRWAY RESISTANCE

In addition to overcoming the elastic resistance of the respiratory system, part of the energy of the respiratory muscles has to be applied to overcome two types of nonelastic resistance: a tissue viscous resistance due to friction, and a resistance to the movement of air in the air passages. This airway resistance may be doubled by bronchial smooth-muscle contraction or reduced to half the normal resistance by bronchodilatation. The airway resistance may also be increased by mucous edema or by intraluminal secretion. The factors causing this bronchoconstriction may be local or they may be a reflex response to inhaled fine, inert particles, smoke, dust, noxious gases, or to the action of the parasympathetic system. The effect of the sympathetic system and epinephrine on bronchial tone is to dilate the airways. The increased sympathicus-tonus during muscular effort thus tends to lower the airway resistance.

In this connection, it should be pointed out that inhalation of the smoke from a cigarette within seconds causes a two- to threefold rise in airway resistance which may last 10 to 30 min (Comroe, 1966b). At rest, this increased airway resistance is not noticeable. In order to give rise to subjective symptoms of distress, the airway resistance has to be increased 4 to 5 times the normal value. However, during muscular effort, with its increased demand on pulmonary ventilation, the effect of tobacco smoking becomes apparent. The causative factor is not nicotine but particles that have a smaller diameter than 1 μm and that affect the sensory receptors in the airway path. The chronic effect of tobacco smoking is an increased secretion in the respiratory tract and a narrowing of the air passages. FEV as well as MVV may be reduced. It is a common observation that athletes involved in events requiring endurance never smoke. This abstention may be explained by the fact that cigarette smoking reduces the respiratory function and increases the amount of carboxyhemoglobin. The latter reduces the oxygen-transporting capacity of the blood. A sprinter, shot-putter, or diver may be unaffected by cigarette smoking, however, since the requirement for aerobic power in these athletic events may be insignificant. It was mentioned above that smoking also reduced the bactericidal effect of the alveolar macrophages.

PULMONARY VENTILATION AT REST AND DURING WORK

The pulmonary ventilation is the mass movement of gas in and out of the lungs. The pulmonary ventilation is mainly regulated so as to provide the gaseous exchange required for the aerobic energy metabolism. Some gaseous exchange does take place through the skin, and some gas is lost in the urine and other secretions, but the volume of gas thus exchanged is negligible.

Methods

The gas volumes are usually measured very accurately with a water-filled spirometer. Room air (or a mixture of gases) is inhaled through a respiratory valve. The expired air is either collected in a bag (Douglas bag) or collected directly in the spirometer.

The volume may also be measured by other means, with the aid of a gas meter or flow meter, for example. The latter is constructed on the principle that the pressure gradient along a rigid tube of uniform cross section is linearly related to the flow of a gas or fluid, as long as the flow is linear. A flow resistance, which may be a fine mesh screen of a dimension which will not affect respiration, is inserted in the tube. With a differential pressure manometer, the pressure difference across the resistance can be measured when gas is flowing. The volume of gas flow is obtained by graphic or electric integration. (For literature references concerning methods, see Mead and Milic-Emili, 1964.)

The amounts of inhaled and exhaled air are usually not exactly equal, since the volume of inspired oxygen in most situations is larger than the volume of carbon dioxide expired. Pulmonary ventilation usually means the volume of air which is *exhaled* per minute. It is exceedingly important that the mouthpiece, respiratory valve, tubes, and stopcocks are so constructed that they cause a minimum of increased airway resistance during heavy physical exertion. Thus, corrugated external breathing tubes should not be used, since they may give rise to turbulence. The diameter of the tubes and all openings should be about 30 mm or *wider.*

Pulmonary Ventilation during Exercise

Figure 7-9 shows how ventilation ($\dot{V}_E$) increases during increasing work loads up to the maximal level. From a resting value of about 6.0 liters $\times$ min^{-1}, the ventilation increases to 100, 150, and in extreme cases, to 200 liters $\times$ min^{-1} (Saltin and P.-O. Åstrand, 1967) (BTPS = gas volume at normal body temperature and ambient barometric pressure, saturated with water vapor.) The increase is semilinear, with a relatively greater increase at the heavier work loads. (The explanation for this is discussed in the section "regulation of respiration.")

Figure 7-10 presents data on maximal pulmonary ventilation for about 225 subjects, from four to about thirty years of age, collected during maximal running for about 5 min. A positive correlation exists between maximal $\dot{V}_E$ and $\dot{V}_{O_2}$, but it is evident that maximal ventilation cannot be used for prediction of maximal oxygen uptake. Maximal pulmonary ventilation is actually not a well-defined parameter. Figure 7-9 illustrates a marked increase in ventilation during heavy exercise without any further increase in oxygen uptake. A well-motivated subject may continue to exercise at very high work loads despite strain (as judged from blood lactic acid concentration), and the person attains a high ventilation; the less motivated one just quits exercise when still submaximal in a physiological sense.

If pulmonary ventilation is expressed in relation to the magnitude of oxygen uptake, it is 20 to 25 liters/liter O_2 at rest and during moderately heavy work, but it increases to 30 to 35 liters/liter O_2 during maximal work. In children under ten years of age, the values are about 30 liters during light work and up to 40 liters/liter O_2 uptake during maximal work (Fig. 7-10).

Figure 7-11 presents mean values for maximal pulmonary ventilation during exercise (running or cycling) in different age groups. The lower ventilation in the older individuals is associated with a reduced maximal oxygen uptake (Fig. 9-11).

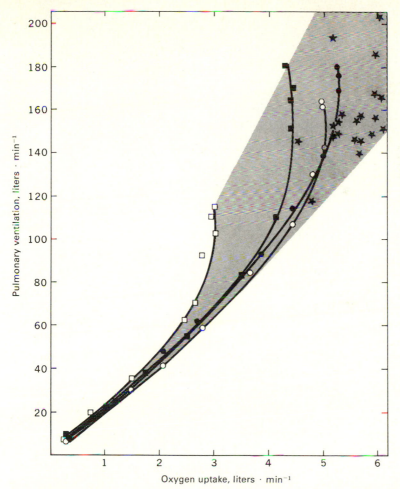

Figure 7-9 Pulmonary ventilation at rest and during exercise (running or cycling). Four individual curves are presented. Several work loads gave the same maximal oxygen uptake. Work time from 2 to 6 min. Stars denote individual values for top athletes measured when maximal oxygen uptake was attained. *(Data from Saltin and P.-O. Åstrand, 1967.)* Individuals with maximal oxygen uptake of 3 liters · min⁻¹ or higher usually fall within the shadowed area. Note the wide scattering at high oxygen uptakes.

Dead Space

Only a part of the inhaled volume of air reaches the alveoles where the gaseous exchange can take place. This part is known as "the effective tidal volume" (V_A). Part of the inspired tidal volume (V_T) occupies the conducting airways. This part is called "dead space" volume (V_D) because it does not take part in the gaseous exchange between alveolar air and blood. During expiration, this dead space component is exhaled first. It has a composition similar to moist inspired air. Then comes the alveolar component, which has a relatively high

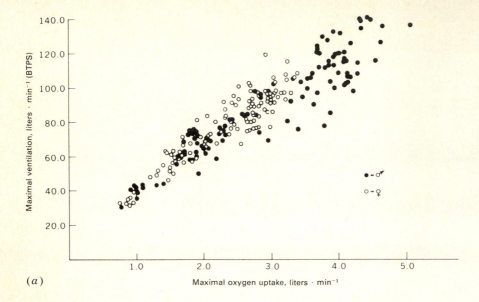

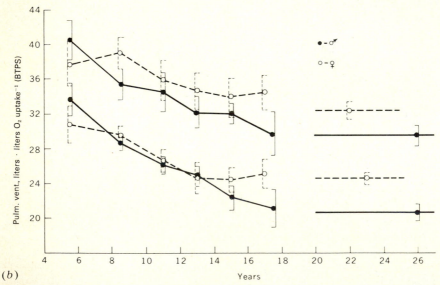

Figure 7-10 Data on 225 subjects from four to about thirty years of age. (*a*) Maximal pulmonary ventilation in relation to maximal oxygen uptake measured during running on a motor-driven treadmill for about 5 min.

(*b*) Average values of ventilation per liter oxygen uptake in relation to age. The upper curves show maximal values (attained during running); and the lower ones, submaximal values during running or cycling, with an oxygen uptake which was 60 to 70 percent of the subject's maximal aerobic power [same subjects as in (*a*)]. Vertical lines denote ± 2 SE (standard error of the mean). *(From P.-O. Åstrand, 1952.)*

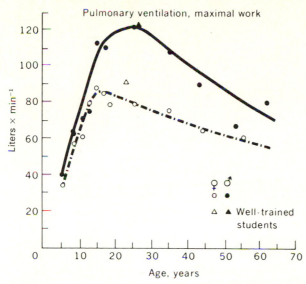

Figure 7-11 Pulmonary ventilation measured after about 5 min exercise with a work load that brought the oxygen uptake to the individual's maximum. Mean values on 350 women and men and about 80 well-trained subjects; exercise on motor-driven treadmill or bicycle ergometer. *(Based mainly on data from P.-O. Åstrand, 1952; I. Åstrand, 1960.)*

concentration of carbon dioxide and a low oxygen concentration. The total expired gas is therefore a mixture of dead space and alveolar gas, or

$$V_T = V_A + V_D$$

From a functional standpoint, this dead space is not merely the result of the anatomic features of the respiratory tract. In addition to the air volume which remains stagnant in the conductive airways, some air eventually reaches alveoles that are not at all, or are poorly, perfused by capillary blood. This reduces the gaseous exchange. In patients suffering from pulmonary disease, an unfavorable relationship between ventilation and perfusion may increase the physiological dead space.

The volume of the dead space may be estimated with the aid of Bohr's formula, which is based on the fact that the expired volume of oxygen at each respiration ($V_T \times FE_{o_2}$*) is equal to the sum of the volume of oxygen contained in the dead space compartment ($V_D \times FI_{o_2}$) and the volume of oxygen coming from the alveolar air ($V_A \times FA_{o_2}$). We therefore arrive at the following formula:

$$V_T \times FE_{o_2} = V_D \times FI_{o_2} + V_A \times FA_{o_2}$$

*F = Fraction of oxygen in the expired, inspired, and alveolar air respectively.

Since $V_A = V_T - V_D$, the formula may be simplified as follows:

$$V_D = V_T \frac{FE_{O_2} - FA_{O_2}}{FI_{O_2} - FA_{O_2}}$$

If the oxygen content of the inspired air is 21 percent, the oxygen content of the expired air 16 percent, the oxygen content of the alveolar air is 14 percent, and the depth of respiration, V_T, is 500 ml:

$$V_D = 500 \ \frac{16 - 14}{21 - 14} = 143 \text{ ml}$$

The same calculation can be made on the basis of CO_2.

With a depth of respiration (tidal volume) of 500 ml at rest, the dead space constitutes approximately 150 ml. The rest of the tidal volume reaches the alveoles. It should be noted, however, that the first portion of the inhaled air is, in reality, the respiratory air which remained in the dead space compartment from the previous respiration. The "fresh" air which is pulled down into the alveoles is diluted into a relatively large volume, i.e., the functional residual capacity. The variations in the gas concentration are therefore relatively small in the alveoles during rest and normal breathing.

Owing to methodological difficulties, it is not easy to measure the exact behavior of the dead space during exercise. Asmussen and Nielsen (1956) estimated that with a tidal volume of 3 liters, the dead space was 300 to 350 ml, whereas Bargeton (1967) concluded from his data that the increase in dead space with increasing tidal volume is very slight and can be taken as a constant for moderate changes in tidal volume. Relatively speaking, the dead space is reduced with increasing tidal volume. If, at a ventilation of 6.0 liter $\times$ min^{-1}, the respiratory frequency is 10, and the dead space 0.15 liter, the alveolar ventilation is

$$6.0 - 0.15 \times 10 = 4.5 \text{ liters} \times \text{min}$$

If the respiratory rate, on the other hand, is 20, and the gross ventilation and dead space are assumed to be unchanged, the alveolar ventilation is only

$$6.0 - 0.15 \times 20 = 3.0 \text{ liters} \times \text{min}$$

Animals which depend on evaporative heat loss from the respiratory tract for their temperature regulation avoid hyperventilation of the alveoles, thanks to a high respiratory rate and a low alveolar ventilation (panting).

From the pulmonary ventilation data which are given in Fig. 7-9, part of the volume does not participate in the gas exchange. The method of conceal-ment, often described in adventure stories, by hiding submerged in water and breathing through a snorkel represents a considerable complication of the gas exchange. The tube (snorkel) represents an extension of the respiratory dead space, and the tidal volume has to be increased by an amount equal to the volume of the tube if the alveolar ventilation is to be maintained unchanged. The breathing may therefore become very laborious.

A second complication of this diving is the increased load on the inspiratory muscles. Within the lungs, there is the same pressure as at the water surface, i.e., atmospheric pressure. The outside of the thorax, however, is subjected to atmospheric pressure *plus* the pressure of the column of water above the diver. At a depth of 1.0 m, this extra pressure will be 0.1 atm, or about 76 mm Hg. The highest pressure the inspiratory muscles can overcome is just above 70 mm Hg, and, therefore, a depth of 1.0 meter would be the maximum that can be tolerated even if the problem of extra dead space could be solved by a system of valves.

Tidal Volume–Respiratory Frequency

By definition, the pulmonary ventilation equals the frequency of breathing multiplied by the mean expired tidal volume, or

$$\dot{V}_E = f \times \overline{V}_T$$

At rest, the respiratory frequency is between 10 and 20. Inspiration occupies less than half the total cycle, the rise in flow being more abrupt than the fall. During physical work of low intensity, it is primarily the tidal volume that is increased. In many types of exercise, this may amount to 50 percent of the vital capacity when the work load is moderately heavy or heavy (Fig. 7-12). The respiratory frequency is also increased, especially in the case of heavy work. Children about five years of age may have a respiratory frequency of about 70 at maximal work, twelve-year-old children about 55, and twenty-five-year-old individuals 40 to 45 (Fig. 7-12) (P.-O. Åstrand, 1952).

The increase in tidal volume is brought about through the utilization of both the inspiratory reserve volume and the expiratory reserve volume (see Fig. 7-6). Inspiration and expiration become more equal in both time and pattern. Naturally the vital capacity limits the tidal volume, but rarely more than 50 percent of the vital capacity is utilized.

When the body was submerged in water, the VC was, in one study, reduced by 10 percent and the expiratory reserve volume was less than 1 liter as compared with 2.5 liters in air. The increase in V_T was affected in water exclusively by utilization of the inspiratory reserve volume (Holmér, 1974). Similarly, the upper limit for the respiratory frequency is determined by the rate at which the neuromuscular system can generate alternating movements. Studies have indicated that an individual spontaneously balances the depth of respiration and respiratory frequency in such a way that a certain ventilation takes place at optimal efficiency, that is, with the utilization of a minimum of energy by the respiratory muscles (Milic-Emili et al., 1960). The greater the

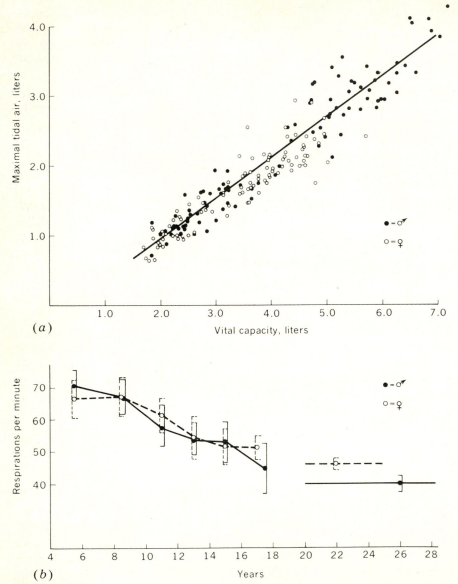

Figure 7-12 (a) Highest tidal volume measured during running at submaximal and maximal speed (work time about 5 min) related to the individual's vital capacity measured in standing position. Altogether, 190 subjects from seven to thirty years of age. On an average, 50 to 55 percent of the vital capacity is used as maximal tidal air.
(b) Respiratory frequency during running at a speed that brings the oxygen uptake up to maximum; average values (± 2 SE) for 225 subjects from four to thirty years of age. *(From P.-O. Åstrand, 1952.)*

pulmonary ventilation, the narrower the range of respiratory frequencies appears to be, yielding minimal energy expenditure (Otis, 1964). In athletic performances, it is therefore advisable to allow the athlete to assume the respiratory pattern which seems natural for him. In many types of physical work, the respiratory frequency tends to become fixed to the work rhythm. Needless to say, this certainly holds for crawl swimming, but it also holds for such activities as bicycle riding, sculling, and running. Therefore, the ventilatory pattern is not exclusively guided by demand for minimal energy expenditure of the respiratory muscles (Flandrois et al., 1961). This is, however, natural since the thoracic cage is also highly affected by muscles other than the true respiratory ones.

In swimming, instruction in the breathing technique may be necessary, but in other cases, the respiratory pattern should be allowed to follow its natural pattern. It should be noted that work on the bicycle ergometer with a pedaling frequency of, for example, 50 usually gives a respiratory frequency related to this pedaling frequency. It often increases stepwise from 12.5 to 16.6, 25.0, 33.0, up to 50 at maximal work. It is important to keep this in mind if one is studying the mechanics and regulation of breathing during work, using only the bicycle ergometer as a means of providing the work load.

Summary The respiratory frequency at which the respiratory work is minimal increases progressively with increasing ventilation. The respiratory frequency at each level of ventilation usually corresponds to the frequency which is spontaneously chosen by the subject. How this regulation is brought about is not known. Probably various receptors, including muscle spindles and the gamma system, also play an important role in the adjustment of the respiratory frequency to the work rhythm.

Respiratory Work (Respiration as a Limiting Factor in Physical Work)

The work of the respiratory muscles consists primarily of overcoming the elastic resistance and the flow-resistive forces. A precise determination of the mechanical efficiency of breathing is not simple, and data in the literature range from a few percent to about 25 percent (Milic-Emili and Petit, 1960). At rest, the respiratory muscles require from 0.5 to 1.0 ml O_2/liter of ventilation. With increasing ventilation, the oxygen cost per unit ventilation becomes progressively greater. It has been estimated that the respiratory muscles during heavy work may tax as much as 10 percent of the total oxygen uptake (Liljestrand, 1918; Nielsen, 1936; Otis, 1964).

A question of considerable importance is whether or not hyperventilation may limit the oxygen uptake capacity. The answer is probably negative for the following reasons: (1) After the maximal oxygen uptake is reached, it is still possible for the subject to continue to work at a higher work load because of the anaerobic processes. At the same time, the pulmonary ventilation is markedly increased, without any distinct ceiling being reached (Fig. 7-9). (2) At an extremely heavy work load which can be tolerated for only a few minutes at the most, the pulmonary ventilation is greater than at a somewhat lower but still maximal load, which may be tolerated for about 6 min. The oxygen uptake is nevertheless the same in both cases (P.-O. Åstrand and Saltin, 1961). (3) At maximal work, it is possible voluntarily to increase the ventilation further, showing that the ability of the respiratory muscles to ventilate the lungs evidently is not exhausted during spontaneous respiration. (4) At heavy work loads, the alveolar oxygen tension increases and the carbon dioxide tension decreases, a fact that indicates an effective gas exchange in the lungs. The oxygen tension of the arterial blood is maintained or only slightly reduced (Fig. 7-13).

These considerations refer to bicycling and running. The situation appears to be comparable for these two types of work.

During running on the treadmill, a group of about 40 male students attained a mean ventilation of 111 liters $\times$ min^{-1} with an oxygen uptake of 4.04 liters $\times$ min^{-1}. During maximal work on the bicycle ergometer, the group's values were 116 and 4.03 liters $\times$ min^{-1} respectively. In a group of about 40 girls, the pulmonary ventilation was 90 and 88 liters $\times$ min^{-1} respectively (P.-O. Åstrand, 1952). In other types of work, the situation may be different, as in strenuous work with the arms, which may hamper free respiration. The same may be true during swimming. During maximal swimming, the mean values for pulmonary ventilation were significantly lower than in maximal running and cycling, according to a study reported by Holmer (1974). Also, in maximal swimming, tidal volume was of the same magnitude as in maximal running, while respiratory rate, apparently governed by the swimming stroke rate, was lower. Despite a reduced maximal pulmonary ventilation in swimming compared with running (in both absolute and relative terms) the oxygen tension and oxygen content in arterial blood were similar in the two types of exercise (but the maximal oxygen uptake was 15 percent lower in swimming) (Holmér, 1974).

To some extent, the pulmonary ventilation may be a limiting factor even though the maximal capacity of the respiratory muscles is not fully taxed. Thus, if the energy demand of the respiratory muscles, in order to increase the pulmonary ventilation, necessitates such a marked increase in the oxygen consumption that all the achieved increase in oxygen content of the alveolar air (and increase in oxygen content of the arterial blood) is entirely utilized by the respiratory muscles themselves, then none of this extra oxygen will benefit the rest of the working muscles of the body. In other words, an increase in pulmonary ventilation beyond a certain point would not be physiologically useful, since all the additional oxygen thus gained would be required for the work of breathing (Otis, 1964). It is even conceivable that the oxygen utilization by the respiratory muscles may become so great that the oxygen supply to other tissues is reduced. However, such a critical limit is probably not reached in normal individuals. It is likely that the blood flow in the vessels of the respiratory muscles is maximal even at a ventilation below the maximal ceiling and that the oxygen content of the blood is more or less completely extracted.

A further increase in ventilation beyond this point is probably met by anaerobic processes. A point in favor of this view is the fact that the oxygen uptake reaches a distinct plateau during extremely heavy work, even if the work load, and thereby also the pulmonary ventilation, is further increased (Fig. 7-9). Even though the full capacity of the respiratory muscles may not be utilized during heavy work, the maximal force which these muscles may develop is limited by the rate at which chemical energy can be transformed into mechanical energy.

DIFFUSION IN LUNG TISSUES, GAS PRESSURES

The role of respiration is to provide the gaseous exchange between the blood and the ambient air. This is accomplished by the flowing of blood through

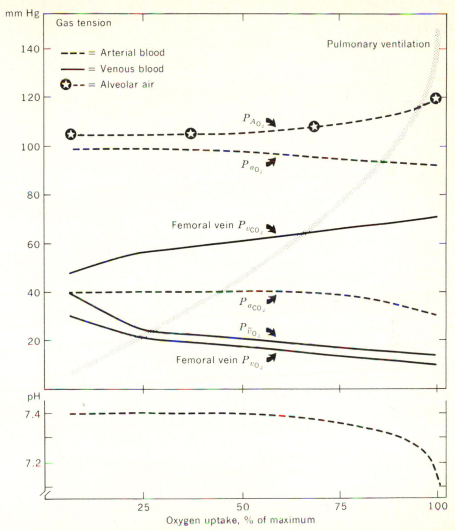

Figure 7-13 Oxygen and carbon dioxide tensions in blood and alveolar air at rest and during various levels of work up to and exceeding the load necessary to reach the individual's maximal oxygen uptake (= 100 percent). At bottom, arterial pH. Curves are based on data from different authors and unpublished studies. (The line denoting pulmonary ventilation refers to an individual with maximal aerobic power of 4.0 liters · min^{-1} if the figures on the ordinate are valid.)

capillaries of extremely small caliber; these are located only a few microns from the alveolar air, which is a derivate of the ambient air (Fig. 7-3). The gas exchange between the capillary blood and the alveolar air is achieved by the process of diffusion (for details, see Forster and Crandell, 1976; West, 1974).

This diffusion takes place as a movement of gas molecules from a region of higher to one of lower chemical activity. The partial pressure of the gas is a measure of this activity. The normal pressure of oxygen, carbon dioxide, and nitrogen in atmospheric air (P_{Bar} = 760 mm Hg), in alveolar air, and in mixed venous blood and arterial blood at rest is given in Fig. 7-14. (If, for example, the oxygen concentration in the alveolar air is 15 percent of the dry gas, its partial pressure is

$$P_{O_2} = {}^{15}/_{100}(760 - 47) = 107 \text{ mm Hg}$$

since the partial pressure of water vapor is 47mmHg.)

Blood flow through a tissue is not always determined by the metabolic activity in the tissue in question. Thus, the oxygen uptake of tissues such as the kidneys and skin is small compared with the magnitude of the blood flow through these tissues. For this reason, the partial pressure of oxygen in the venous blood remains high and the CO_2 pressure relatively low. Comparatively more O_2 is utilized in the muscle, and here a greater amount of CO_2 is produced, so that the partial pressures of these gases in the venous blood are different from those of the above-mentioned organs.

It should be noted that the total gas pressure in venous blood is considerably lower than in arterial blood (706 mm Hg as against 760 mm Hg; see Fig. 7-14). In this way, accumulation of gas in the intrapleural space in the thorax is avoided, despite the opposed recoil of the lungs and chest wall. If gas is trapped behind an occlusion of an airway, it becomes absorbed into the pulmonary circulation because of this subatmospheric gas pressure of venous blood. Under normal conditions, the diffusion processes are so rapid that the gases in the blood leaving the pulmonary capillaries are approximately in equilibrium with the gases in the alveoli. The gas exchange between pulmonary air and blood is achieved entirely by the process of diffusion. No other processes, such as secretion, are involved. An analysis of the gas concentrations and pressures in the expired air at the end of the expiration gives an approximate idea of the gas pressure in the arterial blood. One may simply collect the last portion of the expiratory air volume, either at the end of a single forced expiration (Haldane-Priestley method) or from several repeated respiratory cycles (end-tidal sampling technique), for an analysis of the gaseous composition of expired air. With the aid of modern analytic and registration techniques, it is also possible to follow variations in gas concentrations and pressures continuously during one or several successive respirations (for example, with the aid of a mass spectrometer). Figure 7-15 gives examples of the variations in CO_2 and O_2 pressures in the air in the trachea and in the alveoli during a single respiration. At rest, the variations in the composition of the gases in the alveoli are small because the inhaled air volume is diluted into a relatively large gas volume, the functional residual volume (FRV). During work, when the depth of respiration is increased, the variations become considerably larger.

Because of the length of the respiratory tract, the gas movement during respiration may be considered as a mass movement of gas flow. For the distribution within the small lung units, a molecular diffusion is the main determinant. Because of the small dimensions of the alveoli, a complete mixing within the alveolus probably occurs in less than 0.01 s. The rapid equalization of the gas pressures during the passage of the blood around the alveoli is evident from Fig. 7-16. At rest, the time it takes for the blood to pass the capillary is somewhat less than 1 s, but after 0.1 s the diffusion of the CO_2 already has reached an equilibrium. After a further few tenths of a second, the O_2 has also reached an equilibrium. Thus, during normal resting conditions, the blood in the pulmonary capillaries is almost completely equilibrated with the alveolar oxygen and carbon dioxide pressures. The size of the CO_2 molecule is larger

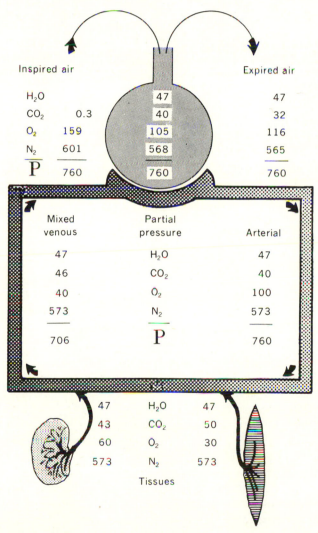

Inspired air		Alveolar	Expired air
H_2O		47	47
CO_2	0.3	40	32
O_2	159	105	116
N_2	601	568	565
P	760	760	760

Mixed venous	Partial pressure	Arterial
47	H_2O	47
46	CO_2	40
40	O_2	100
573	N_2	573
706	P	760

47	H_2O	47
43	CO_2	50
60	O_2	30
573	N_2	573
	Tissues	

Figure 7-14 Typical values of gas tensions in inspired air, alveolar air (encircled), expired air, and blood, at rest. Barometric pressure, 760 mm Hg; for simplicity, the inspired air is considered free from water (dry). Tension of oxygen and carbon dioxide varies markedly in venous blood from different organs. In this figure, gas tensions in venous blood from the kidney and muscle are presented.

than that of the O_2 molecule, which actually slows the rate of diffusion. On the other hand, the CO_2 is about 25 times more soluble in liquids than the O_2, so that the net effect is that the CO_2 diffuses about 20 times more rapidly in aqueous liquids than does oxygen. Both CO_2 and O_2 are carried by the blood mainly in reversible chemical combinations. The hemoglobin plays an overwhelming role in this transportation. In the exchange of the respiratory gases with the blood in the lungs, the primary chemical reactions of these gases occur within the red cell. The "barriers" which have to be passed are the red cell membrane, the plasma, the capillary endothelium, the basement membrane, the interstitial tissue, and the alveolar basement membrane and epithelium (Fig. 7-3). However, the process is very rapid, as shown in Fig. 7-16. Even during heavy work, when the transit time in the capillaries may be only 0.5 s or even less, it may be assumed that a gaseous equilibrium has been reached. In the narrow capillaries, there is hardly any plasma between the red cell and the endothelium, a situation that facilitates the diffusion.

Even though the diffusion rate is sufficiently high, chemical processes are still essential if an adequate volume of gas is to pass the pulmonary membrane. Carbonic anhydrase plays an important role in the exchange of CO_2. If this were not present, the blood would have to remain in the capillaries for almost 4 min for all the CO_2 to be given off. Furthermore, when the CO_2 diffuses into the red cell and then forms protons acting on the hemoglobin to displace the O_2, the exchange of O_2 is thus affected even by the CO_2, as well as by the hemoglobin itself.

The *diffusing capacity of the lung* (D_L) is defined as the number of milliliters of a gas at STPD (Standard Temperature, 0°C, Pressure 760 mm Hg, Dry) diffusing across the pulmonary membrane per minute and per millimeter of mercury of partial pressure difference between

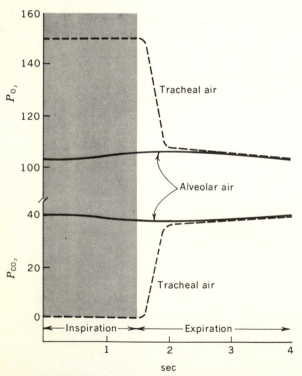

Figure 7-15 Variations in oxygen and carbon dioxide tensions in tracheal air and alveolar air during one single breath at rest. Note the very small fluctuations in gas tensions of the alveolar air.

the alveolar air and the pulmonary capillary blood. It should be emphasized that the diffusion path includes the blood. The diffusing capacity of the lung for the respiratory gases provides an index of the dimensions of the pulmonary capillary bed, the pulmonary membrane, and the overall efficiency of the system in the exchange of respiratory gases. Certain technical difficulties limit the possibility for an exact estimation. Usually the $D_{L_{O_2}}$ is calculated from studies using CO. From a resting value of 20 to 30 ml $O_2 \times min^{-1}$ and a millimeter of mercury mean pressure difference between the alveolar O_2 pressure and the pulmonary capillary O_2 pressure, the $D_{L_{O_2}}$ increases toward 75 ml in individuals with an O_2-uptake capacity of about 5 liters $\times min^{-1}$. (See West, 1974.)

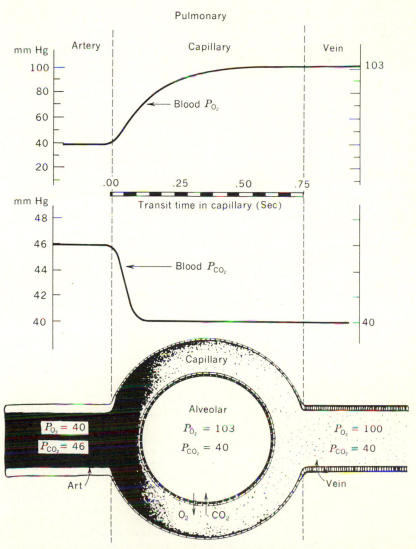

Figure 7-16 Change in the partial pressures of oxygen and carbon dioxide as blood passes along the pulmonary capillary. Note that in the first part of the capillary, the blood is already equilibrated with the alveolar gas. *(From Cherniack and Cherniack, 1961.)*

VENTILATION AND PERFUSION

The difference in P_{O_2} between the alveolar air and arterial blood depends on several factors: the membrane component plays a role; a certain amount of admixture of bronchial and cardiac venous blood occurs; and finally, there is the effect of the passage of some blood through poorly ventilated alveoli.

Since CO_2 diffuses about 20 times more rapidly than does O_2, one cannot speak of any diffusion obstacle for CO_2. The inhaled air is not equally distributed to all the alveoli, and the composition of the gases is therefore not uniform throughout the lungs. The pulmonary capillary bed has a common blood supply, the mixed venous blood, but different areas of the lungs have an uneven perfusion. The composition of the gas in various parts of the alveolar space depends on the ventilation as well as on the blood flow, or the ratio $\dot{V}_A/\dot{Q}$.

Under extreme conditions, the $\dot{V}_A/\dot{Q}$ ratio may vary from zero (when there is perfusion but no ventilation) to infinity (when there is ventilation but no perfusion). When the ratio is zero, the tensions of O_2 and CO_2 of the arterial blood are equal to those in mixed venous blood, since there is no net gas exchange in the capillaries. In the latter case, no modification of the inspired air takes place. Although these extreme situations rarely occur under normal conditions, the various parts of the lung have a wide range of ventilation-perfusion ratio. The "alveolar air" actually represents various contributions from several hundred million alveoli, each possibly having slightly different exchange ratios and gas composition.

In other words, the alveolar gas tensions vary from moment to moment and from place to place within the lungs because of regional inhomogeneity of ventilation and blood perfusion; the supply of air and blood is not perfectly matched in the lungs, even in a healthy individual in any posture (Rahn and Farhi, 1964; West, 1974).

There are mechanisms which to some extent compensate for an uneven ventilation in relation to the blood flow in the capillary bed of the alveoli: (1) inadequately ventilated alveoles have a low P_{O_2}, which in turn causes an alveolar vasoconstriction and reduced blood flow; (2) reduced blood flow produces a reduction in the alveolar P_{CO_2}, which causes a constriction of the bronchioles and, therefore, reduced gas flow. It appears, however, that there is no effective regional variation in the vasomotor tone, and it is apparent that the matching of ventilation against flow is not perfect. The alveolar-arterial oxygen pressure difference of a few millimeters of mercury which actually exists even in normal individuals is primarily a consequence of this unequal distribution. In the lungs, the blood flow is greatly affected by body position. In the discussion of the lung volumes, mention was made of the fact that the vital capacity increases in the standing position, compared with the lying position, owing to the reduction in blood volume in the thorax in the upright position. The effect of gravity on the distribution of the blood, as well as on the perfusion, is such that in the upright position, the perfusion per unit lung volume is about 5 times greater at the base than at the apex of the lung. It is true that the ventilation per unit lung volume changes in the same direction, but only slightly. As a result the $\dot{V}_A/\dot{Q}$ ratio becomes much higher in the upper lobes than in the lower lobes in the erect posture, or above 3 at the top and below 1 at the bottom of the lungs (West, 1962). This is illustrated by Fig. 7-17. As indicated, a high $\dot{V}_A/\dot{Q}$ ratio means either an overventilation or an underperfusion. The result is a highly varying composition of the alveolar air in the different parts of the lung. It has been calculated that the P_{O_2} of the uppermost alveoli may be as high as above 130 mm Hg and of the lowest alveoli below 90 mm Hg. Once again it is apparent that the lung cannot be considered as a homogeneous unit. It should be pointed out that overventilated alveoli cannot completely compensate for a desaturation of arterial blood from underventilated alveoli. The shape of the O_2 dissociation curve is such that an increase in arterial P_{O_2} above the "normal" 100 mm Hg adds little to the oxygen content of the blood, but a reduction in P_{O_2} will effect the oxygen content relatively more.

When the position of the body is changed from upright to supine, the perfusion of the upper lung zone increases markedly at the expense of the lower zone. In the recumbent position, the calculated $\dot{V}_A/\dot{Q}$ therefore becomes quite uniform in the different pulmonary lobes.

Even during light work in the sitting or standing position, the $\dot{V}_A/\dot{Q}$ ratio also becomes more uniform throughout the lungs. Certainly both upper and lower zone blood flows increase, but the former increases relatively more. The slight increase in pulmonary artery pressure that accompanies exercise is one factor that changes the balance between arterial, capillary, and venous pressures on one side and the pressure outside the vessels on the other, favoring the perfusion of the upper zones of the lungs. The pressure within the pulmonary artery may vary between 7 and 20 mm Hg during cardiac cycle at rest but between 15 and 35 mm Hg during exercise with an oxygen uptake of 2.0 liters × min⁻¹. The systolic pressure may exceed 50 mm Hg during maximal work. These pressure variations will actually mean that the distribution of the blood flow to different parts of the lungs will vary not only with body position but also with the cardiac cycle. Even during heavy work, some parts of the lungs may be unperfused during part of the diastole.

What is the effect of this difference in the topographical distribution of air and blood flow? In the final analysis, this interference with the overall gas exchange is, surprisingly, rather insignificant. West (1965) states that the lung, with its uneven distribution (Fig. 7-17), wastes only some 3 percent of its ventilation and 1 percent of its blood flow, compared with an ideal lung. Wasted ventilation, including the effect of dead space, interferes particularly with CO_2 exchange, whereas wasted blood flow affects mainly the oxygen, and the effect may be calculated to be an impairment of the exchange of these gases by a few percentage points.

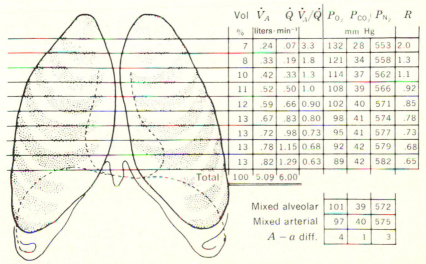

Vol	$\dot{V}_A$	$\dot{Q}$	$\dot{V}_A/\dot{Q}$	P_{O_2}	P_{CO_2}	P_{N_2}	R
%	liters·min⁻¹			mm Hg			
7	.24	.07	3.3	132	28	553	2.0
8	.33	.19	1.8	121	34	558	1.3
10	.42	.33	1.3	114	37	562	1.1
11	.52	.50	1.0	108	39	566	.92
12	.59	.66	0.90	102	40	571	.85
13	.67	.83	0.80	98	41	574	.78
13	.72	.98	0.73	95	41	577	.73
13	.78	1.15	0.68	92	42	579	.68
13	.82	1.29	0.63	89	42	582	.65
Total	100	5.09	6.00				

Mixed alveolar	101	39	572
Mixed arterial	97	40	575
$A - a$ diff.	4	1	3

Figure 7-17 Effects of observed distribution of ventilation and perfusion on regional gas tension within the lung of a normal man in sitting position. The lung is divided into nine horizontal slices, and the position of each slice is shown by its anterior rib marking. Table shows relative lung volume (Vol), ventilation ($\dot{V}_A$), perfusion ($P\dot{Q}$), ventilation-perfusion ratio ($\dot{V}_A/\dot{Q}$), gas tensions (P_{O_2}, P_{CO_2}, P_{N_2}), and respiratory exchange ratio (R) of each slice. Lower table shows differences between mixed-alveolar and mixed-arterial gas tensions which would result from this degree of nonuniformity of $\dot{V}/\dot{Q}$ ratios. *(From West, 1962.)*

It is not clear to what extent the oxygen uptake may be affected by these factors during maximal work. It is evident, however, that an increased gravitational field as well as a drop in the blood pressure regardless of the cause, i.e., a vasovagal syncope, would more seriously affect the $\dot{V}_A/\dot{Q}$ ratio and therefore the gas exchange in the lungs (Bjurstedt et al., 1968).

Summary Considerable regional inequality in ventilation of the alveoli and in blood perfusion of the capillaries exists in the lungs, causing differences in the gas exchange in different parts of the lungs. At rest in the supine or prone position, however, the ventilation distribution is rather well adjusted to follow this perfusion. In the erect posture, hydrostatic forces cause a progressive decrease in perfusion from the bottom to the top of the lung without corresponding variations in the ventilation. Therefore, the upper lobes are relatively underperfused. During exercise, the $\dot{V}_A/\dot{Q}$ ratio becomes more uniform, and the increase in pulmonary arterial pressure is at least one important contributor to this change.

OXYGEN PRESSURE AND OXYGEN-BINDING CAPACITY OF THE BLOOD

Figure 6-16 illustrates how the oxygen-binding capacity of the blood is affected by the partial pressure of oxygen in the blood. At $P_{O_2} = 100$ mm Hg, 98 percent of the hemoglobin is normally saturated with oxygen. The oxygen saturation curve is such that about half the hemoglobin is in the form of HbO_2 and half in the form of reduced Hb at P_{O_2} in the order of 26 mm Hg ($37.0°C$, pH 7.40). The HbO_2 dissociation curve of the blood is affected by the CO_2 pressure, the pH, and blood temperature. Also, salts have their effects, a fact that at least partially explains the difference in curves for various species of animals. P_{CO_2} affects pH and thereby the oxygen saturation curve, but it also has a specific effect in that CO_2 combines with Hb to form carbamino compounds ($HbNH_2 + CO_2 \rightleftharpoons HbNHCOOH$), thereby reducing the capacity of Hb to bind oxygen (Fig. 5-3). During muscular work the CO_2 increases, as does the temperature locally in the muscle, while at the same time the pH is lowered. This causes the liberation of O_2 to increase at a given O_2 tension. Because of this feature of hemoglobin, an effective O_2 diffusion gradient may be maintained between the capillaries and the O_2-utilizing cell. With regard to the effect of an increase in temperature, the diffusion increases about 2 percent/$°C$. These factors aid in the unloading of oxygen to the active muscles, but are, however, of less quantitative importance than the opening of additional capillaries in the muscles. In this manner the distance between the capillary and the muscle cell is reduced, which shortens the diffusion distance. Owing to the shape of the oxygen saturation curve, a change in pH, P_{CO_2}, and blood temperature plays a relatively small role at a P_{O_2} around 100 mm Hg. During very strenuous exercise, the oxygen saturation of arterial blood may, however, be reduced below 95 percent without a corresponding decrease in P_{O_2} (Fig. 6-17). At high altitude, where the alveolar P_{O_2} is lower, the arterial saturation is still more affected in a negative direction by a lowered pH and increased P_{CO_2} and temperature.

An organic phosphate compound, diphosphoglycerate or 2,3-DPG, is normally present in the red cells. Its effect is a shift of the O_2 dissociation curve to the right, thus assisting the unloading of O_2 to peripheral tissues. An increase in 2,3-DPG occurs in chronic hypoxia (Martin et al., 1975). (For ref. see Adamson and Finch, 1975.)

Each Hb molecule has four iron atoms. In reality, the ratio between Hb and HbO_2 may vary as follows: Hb_4, Hb_4O_2, Hb_4O_4, Hb_4O_6, Hb_4O_8, in which Hb_4 is the completely reduced hemoglobin. Only in the case of Hb_4O_8 is it 100 percent saturated. The structure is such that the heme groups do not influence one another's activity, and there is no interaction between them. This means that under normal conditions, these combinations are mixed in proportions which are determined by such factors as P_{O_2}. As far as the volume CO_2 bound to Hb in the carbaminohemoglobin form is concerned, it varies with the degree of O_2 saturation and is greater at low O_2 saturation and less at high O_2 saturation. This may seem like a fortunate coincidence: when the O_2 uptake is great, the O_2 saturation becomes low. At the same time, CO_2 is formed and the capacity to transport CO_2 from the tissues to the lungs is increased. The transport of CO_2 and O_2 was also discussed in Chap. 5. In this connection, it should be recalled that reduced hemoglobin is a weaker acid than oxyhemoglobin, and when reduced, it mops up H^+ ions more readily and thereby helps to prevent too large a reduction in pH through the formation of acid metabolites.

The lower part of the oxygen dissociation curve is, so to speak, a reserve which is utilized during muscular work or pathological conditions. There are very few reported measurements of oxygen content and partial pressure of oxygen in a vein draining a muscle which is working at maximum. Saltin et al. (1968) report from studies on four subjects working at maximum for about 5 min on a treadmill a P_{O_2} that averaged 12 mm Hg in blood collected from the femoral vein (O_2 content, 1.4 vol percent; pH, 7.09; and P_{CO_2}, 70 mm Hg).

Myoglobin in the muscle cell, particulary rich in the slow twitch fibers (Type I fibers) facilitates the O_2 diffusion within the muscle, and this aid to the oxygen transport is most significant at low oxygen pressures near the capillary (Scholander, 1960; Forster, 1967; Wittenberg et al., 1975).

From an O_2 content of 20 vol percent in arterial blood, it may eventually drop during heavy work to below 1 vol percent because of the abundance of capillaries of the muscle tissue with short diffusion distances, a pH around 7.0, and a temperature exceeding 40°C. The mitochondria are apparently working efficiently even if the intracellular P_{O_2} is less than 1 mm Hg; at the end of a capillary in a working muscle, the O_2 tension may probably be less than 10 mm Hg (Chance et al., 1964; Forster, 1967; Stainsby and Otis, 1964).

REGULATION OF BREATHING

The object of breathing is to ensure the exchange of oxygen and carbon dioxide between the blood and atmospheric air. It would therefore appear logical if these gases were to partake in the regulation of the breathing. Such is actually the case. A change in the P_{O_2}, P_{CO_2}, and H^+ concentration in the arterial blood results in a change in ventilation in such a manner as to moderate the primary change (negative feedback). Stretch receptors in the lungs and muscles affect respiration as well as a number of other factors. As in the case of changes in the circulation under different conditions, the question is often: What is the regulating, and what is the disturbing, effect? In respect to P_{O_2}, P_{CO_2} and H^+ concentration in the blood, a change in pulmonary ventilation may influence these factors. The temperature of the blood affects respiration, but a change in pulmonary ventilation does not cause a substantial change in the body temperature of the human. In some animals, such as the dog, respiration also plays a part in temperature regulation. The effect of an increased temperature of the blood in their case also has a much more pronounced effect on respiration than in the case of human beings. Adrenalin affects respiration, but a change in ventilation certainly does not affect the content of adrenalin in the blood. Emotion affects respiration (as in the sighing of a mourning person, the rapid breathing of an excited person). A number of different theories have been advanced concerning the regulation of breathing, but none of them has as yet fully explained how the respiratory volume is adjusted to meet the demand at rest and during physical work. The interested student will find a series of review articles in Fenn and Rahn, 1964, 1965; Kao, 1972; Widdicombe, 1974.

Rest

In the central nervous system, a large number of physical, chemical, and nervous variables are integrated. P_{O_2}, P_{CO_2}, and H^+ concentration, or chemical changes related to them, appear to be the most prominently controlled chemical variables. A lowering of the arterial P_{O_2} stimulates the breathing via *peripheral chemoreceptors* in the carotid and aortic bodies whence impulses go to respiratory centers in the medulla. An increase in P_{CO_2} and H^+ concentration also represents a stimulation leading to an increased ventilation, but this effect is primarily elicited from *medullary chemosensitive receptors*, located on the ventral surface of the medulla, bathing in the cerebrospinal fluid (CSF) separated from the blood by the blood-brain barrier.

In the Medulla oblongata there are separate clusters of neurons widely dispersed but it has not been demonstrated that there are automatically well localized concentrations of expiratory and inspiratory neurons respectively. However, there are respiratory neurons with an inherent rhythmicality, but so far nobody knows where this rhythm originates. An increased activity in the inspiratory generator stimulates the inspiratory muscles and simultaneously causes an inhibition of the neurons to the expiratory muscles. This activity also projects to pools of neurons which command an "off-switch" signal to the inspiratory generator and the inspiration terminates (see Bradley et al., 1975). The relaxing inspiratory muscles permit an expiration (passive at rest). With stronger stimuli of the respiratory centers, such as during exercise, there is a more frequent and forcible alternate activation of the inspiratory and expiratory muscles.

The inspiratory generator inhibits itself by discharging impulses to a *pneumotaxic center* in the pons. This, then, dampens the activity in an *apneustic center*, also located in the pons, which otherwise stimulates the inspiratory generator by spontaneous activity. There are other sources of inhibition of this apneustic center, with a reduced or absent stimulus of the inspiratory neurons as a consequence. When the lung tissues are stretched in the course of inspiration, stretch receptors in the inspiratory bronchioles are stimulated and, via afferent vagus fibers, inhibit the apneustic center, and the inspiration is stopped (the Hering-Breuer reflex). If the vagus is cut in an animal, the respiratory frequency becomes slower and the respiratory depth is increased. The same occurs if the connection between the pneumotaxic center and the lower areas is severed.

Numerous neurons discharging with respiratory modulations can be found throughout the brain stem from caudal medulla to rostral pons.

The respiratory muscles, especially the intercostal muscles, are amply supplied with muscle spindles. Central respiratory drive descends by a common path to the spinal level, being then distributed to both α and γ motoneurons. Thereby we have the tool for a coactivation of the α–γ system. A second and separate input to the respiratory motoneurons originates from the pyramidal and extrapyramidal systems (see Chap. 4). It should be emphasized that the respiratory muscles can be volitionally controlled and they are engaged in many "nonrespiratory" activities. On the spinal level different descending and afferent inputs interact and the interneuronal networks can exert reciprocal effects on the antagonistic muscles.

Central Chemoreceptors

The pulmonary ventilation at rest is chiefly regulated by the chemical state of the blood, particularly its CO_2 tension. An increase in the H^+ and CO_2 concentration (hypercapnia) in the arterial blood causes an increased ventilation.

The chemoreceptors on the surface of the medulla are apparently activated by the extracellular concentration of H^+, i.e., the pH of the CSF (see Loeschke, 1974). (CO_2 diffuses rapidly between blood and CSF, and an increase of P_{CO_2} in the arterial blood therefore causes a lowering of pH in the CSF as well. The barrier between blood and CSF is, however, much less permeable to ions. An administration of fixed acids to the blood will elicit an increased ventilation and a secondary decrease in P_{CO_2} of the blood and CSF. Since the acids do not easily penetrate to the CSF, the effect will actually be a transient alkalosis of the CSF and a simultaneous acidosis of the blood.)

There may be just one type of central chemoreceptor which can transmit CO_2 and fixed acid effects via afferent nerve impulses to the medullary respiratory generators. These central receptors are essential for rhythmic ventilation. There are two primary responses of the respiratory control system to an increase in CO_2 (via the CSF pH): (1) an increased rise of centrally generated inspiratory activity (that will project to the inspiratory motoneurons but also to the "off-switch" mechanism; (2) an increase in the threshold of the "off-switch" neurons (with the consequence that the inspiration is not terminated until a larger tidal volume is reached) (Bradley et al., 1975).

Normally the inspired air is, practically speaking, free of CO_2 (0.03 percent). The inhalation of small amounts of CO_2 causes an increase in ventilation because the alveolar P_{CO_2}, and thereby the arterial P_{CO_2}, is increased. As long as CO_2 is being inhaled, the ventilation remains elevated, since a reduced ventilation would once more increase P_{ACO_2} above the normal level of about 40 mm Hg. At a high concentration of inspired CO_2, mental confusion occurs, and even narcotic effects with concomitant reduced ventilation may be the result. A denervation of the peripheral chemoreceptors in laboratory animals under normoxic conditions hardly alters the response of the breathing to CO_2 excess, which indicates that the effect is elicited centrally. (Wasserman et al. (1975) point out that the ventilatory response to inhaled CO_2 differs from that in which CO_2 is delivered to the lungs via the circulation. Consequently, the ventilatory response resulting from inhaled CO_2 might prove misleading when used to describe mechanisms of respiratory control.)

Peripheral Chemoreceptors

The direct effect of lack of oxygen on the central nervous system is a reduction in its function. For the stimulation of respiration by hypoxia, a stimulation of the peripheral chemoreceptors is required.

The chemoreceptors in the carotid and aortic bodies consist of epithelioid cells with a rich innervation and blood supply; their oxygen usage is exceptionally high (see Biscoe 1971).

The function of the carotid (and aortic) body can be studied by recording the impulse activity electrically in the afferent nerve connecting the chemoreceptors with the brain stem. Figure 7-18 presents an example of such a recording from a few-fiber preparation of the afferent carotid sinus nerve in a cat. It is evident that the discharge of scattered impulses increases with increasing degree of hypoxia. In the upper (a) and lower (g) tracing, the animal is exposed to an hypoxia corresponding to an altitude of 4,000 m. The discharge is relatively strong, about 90 "spikes" of a certain minimal amplitude. When the oxygen tension of the air is made equivalent to sea-level conditions, the activity in the chemoreceptors is quickly reduced to about 25 spikes, and the pulmonary ventilation decreases from 0.85 liter $\times$ min^{-1} to 0.66 liter $\times$ min^{-1} (c). At an altitude of 6,000 m, the discharge is increased, as is the pulmonary ventilation (d). After a change to oxygen breathing, the chemoreceptors almost completely cease to fire (a single spike is seen in the tracing) (f).

It has been shown conclusively that the decisive factor in the hypoxic stimulation of the chemoreceptors is a lowering of P_{O_2}, not a reduced oxygen content as such. Thus, in laboratory animals, up to 80 percent of the hemoglobin may be bound to CO without the respiration being affected, providing the P_{O_2} is kept at a normal level (Heymans and Neil, 1958).

The discharge of the chemoreceptors increases if the blood flow through them is reduced. Stimulation of sympathetic fibers to the carotid bodies as well as blood-borne epinephrine and norepinephrine may reduce the blood flow to the chemoreceptor tissue (Lee et al., 1964; Neil and Joels, 1963). Therefore, chemoreceptors may send an increasing number of nerve impulses centrally, despite an unaltered oxygen tension of the arterial blood, if there is an increased sympathetic activity, as after hemorrhage or during exercise (see below). On the other hand, there are also depressant sinus nerve efferents which may increase the blood flow and diminish the chemoreceptor activity (Biscoe, 1971).

There are considerable individual variations in the reaction to hypoxia.

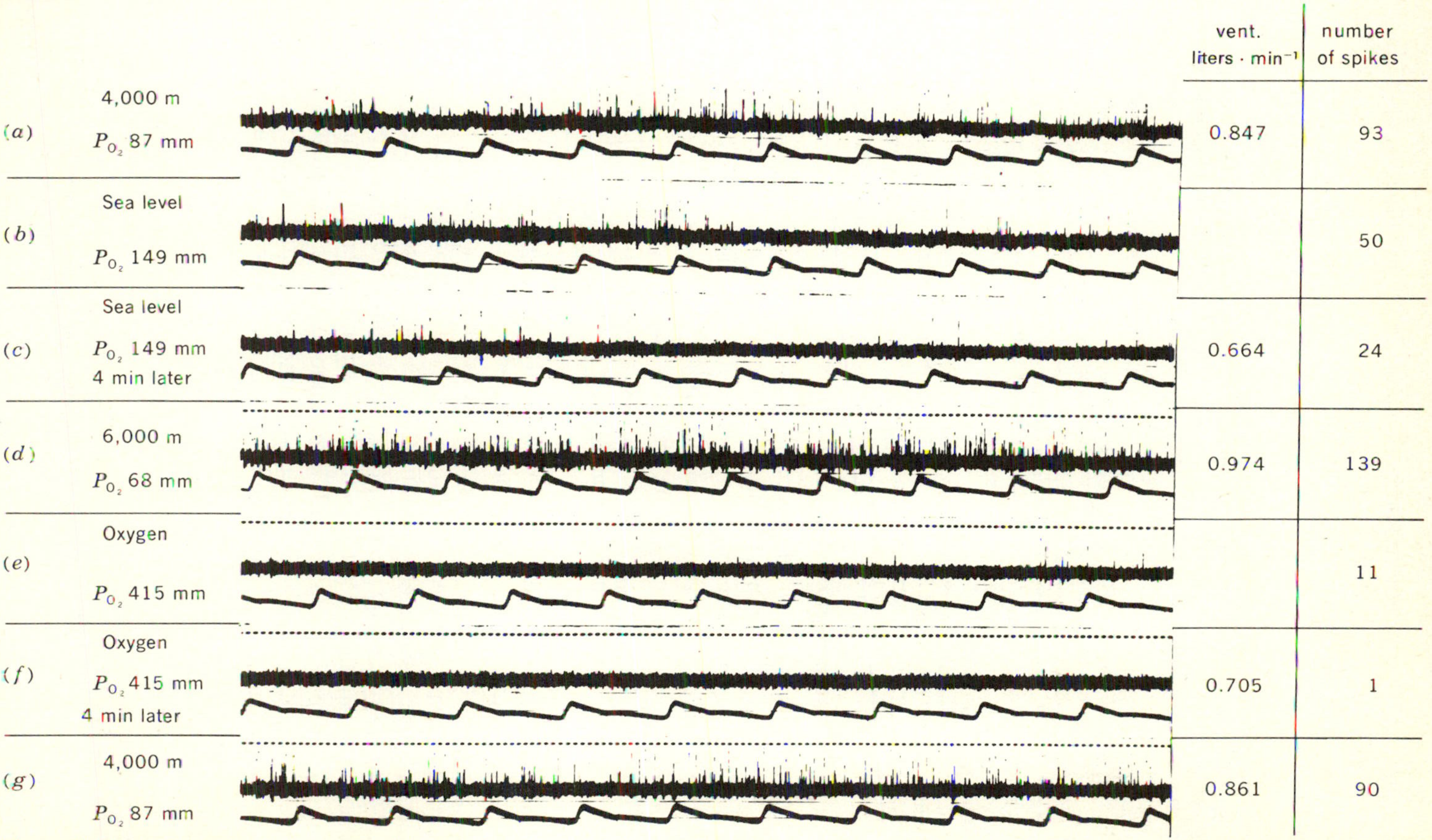

Figure 7-18 Action potentials recorded from a few fibers in the carotid sinus nerve in a cat subjected to different altitudes in a low-pressure chamber and oxygen breathing at 4,000-m simulated altitude. Tracheal oxygen tension to the left. Records on each film strip from top downward: time (50 Hz), electroneurogram and arterial blood pressure (calibration lines at 150 and 100 mm Hg respectively). Right column gives pulmonary ventilation and number of action-potential spikes of an arbitrarily chosen minimal height. Note the relation between chemoreceptor activity and ventilation. (From P.-O. Åstrand, 1954.)

		vent. liters · min⁻¹	number of spikes
(a)	4,000 m P_{O_2} 87 mm	0.847	93
(b)	Sea level P_{O_2} 149 mm		50
(c)	Sea level P_{O_2} 149 mm 4 min later	0.664	24
(d)	6,000 m P_{O_2} 68 mm	0.974	139
(e)	Oxygen P_{O_2} 415 mm		11
(f)	Oxygen P_{O_2} 415 mm 4 min later	0.705	1
(g)	4,000 m P_{O_2} 87 mm	0.861	90

Some persons may be brought to unconsciousness without pulmonary ventilation being significantly affected. In others, the pulmonary ventilation may be increased to double or more during hypoxia.

What it is that actually stimulates these peripheral chemoreceptors is not clear. If the oxygen tension in the arterial blood is high, as is the case when breathing 95 percent O_2, the arterial P_{CO_2} (by breathing CO_2) may increase and its pH drop markedly without the chemoreceptors being stimulated. So, if an hypoxic drive does not exist but the oxygen saturation is adequate, these chemoreceptors then seem rather insensitive to changes in P_{CO_2} and H^+ concentration. At low oxygen tension, the discharge from the chemoreceptors is increased further, however, if P_{CO_2} at the same time is high and the pH low (Hornbein and Roos, 1963; Neil and Joels, 1963). Thus, hypercapnia potentiates the effect of hypoxia, possibly by a change in the intracellular H^+ concentration or reduction in the blood flow to the area where the chemoreceptors are located (see Widdicombe, 1974, p. 247).

Normally, the P_{O_2} in persons breathing ordinary air at sea level is low enough to contribute a small but significant tonic ventilatory stimulus from the peripheral chemoreceptors (Dejours, 1975). Most subjects exhibit a sustained increase in pulmonary ventilation only when breathing 16 percent O_2 and a marked increase in ventilation only when the inspired O_2 has dropped to about 10 percent or less. Thus, the chemoreceptors have different thresholds, some of which are not stimulated markedly until the P_{O_2} has fallen to 40 or 50 mm Hg (8 to 10 percent O_2 inhaled). The chemoreceptors are very resistant to anoxia and can therefore maintain breathing reflexively for long periods of time.

Simultaneous Effect of Hypoxia and Variations in P_{CO_2}

An hypoxic drive via the chemoreceptors increases ventilation. The result is that more CO_2 is expired than is produced. The result of this hyperventilation is that the arterial P_{CO_2} sinks (hypocapnia) and the pH rises. This means a reduced respiratory drive via Pa_{CO_2} and H^+ concentration. If this hypoxic stimulus is now eliminated by the breathing of oxygen-rich air, the ventilation is immediately reduced. The CO_2 is thereby accumulated, which tends to offset the effect of interrupting the P_{O_2}-dependent stimulation.

Summary   The central and peripheral inputs are integrated in the respiratory centers so that P_{O_2}, P_{CO_2}, the acid-base characteristics of the blood, and the internal environment of the body are maintained at a constant level. The inspiratory and expiratory generators in the medulla oblongata are inherently rhythmic and are reciprocally coordinating the inspiratory and expiratory muscles that vary the rate and depth of breathing. Normal respiration is maintained by an interaction between these generators and neurons in the pons, including an apneustic center and a pneumotaxic center. (Inspiration is interrupted by a dampening or cessation of the stimulating effect from the apneustic center on the inspiratory generator and an activation by a "switch-off" mechanism in the medulla.) Stretch receptors in the lungs and muscle spindles in the respiratory muscles may participate in this switch from inspiration to expiration.

The H^+ ion concentration and, especially, P_{CO_2} of the blood and CSF are continuously affecting the respiratory activity in that cells in the proximity of the respiratory neurons are affected. Deviation from a normal Pa_{CO_2} of 40 mm Hg results in a ventilatory response minimizing the deviation. During hypoxia, an increase in ventilation occurs because of discharge from chemoreceptors in the carotic and aortic bodies. A number of structures in the medulla oblongata may influence respiration. Activity of the reticular formation also increases ventilation. This effect forms a link between the proprioceptors in the muscles and joints, the cortex of the brain, and different centers of importance for the motor function, as well as the respiratory neurons. Emotion, voluntary actions, and reflexes such as swallowing can easily alter the pattern of respiration.

Exercise

This topic has attracted respiratory physiologists for almost a century, and one has been searching for a "work factor" that could explain exercise hyperpnea. Efforts have been made to quantitate various factors of chemical and nervous nature with regard to their participation in the regulation of breathing during work. So far, these efforts have not been too promising. Here we shall present a somewhat different approach compared with the classical concepts, and we shall discuss a neurogenic factor as the primary activator of the respiratory muscles during exercise, with a secondary feedback mechanism of chemical nature which will regulate and adjust the respiratory volume mainly according to the composition of the arterial blood.

As is evident from Figs. 7-9 and 7-13, the ventilation increases during muscular work almost rectilinearly with the increase in O_2 uptake up to a certain level, after which the increase in ventilation becomes steeper. As we have already stated, there is a great deal of controversy concerning the mechanism underlying this increase in ventilation during muscular work. During submaximal work, the arterial P_{CO_2}, P_{O_2}, and H^+ concentrations are roughly at the same level as during rest (Fig. 7-13). During very heavy work, the H^+ concentration increases, largely as the result of lactic acid production (HLa) in the working muscles. (HLa + $NaHCO_2 \rightleftharpoons NaLa + H^+ + HCO_3^-$.) Thus pH decreases and may be as low as 7.0. The relative hyperventilation which follows gives an elevated alveolar P_{O_2}, but the arterial P_{O_2} actually drops toward 90 to 85 mm Hg as compared with the normal value of 95 to 100 mm Hg. Pa_{CO_2} drops toward 35 mm Hg or even lower values (Fig. 7-13). A similar drop in Pa_{O_2} and pH at rest should cause a rather moderate increase in ventilation, and a lowering of Pa_{CO_2} would in itself cause a reduced respiratory activity. *The chemical changes in the composition of the arterial blood cannot, then, explain per se the ventilation of 100 to 250 liters $\times$ min^{-1} observed during heavy muscular work.*

The threshold at which the ventilation increases proportionally more than the O_2 uptake does actually vary from person to person. The individual's capacity to supply the working muscles with oxygen is of decisive importance. A person who has a low maximal aerobic power reaches this threshold value at a lower O_2 uptake than does a person who has a high maximal $\dot{V}_{O_2}$ (Fig. 7-9). During work with small muscle groups (the arms, for instance), the ventilation at a given O_2 uptake is greater than it is when larger muscle groups are engaged, for instance, with work involving the leg muscles (Stenberg et al., 1967).

(During bicycle ergometer work with the arms, the ventilation was 89 liters $\times$ min^{-1} at an oxygen uptake of 2.8 liters $\times$ min^{-1}, but only 67 liters $\times$ min^{-1} when the work was performed with the legs. The lactic acid concentrations of the blood were 11 and 4 mmol $\times$ liter^{-1} respectively. The heart rate was 176 and 154 respectively.)

The possibility of impulses from motor centers in the cortex or from the working muscles and joints involved in the work to the respiratory generators has been discussed (Krogh and Lindhard, 1913; Asmussen, 1967; Kao, 1972; Cunningham, 1975). Such impulses might conceivably bring about an alteration in the threshold or set point for the sensitivity of the respiratory neuron pools for CO_2, pH, and O_2.

It should be emphasized that the cerebral hemispheres normally contribute an important component to the volume of breathing during wakefulness, and this cerebral drive will maintain the rhythm of respiration even when metabolic stimuli are temporarily removed (e.g., after a period of hyperventilation). There is a voluntary and behavioral system connected with the act of breathing located in the somatomotor and limbic forebrain structures, which adapts man to vocalization and conditions him for the expected metabolic demands of exercise (see Plum, 1970). As mentioned, the efferent impulses can descend directly to the motoneurons of the respiratory muscles, but they can also make connections with the medullary reticulum. On this lower level of the brain, there are the neural systems subserving the metabolic needs of the body. As a third system we could consider the effects on the respiratory muscles by the postural demands governed by the cerebellum.

Campbell (1964), von Euler, 1974; Granit, 1975, and other investigators have emphasized the importance of the muscle spindles for the activity of the respiratory muscles. Campbell points out that the respiratory muscles are voluntary muscles subject to all the spinal and supraspinal mechanisms that affect tone, posture, and movement. In the anterior horn cells, there is an integration of respiratory and nonrespiratory drives. Muscle spindles are numerous in the intercostal muscles but much less numerous and less evenly distributed in the diaphragm. It should be recalled (from Chap. 4) that an increased activity in the fusimotor γ fibers produces a contraction of the muscle spindle's (intrafusal) muscle, and if the parent muscle (the extrafusal muscle) at the same time is not activated via its α fibers, the receptors in the muscle spindles become deformed and produce an afferent impulse which, via the dorsal root, reaches the spinal cord. This afferent impulse stimulates the α motoneuron, the entire muscle is activated, and the difference in length between the receptor muscle and the muscle as a whole is reduced sufficiently so that the afferent signals cease. This γ motor spindle system represents a "follow-up-length-servo" system. Thus, any difference in length change between intra- and extrafusal muscles is perceived by this receptor of the α motoneuron through excitatory synapses so that the misalignment will be reduced and the demanded length, and therefore the demanded volume, will be assumed. The importance of this mechanism for ventilation is shown by the fact that adding a resistance on the inspiratory side increases the force of contraction of the inspiratory muscles, thanks to an augmentation of their motoneuron discharge by the stretch reflex. In other words, the afferent nerves are a feedback design leading from the muscle spindles which provides information as to how successfully the respiratory muscles have accomplished their task.

It has been shown that the α and γ motoneurons of the inspiratory and expiratory muscles are driven from the respiratory integrating mechanisms in the medulla in a reciprocal fashion (Eklund et al., 1964). Changes in chemical respiratory drive exert reciprocal effects on the expiratory and the inspiratory α and γ excitability. In reality, the γ activity in the respiratory muscles is more dominating than the α activity, so that the intrafusal muscles shorten relatively more than the extrafusal muscles, and thereby the muscle spindles exert a reflex drive on the α motoneurons to achieve the demanded change of volume and pulmonary ventilation. During breathing at rest, the muscle spindles in the intercostal expiratory muscles are passive during expiration but discharge moderately when they are stretched during inspiration. During forced respiration, as during work, the muscle spindles are "called to life" even during expiration and stimulate, by reflex, the motoneurons of the expiratory muscles, at the same time inhibiting the antagonists, that is, the intercostal inspiratory muscles and actually also the diaphragm.

This system of γ and α motoneurons, linked together via the afferent nerves of the muscle spindles and affected by supraspinal mechanisms, especially the reticular formation, may form the basis for the regulation of the ventilation of the lungs. In connection with work, the activity in the reticular formation is increased, partially through impulses from cortex cerebri and other higher centers, and partially through afferent impulses from the muscle spindles in the working muscles, etc. Since the reticular formation activity is unaffected by the conditions in respiratory generators, it is conceivable that this may increase the activity also in the γ and α motoneurons to the respiratory muscles, with an increased volume change in the thorax as a consequence. It would be natural if the respiratory movements and respiratory frequency were adjusted according to the work rhythm, which usually is the case: the respiratory frequency is adjusted in accordance with stride frequency, pedal frequency in cycling, rowing, etc. A change in posture and work rhythm may result in a new pattern of breathing with reference to respiratory frequency and depth. The momentary increased ventilation noticed in connection with the onset of work must then be adjusted according to the requirement for the elimination of CO_2 and uptake of O_2. Here, chemical regulation enters into the picture. The α-γ system has a possible tendency to produce a hyperventilation, especially during the beginning of the work until the produced CO_2 has reached the lungs. Through a negative feedback elicited from the respiratory centers, if Pa_{CO_2} tends to drop, the γ and α activity driving the respiratory muscles may then be inhibited from the respiratory centers. When the CO_2 reaches the lungs without being eliminated in sufficient quantity, the P_{CO_2} of the arterial blood will rise and the inhibition becomes diminished.

This is to some extent a hypothetical discussion which postulates that in connection with muscular work, there is also an increased α–γ activity to the respiratory muscles. The system has, however, every possibility to produce a rhythmic, coordinated switching between inspiration and expiration, partly determined by the work rhythm. It should be recalled that the respiratory muscles are involved in most activities not only to provide adequate pulmonary ventilation, but also to stabilize the trunk in posture or movements. A synchronization between the respiratory movements and work rhythm is therefore important. The α–γ system provides the possibility for such a coordination. [Actually there are two types of intercostal fusimotor neurons driven from different central nervous areas: rhythmic ones, with a functional link with the respiratory drive, and in addition, tonically firing ones, much more readily influenced by various reflexes from postural influence and affected by the cerebellum and other structures (Euler, 1966). This combination of systems makes it possible to integrate respiratory and postural movements at the spinal level and to correct "actual" length of the intercostal muscles to "wanted" length in accordance with the demands of breathing and also with a view to postural engagement.]

During exercise the γ-α system may create a tendency to hyperventilation;

but a negative feedback from the respiratory generators may regulate the ventilation according to the chemical composition of the arterial blood. This feedback may directly affect the motoneurons or operate via the reticular formation.

Figure 7-19 is a simplified diagram presenting components involved in exercise hyperpnea. *The neurogenic factors cannot be considered as a regulating stimulus, but act as an activator. The chemical stimuli are of decisive importance for the finer adjustment of the ventilation.* If one works very hard for 5 s, rests for 5 s, and so forth, the ventilation may eventually rise to 100 liters $\times$ min^{-1}, but no difference in ventilation is detected between the periods of rest and work (Christensen et al., 1960). A variation in afferent impulses from the working muscles is in this case of less importance in that the input of CO_2 to the lungs is continuously high. The greater the CO_2 production, the greater the "independence" of the respiratory muscles.

There are many intriguing experimental findings to consider when analyzing the neurogenic respiratory drive during exercise (Asmussen, 1967). In negative exercise, the mechanical tension developed by the contracting muscles may be 5 to 7 times greater than in positive work, but the pulmonary ventilation, per liter oxygen uptake, is similar in the two types of work. Prolonged static effort with maintained high muscular tension does not represent an especially strong respiratory drive. The very moderate ventilatory

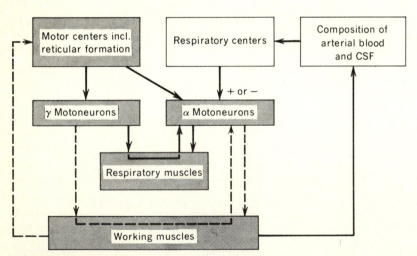

Figure 7-19 Schematic summary of the discussion on regulation of breathing during exercise. The respiratory muscles are activated via their gamma (γ) and alpha (α) motoneurons (filled-line arrows). In a similar way, the other working muscles are activated (dotted-line arrows). The different respiratory centers are influenced directly or indirectly by the chemical composition of the arterial blood and cerebrospinal fluid (CSF), mainly their P_{CO_2}, P_{O_2}, and pH. These centers can then facilitate or inhibit the motoneurons of the respiratory muscles, depending on the effectiveness of the gas exchange in the lungs. Particularly critical is the CO_2 refill from the muscles and CO_2 output from the lungs. (The afferent nerve impulses to motor centers include impulses from receptors located in tendons and joints.)

response to the first seconds of maximal effort, e.g., a sprint, or when running up a flight of stairs at maximal speed should be mentioned.

However, the afferent impulses from exercising limbs and the stimulation from various parts of the brain due to the increased motor activity must be considered as coordinated activators of the respiratory muscles, but their motoneurons are subjected to various degrees of inhibition from the respiratory generators, depending on the chemical composition of the blood.

It is noteworthy that any change in respiratory frequency which adapts to the rhythm of movements is, within limits, automatically followed by a change in tidal air to provide an adequate alveolar ventilation. (We have a similar situation in the regulation of cardiac output: at rest and during not too severe exercise, a change in heart rate is matched by a variation in stroke volume so that the cardiac output is maintained constant.)

Hypoxic Drive during Exercise

If during exercise the normal air is replaced by oxygen, the ventilation drops within seconds, and this effect is particularly marked during heavy work (Asmussen and Nielsen, 1946; Bannister and Cunningham, 1954; Dejours, 1964; Cunningham, 1974). The reaction is so rapid that an effect via the chemoreceptors in the carotid and aortic bodies has to be assumed. This ventilation-reducing effect of O_2 breathing is marked, however, even if the arterial P_{O_2} is at a normal level, and this fact has confused respiratory physiologists.

It has already been mentioned that the chemoreceptors have a relatively large blood flow and high oxygen uptake. They are innervated by sympathetic nerve fibers which supply the arterioles of the carotid and aortic bodies with vasoconstrictor fibers. When these fibers are activated, there is a reduction in the blood flow through the chemoreceptor areas, possibly by the blood being diverted through adjacent arteriovenous anastomoses. It is possible that a change in the P_{CO_2} and H^+ concentration of the arterial blood may contribute to modify the blood flow to the epithelioid cells of the chemosensitive areas.

During work, the sympaticus activity is increased and the concentration of epinephrine and norepinephrine in the blood increases. In fact, Biscoe and Purves (1965) report immediate increase in nerve activity of the cervical sympathetic and the postganglionic nerve to the carotid body by exercise (in the cat), and these changes were abolished by femoral and sciatic nerve section. It is conceivable that the blood flow to the chemoreceptor cells is reduced to such an extent that the O_2 supply to the metabolically very active cells becomes unsatisfactory or that the removal of products of tissue metabolism becomes insufficient. This might then produce an excitation of the chemoreceptors and a stimulation of respiration despite a normal or only slightly different from normal systemic arterial P_{O_2} (Hornbein and Roos, 1962; Neil and Joels, 1963; Lee et al., 1964; Whalen and Nair, 1975).

The heavier the work in relation to the performance capacity of the individual, the greater the sympaticus activity. Thereby, the chemoreceptor drive may also increase, despite the elevated perfusion pressure. This mechanism may also contribute to the increase in ventilation during pronounced emotion as well as during work involving small muscle groups, when the ventilation is relatively high at a given oxygen uptake.

This hypothesis is not supported by Severinghaus et al. (Eisele et al., 1967). They found no difference in end-tidal P_{CO_2} and arterial pH in subjects working after bilateral blockade of the stellate ganglia with Xylocaine compared with normalcy. They conclude that "sympathetic innervation of the carotid body does not appear to contribute to the hyperpnea of exercise." The authors do not state, however, how heavy the work was in relation to the subject's maximal aerobic power. It may have been too mild. Furthermore, all sympathetic fibers to the chemoreceptor areas may not have been blocked; an eventual effect of circulating epinephrine and norepinephrine is difficult to exclude in these experiments.

Summary   During work, the ventilation increases relatively rectilinearly, but this increase is relatively steeper during very heavy work. How this increase in ventilation is elicited is unknown. The comparatively small changes which are observed in the P_{O_2}, P_{CO_2}, and H^+ concentration in the arterial blood cannot explain the increase in ventilation. It is suggested as a hypothesis that muscular work as such, through afferent impulses from the engaged muscle spindles and/or from the central nervous system, increases the activity in the γ and α motoneurons of the respiratory muscles and through spinal and supraspinal reflex centers, and thus in a closely coordinated manner produces an increase in the frequency and depth of respiration, often in pace with the muscular movements. The actual regulation of the respiratory volume then takes place through a negative feedback mechanism, primarily determined by the CO_2 production in relation to CO_2 elimination during expiration. In this manner, the P_{CO_2} of the arterial blood will, via respiratory generators, determine the magnitude of the ventilation. During anaerobic work, the H^+ concentration of the blood will increase, which represents a further stimulation of respiration. During heavier work, the peripheral chemoreceptors in the carotid and aortic bodies will also stimulate respiration, possibly because an increased sympaticus activity will reduce the blood flow to the chemoreceptor areas so that the local P_{O_2} drops in spite of an almost normal value for P_{O_2} in the arterial blood. During maximal work, there is a reduction in Pa_{O_2} which should further increase ventilation. The hyperventilation which follows may even produce a drop in Pa_{CO_2}, but the arterial pH inevitably drops (Fig. 7-13).

BREATHLESSNESS (DYSPNEA)

Dyspnea is difficult, labored, uncomfortable breathing (Comroe, 1966a). The reason why respiration may become consciously troublesome is not clear. The magnitude of the ventilation is not the determining factor: a patient may experience a pulmonary ventilation of 10 liters × min^{-1} as extremely disturbing, whereas the athlete is not consciously troubled by a ventilation as high as

200 liters $\times$ min^{-1}. In certain cases, afferent vagus impulses may be the cause of the dyspnea (for example, through the collapse of some alveoli), but, more commonly, impulses from muscle spindles and thoracic joint receptors appear to give rise to conscious awareness and distress. It may be a question of length/tension, that is, tension appropriateness (Campbell, 1966). Altered afferent signals of the muscle spindles and receptors in the chest wall reaching subcortical and cortical levels may cause unusual sensations. From experience we learn what to expect and how it should be felt or experienced in a certain situation, and the respiration itself, which normally proceeds unconsciously, may give rise to distress when it requires a conscious modification. Light activity at high altitude requires a ventilation which at sea level is associated with heavy work, and this difference may be experienced as distress. A person unaccustomed to muscular work may experience a ventilation of 75 liters $\times$ min^{-1} as unpleasant, but after suitable training this may even appear as a pleasant sensation! (See Pengelly et al., 1974.)

SECOND WIND

During the first minutes of work, the load may appear very strenuous. One may experience dyspnea, but this distress eventually subsides; one experiences a "second wind." The factors eliciting the distress may be an accumulation of metabolites in the working muscles and in the blood because the O_2 transport is inadequate to satisfy the requirement.

By what mechanism this changed environment is brought to consciousness is not known. During heavy work, there is actually at the commencement of work a hypoventilation due to the fact that there is a time lag in the chemical regulation of the respiration. It is then actually a matter of a length/tension inappropriateness in the intercostal muscles. When the second wind occurs, the respiration is increased and adjusted according to requirement.

It appears that the respiratory muscles are forced to work anaerobically during the initial phases of the work if there is a time lag in the redistribution of blood. A stitch in the side may then develop. This is probably the result of hypoxia in the diaphragm. It is most common in untrained persons and is particularly apt to occur if heavy work is performed shortly after a large meal, when the circulatory adjustment at the commencement of work is slower. As the blood supply to the respiratory muscles is improved, the pain disappears. (This theory is not entirely satisfactory. During maximal work, the oxygen tension is often somewhat reduced. It is then possible to increase the ventilation at will. This causes no further increase in oxygen uptake, so that the additional work is probably covered by anaerobic processes. In spite of this, chest pain seldom occurs. It was previously believed that the pain was caused by an emptying of the blood depots in the spleen and the contractions taking place in the spleen. In humans, the spleen serves no such depot function, however. Furthermore, persons who have had their spleen removed may still experience such pain.)

Well-trained athletes who have warmed up adequately prior to a muscular effort seldom experience such pain.

HIGH AIR PRESSURES, BREATH HOLDING, DIVING

High Air Pressures

Although human beings can become acclimatized to low air pressures, there is no way to become acclimatized to high air pressures such as are encountered in deep sea diving and during escape from a submarine when the survivor attempts to get from the inside of the craft where the pressure is normal to the surface through the sea where the air pressure is higher. For every 10 m (33 ft) of sea water the diver descends, an additional pressure of 1 atm is acting upon the body. As the pressure increases, more gases can be taken up by the body and dissolved in the various tissues. At a depth of about 10 m, twice as much gas will be dissolved in the diver's blood and tissues as at sea surface. This is apt to give the diver trouble, mainly because of the nitrogen. The trouble with nitrogen is that it diffuses into various tissues of the body very slowly, and once dissolved, it also leaves the body very slowly when the pressure once more is reduced to the normal atmospheric pressure. This is especially bad when the pressure is suddenly reduced from several atmospheres, as may be the case during submarine escape or deep sea diving. Then the nitrogen is released from the tissues in the form of insoluble gas bubbles. These bubbles congregate in the small blood vessels where they obstruct the flow of blood. This then gives rise to symptoms such as pains in the muscles and joints, and even paralysis may develop if the bubbles become trapped in the central nervous system. These symptoms are known as "the bends." Obviously, the severity of the symptoms depends on the magnitude of the pressure, which means the depth to which the person has descended under water, the length of time spent at that depth, and the speed of ascent to the surface.

The bends can be avoided to a large extent by a slow return to normal pressure so as to allow time for the tissues to get rid of their excess nitrogen without the formation of bubbles. Another way to avoid the bends is to prevent the formation of these bubbles by replacing atmospheric nitrogen with helium, which is less easily dissolved in the body. This is done by having the diver breathe a helium-oxygen gas mixture. Another advantage with this is that it is more apt to prevent the so-called nitrogen narcosis which occurs when air is breathed at 3 atm or more and which results in an onset of euphoria and impaired mental activity with lack of ability to concentrate. With increasing pressures, the individual is progressively handicapped and may be rendered helpless at 10 atm. Pilots of high-flying aircraft may also suffer from bends if there is a sudden loss of pressure in the pressurized cabin, but the symptoms in these cases are usually not so severe as in the divers. In any case, they usually do not occur at altitudes lower than 10,000 m (30,000 ft).

Prolonged breathing of 100 percent oxygen may be quite harmful, for irritation of the respiratory tract may occur after 12 hr, and frank bronchopneumonia after 24 hr. In most individuals, no harmful effects result from breathing mixtures with less than 60 percent oxygen, while newborn infants are particularly susceptible to oxygen poisoning and may suffer harmful effects with oxygen concentrations over 40 percent. The remarkable thing is that oxygen poisoning apparently is no problem when breathing 100 percent oxygen at altitudes over 5,500 m, no matter for how long. Oxygen poisoning, therefore, is not much of a problem in aviation medicine, but it is indeed an important problem in deep sea diving where it may even affect the brain function when pure oxygen is used at depths greater than about 10 m (33 ft), but there are great individual variations in sensitivity to 100 percent oxygen. The onset of symptoms may be hastened by vigorous physical activity at great depths; it starts with muscular twitchings and a jerking type of breathing, and it ends in unconsciousness and convulsions. The exact cause is unknown, but it is assumed that it is a matter of interference with certain enzyme systems in the tissues.

Furthermore, when breathing pure oxygen at a pressure of 3 atm or more, the oxygen dissolved in the blood covers the oxygen need of the body at rest. Therefore, at rest, the hemoglobin of the venous blood is still saturated with oxygen. This interferes with the CO_2 transportation (Chap. 5). The result will be gradual CO_2 retention in the tissue and decrease in pH. (For further details, see Lambertsen, 1971; West, 1974.)

Breath Holding—Diving

Well-nourished individuals may survive without difficulty for weeks without food and for days without water. But they can live only a few minutes without oxygen. The body's ability to store oxygen is extremely limited. The blood may contain up to about 1.0 liter. In the lungs after a normal inspiration, there may be about 0.5 liter oxygen; after a maximal inspiration, it may be about 1.0 liter. (A total capacity of 7.0 liters and a concentration of O_2 of 15 percent = 15/100 $\times$ 7.0 = 1.05 liters.)

The O_2 bound to myoglobin may amount to as much as 0.5 liter, but it can be considered only as a local store and cannot be utilized by other tissues. The hyperbolic slope of the O_2-myoglobin dissociation curve binds the oxygen effectively until its partial pressure drops to extremely low values.

When one holds one's breath at rest, a total of about 600 ml O_2 is available and can be utilized. This is enough to last about 2 min. "Normal" maximal breath-holding time is 30 to 60 s. The arterial P_{O_2} then drops to about 75 to 50 mm Hg and the P_{CO_2} rises to 45 to 50 mm Hg. This elevated CO_2 pressure plays a greater role in forcing the individual to discontinue the breath holding than does the reduced O_2 pressure. The hypoxic drive is also an important factor, as demonstrated by the fact that breath holding may be extended if it is preceded by O_2 breathing. Under these conditions, a further increase of Pa_{CO_2} of 5 to 10 mm Hg may be tolerated. Other factors also affect the capacity for breath holding: (1) if one holds one's breath with a lung volume near the total lung capacity (maximally inflated), the respiratory standstill may be extended until the P_{CO_2} has reached a value about 10 mm Hg higher than with a lung volume near the residual level (this is even the case with O_2 breathing; for this reason, a different hypoxic drive may be excluded). The reason is probably that afferent impulses from the receptors in the lungs and thoracic cage produce a greater stimulus to inspiration in the expiratory position. (2) Breath holding may be prolonged by swallowing; swallowing constitutes a momentary inhibition in inspiration. (3) Single or repeated breaths after the breaking point, without change in alveolar air composition, make a new breath holding possible, and higher Pa_{CO_2} and lower Pa_{O_2} are obtained compared with the first trial. (4) Total paralyses of the muscles by curare prolonged the breath-holding time, and it totally abolished any sensation of breathlessness whatever in the subjects despite a highly elevated arterial CO_2 pressure (Godfrey and Campbell, 1970). Apparently a disproportion between the tension developed in the respiratory muscles and a motor effect is in some way transmitted to sensation by afferent impulses from the muscles and the chest wall. (See also Mithoefer, 1965.)

If a subject voluntarily hyperventilates forcefully for about a minute, the pulmonary air is exchanged more often and the composition of its gases more closely approaches that of the inspired air. In this manner, the alveolar P_{O_2} may increase to about 135 mm Hg, the P_{CO_2} may drop below 20 mm Hg, and the arterial blood will assume similar gas pressure. Because of the shape of the O_2 saturation curve (Fig. 6-16), the O_2 content of the arterial blood is hardly affected, however, and the pulmonary air will receive an addition of only 40 ml $\times$ liter^{-1} pulmonary air (19 vol percent O_2 instead of 15).

A more important effect of hyperventilation in this connection is the reduction in the CO_2 content of the body. Thus the breath holding may be extended to several minutes by providing more room for CO_2. The breaking point occurs probably at a P_{CO_2} of about 40 mm Hg and a P_{O_2} of 45 mm Hg. This is still within the safe range as far as the critical O_2 pressure for the function of the nervous system is concerned, which in the case of the O_2 pressure of the arterial blood is in the order of 25 to 30 mm Hg.

If a maximal breath holding is performed during muscular work (P.-O. Åstrand, 1960), the breath-holding time is shorter than at rest, but owing to the fact that O_2 uptake and CO_2 production occur at a higher rate, the composition of the arterial blood and the pulmonary air will change more rapidly. At the breaking point, the alveolar P_{CO_2} may be as high as 75 mm Hg and the P_{O_2} may drop toward 40 mm Hg (the arterial gas pressures are probably at the same level). If a person breathes pure oxygen prior to the breath holding, the $P_{A_{CO_2}}$ may exceed 90 mm Hg.

If a forceful hyperventilation precedes a breath holding during heavy work, breath holding may be prolonged until convulsions or even fainting occurs (P.-O. Åstrand, 1960). The alveolar P_{O_2} may then drop toward 20 mm Hg, which explains the symptoms. (In experiments of this type in the laboratory, the subjects were secured in a harness in order to prevent their falling off the bicycle ergometer!)

The various possible explanations will not be discussed here. However, on the basis of these experiments, the risk involved in prolonged diving and deep diving without equipment will be emphasized (Lanphier and Rahn, 1963).

If such a dive is performed following a marked hyperventilation, one may apparently hold the breath until unconsciousness occurs. It should be emphasized that the total pressure of the alveolar air at sea level is about 760 mm Hg, but at a depth of 10 m it is doubled (to twice the atmospheric pressure) by the pressure of the water on the thorax. An O_2 percentage which at sea level corresponds to an alveolar O_2 tension of about 25 mm Hg (3.5 vol percent) produces at a depth of 2 m (total pressure = 912 mm Hg) a pressure of 30 mm Hg [$3.5 \times 100^{-1} \times (912 - 47) = 30$]. As long as the diver remains at a depth of 2 m, the O_2 tension may thus be adequate to meet the requirement of the nerve cells. But when the diver approaches the surface, the oxygen tension may drop below the critical level. The danger is greater if the ascent is slow, in that the O_2 utilization then causes the O_2 pressure to drop further. Several cases have been described when divers who were attempting to beat a record, or who for various other reasons have taken their time during prolonged diving, have lost their lives or have been rescued in the nick of time (Craig, 1961; Davis, 1961). Often, hyperventilation had preceded the diving; the diver then exhibited a "strange" behavior or simply ceased to swim and sank. Cases have been described when the swimmer had had no major difficulty holding his breath but then suddenly had lost consciousness.

It is thus justifiable to warn against extremely deep diving and prolonged diving without effective supervision. The diver should avoid hyperventilation prior to the dive (taking, at the most, five deep respirations). During diving without equipment, oxygen breathing prior to the dive may definitely improve the performance. It represents an improvement if the diver's lungs contain 5 liters O_2 instead of barely 1 liter during the dive. In other types of work, the effect is negligible, however, unless the oxygen is inspired during the actual performance. In a few respirations, the extra volume of oxygen is washed out, since no extra stores of oxygen can be deposited in the body. In this connection, another effect of hyperventilation should be stressed. The washing out of CO_2 in connection with the hyperventilation increases the pH. The effect of this alkalosis and reduced Pa_{CO_2} is a vasoconstriction, including the vessels in the brain. Dizziness and cramps may be the result. If one holds one's breath after a hyperventilation against a closed glottis and at the same time contracts the abdominal muscles (Valsalva maneuver), the cardiac output is reduced. This in combination with the vasoconstriction in the cerebral blood vessels may produce an oxygen deprivation in the CNS sufficient to cause a transitory loss of consciousness.

Summary   During breath holding while working, a greater CO_2 content of the blood is tolerated. Since oxygen is rapidly utilized during work, there is a risk that the breath holding will end in unconsciousness. This is particularly true if the breath holding is preceded by marked hyperventilation. The possibility of unconsciousness represents a risk during prolonged and deep diving, especially if the ascent from considerable depths takes place slowly. The intrathoracic pressure increases with the depth of the water. A given oxygen pressure in the alveoli and in the blood may therefore be adequate for the normal function of the central nervous system at a depth of a few meters, but it may become critically low as the diver approaches the surface.

REFERENCES

Adamson, J. W., and C. A. Finch: Hemoglobin Function, Oxygen Affinity, and Erythropoietin, *Ann. Rev. Physiol.*, **37**:351, 1975.

Angus, G. E., and W. M. Thurlbeck: Number of Alveoli in the Human Lung, *J. Appl. Physiol.*, **32**:483, 1972.

Asmussen, E.: Exercise and Regulation of Ventilation, *Circulation Res.*, **20**:1–132, 1967.

Asmussen, E., and M. Nielsen: Studies on the Regulation of Respiration in Heavy Work, *Acta Physiol. Scand.*, **12**:171, 1946.

Asmussen, E., and M. Nielsen: Physiological Dead Space and Alveolar Gas Pressures at Rest and during Muscular Exercise, *Acta Physiol. Scand.*, **38**:1, 1956.

Åstrand, I.: Aerobic Work Capacity in Men and Women with Special Reference to Age, *Acta Physiol. Scand.*, **49**(Suppl. 169):1960.

Åstrand, I., P.-O.: Åstrand, I. Hallbäck, and Å. Kilbom: Reduction in Maximal Oxygen Uptake with Age, *J. Appl. Physiol.*, **35**:649, 1973.

Åstrand, P.-O.: "Experimental Studies of Physical Working Capacity in Relation to Sex and Age," Munksgaard, Copenhagen, 1952.

Åstrand, P.-O.: A Study of Chemoceptor Activity in Animals Exposed to Prolonged Hypoxia, *Acta Physiol. Scand.*, **30**:335, 1954.

Åstrand, P.-O.: Breath Holding during and after Muscular Exercise, *J. Appl. Physiol.*, **15**:220, 1960.

Åstrand, P.-O., and B. Saltin: Oxygen Uptake during the First Minutes of Heavy Muscular Exercise, *J. Appl. Physiol.*, **16**:971, 1961.

Bannister, R. G., and C. J. C. Cunningham: The Effects on the Respiration and Performance during Exercise of Adding Oxygen to the Inspired Air, *J. Physiol. (London)*, **125**:118, 1954.

Bargeton, D.: Analysis of Capnigram and Oxygram in Man, *Bull. de Physio-Pathologie Respiratoire*, **3**:503, 1967.

Biscoe, T. J.: Carotid Body: Structure and Function, *Physiol. Rev.*, **51**:437, 1971.

Biscoe, T. J., and M. J. Purves: Carotid Chemoreceptor and Cervical Sympathetic Activity during Passive Third Limb Exercise in the Anaesthetized Cat, *J. Physiol. (London)*, **178**:43P, 1965.

Bjurstedt, H., G. Rosenhamer, and O. Wigertz: High-G Environment and Responses to Graded Exercise, *J. Appl. Physiol.*, **25**:713, 1968.

Bradley, G. W., C. von Euler, I. Marttila, and B. Roos: A Model of the Central and Reflex Inhibition of Inspiration in the Cat, *Biol. Cybernetics*, **19**:105, 1975.

Braus, H.: "Anatomie des Menschen," Bd. 2, Aufl. 3, p. 134, Springer-Verlag OHG, Berlin, 1956.

Campbell, E. J. M.: Motor Pathways, in W. O. Fenn and H. Rahn (eds.), "Handbook of Physiology," sec. 3, Respiration, vol. I, p. 535, American Physiological Society, Washington, D.C., 1964.

Campbell, E. J. M.: The Relationship of the Sensation of Breathlessness to the Act of

Breathing, in J. B. L. Howell and E. J. M. Campbell (eds.), "Breathlessness," p. 55, Blackwell Scientific Publications, Ltd., Oxford, 1966.

Chance, B., B. Schoener, and F. Schindler: The Intracellular Oxidation-Reduction State, in F. Dickens and E. Neil (eds.), "Oxygen in the Animal Organism," p. 367, Pergamon Press, New York, 1964.

Cherniak, R. M., L. Cherniak, A. Haimark, and V. Cherniack: "Respiration in Health and Disease," W. B. Saunders Company, Philadelphia, 1972.

Christensen, E. H., R. Hedman, and B. Saltin: Intermittent and Continuous Running, *Acta Physiol. Scand.*, **50:**269, 1960.

Cole, P.: Respiratory Mucosal Vascular Responses, Air Conditioning and Thermo Regulation, *J. Laryngol. Otol.*, **68:**613, 1954.

Comroe, J. H., Jr.: "Physiology of Respiration," Year Book Medical Publishers, Inc., Chicago, Ill., 1965.

Comroe, J. H., Jr.: Some Theories of the Mechanism of Dyspnoea, in J. B. L. Howell and E. J. M. Campbell (eds.), "Breathlessness," p. 1, Blackwell Scientific Publications, Ltd., Oxford, 1966a.

Comroe, J. H., Jr.: The Lung, *Sci. Am.*, **214:**(2):56, 1966b.

Craig, A. B., Jr.: Causes of Loss of Consciousness during Underwater Swimming, *J. Appl. Physiol.*, **16:**583, 1961.

Cunningham, D. J. C.: Integrative Aspects of the Regulation of Breathing: a Personal View, in J. G. Widdicombe (ed.), "MTP International Reviews of Science Physiology, Series One, Respiratory Physiology," p. 303, Butterworths, London, 1974.

Davis, J. H.: Fatal Underwater Breath Holding in Trained Swimmers, *J. Forensic Sci.*, **6:**301, 1961.

Dejours, P.: Control of Respiration in Muscular Exercise, in W. O. Fenn and H. Rahn (eds.), "Handbook of Physiology," sec. 3, Respiration, vol. I, p. 631, American Physiological Society, Washington, D.C., 1964.

Dejours, P.: "Principles of Comparative Respiratory Physiology," American Elsevier Publ. Company, New York, 1975.

Eisele, J. H., B. C. Ritchie, and J. W. Severinghaus: Effect of Stellate Ganglion Blockade upon the Hyperpnea of Exercise, *J. Appl. Physiol.*, **22:**966, 1967.

Eklund, G., C. von Euler, and S. Rutkowski: Spontaneous and Reflex Activity of Intercostal Gamma Motoneurons, *J. Physiol.*, **171:**139, 1964.

Eulor, C. von: Proprioceptive Control on Respiration, in R. Granit (ed.), "Muscular Afferents and Motor Control," p. 197, Nobel Symposium I, John Wiley & Sons Inc., New York, 1966.

Euler, C. von: On the Role of Proprioceptors in Perception and Execution of Motor Acts with Special Reference to Breathing," in L. D. Pengelly, A. S. Rebuck, and E. J. M. Campbell (eds.):, "Loaded Breathing," p. 139, Longman Canada Limited, Ontario, 1974.

Fenn, W. O., and H. Rahn (eds.): "Handbook of Physiology," sec. 3, Respiration, vols. I and II, American Physiological Society, Washington, D.C., 1964–1965.

Flandrois, R., R. LeFrançois, and A. Teillac: Comparaison de Plusieurs Grandeurs Ventilatoires dans Deux Types d'Exercise Musculaire, *Biotypologie*, **22:**66, 1961.

Forster, R. E.: Oxygenation of the Muscle Cell, *Circulation Res.*, **20:**1–115, 1967.

Forster, R. E., and E. D. Crandell: Pulmonary Gas Exchange, *Ann. Rev. Physiol.*, **38:**69, 1976.

Godfrey, S., and E. J. M. Campbell: The Role of Afferent Impulses from the Lung and

Chest Wall in Respiratory Control and Sensation, in R. Porter (ed.), "Breathing: Hering-Breuer Centenary Symposium," p. 219, Churchill, London, 1970.

Granit, R.: The Functional Role of the Muscle Spindles—Facts and Hypotheses, *Brain*, **98**:531, 1975.

Heymans, C., and E. Neil: "Reflexogenic Areas of the Cardiovascular System," Churchill, London, 1958.

Holmér, I.: Physiology of Swimming Man, *Acta Physiol. Scand.* (Suppl. 407)1974.

Hornbein, T. F., and A. Roos: Effect of Mild Hypoxia on Ventilation during Exercise, *J. Appl. Physiol.*, **17**:239, 1962.

Hornbein, T. F., and A. Roos: Specificity of H Ion Concentration as a Carotid Chemoreceptor Stimulus, *J. Appl. Physiol.*, **18**:580, 1963.

Kao, F.: "An Introduction to Respiratory Physiology," Excerpta Medica, Amsterdam, 1972.

Krahl, V. E.: Anatomy of the Mammalian Lung, in W. O. Fenn and H. Rahn (eds.), "Handbook of Physiology," sec. 3, Respiration, vol. I, p. 213, American Physiological Society, Washington, D.C., 1964.

Krogh, A.: "The Comparative Physiology of Respiratory Mechanisms," University of Pennsylvania Press, Philadelphia, 1941.

Krogh, A., and J. Lindhard: Regulation of Respiration and Circulation during the Initial Stages of Muscular Work, *J. Physiol.* (*London*), **47**:112, 1913.

Lambertsen, C. J. (ed.): "Proceedings of the Fourth Symposium on Underwater Physiology," Academic Press, Inc., New York, 1971.

Lanphier, E. H., and H. Rahn: Alveolar Gas Exchange during Breath-hold Diving, *J. Appl. Physiol.*, **81**:471, 1963.

Lee, K. D., R. A. Mayou, and R. W. Torrance: The Effect of Blood Pressure upon Chemoreceptor Discharge to Hypoxia and the Modifications of This Effect by the Sympathetic-adrenal System, *J. Quart. Exp. Physiol.*, **49**:171, 1964.

Liljestrand, G.: Untersuchungen über die Atmungsarbeit, *Skand. Arch. Physiol.*, **35**:199, 1918.

Loeschke, H. H.: Central Nervous Chemoreceptors, in J. G. Widdicombe (ed.), "MTP International Review of Science Physiology, Series One, Respiratory Physiology," p. 167, Butterworths, London, 1974.

Martin, L. G., J. M. Connors, J. J. McGrath, and J. Freeman: Altitude-induced Erythrocytic 2-3-DPG and Hemoglobin Changes in Rats of Various Ages, *J. Appl. Physiol.*, **39**:258, 1975.

Mead, J., and G. Milic-Emili: Theory and Methodology in Respiratory Mechanics with Glossary Symbols, in W. O. Fenn and H. Rahn (eds.), "Handbook of Physiology," sec. 3, Respiration, vol. I, p. 363, American Physiological Society, Washington D.C., 1964.

Milic-Emili, G., and J. M. Petit: Mechanical Efficiency of Breathing, *J. Appl. Physiol.*, **15**:359, 1960.

Milic-Emili, G., J. M. Petit, and R. Deroanne: The Effects of Respiratory Rate on the Mechanical Work of Breathing during Muscular Exercise, *Intern. Z. Angew. Physiol.*, **18**:330, 1960.

Mithoefer, J. C.: Breath Holding, in W. O. Fenn and H. Rahn (eds.), "Handbook of Physiology," sec. 3, Respiration, vol. II, p. 1011, American Physiological Society, Washington, D.C., 1965.

Moritz, A. R., F. C. Henriques, Jr., and R. McLean: The Effect of Inhaled Heat on the

Air Passages and Lungs: An Experimental Investigation, *Am. J. Pathol.*, **21**:311, 1945.

Moritz, A. R., and J. R. Weisiger: Effects of Cold Air on the Air Passages and Lungs, *Arch. Intern. Med.*, **75**:233, 1945.

Neil, E., and N. Joels: The Carotid Glomus Sensory Mechanism, in D. J. C. Cunningham and B. B. Lloyd (eds.), "Regulation of Human Respiration," p. 163, Blackwell Scientific Publications, Ltd., Oxford, 1963.

Nielsen, M.: Die Respirationsarbeit bei Körperruhe und bei Muskelarbeit, *Skand. Arch. Physiol.*, **74**:299, 1936.

Otis, A. B.: The Work of Breathing, in W. O. Fenn and H. Rahn (eds.), "Handbook of Physiology," sec. 3, Respiration, vol. I, p. 463, American Physiological Society, Washington, D.C., 1964.

Pappenheimer, J. R.: Standardization of Definitions and Symbols in Respiratory Physiology, *Fed. Proc.*, **9**:602, 1950.

Pattle, R. E.: Surface Lining of Lung Alveoli, *Physiol. Rev.*, **45**:48, 1965.

Pengelly, L. D., A. S. Rebuck, and E. J. M. Campbell (eds.): "Loaded Breathing," Longman Canada Limited, Ontario, 1974.

Plum, F.: Neurological Integration of Behavioural and Metabolic Control of Breathing, in R. Porter (ed.), "Breathing: Hering-Breuer Centenary Symposium," p. 159, Churchill, London, 1970.

Porter, R. (ed.): "Breathing: Hering-Breuer Centenary Symposium. Ciba Foundation Symposium," Churchill, London, 1970.

Rahn, H., and L. E. Farhi: Ventilation, Perfusion and Gas Exchange: The $\dot{V}_A/\dot{Q}$ Concept, in W. O. Fenn and H. Rahn (eds.), "Handbook of Physiology," sec. 3, Respiration, vol. I, p. 735, American Physiological Society, Washington, D.C., 1964.

Saltin, B., G. Blomqvist, J. H. Mitchell, R. L. Johnson, Jr., K. Wildenthal, and C. B. Chapman: Response to Submaximal and Maximal Exercise after Bedrest and Training, *Circulation*, **38**(Suppl. 7):1968.

Saltin, B., and P.-O. Åstrand: Maximal Oxygen Uptake in Athletes, *J. Appl. Physiol.*, **23**:353, 1967.

Scholander, P. F.: Oxygen Transport through Hemoglobin Solution, *Science*, **131**:585, 1960.

Stainsby, W. N., and A. B. Otis: Blood Flow, Blood Oxygen Tension, Oxygen Uptake and Oxygen Transport in Skeletal Muscle, *Am. J. Physiol.*, **206**:858, 1964.

Stenberg, J., P.-O. Åstrand, B. Ekblom, J. Royce, and B. Saltin: Hemodynamic Response to Work with Different Muscle Groups, Sitting and Supine, *J. Appl. Physiol.*, **22**:61, 1967.

Turner, J. M., J. Mead, and M. E. Wohl: Elasticity of Human Lungs in Relation to Age, *J. Appl. Physiol.*, **25**:664, 1968.

Wasserman, K., B. J. Whipp, R. Casaburi, D. J. Huntsman, J. Castagna, and R. Lugliani: Regulation of Arterial P_{CO_2} During Intravenous CO_2 Loading, *J. Appl. Physiol.*, **38**:651, 1975.

Weibel, E. R.: "Morphometry of the Human Lung," Springer-Verlag OHG, Berlin, 1963.

Weibel, E. R.: Morphometrics of the Lung, in W. O. Fenn and H. Rahn (eds.), "Handbook of Physiology," sec. 3, Respiration, vol. I, p. 285, American Physiological Society, Washington, D.C., 1964.

Weibel, E. R.: Morphological Basis of Alveolar-Capillary Gas Exchange, *Physiol. Rev.*, **53**:419, 1972.

West, J.: Regional Differences in Gas Exchange in the Lung of Erect Man, *J. Appl. Physiol.*, **17**:893, 1962.

West, J. B.: "Respiratory Physiology—the Essentials," The Williams & Wilkins Company, Baltimore, 1974.

Whalen, W. J. and P. Nair: Some Factors Affecting Tissue P_{O_2} in the Carotid Body, *J. Appl. Physiol.*, **39**:562, 1975.

Widdicombe, J. G. (ed.): "MTP International Reviews of Science Physiology, Series One, Respiratory Physiology," Butterworths, London, 1974.

Wittenberg, B. A., J. B. Wittenberg, and P. R. B. Caldwell: Role of Myoglobin in the Oxygen Supply to Red Skeletal Muscle, *J. Biol. Chem.*, **250**:9038, 1975.

Skeletal System

CONTENTS

Chapter 8

Skeletal System

THE MANY FUNCTIONS OF BONE

The skeletal system provides the mechanical levers for the muscles so that their contraction may cause the body to move. It is the supporting framework that prevents the entire body from collapsing into a heap of soft tissue; it is the protecting shell or casing for such vital, viable organs as the brain, lungs, heart, and pelvic organs; and it contains within its structures the factories for formed elements of the blood. In addition, it is the great calcium and phosphorus reserve of the body, constantly being drawn upon or added to. Bone is a vital living tissue, continuously undergoing changes of building up and tearing down.

For a comprehensive review of the osseous system, the reader is referred to such volumes as: *Bone as a Tissue*, edited by Rodahl, Nicholson, and Brown (1960); *The Biochemistry and Physiology of Bone*, vols. 1–3, edited by G. H. Bourne (1971), and *Biology of Bone*, by N. M. Hancox (1972). The purpose of this chapter is to present a summary of the structure and functions of bone as they pertain to physical performance, work, stress, and exercise.

EVOLUTION OF THE SKELETAL SYSTEM

In the history of vertebrates, bones and muscles are intimately related in the embryonic development and in physical activity. Primitive multicelled animals probably had only two body layers: one inside layer and one outside layer. With growth in size, a third layer developed in between; muscles and bones are products of this middle layer, from which the circulatory system also originates.

The need for a rigid supporting framework developed early in many of the multicelled animals, especially as they moved from sea to land (Romer, 1957). In most animals, there is some kind of connective tissue filling the spaces between different organs, helping to bind them together. Generally such tissues contain numerous fibers felted together into a supporting structure. With the increase in size and mobility, characteristic for most chordates, came the need for more rigid support, such as the skeletal structures. Many of the nonchordates solved the problem with the development of an outside supporting structure, such as the shell of the mollusk or the hardened superficial armor of the lobster, whereas, in vertebrates, the major skeletal structures are internal, covered by muscles and skin. This arrangement not only facilitates mobility, but also enhances the accessibility of the stored mineral reserves in the bones for participation in metabolic processes and chemical reactions in the body.

The first type of skeletal material which developed in the vertebrates was the *notochord*, already typically developed in such primitive forms as the amphioxus; it consists of a tough but flexible rod running the length of the body, along the back, and below the nerve cord (Romer, 1957). This notochord not only helps to stiffen the body and support the various organs but also serves as a point of attachment for the muscles of the trunk. Although the notochord long persisted as a functional element in the vertebrates, there appeared new skeletal materials, in the form of cartilage and bone, which have assumed and expanded the role of support originally provided by this notochord. Both these substances are formed from the connective tissue, but they differ considerably in their nature and appearance.

STRUCTURE

The *cartilage* is a translucent material formed by round cells (cartilage cells) embedded in a substance which binds them together. It is firm yet elastic, flexible, and capable of rapid growth. It is therefore a useful supporting structure as long as the stress is moderate, and it is a common supporting structure in lower vetebrates. In the shark, for example, it is the only skeletal material in addition to the notochord. On the other hand, in land vertebrates, such as the adult human being, it plays only a limited part in the skeletal makeup and is found only where elasticity is required, such as at the ends of the ribs, as disks between the joints of the backbone, and at the joint surfaces of the limbs.

Bone, like cartilage, is a derivative of the original connective tissue, but it is a highly specialized form and is different from connective tissue proper in that it is very hard. It consists of *osteocytes*, which are cells with long, branching processes that occupy cavities called *lacunas* in a dense matrix made up of collagenous fibers laid down in an amorphous ground substance called *cement*, which is impregnated with calcium phosphate complexes (Fig. 8-1). The ratio between inorganic and organic substances in the bone varies during the life span. The ratio is about 1:1 in the child. In young adults, there is about 4 times more inorganic material than organic, and in the old individual the ratio is about 7:1. Therefore the bone is more fragile in the old than in the young person.

A series of in vitro studies have shown that the development of bone and cartilage cells may be affected by extrinsic factors in the environment. Thus, an experiment by C. A. I. Basset (1962) showed that the formation of normal bone from a tissue culture of primitive fibroblasts depended on adequate oxygenation. When high oxygen pressure was introduced in addition to compaction of the culture, bone formation resulted. With low oxygenation, cartilage was formed from the primitive fibroblasts.

FUNCTION

The cellular components of bone are associated with specific functions. The *osteoblasts* are involved in the formation of bone, the *osteocytes* with the maintenance of bone as a living tissue, and the *osteoclasts* with the destruction and resorption of bone. These cells are closely interrelated, and transformation may occur from one form to the other. Thus, an osteocyte may actually be an osteoblast that has been surrounded by calcified interstitial substance. It may undergo a further transformation and assume the form of an osteoclast, thought to be associated with the resorption of bone.

Many bones, or parts of bones, are compact, solid structures, nourished

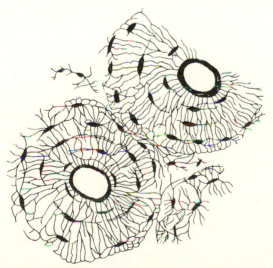

Figure 8-1 Schematic drawing of the structure of bone, showing osteocytes with long branching processes that occupy cavities in the dense bone matrix.

through small canals carrying blood vessels and through tiny tubules connecting cell spaces with one another and with the canals. However, if all bones were solid and compact, they would be unnecessarily heavy in proportion to the strength requirements. The large bones are therefore hollow. They are solid only at the surface, and sufficient bone bars and braces extend from the solid bone exterior into the hollow interior to provide reinforcement in ways more or less similar to the manner in which many bridges are engineered (Fig. 8-2). So wisely is the order of nature arranged that this space in the hollow bone shafts is not wasted, but filled with bone marrow which is used as a factory for red blood cells and capable of producing about 2 to 3 million erythrocytes per second.

Under normal conditions, a state of equilibrium exists between the calcium of the blood and that of the bones, but in disease this balance may be seriously altered. The calcium is absorbed from the food in the small intestine and is excreted by the large intestine and, to a lesser extent, by the kidney. The regulators of calcium metabolism are the parathyroid glands and vitamin D; the former regulate the interchange of calcium between the blood and the bones.

Excess parathyroid hormone results in a rise in the blood calcium with a corresponding fall in the calcium of the bones and a loss of calcium from the body by increased excretion. Vitamin D governs the absorption of calcium from the intestine, but excess intakes of this vitamin cause a mobilization of calcium from the bones and a deposition of calcium in soft tissues.

The maintenance of the normal mineral metabolism of the bones also depends upon the longitudinal pressure on the long bones, brought about by the

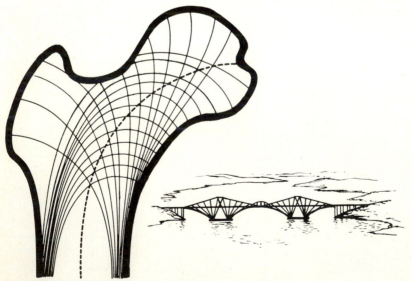

Figure 8-2 Schematic drawing showing reinforcing bars and braces, similar to those of a bridge, of the large bones.

stress of gravity on the upright, ambulatory human frame (Jansen, 1920; Rodahl et al., 1966). In fact, all changes in function of a bone are attended by alterations in its internal structure (Wolff's law). Pressure will stimulate the appositional bone growth. Increased weight bearing will result in increased thickness of the bone and density of the shaft.

The examinations of Atkinson, Weatherell, and Weidmann (1962) and Eisenberg and Gordan (1961) confirm this fact. Atkinson and his coworkers examined the density of the femur in individuals with active and with sedentary habits after fifty years of age. The bone density was slightly higher in the active group. Eisenberg and Gordan found, by measuring the dynamics of growth in the human with nonradioactive strontium, that muscular exercise accelerates the rate of bone deposition. Conversely, elimination of the "normal effects of stress and strain" leads to the familiar picture of disuse osteoporosis.

The pumping action of the muscles on the circulation and the rate of blood flow through the extremities can be estimated from denervation and devascularization experiments (Troupp, 1961). A study on the longitudinal growth of the femur and tibia of rabbits following interruption of vascular and nervous supply demonstrated that retardation of growth is unequivocally more significant following vascular blocking than following nerve paralysis. It was suggested that vascular insufficiency causes hypoxia in the cartilage cells of the epiphysis, thus contributing to growth inhibition by inhibiting the transformation of bone from cartilage cells.

Studies of the effects of prolonged bed rest on calcium metabolism (Rodahl et al., 1966) have shown that the increased urinary calcium excretion that occurs during prolonged confinement to bed is not due to inactivity but to the absence of longitudinal pressure on the long bones. The increased urinary calcium excretion was unaffected by heavy bicycle ergometer work performed in the supine position for 1 to 4 hours daily and by 8 hours of inactive sitting in a wheelchair. But 3 hours' standing per day in addition to recumbent bed rest caused the urinary calcium excretion to return toward normal values (Fig. 8-3). When a pressure equivalent to the body weight of the individual confined to horizontal bed rest was exerted along the longitudinal axis of his body by heavy springs fixed to a shoulder harness and the foot of the bed for 3 hours a day, the urinary calcium elimination also returned toward normal values in one of two subjects.

If the urinary calcium elimination may be taken as an indication of bone mineralization, it appears that gravitational stress on the long bones is essential for normal bone growth. This is supported by recent studies of calcium metabolism in astronauts during prolonged space flights. Thus Whedon et al. (1975), studying mineral and nitrogen metabolism during the United States Skylab flights lasting from 28 to 84 days, observed that urinary calcium excretion increased as in bed rest studies by 80 to 100 percent. The urinary calcium excretion showed a gradual rise during the first 3 weeks of the flight, after which it remained constant at the elevated level for the duration of the space flight, whether it lasted 28 or 84 days.

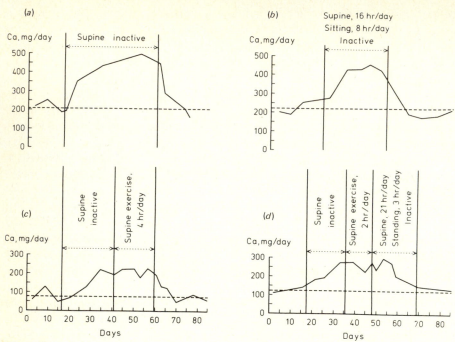

Figure 8-3 Urinary output of calcium during (a) prolonged inactive bed rest; (c) bed rest combined with 4 supine exercise; (b) bed rest combined with 8 hr · day⁻¹ · quiet sitting; and (d) bed rest combined with 3 hr/day standing. *(From Rodahl et al., 1966.)*

It is thus clear that the development of greater sturdiness and strength of bone is brought about by subjecting it to increased pressure. This process may be slower than the increase in muscle strength resulting from a program of weight lifting. In such training, the training intensity or the increase in training load should be sufficiently gradual to allow for the development of skeletal strength to keep pace with the increase in muscle strength.

GROWTH

Bone growth is a unique physiological process, consisting of a constant construction and reconstruction which continue beyond the point of actual growth and development, so that nothing is stable and final in bone except the external shape. A typical long bone, such as the femur, consists of a shaft (the *diaphysis*) and two enlarged ends (the *epiphyses*). The part of the shaft which is next to the epiphysis is called the *metaphysis*. The surface of the long bone is covered by a sheet of connective tissue (the *periosteum*). The shaft of the bone consists of a dense outer layer (the *cortex*), a middle layer of spongy bone, and the inner portion, which is occupied by the marrow cavity. The joint surface of the epiphysis is covered with smooth, glistening white hyaline cartilage. The articular cartilage has to absorb the shocks transmitted to the joint during motion and reflects the wear and tear of the joint.

In providing rigidity, the bone has to sacrifice the flexibility and the expandable traits enjoyed by the cartilage. This naturally presents a problem of growth. Thus, for example, the thigh bone of a two-year-old child has to double its length by adulthood. Although the bone may become stouter simply by adding more bone to its surface layers, it cannot stretch, nor can new bone be added to its ends, for such changes would interfere with its articulate surfaces that form part of the hip joint and the knee joint. This problem is solved by the insertion of a special growth section, the epiphyseal plate, close to both ends of the shaft. On the surface of this plate are layers of actively growing cells. It is here that the increase in length of the bone occurs without interference with the joint surfaces. When the individual is fully grown, the epiphyses fuse with the shaft, and growth terminates.

All the elements of the internal skeleton are first formed in cartilage in the human embryo. Later, the cartilage is destroyed and replaced by bone. It is in this replacement process that a thin film of cartilage is left, separating the ends from the shaft of the bone in the form of the epiphyses, described earlier.

REPAIR-ADAPTATION

Following a fracture of a bone which was initially preformed in cartilage, the first attempt at repair, starting within the first few weeks following the fracture, results in the formation of a mass of fibrocartilage known as *callus*. The bridging of a fracture by new bone occurs in a fashion resembling that of a fixed-arch bridge, according to the principle of cantilevering frequently seen in contemporary architecture (Fig. 8-4). The new bone grows out upon the surface of the cortex on each side of the fracture line and envelops the fibrocartilaginous callus to form an arch of new bone over the fracture gap. The new bone comes down like ribs let down from the arch of the bridge to suspend the deck. This bone gradually replaces the cartilage toward the fracture gap. Finally the "deck" is laid down between the fracture ends and provides permanent union. The superstructure disappears, leaving only the bone required for union of the fracture ends. In the case of a fracture of a long bone in a young individual, this process is usually completed within a period of about 3 to 6 months.

Owing to the readiness with which bone formation and bone absorption can occur, a remodeling of bone is continually taking place. Isotope studies using P^{32} or Ca^{45} indicate that over 20 percent of the calcium and phosphate ions of bone are involved in a fairly rapid exchange in adults (LeBlond and

Figure 8-4 Schematic drawing showing similarity between the process of bone repair and the construction of a fixed-arch bridge. *(Modified from McLean and Urist, 1955.)*

Greulich, 1956). As mentioned, bone tissue is surprisingly sensitive to demands made upon it and responds readily to these demands, so that every change in the function of a bone is followed more or less by a definite change in the internal architecture. Bone bars not stressed will disappear and new ones will be created where altered mechanical forces increase the demand for sturdiness. Thus, as a result of training or a fracture, the lines of stress in a bone may change with the altered direction of mechanical forces, as if it were the most plastic of structures.

Absorption of bone may occur from many causes and leads to bone rarefaction, or *osteoporosis*. It occurs commonly in old age, from disuse, and as a local phenomenon caused by such processes as inflammation, tumors, and aneurysms.

JOINTS

Joints are formed where two or more bones of the skeleton meet one another (Fig. 8-5). The function of the joint determines its character and structure. In areas like the skull, it is important that no movement should be permitted between contiguous bones; in the vertebral column, on the other hand, a slight degree of mobility is desirable, provided that it can be obtained without loss of strength or sturdiness. In other situations, the provision of a more or less wide range of movement is essential. In the first case, *fibrous joints* are provided; these are articulations in which the surfaces of the bones are fastened together by intervening fibrous tissue and in which there is no appreciable motion. In the second case, the connection medium between the bones concerned is white fibrocartilage; such joints are capable of a limited range of movement and are termed *cartilaginous joints* (Fig. 8-5a). In the third case, the opposed bones are separated from one another by a space lined by a special membrane, which is termed *synovial membrane*; such a joint possesses a more or less wide range of movement and is known as a *synovial joint* (Fig. 8-5b, c).

In the cartilaginous joints of the spinal column, the opposed bony surfaces are covered with hyaline cartilage and are connected to one another by flattened disks of fibrocartilage of a more or less complex structure. The bones are also connected by bands of white fibrous tissue known as *ligaments*, which do not form a complete capsule around the joint. A limited degree of movement is made possible by the compressibility of the cartilaginous disk and the degree of leverage which is available (Fig. 8-5d).

Most of the joints of the body, including all joints of the limbs, belong to the synovial group. In these joints (Fig. 8-5), the contiguous bony surfaces are covered with articular cartilage separated by a joint cavity, which in a healthy individual is merely a tiny space. The joint is completely surrounded by an articular capsule consisting of a capsular ligament lined with a synovial membrane. The synovial membrane lines the whole of the interior of the joint, with the exception of the cartilage-covered ends of the articulating bones. The bones are usually connected by ligaments which are added to the capsular

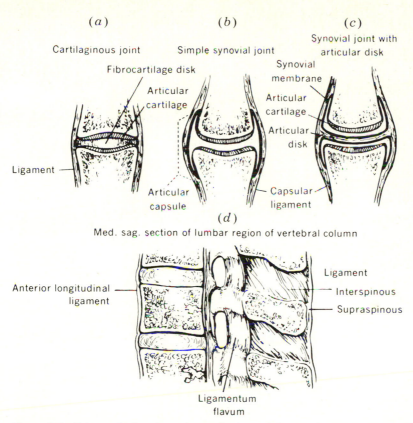

Figure 8-5 Schematic drawings of a cartilaginous joint (*a*), simple synovial joint (*b*), synovial joint with articular disk (*c*), and a sagittal section at the lumbar region of the vertebral column (*d*).

ligaments and situated superficially to the latter. Movements in such a joint may vary from a simple, limited gliding movement to a wide range of movements, such as in the shoulder joint. The joint cavity may be divided by an articular disk of fibrocartilage, as in the knee joint. These structures act as shock-reducing agents and serve to ensure perfect contact between the moving surfaces in any position of the joint. The synovial membrane secretes a small quantity of a viscid fluid termed *synovia*. This fluid acts as a lubricant.

Since the articular cartilage has no blood vessels, it has to be nourished through other channels. There is direct communication between the medullary cavities of the epiphyses and the basal portions of the articular cartilage. An exchange of fluid can also take place between the cartilage and the synovial fluid of the articular space (Holmdal and Ingelmark, 1951; Ekholm, 1951).

Holmdal and Ingelmark (1948) have shown that in trained animals, the articular cartilage is thicker than in untrained ones. The active animals had an increase in both the cellular and intercellular components of the cartilage. Animals living in small cages restricting their opportunities to move around had, in general, thinner cartilages of the knee joints than animals provided with spacious cages.

Figure 8-6 illustrates how the articular cartilage in the knee joint in a matter of minutes can vary in thickness depending on whether the animal is active or inactive. After 10 min of running, the increase in thickness was 12 to 13 percent compared with what it was after 60 min of immobilization (Ingelmark and Ekholm, 1948). Similar results have been obtained on humans. The explanation given for the rapid increase in thickness of the cartilage of an activated joint is that fluid seeps into the cartilage from the underlying bone marrow cavity, when the cartilage is alternately compressed and decompressed.

One consequence of the increased fluid content of the cartilage is a change in its compressibility, diminishing the incongruence between condyle and socket. This change will increase the area of the contact surface of the joint in question and produce a reduced pressure per unit of area of the articular surface during compression with a given force (Ingelmark and Ekholm, 1948). Warm-up activities before vigorous exercise should therefore include those joints that may be stressed during the activity in order to make them less susceptible to trauma.

Another advantage in regular dynamic motion of the joints is that the associated increase in supply of fluid to the articular cartilage also will provide nutriments to the cartilage. It is a long-established clinical observation that a joint suffers from nutritional disturbances when kept inactive. It is not known how frequently such activities should be performed to provide an optimal nutritional situation for articular cartilages.

The movements possible in joints are usually classified as gliding angular movements (flexion-extension, abduction-adduction), circumduction (shoulder and hip joints), and rotation (as in the rotation of the humerus at the shoulder joint, or in the movement of the radius on the ulna during pronation and supination of the hand). Limitation of movements is affected by a number of factors, such as the tension of ligaments or the tension of the muscles that are antagonistic to the movement. In fact, it appears that the tension of the antagonist muscles may never permit a ligament of a joint to be fully stretched. Finally, the movements of some joints are limited by the soft tissues, as in the flexion of the elbow, hip, and knee. A flexion in the hip joint with extended knee is limited by the length of the muscles on the back of the thigh. With a simultaneous flexion in the knee joint, the flexion of the hip can be greatly extended. If, in addition, the flexion of the hip joint is aided by external forces, the flexion may be further increased until it is stopped by the thigh pressing against the abdomen. In other words, the muscles moving a joint cannot, even with maximal force, produce a movement over the full range which the joint actually permits. However, a movement in which external forces are involved may be so extreme, especially when a great force is applied suddenly, that the adjacent articular cartilages may be separated (luxation). At the same time,

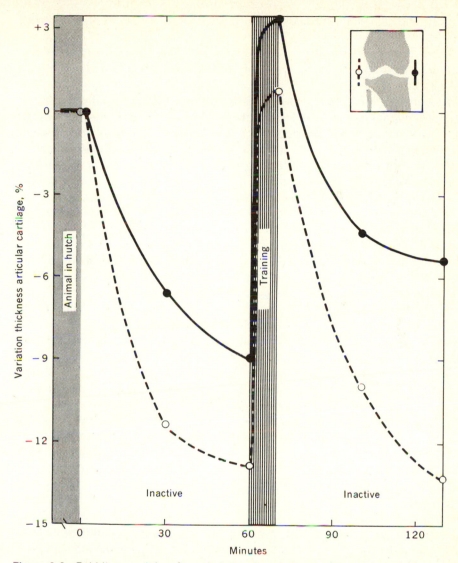

Figure 8-6 Rabbits were taken from their spacious hutches where they could run freely. At "0" minutes, x-ray photographs were taken for measurements of the thickness of the articular cartilage of the fibular and tibial ends of the knee joints (for symbols, see inserted figure). Measurements were repeated after 30 and 60 min of rest in a position which did not stress the rabbits' knees with their body weight. Then the rabbits were exercised on a motor-driven treadmill for 10 min, speed 40 m · min⁻¹. Measurements were repeated after exercise and further periods of rest. Note a significant decrease in cartilage thickness after inactivity and a rapid increase during exercise. (Figures represent the mean of about 50 animals.) *(Modified from Ingelmark and Ekholm, 1948.)*

bone, ligaments, joint capsules, soft tissues, and blood vessels may be damaged.

Since the limiting factor for flexibility is often the length of the muscles, training that produces a lengthening of these muscles will increase the joint flexibility.

In synovial joints where the bones are connected only by ligaments and muscles, the articular surfaces are in constant apposition in all positions of the joint. The maintenance of this apposition is facilitated by atmospheric pressure and cohesion, but the muscles play a far more important role. The balance in tone between the different muscle groups which act on the joint is responsible for maintaining the articular surfaces in constant apposition, so that the stability of any joint depends on the tonus of the muscles that act on it.

Movable joints are innervated by the nerves which supply the muscles that act on them, and it is reasonable to assume that this arrangement establishes local reflex arcs which ensure stability. The part of the articular capsule which is rendered taut in the contraction of a given muscle or group of muscles is innervated by the nerve or nerves supplying their antagonists. This condition may serve to prevent overstretching or tearing of the ligaments.

Studies of different joints both in humans and animals have shown that nerve fibers to a joint arrive both via special branches from motor nerves, periostal nerves, and often cutaneous nerves (Brodal, 1972). The contribution of these various fibers to the overall nerve supply of a joint varies greatly from one joint to another. Thus, the anterior aspects of the knee joints are mainly supplied by nerve fibers coming from nerve branches supplying the adjacent muscles, whereas the ankle joint receives relatively few such fibers. Similarly, the total number of nerve fibers supplying a joint may vary greatly. Thus, the nerve supply to the knee joint is much greater than to the shoulder joint. It has been suggested that the reason for this fact may be that the muscle mass surrounding the shoulder joint may play a role in the perception of the joint position, thus augmenting or partially replacing the function of the afferent nerve fibers of the joint (Brodal, 1972).

In general, there are four types of nerve endings in the joints (Fig. 8-7). Three of them are nerves ending in specialized end organs, as in skin receptors, often surrounded by a capsule of connective tissue, while one of them ends as free-branching nerve endings. Of the former, one type is primarily specialized to give information about changes in joint position (type 1), and a second type to give information about the speed of movement in the joint (type 2). Both types are located in the parts of the joint capsule where bending and stretching take place. A third type of receptor (type 3) appears to be particularly adjusted to register the actual position of the joint, and it is located in the ligaments of the joints. The fourth type of receptor consists of free-branching nerve endings which are pain-sensitive fibers similar to those found elsewhere in the body. The synovial membrane has no nerve fibers of its own, as is also true of the joint cartilage.

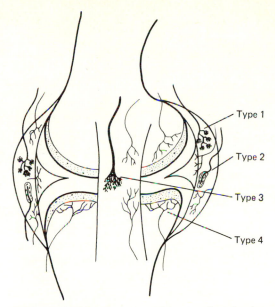

Figure 8-7 Schematic drawing showing the four different types of nerve endings in the joints. *(By courtesy of P. Brodal.)*

Afferent impulses from the joints reach the somato-sensoric region of the cortex. For this reason, it has been postulated that the joint innervation is of major importance for the conscious perception of the position and motions of the joint, whereas afferent impulses from the muscles, which are not consciously perceived, are significant for the subconscious, reflectoric control of motor activity (Brodal, 1972). Skoglund (1956) has shown that afferent impulses from the joints affect the activity of the alpha-motoneurons. This effect is reciprocal on the muscles which act as antagonists on a joint; that is, during passive extension, the extensors are inhibited while the flexors are facilitated. Skoglund (1956) has also provided evidence suggesting that the joint receptors are able to provide information pertaining to the resistance against a movement.

LIGAMENTS AND TENDONS

Tipton et al. (1975) have shown that the mechanical stress produced by regular, prolonged exercise or training causes increased strength of the junctions between ligaments or tendons and bones, and of ligaments which have been repaired. Regular physical exercise of the endurance type may therefore be an important preventive measure, since numerous investigators have demonstrated that the usual site of separation is in the transitional zone between the ligament or the tendon and the bone.

PATHOPHYSIOLOGY OF THE BACK

It is not within the scope of this book to discuss pathological conditions. However, there are diseases that may be influenced by activity or by inactivity. One example is backache. Muscle training, some caution when lifting and carrying loads, proper posture, and application of physiological principles when planning work places and when performing work may prevent or alleviate symptoms from an inadequate back. The problem of backache will therefore be discussed here.

Of all the bony structures of the human body, the spinal column plays a unique role in that it serves as a sustaining rod for the maintenance of the upright position. As such, it is subjected to a complex system of forces and stresses of many types. Frequently these forces are amazingly large. Because of the unique nature of the structure and function of the spine, it is often the site of aches and pains as a result of numerous processes associated with wear and tear. It has often been said that lower backache is the price humans must pay for their upright, two-legged existence. This, however, is not necessarily the case, for similar afflictions may occur in many four-legged animals.

There are few ailments which can hinder muscular activity more dramatically than backache. Apart from being painful, it is indeed a costly condition for the individual and society in terms of lost work output, sick days, and reduced earning power. From a health standpoint, lower backache ranks as one of the most common medical complaints. As a matter of fact, back pain, in one form or another, is experienced by almost all individuals at one time or another. Low backache is also quite common in athletes, especially those whose training includes heavy weight lifting.

Hult (1954) has presented results of examinations of about 1,200 individuals representing different professions. They were divided into two main groups, those engaged in physically light work (such as white-collar workers, workers in light industry, and retail trade employees) and in physically heavy work (including longshoremen, construction workers, and employees in heavy industry). The subjects ranged from twenty-five to sixty years of age. Subjective symptoms revealed by careful notations of case histories could, on the whole, be classified as belonging to one of three syndromes: two common ones, the stiff neck-brachialgia syndrome (in 51 percent of the individuals) and the lumbar insufficiency-lumbago-sciatica syndrome (in 60 percent), and a less common one, the dorsal spine syndrome (in 5 percent of the subjects).

Figure 8-8 (upper part) shows (1) the marked increase in symptoms with age; and (2) a slightly higher percentage of individuals with symptoms of lower back troubles among heavy workers (64 percent) compared with those involved in light work (53 percent). The incidence of symptoms in the upper back was the same in those doing light and heavy work (51 percent).

The results of clinical and roentgenologic examinations are also summarized in Fig. 8-8 (lower part). The trend is the same as for the subjective symptoms, with the age factor being of decisive importance for the occurrence of roentgen signs of disk degeneration, which was as high as about 90 percent in the fifty-five- to fifty-nine-year age group. Such signs appear with consistently higher frequency in those subjects with heavy occupations, even if the difference between the two groups is moderate. Disk degeneration should be interpreted as a more or less physiological process which begins early in some individuals but eventually develops in all persons regardless of occupation.

In patients with a history of lumbago or sciatica, the symptoms were provoked by an accident in only 20 percent of the cases. In an additional 15 to 20 percent, they appeared in connection with heavy lifting or a similar strain. This means that in 60 to 65 percent of those who had had attacks of lumbago or sciatica, the symptoms appeared without such external causative factors.

Static deformities, such as differences in leg length, kyphosis, lordosis, and scoliosis, had no demonstrable significance in the origin of low back trouble in these groups. Nor was there any correlation between low back trouble and body type (evaluated according to Rohrer's index), flat feet, or foot fatigue.

There was some, but not a striking, relation between subjective symptoms and clinical-roentgenologic findings. Thus, disk degeneration could develop and even attain an advanced stage without ever having given rise to pain.

Weight lifters did not in any respect differ significantly from the other subjects.

It is estimated that in Sweden, approximately 2 million working days annually are lost because of back trouble. The occurrence of such a high frequency of symptoms from the back is not at all surprising in view of the complexity of the spinal column, its complicated joint systems, and the very heavy stress to which it may be exposed. An interesting analysis of the tolerance of the spinal column has been published by Morris et al. (1961). They point out that the spinal column may be considered to have both an intrinsic and extrinsic stability, the former being provided by the alternating rigid and elastic components of the spine bound together by a system of ligaments, while extrinsic stability is provided by the paraspinal and other trunk muscles. The stability of the ligamentous spine, which may be considered as a modified elastic rod, depends largely on the action of the extrinsic support provided by the trunk muscles. It has been shown that the critical load value at which buckling of the isolated

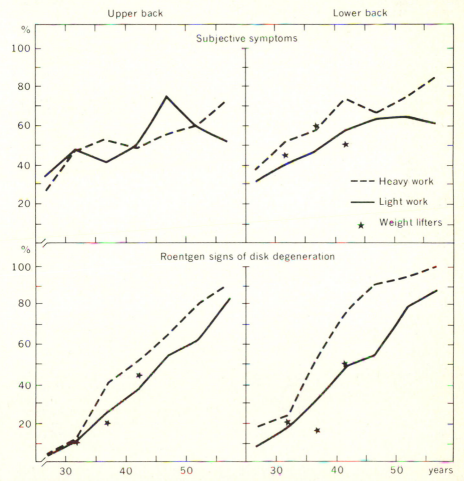

Figure 8-8 Data on approximately 1,200 individuals divided into two main groups according to occupation. Included are data from 56 weight lifters. The upper part of the figure shows results of case histories dealing with symptoms emanating from the upper or lower extremities and the spine. The lower part of the figure shows frequencies of objective incidence of disk degeneration and x-ray–anatomic anomalies. The graphs to the left present cases of stiff neck–brachialgia syndrome ("upper back"); the graphs to the right, lumbar insufficiency–lumbago–sciatica syndrome ("lower back"). *(Data from Hult, 1954.)*

ligamentous spine occurs, fixed at the base, is much less than the weight of the body above the pelvis (Lucas and Bresler, 1960).

In order to explain the discrepancy between the force which the spine as such can tolerate and the much larger forces to which the back is subjected in actual life situations, Morris and his coworkers investigated the role of the compartments of the trunk, i.e., thorax and abdomen, in helping to provide stability of the spine. They then showed that the discrepancy apparently can be explained by the role played by the trunk.

In providing extrinsic stability of the spine, Morris et al. (1961) considered that if the nucleus pulposus of the fifth lumbar disk is considered as the fulcrum of movement and if a heavy weight is lifted with the hands, the arms and trunk form a long anterior lever (Fig. 8-9). The weight being lifted and the weight of the head, arms, and upper part of the trunk are counterbalanced by the contraction of the deep muscles of the back acting through a much shorter lever arm, that is, the distance from the center of the disk to the center of the spinous process. With these factors in mind, they computed the force that results when a 77 kg man lifts a 90 kg weight, and they concluded that the force on the lumbosacral disk is about 8,800 N, if the role of the trunk is omitted. This force is considerably more than segments of the isolated ligamentous spine can withstand without structural failure; compression tests of two vertebral bodies with their intervening disk from subjects under forty years of age resulted in failure of the segments of the spine under compressive forces ranging from 4,400 to 7,600 N. In older subjects, this figure was sometimes as low as 1,300 N.

In a series of experiments, Morris et al. were able to show that since the spinal column is attached to the sides of, and within, the two chambers of the abdominal and thoracic cavities, the action of the trunk muscles converts these chambers into nearly rigid cylinders containing air, liquid, and semisolid material. Thus, these cylinders are capable of transmitting part of the forces generated in loading the trunk, thereby relieving the load on the spine itself. When large forces are applied to the spine, such as when lifting a weight of 90 kg, there is generalized contraction of the trunk muscles, including the intercostals, the muscles of the

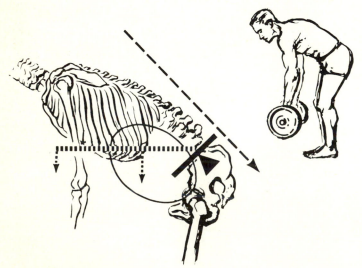

Figure 8-9 Schematic drawing of a 77-kg man lifting 90 kg. *(From Morris et al., 1961.)* The nucleus pulposus of the fifth lumbar disk is considered as the fulcrum of movement. The arms and trunk form a long anterior lever. The weight being lifted is counterbalanced by the contraction of the deep muscles of the back acting on a much shorter lever (the distance from the center of the disk to the center of the spinous process). If the role of the trunk is omitted, the force on the lumbosacral disk would be about 9000 N, which is considerably more than segments of the isolated ligamentous spine can withstand without structural failure. When this does not happen, it is because the contracted trunk muscles convert the abdominal and thoracic cavities into semirigid cylinders which relieve the load on the spine itself.

abdominal wall, and the diaphragm. The action of the intercostals and the muscles of the shoulder girdle renders the thoracic cage a rigid structure firmly bound to the thoracic part of the spine. When inspiration increases intrathoracic pressure, the thoracic cage and spine become a solid, sturdy unit capable of transmitting large forces. By the contraction of the diaphragm, attached at the lower margin of the thorax and overlying the abdominal wall, especially the transversus abdominis, the abdominal contents are also compressed into a semirigid cylinder. The force of weights lifted by the arms is thus transmitted to the spinal column by the shoulder girdle muscles, principally the trapezius, and then to the abdominal cylinder and to the pelvis, partly through the spinal column but also through the rib cage. When larger forces are involved, increased rigidity of the rib cage and increased compression of the abdominal contents are achieved by increased activity of the trunk muscles, resulting in increased intracavitary pressures.

This is all brought about by a reflex mechanism: When a load is placed on the spine, the trunk muscles are involuntarily called into action to fix the rib cage and to compress the abdominal contents. The intracavitary pressures are thereby increased, aiding in the support of the spine. Morris et al. (1961) concluded from their calculations that the actual force on the spine is much less than that considered to be present when this support by the trunk is omitted. The calculated force on the lumbosacral disk is about 30 percent less, and that of the lower thoracic portion of the spine is about 50 percent less than it would be without support by the trunk.

Thus, the study by Morris and his coworkers emphasized the important role of the trunk muscles in the support of the spine. From this it follows that well-developed trunk muscles, including the abdominal muscles, play a significant role in sparing the spine and thus avoiding strain and damage. Although flabby abdominal muscles may expose the spine to injurious stress, well-developed abdominal muscles, on the other hand, are a valuable protective device.

The role of the abdominal muscles in protecting the spine when lifting, especially during the very early part of the lift, is also evident from the calculations of Farfan (1973). In addition, he emphasizes the additional support mechanism for the spine offered by the posterior assembly of the intervertebral joints.

Human beings have had to carry loads from the beginning of their existence and have had time to learn how to perform this task in keeping with the most efficient bioengineering principles. The methods used by primitive native tribes in carrying loads are well known (Fig. 8-10 shows one method).

It has been generally advocated that in lifting and carrying loads, one should bring the center of gravity of the load as close as possible to the axis of the body to minimize the force movement. Although this principle should be followed more often than is commonly done, it should be borne in mind that, in many real-life situations, it is not always practicable to do so.

Often a series of motions are involved in the execution of a given task, including bending, lifting, turning, and walking with the load. However, experience has shown that symptoms of lower backache are most apt to occur when lifting is combined with a twisting

Figure 8-10 A physiological method of carrying a load, practiced by many peoples.

or turning motion. It is therefore wise to arrange the situation so that the individual is facing the direction of movement before picking up the load. It is equally important to have a secure foundation for the feet during such motions, since lower backache very commonly occurs during sudden bodily movements associated with slipping of the foothold. In view of the findings of Morris and his coworkers (1961), it appears reasonable to assume that under such sudden motions, there may not be sufficient time for a reflex contraction of the trunk muscles to occur, thus leaving the spinal column more vulnerable.

These considerations underline the significance of well-trained trunk muscles, including the abdominal muscles, in the protection of the spine. It cannot be overemphasized that in the case of backaches or damage to the back, prevention is the most important aspect, since the back is an inherent weakness in the human body. This fact brings out the need for intelligent indoctrination, starting with children of school age, that will induce a complete understanding of the mechanical aspects of the spinal column and methodical instruction in the proper techniques involving the use of the back in all kinds of daily activities as well as in special industrial tasks. By such early indoctrination, it may well be possible to avoid some of the lower back pain common in adults.

Nachemson (1976) and his group are measuring the intradiscal pressure (in the third lumbar disc) together with EMG and intra-abdominal pressure in various exercises and occupational activities. As expected, good lumbar and also arm support are important factors in decreasing pressures as is avoidance of lifting heavy objects with great distance between the load and the body.

Similar principles are also involved in other joints of the body, such as the neck and the knee. In the case of the knee joint, for example, the importance of the quadriceps muscle in stabilizing the joint when nearly extended is well recognized. In the treatment of hydrops in the knee joint, it is often overlooked that although bed rest and immobilization cause a resorption of the fluid, the concomitant atrophy of the quadriceps muscle caused by immobilization may result in recurrence of the hydrops, at times even more severe than before, as the result of damage caused by inadequate stabilization by the atrophied quadriceps muscle. The correct treatment would be to put the patient to bed in order to relieve the joint of the burden of the body weight, but at the same time to train the leg muscles. In this connection, it should again be noted that the nutrition of the cartilage may be enhanced by allowing some activity of the joint in question.

Not infrequently, lumbar scoliosis is seen as a compensatory measure when the pelvis is tilted to one side, the lower limbs' being of unequal length. To compensate for this, it is only by curving the lumbar spine through an angle equal to the pelvic tilt that the trunk can be held vertical. Usually, this compensatory scoliosis disappears automatically when the pelvic tilt is corrected by the use of an insole of proper thickness.

Observations among young Scandinavian girls engaged in elite calisthenics have indicated that extreme bending of the spine may cause damage, and that such exercises, therefore, should be omitted.

REFERENCES

Atkinson, P. J., J. A. Weatherell, and S. M. Weidmann: Changes in Density of Human Femoral Cortex with Age, *J. Bone Joint Surg.*, **44B:**496, 1962.

Basset, C. A. I.: Current Concepts of Bone Formation, *J. Bone Joint Surg.*, **44A:**1217, 1962.

Bourne, G. H. (ed.): "The Biochemistry and Physiology of Bone," vols. 1–3, Academic Press, Inc., New York, 1971.

Brodal, P.: Leddinnervasjon—et forsömt kapittel? *Fysioterapeuten*, **39:**65, 1972.

Ekholm, R.: Articular Cartilage Nutrition, *Acta Anat.*, **11**(Suppl. 15–2):1951.

Eisenberg, E., and G. S. Gordan: Skeletal Dynamics in Man Measured by Nonradioactive Strontium. *J. Clin. Invest.*, **40:**1809, 1961.

Engström, A., and R. Amprino: Studies of Immobilized Bones, *Experientia,* **6:**267, 1961.

Farfan, H. F.: "Mechanical Disorders of the Low Back," Lea & Febiger, Philadelphia, 1973.

Hancox, N. M.: "Biology of Bone," Cambridge University Press, London, 1972.

Holmdahl, D. E., and B. E. Ingelmark: Der Bau des Gelenkknorpels unter verschiedenen funktionellen Verhältnissen, *Acta Anat.*, **6**:309, 1948.

Holmdahl, D. E., and B. E. Ingelmark: The Contact between the Articular Cartilage and the Medullary Cavities of the Bones, *Acta Anat.*, **12**:341, 1951.

Hult, L.: Cervical, Dorsal and Lumbar Spinal Syndromes, *Acta Orthop. Scand.* (Suppl. 17), 1954.

Ingelmark, B. E., and R. Ekholm: A Study on Variations in the Thickness of Articular Cartilage in Association with Rest and Periodical Load, *Uppsala Läkareförenings Förhandlingar*, **53**:61, 1948.

Jansen, M.: "On Bone Formation: Its Relation to Tension and Pressure," Longmans, Green & Co., Inc., New York, 1920.

LeBlond, C. P., and R. C. Greulich: Autoradiographic Studies of Bone Formation and Growth, in G. H. Bourne (ed.), "The Biochemistry and Physiology of Bone," Academic Press, Inc., New York, 1956.

Lucas, D. B., and B. Bresler: "Stability of the Ligamentous Spine," *Technical Report Series II*, no. 40, University of California, Biomechanics Laboratory, Berkeley, December 1960.

McLean, F. C. and M. R. Urist: "Bone: An Introduction to the Physiology of Skeletal Tissue," The University of Chicago Press, Chicago, 1955.

Morris, J. M., D. R. Lucas and B. Bresler: Role of the Trunk in Stability of the Spine, *J. Bone Joint Surg.*, **43A**:327, 1961.

Nachemson, A. L.: The Lumbar Spine an Orthopedic Challenge, *Spine*, **1**:59, 1976.

Rodahl, K., J. T. Nicholson, and E. M. Brown (eds.): "Bone as a Tissue," McGraw-Hill Book Company, New York, 1960.

Rodahl, K., N. C. Birkhead, J. J. Blizzard, B. Issekutz, Jr., and E. D. R. Pruett: Fysiologiske forandringer under langvarig sengeleie, *Nord. Med.*, **75**:182, 1966.

Romer, A. S.: "Man and the Vertebrates," The University of Chicago Press, Chicago, 1957.

Skoglund, S.: Anatomical and Physiological Studies of Knee-joint innervation in the Cat, *Acta Physiol. Scand.*, **36**(Suppl. 124):1956.

Tipton, C. M., R. D. Matthes, J. A. Maynard, and R. A. Carey: The Influence of Physical Activity on Ligaments and Tendons, *Med. and Sci. in Sports*, **7**:165, 1975.

Troupp, H.: Nervous and Vascular Influence on Longitudinal Growth of Bone; an Experimental Study on Rabbits, *Acta Ortho. Scand.*, **51**(Suppl.):1, 1961.

Whedon, G. D., L. Lutwak, P. Rambaut, M. Whittle, C. Leach, J. Reid, and M. Smith: Mineral and Nitrogen Metabolic Studies on Skylab Flights and Comparison with Effects of Earth Long-term Recumbency, *Proc. COSPAR Symposium on Gravitational Physiology*, Varna, Bulgaria, May 30–31, Bulgarian Academy of Sciences Press, Sofia, 1975.

Physical Work Capacity

CONTENTS

Chapter 9

Physical Work Capacity

DEMAND—CAPABILITY

Athletic competition represents the classical test of physical fitness or performance capacity. Under such conditions the performance may be measured objectively in centimeters or seconds, or it may be judged subjectively, as in gymnastics, figure skating, or diving. The individual's performance is the combined result of the coordinated exertion and integration of a variety of functions. The demands of the actual event must be perfectly matched by the individual's capabilities in order to achieve top performance and championship. It is impossible to present one formula that takes into account all aspects of a person's maximal work power and capacity, since the demands set by different types of activities vary greatly. However, the following factors may serve as a frame of reference for our discussion.

Natural endowment (genetic factors) probably plays a major role in a person's performance capacity, at least for those persons aspiring to the levels required for the attainment of Olympic medals. Since the possible genetic combinations are astronomical in number, it is an interesting question whether a country must have a population of 100,000, 1 million, 10 million, or more, to "breed" an individual with proper endowment for top results. The more popular an event, the greater is the chance that an individual with the suitable

```
Physical Performance

  Energy output

     Aerobic processes

     Anaerobic processes

  Neuromuscular function

     Strength

     Technique

  Psychological factors

     Motivation

     Tactics
```

constitution will participate and thus discover his or her ability. Obviously, the environment and geographic location are also important. If an individual with the perfect endowment for skiing grows up in a place where skiing is impossible, this endowment may be wasted from an athletic standpoint. The fact that an increasingly large number of naturally endowed persons enter the ranks of competitive athletes may in part explain the gradual improvement of athletic records.

Granted the endowment, however, definite improvement in performance may be achieved by *training*, and all factors listed in Fig. 9-1 as contributing to physical performance capacity may be modified. The very intense training programs currently employed in many fields of athletic performance contribute greatly to the improved results. Another factor explaining the gradual improvement in work output and athletic achievement over the years is the better techniques applied and the superior equipment which is becoming available through technical progress.

The athletes themselves are mainly concerned with improving their ability to cut off seconds or add centimeters to their records. The scientist is interested in analyzing why the results improve or vary from time to time. Therefore, the scientific objective is (1) to evaluate quantitatively the influence of the various factors upon the performance capacity in different tasks (performance requirements); (2) to examine how these factors vary with sex, age, and body size (capacity profile); and (3) to study the effect of such factors as training and environment. It is realistic to conclude that scientists have merely begun a

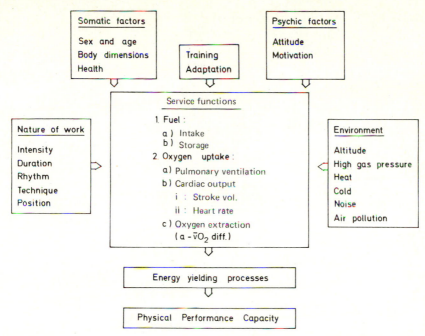

Figure 9-1 Factors influencing the capacity for aerobic muscular activity.

systematic research on the performance capacity and the many factors involved. The most advanced information concerns the energy output by aerobic processes. This may be explained by the fact that methods for quantitative measurements of energy output by the human combustion engine have long been available, actually ever since Lavoisier's discovery of the oxygen utilization by living animals. We shall therefore begin the more detailed analysis of physical work capacity with a discussion of the oxygen uptake during submaximal and maximal exercise and the maximal aerobic power (the individual's maximal oxygen uptake).

(It should be emphasized that *capacity* denotes total energy available, and that *power* means energy per unit of time.)

AEROBIC PROCESSES

The complexity of the capacity for aerobic muscular exercise is illustrated by Fig. 9-1. For each liter of O_2 consumed, about 20 kJ, range 19.7 to 21.2 (5 kcal, range 4.7 to 5.05) will be delivered; hence, the higher the oxygen uptake, the higher the energy output. The oxygen uptake during exercise may be measured with an accuracy of ± 0.04 liter $\cdot$ min^{-1} ($\dot{V}_{O_2} > 1$ liter $\cdot$ min^{-1}). Figure 9-2 gives examples of how the classical Douglas bag method can be applied when studying the aerobic energy output during work or exercise.

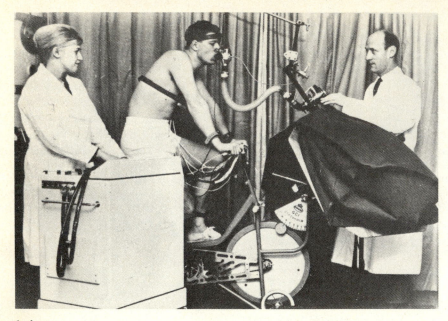

(a)

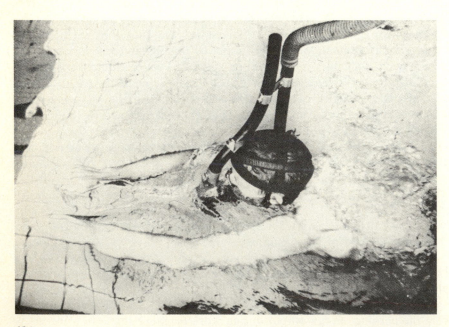

(b)

Figure 9-2 Application of the Douglas bag method for measuring aerobic energy output during different types of exercise. The skier shown in (c) carries a three-way stopcock and a stopwatch on his chest for the recording of time during which the expired air is collected in the Douglas bag. The stopwatch automatically starts and stops when the stopcock is turned.

Work Load and Duration of Work

Figure 9-3a shows how the oxygen uptake increases during the first minutes of exercise to a "steady state" where the oxygen uptake corresponds to the demands of the tissues. When the exercise stops, the oxygen uptake gradually decreases to the resting level; the oxygen debt is paid off.

The slow increase in oxygen uptake at the beginning of exercise is explained by the sluggish adjustment of respiration and circulation, i.e., the sluggish adjustment of the oxygen-transporting systems to work. The attainment of this steady state coincides roughly with the adaptation of cardiac output, heart rate, and pulmonary ventilation. A *steady-state* condition denotes a work situation where oxygen uptake equals the oxygen requirement of the tissues; consequently, there is no accumulation of lactic acid in the body. Heart rate, cardiac output, and pulmonary ventilation have attained fairly constant levels. In light exercise, the energy output during the first minutes of exercise can be delivered aerobically, since oxygen is stored in the muscles bound to myoglobin and in the blood perfusing the muscles. During more severe exercise, anaerobic processes must supply part of the energy during the early phase of exercise, and lactic acid will be produced. [Anaerobic energy is provided by not only the glycogenolysis or glycolysis but also the breakdown

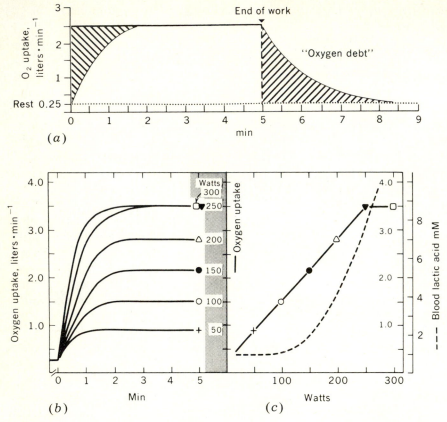

Figure 9-3 (a) During the first minutes of exercise the oxygen uptake increases, then levels off as the oxygen uptake reaches a level adequate to meet the demand of the tissues. At the cessation of exercise, there is a gradual decrease in the oxygen uptake, as the "oxygen debt" is being paid off.
(b) Schematic demonstration of increase in oxygen uptake during exercise on bicycle ergometer with different work loads (noted within shadowed area) performed during 5 to 6 min.
(c) Oxygen uptake in the above-mentioned experiments, measured after 5 min and plotted in relation to work load. Note that 250 watts (1500 kpm · min^{-1}) brought the oxygen uptake up to this subject's maximum and that 300 watts did not further increase the oxygen uptake; the increased work load was possible thanks to anaerobic processes. Maximal aerobic power = 3.5 liters · min^{-1}. (For simplicity, the work load which is sufficient to bring the oxygen uptake to the subject's maximum, in this case 250 watts, may be written $WL_{max_{O_2}}$.) Peak lactic acid concentrations in the blood at each experiment have been included.

of ATP and phosphocreatine (Chap. 2).] With a work load (leg work) that demands an oxygen uptake higher than 50 percent of the individual's maximal uptake and which is performed for some minutes, lactic acid (lactate) appears in the blood in a concentration that can be measured even in the arterial blood. The heavier the work load, the more important is the anaerobic energy contribution. The blood lactate concentration increases, the work becomes subjectively more strenuous, and a decrease in the body's pH affects muscular tissue, respiration, and other functions.

Figure 9-3*b* illustrates the linear increase in oxygen uptake, measured after about 5 min of exercise with different work loads up to a point where the maximum for oxygen transportation appears to be reached. In this case, the maximal oxygen uptake is 3.5 liters · min⁻¹, and this is the *maximal aerobic power* of this subject. There are two main criteria showing that this actually represents the subject's maximal effort: (1) There is no further increase in oxygen uptake despite further increase in work load; and (2) the blood lactate concentration is above 70 to 80 mg · 100 ml⁻¹ of blood, or 8 to 9 mM (P.-O. Åstrand, 1952; I. Åstrand, 1960; Rodahl and Issekutz, 1962). The assumption is that large muscle groups are involved in the exercise and that the work time exceeds 3 min.

From a methodological viewpoint, it is important to emphasize that maximal oxygen uptake is attained at a work load that is not necessarily maximal. From Fig. 9-3*c*, it is obvious that 250 watts were enough to reveal the subject's maximum, 3.5 liters · min⁻¹, but the subject was not exhausted by this work load. The individual could work at a rate of 300 watts for the same period of time. *An all-out test is not necessary for the assessment of an individual's maximal aerobic power.*

There are several reasons for the delayed return of oxygen uptake to resting level after the cessation of exercise (repayment of oxygen debt): (1) refilling of the oxygen content of the body; (2) aerobic removal of anaerobic metabolites; (3) elevated metabolism due to an increase in tissue temperature and a possible increased output of adrenalin (about 13 percent elevation in metabolic rate per degree centigrade); and (4) increased oxygen demand of the activated respiratory muscles and heart. Evidently only part of the excessive oxygen uptake during recovery is used to pay off the energy debt incurred by anaerobic processes (see below). However, the heavier the work load, the more dominating is the anaerobic fraction of this excessive oxygen uptake.

The heavier the work load, the steeper is the increase in the oxygen uptake (and heart rate). This is illustrated by Fig. 9-4. After a 10-min period of work at 50 percent of maximal oxygen uptake, work loads of 300 to 450 watts were applied until exhaustion. The tolerated work time varied from 6 min (300 watts) to less than 2 min (heaviest load). The oxygen uptake at the end was the same in all experiments, or about 4.1 liters · min⁻¹. However, after 1 min of extremely heavy exercise, the oxygen uptake was 4.0 liters · min⁻¹ at the "supermaximal" load but only 3.0 liters · min⁻¹ during the less extreme but still heavy work load of 300 watts, which could not be tolerated for more than 6 min (P.-O. Åstrand and Saltin, 1961a).

There are several implications from these experiments: (1) In studies where maximal oxygen uptake is to be measured, the collection of expired air or other measurements may start after about 1 min of exercise, provided that the work load is extremely heavy (supermaximal) and is preceded by a warming-up period. For many reasons, however, it is wise to aim at a work

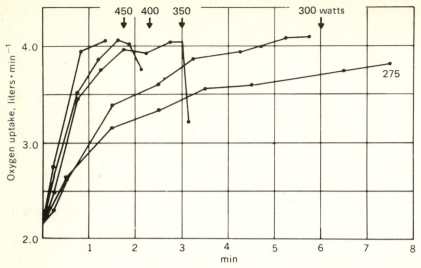

Figure 9-4 Curves showing increase in oxygen uptake during heavy exercise following a 10-min warm-up period. Arrows indicate time when the subject had to stop because of exhaustion. Figures indicate work load on the bicycle ergometer. The subject could continue the load of 275 watts (1650 kpm · min⁻¹) for more than 8 min. *(From P.-O. Åstrand and Saltin, 1961a.)*

period of about 5 min. (2) A work time of 1 min or even less may maximally load the oxygen-transporting system. (3) It is an interesting question why the ability to increase the oxygen uptake maximally within 1 min is not utilized in exercise which can be prolonged to 5 to 10 min and which has a marked oxygen deficit during the first minutes. Such an increase would be an advantage, for the sooner the aerobic processes could come into full swing, the less would be the demand on the anaerobic processes, and less lactic acid would accumulate.

In repeated determinations of maximal oxygen uptake on the same subject, the standard deviation is 3 percent, which includes biological and methodological variables (P.-O. Åstrand, 1952; Taylor et al., 1955; Mitchell et al., 1958; P.-O. Åstrand and Saltin, 1961a).

Summary In many types of muscular exercise, the oxygen uptake increases roughly linearly with an increase in work load. The maximal oxygen uptake or *maximal aerobic power* is defined as the highest oxygen uptake the individual can attain during physical work while breathing air at sea level (work time 2 to 6 min, depending on the work load).

During heavy exercise, anaerobic processes contribute to the energy yield not only at the beginning of work but continuously throughout the exercise period. An accumulation of metabolites will eventually necessitate the termination of the work.

In very heavy exercise, the maximal oxygen uptake and heart rate may be attained within 1 min, provided a sufficient warm-up period precedes the maximal effort.

Intermittent Work

Muscular work in industrial or recreational activities is very seldom maintained for very long at a steady rate. For this reason, a steady state, as discussed earlier, is rarely attained. The classical laboratory studies, with subjects working continuously for 5 min or longer on the treadmill or bicycle ergometer, in many ways, represent artificial situations. Nevertheless, such procedures have distinct advantages when one is studying the physiology of exercise or studying patients, for they provide standardized conditions and permit comparisons to be made on repeated occasions. They may also simulate the demands placed on the body in many sport events. However, from both a practical and a theoretical point of view, it is equally important to study the effect of intermittent work, which better mirrors the type of muscular activities encountered in industry or at home and in most types of ordinary exercise or recreational activity.

Some of the most important principles will be discussed by presenting some experiments concerning intermittent work (I. Åstrand et al., 1960a,b; Christensen et al., 1960; Saltin et al., 1976).

1   A subject whose maximal oxygen uptake was 4.6 liters $\cdot$ min^{-1} could work at 350 watts for about 8 min. Since the oxygen need was approximately 5.2 liters $\cdot$ min^{-1}, the anaerobic processes had to provide part of the energy. When the work load was reduced to 175 watts the work could easily be prolonged to 60 min, the final heart rate was 135 and oxygen uptake 2.45 liters $\cdot$ min^{-1}, and the blood lactate concentration did not increase above resting level. The total oxygen uptake during the hour was 145 liters.
2   In another experiment with the same subject, the work load was again 350 watts, but now work periods of 3 min were alternated with 3-min rest periods. The subject could proceed with great difficulty for 1 hr, and the same total amount of work was performed as in Experiment 1. The oxygen uptake and heart rate were now maximal, as was the peak blood lactate concentration (120 mg $\cdot$ 100 ml^{-1}; 10 mg $\cdot$ 100 ml^{-1} = 1.1 mM). The total energy output during the second experiment was about 10 percent higher than in the first one.
3   When the heavy work periods were shortened by introducing the rest periods more frequently, the *total* oxygen uptake over the hour was not markedly reduced. The subjective feeling of strain was less severe, however, and peak oxygen uptake, heart rate, and blood lactate concentration were now lower. Hence, with intermittent work and rest for 30 s respectively, the heart rate did not exceed 150, the blood lactate was only 20 mg $\cdot$ 100 ml^{-1}, and the total oxygen uptake was 154 liters during the hour. (The subject's maximal heart rate was 190.)

Figure 9-5a illustrates another set of experiments with the same subject. He worked on the bicycle ergometer with an extremely heavy work load of 412 watts. When working continuously at this work load, he became exhausted within about 3 min. When working intermittently for 1 min and resting for 2 min, he could continue for 24 min before being totally exhausted, and the blood lactate concentration rose to 150 mg $\cdot$ 100 ml^{-1}. In another experiment, the periods of work were reduced to 10 s and the rest periods to 20 s. Now he could

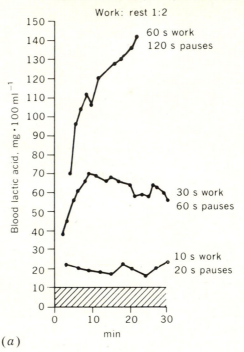

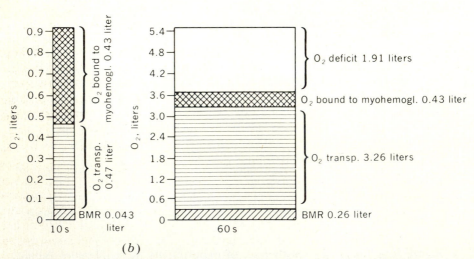

Figure 9-5 (a) The blood lactic acid concentration in a total work production of 247 kJ (25,200 kpm) in 30 min. The work was accomplished with a load of 412 watts (2520 kpm · min⁻¹), the work periods being 10, 30, and 60 s, and the corresponding rest periods 20, 60, and 120 s respectively. *(From I. Åstrand et al., 1960b.)*
(b) The oxygen requirement for 10- and 60 s work at a load of 412 watts (2520 kpm · min⁻¹). The schematic drawing indicates the basal metabolic rate (BMR), the calculated fraction of O₂ bound to myoglobin, transported by the blood, and the O₂ deficit. *(From I. Åstrand et al., 1960b.)*

complete the intended production of 247 kJ within 30 min with no severe feeling of strain, and his blood lactate concentration did not exceed 20 mg · 100 ml^{-1}, indicating an almost balanced oxygen supply to his heavily stressed muscles. With periods of work and rest of 30 and 60 s respectively, intermediate results were obtained.

A prolongation of the rest periods, with the ratio between work and rest changed to 1:4, gave, of course, a decreased total work output but had scarcely any beneficial effect on the subject's fatigue. The critical factor was the length of the work periods, and the duration of the rest pauses and the total time spent resting during the 30-min period were only of secondary importance.

Figure 9-5b attempts to explain the findings presented above. When a person works for short periods at an extremely high energy output, the aerobic supply is apparently adequate despite an insufficient transport of oxygen during the burst of activity. There is, at least, no continuous increase in blood lactate concentration. It is perhaps not very likely that a marked production of lactic acid could be balanced by a simultaneously increased disappearance rate by oxidation and reconversion to glycogen. A possible explanation for a predominantly aerobic oxidation might be that at the beginning of every work period, the muscles have a certain volume of oxygen at their disposal. We may assume that oxygen bound to myoglobin constitutes such an oxygen store, which is consumed during the initial phase of exercise before circulation and respiration can provide an additional supply which may or may not be adequate. During the rest period these depots are refilled with oxygen. Consequently, during severe exercise it is essential that the work periods are kept sufficiently brief to prevent the oxygen supply from being exhausted and the anaerobic lactic acid production from being too great. By spacing the work so that running periods lasted for 10 s and resting for 5 s, a subject could prolong the total work and rest period to 30 min without undue fatigue at a speed that normally exhausted him after about 4 min continuous running.

It has been noticed that in intermittent exercise with high intensity there is a reduction in the phosphocreatine concentration in the working muscles and it is not restored during the resting periods. That means that this phosphohagen does not serve as an energy reserve which oscillates between being uncharged in the exercise period and than charged again during rest (Saltin et al., 1976). It has also been established that the muscle lactate concentration follows similar courses as the blood lactate (Fig. 9-5a). The levels were slightly higher and there was a release of lactate from the working muscles throughout the exercise period. However, the lactate produced with short exercise periods (5 to 20 s) was rather small and balanced by a similar lactate disappearance in the body (Saltin et al., 1976). The recruited muscle fibers had also a high oxidative potential (mainly type I and type IIa).

Dynamic exercise is certainly an intermittent type of work, and its superiority over static work for endurance exercise may be explained partly on the basis of the muscle pump and the alternating emptying and filling of the oxygen store during alternating muscle contraction and relaxation.

Summary   The buffering effect of a hypothetical oxygen store may mean, practically speaking, that a great amount of work can be performed at an extremely heavy work load, with a relatively low peak load on the circulation and the respiration, by the introduction of properly spaced, short work and rest periods ("micropauses"). The heavier the work load, the shorter should be the work periods. This physiological concept has at least two important applications:

1   It may explain why older or physically disabled individuals, in spite of a reduced maximal aerobic power, can remain in jobs involving heavy work, such as forestry, farming, and construction, or enjoy physically demanding hobbies. As long as they are free to choose the optimal length of the work and rest periods, the acute loads on respiration and circulation may not exceed the limits of their reduced capacity. It should be emphasized, however, that if the work pace is determined by a machine, even a less heavy peak load, but with relatively long work periods, may overtax the capacity of the workers whose physical work capacity is limited.

2   If the aim of a training program is to increase muscle strength, a given period of time would permit the highest load on the muscle fibers if periods of rest were frequently introduced between activity periods of 5 to 10 s. On the other hand, a training of the oxygen-transporting system will be more effective if the exercise periods are prolonged to at least 2 to 3 min. This type of work would also adapt the tissues to high lactate concentrations, provided the exercise is severe.

Prolonged Work

When the work time is extended to about 1 hr, the oxygen uptake, heart rate, and cardiac output are maintained at the same level as attained after about 5 min of exercise, provided that the oxygen uptake is not higher than about 50 percent of the maximum. The lactic acid concentration of the arterial blood is not elevated, indicating a steady state (I. Åstrand et al., 1959; I. Åstrand, 1960) (Fig. 9-6a). The well-trained individual can maintain steady state or equilibrium at a still higher relative work load, indicating a more efficient oxygen transportation and oxygen utilization in the working muscles. Elite cross-country skiers can work at 85 percent of their maximal aerobic power at least for 1 hr, oxygen uptake being 4.5 liters · min^{-1} or even higher (P.-O. Åstrand et al., 1963). Marathon runners had energy expenditures requiring from 68 to 100 percent of their maximal oxygen uptake (Maron et al., 1976). Well-trained subjects can work for hours with an oxygen uptake around 70 to 80 percent of their maximum with little or no increase in blood lactate concentration (see Costill, 1970; Hedman, 1957).

When work time is further prolonged, there is a progressive increase in oxygen intake and heart rate, and the subject becomes more or less fatigued. Figure 9-6b illustrates an experiment in which exercise was performed continuously for seven 50-min periods at an oxygen uptake of 50 percent of the subject's maximal oxygen uptake. The subjects rested for 10 min in between, and after 4 hr of work they had a break of 1 hr for lunch. The most fit subject, with a maximal oxygen uptake of 5.60 liters · min^{-1}, worked with an average

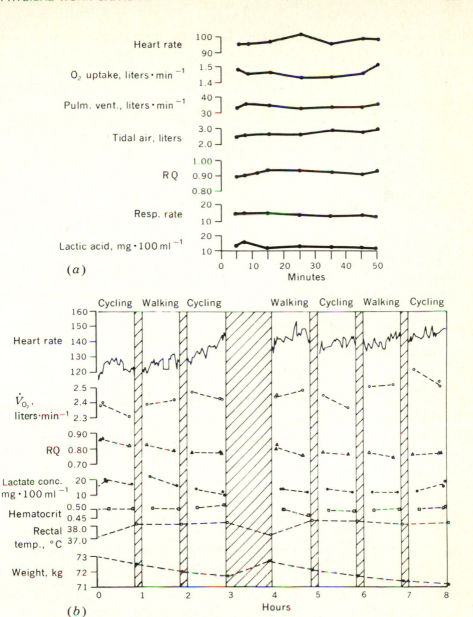

Figure 9-6 (a) Metabolic parameters during 1-hr work in a subject working at a load (1.5 liters $O_2 \cdot min^{-1}$) close to 50 percent of his maximal aerobic power (2.94 liters $O_2 \cdot min^{-1}$). *(Data from I. Åstrand et al., 1959.)*
(b) Metabolic parameters in one subject during an experiment consisting of seven work periods of 50 min each. The shaded columns represent rest periods. The subject's maximal aerobic power was 4.6 liters $O_2 \cdot min^{-1}$. *(From I. Åstrand, 1960.)*

oxygen uptake of 2.75 liters · min^{-1}; one subject with a maximal aerobic power of 2.25 liters · min^{-1} worked with a work load requiring an oxygen uptake of 1.15 liters · min^{-1}. The four subjects participating in the experiments could fulfill the task, but they were fatigued. It appears that a 50 percent load is too high for a steady state if the physical activity is continuous for a whole working day.

During prolonged heavy exercise, the water balance may be disturbed and the stores of available energy, particularly glycogen, may be critically low. Therefore the individual's ability to transport oxygen from the air to the working muscles may not always be the limiting factor. It has been found that the subjective feeling of fatigue during heavy work usually coincides with a drop in blood glucose in the fasting subject (Christensen and Hansen, 1939; Rodahl et al., 1964), and/or a depletion of the glycogen depots in the working muscles (Hultman, 1967). An increase in heart rate with reduction in stroke volume as work proceeds is often observed during prolonged exercise, particularly in a hot environment (see Saltin, 1964; Rowell, 1974). If a dehydration and the fall in blood sugar are prevented by proper supply of fluid and sugar, performance capacity is better maintained during prolonged exercise. (Further discussions appear in Chaps. 14 and 15.)

There are still many unsolved problems, and the limiting factor in prolonged exercise may vary from individual to individual. Training and environment may modify the work performance at a level which cannot be analyzed by the methods presently available. Thus, it is conceivable that the electrolyte balance, the K^+/Na^+ ratios for example, across the muscular cell membrane may be disturbed during prolonged exercise, and that the enzyme systems may be altered. As a matter of fact, in some experiments involving prolonged severe exercise, none of the physiological parameters studied correlates well with the subject's feeling of fatigue or reduction in performance capacity, e.g., blood sugar concentration, maximal oxygen uptake and cardiac output, and blood lactic acid level (Hedman, 1957; Rowell, 1974; Saltin, 1964). From studies of 14 experienced participants in a 100-km run, Oberholzer et al. (1976) concluded that the exhausting run, lasting about 10 hours, caused no distinguishable destructions in mitochondrial or cellular ultrastructure. A decrease in the activities of some extra- and intramitochondrial enzymes could be attributed to a reduced synthesis and/or enhanced loss of enzyme proteins (analysed on biopsies taken out of the vastus lateralis muscle; however, in running this muscle is less active than, e.g., the gastrocnemius).

Motivation is undoubtedly an important factor determining the endurance during heavy exercise. Well-trained, highly motivated subjects may maintain the oxygen uptake at a maximal level for at least 15 min, although most individuals feel forced to stop after 4 to 5 min at a work load which taxes the oxygen-transporting systems to a maximum.

Summary   It is obvious that the individual's maximal aerobic power plays a decisive role in his or her work capacity. If a given work task demands an oxygen uptake of 2.0 liters $\cdot$ min^{-1}, the man with a maximal O_2 uptake of 4.0 liters $\cdot$ min^{-1} has a satisfactory safety margin, but the 2.5-liter man must work close to his maximum, and consequently his internal equilibrium becomes much more disturbed. In prolonged exercise, motivation, state of training, water balance, and depots of available energy are important for the performance capacity.

Muscular Mass Involved in Exercise

The demand on the oxygen-transporting functions varies with the size of the active muscles. Since isometric work hinders the local blood flow and dynamic exercise facilitates the circulation, it follows that a greater oxygen uptake can be obtained during dynamic exercise. Usually, exercise involves both static and dynamic muscle contractions. Static work produces a relatively high heart rate and arterial blood pressure; this result may complicate a work evaluation based on the measurements of heart rate and blood pressure (Chap. 6).

For all practical purposes, the maximal oxygen uptake is approximately the same whether it is measured while running on a treadmill or during cross-country skiing or bicycling (P.-O. Åstrand and Saltin, 1961b). In maximal work on a bicycle ergometer in the supine position, the oxygen uptake is, however, only about 85 percent of the value obtained in the sitting position. But, if the subject works with both legs and arms simultaneously in the supine position, the oxygen uptake, cardiac output, and heart rate go up to the values typical for maximal exercise in the upright position (Stenberg et al., 1967). One plausible explanation for the lower work capacity for cycling in the supine position, despite an optimal venous return to the heart, is the less favorable work position, since the body weight cannot be utilized during the critical stages of the pedaling. Secondly, the blood perfusion of the working leg muscles is enhanced when working in the upright position (Chap. 6, p. 193). (See Fig. 6-23).

In a group of about 70 fairly well-trained women and men, no difference was noted in maximal oxygen uptake in the two types of exercise: work on the bicycle ergometer and uphill running on a motor-driven treadmill at an inclination of 1.75 percent (P.-O. Åstrand, 1952). Kasch et al. (1976) established maximal oxygen uptake on 12 well-trained students using two treadmill protocols: horizontal and inclined (>4 percent). No significant differences were found in maximal oxygen uptake, heart rate, and pulmonary ventilation. Rowell (1974) reports a significantly higher maximal oxygen uptake during running (grade 5 to 15 percent), and P.-O. Åstrand and Saltin (1961b) noticed a 5 percent difference, maximal $\dot{V}_{O_2}$ being higher during running than during cycling. Hermansen (1973) found, on the average, 7 percent higher maximal oxygen uptake in 3° uphill treadmill running than in bicycling. The difference was greatest in well-trained individuals. (See also Fig. 10-1.) It is possible that running at high speed on the treadmill with a low grade is technically so difficult that some subjects fail before reaching maximal oxygen uptake. Another explanation for a higher oxygen uptake when running at high grades and at low to moderate speeds may be that more muscles are active in the uphill running. However, the combined arm-plus-leg work should necessarily include more muscles than leg work only; yet oxygen uptake is not further increased, compared with leg work. On the other hand, slightly higher maximal oxygen uptakes are obtained by "ski-walking," that is, using ski poles as in skiing, while walking uphill at an inclination of 12° (21 percent) at a speed of 60 to 160 m · min⁻¹ (Hermansen, 1973).

In our experience, the maximal oxygen uptake is not consistently higher in running than in cycling, and the small difference sometimes noticed has scarcely any practical consequence. To obtain maximal values when using the bicycle ergometer, motivation and "stimulation" may be particularly important, owing to more pronounced local fatigue in the legs (knee region) when cycling. Furthermore, the work position is critical when using the bicycle ergometer as a tool. The bicycle seat should be high enough and the subject should be positioned almost vertically above the pedals. Otherwise the working position will be more or less similar to work in the supine position. On the whole, there are many reports that treadmill running may give slightly higher maximal oxygen uptakes than bicycling when the inclination is 3° or higher, in which case the oxygen-transporting capacity may be fully taxed without increasing the speed too much.

In arm exercise, the maximal oxygen uptake is about 70 percent of what is attained in leg exercise. The intra-arterial blood pressure during arm work is higher than in leg work at a given oxygen uptake or cardiac output (Fig. 6-22), and the heart rate is also higher. The consequence is a heavier load on the heart. For patients with heart disease or for completely untrained older individuals, heavy work with the arms (such as digging, shoveling snow) may therefore be hazardous. This may in part be due to the Valsalva effect during such maneuvers. The subject is apt to hold his or her breath while lifting the load, increasing the intrathoracic pressure which, in turn, may hinder the normal venous return to the heart.

When combining arm and leg exercise (cranking and bicycling) the highest oxygen uptake that can be attained depends upon the relative load on the arms. In a recent study by Bergh et al. (1976) it was noticed that the oxygen uptake was the same in maximal running as in arm plus leg exercise when the arm work load was 20 to 30 percent of the total rate of work (the total oxygen requirement exceeded the subject's maximal oxygen uptake). Subjects with strong arm and shoulder muscles could be submitted to relatively heavier arm work and still reach the maximum attained during uphill running. Otherwise, a typical finding was that the maximal oxygen uptake became reduced to 90 percent or below the maximum during running, when the arm work load was 40 percent of the total rate of work. However, the difference in oxygen uptake in leg exercise and arm plus leg exercise is much smaller than expected from the difference in mass of working muscles in the two procedures (see discussion in Chap. 6, p. 186). The central circulation may in one way or the other impose a limiting factor for the aerobic power. There are advantages when a large muscle mass is activated. Let us analyse Fig. 9-7. A work load of 350 watts could be tolerated

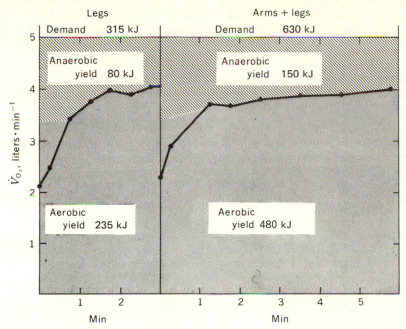

Figure 9-7 Increase in oxygen uptake at the start of exhausting work, following a 10-min warm-up period. Left: illustrates leg work only; right: exercise with the same work load but with both arms and legs involved. Load could be tolerated twice as long with arms and legs working. Calculations of energy demand and yield are explained in text. *(From data presented by P.-O. Åstrand and Saltin, 1961b; subject 1-P.-O.Å.)*

for about 3 min if only the leg muscles were involved. However, with 100 watts for the arms and 250 watts for the legs (=350 watts), the work time could be prolonged to 6 min, even if the oxygen uptake (and cardiac output) did not increase further (P.-O. Åstrand and Saltin, 1961b; Stenberg et al., 1967). Evidently the organism (inclusive heart) could tolerate a prolongation of the work period when a larger mass of skeletal muscles were activated. The subjective feeling of strain is related more to the metabolic rate per square area of muscle than to the total metabolism. Therefore, a training of the oxygen-transporting system is more efficient and is psychologically less strenuous, the larger the muscular mass involved in dynamic work.

It may be concluded that the assessment of the individual's maximal oxygen uptake, i.e., maximal aerobic power, should be made with the subject working in the upright position (running or bicycling) with or without arm exercise added. If the subject is tested when working in the supine position, both arms and legs should be involved in the work.

ANAEROBIC PROCESSES

During light work, the required energy may be produced almost exclusively by aerobic processes, as mentioned, but during more severe work, anaerobic processes are brought into play as well. Anaerobic energy-yielding metabolic processes play an increasingly greater role as the severity of the work load increases. These processes have been briefly discussed in Chap. 2. Here, we shall merely summarize the most important points and present some additional comments. We take an increase in lactic acid concentration in the blood as the main indication of the involvement of anaerobic processes, since the lactic acid concentration is easy to analyze. Furthermore, it should be recalled that the energy yield from the breakdown of ATP and phosphocreatine is indispensable, but that, quantitatively, the available stores of these high-energy phosphates alone can cover only the energy requirement for less than 1 min during maximal exercise. See Table 2-2.

Oxygen Deficit—Lactic Acid Production

1 During light exercise, the oxygen store in the muscle plus the oxygen supplied as the respiration and the circulation adapt to the work will completely cover the oxygen need. Most of the ordinary daily occupations belong to this category of work.

2 During exercise of moderate intensity, anaerobic processes contribute to the energy output at the beginning of the exercise until the aerobic oxidation can take over and completely cover the energy demand. Any produced lactic acid diffuses into the blood and can be traced in the venous blood draining the muscle and, eventually, in the arterial blood if the quantity of lactic acid produced is high enough. As the work proceeds, the blood lactate concentration falls again to the resting level and the work can be continued for hours (Bang, 1936).

3 During heavier exercise, the lactic acid production and, therefore, the rise in blood lactate concentration are higher and remain high throughout the work period. The length of time which the work load may be endured will, to some extent, depend on the subject's motivation.

4 During very severe exercise, there is a continuously growing oxygen deficit and an increase in the lactate content of the blood because of the predominantly anaerobic metabolism. The work cannot be continued for more than a few minutes, as a rule, because the subject's muscles can no longer function.

Figure 9-8 illustrates how the arterial lactic acid concentration increases during and after severe exercise, followed by a slow decline back to the resting level. The lactic acid is produced in the muscles during the actual work, but there is a time lag for the diffusion from the working muscles and redistribution within the body. For a determination of peak lactic acid in the blood, samples must be taken at intervals during the first 5 to 10 min of the recovery period. It should also be noted that it takes up to 60 min or even longer before the resting level is again reached, so that if the effect of a stepwise increasing work load is studied, the samples secured at the end of the last work period not only reflect the anaerobic component of this load, but they are also affected by the preceding work loads. Furthermore, in competitive events where the lactic acid

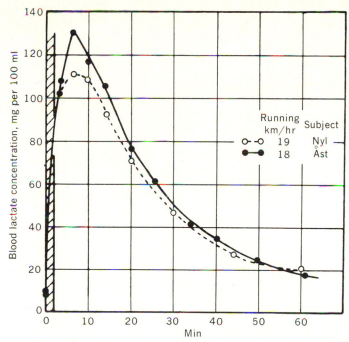

Figure 9-8 Blood lactate concentration after severe work of 2 min duration (shaded column) in two subjects. Peak values occur several minutes following the cessation of work. *(From I. Åstrand, 1960.)*

production is high, the time between heats should be at least 1 hr to allow time for the blood lactate to return to resting values. Obviously, this decision should be based on well-established physiological principles, and it should not be left to the organizing committee's arbitrary judging of whether or not the competitor should have to start a competition with a high tissue lactic acid concentration.

Lactic acid, partly buffered by the bicarbonates in the blood, lowers the pH of the blood, and the respiration is stimulated. An arterial pH as low as 7.0 has been observed after severe exercise (Chap. 7). Therefore, heavy exercise causes a hyperpnea and eventually a dyspnea. In blood samples taken at rest and during and after exercise (steady state as well as maximal effort), Keul et al. (1967) found a very high correlation between lactate concentration and pH values, and also between the sum of lactate and pyruvate concentrations and standard bicarbonate. Of the decrease in standard bicarbonate, about 95 percent was ascribed to the rise in lactate and pyruvate concentration. The remaining 5 percent was due to an increase in free fatty acids in the blood. Thanks to the buffer systems of the blood, a 10-fold increase in lactate concentration caused only a 1.42-fold increase in the H^+ concentration. The blood lactate concentration in men and women is, on the average, the same after maximal exercise and within the range of 11 to 14 mM for twenty- to forty-year-old trained individuals. Children and older individuals usually do not attain such high values (Robinson, 1938; P.-O. Åstrand, 1952; I. Åstrand, 1960). During training, the blood lactate concentration for a given work load is lowered, but the values attained during maximal physical effort are usually higher. The lactate concentration may exceed 20 mM in the blood and rise up to 30 mmoles · kg⁻¹ (wet weight) in working muscles. The highest values reported so far are in samples drawn from well-trained athletes at the end of competitive events of 1- to 2-min duration.

Key enzymes in the glucogenolyses are phosphofructokinase (PFK) and lactate dehydrogen-ase (LDH); see Chap. 2. In children the PFK concentration in skeletal muscles is lower than in adults (Eriksson, 1972). That may be one factor behind the children's lower peak lactate after exhausting exercise. After a period of training, the PFK concentration increased and so did the lactate concentration after maximal exercise. Fast twitch fibers have a 2 to 2.5 times higher enzyme activity of LDH than slow twitch fibers, with a predominance of the muscle type LDH (see Sjödin, 1976). That is one factor behind the emphasis on the fast twitch fibers' potential for anaerobic energy yield. However, 8 elite cross-country skiers characterized by a high maximal oxygen uptake (80 ml · kg⁻¹ · min⁻¹) and a very high percentage of slow twitch fibers (75 percent), typical for endurance athletes, attained very high lactate concentrations in a 5- to 10-min maximal exercise on the treadmill, or in average 14.7 mM (Bergh, personal communication). Apparently, the slow twitch fibers can also work anaerobically quite effectively, and the LDH activity in the skeletal muscles does not directly correlate with their anaerobic power.

Figure 9-7 summarizes experiments performed on bicycle ergometers. From work load and a mechanical efficiency of 22 percent for aerobic work, the energy demand can be calculated. During 3-min exercise, at a rate representing the maximum in leg work for this subject, the energy demand was about 315 kJ (1 kJ = 0.24 kcal). The oxygen uptake was measured continuously and was 10.7 liters. It is calculated that an additional 0.5 liter was utilized from stores bound to myoglobin and hemoglobin, refilled after the exercise. Therefore, the aerobic energy yield can be estimated to be 235 kJ (11.2 × 21). The deficit was then 315 − 235 = 80 kJ, and this energy must have been derived anaerobically. A breakdown of ATP and phosphocreatine may yield 20 to 30 kJ, i.e. it may substitute 1 to 1.5 liters of oxygen. The remaining deficit of a minimum of about 50 kJ must have been yielded by glycogenolysis and glycolysis with a formation of lactic acid.

Since about 220 kJ are released for each 6-carbon unit of glycogen which is converted into lactic acid (Chap. 2), a production of 2 moles or 180 g of lactic acid should yield 220 kJ or about 1.2 kJ · g⁻¹ of lactic acid (220 · 180⁻¹ = 1.22). For a release of 50 kJ, the lactic acid production must then be about 40 g.

As mentioned before, the subject also performed the same work load with both arms and legs, and under these conditions, the work could be prolonged to 6 min before exhaustion (Fig. 9-7). At submaximal exercise, the mechanical efficiency is not significantly different from ordinary cycling. (If anything, the oxygen uptake tends to be higher in work performed with both arms and legs than in leg work alone; it is also conceivable that the mechanical efficiency becomes lower at very heavy exercise since muscles which are at a mechanical disadvantage have to contribute. Therefore, the calculated energy demand is probably a minimal figure.) The energy requirement is therefore assumed to be 105 kJ · min⁻¹, or 630 kJ altogether, during the 6 min. The measured oxygen uptake of 22.3 liters, complemented with 0.5 liter from oxygen stores within the body, covers 480 kJ, leaving 150 kJ for the anaerobic processes. A subtraction of 30 kJ as a contribution from high-energy phosphate compounds leaves 120 kJ from glycogenolysis, i.e., a formation of 100 g lactic acid. (Since more muscles were working, the oxygen from myoglobin and the energy yield from ATP and phosphocreatine may have been larger. Quantitatively, however, this does not alter these calculations significantly.)

Figure 9-9 presents data from a subject who worked for 2.63 min with a work load close to 400 watts with a calculated energy demand of 295 kJ. The total oxygen uptake during activity was 7.1 liters. Adding 0.5 liter from oxygen stores, the aerobic energy yield will be 160 kJ. Therefore, the anaerobic contribution was 135 kJ. Assuming that a breakdown of glycogen yielded 105 kJ, the production of lactic acid would have been about 85 g.

Theoretically, these productions of lactic acid are possible since the glycogen content in the muscles is normally about 15 g · kg⁻¹ wet weight; thus, in 20 kg of muscles, there is 300 g of glycogen.

In Summary

Under very standardized conditions, as in the experiments just discussed, one can estimate the total energy output. With the aerobic energy yield measured,

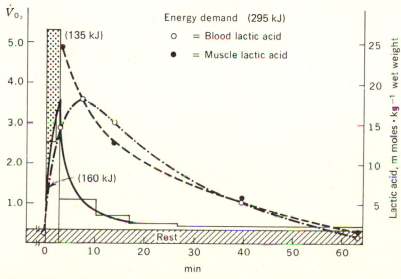

Figure 9-9 Calculated energy requirement for a 2.63-min exercise on a bicycle ergometer (column represents 295 kJ, 70 kcal) and measured oxygen uptake during exercise and during 60 min recovery (dotted area). Horizontal lines denote the level of oxygen uptake measured at rest before exercise. Calculated aerobic energy yield during exercise: 160 kJ (38 kcal); anaerobic energy yield: 135 kJ (32 kcal). Lactic acid concentration was analyzed in blood samples and pieces of skeletal muscle obtained by needle biopsy. *(Data by courtesy of B. Diamant, K. Karlsson, and B. Saltin.)*

the anaerobic contribution can be calculated. In most exercises it is, however, impossible to estimate energy demand. Di Prampero et al. (1973) have postulated that the quantity of lactate accumulating may be calculated from the blood lactate concentration. The weak points in this reasoning are the lack of information about the total water in the body available for uptake of lactate and the uneven distribution of lactate in the various water compartments in the body. Other hidden information is the rate by which lactate is chemically removed during exercise and recovery. In other words, at the present, we have no good method available for a measurement of the anaerobic power and the rate of glycogenolysis.

It should be emphasized that if a standardized exhaustive exercise is performed with blood and muscle lactate concentrations elevated in advance (because other muscle groups were previously exercised to exhaustion) then the performance time will be reduced (see Karlsson et al., 1975). This must be considered when one is warming up before an athletic performance!

Lactic Acid Distribution and Disappearance

We have seen that about 100 g of lactic acid may be produced within a few minutes; in a well-trained top athlete, it may be still higher. Lactic acid is assumed to diffuse freely into all water compartments of the body. With a water content of 40 liters, 4 g of lactic acid would give a concentration of 10 mg $\cdot$ 100 ml^{-1} of water, or roughly 10 mg $\cdot$ 100 ml^{-1} or 1.1 mM in of blood. This concentration is actually the resting level of blood lactate. An additional 4 g of lactic acid would double the concentration.

A production of 40 and 100 g of lactic acid (Fig. 9-8) would increase the blood lactate concentration to approximately 12 and 28 mM respectively. The noticed peak concentration was, however, 17.5 mM blood in both experiments. The discrepancy can be explained as follows: A delay in diffusion into the various water compartments of the body would give a higher blood concentration than expected; a continuous removal by resynthesis and oxidation will cause a lower blood concentration than the expected one. Muscle biopsies on men after maximal exercise have revealed a content of about 30 mmoles lactic acid per kilogram wet muscles, but the blood concentration was just above 20 mmoles/liter (J. Karlsson, 1971). During short-term work the lactic acid concentration in the working muscle group may be 3 to 5 times higher than that in blood (Hultman, 1967). For these and other reasons, it is impossible to calculate the total production of lactic acid, that is, the anaerobic energy yield from glycolysis from the concentration of lactic acid in the blood. A rise merely indicates that an increased formation has occurred.

In Fig. 9-9, data on lactic acid concentration are included, determined in pieces of muscles obtained by biopsies and the lactic acid concentration in blood. During early recovery, it is significantly higher in the exercising muscles than in blood.

At rest, it is generally assumed that most of the lactic acid produced during work is resynthesized back to glycogen, mainly in the liver, but also in the kidneys (Cohen and Little, 1976; Levy, 1962). To what extent gluconeogenesis from lactic acid may take place in mammalian muscles is still an open question (Krebs, 1964a). As pointed out by Keech and Utter (1963), three possible pathways do exist for the formation of phospho-enolpyruvate from pyruvate in all tissues. Previously, several of the essential enzymes for these pathways were believed not to exist in the muscle. However, recent investigations have shown that a set of key enzymes for one of the three possible pathways does exist in the required amounts in the muscle (Newsholme et al., 1976). Lactate is oxidized to CO_2 and H_2O, and this oxidation can occur in the heart muscle (Newman et al., 1937; Carlsten et al., 1961; Keul et al., 1966) and also in skeletal muscles. The degree to which lactic acid replaces the substrates ordinarily oxidized is uncertain (Issekutz et al., 1965). There is a removal of lactate by nonexercising muscles and most likely a subsequent metabolic disposal of the lactate by the resting muscle tissue with an increase in the oxygen uptake in this tissue (see Ahlborg et al., 1976).

Newman et al. (1937) noticed that the removal of lactic acid, accumulated in the body after exhausting exercise, was enhanced if during recovery the subject continued to exercise, but at a lower intensity which normally did not produce any lactic acid. Similar observations are reported by Belcastro and Bonen (1975) and Gollnick and Hermansen (1973). In the 1975 study the optimal removal rate occurred when the "recovery" exercise gave an oxygen uptake of less than 50 percent of the maximum (bicycling), and in the 1973 study at an oxygen uptake of about 65 percent of the maximal one (running). The type of exercise may explain the different optimal levels.

Brooks et al. (1973) have, from their studies on rats, concluded that the "primary fate of lactic acid after exercise appears to be oxidative." There is, however, indirect evidence that a large part of lactate is "lifted up" to glycogen. As mentioned, 1 mole of lactate (90 g) may be produced, and an aerobic utilization of 1 mole lactate demands 3 moles of oxygen, or about 67 liters (Chap. 2). After exhausting exercise, blood and muscle lactate can return to or close to resting level within 1 hr. The total oxygen uptake during that hour can, however, hardly exceed 40 liters, and space must be given for oxidation of free fatty acids during recovery. The hypothesis that most of the lactates are oxidized is not tenable. We can conclude that the fate of the lactic acid produced during heavy exercise is still an unsettled question. The loss of lactic acid with sweat and urine is negligible (Newman et al., 1937).

During steady-state work of moderate intensity, with blood lactate concentrations up to about 4 mM, the uptake of lactates in the liver has been determined (Hultman, 1967; Rowell, 1974). This uptake was 100 to 150 mg · min^{-1}; i.e., it would take 6 to 10 min to remove 1 g of lactic acid. However, the rate of lactate uptake by splanchnic tissues appears to be proportional to its blood concentration. With a production of some 100 g of lactic acid, it is necessary to assume a high disappearance rate, since the peak lactic acid concentration is down to resting level after about 1 hr.

Oxygen Debt

To some extent, lactic acid is eliminated during exercise (Bang, 1936; Newman et al., 1937; see Ahlborg et al., 1976), and part of the aerobic energy deficit during the early stage of exercise can soon be repaid. On the other hand, there is always an oxygen debt after exercise (Fig. 9-3a), which is the total oxygen uptake minus the resting oxygen uptake. If measured after a steady-state exercise, it is of the same magnitude whether the work time is 10 or 60 min.

As discussed at the beginning of this chapter, various factors are involved in the delayed return of oxygen uptake during a recovery to the basal level. After maximal exercise of a few minutes' duration, this excess oxygen uptake, measured during 60 min, may reach values as high as about 20 liters. A refill of the oxygen stores (blood, myoglobin) will demand less than 1 liter of oxygen. At the same elevated tissue temperature and epinephrine concentration at rest that are attained during heavy exercise, up to 1 liter extra oxygen may be consumed. Increased cardiac and respiratory functions may require some 0.5 liter extra oxygen, giving a total of about 2.0 to 2.5 liters of oxygen uptake during recovery, which has nothing to do with the energy transfer in handling anaerobic end products. A breakdown of ATP and phosphocreatine may yield 20 to 30 kJ; that is, it may substitute 1 to 1.5 liters of oxygen. Therefore, up to 4 liters of the oxygen debt may be *alactacid*, i.e., it is not involved in the handling of lactic acid, or the *lactacid* oxygen debt (Margaria et al., 1933; Margaria, 1967). The repayment of the alactacid debt forms a fast component in the decline of the oxygen uptake during recovery from exercise. The lactacid oxygen debt is the main (but not only) factor behind a slower component in the return of oxygen uptake to the pre-exercise level (Fig. 9-9). There is a close relationship between the reduction in ATP and phosphocreatine on one hand, and both oxygen deficit and the fast component of the oxygen debt on the other (see Knuttgen and Saltin, 1973).

According to Krebs (1964), 1 mole of oxygen can remove maximally about 2 moles of lactate, of which 1.7 are resynthesized and 0.3 oxidized to CO_2 and H_2O. Therefore, 22.4 liters of oxygen can remove about 180 g of lactic acid. (It takes 7 moles of ATP to resynthesize 1 mole of glycogen from lactate, and 1 mole of oxygen yields, on an average, 6 of ATP.) As mentioned, a production of 180 g of lactic acid from glycogen yields 220 kJ. However, an aerobic oxidation of glycogen would provide 220 kJ with only about 11 liters of oxygen used. Therefore, the body has to pay about 100 percent interest in the currency of oxygen for the energy borrowed from this anaerobic bank. It is apparently important to keep the glycogen store as high as possible and, since we live in an "ocean of oxygen," the payment is not a problem. Evidently we should expect the oxygen debt to be at least twice as great as the oxygen deficit (Fig. 9-3a), which indeed it is.

The calculated production of 100 g of lactic acid (Fig. 9-7) would demand, for its elimination, about 12 liters of oxygen; the total oxygen debt would be, at the most, 16 liters.

For the experiment presented in Fig. 9-9, the lactacid oxygen debt can be calculated to 10.2 liters ($82 \cdot 180^{-1} \cdot 22.4$); with an alactacid oxygen debt of 4.0 liters, the total oxygen debt would be about 14 liters. The measured oxygen uptake during recovery above the resting demand was 13 liters, a figure which is in good agreement with the one calculated from theoretical considerations.

The oxygen uptake at rest before the experiment was 0.35 liter $\cdot$ min^{-1}, which is quite high. Another time, 0.32 liter $\cdot$ min^{-1} was measured, and with this metabolic rate for 60 min, the oxygen debt would have been calculated to 14.2 liters. With an oxygen uptake of 0.28 liter $\cdot$ min^{-1}, estimated basal metabolic rate, the oxygen debt for a 60-min period would have been 17.1 liters. It should therefore be emphasized that it is very difficult to separate accurately the oxygen debt from resting oxygen uptake according to the definition.

Anyway, considering the potential energy yield for each volume of oxygen consumed, the efficiency of the anaerobic processes is apparently about 50 percent of the aerobic ones. A similar conclusion was drawn by Asmussen (1946) and Christensen and Högberg (1950). The more of the formed lactic acid that is combusted to CO_2 and H_2O, the higher will be the net mechanical efficiency of the anaerobic processes if estimated from the total oxygen uptake during work *and* recovery. Again, it should be emphasized that the measured oxygen uptake during recovery from exhausting exercise is not large enough for a removal of large amounts of lactic acid by oxidation. It is also important to build up the glycogen depot again for future use!

Summary At the beginning of work and during heavy work, there is a discrepancy between energy demand and the energy available by aerobic processes. The anaerobic energy yield must therefore contribute, and a breakdown of glycogen to pyruvic acid and lactic acid is quantitatively the most important step in this anaerobic process.

The heavier the work in relation to the individual's maximal aerobic power, the larger is the oxygen deficit and the more important is the anaerobic energy yield. Under hypoxic conditions, such as at high altitude, it is noticed that the oxygen debt and blood lactate concentration are higher at a given work load compared with sea-level values (Lundin and Ström, 1947; P.-O. Åstrand, 1954; Hermansen and Saltin, 1967; Knuttgen and Saltin, 1973).

It is calculated that about 200 kJ can be provided by the glycogen–lactic acid mechanism in an athlete successful in events of 1- to 2-min duration. This covers an oxygen deficit of about 9 liters. ATP and phosphocreatine may cover an additional oxygen deficit of 1 to 1.5 liters.

Mainly during the recovery period after the exercise, most of the lactic acid formed (up to 85 percent?) is resynthesized to glycogen and the rest is oxidized to CO_2 and H_2O. For these processes extra oxygen is taken up, and this *lactacid* oxygen debt is about twice as high as the oxygen deficit. In the example given, the extra oxygen uptake due to the previous lactic acid formation will be about 16 liters. In addition, a few liters extra oxygen is taken up from the inspired air for the resynthesis of high-energy phosphate compounds, to refill oxygen stores in the body which have been depleted during heavy exercise, and to cover aerobically the increased metabolism due to elevated tissue temperature, the effect of epinephrine, and cardiac-respiratory activity above the basal level. Altogether, the oxygen uptake above the resting basal level after an all-out effort of a few minutes' duration can exceed 20 liters.

INTERRELATION BETWEEN AEROBIC AND ANAEROBIC ENERGY YIELD

Table 9-1 presents the contribution to energy output from aerobic and anaerobic processes respectively in maximal efforts in events involving large

Table 9-1

Process	Work time, maximal effort							
	10 sec	1 min	2 min	4 min	10 min	30 min	60 min	120 min
Anaerobic								
kJ	100	170	200	200	150	125	80	65
kcal	25	40	45	45	35	30	20	15
percent	85	65–70	50	30	10–15	5	2	1
Aerobic								
kJ	20	80	200	420	1000	3000	5500	10,000
kcal	5	20	45	100	250	700	1300	2400
percent	15	30–35	50	70	85–90	95	98	99
Total								
kJ	120	250	400	620	1150	3125	5580	10,065
kcal	30	60	90	145	285	730	1320	2415

muscle groups. The individual's maximal aerobic power is set to 5 liters $\cdot$ min^{-1} = about 100 kJ $\cdot$ min^{-1} and maximal anaerobic capacity to 200 kJ $\cdot$ min^{-1}, equivalent to 9 liters of oxygen uptake in aerobic work. It is assumed that 100 percent of maximal oxygen uptake can be maintained during 10 min, 95 percent during 30 min, 85 percent during 60 min, and 80 percent during 120 min.

With a work time of up to 2 min, the anaerobic power is more important than the aerobic one; at about 2 min there is a 50:50 ratio, and with longer work time the aerobic power becomes gradually more dominating. This is graphically illustrated in Fig. 9-10.

It is very rare that an individual possesses top power for both aerobic and anaerobic processes. Therefore, the analysis in Table 9-1 should not be interpreted and applied too literally. A maximal aerobic power of 100 kJ $\cdot$ min^{-1} may be coupled with a maximal anaerobic yield of 100 kJ. For this individual, the proportional participation of anaerobic and aerobic processes will be different compared with the tabulated data. (For example, a man should be advised to compete over longer distances since his relatively low maximal anaerobic power is then less of a handicap, and he should try to get rid of his rivals before starting the finish!)

An analysis of the energetic demands of different sport events and the athlete's capabilities to fulfill these requirements may help him or her in both training and the selection of suitable events. One factor to consider is that endurance athletes have skeletal muscles with a high proportion of slow twitch fibers and sprinters are characterized by a dominance of fast twitch fibers. However, in many events there is not such a strict pattern in fiber composition (see Costill et al., 1976; Saltin et al., 1977). So far a fiber typing is not a good selective instrument for picking potential athletes.

A trained individual can work at a relatively high oxygen uptake in relation to his or her maximum (up to 60 to 70 percent) without any elevation in blood lactate concentration. In the untrained person, a rise is noted at about 50

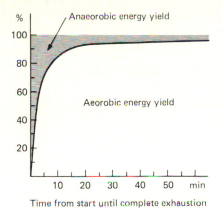

Figure 9-10 Relative contribution in percentage of total energy yield from aerobic and anaerobic processes respectively during maximal efforts of up to 60 min duration for an individual with high maximal efforts of up to 60 min duration for an individual with high maximal power for both processes. Note that a 2-min maximal effort hits the 50 percent mark, meaning that both processes are equally important for success.

percent of maximal aerobic power (P.-O. Åstrand, 1952; Hermansen and Saltin, 1967; Williams et al., 1967).

It is unknown which factor or factors decide the pathway of the energy-yielding processes, i.e., whether the anaerobic or aerobic processes will be preferred. From studies on glycolysis in working frog muscles, Karpatkin et al. (1964) conclude that it is difficult to ascribe activation of some key enzymes during stimulation to changes in the concentration of substrates, activators, and inhibitors. Regulatory factors for some of the steps in the energy-yielding processes were discussed in Chap. 2. With oxygen available, an aerobic utilization of free fatty acids is enhanced. This will increase the concentration of citrate which can inhibit PFK and thereby the glycogenolysis.

In acute hypoxia there is—during a given submaximal exercise—a decreased acceleration from resting oxygen uptake, an increase in the oxygen deficit, and at the higher insities, increased lactate formation compared with control experiments at normal oxygen tension (Knuttgen and Saltin, 1973). In hyperoxic condition the opposite reactions are noticed (Fagraeus, 1974). So, in one way of another, the availability of oxygen has a key role in the interrelation between aerobic and anaerobic energy yield.

It has been pointed out (Chap. 6, p. 186) that the highest volume of oxygen that can be *offered* to exercising skeletal muscles may be decisive for their aerobic power. From a teleological viewpoint it is very efficient that low-threshold slow twitch fibers are recruited during exercise of moderate intensity. They have the enzyme potentials for aerobic work, and the myoglobin enhances the oxygen diffusion within the cells (see Chap. 7, p. 245). During very heavy exercise, the high-threshold fast twitch fibers also become activated, but the oxygen supply is deficient. The myoglobin in the slow twitch fibers directs the oxygen to those fibers preventing the fast twitch fibers from "stealing" oxygen, but the fast twitch fibers are well equipped for anaerobic work and from a mechanical point of view they are the specialists for very intensive exercise. In other words, when the fast twitch fibers are required a limited volume of oxygen is available for them (beginning of exercise, very vigorous exercise). It would be something of a waste of resources to have their metabolic repertoire developed as replicas of the slow twitch fibers. However, with training the potential of the central circulation to transport oxygen out to the tissue is developed. With more oxygen available, it makes sense that fast twitch fibers can improve their aerobic potential; they will not be the same rivals to the slow twitch fibers, and in fact the maximal oxygen uptake will increase (for a more detailed discussion, see Chap. 12).

Summary   In maximal efforts of about 2-min duration, the aerobic energy yield equals approximately the anaerobic yield. With shorter work times, the anaerobic processes dominate; with longer work times, the anaerobic energy yield is more important from a quantitative viewpoint.

MAXIMAL AEROBIC POWER—AGE AND SEX

It should be recalled that the maximal aerobic power is defined as the highest oxygen uptake the individual can attain during physical work while breathing air at sea level. To evaluate whether or not the subject's maximal oxygen uptake has been attained, objective criteria should be used, such as measured oxygen uptake lower than expected from the work load, and/or blood lactic acid concentration higher than about 8 mM.

The information provided by the assessment of maximal oxygen uptake is a measure of (1) the maximal energy output by aerobic processes, and (2) the functional capacity of the circulation, since there is a high correlation between the maximal cardiac output and the maximal aerobic power (see Fig. 6-25).

Direct measurements of the maximal oxygen uptake on 350 individuals ranging in age from four to sixty-five are presented in Fig. 9-11a. All subjects were healthy and moderately well trained; none of them was an athlete. It should be emphasized that it is almost impossible to present "normal material" since it is very difficult to define what is normal. This material is selected, but the age and sex factors which modify maximal aerobic power should be fairly evident in this homogeneous group of subjects.

Before puberty, girls and boys show no significant difference in maximal aerobic power. Thereafter, the women's power is, on an average, 70 to 75 percent of that of the men. In both sexes, there is a peak at eighteen to twenty years of age, followed by a gradual decline in the maximal oxygen uptake. At the age of sixty-five, the mean value is about 70 percent of what it is for a twenty-five-year-old individual. The maximal oxygen uptake for the sixty-five-year-old man (average) is the same as that typical for a twenty-five-year-old woman.

The individual variation should be noticed. Many old subjects have a maximal power that is higher than that found in many much younger individuals. In Fig. 9-11, the "−2 standard deviation line" for male subjects coincides closely with the average values for women, and the 95 percent range is actually ±20 to 30 percent of the mean value at a given age.

Since regular training in previously trained persons can, in most cases, increase the maximal oxygen uptake not more than 10 to 20 percent, it is

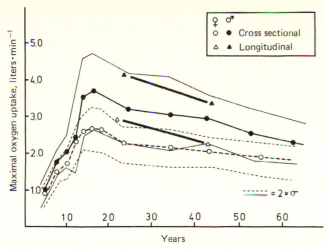

Figure 9-11 Mean values for maximal oxygen uptake (maximal aerobic power) measured during exercise on treadmill or bicycle ergometer in 350 female and male subjects four to sixty-five years of age. Included are values from a group of 86 trained students in physical education *(from P.-O. Åstrand and Christensen, 1964)*, and measured maximal oxygen uptakes in a longitudinal study (21 years) of 35 female and 31 male subjects of the former students *(I. Åstrand et al., 1973).*

evident that the natural endowment is the most important factor determining the individual's maximum. In the 1940s, Gunder Hägg held many world records in middle- and long-distance running. No physiological data pertaining to him are available from this period. His body weight at that time was 69 to 70 kg. In 1963, at the age of forty-five his heart rate and oxygen uptake were recorded while he was working on a bicycle ergometer. His maximal O_2 was 4.0 liters $\cdot$ min^{-1} and his maximal heart rate 181 beats $\cdot$ min^{-1}. His blood lactate concentration was 13.8 mM. His body weight was 94.5 kg. Although he had not trained since 1946, he had a very high maximal aerobic power (see Fig. 9-11). This is an example of how an untrained individual, given a favorable endowment, may have a very high aerobic power. From their studies on dizygous and monozygous twins, Klissouras et al. (1973) conclude that "regardless of age, existing individual differences in functional adaptability of man can be attributed to heredity." Sjödin (1976) reports that monozygous twins are relatively indentical in their fiber composition in skeletal muscles, but in the dizygous pairs there was a wide variation.

The maximal aerobic power does not in itself reveal whether an individual has been physically active in the preceding years.

Figure 9-11 includes data obtained from a group of 86 students in physical education. The mean values are definitely higher than for the average women and men, but the difference between the female and male students in maximal aerobic power is of the same magnitude as in the other material. The highest figures obtained so far on athletes: 7.4 liters $\cdot$ min^{-1} for a male cross-country skier, and 4.5 liters $\cdot$ min^{-1} for a female skier (see Fig. 12-5).

The absolute values for maximal oxygen uptake will inevitably vary for different groups and populations. Selection of subjects is critical and a random sample is difficult to study successfully. For subjects with different occupations, there is a definite trend that the mean values vary to some degree with the nature of the occupation (see I. Åstrand 1967b).

These differences in maximal aerobic power, as well as in many other parameters, are probably partly due to a selection, since those with a strong constitution are overrepresented in occupations with physically demanding tasks. Furthermore, such jobs may in themselves train the oxygen-transport system. The more mechanized the society, the less will such differences probably become among personnel in different occupations. On the other hand, conclusions concerning the general physical standard in a country from physiological data must be drawn with caution.

An interesting question is why the best performance in endurance events is usually obtained by athletes twenty-five to thirty years of age, when the highest maximal oxygen uptake is usually reached at the age of eighteen to twenty. However, there are several factors to be considered. Generally, physical activity is more regular and vigorous for those below than for those above twenty years of age, at least if physical education is compulsory in school. This may explain the results presented in Fig. 9-11. On the other hand, if training is continued, the maximal aerobic power can certainly be maintained or even further increased for another 10-year period. Finally, the performance also depends on technique, tactics, motivation, and other factors, and intensive training and experience over the years make gradual improvement possible.

Figure 9-12 shows how the performance capacity is related to the maximal

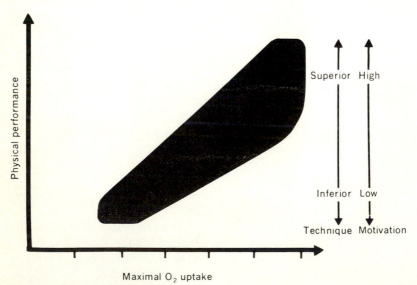

Figure 9-12 Schematic representation of the importance of maximal aerobic power in physical performance involving large muscle groups for more than a minute, and the role played by technique and motivation in modifying top performance.

oxygen uptake in exercises with large muscle groups vigorously involved for 1 min or longer. No one can attain top results in such exercises without a high aerobic power. On the other hand, a high power does not guarantee a good performance, since technique and psychological factors may have a modifying influence in a positive or negative direction.

In work and exercise where the body is lifted (as in walking or running), the oxygen uptake should be related to the body weight. Figure 9-13 presents the same material as in Fig. 9-11, but the maximal aerobic power is expressed in milliliters of O_2 per kilogram gross body weight. The values for young boys and girls are about the same, and the difference between average women and men is now reduced to 15 to 20 percent. The highest values so far recorded are 94 ml $\cdot$ kg^{-1} $\cdot$ min^{-1} for a male cross-country skier and 77 ml $\cdot$ kg^{-1} for a female cross-country skier.

Another way to express the individual's maximal oxygen uptake is to relate it to the dimensions of the body or to various organs. It may be of both theoretical and clinical value to examine whether the maximal oxygen uptake is proportional to heart size, muscular mass, lung volume, etc. Since fatty tissue is metabolically fairly inert but can constitute a large proportion of the body weight, it may be important to exclude it when evaluating the oxygen-transporting capacity.

When the weight of adipose tissue, estimated on the basis of hydrostatic weighing, is subtracted from the gross body weight of the well-trained students in Fig. 9-11 (12 kg or 20.3 percent of the body weight for the women and 8 kg or 10.6 percent for the men, Döbeln, 1956), the maximal oxygen uptake per kilogram fat-free body weight (lean body mass) can be calculated. The average

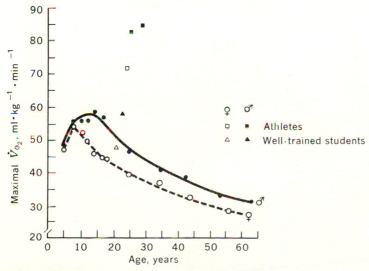

Figure 9-13 Mean values for maximal oxygen uptake expressed in ml O_2 $\cdot$ kg^{-1} $\cdot$ min^{-1}. Same subjects as presented in Fig. 9-11. The standard deviation is between 2.5 and 5 ml O_2 $\cdot$ kg^{-1} $\cdot$ min^{-1}.

figure was the same for both groups: 71 ml. However, since the metabolic rate at rest or during maximal muscular exercise should vary with the body weight raised to the $^2/_3$ power (Chap. 11), it is evident that the woman should have a higher aerobic power per kilogram lean body mass than the man. The explanation for a lower maximal oxygen uptake than expected may be the lower hemoglobin concentration in women. The lower maximal aerobic power of women may therefore be natural, since their dimensions are different from those of men, and the oxygen-binding capacity of the blood is lower. The relative increase in body-fat content in women starts at puberty.

The decrease in maximal oxygen uptake with age is not so simple to explain. Figure 9-14 shows how the maximal heart rate decreases with age from 195 for the twenty-five-year-old to about 165 for the sixty-five-year-old. The lower maximal heart rate with higher age certainly must reduce the maximal cardiac output and hence the oxygen-transporting capacity. Sixty-six of the former students in physical education were reexamined 21 years later and, without exception, there was a decline in oxygen uptake, in average by about 20 percent (Fig. 9-11). In this group there was, however, no correlation between the decline in maximal oxygen uptake and the change in the individual's maximal heart rate.

There is another consequence of a lower maximal heart rate. Studies on 33 building workers (bricklayers, carpenters, laborers) from thirty to seventy years of age showed that the mean heart rate during occupational activity was correlated with the individual's maximal heart rate (Fig. 9-15). For subjects with a maximal heart rate of 185 beats/min, mostly the younger workers, the mean heart rate during occupational activity was 110, and those with a maximum of 150 had a mean of 90 beats · min⁻¹. The maximal oxygen uptake ranged from 2.2 to 3.6 liters · min⁻¹. In general, the worker utilized the same percentage of his maximum in the work operations irrespective of his maximal oxygen uptake. In other words, the older worker with a lower maximal aerobic power keeps a slower tempo than the younger one, but the relative load is the same for the two workers, or about 40 percent. The person with a high maximal heart rate can do a day's work at a higher mean heart rate than a person with a low maximal heart rate, but the relative strain may be the same on the two persons (I. Åstrand, 1967a).

The stroke volume is lower in the older individual, but systolic and mean arterial pressures are higher; the pressures in the pulmonary artery are also

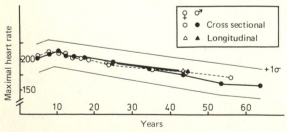

Figure 9-14 Heart rate during maximal exercise in the same 350 subjects represented in Fig. 9-11 *(from P.-O Åstrand and Christensen, 1964)*, and in a longitudinal study of 35 female and 31 male subjects *(from I. Åstrand et al., 1973)*.

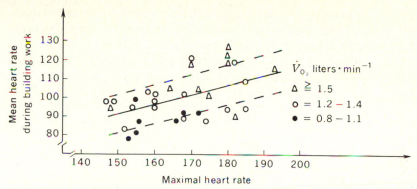

Figure 9-15 Individual values for the relationship between mean heart rate during occupational work (building) and maximal heart rate attained during work on bicycle ergometer. The heart rate was recorded by telemetry. The estimated oxygen uptakes during occupational work are presented on the right, together with their symbols. *(From I. Åstrand, 1967b.)*

higher in the aged (see Chap. 6). The blood lactate concentration during submaximal work is higher in older individuals but lower during maximal exertion (I. Åstrand, 1960).

As individuals grow older, they usually become less physically active. Therefore part of the decrease in maximal oxygen uptake and performance is an effect of inactivity. Saltin and Grimby (1968) have studied middle-aged and old athletes and compared the data obtained with those from former athletes of the same ages who now live a sedentary life. The two groups were about equal as far as their performance in orienteering (including cross-country running) was concerned. It was therefore concluded that there was, at the time, no significant difference in maximal oxygen uptake. As can be seen from Table 9-2, the nonactive former athletes have now fallen behind as far as their maximal aerobic power is concerned, particularly in relation to body weight. In all groups, the mean values are higher than normally found for the same age groups (Figs. 9-11 and 9-13). In the sedentary group, this is due to natural endowment (a highly selected group), whereas in those who are still active, this endowment is further developed by regular physical training. In both groups, there is a decline in maximal heart rate with age.

The heart volume in relation to maximal oxygen uptake was significantly larger in the older athletes and former athletes compared with healthy young males. The cholesterol level in the serum was not different from normal values, but the concentration of the serum neutral fat was lower in most athletes than the normal mean values.

In a separate study with nine well-trained athletes between the ages of forty-five to fifty-five, cardiac output was also measured during exercise (Grimby et al., 1966). The maximal cardiac output was very high, or on an average 26.8 liters · min^{-1} at an oxygen uptake of 3.56 liters · min^{-1}. The stroke volume was also high (average 163 ml), considering the age of the men. A low maximal arteriovenous oxygen difference seemed to be the main limiting factor for the oxygen uptake. This was partially due to a relatively low hemoglobin concentration, but probably also to peripheral factors, possibly including an increased diffusing distance from the capillaries in the aged skeletal muscle.

Table 9-2 Data on performance of present and former athletes in orientation racing. The youngest group is currently competing. The active older athletes are still training and competing regularly. The inactive subjects had, when young, competed successfully with those of the same age who are still active; they discontinued their training more than 10 years ago because of lack of time. N denotes the number of subjects in each group.

	Age						
	20–20	40–49		50–59		60–69	
Function	active N = 9	active N = 15	non-active N = 10	active N = 14	non-active N = 14	active N = 4	non-active N = 5
Max O$_2$ uptake							
liters · min^{-1}	5.4	4.0	3.3	3.4	2.9	2.7	2.6
ml · kg^{-1} · min^{-1}	77	57	44	38	38	43	37
Heart volume, ml		1,050	835	940	915	830	865
Max heart rate		175	182	176	175	165	170
Cholesterol, mg · 100 ml^{-1}		222	231	251	277	286	266
Neutral fat serum, mM		0.85	1.56	0.95	1.44	1.10	1.85
Blood pressure, mm Hg		135/83	128/82	137/81	133/82	138/83	123/86
ECG IV: 1–3*		2	1	4	2	1	0

*Classified according to the Minnesota Code (see Chap. 10, References, I. Åstrand et al., 1967).
Source: Data from Saltin and Grimby, 1968.

In a longitudinal study, Robinson et al. (1976) reexamined former champion runners 25 to 43 years after their competitive careers in track. The mean maximal oxygen uptake declined from 71.4 ml · kg^{-1} · min^{-1} in youth to 41.8 ml at a mean age of 57 years. This should be compared with mean values of 50.6 and 36.6 ml · kg^{-1} · min^{-1} respectively in nonathletes at corresponding ages.

During prolonged heavy physical work, the individual's performance capacity depends largely upon the ability to take up, transport, and deliver oxygen to the working muscle. Consequently, the maximal oxygen uptake is probably the best laboratory measure of a person's physical fitness, provided the definition of physical fitness is restricted to the capacity of the individual for prolonged heavy work (Herbst, 1928; Dill, 1933; P.-O. Åstrand, 1952, 1956).

For decades a discussion has been going on concerning the "limiting factors" in maximal oxygen uptake (P.-O. Åstrand, 1952, 1956). Do they include the oxygen content of the inspired air, the pulmonary ventilation, the diffusion of oxygen from alveolar space to hemoglobin, the hemoglobin content, the blood volume, the ability of the heart to pump blood, the distribution of blood flow, the ability of muscle tissues to receive the offered blood, the diffusion from capillaries to the working cells, the venous blood return, the efficiency of the mitochondria to transfer aerobic energy to the ATP-ADP machinery, access to fuel, the function of the neuromuscular system, or motivation?

We still do not know how critical the various factors are. It should be emphasized that the situation is different in a 5-min effort as compared with a 3-hr performance.

In other sections of the book, many of the listed factors are discussed and analyzed from a viewpoint of oxygen transport (see Chap. 6, p. 184).

Summary The individual's maximal oxygen uptake gives a measure of the "motor power" of the aerobic processes, i.e., of the person's maximal aerobic power. When related to body weight, the ability to move the body can be evaluated. A calculation of the maximal oxygen uptake per kilogram fat-free body weight, or related to muscle mass, blood volume, or other such parameters, makes it possible to analyze dimensions versus function.

In prolonged exercise, there is a high correlation between maximal oxygen uptake and total work output (maximal aerobic capacity). The actual oxygen uptake that can be tolerated is at a certain percentage of the maximum, this percentage being lower the longer the work time.

The maximal oxygen uptake (maximal aerobic power) increases with age up to twenty years. Beyond this age, there is a gradual decline so that the sixty-year-old individual attains about 70 percent of the maximum at twenty-five years. Before the age of twelve, there is no significant difference between girls and boys; thereafter, the average difference in maximal oxygen uptake between women and men amounts to 25 to 30 percent. Related to the body weight, the sex difference in aerobic power after puberty is 15 to 20 percent.

Top athletes in endurance events have a maximal oxygen uptake that is about twice as high as in the average person.

The gradual decline in maximal oxygen uptake with age beyond twenty is at least partially due to a decrease in maximal heart rate. Inactivity is another factor that decreases the functional range of the oxygen transporting system. Inactivity reduces the stroke volume and perhaps the efficiency of the regulation of the circulation during exercise.

It is presently impossible to point at decisive limiting factor(s) for maximal oxygen uptake.

PSYCHOLOGICAL FACTORS

Motivation or drive is the neural process which impels the individual to certain actions in pursuit of specific objectives. It plays an important role in human performance and may be the most important key to success, for abilities and physical capacities alone may be of little use unless the individual is motivated to devote all his or her endowment and capacity to their full limits in the attainment of specific goals. Superior performance may, on the other hand, be impossible to attain if the physical capacity of the body is limited, regardless of motivation.

It is beyond the scope of this book to analyze this exceedingly complex question, and the reader is referred to the psychological literature for a discussion of the psychological factors affecting performance.

REFERENCES

Ahlborg, G., L. Hagenfeldt, and J. Wahren: Influence of Lactate Infusion on Glucose and FFA Metabolism in Man, *Scand. J. Clin. Lab. Invest.*, **36**:193, 1976.

Asmussen, E.: Aerobic Recovery after Anaerobiosis in Rest and Work, *Acta Physiol. Scand.*, **11**:197, 1946.

Åstrand, I.: Aerobic Work Capacity in Men and Women with Special Reference to Age, *Acta Physiol. Scand.*, **49**(Suppl. 169):1960.

Åstrand, I.: Degree of Strain during Building Work as Related to Individual Aerobic Work Capacity, *Ergonomics*, **10**:293, 1967a.

Åstrand, I.: Aerobic Working Capacity in Men and Women in Some Professions, *Försvarsmedicin*, **3**:163, 1967b.

Åstrand, I., P.-O. Åstrand, and K. Rodahl: Maximal Heart Rate during Work in Older Men, *J. Appl. Physiol.*, **14**:562, 1959.

Åstrand, I., P.-O. Åstrand, E. H. Christensen, and R. Hedman: Intermittent Muscular Work, *Acta Physiol. Scand.*, **48**:443, 1960a.

Åstrand, I., P.-O. Åstrand, E. H. Christensen, and R. Hedman: Myohemoglobin as an Oxygen-store in Man, *Acta Physiol. Scand.*, **48**:454, 1960b.

Åstrand, I., P.-O. Åstrand, I. Hallbäck, and A. Kilbom: Reduction in Maximal Oxygen Uptake with Age, *J. Appl. Physiol.*, **35**:649, 1973.

Åstrand, P.-O.: "Experimental Studies of Physical Working Capacity in Relation to Sex and Age," Ejnar Munksgaard, Copenhagen, 1952.

Åstrand, P.-O.: The Respiratory Activity in Man Exposed to Prolonged Hypoxia, *Acta Physiol. Scand.*, **30**:343, 1954.

Åstrand, P.-O.: Human Physical Fitness with Special Reference to Sex and Age, *Physiol. Rev.*, **36**:307, 1956.

Åstrand, P.-O., and B. Saltin: Oxygen Uptake during the First Minutes of Heavy Muscular Exercise, *J. Appl. Physiol.*, **16**:971, 1961a.

Åstrand, P.-O., and B. Saltin: Maximal Oxygen Uptake and Heart Rate in Various Types of Muscular Activity, *J. Appl. Physiol.*, **16**:977, 1961b.

Åstrand, P.-O., J. Hallbäck, R. Hedman, and B. Saltin: Blood Lactates after Prolonged Severe Exercise, *J. Appl. Physiol.*, **18**:619, 1963.

Åstrand, P.-O., and E. H. Christensen: Aerobic Work Capacity, in F. Dickens, E. Neil, and W. F. Widdas (eds.), "Oxygen in the Animal Organism," p. 295, Pergamon Press, New York, 1964.

Bang, O.: The Lactate Content of the Blood during and after Muscular Exercise in Man, *Skand. Arch. Physiol.*, **74**(Suppl. 10):51, 1936.

Belcastro, A. N., and A. Bonen: Lactic Acid Removal Rates during Controlled and Uncontrolled Recovery Exercise, *J. Appl. Physiol.*, **39**:932, 1975.

Bergh, U., I.-L. Kanstrup, and B. Ekblom: Maximal Oxygen Uptake during Exercise with Various Combinations of Arm and Leg Work, *J. Appl. Physiol.*, **41**:191, 1976.

Brooks, G. A., K. E. Brauner, and R. G. Cassens: Glycogen Synthesis and Metabolism of Lactic Acid after Exercise, *Am. J. Physiol.*, **224**:1162, 1973.

Carlsten, A., B. Hallgren, R. Jagenburg, A. Svanborg, and L. Werkö: Myocardial Metabolism of Glucose, Lactic Acid, Amino Acids and Fatty Acids in Healthy Human Individuals at Rest and at Different Work Loads, *Scand. J. Clin. Lab. Invest.*, **13**:418, 1961.

Christensen, E. H., and O. Hansen: Arbeitsfähigkeit und Ehrnährung, *Skand, Arch. Physiol.*, **81**:160, 1939.

Christensen, E. H., and P. Högberg: Physiology of Skiing, *Arbeitsphysiol.*, **14**:292, 1950.

Christensen, E. H., R. Hedman, and B. Saltin: Intermittent and Continuous Running, *Acta Physiol. Scand.*, **50**:269, 1960.

Cohen, J., and J. R. Little: Lactate Metabolism in the Isolated Perfused Rat Kidney: Relations to Renal Function and Gluconeogenesis, *J. Physiol.*, **255**:399, 1976.

Costill, D. L.: Metabolic Responses during Distance Running, *J. Appl. Physiol.*, **28**:251, 1970.

Costill, D. L., J. Daniels, W. Evans, W. Fink, G. Krahenbuhl, and B. Saltin: Skeletal Muscle Enzymes and Fiber Composition in Male and Female Track Athletes, *J. Appl. Physiol.*, **40**:149, 1976.

Dill, D. B.: The Nature of Fatigue, *Personnel*, **9**:113, 1933.

Di Prampero, P. E., L. Peeters, and R. Margaria: Alactic O_2 Debt and Lactic Acid Production after Exhausting Exercise in Man, *J. Appl. Physiol.*, **34**:628, 1973.

Döbeln, W. von: Human Standard and Maximal Metabolic Rate in Relation to Fat-free Body Mass, *Acta Physiol. Scand.*, **37**(Suppl. 126):1956.

Eriksson, B. O.: Physical Training, Oxygen Supply and Muscle Metabolism in 11–13 year old Boys, *Acta Physiol. Scand.* (Suppl. 384), 1972.

Fagraeus, L.: Cardiorespiratory and Metabolic Functions during Exercise in the Hyperbaric Environment, *Acta Physiol. Scand.* (Suppl. 414), 1974.

Gollnick, P. D., and L. Hermansen: Biochemical Adaptations to Exercise: Anaerobic Metabolism, in J. H. Wilmore (ed.), "Exercise and Sport Sciences Reviews," vol. 1, Academic Press, Inc., New York, 1973.

Grimby, G., N. J. Nilsson, and B. Saltin: Cardiac Output during Submaximal and Maximal Exercise in Active Middle-aged Athletes, *J. Appl. Physiol.*, **21**:1150, 1966.

Hedman, R.: The Available Glycogen in Man and the Connection between Rate of Oxygen Intake and Carbohydrate Usage, *Acta Physiol. Scand.*, **40**:305, 1957.

Herbst, R.: Der Gasstoffwechsel als Mass der körperlichen Leistungs-fähigkeit, I, Mitteilung, Die Bestimmung des Sauerstoffaufnahmevermögens beim Gesunden, *Deutsch. Arch. Klin. Med.*, **162**:33, 1928.

Hermansen, L.: Oxygen Transport during Exercise in Human Subjects, *Acta Physiol. Scand.* (Suppl. 399):1973.

Hermansen, L., and B. Saltin: Blood Lactate Concentration during Exercise at Acute Exposure to Altitude, in R. Margaria (ed.), "Exercise at Altitude," Exerpta Medica Foundation, New York, 1967.

Hultman, E.: Studies on Muscle Metabolism of Glycogen and Active Phosphate in Man with Special Reference to Exercise and Diet, *Scand. J. Clin. Lab. Invest.*, **19**(Suppl. 94):1967.

Issekutz, B., Jr., H. I. Miller, P. Paul, and K. Rodahl: Effect of Lactic Acids and Glucose Oxidation in Dogs, *Am. J. Physiol.*, **209**:1137, 1965.

Kamon, E., and K. B. Pandolf: Maximal Aerobic Power during Laddermill Climbing, Uphill Running and Cycling, *J. Appl. Physiol.*, **32**:467, 1972.

Karlsson, J.: Lactate and Phosphagen Concentrations in Working Muscle in Man, *Acta Physiol. Scand.* (Suppl. 358), 1971.

Karlsson, J., F. Bonde-Petersen, J. Henriksson, and H. G. Knuttgen: Effects of Previous Exercise with Arms or Legs on Metabolism and Performance in Exhaustive Exercise, *J. Appl. Physiol.*, **38**:763, 1975.

Karpatkin, S., E. Helmreich, and C. F. Cori: Regulation of Glycolysis in Muscle, *J. Biol. Chem.*, **239**:3139, 1964.

Kasch, F., J. P. Wallace, R. R. Huhn, L. A. Krogh, and P. M. Hurl: $\dot{V}_{O_2}$ max during Horizontal and Inclined Treadmill Running, *J. Appl. Physiol.* **40**:982, 1976.

Keech, D. B., and M. F. Utter: Pyruvate Carboxylase II Properties, *J. Biol. Chem.*, **238**:2609, 1963.

Keul, J., E. Doll, H. Steim, U. Fleer, and H. Reindell: Über den Stoffwechsel des Herzens bei Hochleistungssportlern III. Der oxydative Stoffwechsel des trainierten menschlichen Herzens unter verschiedenen Arbeitsbedingungen, *Ztschr. Kreislaufforsch.*, **55**:477, 1966.

Keul, I., D. Keppler, and E. Doll: Standard Bicarbonate, pH, Lactate and Pyruvate Concentrations during and after Muscular Exercise, *German Medical Monthly*, **12**:156, 1967.

Klissouras, V., F. Pirnay, and J.-M. Petit: Adapation to Maximal Effort: Genetics and Age, *J. Appl. Physiol.*, **35**:288, 1973.

Knuttgen, H. G., and B. Saltin: Oxygen Uptake, Muscle High-Energy Phosphates, and Lactate in Exercise under Acute Hypoxic Conditions in Man, *Acta Physiol. Scand.*, **87**:368, 1973.

Krebs, H.: in F. Dickens, E. Neil, and W. F. Widdas (eds.), "Oxygen in the Animal Organism," p. 304, Pergamon Press, New York, 1964a.

Krebs, H.: Gluconeogenesis, The Croonian Lecture, 1963, *Proc. Roy. Soc. London*, **159**:545, 1964b.

Levy, M. N.: Uptake of Lactate and Pyruvate by Intact Kidney of the Dog, *Am. J. Physiol.*, **202**:302, 1962.

Lundin, G., and G. Ström: The Concentration of Blood Lactic Acid in Man during Muscular Work in Relation to the Partial Pressure of Oxygen of the Inspired Air, *Acta Physiol. Scand.*, **13**:253, 1947.

Margaria, R.: Aerobic and Anaerobic Energy Sources in Muscular Exercise, in R. Margaria (ed.), "Exercise at Altitude," Excerpta Medica Foundation, New York, 1967.

Margaria, R., H. T. Edwards, and D. B. Dill: The Possible Mechanism of Contracting and Paying the Oxygen Debt and the Role of Lactic Acid in Muscular Contraction, *Am. J. Physiol.*, **106**:689, 1933.

Maron, M., S. M. Horvath, J. E. Wilkerson, and J. A. Gliner: Oxygen Uptake Measurements during Competitive Marathon Running, *J. Appl. Physiol.*, **40**:836, 1976.

Mitchell, J. H., B. J. Sproule, and C. B. Chapman: Physiological Meaning of the Maximal Oxygen Intake Test, *J. Clin. Invest.*, **37**:538, 1958.

Newman, E. V., D. B. Dill, H. T. Edwards, and F. A. Webster: The Rate of Lactic Acid Removal in Exercise, *Amer. J. Physiol.*, **118**:457, 1937.

Newsholme, E., L. Hermansen, and O. Vaage: Unpublished results, 1976.

Oberholzer, F., H. Claassen, H. Moesch, and H. Howald: Ultrastrukturelle, biochemische und energetische Analyse einer extremen Dauerleistung (100 km-Lauf), *Schw. Z. Sportmedizin*, **24**:71, 1976.

Robinson, S.: Experimental Studies of Physical Fitness in Relation to Age, *Arbeitsphysiol.*, **10**:251, 1938.

Robinson, S., D. B. Dill, R. D. Robinson S. P. Tzankoff, and J. A. Wagner: Physiological Aging of Champion Runners, *J. Appl. Physiol.*, **41**:6, 1976.

Rodahl, K., and B. Issekutz, Jr.: Physical Performance Capacity in the Older Individual, in "Muscle as a Tissue," chap. 15, McGraw-Hill Book Company, New York, 1962.

Rodahl, K., H. I. Miller, and B. Issekutz, Jr.: Plasma Free Fatty Acids in Exercise, *J. Appl. Physiol.*, **19**:489, 1964.

Rowell, L. B.: Human Cardiovascular Adjustments to Exercise and Thermal Stress, *Physiol. Rev.*, **54**:75, 1974.

Saltin, B.: Aerobic Work Capacity and Circulation at Exercise in Man, *Acta Physiol. Scand.*, **62**(Suppl. 230):1964.

Saltin, B., J. Henriksson, E. Nygaard, P. Andersen, and E. Jansson: Fiber Types and Metabolic Potentials of Skeletal Muscles in Sedentary Man and Endurance Runners, *Annals N.Y. Acad. Sci.*, in press, 1977.

Saltin, B., and G. Grimby: Physiological Analysis of Middle-aged and Old Former Athletes: Comparison with Still Active Athletes of the Same Ages, *Circulation*, **38**:1104, 1968.

Saltin, B., B. Essén, and P. K. Pedersen: Intermittant Exercise: its Physiology and Some Practical Application, in E. Jokl, R. L. Anand, and H. Stoboy (eds.), "Advances in Exercise Physiology," p. 23, S. Karger, Basel, 1976.

Sjödin, B.: Lactate Dehydrogenase in Human Skeletal Muscle, *Acta Physiol. Scand.*, Suppl. 436, 1976.

Stenberg, J., P.-O. Åstrand, B. Ekblom, J. Royce, and B. Saltin: Hemodynamic Response to Work with Different Muscle Groups, Sitting and Supine, *J. Appl. Physiol.*, **22**:61, 1967.

Taylor, H. L., E. Buskirk, and A. Henschel: Maximal Oxygen Uptake as an Objective Measure of Cardiorespiratory Performance, *J. Appl. Physiol.*, **8**:73, 1955.

Williams, C. G., C. H. Wyndham, R. Kok, and M. J. E. von Rahden: Effect of Training on Maximum Oxygen Intake and on Anaerobic Metabolism in Man, *Int. Z. angew. Physiol. einschl. Arbeitsphysiol.*, **24**:18, 1967.

Evaluation of Physical Work Capacity on the Basis of Tests

10

CONTENTS

Chapter 10

Evaluation of Physical Work Capacity on the Basis of Tests

Generally speaking, there have been two main approaches to the assessment of physical performance: (1) physical fitness tests with scoring of actual performance in situations which represent basic performance demands, and (2) studies of cardiopulmonary function at rest and/or during exercise.

PHYSICAL FITNESS TESTS

Since most of the so-called fitness tests, including evaluation of flexibility, skill, strength, etc., are related to special gymnastic or athletic performance, they are really not suitable for an analysis of basic physiological functions. Practice and training in the performance of the actual test may greatly influence the results.

The fact that there may be significant correlation between the results from complicated test batteries applied to a group of individuals, or that the scores are related to certain parameters characteristic of the subjects, does not necessarily mean a direct relationship. Such data may cause confusion rather than solve problems. From a physiological and medical viewpoint, any test battery for the evaluation of physical fitness is rather meaningless unless it is based on sound physiological considerations. The widespread use of such test batteries in physical education can be justified from a pedagogic and psychological viewpoint. It may help the teacher or coach to stimulate the athlete's interest in training. Furthermore, any progress can be evaluated objectively. The selection of such activities and tests should therefore be based on pedagogic and psychological considerations with adaptation to local facilities. If they cannot be justified from these viewpoints, it is better to exclude them from the curriculum altogether. Too often the tests are incorrectly claimed to serve a physiological purpose. Actually, from a physiological viewpoint, application of a test battery may sometimes be unsuitable, since the performance of the tests usually demands maximal exertion of a subject who may be completely untrained.

For a review of the commonly used physical fitness tests, the reader is referred to the fitness test manual published by the American Association for Health, Physical Education and Recreation (1965), and to Mathews (1974).

TESTS OF MAXIMAL AEROBIC POWER

Direct Determination

In laboratory experiments, three methods of producing standard work loads have been mainly applied: running on a treadmill, working on a bicycle ergometer, and using a step test. In Chap. 9 the general methodological criteria were discussed in some detail: the work should involve large muscle groups; and the measurement of the O_2 uptake should be started when the work has lasted a few minutes, to allow the oxygen uptake to reach its maximum. Preferably, several submaximal and maximal work tests should be performed in experiments extended over several days (Fig. 9-3b). Ideally, a definite plateau should be reached when relating O_2 uptake to speed or work load.

The critical question is whether or not the different types of work mentioned give the same maximal O_2 uptake. A number of studies have been made to clarify this question.

It appears that by running on the treadmill uphill ($\geqq 3°$ inclination), the O_2 uptake may be brought to a maximum whereas running horizontally or at a slight inclination may result in a somewhat lower maximal O_2 uptake (Taylor et al., 1955). "Ski-walking" on a motor-driven treadmill with an inclination of 12° at speeds between 60 and 160 m $\cdot$ min^{-1} (the subject walking with slightly bent knees and using ski-poles as in skiing) produced a significantly higher maximal oxygen uptake than in maximal uphill treadmill running (Hermansen, 1973).

Bicycling produces, on the average, a lower O_2 uptake, at least compared with running uphill. In studies in which objective criteria have been used to determine whether the maximal O_2 uptake had been reached for the type of work in question, the values for running are on an average 5 to 8 percent higher than for bicycling.

Table 10-1 summarizes the mean values from several studies.

It is not possible at present to explain the reason for the somewhat higher oxygen uptake when running uphill compared with work on the bicycle ergometer. It can scarcely be caused by the activation of a larger muscle mass

Table 10-1   Mean values for maximal oxygen uptake attained in various types of exercise, compiled from the literature. Maximal oxygen uptake recorded during uphill running is termed 100 percent.

Type of exercise	$\dot{V}_{O_2}$ max, in percent	Some pertinent references
Running, uphill ($\geqq 3°$ incline)	100	Chap. 9; Hermansen, Ekblom, and Saltin, 1970; Hermansen and Saltin, 1969; Kamon and Pandolf, 1972
Running, horizontal	95–100	Hermansen and Saltin, 1969; Kasch et al., 1976
Bicycling, upright (60 rpm)	92–96	Chap. 9; Glassford et al., 1965; Hermansen, Ekblom, and Saltin, 1970; Hermansen and Saltin, 1969; Kamon and Pandolf, 1972; McKay and Banister, 1976
Bicycling, supine	82–85	Chap. 9
Bicycling, upright, one leg	65–70	Davies and Sargent, 1974
Arm cranking	65–75	Chap. 9
Arm and leg (with 10–20% of the total load on the arms)	100	Chap. 9
Step test	97	Kasch et al., 1966; Shepard et al., 1968

during running uphill, since simultaneous work with both arms and legs does not increase the maximal aerobic effect compared with work with the legs only (Table 10-1). The higher work tempo during running may enhance the venous return, but if this is true, running uphill should be no more advantageous than running horizontally.

During bicycling, the subject may often experience a feeling of local fatigue or a sensation of pain in the thighs or knees, which may be disturbing. This discomfort may cause the work effort to be interrupted before the oxygen-transporting organs have been fully taxed. The work position is of critical importance. The subject should be sitting almost vertically over the pedals. The seat should be high enough so that the leg is almost completely stretched when the pedal is in its lowest position. In the Swedish studies (Hermansen and Saltin, Table 10-1), there was no difference between trained and untrained subjects when comparing bicycling and running. However, in this material even the untrained subjects were familiar with bicycling. In persons who have never ridden a bicycle before, a *maximal test* on the bicycle may be undesirable as a method to assess maximal oxygen uptake. Motivation and stimulation of the subject are especially important in the case of bicycling. When a person runs on the treadmill, it is, so to speak, a matter of all or nothing: the subject is forced to follow the speed of the belt or jump off. On the bicycle, it is possible to continue to work at a reduced rate in most types of bicycle ergometers. Figure 10-1 summarizes results of studies by Hermansen

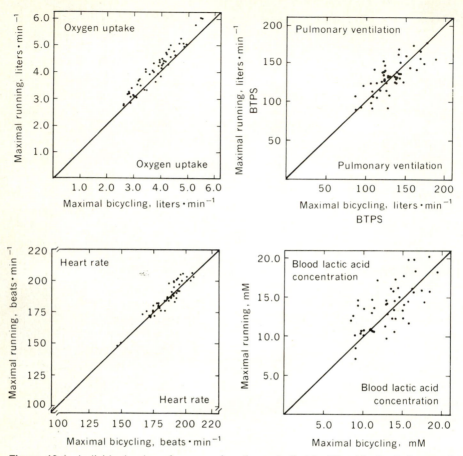

Figure 10-1 Individual values for some functions studied in 55 subjects during maximal running uphill on a treadmill ($\geqq$ 3°) and work on a bicycle ergometer (50 rpm) in a sitting position for about 5 min. Line of identity is drawn. A trend is noted for a somewhat higher maximal oxygen uptake during running compared with cycling, but pulmonary ventilation, heart rate, and blood lactic acid concentration were not different. It should be emphasized that a pedaling rate of 60 rpm may give a higher maximal oxygen uptake than may occur with a pedaling rate of 50 rpm. *(From Hermansen and Saltin, 1968.)*

and Saltin (1969). Despite the 7 percent difference in maximal oxygen uptake between running and bicycling, the maximal pulmonary ventilation, heart rate, and blood lactate concentration were not significantly different in the two procedures.

Normally, a test of maximal aerobic power starts with a submaximal load which also serves as a warming-up activity. After this the load may be increased in one of several ways: (1) The load may be immediately increased to a level which in preliminary experiments has been found to represent the predicted maximal load for the subject. (2) The load may be increased stepwise with several submaximal, maximal, or "supermaximal" loads, the subject working 5 to 6 min at each load, with or without resting periods between each load. (3) The load may be increased stepwise every or every other minute until exhaustion. When any one of these procedures is carefully conducted, they give the same maximal oxygen uptake (authors' observations; Binkhorst and Leeuwen, 1963; McArdle et al., 1973; Stamford, 1976). From a physiological viewpoint, the second method (2) is preferable (see Fig. 9-3b). It is often of interest to obtain steady-state conditions when measuring oxygen uptake, pulse rate, ventilation, etc., at submaximal work loads. This requires a work period of at least 5 min. The more or less continuously increasing work load (procedure 3) is a quick method which may reveal the subject's maximal oxygen uptake. However, since steady-state conditions are not attained at submaximal work, this procedure does not provide reliable information as to how the oxygen-transport problem is solved at different levels of physical effort, a type of information which may be of considerable interest. The oxygen uptake measured by the steady-state method is not the same as that measured by the progressive method for identical work loads (Fernandez et al., 1974).

Type of Exercise   In many cases, the preferable instrument for routine tests or studies of physical work capacity is, in our opinion, the *bicycle ergometer*. The technique involved is simple. The energy output or the oxygen uptake can be predicted with greater accuracy than for any other type of exercise. Within limits, the mechanical efficiency is independent of body weight. This is a definite advantage in studies which require repeated examinations over the years. The work load can, however, simply be selected according to the subject's gross body weight, calculated lean body mass, etc. (for example, 1 or 2 watts $\cdot$ kg^{-1}). The bicycle ergometer operated with a mechanical brake is inexpensive (e.g., "Monark" bicycle ergometer). It is easy to move from place to place, and is not dependent on the availability of electrical power. Since the subject on the bicycle ergometer exercises in a sitting or lying position with arms and chest relatively immobile, it is quite simple to obtain good ECG tracings and to perform studies with indwelling catheters. During *submaximal* work, a pedal frequency of from 40 to 50 revolutions/min produces the lowest O_2 uptake, i.e., the greatest mechanical efficiency, and therefore also a relatively low pulse rate (Grosse-Lordemann and Müller, 1937; Eckermann and Millahn, 1967).

The variation in O_2 uptake with different pedal frequencies at a standard work load should be kept in mind if the O_2 uptake is not measured but merely calculated from the work load used. Bicycle ergometers producing a constant load, even with relatively large variations in pedal frequency, have certain advantages. Nevertheless, it should be clearly realized that the O_2 uptake is not strictly determined by the load (watt) but varies with the pedal frequency. Respiration is also affected by the pedal frequency (Chap. 7). Usually the subject is asked to try to maintain a certain pedal frequency, such as 60 rpm. In our opinion, however, it is best to insist on a fixed pedal frequency such as 50 rpm, since this frequency produces an optimal mechanical efficiency. (Acoustical signals, such as those produced by a metronome, are easier to follow than visual signals.) It should be emphasized, however, that it is by no means essential that the chosen pedal frequency is optimal, as long as the mechanical efficiency is known. Most of the data concerning O_2 uptake at different loads on the bicycle ergometer are obtained from experiments using a pedal frequency of 50 rpm.

In *heavy* or *maximal* work, the optimal pedal frequency is higher, e.g., the use of lower gear when traveling uphill during bicycle racing (P.-O. Åstrand, 1953). 60 rpm may produce a somewhat higher maximal O_2 uptake than 50 rpm (Hermansen and Saltin, 1969; McKay and Banister, 1976). Thus, one should change from 50 rpm to 60 rpm in the case of heavy work, if the objective is to measure maximal oxygen uptake. The type of bicycle ergometer which produces a constant load regardless of pedal frequency has the drawback, during maximal work when the subject is tired and is unable to maintain the tempo, that the load increases for each pedal revolution. The chances are that the subject may be forced to discontinue work before it is possible to complete a critical measurement. With the use of a bicycle ergometer in which the work load varies with the pedal frequency (as on the Krogh or Monark), the load admittedly drops when the tempo no longer can be maintained, but the subject can in any case continue. (It is important to be able to record the pedal frequency continuously if the work load is critical for the experiment.)

The *treadmill*, however, is preferable in studies of young children below the age of about ten. The work load is dependent on the body weight, a fact that may be a disadvantage in longitudinal studies. Since the energy output per kilogram and kilometers per hour is more variable than it is at a given work load on the bicycle ergometer, the oxygen uptake should be measured during walking and running on the treadmill (coefficient of variation is about 15 percent) (Mahadeva et al., 1953; I. Åstrand, 1960). This requirement may limit its application. Older individuals may have some difficulty in walking on the treadmill. The provision of a handrail for support will make the work load still more unpredictable. It should be emphasized that the energy cost of running (jogging) at a low speed is much higher than when walking (Fig. 16-5). Therefore it can be a disadvantage to change the speed more or less gradually from a low to a high one. There will be a speed at which some subjects prefer to walk (e.g., those with long legs) but others will run. The energy demand will be quite different and it cannot be predicted from the speed alone. One example: at a speed of 5.5 km $\cdot$ hr^{-1}, inclination 8° (3.4 mph, 14 percent) the oxygen uptake was 2.9 liters $\cdot$ min^{-1} when the subject was walking. It increased to 3.2 liters $\cdot$ min^{-1} when jogging. When holding onto the railing, the oxygen uptake dropped from 3.2 to 2.9 liters $\cdot$ min^{-1} (jogging) and to 2.2 when the subject was walking. In other words, the oxygen uptake could at this standard speed and slope be anything from 2.2 up to 3.2 liters $\cdot$ min^{-1}, depending on the subject's technique. The treadmill is expensive and immobile. Recordings of ECG and other measurements may be more complicated to make during walking or running than during bicycle-ergometer riding.

Formulas are available for calculating the energy cost of walking or running (Passmore and Durnin, 1955; Bobbert, 1960b; Margaria et al., 1963; Van der Walt and Wyndham, 1973). The *step test* has a more limited application, since it is poorly standardized and offers limited provisions for varying the load on the oxygen-transporting system. However, methods have been developed for gradational step tests (Maritz et al., 1961; Nagle et al., 1965; Kasch et al., 1966; Shephard et al., 1966). Furthermore, a step test with adjustable steps has been proposed by Hettinger and Rodahl (1960). Step tests are particularly useful in field studies and in studies of large numbers of subjects.

At submaximal work loads, there is no difference in the pulse rate at a given oxygen uptake, whether the subject is using a step test or a bicycle ergometer test (Ryhming, 1953). Nor does a comparison between walking, running, and bicycling reveal any significant difference in the pulse rate at a given O_2 uptake (Berggren and Christensen, 1950; Bobbert, 1960a; Hermansen and Saltin, 1969).

We have good experience with the use of the bicycle ergometer for the submaximal exercises but prefer the treadmill test for the maximal exercise. Various aspects of stress testing are discussed in Kattus, 1972; Naughton et al., 1973; Zohman and Phillips, 1973; Åstrand, 1976; Sheffield and Roitman, 1976.

The Douglas Bag Method

The classical method for the determination of oxygen uptake, the Douglas bag method, rests on a very secure foundation. It is theoretically sound, and it is well tested under a wide variety of circumstances. In all its relative simplicity, it is unsurpassable in accuracy.

A disadvantage with the method is that the subject is somewhat hampered by the equipment required for the collection of the expired air. This limits the subject's freedom of movement. Furthermore, it merely provides a mean figure for the oxygen uptake of, say, 30 s, depending on the length of the time in which the expired air is collected.

Figure 10-2 shows the adaptation of this method for experiments using the bicycle ergometer (right) and the treadmill (left). The inner area of the mouthpiece is 400 mm², the inner diameters of the tubes in the valve, stopcock, and bag are 28 mm, and of the connecting tube (smooth, not corrugated), 35 mm. The tubes should be as short as possible. The resistance in this system with a connecting tube of 0.5-m length with a flow rate of 100 liters · min^{-1} is 1 cm H_2O; of 200 liters · min^{-1}, 3 cm H_2O; of 300 liters · min^{-1}, and 6 cm H_2O; and of 400 liters · min^{-1}, 10 cm H_2O. The room temperature should be between 19 and 21°C, and the relative humidity between 40 and 60 percent. The oxygen content of the room air should not be below 20.90 percent.

Respiratory rate is registered via a Marey capsule. Two Douglas bags are connected to the four-way stopcock to enable continuous collection of air. The subject is connected to the bag during an inspiration. The turning of the stopcock starts the stopwatch, which can be read down to .01 min (photograph at the center of Fig. 10-2). When no less than about 50 liters of expired air has been collected in the bag, the stopcock is turned to the second bag or back to room air. The stopcock is always turned during inspiration; the second turning of the stopcock automatically stops the stopwatch. The volume of expired air is measured in a balanced spirometer and the composition of the air is analyzed by the Haldane or Micro-Scholander or by electronic gas analyzers checked against the aforementioned manometric methods. When using electronic gas analysers, care must be taken to correct for water pressure in the expired gas in relation to the gas used for calibration. Otherwise, the error may be as large as 25 percent in the calculation of oxygen uptake (Beaver, 1973). Figure 9-2 gives examples of the adaptation of the same method in experiments carried out in the field.

Work Test Procedures

The current test methods include the upright bicycle ergometer, the step test, and the treadmill. The procedure may vary from a single-level load to an intermittent series of increasing loads with intermittent rest periods, an almost continuous increase in load, or a

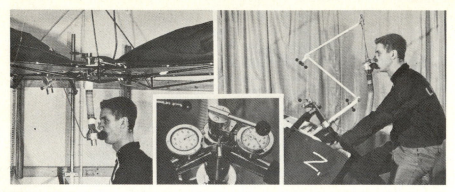

Figure 10-2 Arrangements of respiratory valves, stopcocks, and Douglas bags for the collection of expired air during experiments using a treadmill (left) and a bicycle ergometer (right). With the aid of the arrangement shown in the center of the figure, the handle opens the stopcock for one of the two Douglas bags. At the same time, the corresponding stopwatch is automatically started. When the handle is moved to close the opening to the first bag, the stopwatch is automatically stopped. With the simultaneous opening of the stopcock aperture, leading the expired air into the second bag, the corresponding stopwatch is automatically started. In this manner, the time for the collection of the expired air is taken automatically.

continuous series of increasing loads with a steady state at each level. This latter procedure is possible only when applying loads below 50 to 70 percent of the subject's maximal oxygen uptake.

In the case of single-level loads, the heart rate should reach a level above 120 beats · min⁻¹ and the work period should be about 6 min. In the case of multiple-level tests, we prefer an intermittent series of test loads, or a continuous increase in load but with work periods at each load level which exceed 5 min. If the oxygen uptake is to be measured and if steady-state or "apparent" steady-state values are to be obtained, the noseclip, mouthpiece, and respiratory valve are placed on the subject approximately 4 min after the beginning of the test load and the collection of the expiratory air is initiated about 5 min after the start of the test. (If the metabolic respiratory quotient is to be determined, a much longer work period is required before the air sample is collected.)

The arterial blood pressure can be measured indirectly with a cuff. The diastolic pressure (at the point where the Korotkow sounds change in character—4th phase) may, however, be difficult to note. The product of heart rate and systolic blood pressure is used as a prediction of myocardial oxygen consumption (see Chap. 6, p. 177; Naughton et al., 1973, p. 150).

When the maximal aerobic power is to be assessed, the tests should preferably be extended to 2 days or more. On the first day, two submaximal and one predicted maximal load are performed. On the second day, an additional one or two submaximal loads are performed, followed by a maximal load, which is determined on the basis of the results from the previous day's test. (The necessary criteria for ascertaining that the maximal oxygen uptake has been reached are discussed in Chap. 9.)

If for any reason only one test can be performed in each subject, the following procedure may prove useful.

Bicycle Ergometer

The subject is exposed to one or two submaximal loads. The load should be adjusted so that the heart rate is at least 140 beats · min⁻¹ for subjects less than fifty years of age, and 120 beats · min⁻¹ for subjects more than fifty years of age. Then a "supermaximal" load should be tried. If the subject can tolerate this supermaximal load for at least 2 min, even with considerable difficulty, measurements carried out at the end of the work period are likely to give an oxygen uptake equal to or close to the individual's maximal oxygen uptake. In any case, this will require a well-motivated subject. Usually this supermaximal load is selected on the basis of the individual's predicted maximal oxygen uptake from his or her heart rate at the submaximal work loads, using the nomogram of Åstrand and Åstrand (Fig. 10-7). A load is then selected which will require an oxygen uptake about 10 to 20 percent higher than the predicted maximal oxygen uptake (see Table 10-2). If the subject at the end of the first min on the selected load has difficulty in keeping up the pedaling rate and starts to hyperventilate markedly, the load is lowered slightly so as to allow the subject to continue for a total of about 3 min. If, on the other hand, the subject, after a minute or two, appears to have more strength left than originally predicted, the load is slightly increased.

The objective is to be able to collect two Douglas bags of expiratory air. This requires a collection time of no less than 1 min. It is a matter of experience to decide when to put the mouthpiece and respiratory valve in place on the subject and when to start the collection of the air sample. It is always better, however, to start too early than too late.

This "quick method" requires a considerable amount of experience on the part of the investigator in order to attain satisfactory results. In our opinion, the peak value of blood lactate mirrors fairly well the severity of the load on the aerobic and anaerobic processes and the degree of exhaustion, as does the respiratory quotient as well. Blood for the determination of lactate concentration can easily be obtained from the finger tip. (The hand should be prewarmed and washed in hot water.) Three to four blood samples should be secured within 1 min after the end of the work and again after about 3, 6, and 10 min. The most suitable pedaling rate has been found to be about 60 rpm.

Treadmill

In order to select a suitable starting speed and incline for a maximal run on the treadmill, the subject's maximal oxygen uptake (ml · kg^{-1}· min^{-1}) is first predicted from the heart rate at a submaximal work load on the bicycle ergometer, as described above. Then the data in Table 10-3 are applied.

1 Starting at this initial speed and incline, the incline is increased by 1.5 degrees (2.67 percent) every third min, keeping the speed constant. Under these conditions, not even top athletes are able to run for more than 7 min. Experience has shown that it is advisable to give females a somewhat lower relative starting load than is given males (Table 10-3). Immediately preceding the maximal run, the subject is given a 10-min warm-up on the treadmill at a load corresponding to 50 percent of the selected starting load. The subjects are not allowed to hold onto the treadmill railing during the run.

2 Another procedure, which in our experience has proved quite satisfactory, is to let the subject run in bouts of 3 min, with 4- to 5-min rest periods between each running bout, keeping the treadmill inclination constant at 3° while increasing the speed by 15 m · min^{-1} each time. It is advisable first to allow the subject to become accustomed to the treadmill running and, in particular, to learn how to jump on and off the running treadmill with ease. The actual test should be started at a running speed which is not far from the subject's

Table 10-2

Work load		Oxygen uptake,
watts	kpm · min^{-1}	liters · min^{-1}
50	300	0.9
100	600	1.5
150	900	2.1
200	1200	2.8
250	1500	3.5
300	1800	4.2
350	2100	5.0
400	2400	5.7

Table 10-3   Starting Work Load Used for the Maximal Run on the Treadmill

Predicted max $\dot{V}_{O_2}$ ml · kg^{-1}· min^{-1}	Starting speed and inclination for max run on treadmill*							
	♂ Speed uphill				♀ Speed uphill			
	km · hr^{-1}	Degree	mph	%	km · hr^{-1}	Degree	mph	%
<40	10.0	3.0	6.2	5.25	10.0	1.5	6.2	2.67
40–50	12.5	3.0	7.8	5.25	10.0	3.0	6.2	5.25
55–75	15.0	3.0	9.3	5.25	12.5	3.0	7.8	5.25
>75								
A	15.0	4.5	9.3	7.0				
B	17.5	3.0	10.9	5.25				
C	20.0	1.5	12.5	2.67				

*A = cross-country skiers, skaters, etc.; B = cross-country runners, orientation runners, etc.; C = track runners.
Source: Saltin and Åstrand: *J. Appl. Physiol.*, **23**:353, 1967.

maximal level, yet no greater than the subject will feel confident of keeping up for several minutes.

When comparing three "traditional" treadmill exercise protocols, Froelicher et al. (1974) noticed a higher maximal oxygen uptake when utilizing the Taylor protocol than for the Bruce (−6.5 percent) and Balke protocol (−9.7 percent) (mean differences for 15 subjects).

It should be emphasized that an evaluation of the cardiac performance should be based on the total oxygen uptake (in liters · min^{-1}), for this is correlated to the cardiac output, the myocardial oxygen consumption, and blood flow. In a group of individuals with different body weights the oxygen uptake corrected for weight (ml · kg^{-1}· min^{-1}) is unrelated to the actual load on their hearts. In longitudinal studies the subject's body weight may vary considerably. The ml · kg^{-1} figure for maximal oxygen uptake is a good predictor of the subject's potential to move and lift his body, but again, it may not mirror the cardiac performance.

Another way to express the oxygen uptake during exercise is to relate it to the energy expenditure at rest (= 1 MET, equivalent to approximately 3.5 ml · kg^{-1}· min^{-1} in oxygen uptake.) It may be useful when evaluating the energy demand of exercise in relative terms, but the MET units do not directly reflect the total load on the oxygen transporting system and are, therefore, not well correlated to the myocardial performance and metabolic demand.

Summary   Ideally, any test of maximal oxygen uptake should meet at least the following general requirements: (1) The work in question must involve large muscle groups; (2) the work load must be measurable and reproducible; (3) the test conditions must be such that the results are comparable and repeatable; (4) the test must be tolerated by all healthy individuals; and (5) the mechanical efficiency (skill) required to perform the task should be as uniform as possible in the population to be tested.

The magnitude of the external work can be expressed exactly, and it may be reproduced with a high degree of accuracy in tests using the bicycle ergometer and the treadmill. For these reasons, the use of the bicycle ergometer or the treadmill is preferable to that of the step test. In the case of comparative studies, the same method of producing the test load should be used. Running on the treadmill should be uphill, with an inclination of 3° or more. When using the bicycle ergometer, the subject should be placed in a sitting position directly above the pedals. The seat should be sufficiently high, and the pedal frequency should be about 60 rpm. By preference, the work intensity should be selected so that the subject can proceed for at least 3 min. It has repeatedly been emphasized that the correct criterion for an attained maximal oxygen uptake should be a final "leveling off" of the oxygen uptake despite an increasing work load, i.e., a failure of a higher work load to increase oxygen uptake significantly.

Prediction from Data Obtained at Rest or Submaximal Test

Although the aerobic power in terms of maximal oxygen uptake can be determined with a reasonable degree of accuracy, the method is rather time-consuming. It requires fairly complicated laboratory procedures and demands a high degree of cooperation from the subject. Although this is the method of choice for any scientific investigation, it is by no means a method which may conveniently be applied routinely in the office of a physician.

The practicing physician (especially the industrial physician), the coach, the physical therapist, or anyone else interested in physical performance is nevertheless often faced with the need to assess a person's circulatory fitness. This requirement has created the need for a simple test for such an evaluation of an individual, based upon submaximal work stress. In treating older individuals as well as certain other patients, the physician may be reluctant to expose the patient to the risk of an exhausting maximal work load. This is true whether one is considering job placement, fitness for continued employment, or retirement.

Rest   No objective measurements made on resting individuals will reveal their capacity for physical work or their maximal aerobic power. Even a simple questionnaire may reveal more useful information than can be obtained from measurements made at rest. A low heart rate at rest, a large heart size, or similar parameters may indicate a high aerobic power, but may, on the other hand, be a symptom of disease.

A significant correlation between any parameter and maximal oxygen uptake indicates a direct or indirect dependence. Figure 11-5 (lower part) illustrates such a relationship. The range of the observed values is of decisive importance for the numerical value of the correlation coefficient, but for an evaluation of the individual case, the *standard deviation* from the regression line is the critical factor. The deviation may be large, i.e., a prediction of one parameter from the other is very uncertain, despite a high correlation between the parameters in question. From the data presented in Fig. 11-5 (lower part), we find that the correlation coefficient between maximal oxygen uptake and total amount of hemoglobin (Hb_T) is as high as 0.970, but the standard deviation of hemoglobin weight is still as high as 10.5 percent at an oxygen uptake of 2.6 liters · min^{-1}. The figure shows that one girl with 350 g Hb$_T$ can transport up to 2.7 liters · min^{-1}, but another girl with the same Hb$_T$ reaches an oxygen uptake of only 1.9 liters · min^{-1}. This is to be expected statistically, since 350 g Hb represents an oxygen uptake of 2.25 liters · min^{-1} with such a standard deviation that 95 out of 100 subjects of a similar group are expected to fall within the range 1.8 to 2.7 liters · min^{-1} and 5 subjects will lie outside these values.

In conclusion, it may be stated that the correlation between parameters is of interest when analyzing biological interactions, but for an evaluation of an individual from indirect methods, the standard deviation from the regression line indicates the accuracy of the prediction.

Heart Rate Response to Standardized Work The simplest and most extensively applied way of testing the circulatory functional capacity is to determine the heart rate during or after exercise (step test, treadmill, or bicycle ergometer test). From the heart rate response the circulatory capacity can be evaluated. In the following material, certain basic principles involved in cardiopulmonary function tests will be discussed.

During the Second World War, the Harvard step test was developed as a screening test to select individuals according to their physical fitness. The height of the bench (50 cm, or 20 inches) and the stepping frequency (30 steps · min^{-1}) were chosen so that, roughly, only one-third of the subjects should be able to perform the test for a 5-min period. The heart rate was counted during recovery from the exercise with the subject sitting on the bench (from 1 to 1^1/$_2$ min). There was also a treadmill version of the Harvard fitness test (Johnson et al., 1942). The lower the number of heartbeats during the recovery and the longer the work time, the higher the score. In general, it was noticed that the subjects with high scores also had a better performance in many activities demanding a high aerobic power than did those with lower scores. However, the results of this test, as well as all other similar tests demanding an all-out effort, depend on a number of factors besides a high maximal aerobic power, such as anaerobic power, technique, and especially motivation, which largely affect the duration of the work period. If objective criteria are not used to control the degree of exhaustion or the circulatory load, it is impossible to conclude whether the test was interrupted because the subject was exhausted or just unwilling to continue trying. This fact tends to render the test less useful in the testing of conscripts.

It should also be emphasized that the recording of pulse rate in connection with a work load should take place *during* work. In the same individual, the correlation coefficient between the pulse rate *during* submaximal work and 1 to 1^1/$_2$ min *after* the work may be high. In one study, it was $r = 0.96$ with a deviation from the regression line of 5 percent (see Fig. 10-3). For a group of subjects, the correlation coefficient for steady-state heart rate and recovery rate was reduced to 0.77 and the deviation increased to 10 percent (I. Ryhming,

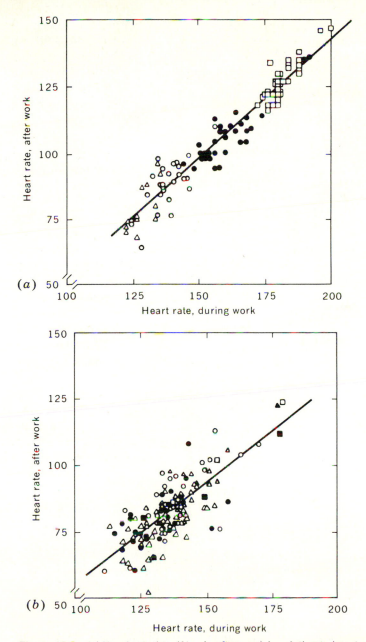

Figure 10-3 (*a*) Heart rate 1 to 1½ min after work in relation to heart rate during work for one subject (female). △ = step test, ○ = 100 watts, ● = 150 watts, □ = 200 watts, (*b*) Heart rate 1 to 1½ min after work in relation to heart rate during work. (Presented are 61 individual average values from step test and 66 from bicycle test.) The figure △ = step test, ● = 150 watts, and ○ = 100 watts. The figures ■ and □ symbolize the average values of 15 different tests with the intensities of 100, 150, and 200 watts for two female subjects. The symbol ▲ shows the average from six step tests done by a male subject with 30 steps · min⁻¹ on a bench 40 cm high. *(From Ryhming, 1953.)*

1953). From these data, it is evident that the recovery heart rate gives only a rough idea of the heart rate attained *during* work if the results are compiled from different subjects.

The development of testing procedures has shown a prominent tendency to analyze separately the various factors of importance for the physical work capacity. In particular tests, procedures have been developed to assess the oxygen-transporting system. Most such modern circulatory exercise tests are based on a linear increase in heart rate with increasing O_2 uptake or work load. Figure 10-4 gives examples of this relationship for two subjects with different maximal aerobic power. If the slope of the heart rate–oxygen uptake line can be determined from measurements made during submaximal exercise and an extrapolation is made to a probable value for maximal heart rate, the individual's maximal oxygen uptake may be predicted (for subject *A* in Fig. 10-4, 3.5 liters $\cdot$ min^{-1}, provided that his maximal heart rate is 195).

One presumption for an assessment of the circulatory capacity based on heart rate is that the cardiac output ($\dot{Q}$) at a given oxygen uptake varies only within reasonable limits. If this is the case, the heart rate (HR) will inversely vary with the individual's stroke volume (SV); that is, the larger the stroke volume, the lower the heart rate, since HR $\cdot$ SV = $\dot{Q}$. The maximal cardiac output and oxygen uptake should then be finally determined by the individual's maximal heart rate. However, several factors exert a decisive influence on the

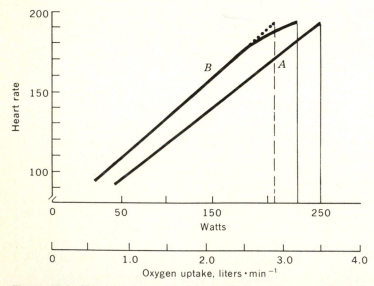

Figure 10-4 The increase in heart rate with increasing work load (and oxygen uptake) is linear within a wide range. In some subjects (*B*), the oxygen uptake may increase relatively more than the heart rate as the work load becomes very heavy. The prediction of maximal oxygen uptake by an extrapolation to the subject's presumed maximal heart rate (195 in this case) suggests a maximum of 2.9 liters $\cdot$ min^{-1} (dotted line), but the actual maximal aerobic power is 3.2 liters $\cdot$ min^{-1}. The individual's maximal heart rate is also a critical factor in an extrapolation.

maximal circulatory capacity and aerobic power, and experimental studies show that there are considerable sources of error in any prediction of the efficiency of the oxygen-transporting system from submaximal tests:

1   *The linear increase in heart rate* with increase in oxygen uptake is a typical feature. There are many exceptions, however. In some cases the oxygen uptake increases relatively more than the heart rate as the work load becomes very heavy (subject *B* in Fig. 10-4). One possible explanation for this phenomenon may be that an efficient redistribution of blood, giving the working muscles an appropriate share of the cardiac output, is not brought about until the very heavy work loads are reached. The consequence is that in this subject, the maximal oxygen uptake will be underestimated by an extrapolation from the heart rate response to submaximal loads.

2   *The maximal heart rate* declines with age (Fig. 9-14). Therefore, if old and young subjects are included in the same study, the circulatory capacity of the older subjects will be consistently overestimated compared with that of the younger subjects. By introducing an age factor, a correction can be made (see below). However, the standard deviation for maximal heart rate within an age group is about ± 10 beats $\cdot$ min^{-1}; thus, 50 percent of the tested subjects will be more or less overestimated and the remainder underestimated. In Fig. 10-4, subject *A* was assumed to have a maximal heart rate of 195, maximal oxygen uptake = 3.5 liters $\cdot$ min^{-1}. If the maximal heart rate was 170, the maximal oxygen uptake will be only 2.9 liters $\cdot$ min^{-1}. An extrapolation of the heart rate to 215 for a subject with the same slope for the relation heart rate to oxygen uptake as subject *A* will reach an oxygen uptake of 4.0 liters $\cdot$ min^{-1}.

3   In cases where the oxygen uptake is predicted from the work load, assuming a fixed *mechanical efficiency*, it should be kept in mind that the mechanical efficiency may vary by ± 6 percent (bicycle ergometer). In Fig. 10-4, 150 watts are indicated with a mean O_2 uptake of 2.1 liters $\cdot$ min^{-1}. An oxygen uptake as low as 1.9 or as high as 2.3 liters $\cdot$ min^{-1} at the same work load would not be unusual, however. The consequence is that in a subject with a low mechanical efficiency (whose oxygen uptake at the submaximal work load is relatively high), maximal oxygen uptake will be predicted to be lower than it really is, since the heart rate is influenced by the extra oxygen transport. It should be noted that the oxygen uptake at a given *submaximal* rate of work on a bicycle ergometer or treadmill is very constant even if the work is performed under different conditions, e.g., with subject exposed to hypoxia or hyperoxia (Chap. 17), hot environment, dehydration (see Rowell, 1974; Saltin, 1964). The variations in mechanical efficiency in the course of a training period are usually small or insignificant (Chap. 12).

4   The last factor to be considered is based on Fig. 6-15. The *cardiac output* is not strictly related to the oxygen uptake but shows individual variations. For the prediction of maximal oxygen uptake from heart rate at a submaximal load, this variation does not matter. The oxygen uptake (measured or predicted) per heartbeat is actually evaluated during the test. When consideration is given to the maximal heart rate, the maximal oxygen uptake is "calculated." However, if the work test is conducted for an evaluation of cardiac performance, e.g., stroke volume, the individual variation in cardiac output and arteriovenous oxygen difference must be taken into consideration.

All these factors must be considered when assessing the efficiency of a person's oxygen-transporting system.

The development of a principle for the prediction of the maximal O_2 uptake will be briefly described. Figure 10-5 gives individual values for the pulse rate at different O_2 uptakes, up to maximal O_2 uptake. The material includes 86 subjects, all of whom were physical education students. As is evident from the figure, the scatter of the data is considerable. At an oxygen uptake of 3 liters · min^{-1}, there are pulse rates from 140 to 220. A pulse rate of 180 beats · min^{-1} represents an oxygen uptake of 2.0 liters · min^{-1} for some female subjects and as much as 5.0 liters · min^{-1} for some male subjects. A closer examination of these data revealed that in the case of men, the pulse rate was on the average 128 at an oxygen uptake representing 50 percent of the maximal O_2 uptake, and 154 at an O_2 uptake representing 70 percent of the maximum. The corresponding values for females were 138 at $\dot{V}_{O_2}$ = 50 percent and 168 at 70 percent (Fig. 10-6). The standard deviation (SD) was in the order of 9 beats/min. This being the case, it may be argued that if a male subject has a pulse rate of 128 at an O_2 uptake of 2.3 liters · min^{-1}, his maximal O_2 uptake ought to be double this value, or about 4.6 liters · min^{-1}. On the basis of Fig. 10-6, a nomogram was constructed for the prediction of maximal O_2 uptake from submaximal pulse rates (120 to 170) (P.-O. Åstrand and Ryhming, 1954). This nomogram has subsequently been modified. The adjusted version is shown in Fig. 10-7. I. Åstrand (1960) has critically examined the nomogram further. She found that the maximal O_2 uptake of persons over twenty-five years of age was consistently overestimated. This could be explained on the basis of the reduction in the maximal pulse rate with age. A correction factor

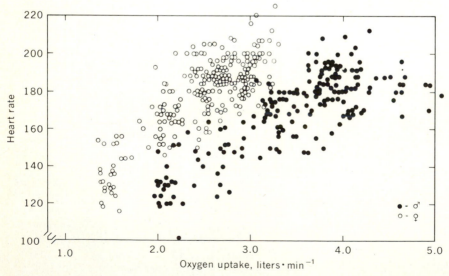

Figure 10-5 Heart rates in relation to oxygen uptake for 86 adult female and male subjects. Maximal as well as submaximal values are represented. *(From P.-O. Åstrand, 1952.)*

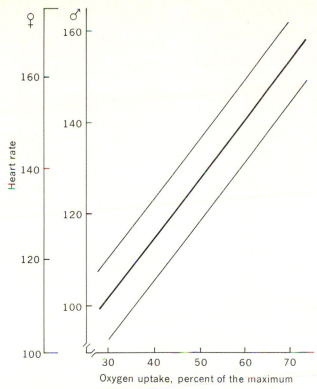

Figure 10-6 Relationship between heart rate during work (bicycle ergometer) and oxygen uptake expressed in percentage of subject's maximal aerobic power. Left of ordinate, heart rates of women; right of ordinate, those of men. Thin lines denote one standard deviation. *(From P.-O. Åstrand and Ryhming, 1954.)*

for age was therefore introduced (Tables 10-4; 10-5). Generally speaking, the application of the nomogram represents an extrapolation to a maximal pulse rate typical for the subject's age, as discussed in connection with Fig. 10-4. In the original group of subjects, the maximal pulse rate was 195. It is therefore understandable that the maximal O_2 of fifty-year-old subjects with a maximal pulse rate of about 170 will be overestimated. Empirically, it was found that the predicted maximal O_2 uptake came closer to the actually measured maximal O_2 uptake if a correction factor of 0.75 was introduced for pulse rates of 170 (see below). If possible, several tests should be performed at different test loads, and the mean figures should be calculated according to the nomogram. However, if need be, the nomogram may be adapted for the use of only one test load.

The standard error of the method for the prediction of maximal oxygen uptake from submaximal exercise tests is about 10 percent in relatively well-trained individuals of the same age, but up to 15 percent in moderately trained individuals of different ages when the age factor for the correction of maximal oxygen uptake is applied (I. Åstrand, 1960). Untrained persons are often

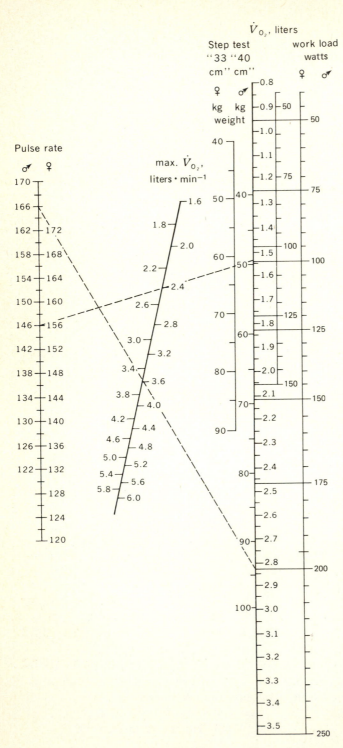

Figure 10-7 The adjusted nomogram for calculation of maximal oxygen uptake from submaximal pulse rate and O_2-uptake values (cycling, running or walking, and step test). In tests without direct O_2-uptake measurement, it can be estimated by reading horizontally from the "body weight" scale (step test) or "work load" scale (cycle test) to the "O_2 uptake" scale. The point on the O_2-uptake scale ($\dot{V}_{O_2}$, liters) shall be connected with the corresponding point on the pulse rate scale, and the predicted maximal O_2 uptake read on the middle scale. A female subject (61 kg) reaches a heart rate of 156 at step test; predicted max. $\dot{V}_{O_2} = 2.4$ liters $\cdot$ min^{-1}. A male subject reaches a heart rate of 166 at cycling test on a work load of 200 watts; predicted max. $\dot{V}_{O_2} = 3.6$ liters $\cdot$ min^{-1} (exemplified by dotted lines). *(From I. Åstrand, 1960.)*

Table 10-4 Prediction of maximal oxygen uptake from heart rate and work load on a bicycle ergometer. The value should be corrected for age, using the factor given in Table 10-2. For the relationship between work load and oxygen uptake, see Table 10-5.

Men

Maximal oxygen uptake, liters · min^{-1}

Heart rate	300 kpm/min 50W	600 kpm/min 100W	900 kpm/min 150W	1200 kpm/min 200W	1500 kpm/min 250W
120	2.2	3.5	4.8		
121	2.2	3.4	4.7		
122	2.2	3.4	4.6		
123	2.1	3.4	4.6		
124	2.1	3.3	4.5	6.0	
125	2.0	3.2	4.4	5.9	
126	2.0	3.2	4.4	5.8	
127	2.0	3.1	4.3	5.7	
128	2.0	3.1	4.2	5.6	
129	1.9	3.0	4.2	5.6	
130	1.9	3.0	4.1	5.5	
131	1.9	2.9	4.0	5.4	
132	1.8	2.9	4.0	5.3	
133	1.8	2.8	3.9	5.3	
134	1.8	2.8	3.9	5.2	
135	1.7	2.8	3.8	5.1	
136	1.7	2.7	3.8	5.0	
137	1.7	2.7	3.7	5.0	
138	1.6	2.7	3.7	4.9	
139	1.6	2.6	3.6	4.8	
140	1.6	2.6	3.6	4.8	6.0
141		2.6	3.5	4.7	5.9
142		2.6	3.5	4.6	5.8
143		2.5	3.4	4.6	5.7
144		2.5	3.4	4.5	5.7
145		2.4	3.4	4.5	5.6
146		2.4	3.3	4.4	5.6
147		2.4	3.3	4.4	5.5
148		2.4	3.2	4.3	5.4
149		2.3	3.2	4.3	5.4
150		2.3	3.2	4.2	5.3
151		2.3	3.1	4.2	5.2
152		2.3	3.1	4.1	5.2
153		2.2	3.0	4.1	5.1
154		2.2	3.0	4.0	5.1
155		2.2	3.0	4.0	5.0
156		2.2	2.9	4.0	5.0
157		2.1	2.9	3.9	4.9
158		2.1	2.8	3.8	4.9
159		2.1	2.8	3.8	4.8
160		2.1	2.8	3.8	4.8
161		2.0	2.8	3.7	4.7
162		2.0	2.8	3.7	4.6
163		2.0	2.8	3.7	4.6
164		2.0	2.7	3.6	4.5
165		2.0	2.7	3.6	4.5
166		1.9	2.7	3.6	4.5
167		1.9	2.6	3.5	4.4
168		1.9	2.6	3.5	4.4
169		1.9	2.6	3.5	4.3
170		1.8	2.6	3.4	4.3

Women

Maximal oxygen uptake, liters · min^{-1}

Heart rate	300 kpm/min 50W	450 kpm/min 75W	600 kpm/min 100W	750 kpm/min 125W	900 kpm/min 150W
120	2.6	3.4	4.1	4.8	
121	2.5	3.3	4.0	4.8	
122	2.5	3.2	3.9	4.7	
123	2.4	3.1	3.9	4.6	
124	2.4	3.1	3.8	4.5	
125	2.3	3.0	3.7	4.4	
126	2.3	3.0	3.6	4.3	
127	2.2	2.9	3.5	4.2	
128	2.2	2.8	3.5	4.2	
129	2.2	2.8	3.4	4.1	
130	2.1	2.7	3.4	4.0	
131	2.1	2.7	3.4	4.0	
132	2.0	2.7	3.3	3.9	
133	2.0	2.6	3.2	3.8	
134	2.0	2.6	3.2	3.8	
135	2.0	2.6	3.1	3.7	
136	1.9	2.5	3.1	3.6	
137	1.9	2.5	3.0	3.6	
138	1.8	2.4	3.0	3.5	
139	1.8	2.4	2.9	3.5	
140	1.8	2.4	2.9	3.4	
141	1.8	2.3	2.8	3.4	
142	1.7	2.3	2.8	3.3	
143	1.7	2.2	2.7	3.3	
144	1.7	2.2	2.7	3.2	
145	1.6	2.2	2.7	3.2	
146	1.6	2.2	2.6	3.2	
147	1.6	2.1	2.6	3.1	
148	1.6	2.1	2.6	3.1	3.6
149		2.1	2.6	3.0	3.5
150		2.0	2.5	3.0	3.5
151		2.0	2.5	3.0	3.4
152		2.0	2.5	2.9	3.4
153		2.0	2.4	2.9	3.3
154		1.9	2.4	2.8	3.3
155		1.9	2.4	2.8	3.2
156		1.9	2.3	2.8	3.2
157		1.9	2.3	2.7	3.2
158		1.8	2.3	2.7	3.1
159		1.8	2.2	2.7	3.1
160		1.8	2.2	2.6	3.0
161		1.8	2.2	2.6	3.0
162		1.8	2.2	2.6	3.0
163		1.7	2.2	2.6	2.9
164		1.7	2.1	2.5	2.9
165		1.7	2.1	2.5	2.9
166		1.7	2.1	2.5	2.8
167		1.6	2.0	2.4	2.8
168		1.6	2.0	2.4	2.8
169		1.6	2.0	2.4	2.8
170		1.6	2.0	2.4	2.7

Source: From a nomogram by I. Åstrand (1960)

Table 10-5   Factor to be used for correction of predicted maximal oxygen uptake
(1) when the subject is over thirty to
thirty-five years of age or (2) when the
subject's maximal heart rate is known.
The actual factor should be multiplied by
the value that is obtained from Table
10-4.

Age	Factor	Max heart rate	Factor
15	1.10	210	1.12
25	1.00	200	1.00
35	0.87	190	0.93
40	0.83	180	0.83
45	0.78	170	0.75
50	0.75	160	0.69
55	0.71	150	0.64
60	0.68		
65	0.65		

underestimated; the extremely well-trained athletes are often overestimated. With a maximal aerobic power predicted to 3.0 liters $\cdot$ min^{-1}, the actual O$_2$ uptake for 5 out of 100 subjects is then less than 2.1 or higher than 3.9 liters $\cdot$ min^{-1} (SD = ± 15 percent). It is important to keep in mind this limitation in accuracy, and this drawback holds true for any submaximal cardiopulmonary test described so far. The validity of the nomogram has been tested in other laboratories. In some cases, there has been good agreement between the actually measured and predicted maximal O$_2$ uptake from the nomogram (Glassford et al., 1965; Teräslinna et al., 1966; Kavanagh and Shephard, 1976). In other studies, the subject's maximal O$_2$ uptake has been found to be underestimated when the nomogram was used (Rowell et al., 1964; Chase et al., 1966). In a study of 28 untrained men between twenty and thirty years of age, there was a statistically significant correlation (at about the 0.1 level) between the mean predicted maximal oxygen uptake (2.62 liters $\cdot$ min^{-1}) and the mean measured maximal oxygen uptake (2.38 liters $\cdot$ min^{-1}). In a further study, an even better correlation was observed (Fig. 10-8) (Hettinger et al., 1961; Rodahl and Issekutz, 1962).

The nomogram discussed is also adapted to a step test (Ryhming, 1953) and running on a treadmill.

In a study of 84 persons, age thirty to seventy years old, von Döbeln et al. (1967) found the nomogram to underestimate the maximal O$_2$ uptake by 0.15 liter $\cdot$ min^{-1}. The standard deviation was 17 percent. Utilizing electronic data-handling techniques, they observed that the introduction of a new correction factor for age eliminated the systematic difference between measured and predicted maximal oxygen uptake. The best prediction was obtained if submaximal heart rate, maximal heart rate, and age were used (SD = 8.4 percent). The precision was only slightly reduced when maximal heart rate was omitted. They also found

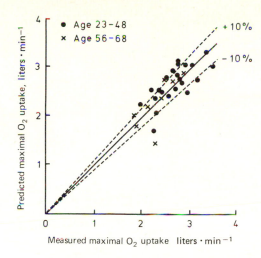

Figure 10-8 Estimated maximal O_2 uptake calculated from the Åstrand nomogram in relation to measured maximal O_2 uptake in 22 male subjects, ages twenty-three to forty-eight (•), and 9 male subjects, ages fifty-six to sixty-eight (x). Broken lines denote a deviation of ±10 per cent from the "ideal line." *(Rodahl and Issekutz, 1962.)*

that the age factor could not be substituted by maximal heart rate without the loss of accuracy in the prediction. Measurement of body size did not contribute to the prediction.

It would be reasonable to assume that some variation in the accuracy may be encountered from one group of subjects to another, with regard to both the difference between predicted and measured maximal oxygen uptake and the scatter of the data. Considering the considerable error of the method and the fact that it is applied, at best, only as a screening test, a consistent difference between measured and predicted maximal O_2 uptake of a few 100 ml · min^{-1} is of no importance.

Maritz et al (1961) and Wyndham et al. (1966) have applied a similar principle in their predictions. Their subjects perform four submaximal work loads (usually step tests). From measurements of heart rate and oxygen uptake the researchers fit a straight line to the four pairs of plots and extrapolate to a mean maximal heart rate for the population in question. Margaria et al. (1965) have also introduced a nomogram based upon similar concepts.

In some test procedures, the physical work capacity (PWC) is evaluated from data on work load, oxygen uptake, or oxygen pulse at a given heart rate, such as 180, 170, or 150 beats · min^{-1} (Sjöstrand, 1947; Wahlund, 1948; Balke, 1954). The methodological error must necessarily be about the same in these tests as in the application of the Åstrand-Åstrand nomogram (I. Åstrand, 1960). The calculated PWC_{170}, etc., is related to the maximal stroke volume of the heart, but it is *no measure of effectiveness or maximal power or rate of work output*. In any such estimate, the maximal heart rate must be considered. If the oxygen uptake is measured during the submaximal test, it is illogical to disregard individual variation in mechanical efficiency by expressing the capacity in watts or kilopond meters per minute at heart rate 170. However, the most important error is introduced when individuals of different ages are compared or evaluated without correcting for the decline in maximal heart rate with age. The mean value for heart rate at a given submaximal oxygen uptake is the same for individuals of the same sex and state of training regardless of age (from twenty-five up to at least seventy years of age) (I. Åstrand, 1960). By definition, therefore, the calculated PWC_{170} or oxygen pulse at a given work load is the same (Fig. 11-7). The real performance capacity, however, declines with age. Furthermore, the subjective feeling of strain is higher at a given heart rate, the older the individual.

Studies have also confirmed that there is a low correlation between the oxygen uptake or work load per minute achieved at a heart rate of 170 or 150 and the measured maximal oxygen uptake, cardiac output, heart size, or blood volume in individuals from twenty to seventy years of age (Strandell, 1964).

The conclusion is that an evaluation of the maximal effect of the oxygen-transporting system, based on studies at submaximal work loads or oxygen uptake, should be done with the utmost caution, especially when persons of different age groups are considered. Figure 11-7 presents mean values for various functional parameters in relation to age. It is evident that the decline in physical performance capacity is not related to a similar change in heart size, blood volume, or heart rate during a standard work load, etc.

There are at least two situations, however, when the submaximal work test has proved to be very useful:

1 In the clinical examination of patients or presumably healthy individuals, it is often important to include an exercise test in order to examine the cardiovascular system under functional stress. Bruce (1971) advocates the use of maximal testing even for cardiac patients (with some exceptions), while others believe it is safer to limit the testing to submaximal rates of work. Blomqvist and others (Blackburn, 1969, p. 281) have reported that the degree of ST-depression, if present in the ECG, will increase with higher O_2 uptake (and cardiac output) up to maximum. In testing patients for diagnostic and prognostic purposes it is, in our opinion, still an open question as to what is gained by a maximal test compared to the disadvantages of such a test. There may actually appear more so-called "false-positive cases" in the maximal test; the submaximal test may end up with some false-negative cases. The Scandinavian Committee on ECG classification (I. Åstrand et al., 1967) recommended that the submaximal exercise testing be terminated at a certain predetermined heart rate: in the age group 20–29 years at 170 beats · min^{-1}; 30–39 years at 160; 40–49 years at 150; 50–59 years at 140; 60–69 years at 130 beats · min^{-1}. The basis for this target heart rate is the fact that the maximal heart rate declines with age and that mean values are in the order of 220 minus the subject's age in years (I. Åstrand, 1960; Naughton et al., 1973).

Considering the wide scatter in maximal heart rates reported in the literature (Blackburn, 1969; Åstrand et al., 1973; Cooper et al., 1977), it is quite meaningless to be too sophisticated in the choice of the target heart rate.

The Scandinavian norms are quite simple to remember, and they have passed the test of time. At present there is hardly any better alternative available for the choice of a submaximal end point. (An analysis of the blood lactate concentration after the end of the exercise will give a hint of whether the subject was working close to his or her maximum or not, that is whether or not the anaerobic energy yield was significant.)

It should be emphasized that in the testing of patients with cardiovascular disease the termination of the test may be necessary due to discomfort or symptoms like angina pectoris, dyspnea, severe leg pains, dizziness, abnormal

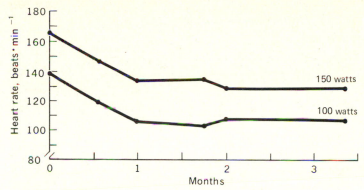

Figure 10-9 Decrease in the heart rate tested at two fixed work loads in the course of $3^{1}/_{2}$ months of training.

ECG changes or ischemic ST changes, or a systolic pressure that fails to rise or even falls as the rate of work increases (See also Naughton et al., 1973; Zohman and Phillips, 1973; Linhart and Turnoff, 1974; ACSM, 1975; Kavanagh and Shephard, 1976; Sheffield and Roitman, 1976).

2 In our experience, the submaximal exercise test is a very useful tool in evaluating whether or not a training program has been effective in improving the individual's circulatory capacity. It has been widely applied in top athletes, in trained and untrained adults, and in children. In this test, the individual is his or her own control; it is a matter of comparing the individual's own performances in repeated tests over months or years. In the simple work test on the bicycle ergometer, a counting of the heart rate is all that is needed for the evaluation. Figures 10-9 and 10-10 present examples of such applications of the test. A gradually decreasing heart rate at a standard load as the training progresses may stimulate the individual's efforts to continue to improve his or her circulatory capacity further. The individual tested is usually interested in knowing the result of the test, and will compare it with other data. In this event, a prediction of the maximal oxygen uptake in absolute figures and per kilogram of body weight is definitely not essential but may be justified if the conductor of the test is aware of the limitations of the method. In these cases, the simple work test is not a research instrument but a training guide, and should not be used to compare one person with another.

It should be noted that the pulse rate at a single submaximal test does not reveal anything about the subject's state of training, since constitutional factors play a greater role than merely the state of training. A person may have a low pulse at a standard load and yet be entirely untrained, and a well-trained individual may show a high pulse rate.

Procedures

The various sources of error associated with every type of prediction have already been emphasized. Having pointed out these sources of error, one may be justified in summarizing a principle which is based on personal experience and which requires a minimum of equipment and facilities. It should always be kept in mind, however, that predicted values

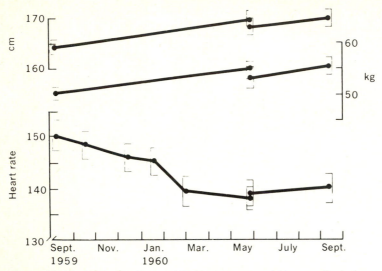

Figure 10-10 Presents data on 163 fourteen-year-old boys collected over a period of 1 year with a work test on a bicycle ergometer (100 watts, oxygen uptake about 1.5 liters · min⁻¹). The decrease in heart rate, observed in steady state of work, suggests an 8 percent increase in maximal oxygen uptake per kilogram body weight from September 1959 until May 1960. After summer vacation for $2^{1}/_{2}$ months, the heart rate for 100 retested boys was somewhat higher as also was the body weight. The maximal oxygen uptake per kilogram weight was now 5 percent lower; the boys apparently did not train as hard when on vacation as they did in school.

cannot be used as a substitute for the actual measurement of precise values when scientific exactness is required.

Under normal conditions, there is in any given individual a roughly linear relationship between oxygen uptake and heart rate during submaximal work (Fig. 10-11). The slope of the line changes with the state of physical training or physical fitness; a fit person is able to transport the same amount of oxygen at a lower heart rate than an unfit person. This relationship in general is independent of sex and age, although females acquire higher heart rates to transport the same amount of oxygen than males do.

On the basis of this oxygen uptake–heart rate relationship, it is possible to predict an individual's maximal oxygen uptake by the heart rate response to two submaximal loads, as is evident from Fig. 10-11, assuming a maximal heart rate of about 190 to 200 beats · min⁻¹. However, since the maximal heart rate declines with age after about age twenty, it is necessary to make corrections for age. It should be pointed out that at a given submaximal work load, a trained sixty-five-year-old subject may have about the same heart rate as a trained thirty-year-old subject. However, the thirty-year-old subject, having a maximal heart rate of about 190, may have a maximal oxygen uptake of about 4.0 liters per min, while the older subject, having a maximal heart rate of about 160, can take up only 3.0 liters of oxygen at a maximum, using the example illustrated in Fig. 10-11.

These considerations form the basis for the nomogram for the estimation of maximal O₂ uptake on the basis of heart rate or pulse response to submaximal work load developed by Åstrand and Åstrand (Table 10-4). This nomogram has been modified by I. Åstrand to include correction for the age factor (Table 10-5).

Performance of the Test

From this discussion it is evident that the pulse rate as such during a work load under standard conditions can be used as an indication of the state of circulatory fitness of an individual. Such tests should be done under steady-state conditions, i.e., during continuous

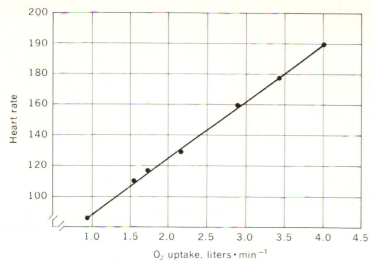

Figure 10-11   The relationship between heart rate and O_2 uptake on the bicycle ergometer.

steady work lasting at least 5 to 6 min. Under such conditions, V_{O_2} equals oxygen demand, so that the metabolic processes are essentially aerobic, with no significant increase in blood lactate. Under these conditions, pulmonary ventilation, heart rate, and cardiodynamic parameters, such as cardiac output, are essentially constant during the last 2 to 3 min of the test lasting 5 to 6 min. The advantage with this is that one may well compare the results from one test to another.

The subject should refrain from energetic physical activity 2 hr preceding the work test. In addition, the test should not be performed earlier than about 1 hr after a light meal, or 2 to 3 hours after a heavier meal has been eaten. Furthermore, the test subject should not smoke for the last hour prior to the commencement of the test.

Experience has shown that the resting heart rate does not normally give any information over and above that provided by the work test. It is therefore not necessary to record the resting heart rate.

Provided that the work is not too heavy, respiration and circulation increase during the first 2 to 3 min of work and then attain a steady state. The increase in pulse rate can easily be established by counting the pulse once every minute. In any case, after 4 to 5 minutes' work the pulse rate has generally reached the steady state. As a rule, a working time of about 5 to 6 min is thus sufficient to adapt the pulse rate to the task being performed. The pulse rate should be taken every minute, the mean value of the pulse rate at the fifth and sixth minutes being designated the working pulse for the work in question. If the difference between these last two pulse rates exceeds 5 beats · min^{-1}, the working time should be prolonged 1 or more minutes until a constant level is reached. The pulse rate is most easily felt over the carotid artery just below the mandible angle. For the inexperienced, it is rather difficult to count the pulse rate: the metronome confuses, the subject is in motion, and the pulse may vary. Practice under experienced supervision is important. The pulse rate may be suitably measured during the last 15 to 20 s of every working minute. The most exact value is obtained by taking the time for 30 pulse beats. A person trained in taking pulse rates, whether it be by palpation or by auscultation, obtains values which are in close agreement with those obtained by ECG recordings, provided the pulse rate is not irregular.

At the end of each exercise stage it has proven quite useful to ask the subject about the *subjective rating of perceived exertion*. Borg (1974) has suggested a scale consisting of 15 grades from 6 to 20 (7 = very, very light; 9 = very light; 11 = fairly light; 13 = somewhat heavy; 15 = heavy; 17 = very heavy; 19 = very, very heavy). In many situations the heart rate mirrors the physical strain experienced subjectively (Ekblom and Goldbarg, 1971).

Choice of Load

For trained, active individuals, the risk of overstraining in connection with a work test is very slight. For female subjects a suitable load is 75 to 100 watts, and for male test subjects, 100 to 150 watts. If the heart rate exceeds about 130 beats per min, the load can be considered adequate and the test can be discontinued after 6 min. If the pulse rate is slower than about 130 beats per min, the load should be increased by 50 watts after 6 min. If time permits testing with several loads, increases of 50 watts should be made in 6-min periods for as long as the pulse rate remains below about 150 beats per min. The next working period should be continued for 6 min even if the pulse rate should then exceed 150 beats per min.

For persons who may be expected to have a lower physical work capacity, persons who are completely untrained, and older individuals, lower loads should be chosen and an initial load of 50 watts might be suitable.

If a physician is not present, work tests on persons over forty years of age should be discontinued if the pulse rate exceeds 150 beats per min. The load for such persons should not be raised above 100 watts for female subjects or 150 watts for male subjects. In the event of pressure or pain in the chest, or marked shortness of breath or distress, the test must be discontinued immediately.

Evaluation of the Work Test

The work test, in the simple form just described, actually gives very limited possibilities of judging the test subject's physical capacity for running, skiing, swimming, etc. In the performance of various kinds of sports and athletics, and in physical work in general, the "motor skill" obviously plays an important part, but other critical factors include technique and motivation. However, the test does give some idea of the aerobic power or the oxygen-transporting capacity, but even here there are sources of error. The maximal pulse rate varies with age, but it can also vary within the same age group. A pulse rate of 150 at work implies an almost maximal effort for a person with a maximal pulse rate of 160, but represents a light load for a person with a pulse ceiling at 200 beats $\cdot$ min^{-1}.

The most valuable application of the work test as described is to test the individual on several different occasions during a period of physical training. In this way it is possible to determine objectively whether the training program has been effective (i.e., whether a given work load is achieved with a lower pulse rate).

Table 10-4 enables the maximal oxygen uptake to be estimated from the heart rate at a certain load. [Example: A male subject has a pulse rate of 147 when working at a rate of 150 watts. According to Table 10-4, his maximal oxygen uptake will be about 3.3 liters $\cdot$ min^{-1}. The maximal oxygen uptake per kilogram of body weight is obtained by dividing the maximal oxygen uptake in ml (in this case, 3300 ml) by the subject's body weight (in this case, 3300 $\cdot$ 74^{-1} = 45 ml $\cdot$ kg $\cdot$ min^{-1}). If different loads have been used, the mean value of the oxygen uptake rate calculated for each load is applied.]

Obviously, the table values are only approximations. Experience has shown that older persons are generally overestimated in regard to predicted maximal oxygen uptake. The value obtained from Table 10-4 must therefore be corrected by multiplication with the age factor given in Table 10-5. (Example: A male subject, weight 79 kg, has a pulse rate of 139 beats per min at 150 watts. If he is fifty years of age, the value will be 3.6 $\cdot$ 0.75 = 2.7 liters $\cdot$ min^{-1}.)

Respiratory Quotient, Lactic Acid   During heavy exercise of short duration (up to 5 min), the respiratory quotient determined on the expired air exceeds 1.0. Issekutz et al. (1962) have shown that the work respiratory quotient (R) under such standardized conditions may be used as a measure of physical work capacity. It was shown that ΔR (work R minus 0.75) increases logarithmically with the work load, and maximal O$_2$ uptake is reached at a ΔR value of 0.40. This observation offers the possibility of predicting the maximal O$_2$ uptake of a person, based on the measurement of R during a single 5-min bicycle ergometer test at a submaximal load.

This proposed method of using the respiratory quotient in the assessment of aerobic work capacity is based on the observation that a part of the expired CO_2 during short, intense work efforts is derived from the body bicarbonate pool as a result of accumulation of lactic acid during exercise. The concentration of lactic acid in the blood starts to increase as soon as the work load exceeds about 50 to 60 percent of the individual's maximal aerobic power. This fact may be utilized as a rough measure to assess the effectiveness of the oxygen-transporting system of an individual during standardized work. This method may be particularly useful when testing the physical work capacity of patients who are kept on drugs which may affect the heart rate, making it impossible to use the Åstrand nomogram for the prediction of the work capacity on the basis of the pulse rate at submaximal loads. The accuracy of this method of predicting maximal oxygen uptake was tested in a group of 32 subjects (Issekutz et al., 1962). The results are given in Fig. 10-12, in which the calculated maximal oxygen uptake is plotted against the actually measured values. The mean difference between measured and predicted values was 0.002 $\pm$ 0.016 liters · min^{-1} with an SD of $\pm$ 0.089. In general, by this method, the estimated maximal O_2 uptake may agree within $\pm$ 5 percent of the directly measured value.

If the blood lactate is not increased, it may indicate that the individual's maximal O_2 uptake is at least twice as high as the O_2 uptake measured during the test load in question.

Limitations   Even when the tests are carried out during strictly standardized conditions, the methodological error in the prediction of the maximal aerobic power is, as previously mentioned, considerable (SD = 10 to 15 percent). By standardized conditions we mean: The subject must be free of any infection. Several hours must have elapsed between the last meal and the test. The subject should not have been engaged in any physical work heavier than the test load the last few hours prior to the test. Smoking should not be allowed

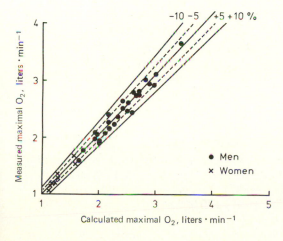

Figure 10-12   Maximal O_2 uptake calculated on the basis of R measurements at submaximal work loads, in relation to the actually measured maximal O_2 uptake in 32 subjects. *(Issekutz et al., 1962.)*

during the last 2 hr before the test. The temperature of the room where the test is to be performed should be 18 to 20°C; the room should be adequately ventilated. An electric fan should be available, especially if the test involves prolonged work. The subject should be relaxed, and there should be no spectators other than the investigators.

Under such standardized conditions, the variation from day to day in heart rate at a given oxygen uptake is less than 5 beats $\cdot$ min^{-1}, provided the state of training is the same. The mean value for a group of subjects undergoing repeated tests remains almost exactly at the same level under these conditions (Ryhming, 1953).

It appears that a number of situations may cause a marked increase in the pulse rate at a submaximal work load, without the maximal oxygen uptake capacity being significantly reduced. The following examples may be mentioned:

1   Dehydration during heavy physical work or during exposure to heat (Saltin, 1964; Chap. 15)

2   Prolonged heavy exercise (Saltin, 1964; Rowell, 1974; Chaps. 14 and 15)

3   Work in a hot environment (Williams et al., 1962; Rowell, 1974; Chap. 15)

4   After pyrogen-induced fever (Grimby and Nilsson, 1963)

5   After ingestion of alcohol (Blomqvist et al., 1970; Chap. 17)

In some of these situations the performance capacity may actually be reduced. Thus, under these conditions the excessive heart-rate response to a given submaximal load is a better criterion for a reduced work capacity than the maximal oxygen uptake. However, the effect is presently impossible to determine quantitatively from such tests. It should again be emphasized that the maximal aerobic power is only one of several factors which determine performance capacity. It should also be stressed that fear, excitement, and related emotional stress may also cause a marked elevation of the heart rate at a submaximal work load without either maximal O_2 uptake or performance capacity being affected. The heavier the work load, however, the less pronounced is this nervous effect on the pulse rate. It is usually recommended that the test load should be sufficiently high so as to bring the pulse rate up to or above 150 beats per min in the case of younger subjects.

In some instances, the pulse rate at a standard work load may be unchanged while the maximal O_2 uptake and the performance capacity are reduced; for example:

1   Following acclimatization for a certain period at high altitude (Christensen, 1937; P.-O. Åstrand, 1954; P.-O. Åstrand and I. Åstrand, 1958; Saltin, 1967)

2   During semistarvation (Keys et al., 1950)

These examples illustrate further the danger of drawing conclusions concerning maximal O_2-uptake capacity and physical performance capacity from data obtained during submaximal tests.

Summary The submaximal exercise test can be applied as a valuable screening test for the evaluation of the functional capacity of the oxygen-transporting system. For research purposes, however, it is not accurate enough to substitute the actual measurements of this capacity. It may be a useful method for selecting the best, the worst, and the average persons from a group. In the examination of patients, the exercise test makes it possible to include observations and studies during physical activity which might simulate the work load of the patient's daily activities. Heart size, electrocardiogram, blood pressure, etc., should be evaluated together with results from work tests. However, any prediction of the work capacity should be avoided if the patient has not been subject to maximal exercise. Repeated submaximal tests on the bicycle ergometer are very useful in controlling the effectiveness of a physical training program. The results may motivate the subject to continue training.

REFERENCES

American College of Sports Medicine (ACSM): "Guidelines for Graded Exercise Testing and Exercise Prescription," Lea & Febiger, Philadelphia, 1975.

American Association for Health, Physical Education and Recreation: The AAHPER Fitness Test Manual, rev. ed., Washington, D.C., 1965.

Åstrand, I.: Aerobic Work Capacity in Men and Women with Special Reference to Age, *Acta Physiol. Scand.*, **49**(Suppl. 169):1960.

Åstrand, I., et al.: The "Minnesota Code" for ECG Classification. Adaptation to CR Leads and Modification of the Code for ECG's Recorded during and after Exercise, *Acta Med. Scand.* (Suppl. 481):1967.

Åstrand, P.-O.: "Experimental Studies of Physical Working Capacity in Relation to Sex and Age," Munksgaard, Copenhagen, 1952.

Åstrand, P.-O.: Study of Bicycle Modifications Using a Motor Driven Treadmill-bicycle Ergometer, *Arbeitsphysiol.*, **15**:23, 1953.

Åstrand, P.-O.: The Respiratory Activity in Man Exposed to Prolonged Hypoxia, *Acta Physiol. Scand.*, **30**:343, 1954.

Åstrand, P.-O.: Quantification of Exercise Capability and Evaluation of Physical Capacity in Man, *Progress in Cardiovascular Diseases*, **19**:51, 1976.

Åstrand, P.-O., and I. Åstrand: Heart Rate during Muscular Work in Man Exposed to Prolonged Hypoxia, *J. Appl. Physiol.*, **13**:75, 1958.

Åstrand, P.-O., and Irma Ryhming: A Nomogram for Calculation of Aerobic Capacity (Physical Fitness) from Pulse Rate during Submaximal Work, *J. Appl. Physiol.*, **7**:218, 1954.

Åstrand, P.-O., and B. Saltin: Maximal Oxygen Uptake and Heart Rate in Various Types of Muscular Activity, *J. Appl. Physiol.*, **16**:977, 1961.

Balke, B.: Optimale Körperliche Leistungsfähigkeit, ihre Messung und Veränderung infolge Arbeitsermüdung, *Arbeitsphysiol.*, **15**:311, 1954.

Beaver, W. L.: Water Vapor Corrections in Oxygen Consumption Calculations, *J. Appl. Physiol.*, **35**:928, 1973.

Berggren, G., and E. H. Christensen: Heart Rate and Body Temperature as Indices of Metabolic Rate during Work, *Arbeitsphysiol.*, **14**:255, 1950.

Binkhorst, R. A., and P. van Leeuwen: A Rapid Method for the Determination of Aerobic Capacity, *Intern. Z. Angew. Physiol.*, **19**:459, 1963.

Blackburn, H. (ed.): "Measurements in Exercise Electrocardiography," Charles C Thomas, Springfield, Ill., 1969.

Blomqvist, G., B. Saltin, and J. H. Mitchell: Acute Effects of Ethanol Ingestion on the Response to Submaximal and Maximal Exercise in Man, *Circulation*, **42**:463, 1970.

Bobbert, A. C.: Physiological Comparison of Three Types of Ergometry, *J. Appl. Physiol.*, **15**:1007, 1960a.

Bobbert, A. C.: Energy Expenditure in Level and Grade Walking, *J. Appl. Physiol.*, **15**:1015, 1960b.

Borg, G.: Psychological Aspects of Physical Activities, in L. A. Larson (ed.), "Fitness, Health and Work Capacity," p. 141, Macmillan, New York, 1974.

Bruce, R. A.: Exercise Testing of Patients with Coronary Heart Disease, *Ann. Clin. Res.*, **3**:323, 1971.

Chase, G. A., C. Grave, and L. B. Rowell: Independence of Changes in Functional and Performance Capacities Attending Prolonged Bed Rest, *Aerospace Med.*, **37**:1232, 1966.

Christensen, E. H.: Sauerstoffaufnahme und respiratorische Funktionen in grossen Höhen, *Skand. Arch. Physiol.*, **76**:88, 1937.

Cooper, K. H., J. G. Purdy, S. R. White, M. L. Pollack, A. C. Linnerud: Age-Fitness Adjusted Maximal Heart Rates, in D. Brunner and E. Jokl (eds.), "The Role of Exercise in Internal Medicine," p. 78, S. Karger, Basel, 1977.

Davies, C. T. M., and A. J. Sargent: Physiological Response to One-and Two-leg Exercise Breathing Air and 45% Oxygen, *J. Appl. Physiol.*, **36**:142, 1974.

Döbeln, W. v., I. Åstrand, and A. Bergström: An Analysis of Age and Other Factors Related to Maximal Oxygen Uptake, *J. Appl. Physiol.*, **22**:934, 1967.

Eckermann, P., and H. P. Millahn: Der Einfluss der Drehzahl auf die Herzfrequenz und die Sauerstoffaufnahme bei konstanter Leistung am Fahrradergometer, *Int. Z. angew. Physiol. einschl. Arbeitsphysiol.*, **23**:340, 1967.

Ekblom, B., and A. N. Goldbarg: The Influence of Physical Training and Other Factors on the Subjective Rating of Perceived Exertion, *Acta Physiol. Scand.*, **83**:399, 1971.

Fernandez, E. A., J. G. Mohler, and J. P. Butler: Comparison of Oxygen Consumption Measured at Steady State and Progressive Rates of Work, *J. Appl. Physiol.*, **37**:982, 1974.

Froelicher, V. F., Jr., H. Brammell, G. Davis, I. Noguera, A. Stewart, and M. C. Lancaster: A Comparison of Three Maximal Treadmill Exercise Protocols, *J. Appl. Physiol.*, **36**:720, 1974.

Glassford, R. G., G. H. I. Baycroft, A. W. Sedgwick, and R. B. J. Macnab: Comparison of Maximal Oxygen Uptake Values Determined by Predicted and Actual Methods, *J. Appl. Physiol.*, **20**:509, 1965.

Grimby, G., and N. J. Nilsson: Cardiac Output during Exercise in Pyrogen-induced Fever, *Scand. J. Clin. Lab. Invest.*, **15**(Suppl. 69):44, 1963.

Grimby, G., and B. Saltin: A Physiological Analysis of Physically Well-Trained Middle-aged and Old Athletes, *Acta Med. Scand.*, **179**:513, 1966.

Grosse-Lordemann, H., and E. A. Müller: Der Einfluss der Tretkurbellänge auf das Arbeitsmaximum und den Wirkungsgrad beim Radfahren, *Arbeitsphysiologie*, **9**:619, 1937.

Hermansen, L.: "Oxygen Transport during Exercise in Human Subjects," *Acta Physiol. Scand.* (Suppl. 399): 1973.

Hermansen, L., and B. Saltin: Oxygen Uptake during Maximal Treadmill and Bicycle Exercise, *J. Appl. Physiol.*, **26:**31, 1969.

Hermansen, L., B. Ekblom, and B. Saltin: Cardiac Output during Submaximal and Maximal Treadmill and Bicycle Exercise, *J. Appl. Physiol.*, **29:**82, 1970.

Hettinger, Th., and K. Rodahl: Ein Modifizierter Stufentest Zur Messung Der Belastungsfähigkeit Des Kreislautes, *Deutsche Medizinische Wochenschrift*, **14:**553, 1960.

Hettinger, Th., N. C. Birkhead, S. M. Horvath, B. Issekutz, Jr., and K. Rodahl: Assessment of Physical Work Capacity, *J. Appl. Physiol.*, **16:**153, 1961.

Issekutz, B., Jr., N. C. Birkhead, and K. Rodahl: Use of Respiratory Quotients in Assessment of Aerobic Work Capacity, *J. Appl. Physiol.*, **17:**47, 1962.

Johnson, R. E., L. Brouha, and R. C. Darling: A Test of Physical Fitness for Strenuous Exertion, *Rev. Canad. Biol.*, **1:**491, 1942.

Kamon, E., and K. B. Pandolf: Maximal Aerobic Power during Laddermill Climbing, Uphill Running and Cycling, *J. Appl. Physiol.*, **32:**467, 1972.

Kasch, F. W., W. H. Phillips, W. D. Ross, J. E. L. Carter, and J. L. Boyer: A Comparison of Maximal Oxygen Uptake by Treadmill and Step-Test Procedures, *J. Appl. Physiol.*, **21:**1387, 1966.

Kasch, F. W., J. P. Wallace, R. R. Hurn, L. A. Krogh, and P. M. Hurl: $\dot{V}_{O_2}$ Max during Horizontal and Inclined Treadmill Running, *J. Appl. Physiol.*, **40:**982, 1976.

Kattus, A. A. (ed.): "Exercise Testing and Training in Apparently Healthy Individuals," Am. Heart Ass., New York, 1972.

Kavanagh, T., and R. J. Shephard: Maximum Exercise Tests on "Postcoronary" Patients, *J. Appl. Physiol.*, **40:**611, 1976.

Keys, A., J. Brozek, A. Henschel, O. Michelsen, and H. L. Taylor: "The Biology of Human Starvation," pp. 675, 735, University of Minnesota Press, Minneapolis, 1950.

Linhart, J. W., and H. B. Turnoff: Maximum Treadmill Exercise Test in Patients with Abnormal Control Electrocardiograms, *Circulation,* **49:**667, 1974.

Mahadeva, K., R. Passmore, and B. Woolf: Individual Variations in the Metabolic Cost of Standardized Exercises: The Effect of Food, Age, Sex, and Race, *J. Physiol.*, **121:**225, 1953.

Margaria, R., P. Cerretelli, P. Aghemo, and G. Sassi: Energy Cost of Running, *J. Appl. Physiol.*, **18:**367, 1963.

Margaria, R., P. Aghemo, and E. Rovell: Indirect Determination of Maximal O_2 Consumption in Man, *J. Appl. Physiol.*, **20:**1070, 1965.

Maritz, J. S., J. F. Morrison, J. Peter, N. B. Strydom, and C. H. Wyndham: A Practical Method of Estimating an Individual's Maximal Oxygen Intake, *Ergonomics*, **4:**97, 1961.

Mathews, D. K.: "Measurement in Physical Education," W. B. Saunders, Philadelphia, 1974.

McArdle, W. D., F. I. Katch, and G. S. Pechar: Comparison of Continuous and Discontinuous Treadmill and Bicycle Tests for Max $\dot{V}_{O_2}$, *Med. Sci. Sports*, **5:**156, 1973.

McKay, G. A., and E. W. Banister: A Comparison of Maximum Oxygen Uptake Determination by Bicycle Ergometry at Various Pedaling Frequencies and by Treadmill Running at Various Speeds, *Europ. J. Appl. Physiol.*, **35:**191, 1976.

Nagle, F. J., B. Balke, and J. P. Naughton: Gradational Step Tests for Assessing Work Capacity, *J. Appl. Physiol.*, **20:**745, 1965.

Naughton, J. P., H. K. Hellerstein, and I. C. Mohler: "Exercise Testing and Exercise Training in Coronary Heart Disease," Academic Press, Inc., New York, 1973.

Passmore, R., and J. V. G. A. Durnin: Human Energy Expenditure, *Physiol. Rev.*, **35**:801, 1955.

Rodahl, K., and B. Issekutz, Jr.: Physical Performance Capacity of the Older Individual, in K. Rodahl and S. M. Horvath (eds.), "Muscle as a Tissue," McGraw-Hill Book Company, New York, 1962.

Rowell, L. B.: Human Cardiovascular Adjustments to Exercise and Thermal Stress, *Physiol. Rev.*, **54**:75, 1974.

Rowell, L. B., H. L. Taylor, and Y. Wang: Limitations to Prediction of Maximal Oxygen Uptake, *J. Appl. Physiol.*, **19**:919, 1964.

Ryhming, I.: A Modified Harvard Step Test for the Evaluation of Physical Fitness, *Arbeitsphysiol.*, **15**:235, 1953.

Saltin, B. : Aerobic Work Capacity and Circulation at Exercise in Man, *Acta Physiol. Scand.*, **62**(Suppl. 230): 1964.

Saltin, B.: Aerobic and Anaerobic Work Capacity at 2300 Meters, *Med. Thorac.*, **24**:205, 1967.

Saltin, B., and P.-O. Åstrand: Maximal Oxygen Uptake in Athletes, *J. Appl. Physiol.*, **23**:353, 1967.

Saltin, B., G. Blomqvist, J. H. Mitchell, R. L. Johnson, Jr., K. Wildenthal, and C. B. Chapman: Response to Submaximal and Maximal Exercise after Bed Rest and Training, *Circulation*, **38**(Suppl. 7): 1968.

Sheffield, L. T., and D. Roitman: Stress Testing Methodology, *Progress in Cardiovascular Diseases*, **19**:33, 1976.

Shephard, R. J.: The Relative Merits of the Step Test, Bicycle Ergometer, and Treadmill in the Assessment of Cardio-Respiratory Fitness, *Intern. Z. Angew. Physiol.*, **23**:219, 1966.

Shephard et al.: Standardization of Submaximal Exercise Tests, *Bull. WHO*, **38**:765, 1968.

Sjöstrand, T.: Changes in the Respiratory Organs of Workmen at an Ore Melting Works, *Acta Med. Scand.*(Suppl. 196):687, 1947.

Stamford, B. A.: Step Increment versus Constant Load Tests for Determination of Maximal Oxygen Uptake, *Europ. J. Appl. Physiol.*, **35**:89, 1976.

Stenberg, J., P.-O. Åstrand, B. Ekblom, J. Royce, and B. Saltin: Hemodynamic Response to Work with Different Muscle Groups, Sitting and Supine, *J. Appl. Physiol.*, **22**:61, 1967.

Strandell, T.: Circulatory Studies on Healthy Old Men with Special Reference to the Limitation of the Maximal Physical Working Capacity, *Acta Med. Scand.*, **175**(Suppl. 414): 1964.

Taylor, H. L., E. Buskirk, and A. Henschel: Maximal Oxygen Intake as an Objective Measure of Cardiorespiratory Performance, *J. Appl. Physiol.*, **8**:73, 1955.

Teräslinna, P., A. H. Ismail, and D. F. MacLeod: Nomogram by Åstrand and Ryhming as a Predictor of Maximum Oxygen Intake, *J. Appl. Physiol.*, **21**:513, 1966.

Van der Walt, W. H., and C. H. Wyndham: An Equation for Prediction of Energy Expenditure of Walking and Running, *J. Appl. Physiol.*, **34**:559, 1973.

Wahlund, H.: Determination of the Physical Working Capacity, *Acta Med. Scand.* (Suppl. 215): 1948.

Williams, C. B., G. A. G. Bredell, C. H. Wyndham, N. B. Strydom, J. F. Morrison, P. W. Flemming, and J. S. Wood: Circulatory and Metabolic Reaction to Work in Heat, *J. Appl. Physiol.*, **17**:625, 1962.

Wyndham, C. H., N. B. Strydom, W. P. Leary, and C. G. Williams: Studies of the Maximum Capacity of Men for Physical Effort, *Intern. Z. Angew. Physiol.*, **22**:285, 1966.

Zohman, L. R., and R. E. Phillips: "Medical Aspects of Exercise Testing and Training," Intercontinental Medical Book Corp., New York, 1973.

Body Dimensions and Muscular Work

11

Chapter 11

Body Dimensions and Muscular Work

The thrill of watching athletic competitions is caused partially by the fact that it is impossible to predict who is going to win.

It is impossible from appearance alone to tell who is an athletic champion. On the other hand, it is often possible to exclude those who obviously *cannot* reach top results in certain sport events, such as shot-putting, rowing, and American football. A tiny individual hardly has a chance in these events.

It is of interest to consider the human resources in relation to muscular work from a biological viewpoint. If we compare animals of different size, it is evident that certain dimensions and functional capacities are determined by fundamental mechanical necessities. In addition, it may be a matter of biological adaptation.

STATICS

If we take two geometrically similar cubes of different size, the relationship between the surface and the volume of the two cubes can easily be calculated when only the scale factor between the sides of the cubes is known. If this length scale is L:1, the surface ratio is L^2:1, and the volume ratio L^3:1.

If we consider two geometrically similar and qualitatively identical individuals, we may expect all linear dimensions (L) to be proportional. The length of the arms, the legs, the trachea, and the individual muscles will have a ratio L:1. If we compare two boys, one 120 cm high, the other 180 cm high, the scale factor will be such that all lengths, levers, ranges of joint motions, and muscular contractions during a specific motion will be related as 120:180 or as 1:1.5 (Fig. 11-1).

Cross sections of, for instance, a muscle, the aorta, a bone, the trachea, the alveolar surface, or the surface of the body are then related as 120^2:180^2, or 1^2:1.5^2, i.e., 1:2.25. Volumes, such as lung volumes, blood volumes, or heart volumes, should similarly be related as 120^3:180^3, or 1^3:1.5^3, i.e., 1:3.375. The same applies to mass measured in units of weight, since the density of biological materials, generally speaking, is independent of size.

The force of gravity acts on the mass (M) of the body. If the limbs supporting this mass are proportional to the area of their cross section, there is a disproportion between the mass (which is proportional to L^3) and the support (which is proportional to L^2). In larger animals an adjustment has taken place: elephants and rhinoceroses have relatively thick, short, and straight legs. In body geometry, these larger animals differ from smaller animals, which have slender limbs compared with their body mass. In relation to total weight, the human has over double as much bone as the mouse (Thompson, 1943). Those of

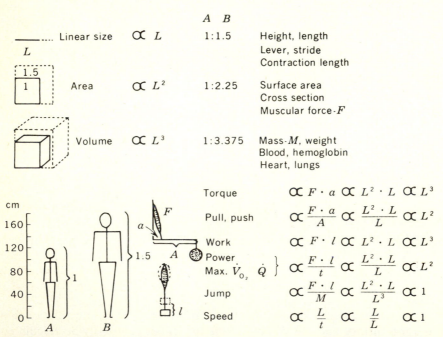

Figure 11-1 Schematic illustration of the influence of dimensions on some static and dynamic functions in geometrically similar individuals. *A* and *B* represent two persons with body height 190 and 180 cm respectively. (α = proportional too.) See text for explanation. *(Partly modified from Asmussen and Christensen, 1967.)*

the larger animals that have relatively slender limbs have "compressed" bodies compared with their length (giraffe, antelope). In large animals, the design of their bodies is dominated by the necessity of supporting their weight. This was pointed out by Galilei in 1638.

DYNAMICS

If we pursue these theoretical considerations further, it is evident that the maximal force a muscle can develop (F), generally speaking, is proportional to ($\propto$) its surface. It would therefore be expected that the 1.5-times taller boy should be able to produce a 2.25-times larger muscular strength, i.e., he should be able to lift a 2.25-times larger weight. The advantage of the 1.5-times longer levers (a) for the taller boy's muscles to work on is offset by the fact that the weight to be lifted also has a 1.5-times longer lever (A). (According to Fig. 11-1:

$$F \cdot a = M \cdot A$$
$$M = \frac{F \cdot a}{A}$$

In accordance with the discussion above, $F \propto L^2$, $a \propto L$; $A \propto L$. The above equation can thus be expressed as follows:

$$M \propto \frac{L^2 \cdot L}{L} \propto L^2.)$$

Force

The magnitude of the work to be performed is determined by the developed force ($\propto L^2$) and the distance the force is applied ($\propto L$). Consequently, $W \propto L^2 \cdot L \propto L^3$. Thus the work which the larger boy in our example should be able to perform is accordingly 3.375 times larger than that which would be expected from the smaller boy.

"Chin-ups"

If it is a matter of lifting one's own body with a mass (M), as in chinning the bar, the formula can be expressed as follows: $F \cdot a = M \cdot A$, where a and A represent the levers for the muscles and the body weight respectively. In other words, the achievement is proportional to the force of the muscles and their levers, but inversely proportional to the body mass and the levers upon which it works. According to the reasoning which we have applied above, it follows that the ability to lift one's own body (i.e., do a chin-up) is proportional to

$$\frac{F \cdot a}{M \cdot A} \text{ or } \frac{L^2 L}{L^3 L} \propto \frac{1}{L} \propto L^{-1}$$

Thus, in the case of the smaller boy the ratio will be 1:1 = 1, but in the case of the larger boy the ratio is 1:1.5 = 0.67, i.e., the larger and stronger boy is actually handicapped by his greater body weight when he has to lift his body, as when chinning the bar.

Time

In many dynamic events an expression of the time scale (t) is necessary. It may be a matter of energy transfer per unit time, the time between two steps or between two heartbeats. It might be of interest to examine this time scale for similar animals of different sizes. If the ratio is expressed as t:1, the relationship between time and acceleration (a) according to the definition of the acceleration is $a \propto L \cdot t^{-2}$. The connection between the time scale and the length scale may then be calculated by the following formula (Döbeln, 1966):

Force = acceleration · mass

$$F = a \cdot M \quad a = \frac{F}{M} \quad a \propto \frac{L^2}{L^3}$$

$$\propto \frac{1}{L} \; (\propto L^{-1})$$

Returning to the formula $a \propto L \cdot t^{-2}$ or $t^2 \propto L \cdot a^{-1}$, we may now replace a with L^{-1} as shown above, and we arrive at the following formula:

$$t^2 \propto \frac{L}{L^{-1}} \quad \text{or} \quad t^2 \propto L^2 \quad t \propto L$$

In other words: *The time scale is proportional to the length scale.*

Acceleration

Since a is proportional to L^{-1}, taller (and heavier) persons are handicapped when it is a matter of accelerating their body mass.

Frequencies

The shorter the time between two steps, heartbeats, etc., the higher the frequency (f), which is another expression of the passing of time. This may be expressed as follows: $f \propto 1 \cdot t^{-1}$, and accordingly, $f \propto 1 \cdot L^{-1} (\propto L^{-1})$.

According to this reasoning, we might expect that the frequency of limb motion should vary as an inverse function of limb length. As pointed out by Hill (1950), this is generally the case. A hummingbird moves its wings about 75 to 100 times · s^{-1} while flying forward, a sparrow some 15 times · s^{-1}, and a stork only 2 or 3 times. These frequencies are roughly in inverse proportion to the linear size of the birds. "If the sparrow's muscles were as slow as the stork's, it would be unable to fly. If the stork's muscles were as fast as the hummingbird's, it would be exhausted very quickly" (Hill, 1950). The maximal force of a contracting voluntary muscle is roughly constant in different animals, being of the order of a few kilograms per square centimeter cross-sectional area (Chap.

4). The speed of contraction varies enormously, however, among different muscles and different animals. The balance between muscle strength, length of levers, and speed of contraction is very delicate. If a human had muscles as fast as those moving wings of a hummingbird, they would soon break the bones and tear the muscles and tendons. It is possible to swing a rod made of fragile material back and forth if it is short and relatively thick. But if it is long and with a thickness just proportional to L^2, the inertia will cause it to break if moved at the same speed. For the same reason, a small motor may run at higher revolutions per minute than a large one, and the strength of the material is an important factor determining the maximal speed. For similar reasons, a smaller creature can tolerate a greater quickness of movement than a larger one. In fact, the margin of safety is quite small, so that it does happen occasionally that muscles tear, tendons break, and bones splinter during unusual strain, such as during strenuous athletic events. *Without altering the general design*, it would be highly hazardous for the athlete's locomotor organs if the muscles could be altered to allow him or her suddenly to run, say, 25 percent faster.

Running Speed

The speed which may be attained in moving the body is determined by the length of stride and the number of movements per unit of time ($\propto 1 \cdot L^{-1}$), among other things. Thus, for similar animals of different size the maximal speed is proportional to $L \cdot L^{-1} = 1$, which means that the speed is the same. Short limbs with short strides move more rapidly and can therefore cover as much ground as do longer ones moving more slowly.

It is well known that "athletic animals" of different size may achieve approximately the same maximal speed. A blue whale of 100 tons and a dolphin of 80 kg attain the same "steady-state" speed of about 15 knots, and a maximal speed of about 20 knots. The speeds of a whippet, a greyhound, and a racehorse, very similar in general design, are nearly the same, or about 65 km $\cdot$ hr^{-1}(40 mph) (Hill, 1950). Gazelles and antelopes with wide variation in size are all able to reach a maximal speed of about 80 km $\cdot$ hr^{-1} (50 mph).

Since Hill (1950) presented his prediction of how performance can be expected to change as a function of body size on the basis of his mathematical model, experimental evidence has shown that it is not quite true that all quadrupeds have the same maximal speed, or that stride frequency is inversely proportional to limb length, or that the rate of oxygen uptake strictly increases with the cube of running speed (Taylor et al., 1970; Schmidt-Nielsen, 1972; Heglund et al., 1974). McMahon (1975), comparing animals ranging in size from mice to horses running at the transition point between trotting and galloping, found that "elastic similarity" (that is, similarity in the structures of animals so that they are similarly threatened by elastic failure under their own weight) provides a better correlation with published data on body and bone proportions, body surface area, resting metabolic rate, and basal heart rate than does geometric similarity.

Jumping

In broad jump and high jump it is also a question of the maximal muscular force which may be developed and the distance which the muscle can shorten before the body leaves the ground. We should therefore expect the performance to be proportional to $L^2 \cdot L \cdot L^{-3} = 1$ (Fig. 11-1), i.e., a small and a large animal should be equally able to lift its center of gravity. In broad jumping, kangaroos, jackrabbits, horses, mule deer, and impala antelopes actually seem to be equal to the record-holding man (Hill, 1950). Borelli drew similar conclusions some 300 years ago (1685). In high jumping, in which the aim is to lift the body as high as possible, the larger animal has an advantage, however, since its center of gravity before the jump is already at a higher level (Hill, 1950).

The outstanding high jumper is without exception a tall individual. This person is usually not geometrically similar to the average individual, but is long-legged, with a body weight that is not proportional to L^3 but lower.

Maximal Running Speed for Children

We might expect that the 180-cm boy would perform better in high jump than the 120-cm-tall boy, which he actually does. Their ability to move their center of gravity vertically, on the other hand, should not be different, nor should their maximal speeds. In an analysis of speed in children of different size, Asmussen has divided them into age groups and plotted maximal speed, calculated from the best time on 50 to 100 m, in relation to body height (Fig. 11-2). From the age

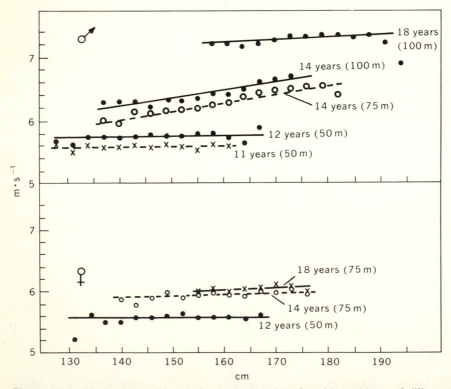

Figure 11-2 Maximal speed in relation to body height for girls and boys of different age; almost 100,000 subjects are included in the statistics. See text for explanation. *(Modified from Asmussen and Christensen, 1967.)*

of about ten the body proportions are about the same. We may therefore consider the children represented in Fig. 11-2 as geometrically similar (Fig. 11-3).

For eleven- to twelve-year old boys there is no significant variation in speed with body size, which would not be expected from the discussion above. The somewhat better performance of the twelve-year-olds over the eleven-year-old boys may be due to maturity of the neuromuscular function, improving the coordination. Even better are fourteen-year-old boys, but one also finds that the taller boys can run faster than the shorter boys. There is a further improvement in coordination with age, but this is also probably due to sexual maturity. Their male sex hormones may have influenced their muscular strength in a positive direction. The smaller fourteen-year-old boys may not have reached puberty, in contrast with the taller boys. In the eighteen-year-old group, there is again hardly any variation in results in spite of a large difference in body height. At this age, all the boys have passed puberty and are sexually mature.

In the girls, there is an increase in maximal speed up to the age of fourteen, but from then on there is no further improvement. The results are not influenced by the size of the girls in any of the age groups, which supports the assumption that the superiority of the taller fourteen-year-old boys is due to the effect of male sex hormones.

This independence of maximal speed with body height is in contrast to the greater muscle strength in taller children. From an anatomic viewpoint, this is just what would be expected: the muscle force should increase in proportion to L^2, but the speed should be independent. As already discussed, the results obtained for boys of different size do not strictly follow the results predicted from body dimensions. Apparently, biological factors may modify muscular

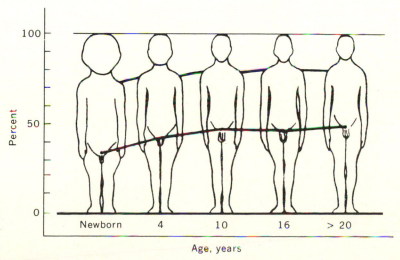

Figure 11-3 Body proportions in different ages. Note that from the age of ten years there is no marked change in proportions. *(From Asmussen and Christensen, 1967.)*

dynamics. We have considered an age factor as well as sexual maturity, which is particularly evident for the boys' performance.

Kinetic Energy

The kinetic energy (KE) developed in a limb depends upon its mass and the square of its velocity, or $KE = M \cdot v^2$; but $v = L \cdot t^{-1}$ and $t \propto L$; therefore, $KE \propto M \cdot L^2 \cdot L^{-2} \propto M$ (or L^3). It follows that the work done during a single movement, calculated per unit of body weight, in producing and utilizing kinetic energy in the limbs should be the same in large and small animals (Hill, 1950).

External Work

Hill (1950) points out that the external work done in overcoming the resistance of air or water is proportional to the square of the linear dimensions, i.e., the surface area. Therefore the effect of this resistance should be the same in similar animals of different size. In running uphill at a given speed, the effect of the slope is inversely proportional to the linear size L. The smaller the animal, the faster it should be when running uphill.

We have concluded that the maximal speed is independent of the animal's size. If one animal is 1,000 times as heavy as another, a movement will be 10 times greater than in the smaller animal. However, the larger animal has to take only one-tenth the number of steps, each step taking 10 times as long, in order to attain the same speed as the smaller animal. Since the work per movement and per unit of body weight is the same in the two animals, it follows that it will take roughly 10 times as long for the larger animal to become exhausted during a maximal run.

Energy Supply

It is obviously important that the power output permitted by the mechanical design be matched by an equivalent supply of chemical energy.

As mentioned above, work (being the product of force and distance moved) is proportional to $L^2 \cdot L = L^3$, or M. The total energy output should therefore be related to the mass of the muscles and the body weight in similar animals. The *power*, or work output per unit of time, must then be proportional to $L^3 \cdot t^{-1}$, that is, $L^3 \cdot L^{-1} \propto L^2$ ($\propto M^{2/3}$).

It is well established that the *basal metabolic rate* in animals with large differences in body size, from the mouse to the elephant, follows this prediction. The resting oxygen uptake is actually proportional to $M^{0.74}$ rather than to $M^{2/3}$, but this difference is surprisingly small considering the wide variation in size, shape, and other factors (Brody, 1945). This relation tells us that smaller animals must be more active metabolically per unit of body weight than larger ones. In proportion to its weight, a mouse has to eat 50 times more than a horse in order to maintain its basic activity.

Theoretically speaking, it would be expected that the *maximal oxygen uptake* should be proportional to L^2 or $M^{2/3}$. This holds extremely well for trained athletes, as is evident from Fig. 11-4. The lower panel of this figure

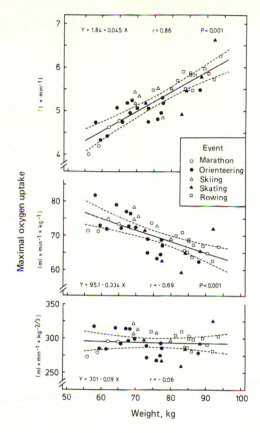

Figure 11-4 Maximal oxygen uptake in a group of Norwegian top athletes trained in different events, expressed as liters · min^{-1}; ml · min^{-1}· kg^{-1}; and as ml · min^{-1}· kg$^{-2/3}$ *(By courtesy of O. Vaage and L. Hermansen.)*

shows that the maximal oxygen uptake, expressed as ml · min^{-1} · kg$^{-2/3}$, is not related to body weight, and may therefore be used as a meaningful fitness index instead of the conventional method of expressing maximal oxygen uptake as ml · min^{-1} · kg^{-1}, which penalizes heavy individuals. Since maximal *cardiac output* and *pulmonary ventilation* are also volumes per unit of time, they should be proportional to $L^3 \cdot L^{-1} = L^2$ or $M^{2/3}$. There is a different approach to analyze this relationship, giving the same result. Cardiac output is the product of the frequency of heartbeats and stroke volume. Frequency is proportional to L^{-1} and stroke volume to L^3, and therefore $\dot{Q} \propto L^{-1} \cdot L^3 \propto L^2$.

Similarly, pulmonary ventilation is the product of respiratory frequency and tidal air: $\dot{V}_E \propto L^{-1} \cdot L^3 \propto L^2$.

With pulmonary ventilation proportional to $\dot{V}_{O_2}$ and with the production of CO_2 proportional to $\dot{V}_{O_2}$ we should expect the alveolar P_{CO_2} to be the same in different mammals, which it really is. Asmussen calculated that the vital capacity measured in subjects seven to thirty years of age (P.-O. Åstrand, 1952) was proportional to $L^{3.1}$ in males and $L^{3.0}$ in females, and thus very close to the expected $L^{3.0}$. It may therefore be concluded that children have lung volumes which are dimensioned to their body size.

In heavy exercise the *heat production* is very great and related to $\dot{V}_{O_2}$ or

L^2, i.e., to the surface of the body from which most of the excess heat is lost, at least in humans. There is also heat loss via the expired air, increasing, within limits, in direct proportion to $\dot{V}_{O_2}$.

Döbeln (1956b) points out that a dimensional analysis should be based on body weight minus adipose tissue, since fat is metabolically inactive. For a group of 65 young female and male subjects, he calculated the value of the exponent b in the equation maximal oxygen uptake = $a \cdot$ (body weight − adipose tissue)^b and found that $b = 0.71$ (maximal oxygen uptake predicted from the Åstrand and Åstrand nomogram; adipose tissue calculated by means of hydrostatic weighing).

Maximal Aerobic Power in Children

There are very few data available on maximal values for oxygen uptake and cardiac output in animals of different size to test these hypotheses. In fully grown humans, the variations in body size are rather limited. For the male subjects eight to eighteen years of age, studied by P.-O. Åstrand (1952), Asmussen has calculated that the maximal attainable oxygen uptake is proportional to $L^{2.9}$ and not, as expected, to $L^{2.0}$.

It thus appears that the children's maximal oxygen uptake is not as high as expected for their size and that, compared with adults, they do not have the aerobic power to handle their weight. It is also significant that the eight-year-old boy could increase his basal metabolic rate only 9.4 times during maximal running for 5 min, but the 17-year-old boy could attain an aerobic power which was 13.5 times the basal power. Therefore the child has less in the way of a power reserve than the adult. It should also be emphasized that the young subjects had a significantly higher oxygen uptake per kilogram of body weight than the older boys and adults when running at a given submaximal speed on a treadmill (P.-O. Åstrand, 1952). These two factors together may explain the fact that children have difficulty in following their parents' speed, even if maximal oxygen uptake per kilogram body weight may be the same. The children's lower efficiency can be explained partially by their high stride frequency which is an expensive utilization of energy per unit of time.

In Chap. 4 we discussed the relatively low muscular strength of children. It is conceivable that their aerobic power may be adapted to their muscular machine. In his analyses, Asmussen found that muscular strength in the eight- to sixteen-year-old boys was proportional to $L^{2.89}$, or exactly the same as the maximal oxygen uptake ($\propto L^{2.90}$).

Since the oxygen is transported by the hemoglobin, it is of interest to compare the children's total amount of hemoglobin (Hb_T) with their body size and maximal oxygen uptake. In most of the subjects just discussed, the total hemoglobin was determined. In similar animals of different size, Hb_T should be proportional to M. However, per kilogram of body weight, the younger boys had only 78 percent of the amount of hemoglobin of the older boys. Thus the amount of hemoglobin was definitely not proportional to body size. Assuming that the maximal oxygen uptake is proportional to Hb_T, we may calculate the maximal $\dot{V}_{O_2}$ in children from the equation $\dot{V}_{O_2} = a \cdot Hb_T^{2/3}$ and use a value for *a* calculated from the data on older boys. It is found that the child's maximal oxygen uptake should be 1.92 liters · min^{-1}, if the child's hemoglobin is as effective in transporting oxygen as is the adult's. This calculated value is not far from the determined 1.75 liters · min^{-1}. Using the exponent 0.74, the calculated maximal aerobic power in the seven- to nine-year-old children will be 1.78 liters · min^{-1}, or very close to real maximum.

The sample of subjects selected for these analyses was limited to 21 individuals, but it was a homogeneous group. They were nonobese and in the same state of training. The results support the assumption that the maximal oxygen uptake in children and young adults is proportional to the muscular strength and to $Hb_T^{0.76}$, or roughly to $Hb_T^{2/3}$, but *not* to $M^{2/3}$.

It may be concluded that children are definitely physically handicapped compared with adults (and fully grown animals of similar size). When related to the child's dimensions, its muscular strength is low and so is its maximal oxygen uptake and other parameters of importance for the oxygen transport. Furthermore, the mechanical efficiency of children is often inferior to that of adults. The introduction of dimensions in the discussion of children's performance clearly indicates that they are not mature as working machines.

Maximal Aerobic Power in Women

For the female subjects eight to sixteen years of age, the maximal $\dot{V}_{O_2}$ is proportional to $L^{2.5}$. In light of the previous discussion, the noted discrepancy from the expected L^2 is not surprising.

Women have approximately the same maximal oxygen uptake per kilogram fat-free body mass as men. However, it should be higher in women because of their smaller size (Döbeln, 1956a and b). The lower Hb concentration in women may explain why they cannot fully utilize their cardiac output for oxygen transport.

Figure 11-5 presents data on 227 children and young adults, stressing these points further. There is a very high correlation between maximal oxygen uptake and body weight for male, nonobese subjects (upper figure). The lower maximal oxygen uptake for female subjects above 40 kg of body weight (age about fourteen years) is largely explained by their higher content of adipose tissue. The lower concentration of hemoglobin in the women's blood also contributes to the observed difference between the sexes. When one relates the total amount of hemoglobin to the maximal oxygen uptake, the difference between regression lines for the female and male subjects is insignificant (lower figure). It should be noted that the exponent *b* in the equation $y = a \cdot x^b$ (i.e., maximal $\dot{V}_{O_2} = a \cdot Hb_T^b$) is 0.76.

Maximal Cardiac Output

From this analysis it is evident that maximal $\dot{V}_{O_2}$ should be proportional to L^2 (or $M^{2/3}$). However, data on children and adults do not support this assumption, since the exponent is closer to 3. There are, nevertheless, reasons to believe that biological factors may have modified the aerobic power in the subjects

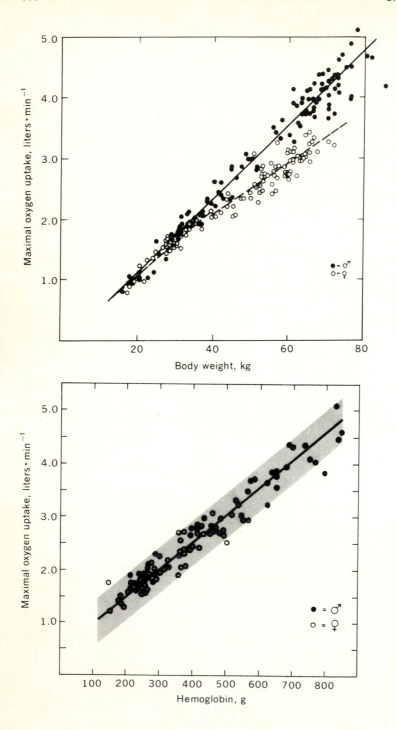

studied. Since $\dot{V}_{O_2}$ must be related to cardiac output, it should be emphasized that if the total output of the heart per minute was proportional to the body weight (L^3), the blood velocity in the aorta would have to be so great in the largest mammals that the heart would be faced with an impossible task (Hoesslin, 1888; Hill, 1950). It is therefore more likely that the cardiac output is proportional, not to the body mass, but to $M^{2/3}$ (or L^2), like the basal metabolism.

The blood pressure is independent of body size, as pointed out by Döbeln, since pressure is force per cross-sectional area, or proportional to $L^2 \cdot L^{-2} = 1$. In hearts working against the same blood pressure and being anatomically uniform in the sense that the coronary blood flow is a given percentage of the total blood flow, the maximal linear velocity of the blood in the aortic ostium during the period of expulsion is the same regardless of the size of the heart. This means that the maximal cardiac output, and consequently the maximal aerobic power of the entire animal, is proportional to the cross-sectional area of the aortic ostium. This area is found to be proportional to L^2 (or actually $M^{0.72}$) (Clark, 1927). Therefore, in uniformly built organisms, the maximal aerobic power is proportional to body weight raised to the $2/3$ power (Döbeln, 1956b).

Heart Weight, Oxygen Pulse

Figure 11-6 illustrates how the heart weight is directly proportional to the body weight in mammals the size of a mouse up to the size of a horse (heart weight = $0.0066 \cdot$ body weight$^{0.98}$; Adolph, 1949). In other words, over the full range of mammalian size, the heart weight is a constant fraction of the body weight. However, an animal capable of severe exercise has a heart ratio greater than 0.6 (heart ratio = heart weight $\cdot$ 100 $\cdot$ body weight^{-1}), while animals incapable of heavy, steady work have ratios less than 0.6 (Clark, 1927; Tenney, 1967). (Compare the data on hare and rabbit in Fig. 11-6.) For the presented parameters, the best correlation and least deviation from the regression line are noticed between heart weight and hemoglobin weight. A similar picture is demonstrated for the oxygen pulse, i.e., oxygen uptake at rest/heart rate, and its relation to body weight (oxygen pulse = $0.061 \cdot$ body weight$^{0.99}$), blood volume, and hemoglobin respectively. Thus, oxygen pulse is a relative measure of the stroke volume.

Figure 11-5 *Upper figure:* Maximal oxygen uptake measured during bicycling or running in 227 female and male subjects four to thirty-three years of age in relation to body weight. For male subjects the exponent $b = 0.76$ in the equation maximal $\dot{V}_{O_2} = a \cdot M$. (For male subjects, $y = -0.108 = 0.060 \, x$; $r = 0.980 \pm 0.004$; deviation from regression line = 7.5 percent.) *Lower figure:* Maximal oxygen uptake for 94 of the same subjects, age seven to thirty years, in relation to total amount of hemoglobin. In the equation $\dot{V}_{O_2} = a \cdot Hb^b$, the exponent $b = 0.76$. (For all subjects, $r = 0.970 \pm 0.006$; 2 $\cdot$ SD within shadowed area. For the determination of total hemoglobin, a CO method was used and the absolute values may be doubtful.) The subjects were all fairly well trained, and none of them was overweight. *(Modified from P.-O. Åstrand, 1952.)*

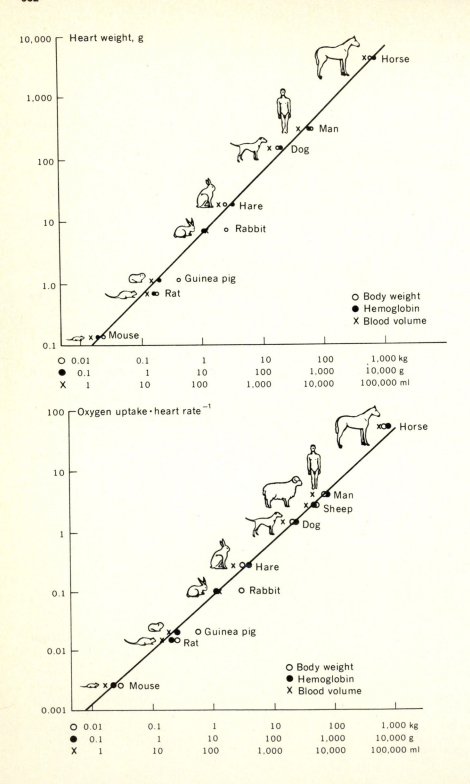

Heart weight, g

10,000

1,000

100

10

1.0

0.1

○ Horse

X ●○ Man

X ●○ Dog

X ●○ Hare

●X ○ Rabbit

X ●○ Guinea pig
X ●○ Rat

X ●○ Mouse

○ Body weight
● Hemoglobin
X Blood volume

○	0.01	0.1	1	10	100	1,000 kg
●	0.1	1	10	100	1,000	10,000 g
X	1	10	100	1,000	10,000	100,000 ml

Oxygen uptake·heart rate^{-1}

100

10

1

0.1

0.01

0.001

X ●○ Horse

X ●○ Man
X ●○ Sheep
X ●○ Dog

X ●○ Hare

●○ Rabbit

X ●○ Guinea pig
●○ Rat

X ●○ Mouse

○ Body weight
● Hemoglobin
X Blood volume

○	0.01	0.1	1	10	100	1,000 kg
●	0.1	1	10	100	1,000	10,000 g
X	1	10	100	1,000	10,000	100,000 ml

Heart Rate

As already mentioned, the heart rate is likely to be proportional to L^{-1} (Lambert and Teissier, 1927). Therefore the larger animal should have a lower heart rate at rest, and possibly also during exercise, than the smaller animal. The resting heart rate of the 25-g mouse is about 700, and of the 3,000-kg elephant, it is 25 beats $\cdot$ min^{-1}. This difference is not a biological adaptation, but a mechanical necessity.

Summary In the case of biological phenomena, we may express the physical basic units of length, mass, and time in one single basic unit: length. This means that all units such as pressure, temperature, and energy may be expressed as derivatives from this basic unit. Table 11-1 is prepared on this basis.

The conclusions which may be drawn from this table obviously do not solve any biological problems. It may, however, serve to facilitate the correct formulation of biological problems. Knowing that a fully grown man of 70 kg on an average has a maximal pulse of 195 beats $\cdot$ min^{-1}, and that a child of 35 kg has a maximal pulse of 210 to 220 beats $\cdot$ min^{-1}, the question is not why this latter value is greater than 195, but why it is less than 245. The latter value of 245 is what might be expected from a purely dimensional consideration.

For a more profound treatment of dimensional analysis and the theory of biological similarity, the reader is referred to a recent review by Günther (1975), who concludes that the body weight of an organism is an adequate reference index for the correlation of morphological and physiological characteristics in comparative physiology. He finds that the statistical analysis of the experimental data can be represented most conveniently by means of the logarithmic equivalent of Huxley's allometric equation: $y = a \cdot W^b$, where y is any function definable in terms of the MLT system (M = mass, L = length, T = time), a is empirical parameter, W is body weight, and b is reduced exponent, the numerical value of which can be obtained from the log-log plot of the experimental data, since the logarithmic expression is a straight line ($\log y = \log a + b \cdot \log W$).

Secular Increase in Dimension

We shall now consider a few additional applications of the effect of dimensions on human performance. Since in many countries man has been growing taller in recent generations, some improvement in athletic performance is to be expected. This steady secular increase in growth is typical of countries with a satisfactory nutritional status. Besides the increase in such bodily dimensions as height and weight at all ages from birth to adulthood, the maturation of certain physiological functions, notably those connected with sexual maturity, is also accelerated. There has been a steady decrease in age of menarche, from about 17 years of age in 1840 to 13$^{1/2}$ years of age in 1960 (Tanner, 1962). A similar trend of earlier maturation of boys is also apparent from the available data; boys now reach their maximal height at an earlier age than they did a generation ago. The influence of sexual maturity on performance is evident

Figure 11-6 Heart weight (upper figure) and oxygen uptake per heartbeat (oxygen pulse) in relation to body weight, total hemoglobin weight, and blood volume, respectively, in various mammals. *(From various sources summarized by Sjöstrand, 1961.)*

Table 11-1   Dimensions in Physics and Physiology

Quantity	Dimension	
	Physical	Physiological
Length	L	L
Mass	M	L^3
Time	t	L
Surface	L^2	L^2
Volume	L^3	L^3
Density	$L^{-3}M$	L^0
Velocity	Lt^{-1}	L^0
Frequence	t^{-1}	L^{-1}
Flow	L^3t^{-1}	L^2
Acceleration	Lt^{-2}	L^{-1}
Force	LMt^{-2}	L^2
Pressure	$L^{-1}Mt^{-2}$	L^0
Temperature*	L^2t^{-2}	L^0
Energy	L^2Mt^{-2}	L^3
Power	L^2Mt^{-3}	L^2

*Physical dimension from J. C. Georgian, The Temperature Scale, *Nature*, **201**:695, 1964.
Source: By courtesy of W. v. Döbeln.

from Fig. 11-2. Asmussen and his collaborators (1955, 1964, 1967) point out that although the height and weight for children of a given age have shifted upward during recent years, the weight-height curves have remained practically unaltered during the last couple of decades. It is also a fact that in spite of the general increase in height and weight of Olympic athletes during the last 30 years, their body proportions are remarkably constant. It may therefore be assumed that even with the increased dimensions, the present-day taller athletes are geometrically no different from those of earlier generations.

If the height of an athlete is 184 cm (the mean height of the participants in decathlon in Rome in 1960), and the height of an athlete 30 years ago was 176 cm (mean height of an athlete about 30 years ago), their heights will compare as 1.06:1, and their muscle strength as 1.13:1 (Asmussen, 1964). This means that, owing to different dimensions, the average top athlete now is 6 percent taller than the top athlete of 30 years ago, but his muscular strength should be 13 percent greater than 30 years ago. Therefore, the maximal work that the muscles of the athlete in decathlon could perform should be 20 percent greater than 30 years ago, for the maximal work a muscle can produce is the product of its maximal force and the distance it can shorten. When the size of the oxygen-transporting organs is the limiting factor, the taller athlete should consequently be able to deliver 13 percent more oxygen to his muscles per unit of time than the smaller athlete.

In such events as throwing the javelin or putting the shot, the increase in bodily dimensions may influence the achievements in two ways: In the first place, the strength of the athlete increases in proportion to the second power of the person's height. This will tend to improve the results, particularly since the weight of the equipment is constant and not varied with the weight of the thrower. Secondly, the greater height from which the javelin and shot start their flight will cause them to travel further. These two factors, and particularly the first one, would result in better records, and may partly account for the improvements in records which have taken place. Anyway, there is clearly a good physiological basis for the selection of tall throwers.

This discussion is included to demonstrate that in some events, part (but only part) of the improvement in results over the years may be due to the athletes' dimensional change. Khosla (1968) finds that the winners in different throwing events in the Olympic Games in Rome and Tokyo were on the average definitely taller and heavier than their competitors. Similarly, the winners in jumping and running, with the exception of 10,000-m and marathon running, were taller. This, he claims, is unfair to people who are less tall, and suggests the classification of the competitors according to height and weight in events in which body size may influence the results.

Old Age

Figure 11-7 summarizes data from the literature on different static and dynamic dimensions in adults from twenty up to sixty years of age. It is evident that a strict interrelation of these functions based on dimensions alone does not occur. The body height is maintained constant, but body weight, heart weight, and heart volume increase with age. Blood volume and total amount of hemoglobin are not markedly changed. Heart rate at a given submaximal work load is the same in the old and the young, i.e., the oxygen transport per heartbeat (oxygen pulse) is constant. However, maximal oxygen uptake, heart rate, stroke volume, pulmonary ventilation, and muscular strength decrease significantly with age. Apparently, an older individual of the same body size as a younger one is in many ways different both in structure and in function.

SUMMARY

We have given a number of examples proving that static and dynamic functions in animals of different sizes in many cases have dimensions which are mechanically meaningful and desirable. The strength of a muscle is adjusted to the strength of bones, tendons, joints, connective tissue, and the muscle itself as a matter of safety. Furthermore, it provides a reasonable efficiency of the movements. There is, in general, a remarkable adjustment of the links involved in the chain of oxygen supply and energy output, so that none of the individual links is much stronger or weaker than necessary. On the other hand, there are examples showing that organisms of different size are not in all respects similar and uniform; there are deviations from the general trend. Deviations may sometimes be of a physical nature; in larger animals, for example, the weight of the skeleton is relatively high. It is also known that training may markedly

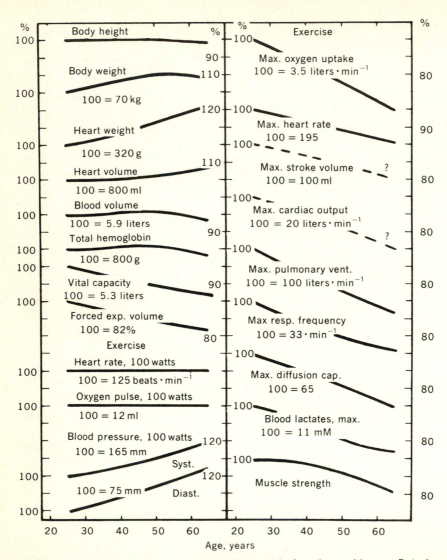

Figure 11-7 Variation in some static and dynamic functions with age. Data have been collected from various studies, including healthy male individuals. For data on the same function, only one study was consulted. The values for the twenty-five-year-old subjects = 100 percent; for the older ages, the mean values are expressed in percentage of the twenty-five-year-old individuals' values. The mean values cannot be considered as "normal values," but their trends illustrate the effect of aging. Note that the heart rate and oxygen pulse at a given work load (100 watts or 600 kpm·min⁻¹, oxygen uptake about 1.5 liters·min⁻¹) are identical throughout the age covered, but the maximal oxygen uptake, heart rate, cardiac output, etc., decline with age. The data on cardiac output and stroke volume are based on relatively few observations and are therefore less certain.

improve physical performance. Sometimes this improvement is accompanied by changes in organic dimensions, but this does not always occur (see Chap. 12). The performance of children is lower than expected from their dimensions, making it evident that biological factors are involved. Women and old individuals have a relatively low maximal oxygen uptake compared with twenty-five-year-old men.

It is very fruitful, however, to consider whether differences in performance of animals of different size, including children and adults, can be explained by purely dimensional factors. If such a consideration should fail to account for the differences, biological adaptations would appear probable.

As stated at the beginning of this chapter, it is usually not possible to tell who might be the best athlete without testing his or her ability. The top skier or runner in endurance events has a maximal aerobic power which is about twice that of ordinary nonathletes of similar age (6.0 liters $\cdot$ min^{-1} compared with 3.0 liters $\cdot$ min^{-1}). Yet the top athlete's body height is not 250 cm, which it would have to be if body dimensions alone were to determine oxygen-transporting ability (maximal $\dot{V}_{O_2} \propto L^2$). In fact, these top athletes are of average height, and some of their dimensions are very similar to those of an average person, while others are different (see Chap. 12).

REFERENCES

Adolph, E. F.: Quantitative Relations in the Physiological Constitution of Mammals, *Science*, **109**:579, 1949.

Asmussen, E.: Growth and Athletic Performance, *FIEP*, **34**(4):22, 1964.

Asmussen, E., and K. Heeböll-Nielsen: A Dimensional Analysis of Physical Performance and Growth in Boys, *J. Appl. Physiol.*, **7**:593, 1955.

Asmussen, E., and E. H. Christensen: "Kompendium i Legemsövelsernes Specielle Teori," Köbenhavns Universitets Fond til Tilvejebringelse af Läremidler, Köbenhavn, 1967.

Åstrand, P.-O.: "Experimental Studies of Physical Working Capacity in Relation to Sex and Age," Ejnar Munksgaard, Copenhagen, 1952.

Borelli, J. A.: De Motu Animalium, *Lugduni in Batavis, Apud Danielem à Guerbeeck* . . . , 1685.

Brody, S.: "Bioenergetics and Growth," Reinhold Book Corporation, New York, 1945.

Clark, A. J.: "Comparative Physiology of the Heart," University Press, Cambridge, England, 1927.

Döbeln, W. v.: Human Standard and Maximal Metabolic Rate in Relation to Fat-free Body Mass, *Acta Physiol. Scand.*, **37**(Suppl. 126):1956a.

Döbeln, W. v.: Maximal Oxygen Intake, Body Size, and Total Hemoglobin in Normal Man, *Acta Physiol. Scand.*, **38**:193, 1956b.

Döbeln, W. v.: Kroppsstorlek, Energiomsättning och Kondition, in G. Luthman, U. Åberg, and N. Lundgren (eds.), "Handbok i Ergonomi," Almqvist & Wiksell, Stockholm, 1966.

Galilei, G.: Discorsi et Dimonstrazioni Matematiche, Interne à due Nueve Scienze, *Elzévir*, 1638.

Günther, B.: Dimensional Analysis and Theory of Biological Similarity, *Phys. Rev.*, **55**:659, 1975.

Heglund, N., T. A. McMahon, and C. R. Taylor: Scaling Stride Frequency and Gait to Animal Size: Mice to Horses, *Science*, **186**:1112, 1974.

Hill, A. V.: The Dimensions of Animals and Their Muscular Dynamics, *Proc. Royal Inst. Great Britain*, **34**:450, 1950.

Hoesslin, H. v.: Ueber die Ursache der scheinbaren Abhängigkeit des Umsatzes von der Grösse der Körperoberfläche, *Arch. f. Anat. Physiol., Physiol. Abt.*, p. 323, 1888.

Khosla, T.: Unfairness of Certain Events in the Olympic Games, *Brit. Med. J.*, **4**:111, 1968.

Lambert, R., and G. Teissier: Théorie de la Similitude Biologique, *Ann. Physiol.*, **3**:212, 1927.

McMahon, T. A.: Using Body Size to Understand the Structural Design of Animals: Quadrupedal Locomotion, *J. Appl. Physiol.*, **39**:619, 1975.

Schmidt-Nielsen, K.: Energetic Cost of Locomotion: Swimming, Running and Flying, *Science*, **177**:222, 1972.

Sjöstrand, T.: Relationen Zwischen Bau und Funktion des Kreislaufsystems Unter Pathologischen Bedingungen, *Forum Cardiologicum*, Boehringer & Soehne, Mannheim-Waldhof, Heft 3, 1961.

Tanner, J. M.: "Growth and Adolescence," Blackwell Scientific Publications, Ltd., Oxford, 1962.

Taylor, C. R., K. Schmidt-Nielsen, and J. L. Raab: Scaling and Energetic Cost of Running to Body Size in Mammals, *Am. J. Physiol.*, **219**:1104, 1970.

Tenney, S. M.: Some Aspects of the Comparative Physiology of Muscular Exercise in Mammals, *Circulation Res.*, **20**:1–7, 1967.

Thompson, D. W.: "On Growth and Form," Cambridge University Press, New York, 1943.

Physical Training

12

CONTENTS

INTRODUCTION

TRAINING PRINCIPLES
Continuous versus intermittent exercise   Training of muscle strength   Training of anaerobic power   Training of aerobic power   Year-round training   Psychological aspects   Tests

BIOLOGICAL LONG-TERM EFFECTS
Locomotive organs   Oxygen-transporting system   Training of cardiac patients   Specificity of training   Recovery from exercise   Mechanical efficiency, technique   Body composition   Blood lipids   Hormones   Psychological changes

Chapter 12

Physical Training

INTRODUCTION

In several sections of this book, it has been shown that different organs or organ systems may be affected by a variety of factors. The effect may be transitory or it may last for a considerable period of time; that is, an adaptation takes place. Figure 9-1, for example, illustrates that the aerobic muscular capacity is affected by training, deconditioning, and acclimatization (altitude, heat, cold). In this chapter we shall discuss how physical training affects the body morphologically as well as functionally.

Conclusions concerning the effect of physical training have often been drawn from studies of well-trained persons, and the data obtained have been compared with similar data from studies on sedentary individuals. The disadvantage of such *cross-sectional studies* is that it may be impossible to determine whether any difference observed depends on constitutional dissimilarities or on the training as such. It is in any case quite obvious that the great maximal aerobic power which is characteristic for the top athlete in endurance largely depends on endowed organic advantages. Thus, a person with a maximal oxygen uptake of 45 ml·kg^{-1}·min^{-1} cannot, under any circumstance, no matter how well trained, attain a maximal oxygen uptake of 80 ml·kg^{-1}·min^{-1}, which is required for Olympic medals in certain sport events

(Fig. 12-15). It may not be quite fair, but it is nevertheless a fact that the "choice of parents" is important for athletic achievement (see Klissouras, 1976; Chap. 4, p. 97).

For these reasons *longitudinal studies* must be designed, in which the same individual is followed for shorter or longer periods of time. However, such studies are rare. Since they are difficult to perform, usually only a few subjects have been included in each study. Even this approach, with long observation periods of the same individuals, is not without objections from a scientific point of view:

1   If the objective is to examine such problems as the effect of physical training, the selection of subjects is critical. If the selection is not done on a strictly random basis, the material may not be representative. If a selection is done from a group of volunteers, it must be remembered that those who volunteer for such studies may do so for special reasons which may be medical, social, psychological, or personal. In any event, such volunteers may not be representative of the population as a whole.

2   Any such study may well represent an intervention of the subject's normal pattern of life. This may in itself bring about different effects: the subjects may perhaps change their diet, smoking habits, or other habits which may elicit effects on the organism which, per se, have nothing to do with the training. It may also be difficult to avoid affecting the control group by the mere fact that the group is being studied and therefore the focus of special attention. In studying patients, arriving at a clear-cut experimental situation is especially complicated in that ethical considerations, in such cases especially, may be brought into the foreground.

3   Experience has shown that the drop-out frequency is rather high among subjects of prolonged training studies.

4   It is quite difficult to establish a person's physical condition objectively prior to the training, and any training effect obviously depends on the initial level at the start of the training. As mentioned on several occasions, at present no method of investigation is available which can definitely reveal and separate the influence of constitutional factors on the one hand, and the effects of training on the other. One is therefore largely forced to rely on the individual's own statements. However, what one person may characterize as a physically active life may, to another, represent a sedentary life. Statements as to occupation and recreational activities may in themselves represent only indications rather than precise, meaningful information regarding degree of physical activity, etc. The duration and intensity of such activities are important factors to be considered.

5   The intensity (tempo) of the actual training is difficult to assess, define, or reproduce. Moreover, it is a time-consuming and complicated task to record this variable, especially in the case of large-scale investigations.

On the basis of these considerations, it is understandable that different investigators arrive at different results concerning the effect of physical activity or inactivity with regard to qualitative as well as quantitative changes.

In this chapter we shall consider (1) the physiological basis for the development of a training program and (2) the biological long-term effects of different levels of physical activity. The discussion will be limited to data which are fairly well documented.

When discussing training and training principles, it is essential to keep in mind the specific purpose of the training. As a rule, the best training is achieved simply by carrying out the activity for which one is training. As is evident from Chap. 1, Table 1-1, physical training may influence a number of the factors which constitute physical performance capacity; that is, it may cause changes not merely in muscle strength and maximal oxygen uptake, but also structural and functional changes in a number of organ systems as well as psychological changes.

Table 12-1 summarizes effects of training on organs and organ functions. Some of the data were obtained on animals, others on humans. There are still many open questions, as indicated in the table. At the end of this chapter, the table will be discussed in some detail.

TRAINING PRINCIPLES

Physical training entails exposing the organism to a training load or work stress of sufficient intensity, duration, and frequency to produce a noticeable or measurable training effect, i.e., an improvement of the functions for which one is training. In order to achieve such a training effect, it is necessary to expose the organism to an overload, that is, to a stress which is greater than the one regularly encountered during everyday life. Generally speaking, it appears that exposure to the training stress is associated with some catabolic processes, such as molecular breakdown of stored fuel and other cellular components, followed by an overshoot or anabolic response that causes an increased deposition of the molecules which were mobilized or broken down during exposure to the training load.

The intensity of the load required to produce an effect increases as the performance is improved in the course of training. The training load is therefore relative to the level of fitness of the individual. The fitter a person is, the more it will take to improve that fitness. Finally, it becomes a matter of time and motivation to continue when the elite athlete has to devote several hours a day to training. The need for a gradual increase in training load with improved performance, in the case of the effect of heart rate, was demonstrated as early as in 1931 by Christensen. He observed that regular training with a given standard work load gradually lowered the heart rate. Further training did not modify this heart rate response. After a period of training on a heavier load, the original standard work load could then be performed with a still lower heart rate. This general principle is apparent during training of a number of functions: *An adaptation to a given load takes place; in order to achieve further improvement, the training intensity has to be increased.* This principle has been elucidated by several studies summarized in the other sections of this chapter.

Table 12-1 Effects of Training on Organs and Organ Functions

Organ or function	Increase	Decrease	No effect	References
Locomotive organs				
Strength of bones and ligaments	x			Ingelmark, 1948, 1957; Viidik, 1966; Tipton et al., 1974; Booth and Gould, 1975
Thickness of articular cartilage	x			Holmdahl and Ingelmark, 1948
Muscle mass (hypertrophy)	x	?	x	Marpurgo, 1897; Siebert, 1929; Vannotti and Pfister, 1934; Man-i et al., 1967; Hollmann and Hettinger, 1976; Jansson and Kaijser, 1977; Saltin et al., 1977
Number of muscle cells	?		x	Gonyea et al., 1977
Muscle strength	x			Clarke, 1973; Hollmann and Hettinger, 1976; Table 12-5
ATP, creatine phosphate, muscle	x			Palladin and Ferdmann, 1928; Yakovlev, 1958
PFK action in muscle	x		x	Gollnick and Hermansen, 1973
SDH action in muscle	x			Holloszy, 1973, 1975; Saltin et al., 1977
Myoglobin	x			Whipple, 1926; Holloszy, 1975
Potassium, muscle	x			Nöcker et al., 1958
Capillary density, muscle	x			Vannotti and Pfister, 1934; Vannotti and Magiday, 1934; Petrén et al., 1936, Fig. 12-7, Brodal et al., 1976; Saltin et al., 1977
Arterial collaterals, muscle	x			Schoop, 1964; 1966
Circulation				
Heart volume	x		x	Roskamm et al., 1966; Reindell et al., 1967; Ekblom, 1969; Fig. 12-8
Heart weight	x			Siebert, 1929; Thörner, 1949; Liere and Northup, 1957
Capillary density, heart	x			Petrén et al., 1936; Fig. 12-7
Coronary collaterals	?			Eckstein, 1957; Tepperman and Pearlman, 1961; Chap. 6.
Blood volume, total	x			Deitrick et al., 1948; Taylor et al., 1949; Hollmann and Venrath, 1963; Miller et al., 1964; Sjöstrand, 1967; Saltin et al., 1968
hemoglobin				
Alkali reserve			x	P.-O. Åstrand, 1956
Hemoglobin concentration			x	P.-O. Åstrand, 1956
Plasma protein concentration			x	P.-O. Åstrand, 1956
Cardiac output, rest			x	P.-O. Åstrand, 1956

				Reference
Submaximal work			?	?	Freedman et al., 1955; Frick et al., 1963; Tabakin et al., 1965;
Maximal work	x		?	?	Andrew et al., 1966; Ekblom et al., 1968; Saltin et al., 1968; Rowell, 1974
Heart rate, rest					Ekblom et al., 1968; Saltin et al., 1968; Rowell, 1974; Barnard, 1975; Fig. 12-11
Submaximal work	x	x	x		Steinhaus, 1933
Maximal work	x	x	x		Christensen, 1931; Clausen, 1976; see "Cardiac output"
Stroke volume, rest	x	?	x		Robinson and Harmon, 1941; Knehr et al., 1942; Ekblom, 1969
Submaximal work	x		?		Ekblom et al., 1968; Saltin et al., 1968
Maximal work			?		See "Cardiac output"
$a-\bar{v}O_2$ difference, rest	?		x		See "Cardiac output"
Submaximal work			?		See "Cardiac output"
Maximal work	x	x	x		See "Cardiac output"; Fig. 12-11
Oxygen uptake, rest			x		P.-O. Åstrand, 1956
Given work load			x		P.-O. Åstrand, 1956
Maximal work	x	x	x		Robinson and Harmon, 1941; Knehr et al., 1942; Taylor et al., 1949; Ekblom et al., 1968; Ekblom, 1969; Saltin et al., 1968; Pollock, 1973; Figs. 12-9, 12-11
Blood lactic acid, rest					
Given work load	?		x		P.-O. Åstrand, 1956; Williams et al., 1967; Ekblom, 1969; Fig. 12-12
Maximal work	x	x			P.-O. Åstrand, 1956; Gollnick and Hermansen, 1973; Fig. 12-12
Local blood flow, working muscle submaximal			x		Elsner and Carlson, 1962; Rohter et al., 1963; Rowell, 1974; Clausen, 1976
Arterial blood pressure, rest	?				P.-O. Åstrand, 1956
Submaximal work		?	x	?	Ekblom, 1969; Tabakin et al., 1965; Frick et al., 1963; Boyer and Kasch, 1970; Kilbom, 1971; Choquette and Ferguson, 1973; Clausen, 1976.
Maximal work			?		Ekblom et al., 1968; Ekblom, 1969
Respiration					
Lung volumes, adults			x	x	P.-O. Åstrand, 1956
Lung volumes, adolescents			?	?	P.-O. Åstrand, 1956
Pulmonary ventilation, rest			?	?	See text
Submaximal work			x	x	See text
Maximal work	x*	x		x	Ekblom et al., 1968
Tidal air, rest	?		x		See text

395

Table 12-1 (continued)

Organ or function	Increase	Decrease	No effect	References
Submaximal work	?		x	
Maximal work	?	?	?	
Respiratory rate, rest		?	x	See text
Submaximal work		?	x	
Maximal work	x			Anderson and Shephard, 1968; Reuschlein et al., 1968; Saltin et al., 1968
Diffusing capacity, rest			x	
Submaximal work				See "Rest"
Maximal work	x*		x	See "Rest"
Miscellaneous				
Body density	x			Skinner et al., 1964; Pářízková, 1973
Serum cholesterol concentration		x	x	Golding, 1961; Altekruse and Wilmore, 1973; Lopezs et al., 1974; Björntorp, 1970; Pyöräiä et al., 1974
Serum triglycerides		x	x	Skinner et al., 1964; Fox and Haskell, 1967; Altekruse, 1973

*Secondary to the increase in maximal oxygen uptake.

However, there is no linear relationship between amount of training and the training effect. Obviously there is a limit to the increase, and the rate and magnitude of the increase vary from one individual to the next.

The exact magnitude of the training load which will produce an optimal training effect is not established in general terms. It varies not only from one individual to another, but also with age. On the whole, it appears that in athletic training, the greater the intensity the better, up to a certain limit. In fitness training for average people, it appears that a training load in excess of about 50 percent of the individual's maximal performance capacity, which roughly corresponds to the work load causing the person to be slightly out of breath, may be sufficient to produce a significant effect.

From these statements it is clear that physical activity is not synonymous with physical training, for the physical activity has to be maintained at a certain level of intensity in order to result in a training effect. In addition to a certain minimal level of intensity, the training stimulus must be of a certain duration in order to produce any training effect. Here again the relationship between training load and duration on the one hand, and training effect on the other, is not clearly established in general terms, and probably never will be, for it depends on which organ systems and what kind of training one is dealing with. Thus, while an increase in isometric muscle strength may be achieved by a few near-maximal contractions of a few seconds' duration once a day, such a brief exposure, even to a very intense physical exercise, may not affect the cardiovascular system sufficiently to improve the oxygen-transporting capacity. The same applies to the required frequency of the training sessions, i.e., how often one has to train in order to attain the desired training effect. From the available evidence, however, it is clear that the same effect may be obtained by relatively shorter daily training sessions as by much longer training sessions two or three times a week.

Older individuals (above fifty years of age) may be less trainable than younger ones. But it should be kept in mind that some effect of training may be noticed even at very old age. It is important for individuals who wish to improve their general state of fitness to ascertain what amount of training may produce the most satisfactory result. One has to weigh the time available for training against the effects achieved by the training. If one has reached such a level of physical fitness that several additional hours of training per week would be necessary in order to attain a further improvement of a few percentage points or less, the small gain would hardly be worth the effort. One has to accept the fact that the daily fluctuations in one's state of fitness may exceed this amount of difference.

It is a general experience that an athlete needs several years to achieve top results. It is not clear what qualitative and quantitative changes in different organ functions may cause the slow, gradual improvement of the results. An analysis of the demands which a particular athletic event places on the body should form the basis for the training program, taking into consideration whatever deficiencies there may be in the athlete's resources or capabilities to meet these demands.

A certain amount of training of the oxygen-transporting organs is necessary for all categories of athletes, regardless of the nature of the athletic event. Thus the individual will be better able to cope with the special training required for the event. Furthermore, even the warm-up prior to the event requires a certain amount of fitness. The general training which all individuals, irrespective of profession, age, and sex, must undergo should include: (1) training of the O_2-transporting function to improve endurance for work; (2) improvement of muscle strength including the abdominal muscles (Chap. 8); (3) training aimed at maintaining joint mobility, the enhancement of the metabolism of the articular cartilage (Chap. 8), and the development of improved coordination (Chap. 4). In addition, (4) low-energy consumers should be stimulated to increase their metabolism through regular exercise so as eventually to become high-energy consumers (Chap. 14). As far as the O_2-transporting function is concerned, distinction should be made between factors involved primarily in the heart and central circulation, and factors involved in the peripheral circulation. In regard to the *central circulation*, the training is effective and less strenuous if as large a muscle mass as possible is engaged in the training. In the case of the *peripheral circulation*, it is a matter of training the muscles that will be engaged in the performance of the type of event or activity in which an improvement is desired.

Continuous versus Intermittent Exercise

It has often been discussed whether physical training is most effective when the work is accomplished continuously or intermittently, i.e., with periods of more intensive muscular work followed by periods of mild exercise or even rest. In the following discussion, we shall present a summary of results from a few studies in which the physiological effects of these types of work are compared (I. Åstrand et al., 1960; Christensen et al., 1960).

One subject was made to accomplish a certain amount of work (635kJ) in the course of 1 hr. This could be done either by uninterrupted work with a load of 175 watts, or by intermittent work with a heavier work load, interrupted by rest periods at regular intervals. The double work load (350 watts) was chosen; thus the required amount of work (635 kJ) could be accomplished by 30 min of work within the span of 1 hr. Working continuously without any rest periods, the subject could tolerate this high work load for only 9 min, at the end of which he was completely exhausted. If, instead, he worked for 30 s, rested for 30 s, worked for 30 s, and so on, he could complete the work with moderate exertion. The longer the work periods, the more exhausting appeared the work, even though the rest periods were correspondingly increased. Some of the results of these studies are summarized in Table 12-2. It appeared that with work periods of 3 min interrupted by 3-min rest periods, the load on the oxygen-transporting organs was maximal (oxygen uptake, 4.60 liters·min^{-1}, heart rate 188), and the degree of exertion was particularly high (blood lactate 13 mM).

Table 12-2 Data on one subject performing 635 kJ (64,800 kpm) on a bicycle ergometer within 1 hr with different procedures

Type of exercise	Oxygen uptake		Pulmonary ventilation, liters · min⁻¹	Heart rate, beats · min⁻¹	Blood lactic acid, mg · 100 ml⁻¹
	liters · hr⁻¹	liters · min⁻¹			
Continuous					
175 watts	146	2.44	49	134	12
350 watts*		4.60	124	190	150
Intermittent					
350 watts					
Work Rest					
½ min ½ min	154	2.90†	63†	150	20
1 min 1 min	152	2.93†	65†	167	45
2 min 2 min	160	4.40	95	178	95
3 min 3 min	163	4.60	107	188	120

*Could be performed for only 9 min.
†Measured during ½ min.
Source: I. Åstrand et al., 1960.

At least one conclusion may be drawn from these experiments: For the purpose of taxing the oxygen-transporting organs maximally, work periods of a few minutes' duration represent an effective type of work. An example of the effect of intermittent work of this type is presented by Vaage (unpublished results, Fig. 12-1). Two athletes trained on the treadmill (inclination, 12°) almost daily for over 100 days. Walking with ski-poles, they started at a speed of about 100 m ·min⁻¹, but later they ran as the treadmill speed was increased to over 140 m · min⁻¹ in order to attain maximal exertion. During the first period, they walked on the treadmill for 4 to 5 min three to five times in succession, with 5- to 9-min rest pauses between each work period. During the subsequent period, they alternated on different days between this very strenuous uphill treadmill walking and running in short bouts of about 20 s, interspersed with about 10-s rests for about 8 min. The whole 8-min sequence was repeated after 5- to 9-min rest periods, and altogether was repeated two to five times each day. This training was replaced by long-distance running or rowing for 40 to 150 min each time on about 30 days, as indicated by ' in Fig. 12-1, showing the effect of this program on the maximal oxygen uptake of the two subjects.

Table 12-3 presents data on one of the subjects obtained in a study by Christensen et al. (1960) of intermittent treadmill running using shorter work periods. It is striking that when the subject ran for 10 s followed by a 5-s pause, he could run for 20 min during the 30-min period at a high speed without undue fatigue and with a low blood lactic acid concentration. At the end of each running period, the load on the oxygen-transporting system was maximal, or 5.6 liters·min⁻¹. On the average, the oxygen uptake when the subject was running was 5.1 liters, but the oxygen requirement can be calculated to be 7.3 liters per work minute. How the deficit of about 46 kJ, (7.3-5.1)·21, is covered is still an open question. The low blood lactate concentration indicates that anaerobic glycogenolysis was not the important energy supplier. The high energy-

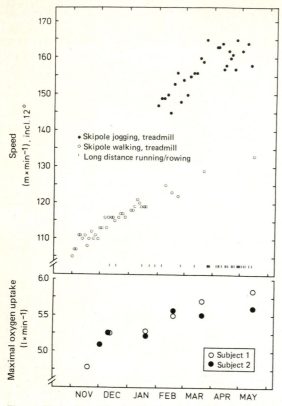

Figure 12-1   The effect of "ski-pole walking" and running on a treadmill (inclination, 12°) on the maximal oxygen uptake in two athletes. *(By courtesy of O. Vaage.)*

containing phosphate compounds may have served as a buffer, supported by aerobic processes utilizing oxygen bound to the myoglobin in addition to the amount of oxygen transported during the running. It should be noted that in all the experiments, the oxygen uptake and pulmonary ventilation were also high during the interspersed resting periods. However, the results indicate that the duration and spacing of exercise and resting periods are rather critical with respect to the peak load on the oxygen-transport system. If the resting period of 5 s is prolonged to 10 s (running for 10 s), the peak oxygen uptake observed will be reduced from 5.6 to 4.7 liters·min^{-1}. Running at the same speed for 15 s, then resting for 15 s, did not bring the oxygen uptake (5.3 liters·min^{-1}) to a maximum.

Figure 12-2 presents another example of how critical the work intensity is for the load on the oxygen-transporting organs in intermittent work with work periods of short duration. During running at a speed of 22.75 km·hr^{-1} for 20 s, 10-s rest, followed by running, and so on, the oxygen uptake becomes maximal. If the speed is reduced to 22.0 km·hr^{-1}, the oxygen uptake is reduced to about 90 percent of the maximum. On the other hand, the subject can continue for

Table 12-3 Data on one subject during intermittent running for 30 min at 20 km/hr on a treadmill. Between the running periods, which were varied, the subject was standing beside the treadmill. During continuous running, he could proceed for 4.0 min, covering a distance of about 1,300 m. Oxygen uptake: 5.6 liters · min⁻¹: pulmonary ventilation: 158 liters · min⁻¹; blood lactic acid concentration: 150 mg · 100 ml⁻¹.

Periods work-rest, s	Distance, m	Oxygen uptake, liters · min⁻¹ Work			Pulmonary ventilation, liters · min⁻¹ Work			Blood lactate, mg · 100 ml⁻¹
		Work			Work			
		Highest	Average	Rest	Highest	Average	Rest	
5–5	5,000	...	4.3	4.5	...	101	101	23
5–10	3,330	...	3.4	3.0	...	81	77	16
10–5	6,670	5.6	5.1	4.9	157	142	140	44
10–10	5,000	4.7	4.4	3.8	109	104	95	20
15–10	6,000	5.3	5.0	4.5	140	139	144	51
15–15	5,000	5.3	4.6	3.8	110	90	95	21
15–30	3,330	3.9	3.6	2.8	96	79	64	16

Source: Data from Christensen et al., 1960.

about 60 min at the lower speed as against only 25 min at the higher speed of 22.75 km·hr⁻¹. It is an important but unsolved question which type of training is most effective: to maintain a level representing 90 percent of the maximal oxygen uptake for 40 min, or to tax 100 percent of the oxygen uptake capacity for about 16 min.

Summary A series of studies have shown that maximal oxygen uptake (and cardiac output) may be attained in connection with repeated periods of work of very high intensity of as short duration as 10 to 15 s, provided the rest periods between each burst of activity are very short (of equal or shorter duration than the work periods). In more prolonged work of several minutes duration, the duration of the rest periods is less critical. If the work periods exceed about 10 min or so, a high level of motivation is required in order to attain maximal oxygen uptakes. In the case of continuous work, the high tempo required for the severe taxation of the oxygen-transporting system must alternate with periods of reduced tempo.

With very short work periods of about 30 s or less, a very severe load may be imposed upon both muscles and oxygen-transporting organs without the engagement of anaerobic processes leading to any significant elevation of the blood lactate. It is thus possible to select the proper work load and work and rest periods in such a manner that the main demand is centered on (1) muscle strength without a major increase in the total oxygen uptake; (2) aerobic processes without significantly mobilizing anaerobic processes; (3) anaerobic processes without maximal taxation of the oxygen-transporting organs; and (4)

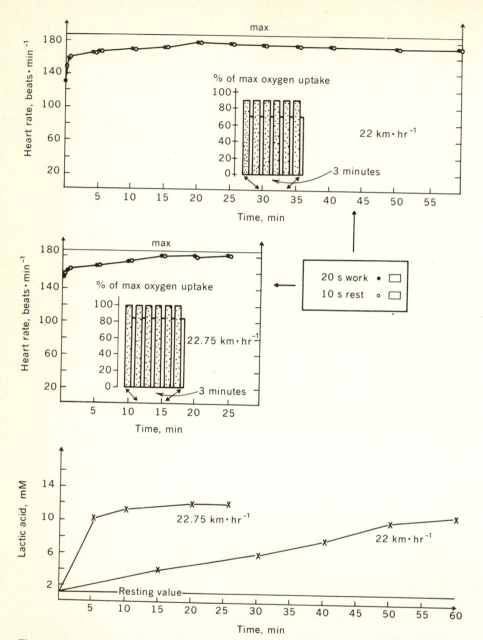

Figure 12-2 Oxygen uptake, heart rate, and blood lactate concentration during a training program, running on the treadmill, with short work and rest periods (20 and 10 s respectively) at speeds of 22.0 km · hr⁻¹ (upper panel) and 22.75 km · hr⁻¹ (lower panel). Note that the heart rate and oxygen uptake are not maximal in the first case (22.0 km · hr⁻¹) but that they are in the second case (22.75 km · hr⁻¹). It should also be noted that the total work time is reduced to half in the latter case. The lactic acid concentration in the blood had reached about the same level in both cases when the work had to be discontinued. *(From Karlsson et al., 1967b.)*

both aerobic and anaerobic processes simultaneously. The alternatives 2 and 4 do not entail maximal taxation of muscle strength; alternative 3 does not necessarily require maximal strength. In the following we shall consider how these principles may be applied.

Training of Muscle Strength

The main point of training is primarily to develop strength and endurance in the type of work in which an improvement is sought, whether it be static or dynamic. Hollmann and Hettinger (1976, p. 217–259) recommend, as an optimal training program for untrained individuals, a load of 50 to 70 percent of maximal isometric strength repeated five times per day, each time maintained for 3 to 6 s. In the more ambitious training, the loads should gradually become maximal. De Lorme and Watkins (1948) developed principles for strength training that are still followed. In their terminology one *repetition maximum* (1 RM) refers to the maximal load that a muscle group can lift one time. Thus, "10 RM" means that a given load can be raised against gravity no more than ten times. A training session may involve three sets of ten repetitions, the first set requiring ten repetitions at a load of $^{1}/_{2}$ 10 RM (e.g., with 25 kg instead of 50 kg weights); the second set with ten repetitions at a load of $^{3}/_{4}$ 10 RM; the third set with 10 RM. The program is usually designed to be progressive, and when the person can perform more than 10 RM with the weight, it is increased in order to bring the number of repetitions back to 10 RM. The number of repetitions, and sets, relative load, and rest intervals between sets are varied depending on the purpose of the training (see O'Shea, 1976). Some weight lifters and body builders train up to four hours daily.

Patients confined to bed rest may avoid muscular atrophy by subjecting the muscles involved to submaximal contractions of a few seconds' duration once a day. The same applies to astronauts during prolonged space flights. Isometric contractions develop isometric strength. Dynamic contractions develop dynamic strength and endurance (see below).

Training of Anaerobic Power

Theoretically, it is conceivable that an improvement of the anaerobic processes in the muscle cell which depend on the high-energy phosphate compounds may be achieved through maximal work of very short duration, up to 10 to 15 s, since the energy for this type of work is delivered primarily through these processes. The rest periods between each maximal effort should probably be at least a few minutes long in order to prevent a major mobilization of glycogenolysis. It is essential that the training involve the same muscle groups that are engaged in the event in which an improvement is sought.

The training of the anaerobic processes, which also involves the splitting of glycogen to lactic acid, may probably be accomplished effectively by periods of maximal effort for about 1 min, followed by 4- to 5-min rest, then a further period of 1 min maximal effort, followed by 4- to 5-min rest, and so on. At the end of four to five such work periods, a highly motivated runner may gradually attain lactic acid concentrations in the blood in excess of 20 mM (Chap. 9) and an arterial pH approaching 7.0 or lower. No doubt this form of training is psychologically very strenuous. If a training of the anaerobic motor power also requires the muscles to be exposed to a high lactic acid concentration, this training is rational provided the proper muscle groups are engaged. In order to produce major changes in the cellular milieu, affecting the respiratory function leading to dyspnea, large muscle groups have to be involved in such activities as running.

A training of the anaerobic motor power is important for many groups of athletes. Since this form of training is psychologically very exhausting, it should preferably not be introduced until a month or two prior to the competitive season. Such strenuous training is not recommended for average people. As pointed out in Chap. 9, there is no reliable method available for accurate measurement of the maximal anaerobic power, including the glyco-genolytic contribution. Therefore, it is difficult to evaluate the effectiveness of different training programs.

Training of Aerobic Power

It has already been pointed out that physical activity ranging from repeated work periods of a few seconds' duration up to hours of continuous work may involve a major load on the oxygen-transporting organs and thereby induce a training effect, provided the work load is sufficiently high. Practical experience has shown that work with large muscle groups for 3 to 5 min, followed by rest or light physical activity for an equal length of time, then a further work period, etc., as required by the individual's ambition and the objective of the training, is an effective method of training. The tempo does not have to be maximal during the work periods. It is not necessary to be exhausted when the work is discontinued (for an explanation, see Fig. 9-3, showing that maximal oxygen uptake may be reached without the subject's being exhausted). Mild exercise, such as jogging, between the heavier bursts of activity may be advantageous, since the elimination of lactic acid is faster than at complete rest (Newman et al., 1937; Gisolfi et al., 1966; Gollnick and Hermansen, 1973). It has been shown experimentally that the cardiac output and the stroke volume attain their highest values at a load which produces the maximal oxygen uptake. During "supermaximal" work, the oxygen uptake as well as the cardiac output and stroke volume may even attain lower values than they do at a slightly lower work load. There is no evidence to support the assumption that it is important to engage the anaerobic processes to any extreme degree in order to train the aerobic motor power.

The justification for a submaximal tempo in the optimal training of the oxygen-transporting system may be further supported by Fig. 12-3. For six subjects, an individual speed was determined which brought them to complete exhaustion at the end of the fourth minute of running. On other days, the speed of the treadmill was decreased stepwise by 0.5 to 1 $km \cdot hr^{-1}$ without changing the total distance of the run. A reduction in speed by as much as 3 $km \cdot hr^{-1}$ for

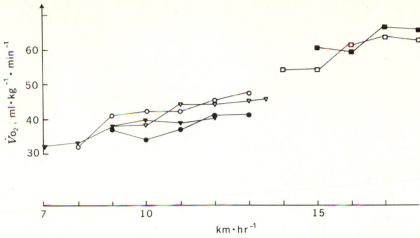

Figure 12-3 Individual data on oxygen uptake in relation to speed for two female (triangles) and four male subjects. Treadmill was set at an angle of 3°. The highest speed for each subject could be maintained for just 4.0 min. *(From Karlsson et al., 1967a.)*

some of the subjects did not reduce the oxygen uptake. Therefore, since maximal oxygen uptake can be attained at a submaximal speed, this lower speed may be sufficient and probably optimal as a training stimulus. A highly motivated individual and someone with a high anaerobic capacity will show a wide plateau; others may just be able, or willing, to push themselves to the point where the maximal oxygen uptake is reached (for example, 250 watts in Fig. 9-3b), or not even that far. It is thus impossible to delineate a border value where the greatest load on the oxygen-transporting organs is attained. In the case of healthy young persons, the speed of running may be reduced to about 80 percent of the maximum, which may be maintained for a period of 3 to 5 min. If, in other words, the distance which may be covered by running, swimming, or bicycling in a matter of, say, 3.0 min is covered, instead, in about 3.5 min, the demand on the aerobic processes remains the same. Therefore the stopwatch, in such cases, should be used to maintain a reduced tempo, not to stimulate the trainee to attain a better achievement in terms of better timing. Figure 12-4 illustrates how this type of training may be arranged, and what effect it has on heart rate and oxygen uptake. (It should be emphasized that in exercise in the upright position the maximal stroke volume is attained *during* and not after the exercise. It is a misconception to believe that the advantage of interval training is that frequent recovery periods as such should elicit an effective training of the central circulation.)

   As an aid to determine whether the load has been maximal or nearly maximal, the heart rate during the work, i.e., the work pulse, may be used. It should not differ more than 10 beats·min^{-1} from the individual's maximal heart rate, assessed during controlled laboratory experiments. A person accustomed to heavy work can also sense when the pulmonary ventilation has reached the

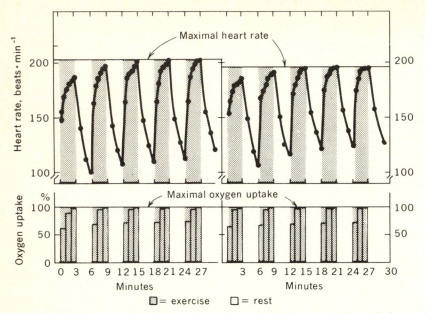

Figure 12-4 Heart rate and oxygen uptake recorded in two subjects during training with alternating 3-min running and 3-min rest. The efforts were not maximal, but the oxygen uptake reached maximal values, as did the heart rate. *(From Saltin et al., 1968.)*

steep part of the slope on the curve relating pulmonary ventilation to oxygen uptake (see Fig. 7-9). This type of training is certainly far more pleasant than when the tempo is higher. Thus, one should definitely distinguish between training of the aerobic and the anaerobic processes. In competitive events in which a superior effect is required in both these processes, the training obviously should also include this combination. This aspect, however, should be postponed until a few months prior to the start of the season; otherwise, the athlete may be unable to keep up the training program for psychological reasons.

The type of training described for the oxygen-transporting system, with a submaximal tempo for periods of 3 to 5 min, may increase the maximal oxygen uptake and is probably also effective in eliciting many of the side effects which have been described earlier in this book. The ability to work for prolonged periods of time, utilizing the largest possible percentage of the maximal oxygen uptake, may probably be primarily developed just by working continuously during long periods of time (endurance training). The capacity to store glycogen in the muscles and the ability to mobilize and to utilize free fatty acids play a major role in prolonged work (see Chap. 14). Endurance athletes may devote a great deal of time to "distance training." They may run or ski up to 250 to 350 km (150 to 220 miles) weekly. At present there are no studies available for an objective evaluation of the effects of such training quantities. It should be emphasized that many top athletes of today are professional in the sense that they are "employed" full time in their sports and are thus free to devote a great deal of time to training.

However, it should again be emphasized that the recruitment of motor units is dependent on the rate of work. At low intensities slow twitch fibers (type I) are activated, and with increased load more and more units become recruited and the rate of fiber contractions increases. At high intensities the fast twitch fibers (type II) are also thrown into action (see Chap. 4, p. 107). Analyses of the exercising muscle groups have revealed that the slow twitch fibers are the first to lose their glycogen content in prolonged submaximal activity. However, when these fibers are depleted of their glycogen stores, fast twitch fibers are apparently recruited, for a reduction of glycogen content has been observed in these fast fibers under such conditions. When the subjects are working at rates exceeding 100 percent of maximal $\dot{V}_{O_2}$, both fiber types appear to be continuously involved (Gollnick et al., 1974). Thus, in heavy interval training all fibers in the activated muscle groups may be working simultaneously. As pointed out in Chap. 9, the fast twitch fibers may be forced to depend primarily on an anaerobic energy yield since the oxygen supply is not sufficient to cover the demand of all the fibers for an aerobic metabolism. In "distance training" the oxygen uptake is, most of the time, below the maximum. With the depletion of the glycogen in the slow twitch fibers, a central nervous system mechanism, eventually assisted by an afferent impulse traffic from muscle spindles, will cause the recruitment of new motor units, the high-threshold fast twitch fibers being the last reserve. The timing of these events depends on initial glycogen stores, rate of work, and on the exercising person's will power. At any rate, the biochemical and morphological adaptations to chronic exercise will occur locally in the training fibers. One can speculate that an intensive interval training provides a balanced adaptation of both slow and fast twitch fibers to a combined aerobic and anaerobic stress situation, as in a 5000 or 10,000 m race. An exclusive "distance training" will promote the fast twitch fibers' aerobic metabolism (e.g., with an increase in their SDH activity). However, the question is whether this submaximal exercise is very effective in improving the maximal aerobic power. If, during training, there is no marked increase in maximal oxygen uptake, it is of questionable value to devote much time to increasing the aerobic potential of *all* muscle fibers. If the objective of the training is to improve the ability to perform prolonged work, it may be advisable to concentrate on a "distance training," for it may be effective in stimulating an increase in capillary density, myoglobin content, and activity in enzymes specialized on the metabolism of free fatty acids. However, for an endurance athlete the optimum must be to combine submaximal, maximal, and "supermaximal" training. (There are experiments on rats showing that running durations of moderately intense exercise longer than 2 hours·day^{-1} do not further increase the oxidative capacity of the working muscles; Terjung, 1976).

It is interesting to note that today's top athletes are not superior in their maximal oxygen uptake in comparison to the middle distance runners of the 1930–1950 era (Åstrand, 1956; Fig. 12-5). The extreme "distance training" typical for many training programs today has not been very effective in improving the maximal aerobic power considering the relatively modest amount of time devoted to training in earlier days.

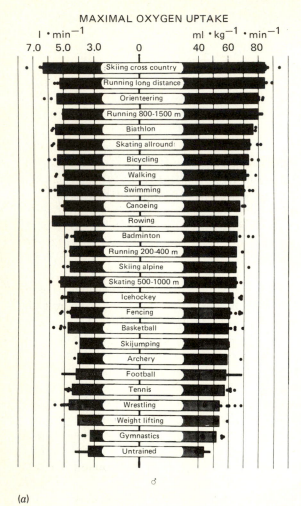

MAXIMAL OXYGEN UPTAKE

Figure 12-5 Average maximal oxygen uptake in liters per minute (left part) milliliters per kilogram body weight times minutes for male (a) and female (b) Swedish national teams in different sports. Dots indicate individual values higher than the mean value. *(Compiled from various sources by Ulf Bergh.)*

(a)

It should be emphasized that it takes a minimum of 48 hours to refill emptied glycogen stores (Piehl, 1974). This must be an additional complication for those who have the ambition to train daily for several hours at high rates of work.

Whatever "special effects" these two types of training may elicit in order to improve the maximal aerobic power and the maximal aerobic capacity, a certain overlapping in the elicited effect is highly probable.

The athlete often needs a certain amount of variation in the training. According to the above, there are considerable possibilities for variation, even though certain programs are more critical than others concerning work time, rest periods, and intensity.

In the case of patients and completely untrained individuals, it is out of the question to prescribe an accelerated tempo in connection with the training of

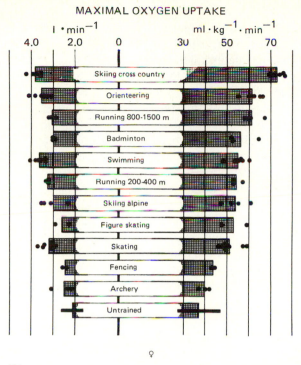

MAXIMAL OXYGEN UPTAKE

(b)

circulation. A previously bedridden patient should be satisfied with a training load which commences by elevating the heart rate by about 30 beats · min^{-1} above the resting value (to about 100 beats · min^{-1}). For the habitually sedentary individual, an elevation of the heart rate by about 60 beats · min^{-1} may be a suitable initial intensity. The principle of intermittent work is also valid for these categories of individuals. With daily training periods of from 15- to 30-min duration, or even with only a few training periods per week, the tempo may gradually be increased. The individual's health, age, and interest may determine how strenuous the training should be. It should be pointed out that in many cases it is not necessary, or in some cases even desirable, to attain maximal oxygen uptake and cardiac output during the training. In working with patients and average individuals, the "endurance training" may conveniently be accomplished by such activities as walking or bicycling. We can conclude that for an untrained individual an exercise that demands an oxygen uptake exceeding 50 percent of his or her maximal will, when repeated two to three half hours a week, gradually increase the maximal oxygen uptake (see Kilbom, 1971; Kearney et al., 1976). Training at an 80 percent level of maximal oxygen uptake may elicit a good effect (e.g., an increase in maximal oxygen uptake of about 15 percent; see Pollock, 1973; Hollmann and Hettinger, 1976). Another rule of thumb is to aim at a heart rate during training of 195 minus the age in years.

It has repeatedly been emphasized that large muscle groups must be engaged when training the central circulation. Figure 12-5 may be of interest in this connection. In a number of top Swedish athletes belonging to the National Team, the maximal oxygen uptake was determined by laboratory experiments. In most cases the treadmill was used. It is evident that the athletes who had the highest maximal oxygen uptakes had selected events that placed heavy demands on their aerobic power. This means that these events also represent an excellent form of training of the oxygen-transporting system. Ball games fall rather low on the scale (data for handball, basketball, and soccer in Chap. 16 support this statement). This may be explained by the fact that each period of activity with a high tempo is frequently interrupted by periods of reduced tempo (Table 12-2). Most competitive calisthenics entail work periods up to 1 min. Needless to say, calisthenics may be performed in such a manner that they entail an excellent training effect of the aerobic power (straddle jump, sequences of movements engaging large muscle groups). Intensive activities involving small muscle groups (chinning the bar, push-ups, weight lifting) may be harmful for untrained individuals and cardiac patients, since these activities produce a high heart rate and high blood pressure at a given cardiac output and oxygen uptake (Chap. 6).

"Circuit training" (Morgan and Adamson, 1962) entails a series of activities performed one after the other. At the end of the last activity, one starts from the beginning again and carries on until the entire series has been repeated several times. By a preliminary test in which many of the activities may be performed at maximal exertion, the number of repetitions of each activity is determined. The advantage with this circuit training is that every individual undergoes a program adjusted to his or her level of fitness. Persons may follow their own improvement by recording the time required for the series of repetitions, and endeavor to shorten the time required. At the end of a few weeks' training, a new test is performed on the number of repetitions of each activity the individual can manage (for example, push-ups). This training produces a high degree of motivation. The program may be accomplished in a limited space. The disadvantage is that an untrained individual is exposed to tests requiring maximal exertion. It has been found that a correctly devised program varying the involvement of large and small muscle groups, mixing static and dynamic work, does not produce maximal oxygen uptake measured on the bicycle ergometer, but only about 80 percent of the maximal O_2 uptake (Hedman, 1960). In spite of this, the heart rate is almost maximal, the lactic acid concentration in the blood is very high, and the degree of exertion is considerable. Circuit training may be included in a training program, not only for athletes (especially those who fall within the lower part of Fig. 12-5) but also for school children for the sake of variation and for experience.

Year-round Training

In all cases, regularity in the performance of training is important. Within a month it is possible to develop a reasonable level of fitness, strength, and so on, but these qualities are lost when the training is discontinued. As already pointed out, less effort is required to maintain a certain level of fitness than to develop this level in the first place. With nonathletes, it may therefore be worthwhile, when time permits, to devote more time to training, and then, later on, to endeavor to maintain the acquired level of fitness by only a few training periods per week.

The athlete often has to train many different functions (aerobic and anaerobic power, strength, endurance, technique). It may be practical and sometimes necessary, because of time limitations, to concentrate on some of these functions at certain periods. This particular type of training has to be continued, however, even though it may be at a reduced intensity, when the athlete concentrates on the next function. Otherwise, the individual will lose the improvement already attained. The problem is to decide how intensively the different functions have to be trained in order to maintain a satisfactory level.

The keen competition of today necessitates year-round training. For a variety of reasons this is necessary, especially with regard to the oxygen-transporting system. This is particularly important for the aging athlete. The reason why many thirty- to thirty-five-year-old athletes may continue to rank among the world elite in endurance events if often due to a relatively hard training during all seasons of the year. The older athlete may be what is known as "hard to train," compared with younger athletes, and the person whose fitness is allowed to deteriorate may have considerable difficulty in regaining the former level of training. The seasonal variations in maximal oxygen uptake may be pronounced. Fig. 12-6 presents data on three top cross-country skiers. It should be noted that they have apparently more or less reached a ceiling, and the intensive daily training they undergo over the years will not markedly affect the maxim. Included are data on two subjects who began an intensive training in 1969. The improvement in maximal oxygen uptake is quite remarkable. On the other hand, they could not compete with the athletes who apparently had a more favorable genetic background.

Psychological Aspects

It appears that humans have a tendency to become inactive after they have reached puberty. Unfortunately, physical activity without a definite purpose becomes, from then on, rather rare. In the past, and in some countries and occupations even today, physical activity was a part of one's daily life and work. Inasmuch as the need for physical activity in connection with most daily occupations is about to be abolished, it will be necessary to devote some of the leisure time to physical activity to maintain an optimal function of the body. In this connection, it should be emphasized that effective physical training does not have to be very stressful. Nevertheless, it is a common misconception that physical training is unpleasant and difficult to arrange. Often the ideal solution is to acquire a hobby which involves some degree of physical activity, even though the training may not be very intensive. As a rule, most people can be

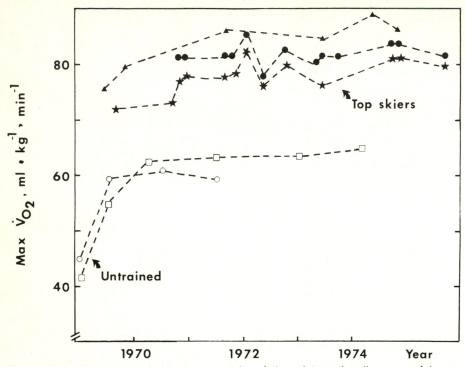

Figure 12-6 Data on maximal oxygen uptake of three internationally successful cross-country skiers, and two "normal" subjects who started an intensive physical training in 1969. *(By courtesy of U. Bergh and B. Ekblom.)*

motivated to exercise, provided proper facilities are available. Simple training tracks can, for example, be constructed in parks. Psychologically speaking, it is far more attractive to walk or run a 5-km course than to cover a 1-km course 5 times, to say nothing of 100 laps on a 50-m course. It is more acceptable to cover a distance which offers a certain amount of variation than to spend an equal amount of energy indoors on a stationary bicycle.

On the whole, it has to be admitted that for nonathletes, physical training may not be sufficiently enjoyable to make them exercise merely for fun, without being motivated by the prospects of the benefits they may obtain from the activity. One more or less has to accept the fact that one has to train and exercise as a matter of necessity in order to become fit, whether one likes it or not. It may therefore be best to get the exercise program over with the first thing in the morning—the sooner the better; then it is done. Eventually, as the benefits become apparent, one will miss the program if one does not keep it up. As an alternative, one may join a fitness club to train in a group; it may be more fun that way, and perhaps easier to continue, as one will be missed if failing to show up.

Tests

Objective tests to determine the effect of a training on different functions are both important and desirable. Such tests may be an aid in the development of the program and they encourage the individual to continue the training. Apart from the bicycle ergometer test, a simple step test may be used (Chap. 11). Figures 11-7 and 11-8 presented examples of the heart rate during a standardized work load during a period of training.

Summary From a practical standpoint, it may be logical to list four components of a rational training program aimed at developing the different types of power:

1 Bursts of intense activity lasting only a few seconds may develop muscle strength and stronger tendons and ligaments.
2 Intense activity lasting for about 1 min, repeated after about 4 min of rest or mild exercise, may develop the anaerobic power.
3 Activity with large muscles involved, less than maximal intensity, for about 3 to 5 min, repeated after rest or mild exercise of similar duration may develop the aerobic power.
4 Activity at submaximal intensity lasting as long as 30 min or more may develop endurance, i.e., the ability to tax a larger percentage of the individual's maximal aerobic power.

BIOLOGICAL LONG-TERM EFFECTS

In this section we will discuss in some detail the various long-term training effects indicated in Table 12-1. It should be emphasized that by long-term effects, we mean changes whose development may require a certain amount of time (weeks, months, or years).

Locomotive Organs

In Chap. 8 it was pointed out that *bones*, *ligaments*, and *joint cartilages* are affected by use as well as by disuse. Bone structures which are not stressed may disappear and new bone trabeculae may be created where altered mechanical forces increase the demand for sturdiness. The hard interstitial substance consists of carbonates and phosphates of calcium, and may constitute up to about 60 percent of the dry, fat-free weight in adult life. After a long period of inactivity, it may be as low as about 40 percent (Ingelmark, 1957). The increased urinary calcium and phosphorus elimination during prolonged bed rest depends on a demineralization of the bones (Deitrick et al., 1948; Rodahl et al., 1967). The thickness of the articular cartilage is greater in trained animals than in untrained ones. The compressibility of the cartilage will therefore increase with training, providing greater possibilities of compensating for any incongruence of the articular surfaces of the cartilages. The contact surface will increase, and when stressed, the force per unit of surface will thus decrease (Holmdahl and Ingelmark, 1948; Ingelmark, 1957). Training causes a hypertrophy of the intercellular substance of connective tissue, increasing the volume of tendons and ligaments, and enhances their tensile strength. The hypertrophy of the tendons is accompanied by a hypertrophy of the muscles attached to them (Ingelmark, 1948). In the bone-ligament-muscle system the attachment of the ligament to the bone is the weakest point, although they tend to become stronger in trained animals (Viidik, 1966; Tipton et al., 1974). (For details see Booth and Gould, 1975.)

During training there is an increase in *muscle mass* by an enlargement of the already existing fibers. Most investigators find no increase in the number of muscle cells with training through a division of already existing cells. Results from animal experiments have been published, however, which indicate that prolonged excessive training may create new muscle fibers, even though the most important factor in the volume change of the muscle is a true hypertrophy of the existing muscle fibers (see Gonyea et al., 1977). The total amount of protein in the muscle increases with training and decreases with inactivity. Disuse atrophy is associated with a decline in myofibrils, and the proportion of sarcoplasm proteins rises (Helander, 1958). There is, in other words, a selective reduction in the contractile myofibril proteins, and this is more pronounced than the reduction which takes place in the total amount of protein. With inactivity there is an increase in the fat content of the muscle, whereas it decreases with training.

The hypertrophy of skeletal muscles in response to muscle training appears to be independent of growth hormone, insulin, testosterone, or thyroid hormone. The increase in muscle weight seems to be the result of greater protein synthesis and reduced protein breakdown. RNA and DNA synthesis also increases. On the basis of in vitro studies of rat muscles, Goldberg et al. (1975) suggest that it is the increased tension, passive or active, which is the critical event in initiating compensatory growth of the skeletal muscle. At any rate, muscle hypertrophy and strength are related to an increase in myofibrillar protein (actin and myosin) and, therefore, increased contractile filaments per myofibril.

In weight lifters the fast twitch fibers (type II) have a proportionally large area, and it can increase by strength training (see Edström and Ekblom, 1972; Thorstensson, 1976). Muscle groups subjected to intensive endurance training have "normal" or relatively small fiber areas, especially for the slow twitch fibers (type I) (Jansson and Kaijser, 1977; Nygaard et al., 1977). Small sizes of the trained muscle fibers together with increased numbers of capillaries surrounding them will reduce the diffusion distance in the endurance trained muscles.

Both training and disuse of a muscle are associated with *biochemical changes* in the muscle (Howald and Poortmans, 1975). According to Nöcker et al. (1958), training causes an increase in the potassium content of the skeletal muscles. During exhausting work, however, the potassium content drops to lower levels in trained than in untrained individuals. Enzyme systems in the muscles are affected by training. Thanks to the development of the needle biopsy technique it is now possible to study in detail human skeletal muscles. Endurance exercise, such as long-distance running, leads to an increase in the capacity for aerobic metabolism of the muscle cell, associated with an increased capacity to oxidize pyruvate and long-chain fatty acids, due to an increase in the level of a number of mitochondrial enzymes. The increased mitochondrial enzyme activity appears to be due to an increase in enzyme protein. From electron-microscopic studies, increased size and number of mitochondria seem to be responsible for the increase in mitochondrial protein (see Howald and Poortmans, 1975, p. 372). As pointed out by Holloszy (1975), the skeletal muscle, as a result of these and other exercise-induced biochemical adaptations, tends to become more like heart muscle in its enzyme pattern.

The effect of training on the activity of glycolytic enzymes in skeletal muscle in humans has been studied by a number of investigators. The results are to some extent conflicting. Gollnick et al. (1972) observed no consistent difference in PFK activity, although there were marked differences in oxidative capacity as judged by succinate dehydrogenase (SDH) activity. In a later study (Gollnick and Hermansen, 1973), however, a doubling of the PFK activity was found after 5 months' training. Eriksson et al. (1971) studied the effects of training on the PFK activity of biopsy samples from the vastus lateralis muscle of ten- and eleven-year-old boys and found an increase of about 40 percent after 2 months' training. Succinate dehydrogenase activity had also increased. Increase in ATP and CP concentration in skeletal muscle has also been reported after training, the real significance of which is not quite clear (Gollnick and Hermansen, 1973).

Total glycogen synthetase activity is reported to be increased in human skeletal muscle in response to exercise training. Glycogen-branching enzyme activity is also increased in muscle with training. These findings indicate an increased capacity for glycogen synthesis in trained muscle (Holloszy and Booth, 1976). In fact, glycogen content is usually somewhat higher in muscles that are trained compared with the untrained muscles, but there is no difference in the glycogen content between slow and fast twitch fibers (see Piehl, 1974).

Costill et al. (1976) examined biopsy samples from the gastrocnemius muscle in male and female athletes and in untrained subjects. Their findings confirm earlier reports suggesting that the athletes' performance in terms of strength, speed, and endurance is in part a matter of genetic endowment. Succinate dehydrogenase was found to correlate significantly with maximal oxygen uptake. While sprint and endurance-trained athletes are characterized by distinct fiber compositions and enzyme activities (more fast and slow twitch fibers respectively), participants in strength events, like throwers, high jumpers, weight lifters, have relatively low muscle enzyme activities and a variety of fiber compositions (Costill et al., 1976; Saltin et al., 1977).

In endurance-trained athletes there is usually a "normal" or slightly reduced activity of glycolytic enzymes. Despite a decrease in total LDH activity, an increase in relative activity of the heart-specific isozymes, the H-form, has been noted. In elite distance runners and swimmers the SDH activity may be 20 to 25 mmoles $\cdot$ kg^{-1} $\cdot$ min^{-1} compared to approximately 7 mmoles in sedentary individuals and 3 to 4 mmoles $\cdot$ kg^{-1} $\cdot$ min^{-1} after prolonged physical inactivity (Saltin et al., 1977). SDH analyses on isolated muscle fibers show a higher activity in slow twitch fibers. However, in highly trained cross-country runners, the SDH activity was identical in the slow and fast twitch fibers (Jansson and Kaijser, 1977). This indicates an improved aerobic potential for the fast twitch fibers. As a consequence there is an increase in the type IIa fibers at the expense of IIb fibers, which may disappear altogether (Nygaard et al., 1977). The observed range in aerobic potential in individuals of different states of physical training is much greater than the variations in maximal oxygen uptake. As pointed out in Chap. 6, there are studies indicating that the enzyme systems do not limit the maximal oxygen uptake. Henriksson and Reitman (1977) noticed a 19 percent increase in maximal aerobic power in their 13 subjects who had trained for endurance for 8 to 10 weeks. The activities of SDH and cytochrome oxidase had increased 32 and 35 percent respectively (muscle samples from vastus lateralis). Within two weeks posttraining, the cytochrome oxidase activity had returned to the pretraining level and after six weeks the SDH activity was back to the control level. However, the maximal oxygen uptake was still high or 16 percent above the pretraining level. The authors conclude that an enhancement of the oxidative potential in skeletal muscle is not a necessity for a high maximal oxygen uptake. As pointed out, changes in the enzyme pattern may be of great importance in the utilization of various substrates in the muscles which may have particular consequences in prolonged excercise (see Chap. 14).

It has been claimed that the serum level of various enzymes is raised after exercise, but this increment is more pronounced in untrained than in trained individuals (Fowler et al., 1962; Garbus et al., 1964). This finding suggests a difference in cellular permeability.

Several investigators have found an increase in the number of capillaries in the muscle as the result of training (Fig. 12-7). The most recent and probably most accurate data on capillarization of muscle tissue in trained and untrained individuals have been obtained by Brodal et al. (1976), using the electron microscope. They counted the number of capillaries per fiber and per mm^2 in biopsy samples from the vastus lateralis of the quadriceps muscle. The mean number of capillaries per fiber was 41 percent greater in the trained than in the untrained group. The difference was statistically highly significant ($P<0.001$), and was of the same order of magnitude as the difference in maximal oxygen uptake between the two groups, i.e., 41 percent. The number of capillaries per mm^2 was 40 percent higher in the trained group than in the untrained group, the

Figure 12-7 Increased vascularization of a muscle as the result of training.
(a) Capillary density in three muscles of guinea pigs in the course of daily training on a treadmill at a speed that was gradually increased to 60 m $\cdot$ min^{-1}, running distance about 1,800 m each time. Animals were analyzed after different training times. Note that in the masseter muscle there was no change in capillary density, in contrast with the systematically exercised muscles. (Modified from Petrén, 1936.)
(b) Diagram showing the relation between the capillary density (capillaries per mm^2) and the fiber area (fibers per mm^2). Each point represents the values from one person. Two diameter values are indicated to show the relationship between the number of fibers per mm^2 and the lesser fiber diameter. It is seen that the number of capillaries decreases with increasing fiber area or fiber diameter. Increasing the fiber diameter with 10μm reduces the number of capillaries per mm^2 by about 150. It is evident that an endurance-trained person (with maximal oxygen uptake of more than 70 ml $\cdot$ kg^{-1} $\cdot$ min^{-1}) may have fewer capillaries per mm^2 than an untrained person (with maximal oxygen uptake of 50 ml $\cdot$ kg^{-1} $\cdot$ min^{-1}), if their fiber diameters are sufficiently different. (By courtesy of P. Brodal, F. Ingjer, and L. Hermansen.)

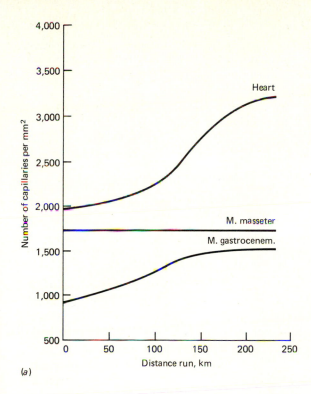

(a)

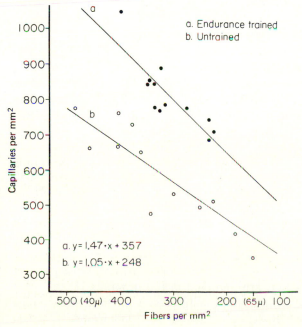

(b)

difference being statistically highly significant ($P<0.001$). Similar observations of a higher capillary density in endurance-trained athletes are reported by Saltin et al., (1977) who also noticed an increased capillary density in a longitudinal training study on humans. An increase in the number of capillaries reduces the tissue cylinder around a capillary, increasing the capillary surface area for an exchange of materials. It would also tend to raise P_{O_2} and lower P_{CO_2} and the concentration of metabolites in the interstitial fluid around the muscle fiber. An increase in the myoglobin content in trained muscles will also enhance the diffusion of oxygen (Whipple, 1926; Holloszy, 1975; Meldon, 1976; Chap. 7).

Training may also develop the arterial tree, probably by opening potential collateral vessels (Schoop, 1964, 1966). The increased blood flow with a vasodilation distal to the arterial tree in question is believed to be an important factor in the development of these collaterals.

In Chap. 4 it was pointed out that the measured *muscle strength* may vary greatly from test to test in the same subject. The explanation is probably that some of the training takes place in the motoneurons associated with the muscles being trained. The less the inhibition of these motoneurons, the more motor units may contract at tetanus frequency. In certain situations (danger, competition, etc.) and as the result of training, this inhibition may itself be inhibited, resulting in the development of a larger muscular strength (Ikai and Steinhaus, 1961). These two factors—a variation in the composition of the muscle tissue and a variation in the number of muscle cells which may be engaged through maximal exertion of the subject's willpower—may explain the fact that the effective area of the cross section of a muscle is not a dependable measure of its maximal strength (McMorris and Elkins, 1954; Rasch and Morehouse, 1957; Ikai and Fukunaga, 1970). Hollmann and Hettinger (1976) stress the importance of motivation for the development of maximal strength.

As discussed in Chap. 4, one may distinguish between maximal static (isometric) and dynamic strength (sometimes incorrectly called isotonic strength). In addition, the endurance is also of interest, with reference to both static and dynamic exercise. Figure 4-30 shows that only when the exerted force is reduced to a level below 15 percent of the maximal strength is it possible to maintain the contraction "indefinitely." During short efforts lasting only for seconds, the anaerobic processes play a dominating role, with high phosphate compounds involved in the energy yield. In more prolonged efforts, the aerobic processes assume a more dominating role, in which case an effective circulation is of major importance. For this reason, and in view of the important role played by the nervous system in the development of muscular strength (with a special nervous circuit for every single movement of the body), it is obvious that the effect of the training is related primarily to those parts of the body and the movements that are involved in the training. The effect is considerably less pronounced in unfamiliar activities, even though the same muscle groups may be engaged (Rasch and Morehouse, 1957; Clarke, 1973). A considerable confusion has resulted from the fact that some investigators have used the same activity both for testing and for training, while others have used a test situation which has not corresponded to the actual training.

It has been shown that maximal isometric training to a great extent increases the isometric strength without significantly affecting the endurance. Compared with the increase in strength, the endurance may actually be reduced. This may be explained by the occurrence of a muscular hypertrophy without a corresponding increase in vascularization. The result may be an increase in the tissue cylinder surrounding the capillary. The muscular hypertrophy is, as mentioned, occurring in the fast twitch fibers. These fibers are also most susceptible to fatigue during repeated fast maximal-voluntary contractions (Thorstensson, 1976).

In Chap. 4, it was pointed out that the structural and metabolic changes in a denervated muscle are not identical with those of an inactivated muscle with an intact nerve supply. In a muscle subject to disuse atrophy caused by a cast, the total nitrogen content is reduced to a lesser extent than is the case after neurotomy (Helander, 1958). A stimulation of a denervated muscle will not lead to a hypertrophy, as it does in a normal animal subjected to training (Gutmann et al., 1961). Apparently the nerve cell has a trophic function essential for the muscle metabolism and protein synthesis. When an extremity is immobilized in a cast, the loss of muscle strength is greater than it is when the individual is immobilized by prolonged bed rest. Müller and Hettinger (1953) have reported experiments in which a muscle that was immobilized by a cast lost 20 percent of its maximal strength within a week.

There is only a slight correlation between static strength and speed of movements. A training of the progressive resistance type improving the strength, however, may also increase the speed of movement, but the change is not large (Clarke and Henry, 1961). A strength training has no influence on reaction time ability (Beers, 1935; Clarke and Henry, 1961). For further references, see Burke and Edgerton (1975), Clarke (1973), Hollmann and Hettinger (1976).

Petersen et al. (1961) have conducted several series of investigations concerning different types of training. The results are summarized in Table 12-4. They show convincingly what has been pointed out in several connections: that the improvement in performance is tied primarily to the function which is being trained. A training of static endurance increases the ability to attain static endurance; a training of dynamic endurance produces a marked improvement in this form of work without improving endurance for static work. Some improvement, although very slight, is observed in the maximal strength through training of endurance with lower tension developed at each contraction. Of importance for the improvement of endurance is a more abundant blood supply to the muscle in question. It is therefore natural that a type of training which produces an increase in the flow rate through the muscle also represents the most effective stimulus for the development of the vascular bed, i.e., for dynamic work. The maximal strength, on the other hand, depends on the neuromuscular function and especially on the ATP-ADP machinery and

Table 12-4 Measurement before and after training of maximal strength, arm flexor, in (1) isometric and (2) dynamic contractions. Endurance was tested with load that was 60 percent of the maximal strength and the number of contractions (kpm) was counted before fatigue made further successful contractions impossible for (3) isometric and (4) dynamic exercise. Training was restricted to one type of muscle work at a time. The figures give the average performance before (e.g., 378 kp · cm) and after (e.g., 400 kp · cm) training as well as the percentage difference. (1 kp = 9.8 N.)

Training	Maximal strength		Endurance 60% of maximum		Study
	(1) Isometric, kp × cm*	(2) Dynamic, kp†	(3) Isometric, No. of contractions	(4) Dynamic, kpm	
Dynamic "60%" 150 contr./day 30 days	378 to 400 +6%	8.8 to 11.4 +29%		16.4 to 843 +5040%	Petersen et al., 1961
Isometric "60%" 150 contr./day 30 days	425 to 441 +4%	10.7 to 11.3 +6%	29 to 336 +1060%	17 to 24 +41%	Hansen, 1961
Dynamic "60%" 100 contr./hour 30 days		12.1 to 13.7 +13%	100 to 98 −2%	60 to 438 +630%	Hansen, 1967

*Torque.
†Weight was lifted 25 cm.

not on the aerobic energy yield. High-threshold fast twitch fibers are particularly important for the development of great strength.

In this connection, the observations by Ikai (1964) are of interest. He examined the number of contractions of arm flexor and knee extensor performed once a second against a load which was one-third the maximal strength, and the exercise was continued to exhaustion. Table 12-5 summarizes some of the results.

It is noteworthy that throwers were stronger than the rest of the subjects, but relatively speaking, they had the lowest endurance, especially in the leg muscles. The male middle- and long-distance runners had no stronger leg muscles than did the average men, but they were characterized by an impressive endurance; the performance of their arm muscles did not differ from the average. If anything, the endurance test was inferior in this case. These results probably reflect differences in training in these groups of subjects.

Electrical stimulation indirectly via motor nerves or directly applied on the muscles may increase the maximal voluntary isometric strength (see Hollmann and Hettinger, 1976, p. 245). The drawback is that coordination is not properly

Table 12-5 Muscle Strength and Endurance

Subjects*	Leg extensor		Arm flexor	
	Max strength, kp	No. of contractions $1/3$ of max strength	Max strength, kp	No. of contractions $1/3$ of max strength
Men				
Average men (5;10)	55	48	17	75
Sprinters (7)	71	52	19	65
Middle- and long-distance runners (6)	55	399	19	48
Hurdlers (3)	61	67	19	46
Jumpers (8)	68	49	21	45
Throwers (8)	88	38	26	51
Women				
Average women (10)			9	70
Sprinters (3)	48	71	12	71
Middle- and long-distance runners (4)	53	68	10	68
Hurdlers (5)	48	67	10	67
Jumpers (2)	56	57	12	58
Throwers (1)	68	43	17	43

*Number of subjects in parentheses.
Source: Ikai, 1964.

trained. As far as the specificity of a training effect is concerned, one is once again reminded of the fact that the nervous control plays a decisive role for the developed strength. An inhibition of the motoneuron may be more or less pronounced. It appears as if a certain strength, developed in the course of a long period of time, is maintained largely by the same motor units (Carlsöö, 1952). An increase in the developed force in the same activity means that the same motor units became engaged with greater frequency and that new motor units are recruited in addition. Only at maximal effort are the motor units representing the "last reserve" thrown into play, especially in emergency situations. It is possible that these reserve units may become more easily engaged as the result of training. In order to train these motor units in the motions of question, maximal exertion is required. Many individuals usually exert themselves far from maximally. The reason why one does not find atrophied muscle cells in sedentary individuals may be that the motor units, which in the mentioned cases are only recruited during maximal effort, are already engaged at moderate exertion in a different motion pattern.

The application of eccentric contractions in the training has no advantage compared with the concentric or isometric contractions (see Clarke, 1973, p. 82). The *isokinetic* exercise, i.e., with the speed of movement being controlled and held constant, is a relatively new design (Thistle et al., 1967). However, it does not simulate natural movements, with the possible exception of swimming, and therefore the value of isokinetic training is difficult to evaluate. For the testing of maximal dynamic strength, the isokinetic testing apparatus is very useful. From Fig. 4-24 it is evident that the speed of contraction is decisive for the maximal strength. It is therefore very important that the speed of muscle contraction can be exactly repeated in a posttraining test. With the force-velocity curve in mind, it should also be pointed out that it may be misleading to express the demand on maximal dynamic strength in running, swimming, jumping, etc. as a percentage of maximal isometric strength. At a high velocity of contraction, "submaximal" force may, in this respect, be maximum or at least closer to maximum. (It should be emphasized that the curves in Fig. 4-24 represent data from in vitro studies. In in vivo experiments, inhibitions at various levels within the CNS may reduce the individual's ability to utilize up to 100 percent of the force potential of the muscles). The loss of muscular strength after a period of training is relatively slow; although there is a significant loss of strength after about a month, it may take several months before it gradually returns to pretraining level. It takes less effort to maintain an established level of strength than to develop strength.

Summary The degree of activity may to a great extent affect the locomotor organs morphologically as well as biochemically. Training is associated with an increased strength of the muscle attachment of the bone and an increase in the contractile force of the muscle. The processes which are the basis for the energy yield are affected in a favorable manner, and the peripheral circulation is improved. In all probability, the central nervous system is also affected, leading to improved coordination and making it possible for more motor units to be engaged simultaneously at tetanus frequency.

A person's potential for development of muscle strength is established at birth, since a given individual is born with a certain fixed number of muscle fibers which, on the whole, remains unaltered throughout life. Training increases the maximal muscular strength and inactivity reduces this strength. This is related to changes in the thickness of each muscle fiber and therefore also to the cross section of the muscle as a whole.

Muscular endurance is related to the exerted force in relation to the maximal strength which the muscle can develop; when the exerted force is below 15 percent of the muscle's maximal strength, the contractions may be continued more or less indefinitely. In the training of a muscle, the objective of the training should be kept in mind in that the physiological processes which are the basis for the development of muscular strength and which are affected through training are different, depending on the duration of the muscular effort. Brief efforts depend primarily on anaerobic processes. In prolonged efforts, aerobic processes play a major role.

Isometric training increases the isometric strength without significantly affecting endurance. A training load taxing from 50 to 100 percent of maximal force (beginners) in the form of one contraction per day, maintained for 3 to 6 s, may be effective. Progressive dynamic exercise will result in an increase in muscular strength and can induce muscular hypertrophy. Programs that apply from 2 RM for one set up to 10 RM for three sets all seem to develop strength, but an optimum may be around 6 RM for three sets, three times a week. Repeated submaximal dynamic contractions primarily increase dynamic endurance. Apparently the nerve cell has an essential trophic function governing the metabolism and protein synthesis associated with the increase in muscular strength through training.

Oxygen-transporting System

Like the skeletal muscle, the heart is adaptable to variations in the individual's physical activity. The changes just described also occur in essence in the heart muscle. It is typical for athletes in endurance events to have a large *heart volume* (Figs. 6-24, 12-8). In relation to body weight, the rabbit has a small heart compared to the hare (Fig. 11-6). With training, the rabbit becomes more like the hare as far as its physical constitution is concerned (Arshavsky, personal communication). Animal experiments have demonstrated an enhanced vascularization of the heart muscle as a consequence of training (Fig. 12-7). An important question which is not as yet fully answered is whether an enhancement of collateral formation in the heart muscles is limited to already vascularly handicapped areas or whether it may also occur in a healthy heart.

There is a high correlation not only between heart volume and maximal O_2 uptake in persons of a certain age but also between *blood volume* or *total hemoglobin* and maximal O_2 uptake (Fig. 11-5). As we have already pointed out, however, there are considerable individual variations in these parameters.

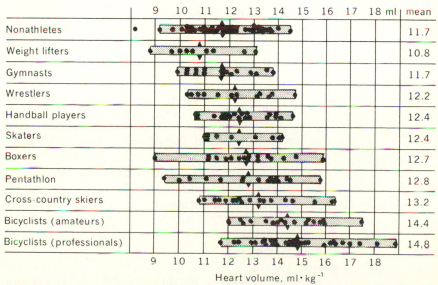

	9 10 11 12 13 14 15 16 17 18 ml	mean
Nonathletes		11.7
Weight lifters		10.8
Gymnasts		11.7
Wrestlers		12.2
Handball players		12.4
Skaters		12.4
Boxers		12.7
Pentathlon		12.8
Cross-country skiers		13.2
Bicyclists (amateurs)		14.4
Bicyclists (professionals)		14.8

Heart volume, ml·kg^{-1}

Figure 12-8 The heart volume per kilogram body weight for members of different national teams of Germany and a group of untrained individuals of the same age. Each dot represents one subject. Note the large individual variations within a group. *(From Roskamm, 1967.)*

Older persons may largely retain their circulatory dimensions although they have a reduced maximal aerobic power (Fig. 11-7). This is true even in the case of older former athletes, especially as regards the heart size (Holmgren and Strandell, 1959; Grimby and Saltin, 1966; Reindell et al., 1967). Training may increase and bed rest decrease these parameters, but the interrelationship between these parameters may not always be consistent (Deitrick et al., 1948; Taylor et al., 1949; Miller et al., 1964; Saltin et al., 1968). Bed rest primarily causes a decrease in the plasma volume, and a reduction in the red cell volume occurs only after prolonged bed rest (Miller et al., 1964; Vogt et al., 1967).

An interesting question is what happened to the large heart volumes of the girl swimmers presented in Fig. 6-24 after the termination of their intensive training. Sixteen of the 30 girls were restudied in 1971; at that time the mean heart volume was 625 ml as compared with 625 ml in 1961 (Eriksson et al., 1975). The maximal oxygen uptake had decreased from 2.80 to 2.17 liters $\cdot$ min^{-1}, and with a 6 kg gain in body weight the weight-corrected maximal oxygen uptake had declined from 51.4 to 36.1 ml $\cdot$ kg^{-1} $\cdot$ min^{-1}. After a 12-week swim-training (in 1971), the maximal oxygen uptake increased by 14 percent or to 2.47 liter $\cdot$ min^{-1}, with no significant change in heart volume (635 ml). (The training included, on the average, 16 sessions, ending with a week's stay in a training camp with relatively intensive swim-training). The high correlation between heart volume and maximal aerobic power noticed in 1961 was lacking in 1971. The 12-week training only partly restored the relationship. The authors concluded that the intense training early in life did not constitute a definite advantage 10 years later in life when the women underwent a training program.

It has long been established that individuals known to have considerable endurance usually have a slow resting *heart rate.* Hoogerwerf (1929) found a mean pulse of 50 beats $\cdot$ min^{-1} in 260 athletes participating in the Amsterdam Olympic Games (1928), the lowest value being 30 beats $\cdot$ min^{-1}. In one cross-country skier, the resting heart rate was repeatedly as low as 28 beats $\cdot$ min^{-1}, while a heart rate of 170 was recorded during heavy exercise (unpublished results). Habitual training enables a person to achieve a certain cardiac output at rest, as well as during work, with a slow heart rate and a large stroke volume. This improves the economy of the heart muscle as far as energy requirement and oxygen demand are concerned (see Chap. 6).

It is assumed that training causes an increased centrogenic vagal cholinergic drive combined with a sympathoinhibitory mechanism. Prolonged inactivity on the other hand causes preponderance of the oxygen-wasting adrenergic system. Herrlich et al. (1960) have shown that rats, after prolonged training in a running case, showed a significant increase of the atrial acetylcholine content. This may be the result of an increased vagal discharge. Trautwein et al. (1960) have presented evidence of a spontaneous release of acetylcholine in the right atrium of the dog's heart. It is conceivable that this production may also be affected by training. (For further discussion of the mechanism which may be involved in the adaptation of heart rate as a result of physical training, see Tipton et al., 1971; Barnard, 1975).

Whether or not the stretch receptors in the dilated and hypertrophic atria might elicit a bradycardia has also been discussed. There is no evidence, however, in support of such a mechanism. A cardiac patient may also have a large heart, but this is not associated with any bradycardia. The increased blood volume caused by training also causes an improved venous return and enhances the filling of the heart during diastole.

Studies by Scheuer and coworkers (1973) in rats subjected to moderate and more severe training programs suggest that the training enhances both the function of the myocardium as a muscle (increased contractility) and the heart as a pump (increased external cardiac work capacity). Dynamic changes were not dependent on an increase in the cardiac weight. There is no convincing evidence that enhanced energy-generating mechanisms are of sufficient quantitative importance to explain this improved myocardial performance. However, there are indications that physical training induces an alteration in the structure or control of the myosin molecule, including an increase in ATPase activity. The isolated working hearts of trained rats also demonstrated an increased resistance to hypoxia. (For further discussion of the training effects on the cardiac function, see Barnard, 1975.)

It may be concluded that the mechanism underlying the effect of training on the work of the heart is by no means clear. It has an effect on the contractile force of the heart muscle and an effect on the relationship between sympaticus and parasympaticus. The final result is a diminished demand on the oxygen consumption and thereby on the blood flow through the heart muscle at a given cardiac output. There are good indications to suggest that the blood supply is enhanced in the trained heart.

At rest, the *pulmonary ventilation* per liter of oxygen consumed is more or less unchanged after training. Possibly the depth of respiration is somewhat increased and the respiratory rate correspondingly reduced. (It should be pointed out that any measurement of pulmonary ventilation at rest is very difficult to achieve without affecting the subject.)

Several reports state that the *cardiac output* at submaximal work and at a given oxygen uptake is not significantly affected by training. Others have found that it is somewhat lowered. Bevegård et al. (1963) found well-trained athletes to have the same cardiac output at a given oxygen uptake as did untrained persons. On the whole, the maximal heart rate appears to be the same at different levels of training. The fact that the mean maximal heart rate is somewhat lower in athletes engaged in endurance events (about 10 beats $\cdot$ min^{-1}, Saltin and P.-O. Åstrand, 1967) may be due to constitutional factors. The possibility cannot be excluded, however, that many years of training, perhaps especially when associated with a marked increase in heart volume, might also bring about a drop in the maximal heart rate (Ekblom, 1969).

An increased stroke volume combined with an unaltered maximal heart rate means an increased maximal cardiac output as a consequence of habitual physical activity. Since, in addition, training improves the possibility for the tissues to utilize the available oxygen volume, there is a dual basis for an *increased maximal oxygen uptake*: (1) an increased maximal cardiac output; (2) an increased a-$\bar{v}O_2$ difference. From a review of the literature, Rowell (1974) concluded that: (1) in young subjects with initially low maximal oxygen uptake, an increase is by equal increments in cardiac output (stroke volume) and a-$\bar{v}O_2$ difference; in older individuals only cardiac output increases; (2) when the preconditioning maximal oxygen uptake is high, the changes are small and due almost exclusively to a rise in maximal cardiac output. There are methodological difficulties in the measurement of regional blood flows, but there are studies indicating that a smaller fraction of the cardiac output is distributed to working muscles during submaximal exercise in the well-conditioned subject. Instead the splanchnic and renal blood flow appears to increase, which at least can partly explain a constant cardiac output (see Rowell, 1974, p. 92; Clausen, 1976). The reader should be reminded that the morphological and biochemical changes in skeletal muscles induced by training may provide an adequate oxygen supply despite a reduced blood flow. In prolonged exercise a more "generous" distribution of the blood flow to the liver, kidneys, skin, etc. would be an advantage.

In a recent review by Pollock (1973) covering a wide selection of studies concerning the effect of endurance training on maximal oxygen uptake, the reported improvements vary from 0 to 93 percent. Again, initial level of fitness (natural endowment, habitual activity), intensity, frequency, and duration of training, not to mention the person's age, are factors that will influence the long-term effects of physical activity. We will present a few representative studies to illustrate effects of training on some parameters.

Saltin et al. (1968) carried out extensive studies on the effect of a 50-day period of physical training following a 20-day period of bed rest in five male subjects, aged nineteen to twenty-one. Three of the subjects had previously been sedentary and two of them had been physically active. The training program was rather intensive and continuously supervised. The weekly schedule included two workouts daily, Monday through Friday; on Saturdays there was only one workout, and Sundays were free. The workouts consisted of both interval and continuous exercise, mainly outdoor running of from 2.5 up to 7 miles. In the interval exercise, the speed was chosen so that the oxygen demand during 2- to 5-min periods of running was at or near the individual's maximal oxygen uptake. During the continuous running, usually for more than 20 min, the oxygen uptake varied from 65 to 90 percent of the subject's maximal value. One of the subjects who had not trained before was covering an average of 40 miles per week (64 km), and one of the previously trained subjects covered about 50 miles per week (80 km). The maximal oxygen uptake dropped from an average of 3.3 in the control study (before bed rest) to 2.4 liters $\cdot$ min^{-1} after bed rest, a 27 percent decrease (see Fig. 12-9). The stroke volume during supine exercise on a bicycle ergometer at 100 watts (oxygen uptake, 1.5 liters $\cdot$ min^{-1}) decreased from 116 to 88 ml, or about 25 percent. The heart rate increased from 129 to 154 beats $\cdot$ min^{-1}. The cardiac output at this standard load fell from 14.4 to 12.4 liters $\cdot$ min^{-1}. Thus the arteriovenous O$_2$ difference became somewhat increased. Also, during upright exercise at submaximal loads, there was a reduction in cardiac output (15 percent) and stroke volume (30 percent) (Fig. 12-10). An oxygen uptake that could normally be attained with a heart rate of 145 required a heart rate of 180 beats $\cdot$ min^{-1} after bed rest. During maximal treadmill exercise, the cardiac output fell from 20.0 to 14.8 liters $\cdot$ min^{-1} (a 26

liters · min⁻¹

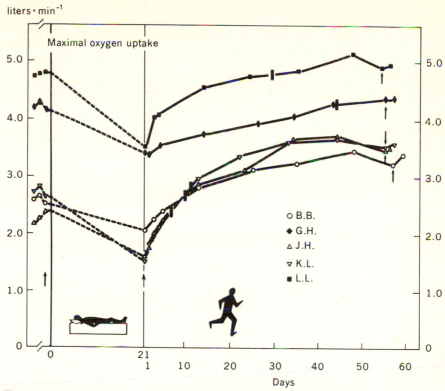

Maximal oxygen uptake

o B.B.
♦ G.H.
△ J.H.
▽ K.L.
■ L.L.

Days

Figure 12-9 Changes in maximal oxygen uptake, measured during running on a motor-driven treadmill, before and after bed rest and at various intervals during training; individual data on five subjects. Arrows indicate circulatory studies. Heavy bars mark the time during the training period at which the maximal oxygen uptake had returned to the control value before bed rest. *(From Saltin et al., 1968.)*

percent reduction). It should be recalled that the maximal oxygen uptake was now 27 percent less. Since the oxygen content of arterial blood did not change, the maximal arteriovenous O_2 difference was therefore not modified by the bed rest. These findings are illustrated in Fig. 12-10. The maximal heart rate was not altered, so the fall in maximal cardiac output was due to a reduction of stroke volume.

The physical training produced an increase in maximal oxygen uptake from 2.52 to 3.41 liters · min⁻¹ in the previously sedentary subjects (an increase of 33 percent), and from 4.48 to 4.65 liters · min⁻¹ in the previously active subjects (Fig. 12-9). The most dramatic improvement in maximal aerobic power was noted with the three usually sedentary subjects when comparing the "after bed rest" values with the posttraining ones. They actually increased their maximal oxygen uptake by 100 percent, or from 1.74 to an average of 3.41 liters · min⁻¹ as mentioned. This study illustrates how critical the level of physical activity is before a training regime for the evaluation of the effective-

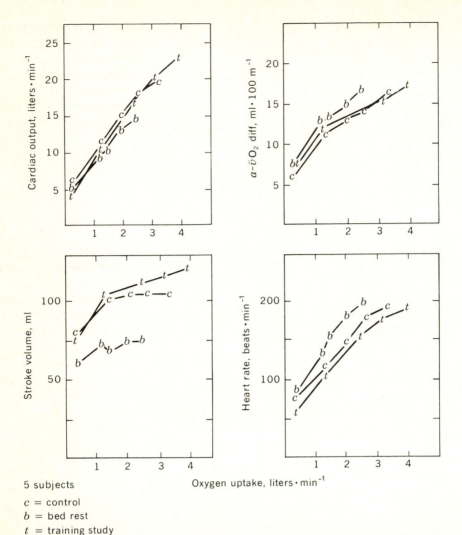

Figure 12-10   Mean values of cardiac output, arteriovenous oxygen difference, stroke volume, and heart rate in relation to oxygen uptake during running at submaximal and maximal intensity before (*c*), after (*b*) a 20-day period of bed rest, and again after a 50-day training period (*t*). *(Modified from Saltin et al., 1968.)*

ness of a training program to improve the maximal aerobic power. In the previously physically active subjects, the improvement was only 4 percent. Starting with the "after bed rest" level, the increase was, however, 34 percent (from 3.48 to 4.65 liters · min^{-1}). For the three previously sedentary subjects, the improvement was 33 and 100 percent respectively. In other words, the training program applied may be said to have caused the maximal oxygen uptake of the participants to increase from 4 to 100 percent, depending on how the initial level is defined. From Fig. 12-9 it is evident that the three sedentary subjects exceeded their control values as early as about 10 days after the commencement of the training. The two previously active subjects required 30 to 40 days to achieve the noted improvement.

Figure 12-11 illustrates the variation in maximal oxygen uptake for the three normally inactive subjects. The higher maximum noticed during their normal sedentary life, compared with the more extreme inactivity when immobilized in bed, was due to a higher cardiac output. The further improvement in maximal oxygen uptake with intensive training was partly due to a still higher cardiac output and partly to an increased arteriovenous O_2 difference by a more complete extraction of oxygen from the blood in the tissues.

Figure 12-10 summarizes the circulatory data for the five subjects. The training did increase the stroke volume and decrease the heart rate at submaximal work loads. The maximal heart rate was, however, not modified by the training. The maximal cardiac output in the sedentary subjects decreased from 17.2 to 12.3 liters $\cdot$ min^{-1} during bed rest; after the training it rose to 20.2 liters $\cdot$ min^{-1}. The arteriovenous oxygen difference, in the same experiments, was 14.7, 14.9, and 17.0 ml $\cdot$ 100 ml^{-1} blood respectively. The stroke volume fell from 90 to 62 ml during bed rest, but after training it increased to 105 ml. The changes in heart rate at various levels of oxygen uptake due to inactivity and caused by activity are obvious. However, when the heart rate was related

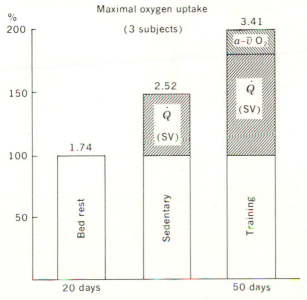

Figure 12-11 Maximal oxygen uptake during treadmill running for three subjects after bed rest (= 100 percent), when they are habitually sedentary, and after intensive training, respectively. The higher oxygen uptake under sedentary conditions compared with bed rest is due to an increased maximal cardiac output ($\dot{Q}$). The further increase after training is possible because of a further increase in maximal cardiac output and arteriovenous O_2 difference (a-$\bar{v}O_2$). The maximal heart rate was the same throughout the experiment; therefore the increased cardiac output was due to a larger stroke volume (SV). *(From data obtained by Saltin et al., 1968.)*

to the oxygen uptake in percentage of the maximum, the heart rate response remained the same in the different conditions.

The mean value for the heart volume in the three sedentary subjects was 740 ml. After bed rest, it was 690 ml and increased to 810 ml at the end of the training, an increase of 17 percent. The increase in stroke volume was 69 percent, however.

Blood volume decreased significantly during bed rest (from 5.06 to 4.70 liters in the 5 subjects, i.e., a 7 percent reduction). The fall in plasma volume was slightly more pronounced than in the red cell mass. During training, the plasma volume and red cell mass increased again, in most subjects above the control values.

The lean body mass changed from 66.3 to 65.3 kg after bed rest and increased to 67.0 kg after training. There were no significant changes in ultrastructural morphology of voluntary muscle (quadriceps) during the experimental period. The basal heart rate recorded during sleep was on the average 51.3 beats/min after bed rest, and it decreased to 39.7 at the end of the training.

There are no data available showing how far the maximal oxygen uptake (and cardiac output) can be increased by intensive training for longer periods of time. Ekblom et al. (1968) studied eight subjects before and after 16 weeks of rather intensive training. The maximal oxygen uptake increased by 16 percent, or from 3.16 to 3.68 liters $\cdot$ min^{-1}, partly owing to an increased arteriovenous oxygen difference (from 13.8 to about 14.5 ml $\cdot$ 100 ml^{-1}) and partly because of an increased cardiac output (from 22.4 to about 24.2 liters $\cdot$ min^{-1}). Since the maximal heart rate was unchanged, the direct cause of the higher cardiac output was the greater stroke volume (from 112 to 127 ml). Two of the subjects continued their training up to about 30 months. The maximal oxygen uptake increased by 42 and 26 percent, the maximal cardiac output increased by 15 and 20 percent, the maximal stroke volume increased by 24 and 20 percent, the maximal $a\text{-}\bar{v}O_2$ difference increased by 23 and 6 percent, and the maximal heart rate dropped 10 percent in one case and was unchanged in the other. (See Fig. 12-6.)

Karvonen et al. (1957), in studies of six subjects (trained by running on a treadmill $^1/_2$ hr daily, four to five times a week over a period of 4 weeks), noticed that "the heart rate during training has to be more than 60 percent of the available range from rest to the maximum attainable by running or above approx. 140 per minute, in order to produce a decrease of the working heart rate."

Kilbom and coworkers (see Kilbom, 1971) submitted 49 female sales personnel, nineteen to sixty-four years of age, to an interval training on bicycle ergometers. They trained two-to-three half hours per week for 6 to 8 weeks, and the oxygen uptake was about 70 percent of their maximal aerobic power. In all age groups there was a mean increase of about 10 percent in the maximal oxygen uptake, a decrease in submaximal, but no change in maximal heart rate. Studies on a smaller group of the subjects indicated that an increase in stroke volume, and therefore cardiac output, was the cause of the improved aerobic power. Many of the subjects observed a reduced swelling of the calf and the foot during a working day after the training period.

Jones et al. (1962) noticed that 5 min of rope skipping daily for a 4-week period reduced the mean heart rate at a fixed ergometer test load of 75 watts from 159 to 141 beats·min^{-1} in a group of seven habitually sedentary women nineteen to forty-two years of age. (This difference is statistically significant at less than the 0.01 level.) The mean heart rate during the rope skipping was 168 beats·min^{-1} at the beginning of the training, but dropped to 145 beats·min^{-1}.

Roskamm (1967) summarizes experimental studies conducted in Freiburg, Germany. Altogether, 80 soldiers took part in the study. They were divided into four groups: Three of the groups trained $1/2$ hour daily, 5 days per week for 4 weeks. The total amount of training, measured in kpm, was the same in all these groups. All the training was done on bicycle ergometers. One group trained continually with a work load which brought the heart rate to a value of about 70 percent of the difference between the heart rate at rest and during maximal work. (Example: If the heart rate at rest was 65 beats/min and the maximum 195, the "70 percent load" should increase the heart rate to about 155 beats·min^{-1}: $195 - 65 = 130$; $130·70·100^{-1} = 91$; $65 + 91 = 156$.) A second group trained at alternating work loads: for 1 min the load was 50 percent higher than this 70 percent load, followed by a 1-min period when the load was 50 percent lower. The third group performed a similar training with varying work loads but each period was extended to $2^{1/2}$ min. The fourth group served as a control group without additional exercise. As an index of the training effect, the maximal load which could be achieved per pulse beat during work on the bicycle ergometer was used (maximal watt-pulse, similar to maximal oxygen pulse). It was found that all groups that trained improved their values significantly, compared with the control group. Almost all the trained soldiers had an increase of more than 10 percent. The group that carried out the $2^{1/2}$-min interval training showed the greatest improvement of the maximal watt-pulse. On the other hand, the decrease in heart rate at the standard load (100 watts) was most pronounced for the subjects training continually at the 70 percent load.

In another study, the Freiburg researchers noticed a 20 percent increase in the maximal watt per pulse beat in a group of 18 younger persons who were training $1/2$ hour daily for 4 weeks. Some of the subjects then stopped their training, and their increased level of physical performance was partially lost within 2 weeks. The other subjects continued to train, but only every third day. This amount of training did not further improve their physical condition but was enough to maintain the attained level by the previous training. This finding is in agreement with the experience that it takes less training effort to maintain a level of physical fitness than it does to attain the approved level in the first place.

The final study by Roskamm et al. (1966) which will be discussed here involved 18 men, aged 16 to 18, 20 to 30, and 50 to 60 years, and 6 women, aged 20 to 30 years. The training and evaluations were done according to the procedure outlined above. The heart rate during training was, in the case of the younger subjects, about 150 beats·min⁻¹, and in the case of the older subjects, about 130 beats·min⁻¹ (lower maximal heart rate). After 4 weeks' training, every group showed a significant increase in the maximal watt-pulse, but the training effect was less pronounced in the fifty- to sixty-year-old subjects than in the younger subjects.

Hollmann and Hettinger (1976, p. 555) present preliminary data from studies on male subjects, fifty-five to seventy years of age, who had been physically inactive for a minimum of 20 years. After 8 weeks of training, three to five times per week in sessions of 1 to 2 hours duration, the average increase in maximal oxygen uptake was 20 percent. A similar training effect is reported by Kasch et al. (1973) on middle-aged men, thirty-nine to sixty years of age. The "normal" decline in maximal oxygen uptake with age beyond twenty years can apparently be modified by regular training, corresponding to the rejuvenation of 15 to 20 years (see Fig. 9-11). It should be pointed out, however, that in the group of former physical education students who were restudied 20 years later, a decline in maximal oxygen uptake was noted without exception, although most of them were still very active physically (Fig. 9-11).

At a given submaximal O₂ uptake, the content of *lactic acid* in the blood is lower in a trained subject than in an untrained one, as shown by a series of studies (Fig. 12-12). This may be interpreted as an expression of a more

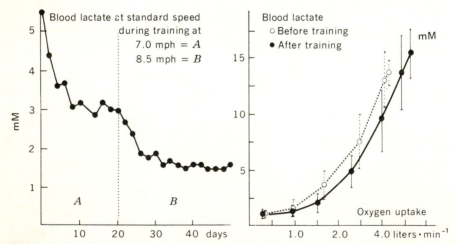

Figure 12-12 *To the left:* Progressive decrease of blood lactate for a standard amount of exercise: running on a treadmill at 7 mph for 10 min. During the first 20 days (*A*), training consisted of running daily on the treadmill for 20 min at 7 mph. A steady level of blood lactic acid is reached around 3 mmoles. During the following 30 days (*B*), training is increased to running at 8.5 mph for 15 min daily. Blood lactic acid decreases further, and a new steady level is reached around 1.5 mM after the standard test. *(From Edwards et al., 1939.)*
To the right: Peak concentration of lactic acid in blood in relation to oxygen uptake up to maximum before (dotted line) and after (solid line) 16 weeks of physical training. Mean values on eight subjects exercising on a bicycle ergometer. Vertical bars indicate ±1 standard deviation. *(From Ekblom et al., 1968.)*

effective oxygen transport during the beginning of the work, leading to a diminished anaerobic energy yield. When a certain increase in blood lactate is observed at an O_2 uptake corresponding to 50 percent of the maximal O_2 uptake in an untrained individual, this percentage may be elevated to 60 or 70 percent in a well-trained individual (see Karlsson et al., 1972; Chap. 9).

During maximal work, the well-trained individual usually achieves a higher concentration of lactic acid in the blood and a lower blood pH than an untrained individual. To what extent this may be due to an increased physiological tolerance for lactic acid with its side effects, or a greater psychic ability to exert oneself, is not clear.

No consistent data are available concerning the blood pressure during work before and after training. In the case of persons without elevated blood pressure, the changes are in any case moderate (Ekblom, 1969; Kilbom, 1971). The effect of an endurance training in middle-aged, healthy persons and hypertensive patients is often a lowering of both systolic and diastolic arterial pressures (Boyer and Kasch, 1970; Choquette and Ferguson, 1973; Clausen, 1976).

During submaximal work of relatively low intensity, the *pulmonary ventilation per liter* O_2 *consumed* does not change materially with training. Apparently the depth of respiration is increased somewhat, associated with a corresponding reduction in the respiratory rate. During heavier work, the ventilation per liter O_2 uptake is reduced, but reaches a higher level during maximal work. A study of Fig. 7-9 facilitates the explanation of these findings. During light work, the level of the pulmonary ventilation is primarily determined by the CO_2 production which is directly related to the O_2 utilization. During heavier work, the pH is also altered, primarily by the increase in blood lactate. This causes a relatively steeper increase in the pulmonary ventilation. Since the blood lactate concentration is generally lower during submaximal work following training, the respiratory drive is reduced, and the result is a lower pulmonary ventilation. The higher maximal ventilation is partly due to the increased maximal aerobic power, leading to an increased CO_2 production, and is partly due to the higher maximal lactic acid level. Thus, the untrained individual and individuals with low maximal aerobic power fall within the left side of the shaded area in Fig. 7-9, while training moves the curve downward to the right. Since the vital capacity does not change with training (in adults), no major changes in the depth of respiration are to be expected during work. It should be recalled that this is up to 50 to 55 percent of the vital capacity in trained as well as in untrained individuals (Chap. 7). At any given level of pulmonary ventilation the mechanical work of breathing is the same for untrained and trained individuals (Milic-Emili et al., 1962).

Another effect of training is that individuals, during prolonged work, may tax a somewhat larger percentage of their maximal O_2 uptake than they do when untrained. The reason is difficult to assess. A more effective O_2 supply to the working muscle due to an enlargement of the vascular bed, and enhanced diffusion, enzymatic potential for an increased utilization of free fatty acids as substrate, an increased glycogen content, and a higher psychic "fatigue threshold" may contribute in a positive direction. With training, individuals become more accustomed and willing to push themselves closer to their limit. It is important to keep in mind, however, that achievements in endurance events may be improved beyond those that are mirrored in changes in the maximal aerobic power.

Training of Cardiac Patients

While it is not the purpose of this book to discuss the training of patients, mention should be made of the increasing interest in prescribing regular exercise for patients with various cardiovascular diseases, notably patients who have had myocardial infarction (for references see Kellermann, 1975; Wilhelmsen et al., 1975; Clausen, 1976; Brunner and Jokl, 1977). In general, most patients who have had myocardial infarctions, and who undergo retraining programs, demonstrate the same type of cardiovascular adaptations as persons with normal hearts.

Specificity of training: When discussing muscular strength, it was pointed out that specific training may not improve the strength in other types of activities. There is a similar lack of transfer in training of the oxygen-transporting system.

In studies by Saltin et al., (1976), subjects were performing endurance training with one leg and sprint training with the other leg on a bicycle ergometer at work loads demanding about 75 and 150 percent respectively of the one-leg maximal. In other groups one leg was trained and the other one served as a control. After 4 weeks of training, the maximal oxygen uptake was increased when testing the trained leg (with an average increase of 11 to 23 percent). The greater the increase in maximal oxygen uptake, the more marked was the reduction in heart rate during submaximal one-leg exercise. When exercising the untrained leg only, a much smaller reduction in the heart rate was observed compared with the pretraining level. When doing maximal two-leg exercise, the oxygen uptake was about 10 percent higher after training. It is interesting to note that in those subjects who had been training one leg by sprint and the other leg by endurance exercises the increase in maximal oxygen uptake was, on the average, 0.32 and 0.54 liters·min^{-1} respectively in the one-leg test; when working with both legs the maximal oxygen uptake was only 0.30 liters·min^{-1} higher after training. One can assume that the one-leg exercise never brought the cardiac output up to the same level as the two-leg exercise before the period of training. However, with a training load that demanded approximately 60 percent of the two-leg maximal oxygen uptake, a 10 percent improvement may be considered a good result of the training. A decrease in lactate response and enhancement of SDH activity were observed only in exercise with the trained leg. Davies and Sargeant (1975) report an increase in maximal oxygen uptake by 0.34 liters·min^{-1} (14 percent) after training with one-leg maximal oxygen uptake. When submitted to a two-leg maximal exercise, the maximal oxygen uptake increased only 0.15 liters·min $^{-1}$ or 5 percent. The conclusion was that the training effect of the one-leg exercise was mainly peripheral in that particular leg, but in the two-leg exercise the central cardiovascular system was limiting the oxygen transport. Applying similar training and testing designs, Henriksson (1977) noted after 2 months of training a 27 percent higher SDH activity in the trained leg than in the untrained one (quadriceps femoris muscle). The maximal oxygen uptake was 11 and 4 percent respectively higher in the posttraining test than before training. At a submaximal two-leg exercise at an average of about 70 percent of the maximal oxygen uptake (for one hour) the subjects apparently worked harder with the trained leg. However, the degree of utilization of free fatty acids was higher in the trained than in the untrained leg indicating a difference in the oxidative capacity.

These studies indicate that irrespective of what factors trigger the adaptations on the cellular levels causing effects mainly in those muscles which are being trained, there are no, or only modest, effects elicited in nontrained muscles.

Clausen (1976) summarizes data from studies on healthy, young subjects in whom the circulatory response to arm-and-leg exercise was assessed after a period of training of either arms or legs. Arm training caused a marked reduction in heart rate during exercise with the trained arm muscles (from 137 to 118 beats·min^{-1}). During exercise with the nontrained leg muscles, a much less pronounced decrease in heart rate was seen (from 132 to 124 beats·min^{-1}). After leg training, however, the decrease in heart rate was about the same in the test with the trained leg muscles (from 135 to 122) as with the untrained arm muscles (from 127 ro 112 beats·min^{-1}). It is logical that there is more of a transfer effect on the central hemodynamics after training with large muscle groups involved as compared with training with small muscle groups.

Finally, Fig. 12-13 gives an example of data on maximal oxygen uptake measured on a champion swimmer when swimming in a swimming flume and also when running on a treadmill. For about four years his maximum was the same when running, but varying oxygen uptake values were measured during maximal swimming, owing to variations in training intensity, illness, etc. The variation was most pronounced in freestyle swimming using only the arms. Peaks were noted when the swimmer was most successful (winning two Olympic gold medals in 200 and 400 m medley, 1972).

Two identical twin sisters reached similar maximal oxygen uptake when running (3.6 liters·min⁻¹), but the sibling who had been training for competitive swimming attained a 30 percent higher oxygen uptake than her non-swim-trained sister. When swimming with arms only, the swim-trained sister could reach a 50 percent higher maximum (Holmér, I., and P.-O. Åstrand, 1972; see also Magle et al., 1975).

Conclusions: (1) A treadmill test is not a good predictor of performance in other types of activities; (2) in order to utilize the aerobic potential in an optimal way in a given activity, one must train in that activity.

Recovery from Exercise

There are very few longitudinal studies concerning the payment of oxygen debt, return to resting level of heart rate, cardiac output, temperature, etc., before and after training. The fast recovery of athletes after muscular work, in contrast, is well established.

It has been observed by Hartley and Saltin (1968) that untrained subjects who (1) work for 6 min at an O_2 uptake of about 40 percent of their maximal aerobic power, then (2) work for 6 min at a load of about 70 percent of their maximum, followed by (3) a 10-min rest, and then (4) repeat the 40 percent load, now have the same O_2 uptake and cardiac output as under

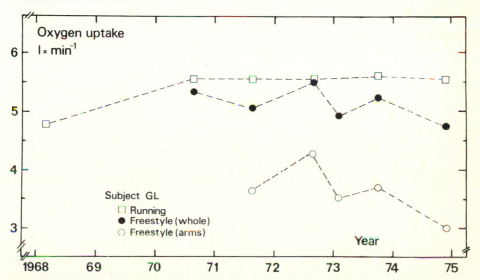

Figure 12-13 Oxygen uptake during swimming and running at maximal intensities by one world-class swimmer over a six-year period. Note the unchanged oxygen uptake during running from 1970, whereas oxygen uptake during swimming with the whole stroke and with only the arms varied considerably during the seasons. *(From Holmer, 1973.)*

(1), but they now have a significantly higher heart rate, in many cases as much as 20 beats/min higher, and consequently also have a smaller stroke volume. Following training, this increase in heart rate is much less, and well-trained athletes may accomplish the entire procedure with the results from (1) and (4) being almost identical. This type of testing appears quite promising as a method of assessing the level of physical fitness or the level of habitual physical exercise.

Mechanical Efficiency, Technique

In activities which are relatively uncomplicated technically, such as walking, running, or bicycling, there is a very slight increase in efficiency with training, but this increase is less than the variability among individuals. There is no definite difference in the consumption of energy per kilogram body weight and per kilometer in differently trained runners (Böje, 1944; Erickson et al., 1946; P.-O. Åstrand, 1956). On a bicycle ergometer, both Olympic medal bicyclists and untrained persons have the same mechanical efficiency at submaximal work loads. In Eskimos totally unfamiliar with the use of a bicycle, however, Vokac and Rodahl (1976) found a distinctly higher oxygen uptake at a given work load on the bicycle ergometer than they found in Caucasians. The more complicated the exercise, the greater are the individual variations in mechanical efficiency and the greater is the improvement with training. It should be pointed out that the mechanical efficiency of a person performing heavy work may be overestimated if the assessment is based on measurement of O_2 uptake, since anaerobic processes may have contributed to the energy yield.

In many achievements, the aim of the training is not primarily to reduce the energy expenditure during the event in question, but to attain an improvement "at all costs," for example, by increasing the developed power. Lauru (1957) gives an example with an athlete doing a broad jump from a "force-platform" before and after training. (With the aid of this force-platform, force phenomena during movement arising vertically, frontally, and transversely can be recorded and analyzed.) Before training, the push against the platform during the jump reached 1000 N. During the training, the push increased to 1400 N. The performance was at least partially improved by an improved coordination and a smoother sequence of motions.

Motor learning is so specific in nature that one cannot speak of a general learning ability with reference to motor coordination in skilled movements (see Chap. 4, p. 93; Albinson and Andrew, 1976; Stelmach, 1976).

Body Composition

During inactivity or habitual training, the body weight may remain relatively constant although inactivity often produces a gradual weight gain. During intensive training, the specific gravity of the body increases while the skin-fold thickness decreases (see Pǎrízková, 1973; Chap. 14). Increased specific gravity (i.e., changed from 1.068 to 1.077 and from 1.058 to 1.063 in the studies just mentioned) indicates a reduction in the fat content. This is also supported by the reduced skin-fold thickness observed. At the same time, there is a certain increase, especially of the muscle mass and the blood volume.

In the case of fat-free body mass, however, it is a matter of variations of only a few kilograms in the body weight with different states of training. It is therefore evident that change in body weight alone is an inadequate index of possible alterations in body composition.

Blood Lipids

A number of studies during the last two decades have examined the possible relationship between physical activity and blood lipid levels. Many of them have investigated the effect of a specific training program on the blood cholesterol level. The results are conflicting: about half the studies show a blood cholesterol-lowering effect of exercise; the other half show no effect. Golding (1961) found a 30 percent drop in blood cholesterol as the result of 25 weeks' training 1 hour five times per week; Altekruse and Wilmore (1973) reported a 10 percent drop at the end of 10 weeks' training of 15 min three times a week; Lopezs et al. (1974) found a 4 percent drop at the end of 7 weeks' training four times a week; Björntorp (1970) and Pyörälä et al. (1971) observed no change.

Hormones

It has been mentioned that training will affect the balance between sympathetic and parasympathetic activities. Plasma concentrations of norepinephrine increase with work intensity, but the increment is less following physical training. However, the concentration at a given percentage of maximal oxygen uptake seems not to be affected (see Hartley, 1975).

Insulin levels are higher but glucagon lower after training (see Hartley et al., 1972; Bloom et al., 1976). There is evidence that increased physical activity and training will increase the thyroxine turnover (Terjung and Winder, 1975). Plasma concentrations of growth hormone increase during exercise, but there are controversial reports on the effect of training in this response (Hartley 1975; Shephard and Sidney, 1975).

We can conclude that physical training does affect many hormone producing systems, but the significance of these responses is far from evident. As mentioned, many of the adaptations on a cellular level triggered by training are unique for the trained muscles, but are not found in "idle" muscles which must be subjected to the same changes in hormone levels.

Psychological Changes

"There is a general assumption that physical fitness has pyschological correlates. The concept of body-mind relationships is age-old. Psychosomatic research has indicated that physical changes result from continued psychological states; it seems logical to assume the reverse: that psychological changes result from physical states, such as fitness. Although there is a general assumption that this is so and considerable claims rather vaguely documented, there are surprisingly few firmly validated data" (Hammett, 1967). There are several studies demonstrating positive correlations between athletic ability on the one side and the general level of social adjustment and many psychological factors on the other. However, as emphasized by Hammett, they cannot be regarded as valid indications of psychological change related to increasing physical activity since it is equally possible that they reflect predilections; that is, persons with certain psychological characteristics may gravitate to physical training programs. There are too few longitudinal studies to date to permit us to draw any firm conclusions. In Kilbom's study (1971) the female sales personnel were questioned five times a day about some emotional dimensions and about subjective well-being. The emotional variables (irritation, interest in work, joy, etc.) were not significantly influenced, either from the morning to the afternoon, or before as compared to after training. However, the physical variables showed that fatigue was significantly more pronounced in the afternoon than in the morning, but it was less pronounced after training.

It is a common observation that a person engaged in physical training eventually experiences a given work load as being lighter, and even work at a given heart rate may appear lighter. Breathlessness is probably largely a question of lack of experience (Chap. 7).

REFERENCES

Albinson, J. G., and G. M. Andrew (eds.): "Child in Sport and Physical Activity," University Park Press, Baltimore, 1976.

Altekruse, E. B., and J. H. Wilmore: Changes in Blood Chemistries Following a Controlled Exercise Program, *J. Occup. Med.*, **15**:110, 1973.

Anderson, T. W., and R. J. Shephard: Physical Training and Exercise Diffusing Capacity, *Intern. Z. Angew. Physiol.*, **25**:198, 1968.

Andrew, G. M., C. A. Guzman, and M. R. Becklake: Effect of Athletic Training on Exercise Cardiac Output, *J. Appl. Physiol.*, **21**:603, 1966.

Åstrand, I., P.-O. Åstrand, E. H. Christensen, and R. Hedman: Intermittent Muscular Work, *Acta Physiol. Scand.*, **48**:448, 1960.

Åstrand, P.-O.: Human Physical Fitness with Special Reference to Sex and Age, *Physiol. Rev.*, **36**:307, 1956.

Barnard, R. J.: Long Term Effects of Exercise on Cardiac Function, in J. H. Wilmore, and J. F. Keogh (eds.), "Exercise and Sport Sciences Reviews," vol. 3, p. 113, Academic Press, Inc., New York, 1975.

Beers, L. B.: The Acute and Chronic Effects of Exercise on the Latent Period of the Gastrocnemius Muscle in Man, *Arbeitsphysiol.*, **8**:539, 1935.

Bevegård, S., A. Holmgren, and B. Jonsson: Circulatory Studies in Well-trained Athletes at Rest and during Heavy Exercise, with Special Reference to Stroke Volume and the Influence of Body Position, *Acta Physiol. Scand.*, **57**:26, 1963.

Björntorp, P.: Metabolism in Patients with Ischemic Heart Disease and Obesity after Training, in B. Pernow and B. Saltin (eds.), p. 493, "Muscle Metabolism during Exercise," Plenum Press, New York, 1970.

Bloom, S. R., R. H. Johnson, D. M. Park, M. J. Rennie, and W. R. Sulaiman: Differences in the Metabolic and Hormonal Response to Exercise between Racing Cyclists and Untrained Individuals, *J. Physiol.*, **258**:1, 1976.

Booth, F. W., and E. W. Gould: Effects of Training and Disuse on Connective Tissue, in

J. H. Wilmore, and J. F. Keogh (eds.), "Exercise and Sport Sciences Reviews," vol. 3, p. 83, Academic Press, Inc., New York, 1975.

Boyer, J., and F. Kasch: Exercise Therapy in Hypertensive Men, *J.A.M.A.*, **211**:1668, 1970.

Brodal, P., F. Ingjer, and L. Hermansen, "Number and Density of Capillaries in the Quadriceps Muscle of Untrained and Endurance-trained Men: A Quantitative Electronmicroscopical Study," *Am. J. Physiol.* (in press), 1977.

Brunner, D., and E. J. Jokl (eds.): "The Role of Exercise in Internal Medicine," S. Karger, Basel, 1977.

Burke, R. E., and V. R. Edgerton: Motor Unit Properties and Selective Involvement in Movement, in J. H. Wilmore and J. E. Keogh (eds.), "Exercise and Sport Sciences Reviews," vol. 3, p. 31, Academic Press, Inc., New York, 1975.

Böje, O.: Energy Production, Pulmonary Ventilation and Length of Steps in Well-trained Runners Working on a Treadmill, *Acta Physiol. Scand.*, **7**:362, 1944.

Carlsöö, S.: Nervous Coordination and Mechanical Function of the Mandibular Elevators, *Acta Odont. Scand.*, **10**(Suppl. 11): 1952.

Choquette, G., and R. J. Ferguson: Blood Pressure Reduction in "Borderline" Hypertensives Following Physical Training, *Can. Med. Ass. J.*, **108**:699, 1973.

Christensen, E. H.: Beiträge zur Physiologie schwerer körperlicher Arbeit, *Arbeitsphysiol.*, **4**:1, 1931.

Christensen, E. H., R. Hedman, and B. Saltin: Intermittent and Continuous Running, *Acta Physiol. Scand.*, **50**:269, 1960.

Clarke, D. H.: Adaptations in Strength and Muscular Endurance Resulting from Exercise, in J. H. Wilmore (ed.), "Exercise and Sport Sciences Reviews," vol. 1, p. 73, Academic Press, Inc., New York, 1973.

Clarke, D. H., and F. M. Henry: Neuromotor Specificity and Increased Speed from Strength Development, *Res. Quart.*, **32**:315, 1961.

Clausen, J.P.: Circulatory Adjustments to Dynamic Exercise and Effect of Physical Training in Normal Subjects and in Patients with Coronary Disease, *Progr. Cardiovasc. Dis.*, **18**:459, 1976.

Costill, D. L., J. Daniels, W. Evans, W. Fink, G. Krahenbuhl, and B. Saltin: Skeletal Muscle Enzymes and Fiber Composition in Male and Female Track Athletes, *J. Appl. Physiol.*, **40**:149, 1976.

Davies, C. T. M., and A. J. Sargeant: Effects of Training on the Physiological Responses to One- and Two-leg Work, *J. Appl. Physiol.*, **38**:377, 1975.

Deitrick, J. E., G. D. Whedon, and E. Shorr: Effects of Immobilization upon Various Metabolic and Physiologic Functions of Normal Men, *Am. J. Med.*, **4**:3, 1948.

DeLorme, T. L., and A. L. Watkins: Techniques of Progressive Resistance Exercise, *Arch. Phys. Med.*, **29**:263, 1948.

Dill, D. B., E. E. Phillips, Jr., and D. MacGregor: Training: Youth and Age, *Am. N.Y. Acad. Sci.*, **134**:760, 1966.

Eckstein, R. W.: Effect of Exercise and Coronary Artery Narrowing on Coronary Collateral Circulation, *Circulation Res.*, **5**:230, 1957.

Edström, L., and B. Ekblom: Differences in Sizes of Red and White Muscle Fibres in Vastus Lateralis of Musculus Quadriceps Femoris of Normal Individuals and Athletes. Relation to Physical Performance, *Scand. J. Clin. Lab. Invest.*, **30**:175, 1972.

Edwards. H. T., L. Brouha, and R. T. Johnson: Effect de l'entrainement sur le taux de l'acide lactique au cours du travail musculaire, *Le Travail Humain*, **8**:1, 1939.

Ekblom, B.: Effect of Physical Training on Oxygen Transport System in Man, *Acta Physiol. Scand.*, Suppl. 328, 1969.

Ekblom, B., P.-O. Åstrand, B. Saltin, J. Stenberg, and B. Wallström: Effect of Training on Circulatory Response to Exercise, *J. Appl. Physiol.*, **24**:518, 1968.

Elsner, R. W., and L. D. Carlson: Postexercise Hyperemia in Trained and Untrained Subjects, *J. Appl. Physiol.*, **17**:436, 1962.

Erickson, L., E. Simonson, H. L. Taylor, H. Alexander, and A. Keys: The Energy Cost of Horizontal and Grade Walking on the Motordriven Treadmill, *Am. J. Physiol.*, **145**:391, 1946.

Eriksson, B. O., A. Lundin, and B. Saltin: Cardiopulmonary Function in Former Girl Swimmers and the Effects of Physical Training, *Scand. J. Clin. Lab. Invest.*, **35**:135, 1975.

Fowler, W. M., S. R. Chowdhury, C. M. Pearson, G. Gardner, and R. Bratton: Changes in Serum Enzyme Levels after Exercise in Trained and Untrained Subjects, *J. Appl. Physiol.*, **17**:943, 1962.

Fox, S. M., III, and W. L. Haskell: Population Studies, *Can. Med. Ass. J.*, **96**:806, 1967.

Freedman, M. E., G. L. Snider, P. Brostoff, S. Kimelblot, and L. N. Katz: Effects of Training on Response of Cardiac Output to Muscular Exercise in Athletes, *J. Appl. Physiol.*, **8**:37, 1955.

Frick, M. H., A. Konttinen, and H. S. S. Sarajas: Effects of Physical Training on Circulation at Rest and during Exercise, *Am. J. Cardiol.*, **12**:142, 1963.

Garbus, J., B. Highman, and P. D. Altland: Serum Enzymes and Lactic Dehydrogenase Isoenzymes after Exercise and Training in Rats, *Am. J. Physiol.*, **207**:467, 1964.

Gisolfi, D., S. Robinson, and E. S. Turrell: Effects of Aerobic Work Performed during Recovery from Exhausting Work, *J. Appl. Physiol.*, **21**:1767, 1966.

Goldberg, A. L., J. D. Etlinger, D. F. Goldspink, and C. Jeblecki: Mechanism of Work-Induced Hypertrophy of Skeletal Muscle, *Med. and Sci. in Sports*, **7**:185, 1975.

Golding, L.: Effects of Physical Training upon Total Serum Cholesterol Levels, *Res. Quart.*, **32**:499, 1961.

Gollnick, P. D., R. B. Armstrong, C. W. Saubert IV, K. Piehl, and B. Saltin: Enzyme Activity and Fiber Composition in Skeletal Muscle of Untrained and Trained Men *J. Appl. Physiol.*, **33**:312, 1972.

Gollnick, P. D., and L. Hermansen: Biochemical Adaptations to Exercise: Anaerobic Metabolism, in J. H. Wilmore (ed.), "Exercise and Sport Sciences Reviews," vol. 1, p. 1, Academic Press, Inc., New York, 1973.

Gollnick, P. D., K. Piehl, and B. Saltin: Selective Glycogen Depletion Pattern in Human Muscle Fibres after Exercise of Varying Intensity and at Varying Pedalling Rates, *J. Physiol.*, **241**:45, 1974.

Gonyea, W., G. C. Ericson, and F. Bonde-Petersen: Skeletal Muscle Fiber Splitting Induced by Weight-Lifting Exercise in Cats, *Acta Physiol. Scand.*, **99**:105, 1977.

Grimby, G., and B. Saltin: Physiological Analysis of Physically Well-trained Middle-aged and Old Athletes, *Acta Med. Scand.*, **179**:513, 1966.

Gutmann, E., R. Beránek, P. Hník, and J. Zelená: Physiology of Neurotrophic Relations, *Proc. 5th Nat. Congr. Czech. Physiol. Soc.*, 1961.

Hall, V. E.: The Relation of Heart Rate to Exercise Fitness: An Attempt at Physiological Interpretation of the Bradycardia of Training, *Pediatrics*, **32**(Suppl.):723, 1963.

Hammett, V. B. O.: Psychological Changes with Physical Fitness Training, *Can. Med. Ass. J.*, **96**:764, 1967.

Hansen, J. W.: The Training Effect of Repeated Isometric Muscle Contractions, *Intern. Z. Angew. Physiol.*, **18**:474, 1961.

Hansen, J. W.: Effect of Dynamic Training on the Isometric Endurance of the Elbow Flexors, *Intern. Z. Angew. Physiol.*, **23**:367, 1967.

Hartley, L. H.: Growth Hormone and Catecholamine Response to Exercise in Relation to Physical Training, *Med. Sci. Sports*, **7**:34, 1975.

Hartley, L. H. et al.: Multiple Hormonal Responses to Graded Exercise in Relation to Physical Training, *J. Appl. Physiol.*, **33**:602, 1972.

Hartley, L. H., and B. Saltin: Reduction of Stroke Volume and Increase in Heart Rate after a Previous Heavier Submaximal Work Load, *Scand. J. Clin. Lab. Invest.*, **22**:217, 1968.

Hedman, R.: Fysiologiska Synpunkter på Cirkelträning, *Tidskrift i Gymnastik*, 87:1, 1960.

Helander, E.: Adaptive Muscular "Allomorphism," *Nature*, **182**:1035, 1958.

Henriksson, J.: Training Induced Adaptation of Skeletal Muscle and Metabolism during Submaximal Exercise, *J. Physiol.*, 1977 (in press).

Henriksson, J., and J. S. Reitman: Time Course of Changes in Human Skeletal Muscle Succinate Dehydrogenase and Cytochrome Oxidase Activities and Maximal Oxygen Uptake with Physical Activity and Inactivity, *Acta Physiol. Scand.*, **99**:91, 1977.

Herrlich, H. C., W. Raab, and W. Gigee: Influence of Muscle Training and of Catecholamines on Cardiac Acetylcholine and Cholinesterase, *Arch. Intern. Pharmacodyn.*, **129**:201, 1960.

Hollmann, W., and Th. Hettinger: "Sportmedizin—Arbeits-und Trainingsgrundlagen," F. K. Schattauer Verlag, Stuttgart, 1976.

Hollmann, W., and H. Venrath: Die Beeinflussung von Herzgrösse, maximaler O_2-Aufnahme und Ausdauergrenze durch ein Ausdauertraining mittlerer und hoher Intensität, *Der Sportarzt*, **14**:189, 1963.

Holloszy, J. O.: Biochemical Adaptations to Exercise in Aerobic Metabolism, in J. H. Wilmore (ed.), "Exercise and Sport Sciences Reviews," vol. 1, p. 45, Academic Press, Inc., New York, 1973.

Holloszy, J. O.: Adaptation of Skeletal Muscle to Endurance Exercise, *Med. Sci. Sports*, **7**:155, 1975.

Holloszy, J. O., and F. W. Booth: Biochemical Adaptations to Endurance Exercise in Muscle, *Ann. Rev. Physiol.*, **18**:273, 1976.

Holmdahl, D. E., and B. E. Ingelmark: Der Bau des Gelenkknorpels unter verschiedenen Funktionellen Verhältnissen, *Acta Anat.*, **6**:309, 1948.

Holmér, I.: Physiology of Swimming Man, *Acta Physiol. Scand.*, Suppl. 407, 1974.

Holmér, I., and P.-O. Åstrand: Swimming Training and Maximal Oxygen Uptake, *J. Appl. Physiol.*, **33**:510, 1972.

Holmgren, A., and T. Strandell: The Relationship between Heart Volume, Total Hemoglobin and Physical Working Capacity in Former Athletes, *Acta Med. Scand.*, **163**:149, 1959.

Hoogerwerf, S.: Elektrokardiographische Untersuchungen der Amsterdamer Olympiakämpfer, *Arbeitsphysiol.*, **2**:61, 1929.

Howald, H. and J. R. Poortmans (eds.): "Metabolic Adaptation to Prolonged Physical Exercise," Birkhäuser Verlag, Basel, 1975.

Ikai, M.: The Effects of Training on Muscular Endurance, *Proc. Int. Congr. Sport Sci*, p. 109, 1964.

Ikai, M., and T. Fukunaga: A Study on Training Effect on Strength per Unit Cross-Sectional Area of Muscle by Means of Ultrasonic Measurement, *Int. Z. angew. Physiol.*, **28:**172, 1970.

Ikai, M., and A. H. Steinhaus: Some Factors Modifying the Expression of Human Strength, *J. Appl. Physiol.*, **16:**157, 1961.

Ingelmark, B. E.: Der Bau der Sehnen während verschiedener Altersperioden und unter wechselnden funktionellen Bedingungen I, *Acta Anat.*, **6:**113, 1948.

Ingelmark, B. E.: Morpho-physiological Aspects of Gymnastic Exercises, *FIEP-Bull.*, **27:**37, 1957.

Jansson, E., and L. Kaijser: Muscle Adaptation to Extreme Endurance Training in Man, *Acta Physiol. Scand.*, **99,** 1977 (in press).

Jones, D. M., C. Squires, and K. Rodahl: Effect of Rope Skipping on Physical Work Capacity, *Res. Quart.*, **33:**236, 1962.

Karlsson, J., P.-O. Åstrand, and B. Ekblom: Training of the Oxygen Transport System in Man, *J. Appl. Physiol.*, **22:**1061, 1967.

Karlsson, J., L. Hermansen, G. Agnevik, and B. Saltin: Energikraven vid löpning, *Idrottsfysiologi*, Rapport 4, Framtiden, Stockholm, 1967b.

Karlsson, J., L.-O. Nordesjö, L. Jorfeldt, and B. Saltin: Muscle Lactate, ATP, and CP Levels during Exercise after Physical Training in Man, *J. Appl. Physiol.*, **33:**199, 1972.

Karvonen, M. J., E. Kentala, and O. Mustala: The Effects of Training on Heart Rate, *Am. Med. Exp. Fenn.*, **35:**307, 1957.

Kasch, F. W., W. H. Phillips, J. E. L. Carter, and J. L. Boyer: Cardiovascular Changes in Middle-aged Men during Two Years of Training, *J. Appl. Physiol.*, **34:**57, 1973.

Kearney, J. T., G. A. Stull, J. L. Ewing, Jr., and J. W. Strein: Cardiorespiratory Responses of Sedentary College Women as a Function of Training Intensity, *J. Appl. Physiol.*, **41:**822, 1976.

Kellermann, J.: Rehabilitation of Patients with Coronary Heart Disease, *Prog. Cardiovasc. Dis.*, **17:**303, 1975.

Kilbom, Å.: Physical Training in Women, *Scand. J. Clin. Lab. Invest.*, **28**(Suppl. 119), 1971.

Klissouras, V.: Prediction of Athletic Performance: Genetic Considerations, *Can. J. Appl. Sport Sci.*, **1:**195, 1976.

Knehr, C. A., D. B. Dill, and W. Neufeld: Training and Its Effects on Man at Rest and at Work, *Am. J. Physiol.*, **136:**148, 1942.

Lautu, L.: Physiological Study of Motion, *Advanced Management*, **22:**17, 1957.

Liere, E. J. van, and D. W. Northup: Cardiac Hypertrophy Produced by Exercise in Albino and in Hooded Rats, *J. Appl. Physiol.*, **11:**91, 1957.

Lopezs, A., R. Vial, L. Balart, and G. Arroyave: Effect of Exercise and Physical Fitness on Serum Lipids and Lipoproteins, *Atherosclerosis*, **20:**1, 1974.

Magel, J. R., G. F. Foglia, W. D. McArdle, B. Gutin, G. S. Pechar, and F. I. Katch: Specificity of Swim Training on Maximum Oxygen Uptake, *J. Appl. Physiol.*, **38:**151, 1975.

Man-i, M., K. Ito, and K. Kikuchi: Histological Studies of Muscular Training, *Res. Physical Education*, **11:**153, 1967.

Marpurgo, P.: Über Aktivitäts-Hypertrophie der willkürlichen Muskein, *Virchows Arch.*, **150:**522, 1897.

McMorris, R. O., and E. C. Elkins: A Study of Production and Evaluation of Muscular Hypertrophy, *Arch. Phys. Med.*, **35:**420, 1954.

Meldon, J. H.: The Theoretical Role of Myoglobin in Steady-state Oxygen Transport to Tissue and Its Impact upon Cardiac Output Requirements, *Acta Physiol. Scand.*, Suppl., 440, p. 93, 1976.

Milic-Emili, G., J. M. Petit, and R. Deroanne: Mechanical Work of Breathing during Exercise in Trained and Untrained Subjects, *J. Appl. Physiol.*, **17**:43, 1962.

Miller, P. B., R. L. Johnson, and L. E. Lamb: Effects of Four Weeks of Absolute Bed Rest on Circulatory Functions in Man, *Aerospace Med.*, **35**:1194, 1964.

Morgan, R. E., and G. T. Adamson: "Circuit Training," G. Bell and Sons, Ltd., London, 1962.

Müller, E. A., and Th. Hettinger: Über Unterschiede der Trainingsgeschwindigkeit atrophierter und normaler Muskeln, *Arbeitsphysiol.*, **15**:223, 1953.

Newman, E. V., D. B. Dill, H. T. Edwards, and F. A. Webster: The Rate of Lactic Acid Removal in Exercise, *Am. J. Physiol.*, **118**:457, 1937.

Nöcker, J., D. Lehmann, and G. Schleusing: Einfluss von Training und Belastung auf den Mineralgehalt von Herz und Skelettmuskel, *Intern. Z. Angew. Physiol.*, **17**:243, 1958.

Nygaard, E., H. Bentzen, M. Houston, H. Larsen, E. Nielsen, and B. Saltin: Capillary Supply and Morphology of Trained Human Skeletal Muscle, paper presented at the 27th International Congress of Physiological Sciences, Paris, 1977.

O'Shea, J. P.: "Scientific Principles and Methods of Strengths Fitness," 2d ed., Addison-Wesley Publ. Company, Reading, Mass., 1976.

Palladin, A., and D. Ferdmann: Über den Einfluss der Trainings der Muskeln auf ihren Kreatingehalt, *Hoppe-Seylers Z. Physiol. Chem.*, **174**:284, 1928.

Pařízková, J.: Composition and Exercise during Growth and Development, in G. L. Rarick (ed.), "Physical Activity, Human Growth and Development," p. 97, Academic Press, Inc., New York, 1973.

Petersen, F. B., H. Graudal, J. W. Hansen, and N. Hvid: The Effect of Varying the Number of Muscle Contractions on Dynamic Muscle Training, *Intern. Z. Angew. Physiol.*, **18**:468, 1961.

Petrén, T.: Die totale Anzahl der Blutkapillaren im Herzen und Skelettmuskulatur bei Ruhe und nach langer Muskelübung, p. 165, *Verhandl. Anatom. Gesellsch* (Suppl. *Anat. Anz., 81*), 1936.

Petrén, T., T. Sjöstrand, and B. Sylvén: Der Einfluss der Trainings auf die Heufigkeit der Capillaren in Herz- und Skelettmuskulatur, *Arbeitsphysiol.*, **9**:376, 1936.

Piehl, K.: Glycogen Storage and Depletion in Human Skeletal Muscle Fibres, *Acta Physiol. Scand.*, Suppl. 402, 1974.

Pollock, M. L.: The Quantification of Endurance Training Program, in J. H. Wilmore (ed.), "Exercise and Sport Sciences Review," vol. 1, p. 155, Academic Press, Inc., New York, 1973.

Pyörälä, K., R. Kärävä, S. Punsar, P. Oja, P. Teräslinna, T. Partanen, M. Jääskeläinen, M-L. Pekkarinen, and A. Koskela: A Controlled Study of the Effects of 18 Months' Physical Training in Sedentary Middle-aged Men with High Indexes of Risk Relative to Coronary Heart Disease, in O. Andrée Larsen and R. O. Malmborg (eds.), "Coronary Heart Disease and Physical Fitness," Munksgaard, Copenhagen, 1971.

Rasch, P. J., and L. E. Morehouse: Effect of Static and Dynamic Exercise on Muscular Strength and Hypertrophy, *J. Appl. Physiol.*, **11**:29, 1957.

Reindell, H., K. König, and H. Roskamm: "Funktionsdiagnostik des gesunden und kranken Herzens," Georg Thieme Verlag, Stuttgart, 1967.

Reuschlein, P. S., W. G. Reddan, J. Burpee, J. B. L. Gee, and J. Rankin: Effect of Physical Training on the Pulmonary Diffusing Capacity during Submaximal Work, *J. Appl. Physiol.*, **24**:152, 1968.

Robinson, S., and P. M. Harmon: The Lactic Acid Mechanism and Certain Properties of the Blood in Relation to Training, *Am. J. Physiol.*, **132**:757, 1941a.

Robinson, S., and P. M. Harmon: The Effects of Training and of Gelatin upon Certain Factors Which Limit Muscular Work, *Am. J. Physiol.*, **133**:161, 1941b.

Rodahl, K., N. C. Birkhead, J. J. Blizzard, B. Issekutz, Jr., and E. D. R. Pruett: Physiological Changes during Prolonged Bed Rest, in G. Blix (ed.), "Nutrition and Physical Activity," p. 107, Almqvist & Wiksell, Stockholm, 1967.

Rohter, F. D., R. H. Rochelle, and C. Hyman: Exercise Blood Flow Changes in the Human Forearm during Physical Training, *J. Appl. Physiol.*, **18**:789, 1963.

Roskamm, H.: Optimum Patterns of Exercise for Healthy Adult, *Can. Med. Ass. J.*, **22**:895, 1967.

Roskamm, H., H. Reindell, and K. König: "Körperliche Aktivität und Herz und Kreislauferkrankungen," Johann Ambrosius Barth, Munich, 1966.

Rowell, L. B.: Human Cardiovascular Adjustments to Exercise and Thermal Stress, *Physiol. Rev.*, **54**:75, 1974.

Saltin, B., and P.-O. Åstrand: Maximal Oxygen Uptake in Athletes, *J. Appl. Physiol.*, **23**:353, 1967.

Saltin, B., B. Blomqvist, J. H. Mitchell, R. L. Johnson, Jr., K. Wildenthal, and C. B. Chapman: Response to Submaximal and Maximal Exercise after Bed Rest and Training, *Circulation*, **38**(Suppl. 7): 1968.

Saltin, B., L. H. Hartley, Å. Kilbom, and I. Åstrand: Physical Training in Sedentary Middle-aged and Older Men, *Scand. J. Clin. Lab. Invest.*, **24**:323, 1969.

Saltin, B., J. Henriksson, E. Nygaard, P. Andersen, and E. Jansson: Fiber Types and Metabolic Potentials of Skeletal Muscles in Sedentary Man and Endurance Runners, *Annuls N.Y. Acad. Sci.*, 1977 (in press).

Saltin, B., K. Nazar, D. L. Costill, E. Stein, E. Jansson, B. Essén, and P. D. Gollnick: The Nature of the Training Response; Peripheral and Central Adaptations to One-Legged Exercise, *Acta Physiol. Scand.*, **96**:289, 1976.

Scheuer, J.: Physical Training and Intrinsic Cardiac Adaptations, *Circulation*, **47**:677, 1973.

Schoop, W.: Bewegungstherapie bei peripheren Durchblutungsstörungen, *Med. Welt*, **10**:502, 1964.

Schoop, W.: Auswirkungen gesteigerter körperlicher Aktivität auf gesunde und krankhaft veränderte Extremitätsarterien, in Roskamm et al. (eds.), "Körperliche Aktivität und Herz und Kreislauferkrankungen," p. 33, Johann Ambrosius Barth, Munich, 1966.

Shephard, R. J., and K. H. Sidney: Effects of Physical Exercise on Plasma Growth Hormone and Cortisol Levels in Human Subjects, in J. H. Wilmore, and J. F. Keogh (eds.), "Exercise and Sport Sciences Reviews," vol. 3, p. 1, Academic Press, Inc., New York, 1975.

Siebert, W. W.: Untersuchungen über Hypertrophie des Skelettmuskels, *Z. Klin. Med.*, **109**:350, 1929.

Sjöstrand, T. (ed.): "Clinical Physiology," Svenska Bokförlaget, Stockholm, 1967.

Skinner, J. S., K. O. Holloszy, and T. K. Cureton: Effects of a Program of Endurance Exercises on Physical Work, *Amer. J. Cardiol.*, **14**:747, 1964.

Steinhaus, A. H.: Chronic Effects of Exercise, *Physiol. Rev.*, **13**:103, 1933.

Stelmach, G. E. (ed.): "Motor Control, Issues and Trends," Academic Press Inc., New York, 1976.

Tabakin, B. S., J. S. Hanson, and A. M. Levy: Effect of Physical Training on the Cardiovascular and Respiratory Response to Graded Upright Exercise in Distance Runners, *Brit. Heart J.*, **27:**205, 1965.

Taylor, H. L., A. Henschel, J. Brozek, and A. Keys: Effects of Bed Rest on Cardiovascular Function and Work Performance, *J. Appl. Physiol.*, **2:**223, 1949.

Tepperman, J., and D. Pearlman: Effects of Exercise and Anemia on Coronary Arteries of Small Animals as Revealed by the Corrosion-Cast Technique, *Circulation Res.*, **9:**576, 1961.

Terjung, R. L.: Muscle Fiber Involvement during Training of Different Intensities and Durations, *Am. J. Physiol.*, **230:**946, 1976.

Terjung, R. L., and W. W. Winder: Exercise and Thyroid Function, *Med. Sci. Sports*, **7:**20, 1975.

Thistle, H. G., H. J. Hislop, M. Moffroid, and E. W. Lowman: Isokinetic Contraction: A New Concept of Resistive Exercise, *Arch. Phys. Med. Rehabil.*, **48:**279, 1967.

Thörner, W.: Neue Beiträge zur Physiologie des Trainings, *Arbeitsphysiol.*, **14:**95, 1949.

Thorstensson, A.: Muscle Strength, Fibre Types and Enzyme Activities in Man, *Acta Physiol. Scand.*, Suppl. 443, 1976.

Tipton, C. M., W. C. Eastin, and R. A. Carey: Evaluation of Training and Detraining Effects in Dogs, *The Physiologist*, **14:**245, 1971.

Tipton, C. M., R. D. Matthes, and D. S. Sandage: In Situ Measurement of Junction Strength and Ligament Elongation in Rats, *J. Appl. Physiol.*, **37:**758, 1974.

Trautwein, W., W. J. Whalen, and E. Grosse-Schulte:Elektrophysiologischer Nachweis spontaner Freisetzung von Acetylcholin in Vorhof des Herzens, *Pflügers Arch.*, **270:**560, 1960.

Vannotti, A., and M. Magiday: Untersuchungen zum Studium des Trainiertseins, *Arbeitsphysiol.*, **7:**615, 1934.

Vannotti, A., and H. Pfister: Untersuchungen zum Studium des Trainiertseins, *Arbeitsphysiol.*, **7:**127, 1934.

Viidik, A.: Biomechanics and Functional Adaptation of Tendons and Joint Ligaments, in F. G. Evans (ed.), "Studies on the Anatomy and Function of Bone and Joints," p. 17, Springer-Verlag OHG, Heidelberg, 1966.

Vogt, F. B., P. B. Mack, P. C. Johnson, and L. Wade, Jr.: Tilt Table Response and Blood Volume Changes Associated with Fourteen Days of Recumbency, *Aerospace Med.*, **38:**43, 1967.

Vokac, Z., and K. Rodahl: Maximal Aerobic Power and Circulatory Strain in Eskimo Hunters in Greenland, Nordic Council for Arctic Medical Research Report no. 1, 1977.

Whipple, G. H.: The Hemoglobin of Striated Muscle, 1, Variations Due to Age and Exercise, *Am. J. Physiol.*, **76:**693, 1926.

Wilhelmsen, L., H. Sanne, D. Elmfeldt, G. Grimby, G. Tibblin, and H. Wedel: A Controlled Trial of Physical Training after Myocardial Infarction, *Preventive Med.*, **4:**491, 1975.

Yakovlev, N. N.: Problem of Biochemical Adaptation of Muscles in Dependence on the Character of Their Activity, *J. Gen. Biol. USSR* (Eng. transl.), **19:**417, 1958.

Applied Work Physiology

CONTENTS

INTRODUCTION

FACTORS AFFECTING THE ABILITY TO PERFORM SUSTAINED PHYSICAL WORK

ASSESSMENT OF WORK LOAD IN RELATION TO WORK CAPACITY
Assessment of the maximal aerobic power   Assessment of the physical work
load   Assessment of the organism's response to the total stress of work

ENERGY EXPENDITURE OF WORK, REST, AND LEISURE
Classification of work   Daily rates of energy expenditure   Energy expenditure during
specific activities

FATIGUE
General physical fatigue   Local muscular fatigue

CIRCADIAN RHYTHMS AND PERFORMANCE
Circadian rhythms in the human   Shift work

EFFECTS OF MENSTRUATION

Chapter 13

Applied Work Physiology

INTRODUCTION

The practical application of the principles of work physiology involves the study of the functions of the organism subjected to the many stresses of muscular work. Since some degree of muscular activity is required in all kinds of work, even in most intellectual occupations, and in all expressions of life, work physiology is of interest not merely to those concerned with manual labor.

The main objective of the work physiologist is to make it possible for individuals to accomplish their tasks without undue fatigue so that at the end of the working day, they are left with sufficient vigor to enjoy their leisure.

With the development of mechanization, automation, and many work-saving devices, modern technology has contributed greatly to the elimination of much heavy physical work. Nevertheless, some heavy physical work, at least occasionally, is still obligatory in a number of occupations, such as commercial fishing, agriculture, forestry, construction, transportation, and many service occupations. In some cases, when the work load appears excessively high, it is evident that the only reason why the task can be accomplished is that it is performed by intermittent work, i.e., brief work periods interspaced with short periods of rest. However, the tendency is for the elimination of physical strain, while at the same time the need for increased output and greater efficiency through rationalization and automation has resulted in an accelerated tempo in most industrial operations. The short working week has resulted, on the whole, in a greater work intensity. The outcome is increasing nervous tension and mounting emotional stress. Consequently, the greatest problem in many industrial operations today is not the physical load, but rather, the mental stress and unfavorable working environment.

Thus, the task facing the work physiologist in the field is to assess the strain imposed on the working organism by the total stress of the work and the working environment. Since practical experience has shown that one cannot tax more than some 30 to 40 percent of one's maximal aerobic power during an 8-hour working day without developing subjective or objective symptoms of fatigue, one of the most obvious problems is to determine the ratio between work load and work capacity. If the burden placed upon the worker is too high in relation to the person's capacity for sustained physical work, fatigue invariably will develop. This is true whether the work in question involves the entire body (large muscle groups) or only part of it (small muscle groups). A basic task for the work physiologist must therefore be that of measuring the rate at which the work is being done, i.e., the work load, and of matching this rate with the worker's ability to perform the work.

FACTORS AFFECTING THE ABILITY TO PERFORM SUSTAINED PHYSICAL WORK*

The relationship between the work load and work capacity is affected, however, by a complicated interplay of many factors, internal as well as external, which must be taken into consideration (Fig. 13-1).

Ability to perform physical work basically depends on the ability of the muscle cell to transform chemically bound energy in the food into mechanical energy for muscular work, that is, into the energy-yielding processes in the

---

*Definition of terms and units: Power = ability to work; the faculty of performing work; capacity for performance; the rate of transfer of energy. Capacity = ability; maximum power output. Load = the burden placed upon the worker; the rate at which work is being done at any time.

Units: In order to facilitate the transition from old units to the new SI system (Le Système International d'Unités), both the conventional and the new units will be used.

The new unit for force (F) is newton (N). $F = m \cdot a$, where m = mass; a = acceleration. 1 N = 1 kg $\cdot$ m $\cdot$ s^{-2} = the force which gives the mass of 1 kg an acceleration of 1 m $\cdot$ s^{-2}. In the old system, the commonly used unit for force is kp (1 kp is the force acting on the mass of 1 kg at normal acceleration of gravity): 1 kp = 9.80665 N, or approximately 10 N.

The unit for work or energy is derived from the equation: $W = F \cdot L$ = force times distance. The unit for force is N and the unit for distance is m (meter); therefore the unit for work or energy is N $\cdot$ m = joule (J). 1000 J = 1 kJ; 1000 kJ = 1 MJ; 1 cal = 4.1868 J; 1 kcal = 4.1868 kJ.

For power (work/time) the following units apply:
  1 watt = 1 joule $\cdot$ s^{-1} = 6.12 kpm $\cdot$ min^{-1}
  9.81 watts = 1 kpm $\cdot$ s^{-1}
  16.35 watts = 100 kpm $\cdot$ s^{-1}
  1 hp (horsepower) = 736 watts = 75 kpm $\cdot$ s^{-1} = 4500 kpm $\cdot$ min^{-1}

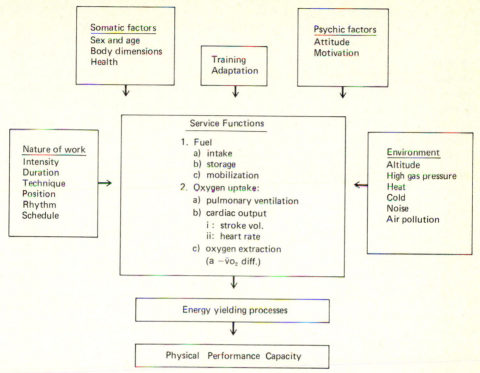

Figure 13-1 Factors affecting physical performance capacity.

muscle cell (see Chap. 2). This in turn depends on the capacity of the service functions that deliver fuel and oxygen to the working muscle fiber, i.e., on the nutritional state, nature and quality of the food ingested, frequency of meals (see Chap. 14), oxygen uptake including pulmonary ventilation (see Chap. 7), cardiac output and oxygen extraction (see Chap. 6), and the nervous and hormonal mechanisms which regulate these functions.

Many of these functions depend on somatic factors which may be partially genetically endowed; others may depend on sex, age, body dimensions, and state of health. In addition, physical performance is to a significant extent a function of psychological factors, notably motivation, attitude to work, and the will to mobilize one's resources for the accomplishment of the task in question. Several of these factors may be affected by training and adaptation.

Physical performance may also, directly or indirectly, be greatly influenced by factors in the external environment. Thus, air pollution may affect physical performance directly by increasing air-way resistance and thereby pulmonary ventilation, and indirectly, by causing ill health. Noise is a stress which not only may damage hearing, but also causes an elevation of heart rate and affects other physiological parameters that reduce physical performance. Cold weather, if severe, may in itself reduce physical performance because of numbness of the hands or lowered body temperature (the opposite effect of a warm-up prior to an athletic competition). But it may also involve the hobbling effect of bulky clothing and the slowing down of ordinary simple functions because of snow and ice. Heat, if intense, may greatly reduce endurance because of the need for more of the circulating blood volume to be devoted to the transportation of heat rather than to the transportation of oxygen (see Chap. 15), and because of the effect of dehydration often accompanying heat exposure as a result of loss of body fluids (sweating). Although high gas pressures, encountered in underwater operations in connection with modern offshore oil exploration, present new and rather unique problems for the work physiologist, the drastic reduction in physical work capacity at high altitudes is one of the best studied problems concerning the environmental effects on physical work capacity.

Finally, the nature of the work to be performed, apart from work intensity and duration, is of decisive importance when considering an individual's capacity to endure prolonged work stress. Since all life functions generally consist of rhythmic, dynamic muscular work in which work and rest, muscle contraction and relaxation, are interspersed at more or less regular, fairly short intervals, the ideal way to perform physical work is to perform it dynamically, with brief work periods interrupted by brief pauses. This routine will provide some rest during the actual work period, so the worker may avoid fatigue and exhaustion and be able to leave the workplace with sufficient vigor left over for the enjoyment of leisure. Similarly, the working position is also important in that working in a standing position may represent a greater circulatory strain than working in a sitting position does. Conversely, working in a standing position, which will permit the worker to move about and thereby vary the load on individual muscle groups and facilitate circulation, may at times be preferable. The working technique may be of major importance in conserving energy and in providing varied use of different muscle groups. The monotony of a working operation may be a stress for some individuals but a relief for others who can carry on the work more or less automatically while thinking about something else. In any case, the tempo of work performance may be extremely important and impose stresses which, in some instances, may be unbearable or harmful to the individual. Finally, the work schedule, including shift work, is a problem requiring increasing attention in modern industry.

ASSESSMENT OF WORK LOAD IN RELATION TO WORK CAPACITY

In view of the fact that maximal oxygen uptake varies greatly from one person to another, a work load that is fairly easy for one worker may be quite exhausting for another. Suppose two men are to perform the same task, such as carrying a heavy load uphill, requiring an energy expenditure of about 2 liters of oxygen per min. One of the individuals has a maximal oxygen uptake of 6.0 liters $\cdot$ min^{-1}, the other 2.0 liters $\cdot$ min^{-1}. In the first case, the individual is merely taxing 30 percent of his aerobic power. Consequently, he can carry on all day without fatigue, as is normally true when the work load is less than about 40 percent of the individual's maximal aerobic power. Furthermore, he can continue to cover more than half his energy expenditure from the oxidation of fat. The second man, on the other hand, is taxing his aerobic power maximally and can carry on for only a few minutes, during which time he is compelled to rely on his carbohydrate stores as a source of metabolic fuel (see Chap. 14).

Expression of the work load, as such, in absolute values (liter oxygen uptake per min) may therefore be quite meaningless. Instead, it should be expressed in percentage of the individual's maximal aerobic power. This means that the ratio between load and power should be assessed individually; that is, the individual's maximal oxygen uptake has to be determined and his or her rate of work has to be assessed. In general, the same principle also applies to the muscle groups which are engaged in the performance of the work in question, since only a certain percentage of the maximal muscle strength can be taxed without developing muscular fatigue (see Chap. 4).

Assessment of the Maximal Aerobic Power

A person's maximal aerobic power may be determined by direct measurement of the individual's maximal oxygen uptake, or estimated on the basis of data obtained from submaximal tests (see Chap. 11).

Assessment of the Physical Work Load

The physical work load may be assessed either by measurement of the oxygen uptake during the actual work operation or by indirect estimation of the oxygen uptake on the basis of the work pulse recorded during the performance of the work.

Measuring the Oxygen Uptake in a Typical Work Situation Since the validity of using oxygen uptake as a basis for measuring energy expenditure has been established, this indirect calorimetry has been used to determine the energy cost of a great variety of human activities. Figure 13-2 presents oxygen uptakes based on average values for various loads on a bicycle ergometer, compared with equivalent activities. With the development of highly portable devices for collecting expired air under field conditions and rapid methods of analyzing the oxygen and carbon dioxide content of the air samples, a vast body of knowledge of the energy cost of physical work has been accumulated.

In field studies, the classical method is to collect the expired air in Douglas bags carried on the subject's back. At present, other methods are available by which the volume of the expired air is measured with flow meters, and aliquot samples of the expired air are collected in a small rubber bladder (the Max Planck respirometer, Müller and Franz, 1952; the integrating motor pneumotachygraph, or IMP, Wolff, 1958). The expired air is then analyzed for O_2 and CO_2, using a conventional gas-analysis technique (Haldane or Micro-Scholander) or electronic O_2 and CO_2 analyzers. If accuracy is not too critical

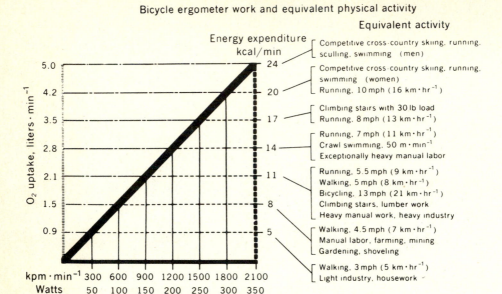

Bicycle ergometer work and equivalent physical activity

Figure 13-2 Bicycle ergometer work and equivalent physical activity.

(± 10 percent), it may be sufficient to analyze only the O_2 content by a portable O_2 analyzer. Various methods are described and discussed by Consolazio et al. (1963), Durnin and Passmore (1967), Banister and Brown (1968), Morehouse (1972), and Kamon (1974).

An example of measured oxygen uptake in typical commercial fishing operations is given in Fig. 13-3. It is evident that the measured oxygen uptake during the work performance represents the energy expenditure only at the time when the expired air sample is collected, and may not be representative for the work performed during the whole working day. Furthermore, it is a common experience that the test subject tends to be affected by the investigation, causing the test situation to be atypical. The test equipment is apt to affect heart rate, pulmonary ventilation, and oxygen uptake, and may hamper the actual work operation.

In contrast to this, the indirect assessment of the work load on the basis of the continuously recorded heart rate reveals a general picture of the overall activity level during the entire working day. Moreover, on the basis of time-activity records for each subject, collected by an observer during the whole working day, it is possible to separate the different activities with respect

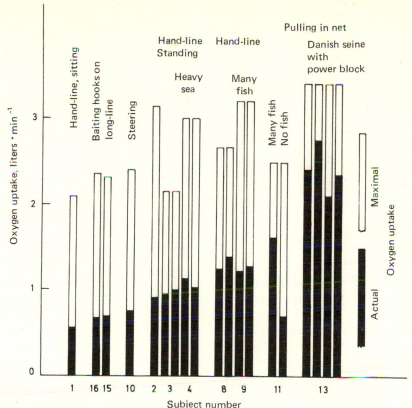

Figure 13-3 Measured oxygen uptake in typical fishing operations. *(From I. Åstrand et al., 1973.)*

to heart rate. Thus, the indirect assessment of work load based on the recorded work pulse may be preferable in many work situations.

Indirect Assessment by Recording the Heart Rate during Work In a given person, there is generally a linear relationship between O_2 uptake and heart rate. Therefore, the heart rate, under certain standardized conditions, may be used to estimate work load, if the work load–heart rate relationship has been established for the individual in question, if roughly the same large muscle groups are engaged in the work in both cases, and if environmental temperature, emotional stress, etc., are the same (see Chap. 6). pg. 189

The individual's circulatory response to work engaging large muscle groups may be measured on a bicycle ergometer. Starting at a low load such as 50 watts, the load is increased stepwise every 6 min, usually by 50 watts, until a heart rate of about 150 beats · min^{-1} is reached (Fig. 13-4). On the basis of the resulting line representing the relationship between the individual's work load

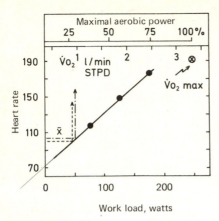

Figure 13-4 The individual relationship between the heart rate at different submaximal work loads and the predicted corresponding oxygen uptake (Åstrand, 1967) is established graphically (scale $\dot{V}_{O_2}$). The measured maximal oxygen uptake ($\dot{V}_{O_2}$ max) is used to construct another parallel scale which shows the load expressed in percentage of the individual's maximal aerobic power. The weighted mean ($\bar{x}$) of the continuous recording of the heart rate is then used to assess the approximate average oxygen uptake during work as well as the load expressed as percentage of the maximal aerobic power.

and heart rate, it is possible to estimate the work load from the heart rate recorded during a specific work situation in the field (Fig. 13-5, 13-6, showing typical examples).

The recording of the heart rate in the field is most conveniently accomplished with the aid of a portable miniature battery-operated tape recorder (such as those produced by Avionics, Philadelphia, U.S.A., or Hellige, Freiburg/Breisgau, Germany), monitoring the heart rate continuously in the form of a simplified electrocardiogram. The recorded heart rate, coupled with time-activity records, shows the degree and variations of the circulatory strain. By comparing the individual's heart rate during work in the field with the heart rate response to known, increasing work loads on a bicycle ergometer, the heart rate can be converted into the approximate oxygen uptake (Fig. 13-5).

The heart rate curves thus obtained may be replayed and transcribed on recording paper and graphically evaluated (I. Åstrand et al., 1973), or evaluated by computer analysis (Rodahl et al., 1974a).

It should be borne in mind that the estimation of oxygen uptake from recorded heart rate may be subject to considerable inaccuracy. However, in a field study by Rodahl et al. (1974a), 24 direct measurements of oxygen uptake with the Douglas bag method were compared with the oxygen uptake calculated from the simultaneously recorded heart rate in six fishermen. The calculated values deviated from the measured ones in both directions by no more than ± 15 percent (Fig. 13-7).

The reproducibility of the results from day to day in such field studies over a 3-day period was examined in connection with a study of fishermen by Rodahl et al. (1974a). They found a remarkable reproducibility of the day-to-day results in the same individual doing the same work. The weighted mean heart rates were practically the same for all 3 days, and the distribution curves, when superimposed, showed the same shape (Fig. 13-6).

It is thus clear that the use of the recorded heart rate in the field, compared with the heart rate at known work loads on the bicycle ergometer, may be used as a basis for the estimation of the work load when the work operation involves

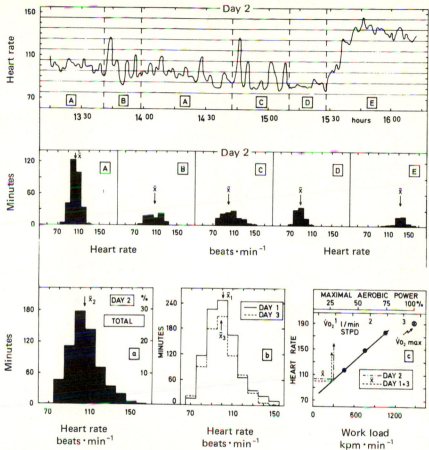

Figure 13-5 Example of analysis of the heart rate recorded at sea in a twenty-one-year-old net-fisherman (T.F.). Upper panel: Heart rate curve in the latter part of the second day observation period. Activities: A—Arranging and putting out net. B—Bleeding and cleaning the fish. C—Other, unspecified activities. D—Resting. E—Unloading the catch at the pier. Middle panel: Heart rate distribution curves and weighted means ($\bar{x}$) of the above listed five types of activity throughout the whole observation period. Lower panel: (a) Heart rate distribution curve and weighted mean ($\bar{x}_2$) for the whole observation period of the second day (06.45–16.50 hr). (b) Corresponding results from the first and third day of observation. (c) Relationship between work load, heart rate, estimated oxygen uptake ($\dot{V}_{O_2}$) and percent of maximal aerobic power in subject T.F. as assessed in the ergometer test. Estimation of the average oxygen uptake and percent of maximal aerobic power during the 3 days of observation on board by using the weighted means ($\bar{x}$) of the heart rate from sections (a) and (b) of the lower panel. *(From Rodahl et al., 1974a.)*

the use of the same large muscle groups as are used in the bicycle work. This comparison may not be feasible in work situations where mostly small muscle groups are involved, such as in arm work, since it is well established that the heart rate is higher than in leg work at the same work load. Although this is indeed the case in prolonged work lasting 5 min or more, Vokac et al. (1975)

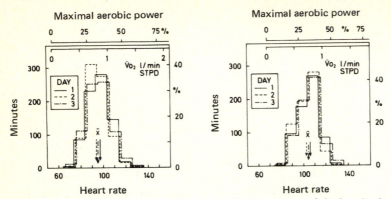

Figure 13-6 Weighted means ($\bar{x}$) and distribution curves of the heart rate and corresponding work load in two subjects during consecutive days of observation.
Left panel: forty-one-year-old long-line fisherman (K.S.).
Right panel: fifty-six-year-old catch handler (M.E.) on land. *(From Rodahl et al. 1974a.)*

have shown that the discrepancy in heart rate between arm and leg work at the same work load is not so marked at the onset of the work, but increases as the work proceeds (Fig. 13-8). Since most ordinary work operations involve a dynamic type of work with a rhythmic alteration between muscular contraction and relaxation, in which each period of work effort is rather brief, it appears that the use of the recorded heart rate as a basis for the estimation of work load may be acceptable even in many work situations involving arm work or the use of small muscle groups.

Summary   The continuous recording of the heart rate permits an uninterrupted collection of data which reflect the work load during the entire day. A quantitative numerical analysis of the recorded data, supplemented by visual analysis of the replayed heart rate curves (Fig. 13-5), permits a comprehensive and dynamic evaluation of the circulatory strain imposed by work loads of varying intensity. The use of a computer makes it possible to analyze large

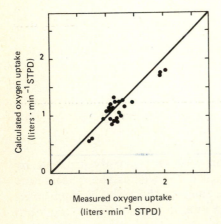

Figure 13-7 Relationship between oxygen uptake measured by the Douglas bag method and calculated from the simultaneously recorded heart rate. Twenty-four observations in six subjects during various fishing operations. *(From Rodahl et al., 1974a.)*

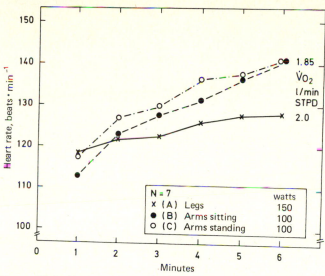

Figure 13-8 Heart rates in the first 6 min of cycling at 900 kpm · min⁻¹ (150 watts) (A) and of arm cranking at 600 kpm · min⁻¹ (100 watts) in sitting (B) and standing (C) positions. *(From Vokac et al., 1975.)*

series of observations with respect to the mean values, peak values, time distribution, and the occurrence and duration of excessively high heart rates, for example, over 50 percent of the heart rate reserve (the maximal heart rate minus the heart rate at rest). Since the heart rate response to one and the same work load varies with individuals, the circulatory strain is best expressed as a percentage of the subject's heart rate reserve (I. Åstrand, 1960). The results can be presented more conveniently and more graphically by converting the recorded heart rate individually into the corresponding estimated oxygen uptake (I. Åstrand, 1967). Estimated oxygen uptake serves, then, as the measure of the work load, and, expressed in percentage of the subject's maximal aerobic power, it indicates the relative degree of the exertion in the same way as the percentage of the heart rate reserve. In most cases, the reliability of the conversion is adequate for all practical purposes of field investigation (Rodahl et al., 1974a).

Assessment of the Organism's Response to the Total Stress of Work

The total stress imposed on the organism by a given work situation (physical as well as psychological) is generally reflected by a certain nervous and hormonal stimulation, more or less proportional to the degree of the stress.

Nervous Response An increased sympathicotonus, brought about by emotional as well as physical stress, will give rise to an accelerated heart rate, which consequently may serve as an index of stress response. It is thus well established that the heart rate increases linearly with increasing physical work load, provided there is no major change in the subject's emotional state. Also, the heart rate may vary markedly as a consequence of emotional stress in an individual who is at rest or subject to a constant light work load. However, in view of the fact that the heart rate in most ordinary life situations more closely reflects physical rather than emotional stress of work, and that it may simultaneously be greatly influenced by a number of other factors as well, such as the size of the muscle groups engaged in the work, venous return, static work, body position, and environmental temperature, heart rate data may be difficult to interpret as an index of total, not merely circulatory, stress response.

Hormonal Response   It is generally agreed that the total stress response of the organism is reflected by the sympatheticoadrenomedullary activity. This may be roughly assessed by measuring urinary excretion of epinephrine and norepinephrine by the method described by von Euler and Lishajko (1961) as modifed by Vaage (1974), using the resting night urine as base value. Expressed in $ng \cdot min^{-1}$, it may serve as a measure of occupational stress. A series of examples of urinary catecholamine excretion in different occupations is given in Fig. 13-9. A nearly tenfold increase in the epinephrine excretion and about a fourfold increase in norepinephrine excretion were observed during the workday, as compared with the excretion of coastal fishermen during the night (I. Åstrand et al., 1973). An even greater increase in norepinephrine excretion was observed in war college cadets during a strenuous battle course including marked sleep deprivation (Holmboe et al., 1975). The marked individual variation in catecholamine elimination under comparable stress situations is shown in Fig. 13-10 (Holmboe et al., 1975).

An increase both in epinephrine and norepinephrine may be due to a

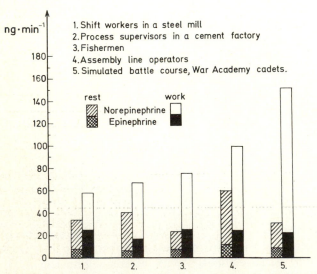

Figure 13-9   Urinary catecholamine elimination in different occupations.

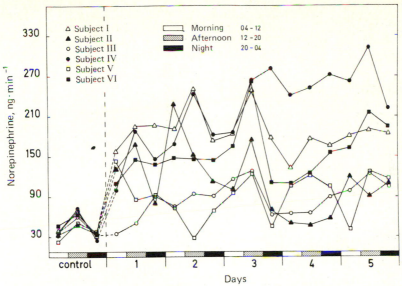

Figure 13-10   Urinary excretion of norepinephrine in war academy cadets during simulated battle course. *(From Holmboe et al., 1975.)*

number of single factors or a combination of them. Generally, both the circadian rhythm (Fröberg et al., 1972) and the change of body posture from the recumbent to the standing position (Sundin, 1956) increase the catecholamine excretion during the day. This increase may be markedly enhanced by the effect of physical exertion (Euler and Hellner, 1952), cold (Lamke et al., 1972), and emotional factors (Euler, 1964; Levi, 1967). The level of plasma catecholamines increases with both the duration and the severity of the muscular exertion (Banister and Griffiths, 1972).

ENERGY EXPENDITURE OF WORK, REST, AND LEISURE

Classification of Work

Evidently, the human being is not ideally suited to be a source of mechanical power, and in this respect cannot compete with modern mechanical devices, such as a bulldozer or a truck. The power output of an average man engaged in prolonged work over an 8-hour working day may amount to little more than .1 horsepower (1 horsepower ~ 750 watts). A horse may yield at least 7 times that amount, and an ordinary farm tractor, 70 horsepower. Yet humans must be physically active, or they will deteriorate. The solution, therefore, is to include some physical activity in the daily work, to provide the opportunity for varied use of the locomotive system, and to select a proper ratio between work and rest in order to supply adequate recuperation *during* work.

   In most instances, at least in the Western world with its advanced technology, excessively heavy work can easily be eliminated with technical aids; it is merely a matter of cost and priority. Establishing limits for permissible physical work loads is therefore of limited practical value. Of far greater importance to the worker today is the manner in which the work is being performed, the opportunity to influence the working situation and to govern one's own rate of work, the safety and the general atmosphere of the working environment, the arrangement of work shifts, etc. In most jobs in modern industry, the worker or operator is able to adjust the rate of work according to one's personal capacity. However, there are some exceptions, as when the work is performed by a team. Here, the weak have to keep up with the strong. In such teamwork, older workers, who are generally slower and who have a reduced physical working capacity, may be hard-pressed to keep pace with the younger members of the team. In any event, great individual differences do exist in physical working capacity, and practical experience has indicated that a work load taxing 30 to 40 percent of the individual's maximal oxygen uptake is a reasonable average upper limit for physical work performed regularly over an 8-hour working day. Similarly, no more than 40 percent of maximal muscle strength should be applied in repetitious muscular work in which the time of each muscular contraction is about one-half the time of each period of relaxation (Banister and Brown, 1968).

The physiological and psychological effects of a given energy output (per min, per 8 hr, per day) are determined by the individual's maximal aerobic power, size of the engaged muscle mass, working position, whether work is intermittent at a high rate or continuous at a lower intensity, and environmental conditions. In general, a person's subjective experience of a particular work load or rate of work is more closely related to heart rate than to oxygen uptake during the performance of the work, since the work pulse, in addition to the actual work load, also reflects emotional factors, heat, the size of the engaged muscle groups, etc.

Bearing these reservations in mind, the following identification of prolonged physical work, classified as to severity of work load and to cardiovascular response, may be of some use. These figures refer to average individuals twenty to thirty years of age, and can be used only as general guide lines in view of the vast individual variations in ability to perform physical work.

In terms of oxygen uptake:

Light work	up to 0.5 liter $\cdot$ min^{-1}
Moderate work	0.5–1.0 liter $\cdot$ min^{-1}
Heavy work	1.0–1.5 liters $\cdot$ min^{-1}
Very heavy work	1.5–2.0 liters $\cdot$ min^{-1}
Extremely heavy work	over 2.0 liters $\cdot$ min^{-1}

In terms of heart rate response:

Light work	up to 90 beats $\cdot$ min^{-1}
Moderate work	90–110 beats $\cdot$ min^{-1}
Heavy work	110–130 beats $\cdot$ min^{-1}
Very heavy work	130–150 beats $\cdot$ min^{-1}
Extremely heavy work	150–170 beats $\cdot$ min^{-1}

Daily Rates of Energy Expenditure

An estimate of daily energy expenditure is important not only for the calculation of energy need. It is also necessary in order to determine fairly accurately the level of physical activity of an individual or a group of individuals, in connection with attempts to ascertain the role of physical activity in health and disease, for example, in the prevention and treatment of ischemic heart disease. Such an estimate of energy output can be made by several methods: (1) the 24-hr recording of heart rate by portable miniature tape-recorders, as described earlier in this chapter; (2) estimation based on time-activity data and measurements of the energy cost of all pertinent activity; and (3) assessment (by food weighing) of daily food intake required to maintain body weight. All three of these methods are about equally accurate and reliable, with an error of no more than about 15 percent (Rodahl, 1960).

As expected, there is a wide individual variation in energy output depending on occupation, leisure activity, and attitude or individual proneness to physical activity in general. The range of daily rates of energy expenditure is from 1340 up to 5000 kcal (5.63 to 21.00 MJ) (Durnin and Passmore, 1967). About 2900 kcal · day^{-1} (12 MJ · day^{-1}) is a reasonable expenditure for a man who is not engaged in heavy manual labor but who is regularly active during leisure time. A reasonable figure for his wife would be about 2100 kcal · day^{-1} (9 MJ · day^{-1}) (Table 13-1), representing energy expenditure by reference man and woman, according to the Food and Nutrition Board (1964). It is possible, however, that these figures may be slightly too high because of the increasing inactivity and the sedentary life led by large segments of the population.

Available evidence shows that the effect of environmental temperature, as such, on the metabolic work is very small (Consolazio, et al., 1963). Gray et al. (1951) reported variations in metabolism up to 4 percent in men working at a fixed bicycle ergometer work load at ambient temperatures of $-15°C$ and $32°C$. This variation, however, is probably within the experimental error. It appears that any increase in the energy cost of work in a cold environment is primarily due to the additional energy cost of bodily movement in bulky clothing through difficult terrain, snow, etc. Nelson et al. (1948) found that metabolic heat production for a given amount of work remained unchanged in three men walking in seven hot environments between $32°C$ and $49°C$.

In the majority of professional activities (including office work), housework, light industry, laboratory and hospital work, and retail and distribution trade, the energy output is less than 5 kcal · min^{-1} (less than 20 kJ · min^{-1}, or less than 1 liter O_2·min^{-1}) (Fig. 13-11). In the building industry, agriculture, the iron and steel industries, and the armed services, there are many jobs which occasionally demand a caloric expenditure of up to 7.5 kcal · min^{-1} (or 30 kJ · min^{-1}) or even higher, particularly if mechanical aids are few and prefabricated materials are utilized to only a small extent. Still higher energy demands are found in fishing, forestry, mining, and dock labor, where figures up to or exceeding 10 kcal · min^{-1} (40 kJ · min^{-1}) have been reported.

Even if her energy output may be relatively low, a housewife, because of

Table 13-1

		Man						Woman			
	Time,	Rate		Total				Rate		Total	
Activity	hr	kcal · min⁻¹	kJ	kcal · min⁻¹	MJ		kcal · min⁻¹	kJ	kcal · min⁻¹	kcal · min⁻¹	MJ
Sleeping and lying*	8	1.1	4.6	540	2.3		1.0	4.2	480	2.0
Sitting†	6	1.5	6.3	540	2.3		1.1	4.6	420	1.7
Standing‡	6	2.5	10.5	900	3.8		1.5	6.3	540	2.3
Walking§	2	3.0	12.5	360	1.5		2.5	10.5	300	1.3
Other‖	2	4.5	18.8	540	2.3		3.0	12.5	360	1.5
Total	24			2880	12.2				2100	8.8

*Essentially basal metabolic rate plus some allowance for turning over or getting up or lying down.

†Includes normal activity carried on while sitting, e.g., reading, driving an automobile, eating, playing cards, and desk or bench work.

‡Includes normal indoor activities while standing and walking spasmodically in limited areas, e.g., performing personal toilet, moving from one room to another.

§Includes purposeful walking, largely outdoors, e.g., home to commuting station to work site, and other comparable activities.

‖Includes spasmodic activities in occasional sport exercises, limited stair climbing, or occupational activities involving light physical work. This category may include weekend swimming, golf, tennis, or picnic using 20 to 85 kJ · min⁻¹ (5 to 20 kcal) for limited time.

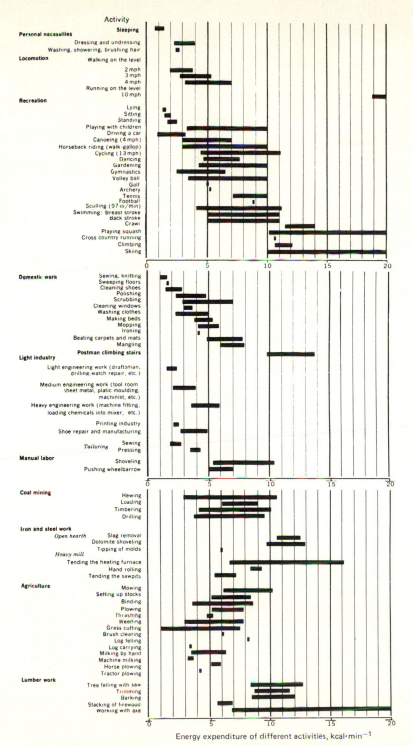

Figure 13-11 Energy expenditure of different activities. Bars denote range of data presented in literature (See also Karvonen, 1974.)

her lower maximal aerobic power, may strain herself as much doing domestic work as a lumberjack, farmer, commerical fisherman, or miner whose work requires a higher energy output, but who also has a higher maximal aerobic power.

The energy expenditure in recreational activity naturally covers the whole range from near resting values up to utilization of the full power of aerobic and anaerobic processes, depending on the type of activity and the degree of vigor with which it is pursued.

Various attempts have been made to establish maximal permissible limits for daily energy output for men working at the same task on a year-round basis (Banister and Brown, 1968). Lehmann (1953) suggested 4800 kcal · day^{-1} (20 MJ · day^{-1}) as the limit. Subtracting about 2300 kcal (9.66 MJ) for basal metabolism, eating and various basic necessities, leisure activity and travel to and from work, 2500 kcal (10.34 MJ) is left for the actual 8-hour work. Banister and Brown (1968) consider 2000 kcal (8.40 MJ) a more suitable load for heavy workers, giving an average rate of energy expenditure of 4.2 kcal · min^{-1} (18 kJ · min^{-1}).

Establishing such norms may be quite meaningless, however, in view of the great individual differences in physical work capacity or fitness. Furthermore, the level of activity in many industrial tasks is actually self-regulatory in that the rate of work and the spacing of rest pauses are set by the individual's level of physical fitness. In fact, in some cases, such as the older commercial fishermen studied by Rodahl et al. (1974a), the only way it is possible for a person to endure work loads close to the permissible physiological limits, day after day, year after year, is by working intermittently, with periods of high work intensity interspersed with frequent, brief rest periods.

In terms of strain imposed upon the worker, the peak load of the task is more important than the mean energy expenditure. A steel worker may expend 10 kcal · min^{-1} (40 kJ · min^{-1}) during 1 hour of shoveling gravel or dolomite, but during the rest of the 8-hour shift, energy output may be only 2 to 2.5 kcal · min^{-1} (8-10 kJ · min^{-1}). The 8-hour energy expenditure is then 600 + 900 = 1500 kcal (6.30 MJ). A worker with a job which demands a consistent rate of work with no peak loads may attain a higher 8-hour energy expenditure (e.g., 480 · 4 = 1920 kcal, (or 8.00 MJ), without requiring as strong a physique as does the steel worker. Heavy work or awkward working positions may often hamper the recruitment for certain types of work, even though these factors may be operating for only short periods of time. The same applies to many types of automated industrial operations where monotony or lack of personal influence on the process may appear boring.

Energy Expenditure during Specific Activities

Data for O_2 uptake, expressed as kilocalories per minute, of a variety of activities taken from various sources but mostly from Passmore and Durnin (1955) and Spitzer and Hettinger (1958), are summarized in Fig. 13-11.

Sleeping   In the past it has been difficult to obtain accurate values for energy expenditure in a sleeping subject because of the discomfort and restriction imposed by noseclip and mouthpiece, or the technical problem of leaks when a face mask is used. The use of the open-circuit method with a plastic hood covering the entire head of the patient and the Noyons basal metabolism apparatus (Issekutz et al., 1963) offers many distinct advantages when making metabolic measurements in a resting or sleeping individual. With this method, repeated samples can be taken throughout the observation period without disturbing the subject in any way.

Generally, one-third of the 24-hr period is spent in bed sleeping or resting. The energy spent in this way accounts for about one-tenth to one-quarter of the daily expenditure.

The metabolism, as a rule, may fall below the basal metabolic rate (BMR) when the fasting subject is asleep. The basal metabolic rate is the rate of energy metabolism in a resting individual 14 to 18 hours after eating. At the start of the night's sleep, however, the metabolism may be above the basal level because of the specific dynamic action (SDA) effect of the evening meal. These two factors thus tend to cancel each other out, so that the energy expenditure throughout the night is not far from that of the BMR value. In any case, a deviation of 10 percent above or below the normal basal level represents an error of less than about 3 percent of the total 24-hr energy expenditure. The BMR value may therefore be taken as a measure of the metabolic rate of a subject in bed, asleep or awake.

Personal Necessities   Under normal circumstances, an individual spends not more than 1 hr of the day in carrying out personal necessities, such as washing, shaving, brushing hair, cleaning teeth, dressing, and undressing. Energy expenditure values for such activities are summarized in Fig. 13-11.

Sedentary Work   For all practical purposes, the energy expenditure of mental work, including office work, etc., is not materially different from that of sitting or standing unless such occupations involve a great deal of physical activity, such as walking, bending, and opening drawers. Passmore and Durnin (1955) list a mean value of 1.6 kcal · min^{-1} (7 kJ · min^{-1}) for miscellaneous office work while sitting, and 1.8 kcal · min^{-1} (8 kJ · min^{-1}) while standing.

Although the brain utilizes a substantial part of the total O_2 uptake of the body at rest, it is well established that mental work requires only an insignificant rise in O_2 uptake, at least as long as the mental effort is not associated with markedly increased muscular tension or emotional stress (Benedict and Carpenter, 1909). Benedict and Benedict (1933) observed no substantial difference in metabolism at rest and during mental effort in six subjects engaged in 15-min periods of arithmetic exercises. Others (Eiff and Göpfert, 1952) have found an average rise in metabolism during mental efforts of as much as 11 percent, but this difference was apparently due to a concomitant rise in muscle tone. However, it is indeed interesting to note that although the central nervous system is extremely sensitive to lack of oxygen and lowered oxygen tension, an increase in intellectual functions is not associated with a significant increase in the overall O_2 uptake.

Housework   From Fig. 13-11 it is apparent that domestic work involves many tasks which may be classified as fairly heavy physical work, although modern equipment has contributed greatly to making life somewhat easier for the person keeping house today. This is all relative, however, since performance of physical work depends not only on the severity of the work load, but also on the physical work capacity of the individual. For this reason, the lighter work load of present-day domestic occupations, due to modernizations, may represent, relatively speaking, as heavy a load on present-day housekeepers as that of their grandmothers, if the latter were more physically fit.

Figure 13-12 presents an example of a continuous recording of heart rate during a day's work in a Swedish housewife, showing that a great deal of physical activity is involved even in modern housekeeping.

Light Industry   A fair amount of data is available regarding the energy cost of different kinds of manual labor and industrial tasks. With the development of automation, the physical work load of the industrial worker on the whole has been greatly reduced. This is evident from an early study by Kagan et al. (1928), who compared energy expenditure by men assembling machinery entirely by hand with those using a conveyer system. In the former case, the energy expenditure varied from 5.2 to 6.4 kcal $\cdot$ min^{-1} (22 to 27 kJ $\cdot$ min^{-1}); in the latter case, it varied between 1.8 and 4.7 kcal $\cdot$ min^{-1} (8 to 20 kJ $\cdot$ min^{-1}). According to Passmore and Durnin (1955) and Durnin and Passmore (1967), a wide variety of industrial activities, classified by them as light industry, demanded energy expenditure rates between 2 to 5 kcal $\cdot$ min^{-1} (8 to 20 kJ $\cdot$ min^{-1}) for men and 1.5 to 4 kcal $\cdot$ min^{-1} (6 to 17 kJ $\cdot$ min^{-1}) for women. The authors pointed out that further subdivision or classification is quite impracticable.

Recent unpublished studies carried out by the Institute of Work Physiology in Norway show that the work load in most light industries is below 30 percent of the worker's maximal oxygen uptake.

Manual Labor   A number of studies have shown that the energy expended during the performance of similar types of work may vary greatly, depending on the technique used in accomplishing the work. This is even true for relatively simple activities, such as carrying a load, depending on how the activity is performed. Bedale (1924) showed that when a person carried a load, the energy expenditure was minimal when a yoke across the shoulder was used; it was maximal when the load was carried on the hip under the arm. When carrying a load, the energy expenditure rises markedly with the increased speed

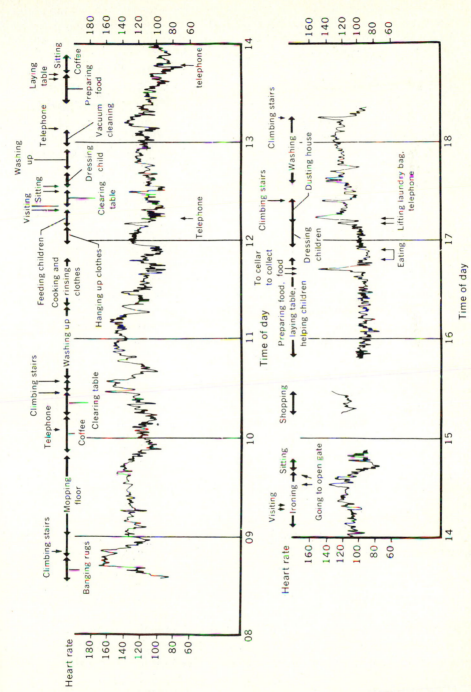

Figure 13-12 Heart rate in a housewife during an average day of housework.

of walking (Brezina and Kolmer, 1912; Cathcart et al., 1923). The energy cost of climbing up and going down stairs with a load is, according to Crowden (1941), 11 times that of walking on the level bearing the same load. In a study of the energy cost of transporting a load with the aid of a wheelbarrow, Hansson (1968) showed that the energy expenditure is higher the smaller and softer the wheel, and that a two-wheel cart is more efficient than a wheelbarrow with a single wheel.

There are few data available to assess the overall energy expenditure of building work. Accurate assessment of the activity is particularly difficult because of the variety of individual operations involved.

A rough estimate of the efficiency of a particular work operation may be achieved by measuring the energy cost based on oxygen uptake or heart rate, or by recording output at a constant O_2 uptake or heart rate. Thus the influence of the choice of tools on the rate of work and work effort was studied in a group of men engaged in the nailing of impregnated paper under standardized conditions (Hansson, 1968). The rate attained in nailing when using a stapler was 3 to 4 times higher than when using an ordinary hammer. There were no significant differences, however, in oxygen uptake per minute, quality of work, or estimated degree of fatigue when using the different tools. I. Åstrand et al. (1968) have further studied the energy cost of hammer nailing at three different heights: at bench level, into a wall at head level, and into a ceiling above the head. The number of hammer strokes did not differ significantly in the three situations. However, the number of nails driven per minute was lower when nailing into the wall (10.6 nails $\cdot$ min^{-1}) than when nailing into the bench (14.6), and still lower for nailing into the ceiling (4.5 nails $\cdot$ min^{-1}), indicating that the strokes became less powerful or were less well aimed when nailing into the wall and ceiling than when nailing into the bench. Nailing in the three positions resulted in an oxygen uptake of about 1.0 liter $\cdot$ min^{-1} in each case. The 11 subjects, all of whom were skilled carpenters, also performed leg work on a bicycle ergometer with the same oxygen uptake. Nailing into the wall or ceiling resulted in a greater elevation of the heart rate than nailing at bench level; the heart rate during bicycle exercise at a comparable oxygen uptake was lower (or 102 beats $\cdot$ min^{-1}) than for all types of nailing (130 beats $\cdot$ min^{-1} when nailing into the ceiling). It is interesting to note that leg exercise with an oxygen uptake of 1.4 liters $\cdot$ min^{-1} gave approximately the same rise in heart rate as nailing with an oxygen uptake of only 1.0 liter $\cdot$ min^{-1}. It should be noted that this finding refers to prolonged continuous work in a more or less steady-state situation, and not to intermittent work as described elsewhere in this chapter. The intra-arterially measured blood pressures during nailing were higher than during leg exercise at a given oxygen uptake (and probably cardiac output), the difference being most pronounced between nailing into the ceiling and bicycle work ($P < 0.01$).

These examples illustrate that (1) the work output may vary with the tools used and with the working positions, even though the energy expenditure is the same; and (2) the physiological effects of a given energy demand may vary considerably, depending on tools and techniques.

Studies of coal miners in different countries have shown a good general agreement for the work with pick and shovel. It appears that the energy expenditure of shoveling ranges, as a rule, from 6 to 7 kcal · min⁻¹ (25 to 29 kJ · min⁻¹). In a German study (Lehmann et al., 1950), the mean gross energy expenditure during actual coal mining was 5 kcal · min⁻¹ (21 kJ · min⁻¹), and the mean expenditure of energy per minute for the total time spent underground was 3.5 (15 kJ). Walking to and from the coal face in a stooping position may, according to Passmore and Durnin (1955), require as much as 10 kcal · min⁻¹ (42 kJ · min⁻¹). It appears that in spite of the increased mechanization, mining is still hard physical work. Studies in the iron and steel industry (Lehmann et al., 1950; Christensen, 1953) show wide variations in energy expenditure, so that generalizations are not justified. Although some of the workers periodically may work very hard, many are doing only light work in most modern plants.

Several studies have verified the general impression that farming is hard work, at least in the busy season, especially where the advantage of mechanization is not available (Durnin and Passmore, 1967). Milking by hand requires about 4.5 kcal · min⁻¹ (18 kJ · min⁻¹) as against 3.5 kcal · min⁻¹ (15 kJ · min⁻¹) for machine milking. Horse ploughing entails 6 kcal · min⁻¹ (25 kJ · min⁻¹), whereas tractor ploughing requires less than 5 kcal · min⁻¹ (21 kJ · min⁻¹) (Hettinger et al., 1953a and b).

In Norwegian coastal fishermen, the average energy expenditure of all activities on board during the whole day at sea amounted to 34 to 39 percent of the fisherman's maximal aerobic power, with occasional peaks up to 80 percent (Rodahl, et al., 1974a). The heart rate exceeded 50 percent of the fisherman's heart rate reserve for 9 to 23 percent of the time. The most strenuous activities were pulling in the seine with a power block (oxygen uptake up to 2.7 liters · min⁻¹) and unloading the catch, taxing the subjects by more than 50 percent of their maximal aerobic power for two-thirds of the duration of these activities. In general, it appears that the energy expenditure of commercial fishermen during the active fishing season may reach levels as high as about 5000 kcal · day⁻¹ (21 MJ · day⁻¹).

It is generally recognized that lumber work involves heavy expenditure of energy. In fact, lumbering is probably the hardest form of physical work, requiring sometimes as much as 6000 kcal · day⁻¹ (25 MJ · day⁻¹) (Lundgren, 1946). Hansson (1965) noted that lumberjacks with very high earnings were characterized by a particularly high maximal aerobic power compared with that of the average earner. With regard to muscle strength and precision in a variety of standardized work operations, there was no significant difference between the two categories. Because of his higher aerobic power, the top worker could attain a higher work output, he did not take as long breaks, and he became less tired at a given energy output, compared with his less productive colleague. It appears that workers involved in manual labor, who are more or less free to set their own pace, normally accept working with an energy output which is less than about 40 percent of their individual maximal aerobic power (I. Åstrand, 1967; Chap. 9).

Recreational Activities A general summary of measurements of energy expenditure of a variety of common recreational activities is given in Fig. 13-11, ranging from light indoor recreation and games to strenuous outdoor sports (see also Chap. 16). It should be pointed out that the listed figures can be taken as approximate values only, in view of the many factors which may have a profound influence upon the energy expenditure of such activities. Individuals pursue their recreations with very different degrees of vigor. Swimming, ball games, dancing, gardening, etc., can be enjoyed with an energy expenditure which ranges from low to very high. Playing volleyball or taking a swim does not involve the same degree of physical activity for all persons. Anyone, even the partially incapacitated person, can find a suitable recreation.

Data for energy expenditure of playing children have been published by Taylor et al. (1948, 1949) and Cullumbine (1950). The reported values range from 1 to 3 kcal · min^{-1} (4 to 13 kJ · min^{-1}). (Walking and running are discussed in Chap. 16.) Data for energy expenditure during different sports activities are given by Banister and Brown (1968) and show variations from 2 kcal · min^{-1} (playing pool) to 23 kcal · min^{-1} (8 to 100 kJ · min^{-1}) (sprinting).

Military Activities Data representing the energy cost of various military activities (Goldman, 1965; Rodahl et al, 1974b) are mostly gathered during training or maneuvers and not under realistic combat conditions. In general, it appears that the energy expenditure under these conditions very seldom exceeds 10 kcal · min^{-1} (40 kJ · min^{-1}), and that energy expenditure in excess of 7.0 kcal (30 kJ) seldom lasts more than 10 min.

FATIGUE

General fatigue may be a symptom of disease. It may also be psychological in nature, often associated with lack of motivation, lack of interest, low reserve capacity, etc. None of these problems will be dealt with here since they fall outside the scope of this book. We shall deal only with fatigue that has a functional physiological basis. Such physiological fatigue, thought to be a warning mechanism preventing overstraining the organism or part of it, may be general and systematic, or it may be local and, as a rule, muscular in nature.

General Physical Fatigue

Christensen (1960) defines physical fatigue as a state of disturbed homeostasis due to work and to work environment. This may give rise to subjective as well as objective symptoms.

It should be emphasized, however, that, so far, very little is known about the nature of this disturbed homeostasis. Thus, all that can be said with certainty at present is that the fall in the blood sugar observed in a fasting subject engaged in prolonged submaximal work lasting several hours causes disturbed homeostasis in the central nervous system leading to a feeling of fatigue as one of the symptoms of hypoglycemia. It is also clear that the accumulation of lactic acid in muscles engaged in intense work involving anaerobic metabolic conditions is a sign of disturbed homeostasis leading to symptoms of local fatigue. A further insight into the physiological nature of fatigue would be of utmost importance to the worker as well as to the athlete.

Subjective Symptoms of Fatigue   These symptoms may range from a slight feeling of tiredness to complete exhaustion. Attempts have been made to relate these subjective feelings to objective physiological criteria such as the accumulation of lactate in the blood. While such a relationship is often observed in connection with prolonged strenuous physical efforts, this relationship is not always present in prolonged light or moderate work. Subjective feelings of fatigue usually occur at the end of an 8-hr workday when the average work load exceeds 30 to 40 percent of the individual's maximal aerobic power, and certainly when the load exceeds 50 percent of the maximal aerobic power.

Objective Symptoms of Fatigue   I. Åstrand (1960) observed a rise in heart rate in subjects working at a load corresponding to about 50 percent of the individual's maximal oxygen uptake during a period of about 8 hours. However, since these experiments were carried out during the day (morning and afternoon), it is still an open question as to whether these changes are in fact due to, or partially due to, the development of fatigue, or may be the result of the normal circadian rhythms. In normal environment a rise in rectal temperature above 38°C is noted when the oxygen uptake exceeds 50 percent of the individual's maximal oxygen uptake (I. Åstrand, 1960; Chap. 15).

Local Muscular Fatigue

This subject is discussed in considerable detail in Chap. 4. The practical application of this discussion may be summarized as follows: Both in static and dynamic muscular work, endurance is related to the developed tension expressed in percentage of the maximal tension which the muscle can develop. Thus, in isometric muscular contraction (see Fig. 4-30), a 50 percent load can be maintained for about 1 minute, while the contraction may be maintained almost indefinitely as long as the load is less than 15 percent of the maximal force of the muscle. Similar relationships exist in the case of rhythmic contractions (see Fig. 4-31). Thus, if the load corresponds to about 80 percent of maximal strength, only 10 contractions per min can be maintained in a steady state, but if the load is reduced to about 60 percent of maximal strength, a contraction rate of $30 \cdot min^{-1}$ can be maintained. Therefore the stronger the muscles, the greater load they can endure without developing muscular fatigue. But, in any situation, the load and the rate of contraction have to be adjusted according to the strength of the muscle in order to avoid muscular fatigue.

CIRCADIAN RHYTHMS AND PERFORMANCE

Circadian Rhythms in the Human

In human beings, a variety of physiological functions, such as heart rate, oxygen uptake, rectal temperature, and urinary excretion of potassium and catecholamines, show distinct rhythmic changes in the course of a 24-hr period, with the values falling to their lowest during the night (low dip around 4 A.M.) and rising during the day, reaching their peak in the afternoon. This phenomenon is known as circadian rhythms, and is thought to be regulated by several separately operating biological clocks. It occurs in most individuals, although there are apparently a few exceptions; some individuals show reversed rhythms, the rectal temperature, for example, being highest at night (Folk, 1974).

These rhythmic changes in physiological functions have been found to be associated with changes in performance. This relationship appears to exist especially in the case of rectal temperatures and performance. In general, the lowest performance is observed early in the morning (about 4 A.M.). Thus, the delay in answering calls by switchboard operators on night shift was twice as long between 2 and 4 A.M. as during the daytime (Colquhoun, 1971). A similar relationship may exist also in the case of athletic performance. Thus, A. Rodahl et al. (1976), studying the performance of top swimmers who competed under comparable conditions early in the morning and late in the evening, found that the swimmers performed significantly better in the evening than in the morning ($P < .001$).

These findings show that circadian rhythms must be considered when interpreting the results from prolonged physiological experiments and when performing fitness tests in athletes at different times of the day.

Shift Work

The fact that human beings are "day-animals" and that some of their basic physiological functions which are associated with their performance capacities are subject to circadian rhythmic changes, suggests that humans may not ideally be suited for night work. Nonetheless, shift work has been practiced for generations in one form or another. Yet, little precise information is available as to what effects shift work has on physiological functions or physical performance, and there is no general agreement as to what type of shift work or work schedule is to be preferred. Most of the available information refers to clinical, social, or psychological aspects of shift work (Aanonsen, 1964). A review of the literature indicates that the health of shift workers in general is good in spite of such complaints as loss of sleep, disturbance of appetite and digestion, and a high rate of ulcers. The social and domestic effects of shift work represent greater problems than do the physiological effects. The results of studies pertaining to the effects on productivity are conflicting, as are results concerning accident rates. Absenteeism because of illness appears to be lower among shift workers than among day workers. It has been suggested that the physiological and biological effects are probably related to circadian rhythms rather than to work schedule. To what extent such circadian rhythms are related to health, performance, and a feeling of well-being is still undetermined.

Systematic studies of men engaged in rotating shift work and in continuous night work (Vokac and Rodahl, 1974, 1975) indicate that shift work does represent a physiological strain on the organism. It causes a desynchronization between functions such as body temperature and the biological clocks governing these functions. These studies show that there are considerable individual differences in the reaction to shift work, supporting the general experience that not everyone is equally suited for such work (some individuals consistently show relatively high values for urinary catecholamine elimination during shift work, whereas others have consistently low values). As judged by the catecholamine excretion, the greatest strain occurs when the worker, after several free days, starts work on night shift. The results of this study indicate that it is preferable, from a physiological standpoint, to distribute the free days more evenly throughout the entire shift cycle; that is, to alternate between work and free time regularly instead of assigning several consecutive free days.

The study of continuous night work shows that at the onset, body temperature and work pulse fell in the course of the night as if the subject were sleeping, although he was working (Fig. 13-13). It takes several weeks for this normal rhythm to be reverted (i.e., for an increase in body temperature in the course of the night work). In view of this, it would appear unrealistic to keep shift workers on continuous night work for prolonged periods in order to obtain the benefit of the reverted physiological reactions, since such a reversion takes too long to occur and is lost when interrupted by a single day.

Disturbances in circadian rhythms may give rise to considerable problems for those who have to travel by air from one continent to another in order to conduct business, to take part in political negotiations, or to participate in athletic competitions. It is an open question whether the indisposition or functional disturbances experienced after such intercontinental flights are in fact due to disturbed circadian rhythms, to loss of sleep, or to both. It is a common experience, however, that by being able to sleep during such travel, if necessary by using sleep-producing drugs, the individual can maintain a reasonable functional capacity in spite of the rapid shift from one time zone to another.

EFFECTS OF MENSTRUATION

Observations concerning the effect of menstruation on physical work capacity have yielded conflicting results (Gamberale et al., 1975). In general it appears that a subjectively perceived exertion may be more pronounced during the menstrual cycle, while no change in heart rate or oxygen uptake is apparent

476

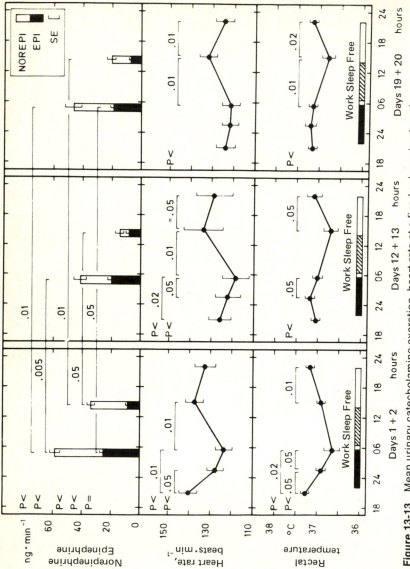

Figure 13-13 Mean urinary catecholamine excretion, heart rate at a fixed submaximal work load, and rectal temperature in four steel mill workers during 3 weeks continuous night shifts. *(From Vokaz and Rodahl, 1975.)*

during the different phases of the menstrual cycle. Furthermore, Olympic gold medals have been attained by menstruating women athletes. (For further references, see Drinkwater, 1973.)

REFERENCES

Aanonsen, A.: "Shift Work and Health," Universitetsforlaget, Oslo, 1964.

Åstrand, I.: Aerobic Work Capacity in Men and Women with Special Reference to Age, *Acta Physiol. Scand.*, **49**(Suppl. 169): 1960.

Åstrand, I.: Degree of Strain during Building Work as Related to Individual Aerobic Work Capacity, *Ergonomics*, **10**:293, 1967.

Åstrand, I., A. Guharay, and J. Wahren: Circulatory Responses to Arm Exercise with Different Arm Positions, *J. Appl. Physiol.*, **25**:528, 1968.

Åstrand, I., P. Fugelli, C. G. Karlsson, K. Rodahl, and Z. Vokac: Energy Output and Work Stress in Coastal Fishing, *Scand. J. Clin. and Lab. Invest.*, **31**:105, 1973.

Banister, E. W., and S. R. Brown: The Relative Energy Requirements of Physical Activity, Chap. 10, in "Exercise Physiology," Academic Press, Inc., New York, 1968.

Banister, E. W., and J. Griffiths: Blood Levels of Adrenergic Amines during Exercise, *J. Appl. Physiol.*, **33**:674–676, 1972.

Bedale, E. M.: "Comparison of the Energy Expenditure of a Woman Carrying Loads in Eight Different Positions," Medical Research Council Industrial Fatigue Research Board, no. 29, 1924.

Benedict, F. G., and T. M. Carpenter: "Influence of Muscular and Mental Work on Metabolism and Efficiency of the Human Body as a Machine," U. S. Department of Agriculture, Office of Experimental Stations, Bul. 208, 1909.

Benedict, F. G., and C. G. Benedict: "Mental Effort in Relation to Gaseous Exchange, Heart Rate and Mechanics of Respiration," Carnegie Institute, Washington, D.C., no. 446, 1933.

Brezina, E., and W. Kolmer: Über den Energieverbrauch bei der Geharbeit unter dem Einfluss verschiedener Geschwindigkeiten und Verschiedener Belastungen, *Biochem. Z.*, **38**:129, 1912.

Cathcart, E. P., D. T. Richardson, and W. Campbell: Maximum Load to be Carried by the Soldier, *J. Roy. Army Med. Corps.*, **40**:435; **41**:12; **87**:161, 1923.

Christensen, E. H.: "Physiological Valuation of Work in the Nykoppa Iron Works," in W. F. Floyd and A. T. Welford (eds.), Ergonomic Society Symposium on Fatigue, p. 93, Lewis, London, 1953.

Christensen, E. H.: Muscular Work and Fatigue, Chap. 9 in K. Rodahl and S. M. Horvath (eds.), "Muscle as a Tissue," McGraw-Hill Book Company, New York, 1960.

Colquhoun, W. P.: Circadian Variations in Mental Efficiency, "Biological Rhythms and Human Performance," pp. 39–108, Academic Press, Inc., London, 1971.

Consolazio, C. F., R. E. Johnson, and L. J. Pecora: "Physiological Measurements of Metabolic Functions in Man," McGraw-Hill Book Company, New York, 1963.

Crowden, G. P.: Stair Climbing by Postmen, *The Post* (London), p. 10, July 26, 1941.

Cullumbine, H.: Heat Production and Energy Requirements of Tropical People, *J. Appl. Physiol.*, **2**:640, 1950.

Drinkwater, B. L.: Physiological Responses of Women to Exercise in J. H. Wilmore (ed.), "Exercise and Sport Sciences Reviews," p. 125, Academic Press, Inc., New York, 1973.

Durnin, J. V. G. A., and R. Passmore: "Energy, Work and Leisure," William Heinemann, Ltd., London, 1967.

Eiff, A. W., and H. Göpfert: Ausmass und Ursachen der Energieumsatz-Veränderungen bei geistiger Arbeit, *Z. Ges. Exp. Med.*, **120**:72, 1952.

Euler, U. S. von,: Quantification of Stress by Catecholamine Analysis, *Clinical Pharmacology and Therapeutics*, **5**:398–404, 1964.

Euler, U. S. von, and S. Hellner: Excretion of Noradrenaline and Adrenaline in Muscular Work, *Acta Physiol. Scand.*, **26**:183–191, 1952.

Euler, U. S. von, and F. Lishajko: Improved Technique for the Fluorimetric Estimation of Catecholamines, *Acta Physiol. Scand.*, **51**:348–356, 1961.

Folk, G. Edgar, Jr.: "Textbook of Environmental Physiology," 2d ed., Lea & Febiger, Philadelphia, 1974.

Food and Nutrition Board: "Recommended Dietary Allowances," Nat. Acad. Sci., Nat. Res. Coun., pub. 1146, Washington, D.C., 1964.

Fröberg, J., C. G. Karlsson, L. Levi, and L. Lidberg: Circadian Variations in Performance, Psychological Ratings, Catecholamine Excretion, and Diuresis during Prolonged Sleep Deprivation, *Intern. J. of Psychobiol.*, **2**:23–36, 1972.

Gamberale, F., L. Strindberg, and I. Wahlberg: Female Work Capacity during the Menstrual Cycle: Physiological and Psychological Reactions, *Scand. J. Work Environ. & Health*, **1**:120, 1975.

Goldman, R. F.: Energy Expenditure of Soldiers Performing Combat Type Activities, *Ergonomics*, **8**:322, 1965.

Gray, E. L., C. F. Consolazio, and R. M. Karl: Nutritional Requirements for Men at Work in Cold, Temperate and Hot Environments, *J. Appl. Physiol.*, **4**:270, 1951.

Hansson, J. E.: The Relationship between Individual Characteristics of the Worker and Output of Work in Logging Operations, *Studia Forestalia Suecia*, no. 29, Skogshögskolan, Stockholm, 1965.

Hansson, J. -E.: Work Physiology as a Tool in Ergonomics and Production Engineering, Al-Rapport 2, *Ergonomi och Produktionsteknik*, National Institute of Occupational Health, Stockholm, 1968.

Hettinger, T., and W. Wirths: Der Energieverbrauch beim Hand und Motorpflügen, *Arbeitsphysiol.*, **15**:41, 1953a.

Hettinger, T., and W. Wirths: Über die körperliche Beanspruchung beim Hand und Maschinemelken, *Arbeitsphysiol.*, **15**:103, 1953b.

Holmboe, J., H. Bell and N. Norman: Urinary Excretion of Catecholamines and Steroids in Military Cadets Exposed to Prolonged Stress, *Försvarsmedicin*, **11**:183, 1975.

Issekutz, B., Jr., N. C. Birkhead, and K. Rodahl: Effect of Diet on Work Metabolism, *J. Nutr.*, **79**:109, 1963.

Kagan, E. M., P. Dolgin, P. M. Kaplan, C. O. Linetzkaja, J. L. Lubarsky, M. F. Neumann, J. J. Semernin, J. S. Starch, and P. Spilberg: Physiologische Vergleichsuntersuchung der Hand-und Fleiss-(Conveyor) Arbeit, *Arch. Hyg. (Berlin)*, **100**:335, 1928.

Kamon, E.: Instrumentation for Work Physiology, *Transact. N.Y. Acad. Sci.*, no. 7, **36**:625, 1974.

Karvonen, M. J.: Work and Activity Classifications, in L. A. Larson (ed.), "Fitness,

Health, and Work Capacity," p. 38, Macmillan Publishing Co., Inc., New York, 1974.

Lamke, L. O., S. Lennquist, S. O. Liljedahl, and B. Wedin: The Influence of Cold Stress on Catecholamine Excretion and Oxygen Uptake of Normal Persons, *Scand. J. Clin. Lab. Invest.*, **30:**57–62, 1972.

Lehmann, G., E. A. Müller, and H. Spitzer: Der Kalorienbedarf bei gewerblicher Arbeit, *Arbeitsphysiol*, **14:**166, 1950.

Lehmann, G.: "Praktische Arbeitsphysiolgie," Thieme, Stuttgart, 1953.

Levi, L.: Sympatho-adrenomedullary Responses to Emotional Stimuli: Methodologic, Physiologic and Pathologic Considerations, in E. Bajusz (ed.), "An Introduction to Clinical Neuroendocrinology," pp. 78–105, S. Karger, New York, 1967.

Lundgren, N.: Physiological Effects of Time Schedule Work on Lumbar Workers, *Acta Physiol. Scand.*, **13**(Suppl. 41): 1946.

Morehouse, L. E.: "Laboratory Manual for Physiology of Exercise," The C. V. Mosby Company, St. Louis, 1972.

Müller, E. A., and H. Franz: Energieverbrauchsmessungen bei beruflicher Arbeit mit einer verbesserten Respirations-Gasuhr, *Arbeitsphysiol*, **14:**499, 1952.

Nelson, N. A., W. B. Shelley, S. M. Horvath, L. W. Eichna, and T. F. Hatch: Influence of Clothing, Work and Air Movement on the Thermal Exchanges of Acclimatized Men in Various Hot Environments, *J. Clin. Invest.*, **27:**209, 1948.

Passmore, R., and J. V. G. A. Durnin: Human Energy Expenditure, *Physiol. Rev.*, **35:**801, 1955.

Rodahl, A., M. O'Brien, and R. G. R. Firth: Diurnal Variation in Performance of Competitive Swimmers, *J. Sports Med. and Physical Fitness*, **16:**72, 1976.

Rodahl, K.: "Nutritional Requirements under Arctic Conditions," Norsk Polarinstitutt Skrifter No. 118, Oslo University Press, Oslo, 1960.

Rodahl, K., Z. Vokac, P. Fugelli, O. Vaage and S. Maehlum: Circulatory Strain, Estimated Energy Output and Catecholamine Excretion in Norwegian Coastal Fishermen, *Ergonomics*, **17:**585–602, 1974a.

Rodahl, K., T. Wessel-Aas, P. O. Huser and T. S. Nilsen: A Physiological Evaluation of a Waterproof, Partially Permeable Protective Suit against Chemical and Bacteriological Warfare, *Försvarsmedicin*, **10:**24, 1974b.

Spitzer, H., and Th. Hettinger: "Tafeln für Kalorienumsatz bei körperlicher Arbeit," REFA publication, Darmstadt, 1964.

Sundin, I.: The Influence of Body Posture on the Urinary Excretion of Adrenalin and Noradrenalin, *Acta Med. Scand.*, **154:**(Suppl. 313): 1956.

Taylor, C. M., M. W. Lamb, M. E. Robertson, and G. MacLeod: Energy Expenditure for Quiet Playing and Cycling of Boys 7 to 15 Years of Age, *J. Nutr.*, **35:**511, 1948.

Taylor, C. M., O. F. Pye, A. B. Caldwell, and E. R. Sostman: Energy Expenditure of Boys and Girls 9 to 11 Years of Age (1) Sitting Listening to the Radio (Phonograph), (2) Sitting Singing and (3) Standing Singing, *J. Nutr.*, **38:**1, 1949.

Vaage, O.: Fluorometric Determination of Epinephrine and Norepinephrine in 1 ml Urine Introducing Dithiotreitol and Boric Acid as Stability and Sensitivity Improving Agents of the Trihydroxyindole Method, *Biochem. Med.*, **9:**41–53, 1974.

Vokac, Z., and K. Rodahl: "A Study of Continuous Night Work at the Norwegian Steel Mill at Mo i Rana," Nordic Council for Arctic Medical Research Report, no. 10, 1974.

Vokac, Z., H. Bell, E. Bautz-Holter, and K. Rodahl: Oxygen Uptake/Heart Rate

Relationship in Leg and Arm Exercise, Sitting and Standing, *J. Appl. Physiol.*, **39**:54, 1975.

Vokac, Z., and K. Rodahl: Field Study of Rotating and Continuous Night Shifts in a Steel Mill, Proceedings of the Third International Symposium on Night- and Shiftwork in Dortmund, Oct. 28–31, 1974, in P. Colquhoun, S. Folkard, P. Knaut, and R. Rutenfranz (eds.), "Experimental Studies in Shiftwork," Westdeutscher Verlag, Opladen, 1975.

Wolff, H. S.: The Integrating Motor Pneumotachograph: A New Instrument for the Measurement of the Energy Expenditure by Indirect Calorimetry, *Quart. J. Exper. Physiol.*, **43**:270, 1958.

Nutrition and Physical Performance

14

CONTENTS

Chapter 14

Nutrition and Physical Performance

INTRODUCTION

To a certain extent we are a product of what we eat. Molecules and atoms in the food we ingest are used to build and maintain the different cells, tissues, and organs in the body. It is therefore fairly essential that the necessary elements of nutrition be included in our daily diet in order to provide the proper building material and supplies for our tissues.

Food is also fuel for the biological machinery of the body. The energy-containing foodstuffs we consume are oxidized in the cells with the aid of the oxygen we inhale. As a result of this process, all the energy we need for our existence and accomplishments is liberated.

It therefore appears reasonable to expect that nutrition may well play a role in physical performance. It is well known that undernutrition definitely does impair performance, but it is still an open question whether more than enough of an essential element of nutrition is better than just enough.

The question of what an athlete should eat in order to achieve superior performance is as old as the recorded history of organized sports. The practice of consuming large quantities of meat to replenish the supposed loss of muscular substances during heavy muscular work was first recorded in Greece during the fifth century B.C. (Christophe and Mayer, 1958). Two athletes, instead of eating the predominantly vegetarian diet of the time, adopted a regimen consisting of large quantities of meat, resulting in increased body bulk and weight. Many of the food taboos of primitive tribes are related to this question. The discovery of vitamins offered a new and promising area of experimentation by the more imaginative minds on the basis of the assumption that if a little of something is good, a great deal of it must be much better. More recently, protein tablets have figured prominently in dietary discussion, especially among weight lifters. Since the muscles consist mainly of protein, it was assumed that ingestion of excess protein might enhance muscle growth and improve strength, as the Greek pioneers believed some 2,000 years ago.

Repeated claims have been made by different workers over the last several decades that physical work performance of average persons can be significantly improved by special diets or dietary supplements (Simonson, 1951), and even greater improvements can thus be made in athletic performance (Mayer and Bullen, 1960). Yet, Mayer and Bullen conclude: "The concept that any well-balanced diet is all that athletes actually require for peak performance has not been superseded." It is the purpose of this chapter to examine whether this, in the light of more recent evidence, is still true. Water intake and the exchange of water will be discussed in Chap. 15.

NUTRITION IN GENERAL

We need food as building blocks for our tissues, and we need food for energy.

The most important building material is protein. It is essential for building new cells and tissues in growth and development, and for replacing parts of old cells which constantly are being broken down. There are different kinds of proteins, depending on their amino acid composition. Plant proteins are not identical with proteins of animal origin. The tissues of our body need some amino acids which we cannot live without because the body is unable to synthesize them. They are therefore vital, and are known as the essential amino acids. Meat, fish, eggs, milk, and cheese contain proteins with about the proper composition of amino acids. They are therefore the most suitable sources of protein.

An adult needs about 1 g protein of proper composition per kg body weight per day. This corresponds to about 70 g for a full-grown man. Athletes may perhaps have a somewhat greater protein requirement during periods of intense training, especially that involving muscle strength. However, the exact protein requirements of different categories of athletes in training have, so far, not been established by scientifically controlled metabolic balance studies. (FAO and WHO have in their recommendations of 1973 suggested that an intake of 0.57 g egg protein per kg body weight should maintain nitrogen balance in young, healthy individuals. See "Energy and Protein Requirements" Wld. Hlth Org. techn. Rep. Ser. 1973, No. 522. However, this recommended protein intake requires a relatively high-energy intake; see Garza et al., 1976.)

Minerals are also necessary for the maintenance of body structures, and we need a number of vitamins for a variety of catalytic processes in our biological machinery. However, food is also the fuel for the body machinery. Some of the ingested food is metabolized and used as soon as it is resorbed. But most of it is stored temporarily, mostly in the form of fat. This stored energy may then be mobilized as needed.

In the development of the ideal energy stores for the mobile animal, such as the human, nature has to meet a number of requirements. For reasons of portability, each molecule should carry a large amount of energy per unit weight. The material should be fitted into various oddly shaped spaces and compartments of the body. It should possess a great storage stability and, at the same time, be readily available and capable of being rapidly converted into oxidizable substrate when needed, without being spontaneously explosive. As in all other efficient operations, the overhead cost should be low, that is, the handling expenses, including costs of storage and transport, should be minimal.

As pointed out by Dole (1964), fat or triacylglycerol (triglycerides) meet these requirements remarkably well. They are high in energy content, and they are stable, yet readily mobilized. The amount of energy held per unit weight of any molecule depends on its content of oxidizable carbon and hydrogen. Oxygen in a molecule of stored energy merely adds dead weight, since oxygen atoms can be obtained from the air as needed. Fat, therefore, approaches maximal storage efficiency since it contains as much as 90 percent carbon and hydrogen and has an energy density of $39 \text{ kJ} \cdot \text{g}^{-1}$ (9.3 kcal). It is much superior in storage efficiency to carbohydrates, which contain only 49 percent carbon and hydrogen and have an energy density of only $17 \text{ kJ} \cdot \text{g}^{-1}$ (4.1 kcal). The difference is even greater when one considers that fat is deposited in droplets, while carbohydrate is deposited together with an appreciable amount of water, in mammals 2.7 g water per g dried glycogen (Weis-Fogh, 1967). In other words, hydration dilutes the energy density of glycogen to about $4 \text{ kJ} \cdot \text{g}^{-1}$ (1 kcal). Thus, carbohydrate is a rather inferior material as an energy store in terms of portability. The energy value of 1 g of adipose tissue, which does not consist of pure fat, is about 25 to 29 kJ (6 to 7 kcal). On the other hand, it should be pointed out that there is an almost 10 percent higher energy yield per liter of oxygen used when carbohydrate is combusted than when protein and fat are burned.

The different animal species have, through adaptation, met their own peculiar needs. In migrating fishes like the salmon and the eel, and in birds, fat constitutes the main source of energy. Weis-Fogh (1967) points out that owing to impaired weight economy, carbohydrate cannot sustain a bird in flight for more than a few hours, and the recorded endurance of 1 to 3 days observed in some typical migrants therefore depends almost exclusively on the utilization of fat mobilized from stored triglycerides.

DIGESTION

The digestion of food starts in the oral cavity where the food is broken up into smaller particles by the teeth during chewing, and is mixed with saliva. The secretion of saliva is brought about by reflexes. It is greatest when the food is dry and gritty. The saliva contains enzymes which split starch into maltose. This process is more complete the longer the food is chewed.

When the chewed food has attained the proper consistency (bolus), it is gathered by the tongue and pushed backward toward the pharynx. The swallowing reflex automatically provides for the swallowing of the bolus, whereby the food is moved by peristalsis down the esophagus into the stomach. The stomach is not a relaxed bag but a slightly constricted muscular container. It encompasses the stomach content by an even, gentle pressure and is expanded as the stomach is filled. This tension of the stomach wall apparently is connected with the sensation of satiety.

As the food is dissolved by the gastric juice which is secreted from numerous glands in the stomach mucosa, it is ejected into the duodenum. The gastric secretion is regulated by both neurogenic and hormonal mechanisms. The secretory glands may be excited by smell and taste, and possibly also by the mere thought of food. Fatty food, on the other hand, suppresses the production. The gastric juice causes the breakdown of proteins and the emulsification of fat.

The digestion of the food (chyme) in the small intestine is accomplished by the intestinal juice, which contains several digestive enzymes capable of breaking down carbohydrates, fat, and proteins into molecules which can be transported through the intestinal wall into the blood. The remainder is moved along by the intestinal peristalsis down into the large intestine and eventually expelled as feces.

The nutrient molecules which are resorbed through the intestinal wall are partially used at once as building blocks for the cells and tissues of the body, and as fuel. The rest is stored, partly in the form of glycogen, but for the greater part as fat in the adipose tissues around the guts, under the skin, and for a small part also in the muscles.

For a more comprehensive review, the reader is referred to the standard textbooks of physiology and nutrition. A simplified schematic presentation of the actions of the different digestive enzymes may be summarized as follows:

Sequence of protein digestion:

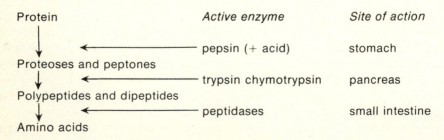

Protein	Active enzyme	Site of action
↓ ←	pepsin (+ acid)	stomach
Proteoses and peptones		
↓ ←	trypsin chymotrypsin	pancreas
Polypeptides and dipeptides		
↓ ←	peptidases	small intestine
Amino acids		

Sequence of fat digestion:

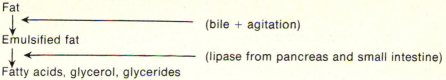

Fat
↓ ← ─────────────────────── (bile + agitation)
Emulsified fat
↓ ← ─────────────────────── (lipase from pancreas and small intestine)
Fatty acids, glycerol, glycerides

Sequence of carbohydrate digestion:

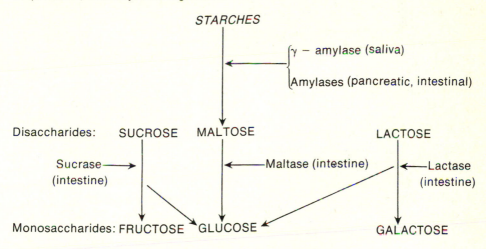

STARCHES

{γ – amylase (saliva)

Amylases (pancreatic, intestinal)

Disaccharides: SUCROSE   MALTOSE   LACTOSE

Sucrase ──→   ←──── Maltase (intestine)   ←── Lactase
(intestine)   (intestine)

Monosaccharides: FRUCTOSE   GLUCOSE ←   GALACTOSE

ENERGY METABOLISM AND THE FACTORS GOVERNING THE SELECTION OF FUEL FOR MUSCULAR WORK

Of the various nutrients in the food we eat, it is only the carbohydrates, fat, and proteins that can yield energy for muscular work. These three sources, however, do not contribute equally to the energy-yielding processes in the muscle cell. The fact that nitrogen excretion is not significantly increased during muscular work in the fed individual (Crittenden, 1904; Cathcart and Burnett, 1926; Hedman, 1957) shows that protein is not used as a fuel to any appreciable extent as long as the energy supply is adequate (Pettenkofer and Voit, 1866; Chaveau, 1896; Krogh and Lindhard, 1920). Under such conditions, proteins are used almost exclusively to replace those parts of the cells which are being broken down. It is therefore only when an individual is undernourished or fasting that proteins may be utilized as a source of metabolic energy at the expense of the cells and tissues of the body. From these facts, it is clear that the choice of fuel for the working muscle is limited to carbohydrate and fat. The percentage participation of these two fuels in the energy metabolism is usually assessed by the determination of the nonprotein respiratory quotient

(R), which is the ratio CO_2 volume produced/O_2 volume utilized. An estimation of the amount of O_2 (ml $\cdot$ min^{-1}) used for the oxidation of fat can be obtained from the formula: $(1-R) \cdot 0.03^{-1} \cdot O_2$. In this calculation, the R is handled as a "nonprotein R." The fact that the energy value of O_2 varies with R is not taken into consideration, but this may represent, at the most, a 7 percent difference.

The percentage participation of these two major fuels in the energy metabolism depends on a variety of factors:

1   Type of muscular work: whether it is (a) continuous or intermittent; (b) brief or prolonged; (c) light or heavy in relation to the maximal aerobic power of the engaged muscle groups
2   State of physical training: whether the individual is untrained or well-trained
3   The diet: whether it is high or low in carbohydrates
4   State of health: certain pathological conditions, such as diabetes, affecting the organism's choice of fuel

The adequacy of the oxygen supply to the working muscle cell is of prime importance because the oxidation of free fatty acids (FFA) depends on oxygen as the hydrogen acceptor. Thus, an inadequate oxygen supply more or less restricts the usable fuel to carbohydrate, of which there are limited stores. In addition, anaerobic oxidation of carbohydrate is about one-twentieth as efficient in yielding energy as aerobic oxidation of carbohydrates (see Chap. 2). When the oxygen supply is lacking, the carbohydrate oxidation proceeds only as far as the formation of lactate. The accumulation of lactic acid in the muscle impairs the function of the muscle cells. Furthermore, the increased lactate in the blood may inhibit the mobilization of FFA (Fredholm, 1969). This in turn suppresses fat metabolism further by limiting the supply of FFA substrate to the muscle cell, although a study by Ahlborg et al. (1976) suggests an augmented removal of FFA from the plasma pool in humans following lactate infusion.

Since the ability to utilize fat as a fuel depends on the oxygen-transporting capacity, the choice of fuel for the working muscle depends on the work load in relation to the individual's maximal oxygen uptake. The greater the maximal oxygen uptake, the greater the percentage contribution of fat to the energy metabolism at a given work load. Since training increases the maximal oxygen uptake, it also increases the facility for utilizing fat as a source of muscular energy during certain types of activity.

In prolonged work, it is a distinct advantage for the fasting individual to be able to utilize fat as a source of muscular energy, since the fat stores are infinitely larger than those of carbohydrates. Stored fat amounts perhaps to some 400 MJ (10^5 kcal) or more in a well-fed man of average size. The available energy in the form of stored ATP and phosphocreatine is only a few kJ, sufficient for less than a minute of strenuous physical effort. Stored glycogen amounts to some 8MJ (2000 kcal); see below.

The fact that physical training which increases the maximal oxygen uptake also increases the individual's facility for fat utilization was shown by Issekutz et al. (1965) in dogs. A trained and an untrained dog performed the same work load on the treadmill, which the untrained dog could endure for only 30 min (Fig. 14-1). In the untrained dog, the blood lactate rose to 75 mg · 100 ml⁻¹, whereas, in the trained dog, it was only 27 mg · 100 ml⁻¹. In the trained dog, the utilization of FFA rose during the experiment, while in the untrained dog it declined. One reason for this difference is that training enhances the O_2 supply to the working muscle cells, so that the work, in the case of the trained dog, may be performed to a greater extent aerobically. Consequently, less lactic acid is formed in the trained dog. The unfit dog, at the same work load, produces more lactic acid, which inhibits the FFA release from the adipose tissue (for reasons which will be explained later). The result is that the plasma FFA drops. A decreased plasma FFA always means decreased turnover rate or decreased rate of utilization of FFA (see below).

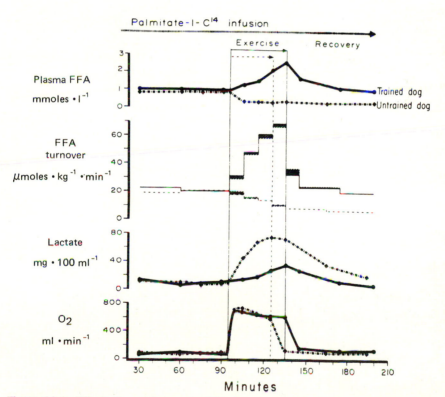

Figure 14-1 Effect of exercise on the FFA (free fatty acid) turnover in a trained dog (body weight 11.4 kg) and in an untrained dog (body weight 10.5 kg). At zero time, a priming dose of radiopalmitate was given intravenously, 220 and 276 mμc/kg, respectively, followed by an infusion from 0 to 200 min of 12 and 15.3 mμc/kg min, respectively. The trained dog ran on the treadmill for 40 min (from 95th to 135th min), and the untrained dog for 30 min (from 95th to 125th min). Shaded area represents differences between the rate of release and rate of uptake of FFA. *(From Issekutz et al., 1965.)*

At rest, fat and carbohydrate contribute about equally to the energy supply, as they also do during light or moderate muscular work in the fasting individual (Christensen and Hansen, 1939). As the work progresses, fat contributes in an increasing amount to the energy yield (Fig. 14-2). During moderately heavy work (in fasting subjects) which may be endured for 4 to 6 hours (including rest pauses), as much as 60 to 70 percent of the energy may be derived from fat at the end of the work period (Pruett, 1971; Table 14-1). Thus, the longer the work lasts, the smaller the percentage contribution of carbohydrate to energy metabolism in the fasting subject.

In spite of the reduced carbohydrate utilization during prolonged moderate work, the availability of glycogen may nevertheless be a factor limiting endurance because of the limited stores.

At work loads up to some 75 percent of the individual's maximal oxygen uptake, performed with 15-min rest during each hour of work, and tolerated for perhaps as long as 3 to 6 hours, the hepatic glucose supply may be the limiting factor causing a drop in the blood sugar. This leads to central nervous system symptoms typical of hypoglycemia (dizziness, partial blackout, nausea, confusion, etc.) in fasting subjects (Pruett, 1971; Fig. 14-3) without depletion of the muscle glycogen depots. The fact that the hepatic sugar output and the blood sugar level indeed are the limiting factor under these conditions was clearly demonstrated by Christensen and Hansen (1939). Their experiments were carried out until the subjects became exhausted. At the point of exhaustion, both subjective symptoms and laboratory findings indicated that the subjects had hypoglycemia. At this point, 200 g of glucose was ingested. Within 15 min the subjective symptoms disappeared. The blood sugar level rose and the subjects could continue the work for another hour.

Similarly, studies at Lankenau Hospital (Rodahl et al., 1964) on the effect of feeding on blood glucose levels in prolonged work (15-min rest per hour of work) showed that fed subjects could complete 6 hours of work at a load corresponding to approximately 60 percent of their maximal oxygen uptake without difficulty, their blood sugar level remaining more or less unchanged; the same subjects, when fasting, could barely complete the work period, and

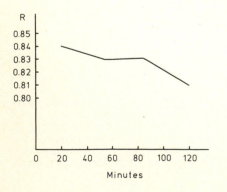

Figure 14-2 Respiratory quotient (R) in a subject working at 175 watts for 120 min while living on a normal diet. *(Modified from Christensen and Hansen, 1939.)*

Table 14-1   Average respiratory quotients at the beginning and end of 4 1/2-hr work (with 15-min rest every hour), and the percentage of the energy utilized, which was derived from carbohydrate at 50 percent and 70 percent max $\dot{v}_{0_2}$ on a standard diet. (*From Pruett, 1971.*)

50% max $\dot{v}_{0_2}$					70% max $\dot{v}_{0_2}$				
R		*% from carbohydrate*			*R*		*% from carbohydrate*		
First hr	Last hr	First hr	Last hr	Average	First hr	Last hr	First hr	Last hr	Average
0.84	0.81	47	37	40	0.89	0.84	64	47	53

had a marked drop in their blood sugar level (Fig. 14-4). Later studies have shown that oral administration of 225 ml of a 5 percent glucose solution, given every 15 min, is sufficient to prevent hypoglycemia in subjects working at 70 percent of their maximal oxygen uptake for 3 hours (Staff and Nilsson, 1971).

The central nervous system with its low glycogen content depends to a very great extent on the blood sugar. It has been calculated that in humans, approximately 60 percent of the hepatic sugar output serves the brain metabolism (Shreeve et al., 1956; Reichard et al., 1961). It would therefore seem essential that some sort of barrier should exist to prevent the blood glucose from freely entering the muscle cells and being used in their metabolism, since this might lead to a too rapid fall in blood glucose levels resulting in severe symptoms of hypoglycemia. In this respect, it should be pointed out that the

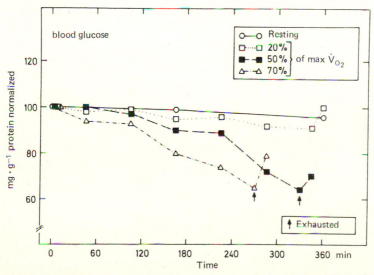

Figure 14-3 Effect of rest and three levels of exercise on blood glucose concentrations in seven subjects living on the standard diet. (*From Pruett, 1971.*)

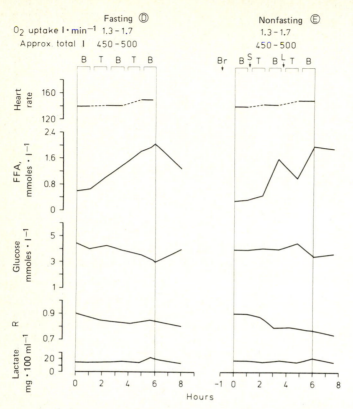

Figure 14-4 Effect of prolonged exercise. D: fasting subjects; B: work on the bicycle ergometer at 100 watts for 60 min; T: walk on the treadmill with a slope 8.6 percent at a speed 5.6 km · hr⁻¹ (3.5 mph) for 60 min; E: nonfasting subjects; Br: breakfast; S: snack; L: lunch. *(From Rodahl et al., 1964.)*

permeability of the cell membrane for glucose depends on the plasma insulin concentration which falls parallel with the blood sugar during prolonged moderately heavy exercise (Pruett, 1971). This may serve to reduce glucose uptake by the working muscle cell. Furthermore, enzymes are necessary for the uptake of glucose across the membrane, and at least one such enzyme, hexokinase, is inhibited by products from the breakdown of glycogen (Hultman, 1967). Therefore, stored muscle glycogen is a more readily available substrate for the energy metabolism in the working muscle cell than exogenous glucose. This fact is an advantage for the central nervous system, which might otherwise be competing with the muscles for blood glucose and suffer from hypoglycemia as a consequence. However, during prolonged exercise there is a significant utilization of glucose in the working muscles. This is partly balanced by an enhanced hepatic gluconeogenesis from various glucose precursors e.g., alanine and glycerol. Eventually glucagon may have a stimulatory effect on the hepatic uptake of glucose precursors and on the gluconeogenesis. Wahren et al. (1975) estimated the total glucose output from the liver during 4 hours of exercise (at 30 percent of the subject's maximal oxygen uptake) to be about 75 g. The glucose production in the liver was estimated to 15 to 20 g. With a total glycogen content in the liver amounting to 75 to 90 g, it is evident that the situation may be critical in prolonged heavy exercise (see Hultman and Nilsson, 1973).

At heavier work loads that are above 75 to 80 percent of the individual's maximal capacity, and that can be tolerated for only about an hour and a half at the most, a significant depletion of the muscle glycogen stores may be the factor limiting endurance. Under these conditions, there is usually no fall in blood sugar, because the duration of the work period is too short to cause a depletion of hepatic glycogen stores to any significant extent, while the muscle glycogen stores may indeed be markedly reduced. Normally, some glucose from the blood enters the muscle cell at all times. However, when the muscle glycogen store is nearly depleted, as it is toward the end of heavy prolonged exercise, evidently an increasing amount of glucose enters the muscle cells from the blood.

By the needle biopsy technique, small pieces of muscle tissue (10 to 20 mg) can be sampled in the human, and their content of glycogen and other substances analyzed (Bergström, 1962). By obtaining samples during various stages of different dietary regimens, during exercise, etc., the variation in glycogen content of the muscles can be followed.

On a normal mixed diet, the glycogen content in the quadriceps femoris muscle in the individual ranges from about 1.0 to 2.0 g · 100 g^{-1} wet muscle (Hultman, 1967). In the deltoid muscle, the glycogen content is significantly lower, or about 1 g · 100 g^{-1} wet muscle. Figure 14-5 presents data from experiments in which biopsies were taken every twentieth min during exercise

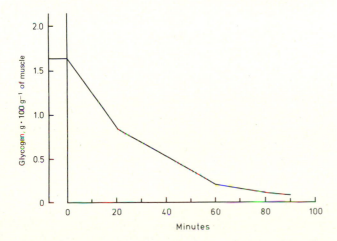

Fig. 14-5 Average values for glycogen content in needle-biopsy specimens from the lateral portion of the quadriceps muscle taken before and at intervals during exercise until exhaustion, in a group of 20 subjects. Oxygen uptake averaged 77 percent of the maximal aerobic power. (Modified from Hermansen et al., 1967.)

on a bicycle ergometer, with an average oxygen uptake of 77 percent of the individual's maximal aerobic power. Two groups of subjects participated in the study; 10 were trained, 10 were untrained and had a lower maximal oxygen uptake. Therefore, the "77 percent load" corresponds to a mean oxygen uptake of 3.4 liters $\cdot$ min^{-1} for the trained and only 2.8 liters $\cdot$ min^{-1} for the untrained subjects. After about 90 min the work had to be terminated because of the subjects' exhaustion. The glycogen content was then in the order of 0.1 g $\cdot$ 100 g^{-1} wet muscle. It should be emphasized, however, that the biopsies were taken only from the lateral and superficial portion of the quadriceps femoris muscle, and the analyses do not necessarily reveal events in other portions of the working muscles. The respiratory quotient was higher in the untrained group (about 0.95) than in the trained subjects (about 0.90), indicating that FFA metabolism plays a larger role in the trained men. The actual combination of glycogen was about 2.8 g $\cdot$ min^{-1} in both groups, despite the difference in energy expenditure (3.4 versus 2.8 liters O_2 per min). At the end of exercise, the blood sugar level was still 80 mg $\cdot$ 100 ml^{-1}.

The available evidence shows that at work loads exceeding about 75 percent of the individual's maximal oxygen uptake, the initial glycogen content in the skeletal muscles is decisive for the individual's ability to sustain such exercise for more than an hour. Thus, Bergström et al. (1967) showed that at a work load of about 75 percent of the maximal oxygen uptake, the larger the initial muscle glycogen stores, the longer the subject could continue to work at this load (Fig. 14-6). After a normal mixed diet giving an initial glycogen content of 1.75 g $\cdot$ 100 g^{-1} wet muscle, the subject could tolerate the 75 percent work load for 115 min. After the subject spent 3 days on an extreme fat and protein diet, the glycogen concentration was reduced to about 0.6 g $\cdot$ 100 g^{-1} wet muscle and the standard load could be performed for only about 60 min. After 3 days on a carbohydrate-rich diet, the subject's glycogen content became higher, 3.5 g $\cdot$ 100 g^{-1} wet muscle, and the time on the 75 percent load could now be prolonged to about 170 min on the average. It was further observed that the most pronounced effect was obtained if the glycogen depots were first emptied by heavy prolonged exercise and then maintained low by giving the subject a diet low in carbohydrate, followed by a few days with a diet rich in carbohydrates (see Fig. 14-7). With this procedure, the glycogen content could exceed 4 g $\cdot$ 100 g^{-1} wet muscle and the heavy load could be tolerated for longer periods, in some subjects for more than 4 hr. The total muscle glycogen content under these conditions could exceed 700 g.

Bergström and Hultman (1966) also performed an experiment in which one subject worked with his left leg and the other subject simultaneously worked with his right leg on the same bicycle ergometer. After several hours' work, the exercising leg of each individual was almost depleted of glycogen while the resting leg still had a normal glycogen content. Feeding the subjects a carbohydrate-rich diet on the following days did not markedly influence the depots of the resting limb, but in the previously exercised leg the glycogen content increased rapidly until the values were about twice as high as those in

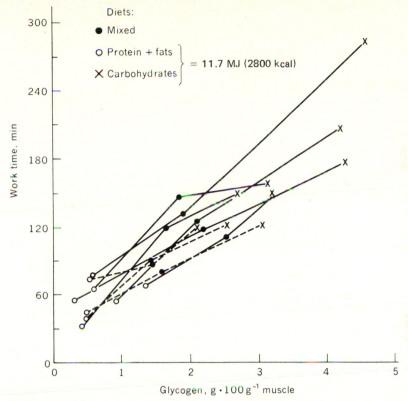

Figure 14-6 Relation between initial glycogen content in the quadriceps muscle in nine subjects who had been on different diets, and maximal work time when working at a given load demanding 75 percent of maximal aerobic power. Dotted lines denote subjects who had been on a carbohydrate diet prior to the fat-plus-protein diet. *(From Bergström et al., 1967.)*

the nonexercised leg (Fig. 14-8). This experiment shows that exercise with glycogen depletion enhances the resynthesis of glycogen. It also shows that the factor must be operating locally in the exercising muscle. (See Bergström et al., 1972.)

In prolonged athletic events in which the work load exceeds about 75 percent of the maximal oxygen uptake, not only endurance, but also speed, is affected by the initial muscle glycogen content. This is schematically illustrated in Fig. 14-9. A group of subjects participated in two 30-km cross-country running races, on the first occasion after their normal mixed diet, and on the second occasion after a few days on an extremely high carbohydrate diet after previously emptying the glycogen depots. At several points on the track, the running time was recorded. It was observed that the lower the initial muscle content, the lower the ability to maintain a high running speed toward the end of the race. It should be noted, however, that even in the subjects with the lowest initial muscle glycogen content, the speed was maintained during the

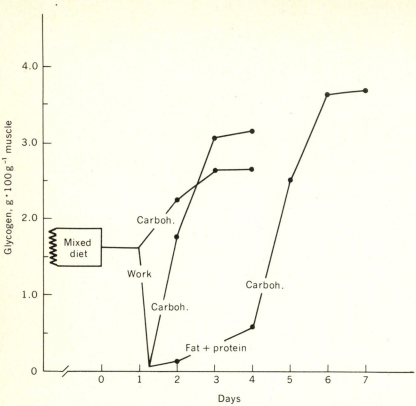

Figure 14-7 Different possibilities of increasing the muscle glycogen content. For further explanation, see text. *(From Saltin and Hermansen, 1967.)*

first hour of the race. This shows that a high glycogen content in the muscles did not enable the subject to attain a higher speed at the beginning of the race any more than a low initial glycogen level did.

Under such conditions of very intense or near maximal effort of sufficient duration, the draining of the muscle glycogen apparently is high enough to cause a significant depletion of the muscle glycogen stores. In this event, then, the cause of exhaustion is located in the working muscle tissue. However, even at the point of such exhaustion, the muscle is still utilizing significant amounts of glycogen or glucose as judged by the respiratory quotient.

It should be pointed out that there is one drawback with the high glycogen storage: it was mentioned that each gram of glycogen is stored together with about 2.7 g of water. With a glycogen storage of 700 g, there is then an increase in body water amounting to about 2 kg. In activities in which the body weight has to be lifted, an excessive glycogen store should therefore be avoided.

It is thus clear that under certain conditions, i.e., during heavy work equivalent to about 75 percent or more of the maximal oxygen uptake, the level of the muscle glycogen stores may significantly affect performance after about 1 hour. This may be influenced by the diet (Fig. 14-7).

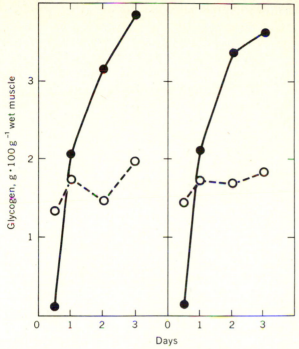

Figure 14-8 Two subjects were exercised on the same bicycle ergometer, one on each side working with one leg, while the other leg rested (dashed line). After working to exhaustion, the subjects' glycogen content was analyzed in specimens from the lateral portion of the quadriceps muscle. Thereafter, a carbohydrate-rich diet was followed for 3 days. Note that the glycogen content increased markedly in the leg that had been previously emptied of its glycogen content. *(Modified from Bergström and Hultman, 1966.)*

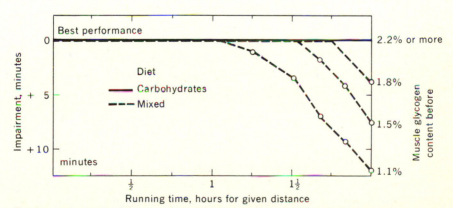

Figure 14-9 Schematic illustration of the importance of a high glycogen content in the muscle before a 30-km race (running). The lower the initial glycogen store, the slower became the speed at the end of the race compared with the race performed when the muscle glycogen content was 2.2 g or more per 100 g muscle at the start of the race. For the first hour, however, no difference in speed was observed. *(By courtesy of B. Saltin.)*

At extremely heavy or near maximal and maximal work loads, glycogen is the major, or almost exclusive, source of energy for muscular work. Under such conditions, the energy metabolism is predominantly anaerobic. This limits the fuel to carbohydrate with the accumulation of lactic acid. The lactic acid suppresses the FFA mobilization, which further reduces the available FFA substrate in the muscle cell.

Following cessation of such intense or near maximal work, however, there is a marked rise in plasma FFA levels lasting for several hours, or actually until the next meal is taken (Fig. 14-10). Such a postexercise rise in FFA levels may also occur following intermittent work (Fig. 14-11).

There are other fuels which may participate as sources of energy, such as amino acids, acetoacetate, and 3-hydroxy-butyrate, but which play an insignificant role, at any rate in well-fed individuals engaged in intense muscular work at high levels of energy expenditure (McGilvery, 1975). During prolonged fasting, however, Owen et al. (1967) have shown that ketone bodies may replace glucose oxidation in the brain.

The metabolic rate of the nervous system is exceedingly high even at rest. Although intellectual work does not seem to cause an appreciable elevation of the overall oxygen uptake of the individual, it appears likely that the rate of energy utilization of the nervous tissues is markedly increased during physical exertion, owing to the vastly increased nervous impulse traffic, the increased rate of the sodium pump, etc. The energy supply to the central nervous system during strenuous physical exercise should therefore be subject to more attention, especially since some unpublished observations have indicated reduced CNS function in athletes during strenuous physical exertion (such as orienteering competitions).

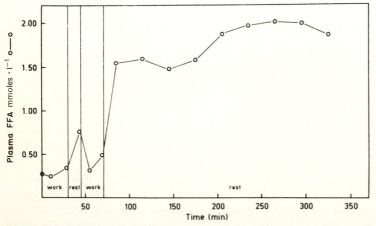

Figure 14-10 The plasma FFA (free fatty acid) response to two exhausting work bouts on the bicycle ergometer at 87 percent max $\dot{V}_{O_2}$. *(From Pruett, 1971.)*

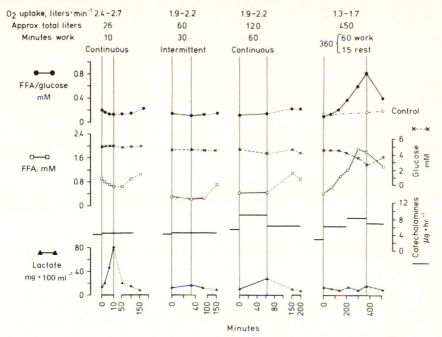

Figure 14-11 Effects of work of different intensity and duration (heavy exercise on the bicycle ergometer for 10 min; intermittent run: 5 s work, 5 s rest, on the treadmill, slope 8.6 percent, speed 12 km · hr⁻¹, 7.5 mph for 30 min; continuous work on the bicycle ergometer at 150 watts for 60 min, and alternating work on the bicycle ergometer at 100 watts for 60 min and work on the treadmill with a slope 8.6 percent at a speed 5.6 km · hr⁻¹, 3.5 mph for 60 min for a total period of 360 min or until exhaustion), in normal fasting subjects in relation to blood lactate levels, urinary catecholamine excretion, plasma FFA, and blood glucose. *(From Rodahl et al., 1964,)* Note the rise in plasma FFA and the drop in blood glucose during prolonged work, and the marked rise in blood lactate level during heavy short exercise, while it is essentially unchanged during the prolonged work. Note also the rise in plasma FFA following the cessation of short, heavy work.

REGULATORY MECHANISMS

The utilization of metabolic fuel is regulated by the interplay of a number of factors, both physiological and biochemical in nature (Fig. 14-12).

First of all, a high carbohydrate diet favors a higher participation of carbohydrate in energy metabolism. This fact was first described by Christensen and Hansen in 1939. It was confirmed by Issekutz et al. in 1963 and by Bergström et al., 1967, and Pruett in 1971. Issekutz et al. (1963)found that it is the carbohydrate intake rather than the amount of fat in the diet which determines whether the preferred fuel is FFA or glucose. Ingestion of 100 g glucose immediately before exercise causes a shift of work metabolism toward carbohydrate and a corresponding reduction in FFA metabolism. The reason probably is that the carbohydrate intake causes an increased production of insulin which inhibits FFA mobilization, and therefore suppresses FFA oxidation.

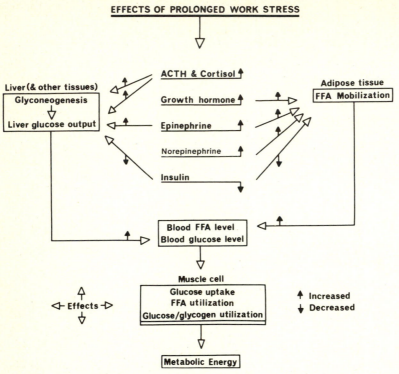

Figure 14-12 Schematic representation of the effect of exercise on hormone secretion and the concomitant effect of those hormones on fat and carbohydrate metabolism. *(From Pruett, 1971.)*

Secondly, the utilization of FFA during muscular work is determined by the level of plasma FFA. An increased plasma FFA concentration always means an increased rate of FFA utilization. At any given plasma FFA level, the FFA turnover rate is about twice as high during exercise as during rest (Fig. 14-13) (Issekutz et al., 1964; Paul, 1975). As pointed out in Chap. 12, one-leg training induced during a standardized two-leg exercise increased utilization of FFA in the trained leg compared with the untrained leg. Certainly the arterial concentration of FFA was identical in the two legs. Apparently local factors affected by training will also influence the FFA utilization.

The plasma FFA level during exercise is in turn determined by the combined effects of several factors:

1 Norepinephrine is a most powerful stimulator of FFA mobilization. Even small increases in norepinephrine cause a marked rise in plasma FFA levels associated with a corresponding increase in FFA turnover (Issekutz, 1964; Rodahl and Issekutz, 1965) (Fig. 14-14). Both during short, intense muscular work and during prolonged, moderately heavy work, there is a considerable increase in catecholamine production (Fig. 14-11) which would increase FFA mobilization and utilization.

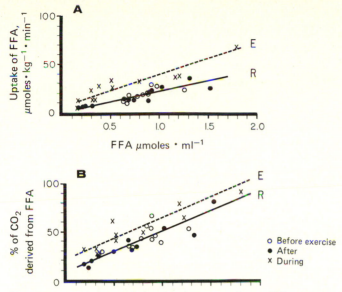

Figure 14-13   A: interrelationship between plasma FFA concentration and uptake of FFA, in rest (R) and during the last 10 min of exercise (E). $Y_R = 23.8 \times -1.5$; SE of regression coefficient: ± 2.5 ($P < .001$). $Y_E = 32.8 \times +6.1$. SE of the regression coefficient: ± 3.1 ($P < .001$). B: interrelationship between plasma FFA level and the contribution of FFA to the fat oxidation in rest (continuous lines), and in the 30th min of exercise (broken line). $Y_R = 41.9 \times +9.3$; SE of the regression coefficient: ± 6.3 ($P < .001$). $Y_E = 42.2 \times +19.8$; SE of the regression coefficient: ± 9.2 ($P < .01$). *(From Issekutz et al., 1964.)*

2   The lactate accumulated during strenuous muscular work will, on the other hand, suppress the mobilization of FFA from the adipose tissues. The possibility that lactate might function as a physiological inhibitor of FFA mobilization in severe exercise was first suggested by Issekutz and Miller (1962), who observed an inverse correlation between FFA and lactate levels in the blood (Fig. 14-15). Miller et al. (1964) showed that lactate infusion decreased the inflow of FFA into the plasma in depancreatized dogs. Fredholm (1969) induced FFA mobilization in isolated adipose tissue preparations by sympathetic nerve stimulation, which could be counteracted by lactate infusion in physiological concentrations. From his experiments, it appears that an increased re-esterification of FFA is the major mechanism underlying the lactate effect. This view is supported by the findings of Issekutz et al. (1975).

In prolonged, moderately heavy muscular work, the lactate level is not significantly increased, but the catecholamine level is (Fig. 14-11). This may explain the observed rise of FFA mobilization under such conditions. In contrast, the accumulation of lactate during short, intense work may prevent a rise in plasma FFA during the work period itself, but permit a subsequent rise after work when the lactate level has dropped (Fig. 14-11)(Rodahl et al., 1964).

3   A number of hormones other than catecholamines are affected by physical exercise. As already mentioned, insulin suppresses FFA mobilization. However, insulin is not increased during prolonged exercise (Fig. 14-16; Pruett, 1971). Insulin, therefore, cannot prevent the observed rise in FFA mobilization

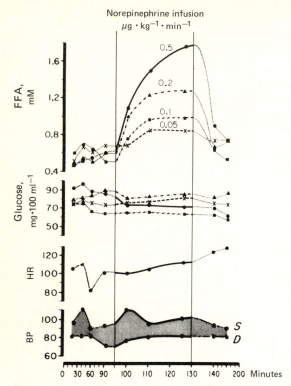

Figure 14-14 Effect of intravenous norepinephrine infusions on the plasma FFA and blood glucose of a resting nonanesthetized dog. Pulse rate (PR) and blood pressure (BP) values obtained with the highest infusion rate (0.5 μg · kg⁻¹ · min⁻¹) are shown on the figure. *(From Issekutz, 1964.)*

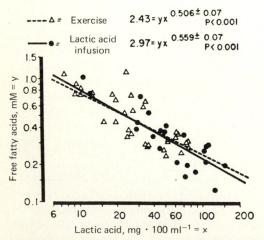

Figure 14-15 Treadmill experiment on a dog (9.5 kg). Interrelationship between plasma FFA level and blood lactic acid concentration. *(From Issekutz and Miller, 1962.)*

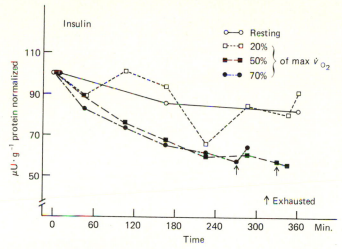

Figure 14-16 The effect of rest and three levels of exercise on plasma immunoreactive insulin concentrations in six subjects living on the standard diet. *(From Pruett, 1971.)*

during prolonged exercise. It is known that growth hormone is increased during exercise (for ref. see Hartley, 1975). This increase may affect both glucose and FFA metabolism. The administration of growth hormone is reported to cause an increase in plasma FFA lasting several hours (Grunt et al., 1967). It therefore appears likely that the long-lasting increase in FFA plasma levels following the cessation of short, intense exercise is due to a growth hormone effect rather than to a norepinephrine effect, since this effect probably is of a fairly short duration.

It is well known that cortisol, which is produced during both physical and mental stress, has a profound influence on carbohydrate metabolism. It increases the catabolic breakdown of proteins and the formation of carbohydrates from protein, which in turn may enter the metabolic pool. Issekutz and Allen (1971) have shown in experiments on dogs that even short-term treatment with metylprednisolone increases the glycogen depots of the body, causing a corresponding increase in endurance. This process, however, takes place at the expense of the protein content of the tissues and may therefore be harmful in the long run. As the result of prolonged cortisol administration, muscular wastage occurs, causing reduced muscle strength and endurance.

From this brief review it is thus clear that the hormonal regulation of fat and carbohydrate metabolism during exercise depends on a very complicated balance between many factors, representing a series of unsolved problems for future research into the regulation of energy supply during heavy exercise.

A very important effect of training is certainly the induced increase in maximal oxygen uptake. Another positive effect is the increase in oxidation of FFA and the reduced energy yield from glycogen. This change in fuel utilization is not only evident at a given metabolic rate but also when working at a given percentage of the maximal aerobic power. This adaptation is certainly a glycogen-saving mechanism. A reduced lactate production in the trained individual may be one factor behind this difference in metabolic pattern. The modification in the mitochondrial enzyme profile induced in trained muscles may also be of importance. A reduction in the diffusion distance between capillaries and the interior of the muscle cells will facilitate the FFA uptake. An increased oxidation of FFA will elevate the concentration of citrate which in turn will inhibit the activity of the enzyme phosphofructokinase retarding the glycogenolysis and lactate formation. All this means that the endurance-trained individual can work closer to his maximal aerobic power for longer periods of time than the untrained person.

FOOD FOR THE ATHLETE

The nutritional requirements in general have been briefly reviewed earlier in this chapter. On the whole, these requirements also hold for the athlete. A varied, well-balanced diet in adequate amounts is all that is necessary from a nutritional point of view for the body to function optimally, and for providing a biological basis for top performance. It is true that the requirements for protein and certain minerals may be somewhat increased in athletes during training in events which require muscular strength. However, the total food intake in athletes undergoing such training is also increased, often by as much as 4 MJ · day^{-1} (1000 kcal) or more in connection with the regular training and participation in competitions. If their diet is balanced, they will then automatically take in more of these nutrients as they increase their food intake. This is true provided that their diet is not too high in fat, i.e., that no more than 35 to 40 percent of the energy is derived from fat, and that the content of refined sugar is relatively low, with the consumption of milk, fish, meat, vegetables, fruit, berries, and grain products being comparatively high. In this way, an adequate intake of proteins, iron, calcium, and vitamins is secured. This diet holds for athletes in general, whether they are engaged in the training of strength or endurance, whether they are engaged in competitions or not. However, the purpose of the training must also be kept in mind. The nutritional body building should be geared to that for which the body is being trained. If we train for participation in long-distance running where the body weight must be carried long distances, the body weight and muscle mass should be no greater than is necessary in order to support the functional needs. All body weight in excess of this requirement is dead weight. The same holds for high jumpers who have to lift the body against the force of gravity. In weight lifting and throwing, on the other hand, these considerations are unimportant.

Nutritional surveys in Norway (K. Solvoll, unpublished results) have verified the general impression that throwers eat more than runners. The throwers are heavier because of their larger muscle mass. Therefore their overall energy requirement is greater. However, the energy intake per kg body weight is about the same. The percentage of the energy from proteins, fat, and carbohydrates is about the same in throwers and runners. Fourteen percent of the energy is taken in the form of proteins, close to 40 percent in the form of fat, and about 46 percent in the form of carbohydrates. The surveys showed that both groups of athletes consumed, as a rule, much more meat and milk than the population as a whole.

When discussing the dietary requirements of athletes, it is necessary to distinguish between events of very short duration, which mainly involve technique and muscular strength and which last for only seconds or a minute or two at the most, and events which last for a long period lasting up to several hours, and which therefore require endurance. In the case of the endurance events, it is necessary from a nutritional standpoint to distinguish between events lasting less than an hour and events of significantly longer duration.

Events Lasting Less than 1 Hour

In very intense physical exertion or athletic events lasting less than 1 hour, the available supply of stored energy fuel is generally ample to cover the need. Under such conditions a special diet is unnecessary. Because the digestion of a meal causes a redistribution of blood from the muscles to the guts, physical exercise shortly after a meal will result in a competition between the guts and the working muscles for the available blood supply. Heavy muscular work should therefore be avoided immediately following a heavy meal. As a general rule, a meal should not be ingested later than $2^1/_2$ hr prior to an athletic event. Furthermore, the last meal prior to the event should be light. It should consist only of ingredients which the individual knows from experience can be tolerated well. Excessive quantities of carbohydrates should not be ingested prior to the event because of a possible subsequent insulin effect, previously explained.

Events Lasting between 1 and 2 Hours

In the case of athletic events involving large muscle groups at very high work loads for periods exceeding an hour or so, the available evidence suggests that it would be advisable for the competitor to ingest ample quantities of carbohydrates several days preceding the events in order to fill the muscle glycogen depots (Fig. 14-7). The individual should avoid heavy muscular work which might deplete the existing glycogen depots prior to the event. On the other hand, it is not advisable to live on a high carbohydrate diet regularly, since this would condition the metabolic processes to a high utilization of carbohydrate fuel, rather than FFA, as already explained.

Events Lasting for Several Hours

In the case of very prolonged physical exertions lasting 3 to 4 hours or more, it might also be an advantage to ingest ample quantities of carbohydrates several days preceding the event, as mentioned above. Here, however, it is more important, relatively speaking, to ingest carbohydrates during the actual event in order to supplement the hepatic sugar output.

If the period of exertion during the event is very long, it might be advantageous to consume moderate amounts of sugar, preferably in the form of a flavored glucose solution, before warm-up. During a 50-km race, cross-country skiers may ingest as much as 1 liter of various sugar solutions corresponding to about 50 to 400 g sugar, divided into seven to eight portions, taken at the various control posts some 5 to 6 km apart. The ideal arrangement would be to issue the sugar solution just before a slack downhill slope so that the skier could drink the solution while gliding downhill. The sweet sugar solution should be rinsed down with water, which would also serve to replace some of the fluid loss. In view of the fact that the sugar solution is rather concentrated, it is important that the competitor become accustomed to it during training in order to avoid any unpleasant reaction during the event itself. This regimen has been well tried and has survived the test of time. It may therefore be recommended for other competitive events such as long-distance bicycle racing, walking, and marathon running.

It should be emphasized that the rate of absorption of glucose, water, and various minerals in the gastrointestinal tract is not affected by exercise, at least not by loads demanding less than about 70 percent of the maximal oxygen uptake. In Fordtran and Saltin's experiments (1967), at least 50 g glucose was emptied from the stomach during 1 hr of heavy exercise. This amount corresponded to one-fourth to one-half the carbohydrates required by the body during this period. The data of Fordtran and Saltin (1967) suggest that gastric emptying and intestinal absorption of saline solutions could be rapid enough to replace all the losses of sweat incurred during heavy exercise, even in hot environments. An addition of glucose to water may cause inhibition of gastric emptying, and high concentration of sugar in orally ingested water may cause large amounts of fluid to be retained in the stomach. This may produce abdominal discomfort during exercise. In our experience with cross-country skiers, we have noticed that some of them prefer a weak (about 10 percent) glucose solution, whereas others tolerate and really prefer a 40 percent glucose solution.

With the exception of the days preceding the event and on the day itself, the athlete should consume a regular, well-balanced diet. The greatly increased energy expenditure automatically increases the appetite, with the result that the competitor ingests more food. If the diet is well balanced to start with, the increased intake will also automatically cover the athlete's increased requirements for protein, vitamins, and minerals. It has been shown that during a 10-day period, maximal oxygen uptake is not affected by a reduction of the protein intake to 4 g $\cdot$ day^{-1} (Rodahl et al., 1962), nor is the performance improved by an increased ingestion of protein, up to 160 g $\cdot$ day^{-1} (Darling et al., 1944). If an extra supply of protein seems advisable in connection with muscle training or "muscle building" during convalescence or during growth, there is no reason to resort to costly preparations, since all the required protein and all the essential amino acids can be obtained from meat, fish, and milk. The old rule of thumb that 1 g of protein $\cdot$ kg body weight^{-1} $\cdot$ day^{-1} covers the demand may well be valid also for the hard-training athlete. However, properly controlled metabolic balance studies to determine the exact protein requirement of the athlete have not as yet been carried out.

With regard to the distribution and the frequency of meals throughout the day, it appears that rather frequent, moderate meals are more effective in yielding maximal performance than fewer, larger meals. Hutchinson (1952) and Mayer and Bullen (1960) recommend that athletes should have at least three meals a day. It is a common experience, however, that athletes in active training often are unable to maintain regular meal schedules, and may incur detrimental effects.

It is well known that vitamin deficiency causes an impairment of the performance capacity, but it takes a very long time to develop such vitamin deficiency in an individual living on a deficient diet (Rodahl, 1960). For this reason, an individual may be without any vitamin intake for a week or so without any detectable detrimental effect on the work capacity. As long as the usual vitamin intake is adequate, additional vitamin intake does not improve the performance (Keys, 1943; Simonson, 1951; Mayer and Bullen, 1960). Particular attention has been focused on the possible beneficial effect of the B-vitamins on performance capacity, but the studies, almost without exception, have been negative. In the case of the water-soluble vitamins, the ingestion of large quantities of vitamin pills is a rather expensive way of increasing the vitamin content of the urine, which serves no useful purpose in the first place.

The iron intake is at present subject to considerable interest because low serum-iron values have been observed in many top athletes. It has been pointed out that inasmuch as the oxygen is transported by hemoglobin, which contains iron, an iron deficiency may cause a reduced capacity for oxygen transport. It is hard to conceive how reduced serum-iron levels may affect the oxygen-transporting system as long as the hemoglobin content of the blood is normal, as it usually is in the athletes in question. However, it is conceivable that the observed reduction in the serum iron may be interpreted as a forerunner for a reduction in the hemoglobin content of the blood. On the basis of the available evidence, it appears that there is no obvious justification for excessive iron intake in the athlete.

In certain sports, such as boxing, wrestling, and horse-racing, athletes at times subject themselves to stringent dietary regimens combined with dehydration in an endeavor to lose weight and thus to obtain a lower weight classification, for which they do not properly qualify. This allows them the advantage of competing with contestants who weigh less than they themselves normally do. Such weight classifications have been established to provide competition on an equitable basis. The violation of these standards by sudden, self-inflicted starvation and dehydration not only defies good sports ethics, but may also perhaps harm the health of the individual concerned. Such practices have been condemned by the American Medical Association (1959, 1967) and the American College of Sports Medicine (1976).

During competitive cross-country orientation, a mean fluid loss of 3.0 liters during the 90-min duration of the competition has been observed (Saltin, 1964). Even greater fluid loss of up to 5 or 6 liters has been observed during ski racing and in bicycle racing in a hot environment. Sweat loss of a similar magnitude has been reported for many industrial operations.

In the army it used to be advocated that during strenuous field maneuvers, the soldier should drink as little as possible in order to reduce the amount of sweating. This advice is in total violation of the physiological fact that sweating is a vital process which, during heavy muscular work or in the heat, cools the body and prevents overheating. It is now generally recognized that the lost water has to be replaced, preferably at the same rate at which it is lost (Adolph, 1947; see Chap. 15).

As part of the dietary advice for the athlete, emphasis should be placed on the fact that weight loss through dehydration should be avoided. If the training is intense, and especially if it occurs in a hot climate, ample fluid should be taken, even on the day preceding the event. It is advisable to drink water a few hours prior to the event. In the case of prolonged efforts in a hot climate, adequate fluid intake is essential. Apparently the death of an athlete during the bicycle race in Rome in 1960 may have been due, at least in part, to extreme dehydration. A dehydration corresponding to a loss of body water in excess of 1 to 2 percent of the body weight should in any case be avoided (Adolph, 1947; Ladell, 1955; Saltin, 1964).

The sweat is hypotonic compared with the body fluid (Robinson and Robinson, 1954), so that relatively more fluid than salt is lost from the body during sweating. Sweating, therefore, causes an increase in the NaCl concentration in the body. Under these conditions, ingestion of extra NaCl is contraindicated (Ladell, 1955). Only in prolonged activity associated with intense sweating lasting more than a week or so is the ingestion of additional salt indicated, provided that the daily diet contains sufficient amounts of salt.

It should be kept in mind that whatever the physiological principles for an optimal diet, the practical considerations dictate that the diet has to be acceptable to the individual. If an athlete believes in a food fad or in a miracle pill, the fad or the pill may result in victory, provided, of course, that the diet otherwise is fully adequate.

PHYSICAL ACTIVITY, FOOD INTAKE, AND BODY WEIGHT

Energy Balance

The energy requirement is essentially a question of energy balance or energy intake versus energy expenditure. Any excess intake of food energy over and above the daily need will be stored as fat. The result is weight gain.

The energy requirement is directly proportional to body size and degree of physical activity. As a rough guide, sedentary, middle-aged people need about 150 kJ (35 kcal) per kg body weight per day. This corresponds to about 10.5 MJ (2500 kcal) per day for a 70-kg man.

As long as such a person adheres to his or her general activity pattern in energy balance, there will be only minor fluctuations in energy expenditure. If the individual regularly takes in roughly this amount of energy, the body weight will remain the same. On growing older, however, the person is apt to be less active and therefore to experience a gradual decline in energy expenditure. Consequently, if food intake continues to be about the same as before, the person will no longer spend what is taken in. The difference will be deposited as stored energy in the fat depots. Thus, if the person takes in some 1.5 MJ (350 kcal) more than is spent each day (the amount of energy contained in a piece of apple pie), 10.5 MJ (3500 kcal) will be stored in 10 days. This means that the individual will have deposited about half a kg of fat in the body, for there are approximately 29 MJ (7000 kcal) in each kg of stored fat. This uptake, if continued day after day, will result in a gain of about 18 kg in body weight in a year. Simply omitting a single piece of apple pie from the daily diet could have prevented this weight gain. Or, if the person had kept up his or her former level of physical activity, the slice of apple pie could have been enjoyed every day without gain in weight.

It is a common observation that athletes, such as runners and javelin throwers engaged in intense training, need about 200 kJ (50 kcal) per kg body weight per day. Thus, a javelin thrower weighing about 90 kg consumes about 19 MJ (4500 kcal) per day. A 70-kg runner consumes between 12.5 and 14.5 MJ (3000 and 3500 kcal) daily. The higher energy intake, compared with that of most sedentary individuals, is caused by the fact that athletes, owing to their daily 1- to 3-hour training activities, expend at least 4 MJ (1000 kcal) more per day than normal persons. Jogging, for example, involves an energy expenditure of more than 1.7 MJ (400 kcal) per hour.

As mentioned previously, about one-third to one-half the total daily energy expenditure is used to maintain the basal metabolic rate (BMR). This BMR may vary greatly from one individual to another. A BMR 10 percent below or above the normal average is not uncommon. In sedentary individuals with a daily total energy expenditure of about 10.5 MJ (2500 kcal), a difference in BMR of 20 percent may mean a difference of 2.1 MJ (500 kcal) per day in overall energy metabolism merely due to a basic difference in BMR. This may well explain why some persons remain slim although they eat more than some heavy persons who are equally active. It should be emphasized that such small differences in BMR, 10 percent below and 10 percent above the average, are clinically classified as normal and that a BMR 10 percent higher than the average will mean an additional energy expenditure of 375 MJ (90,000 kcal), or the equivalent of more than 10 kg body fat, in a year.

On a short-term basis, on the other hand, only muscular work can cause major differences in energy expenditure, in that such work easily may cause a marked increase in energy expenditure over the resting value. However, physical activity of a more moderate degree, spread out over the whole day, may also count in the long run. There are great individual differences in an individual's attitude to activity. Some instinctively seek it, whereas others avoid it. Some remain seated while others get up and move about. These are indeed small differences but in the course of the day, they may add up to several hundred kilojoules, for it takes only some 15 steps to expend 4 kJ (1 kcal).

There is no real basis for the attitude that one should refrain from unnecessary activity that spends extra energy when large segments of the world population are starving. Physical exercise increases the utilization only of fat and carbohydrate, not of protein, which is the critical nutrient for poorly fed people. Therefore, exercise does not deprive the needy people of their essential proteins.

The relationship between food intake, body weight, and physical activity has been systematically studied in laboratory animals. Mayer et al. (1954) exercised mature rats, accustomed to a "caged" sedentary existence, on a treadmill for increasing daily periods. They observed that for moderate exercise of 20- to 60-min duration, there was no corresponding increase in food intake. In fact, there was a significant decline in food intake, and consequently body weight also decreased. When the exercise was extended from 1 to 6 hr, food intake increased linearly with energy expenditure, and body weight was maintained. When the daily exercise was extended beyond 6 hr, the animals lost weight, their food intake decreased, and their appearance deteriorated. It has also been shown that rats that are overfed voluntarily reduce their activity. This, in turn, increases the weight gain, and a vicious circle is created. On the other hand, when fat rats are fed a calorically restricted diet, their spontaneous activity increases as the weight loss progresses. Thus, both slight and exhausting activities do not appear to be directly related to corresponding changes in food intake.

It has been observed that animals whose activities are restricted by confinement in small cages consume more food than they require and therefore accumulate fat (Gasnier and Mayer, 1939; Ingle and Nezamis, 1947). This is the basis for the practice of fattening cattle, pigs, and geese by restricting their activity.

Mayer et al. (1956) made a study in an industrial population in India engaged in a very wide range of physical activity, from tailors and clerks to coolies carrying heavy loads for several hours a day. The diet was quite uniform for each individual and showed little variety within groups and from group to group. It was observed that the sedentary individuals had the highest energy intakes and the highest body weight. The light workers had lower energy intakes and the highest body weight. The groups engaged in medium-heavy and very heavy work had increasing energy intakes, but their body weights were the same. It thus appears that sedentary individuals, like the animals, are apt to eat more than they need and to become obese. The remarkable thing is that, similar to the rats on light treadmill work, the group engaged in light work did not eat more than the sedentary workers; they did, in fact, eat less.

Summary Within a wide range of energy expenditures through various degrees of physical activity, there is an accurate balance between energy output and intake so that the body weight is maintained constant. If the daily activity is very intensive and prolonged, the spontaneous energy intake is often less than the output, with a reduction in body weight as a result. A daily energy expenditure below a threshold level often leads to an energy surplus and consequent obesity. In this case, satiety is not reached until more energy has been taken in than has been expended.

Regulation of Food Intake

The exact mechanism by which persons regulate the amount and kind of food they need is not completely understood. As in many control systems, one has to extrapolate from animal studies (Hamilton, 1965; Mayer and Thomas, 1967). The hypothalamus contains a feeding center responsible for the urge to eat or the initiation of feeding, and a satiety center capable of exerting inhibitory control over this feeding center.

Mayer and Thomas (1967) point out that there is some sort of short-term feedback: (1) Information concerning the nutritive value of ingested foods is relayed from gastric sensors to the hypothalamic regulation system by way of neural or humoral pathways, or both. (2) Food intake is also intimately related to the regulation of blood glucose. The information concerning the availability of glucose to the cells mediated via glucose receptors is more important than the actual blood glucose concentration. Hypoglycemia is, however, almost always followed by hunger sensations. (3) The control of heat exchange is connected in a complex manner with the center regulating food intake. An elevation of the body temperature inhibits the sensation of hunger (Andersson, 1967). This may explain the poor appetite experienced during periods of intense physical activity and the poor appetite often noticed in patients with fever. All this information is integrated with myriads of other exteroceptive and interoceptive inputs at a particular moment, determining the balance of appetite and satiety and the initiation, continuance, or termination of the feeding response.

Cumulative errors in this system could, in turn, be corrected through a long-term feedback, or a lipostatic regulation, i.e., an inhibition of food intake whenever sufficient energy is derived from the mobilization of surplus body fat. How the adjustment of feeding is modified by the state of the fat stores is not clear. It is noticed, however, that after a period of weight gain, food intake is often reduced and the body weight becomes stabilized at a new level.

As pointed out by Garrow (1974) at a symposium on energy balance, humans are not rats. He makes the following observations. The rat has an astonishing ability to regulate its energy intake over long periods of time, but it applies only if the food available is monotonous, if the rat has been fed *ad libitum* with this diet, and if flavoring agents, especially sweeteners, are excluded. These conditions are not relevent for the human species. For primitive man there was available a relatively small selection of naturally occurring animal and plant foods. Appetite,which may at one time have been a reliable guide to a correctly balanced diet, is now merely a sensation which can be manipulated in many ways by food manufacturers. In experiments, subjects have been over- or underfed by various methods (e.g., by supplement of food by an intragastric tube, or by meals with differing energy density but indistinguishable in taste and volume). The correlation between the state of energy balance and the voluntary energy intake is very low, and the individual's ability to regulate the food intake, at least over a short term, is very poor.

"Ideal" Body Weight

The body size and shape are largely determined by the skeletal size, for a certain amount of muscle and other tissues usually go along with a certain amount of bone. Therefore, the "ideal" body weight, the body weight which includes only a minimal amount of body fat, depends largely on the skeletal size. The "ideal" body weight may be modified to some extent by enlargement of the muscles by training, especially such training as weight lifting; a person may therefore be overweight without being obese. However, for all practical purposes the excess weight, or the weight over and above the ideal body weight, represents accumulated body fat.

There are graphs and tables, usually based on height and sex, giving the so-called ideal body weight. One of the best known of these norms is the Metropolitan Life Insurance table (1963). Such norms are, however, rather inaccurate and meant only as a general guide. Thus the need for a relatively simple but meaningful method of assessing body composition, including the amount of adipose tissue, is apparent. There are several methods for a quantitative classification of body build based on a series of anthropometric measurements, including various diameters (e.g., the radial-ulnar diameter), skin-fold thickness, body densitometry, soft tissue radiography, methods using fat-soluble gases, and total body potassium determination (von Döbeln, 1959; Mayer and Thomas, 1967; Wilmore and Behnke, 1969; Björntorp, 1974; Garrow, 1974; Pollock et al., 1976). Body composition, particularly in athletes, is a better guide for determining the desirable weight than the standard height-weight-age tables because of the high proportion of muscular content in their total body composition. However, for practical purposes, obesity is most conveniently diagnosed and judged by the application of simple height-weight methods. These measurements are at least quite accurate.

There are observations of animals as well as of humans showing that an increase in physical activity may have a profound effect on the body composition, increasing the protein/fat ratio, but the body weight may remain the same (Larsson, 1967; see Chap. 13).

As a general guide, the body weight at the age of about twenty is usually not far from the "ideal" body weight for most adults later on in life.

Obesity

If energy intake exceeds energy output, the excess energy will be stored mainly as adipose tissue. If this state of affairs is maintained over a period of time, it will lead to obesity. As Mayer and Thomas (1967) have pointed out, obesity is often the result of too little physical activity rather than of overeating.

Greene (1939) studied more than 200 overweight adult patients in whom the onset of obesity could be traced to a sudden decrease in activity. In a study concerning the relationship of body weight and physical activity in children, Bruch (1940) found that inactivity was characteristic of the majority of 160 obese children examined; 76 percent of the boys and 88 percent of the girls were physically inactive. It has been suggested by Rony (1940) that laziness or a decreased tendency to muscular activity is a primary characteristic of obese subjects, and Bronstein et al. (1942) found that the majority of the obese children they studied spent most of their leisure time in sedentary activities. Similar observations have been made by Simonson (1951), Juel-Nielsen (1953), and others. Other studies (Peckos, 1953; Fry, 1953) indicate that obese children do not have higher average energy intakes than do control children of the same height and age.

In a study by Johnson et al. (1956), energy intake and activity were systematically compared in paired groups of obese and normal-weight school girls. Their findings were that suburban high school girls were generally not very active, but nevertheless there was a marked difference between the groups in that the obese groups were much less active than the nonobese. Generally speaking, the time spent by the obese groups in sports or any other sort of exercise was less than half that spent by the lean girls. Energy intakes were generally larger in the nonobese girls than in the obese, and it was concluded that inactivity was more important than overeating in the development of obesity. It is interesting to note that when these school girls attended summer camp, they all, both obese and nonobese, almost without exception lost weight under a program of enforced strenuous activity in spite of simultaneous increased food intake.

Stefanik et al (1959), in a summer camp study, found that obese boys had significantly smaller energy intakes both during the school year and at the summer camp than the nonobese controls. Similar observations have been made by Bullen et al. (1964).

Chirico and Stunkard (1960) made a rough estimate of the degree of physical activity of obese and normal subjects of similar occupation and social status. They were asked to wear a pedometer for recording the number of steps throughout the day. The nonobese subjects were about twice as active as the obese ones. Durnin (1967) has also noticed that overweight individuals are less physically active than those who are nonobese.

Mayer and Thomas (1967) remark that comparison of the hunger and satiety pictures in obese and nonobese individuals suggests that abnormalities of satiety may be more prevalent in the obese than abnormalities of hunger. Nisbett (1968) made observations which fit this assumption. He told his subjects, normal as well as overweight, to eat as many sandwiches as they wanted, but a limited number were served at the table, the rest being left in a refrigerator in the same room. He found that the overweight individuals confronted with three sandwiches ate 57 percent more than those confronted with only one sandwich. In contrast, normal and underweight subjects were completely unaffected by the difference between experimental conditions; both groups ate as many sandwiches whether they were initially offered one or three. Obese individuals would habitually eat everything they were served in a typical meal, but they might not necessarily refill from the refrigerator. Schachter et al. (quoted from Nisbett, 1968) found that obese subjects ate no more food after being deprived for several hours than they did after being recently fed, while normal subjects ate much more food after they had been deprived.

Recent studies have distinguished between two types of obesity: One type in which there is an increased number of fat cells (hyperplasia-obesity), and another in which there is a normal number of fat cells, but each fat cell has an increased content of triglycerides (hypertrophy-obesity). (For a review see Björntorp, 1974.) In patients with the latter type, the prognosis for weight reduction is better than in the case of hyperplasia-obesity (Rognum and Kind, 1973). It appears that the number of fat cells is largely determined during the period from about 30 weeks' gestation to the age of about one year (Brook, 1973). According to Salans et al. (1973) there may be a second critical period for such a fat-cell hyperplasia between 9 and 13 years of age. (For additional reading on the subject of obesity, see Mayer, 1968 and Garrow, 1974.)

Summary It appears conclusively that obesity is to a large extent the result of reduced activity with a maintenance of an "old-fashioned" appetite center set for an energy expenditure well above the one typical for a sedentary individual. This is true for children as well as for adults. The reason for their different attitudes toward physical exertion is not clear. When studying very obese individuals on the bicycle ergometer, the authors noted that obese subjects complained of fatigue and felt exhausted at low work loads, when the blood lactic acid levels were not significantly elevated and would be easily tolerated by normal individuals. Obese individuals are characterized by their response to external rather than to internal cues in their eating.

It is particularly important to stimulate young individuals to regular physical activity, for such activity will in the long run effectively counteract obesity by keeping the individual within the range where spontaneous energy intake is properly regulated by the energy output. Obesity in infancy may increase the number of fat cells, causing predisposition for subsequent overweight. The treatment of obesity is particularly difficult in patients with an increased number of fat cells.

Slimming Diets

There are numerous slimming diets available, but most of them are not based on physiological principles. It should be emphasized that the most lenient way to reduce weight involves allowing adequate time for the measures to take effect. A weight loss of more than 0.5 to 1.0 kg per week is not to be recommended. The obese person's diet should be critically examined and an attempt made to eliminate about a thousand kilojoules per day, for example, by substituting artificial sweetenings for sugar and low-fat milk for whole milk, and by eliminating butter and all visible fat. Furthermore, a 2-km walk per day would add 400 kJ (100 kcal) to the expenditure, the result being a total net reduction in fatty tissue equivalent to about 1.5 MJ $\cdot$ day^{-1}. A combination of a small dietary restriction and an increased energy expenditure equivalent to a total of some 2 MJ (500 kcal) per day will mean the loss of half a kilo in body weight per week.

It is a common experience that when a slimming diet is instituted, there is often a sudden drop in body weight, followed by a more gradual decline. This initial drop, which may indeed have a gratifying and encouraging effect, may be due largely to loss of body water, which is regained later on. During the first day or two of energy restriction, stored glycogen is mobilized to cover the energy requirement. Since each gram of glycogen is stored together with almost 3 g of water, the mobilized glycogen liberates this water, which is eliminated and may thus account for the initial drop in body weight. When returning to an adequate diet, glycogen is again stored, together with the required amount of water. This, then, would account for the rapid transient weight gain when changing from energy restriction to adequate diet.

It is a mistake to avoid carbohydrates competely, as muscle and nerve cells need them in their metabolism. They are particularly important for anyone who is physically active.

Optimal Supply of Nutrients

With the exception of children during growth, of women during pregnancy, and sometimes of convalescents, the energy intake should not exceed the energy expenditure. The energy requirements vary naturally with the individual's physical activity. On the other hand, the need for most of the nutrients is comparatively independent of the individual's activity level; therefore the less active the individual, the higher is the content of the essential nutrients required per energy unit in order to obtain the desired optimal nutritional level (Wretlind, 1967). In a homogeneous population, a dietary tradition is usually quite similar. Blix (1965) found that a linear relation exists between daily energy supply and supply of many nutrients (protein, calcium, vitamin A, thiamin, iron). This is illustrated in Fig. 14-17. Through the centuries people obtained their choice of food geared to an energy output of 12.5 MJ (3000 kcal) or more, which also gave all the nutrients needed. For many of the nutrients mentioned, the energy intake should actually exceed about 10.5 MJ (2500 kcal) to ensure an adequate supply. In other words, the diet in Sweden, and no doubt in many other countries, seems to be adjusted to persons with a caloric requirement of at least 10.5 to 12.5 MJ (2500 to 3000 kcal) (Wretlind, 1967). This diet, however, is not suitable for the large number of "low energy consumers" who actually do exist, as exemplified in Fig. 14-17. This unsuitability may explain the rather common disturbances in the state of health and well-being associated with malnutrition even in countries with plenty of food available. It has been noted that in Sweden, as well as in many other countries, the percentage of energy from protein has remained remarkably constant from the

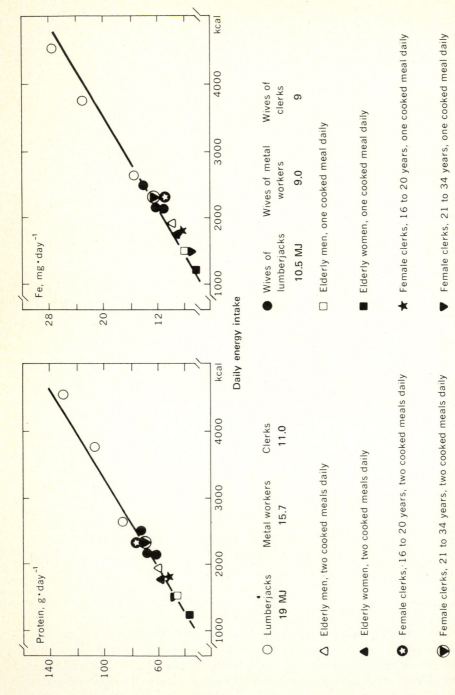

Figure 14-17 Relationship between total energy intake per day and the supply of protein and iron respectively. With the traditional food composition, some 10.5 MJ (2500 kcal) should be consumed in order to ensure a sufficient supply of these and other nutrients. *(From Blix, 1965.)*

end of the nineteenth century to the present day, or between 11 and 12 percent; the proportion of fat in the energy supply, however, has greatly increased from about 20 to 40 percent. Per capita, the amounts of energy of protein and carbohydrates have decreased, but the amount of fat has increased during this period of time (Wretlind, 1967).

There are two general ways of improving the nutritional conditions of the low energy consumers: (1) Change their food habits so that their diet consists of a higher content of essential nutrients per energy unit than it now does. A dietary habit should be developed so that the requirements of those who consume only 6.3 to 8.4 MJ (1500 to 2000 kcal) per day can be satisfied. Wretlind (1967) has pointed out, as an example, that a 20 percent increase in the iron content of the diet (from 10 to 12 mg · day^{-1}) would reduce the risk of anemia from 25 to 3 percent. (2) Low energy consumers should be stimulated to become high energy consumers by taking part in regular physical activity in one form or another. By an increase in energy output, they can, without the risk of obesity, eat more and automatically get a greater supply of essential nutrients.

Summary An improved diet in a modern society should include the following: (1) The fat content should be between 25 and no more than 35 percent of the energy intake, partially in the form of polyunsaturated fatty acids. (2) The amount of refined sugar in the diet should be low. (3) The consumption of vegetables, fruit, berries, low-fat milk, fish, lean meat, and cereal products should be relatively high. The content of protein, iron, calcium, and certain vitamins in the diet will then be high enough to satisfy the requirements of low energy consumers.

Low energy consumers should be stimulated to become more physically active. Such activity will allow them to eat more, which will automatically furnish them with more of the essential nutrients. The need for such nutrients does not vary significantly with the level of physical activity.

However, a high activity level and high energy intake do not necessarily guarantee that the individual is on the "safe side" from a nutritional point of view. Studies on girl swimmers with an energy balance of about 17 MJ (4000 kcal) per day have shown that some of them had an intake of some essential nutrients just at or even below the recommended values. Too much of their energy intake was based on sweetened soft drinks, cakes, candy, etc. (Hultén, unpublished data). It may be devastating if they suddenly give up their intense physical training but maintain their poor dietary habits.

REFERENCES

Adolph, E. F.: "Physiology of Man in the Desert," Interscience Publishers, Inc., New York, 1947.

Ahlborg, G., L. Hagenfeldt, and J. Wahren: Influence of Lactate Infusion on Glucose and FFA Metabolism in Man, *Scan. J. Clin. Lab. Invest.*, **36**:193, 1976.

AMA committee on the Medical Aspects of Sports, Wrestling and Weight Control, *JAMA*, **201**:541, 1967.

American College of Sports Medicine: Position Stand on Weight Loss in Wrestlers, *Med. Sci. Sports,* **8**(2):xi, 1976.

American Medical Association (editorial): Crash Diets for Athletes Termed Dangerous, Unfair, *A.M.A. News,* **2**:2, 1959.

Andersson, B.: The Thirst Mechanism as a Link in the Regulation of the "Milieu Intérieur," in "Les Concepts de Claude Bernard sur le Milieu Intérieur," p. 13, Masson et Cie, Paris, 1967.

Bergström, J.: Muscle Electrolytes in Man, Determined by Neutron Activation Analysis on Needle Biopsy Specimens, A Study on Normal Subjects, Kidney Patients, and Patients with Chronic Diarrhoea, *Scand. J. Clin. Lab. Invest.,* **14**(Suppl. 68):1962.

Bergström, J., and E. Hultman: Muscle Glycogen Synthesis after Exercise: An Enhancing Factor Localized to the Muscle Cells in Man, *Nature,* **210**:309, 1966.

Bergström, J., L. Hermansen, E. Hultman, and B. Saltin: Diet, Muscle Glycogen and Physical Performance, *Acta Physiol. Scand.,* **71**:140, 1967.

Bergström, J., E. Hultman, and A. F. Roch-Norlund: Muscle Glycogen Synthetase in Normal Subjects, *Scand. J. Clin. Lab. Invest.,* **29**:231, 1972.

Björntorp, P.: Effects of Age, Sex and Clinical Conditions on Adipose Tissue Cellularity in Man, *Metabolism,* **23**:1091, 1974.

Blix, G.: A Study on the Relation between Total Calories and Single Nutrients in Swedish Food, *Acta Soc. Med. Upsal.,* **70**:117, 1965.

Bronstein, I. P., S. Wexler, A. W. Brown, and L. J. Halpern: Obesity in Childhood, *Am. J. Diseases of Children,* **63**:238, 1942.

Bruch, H.: Energy Expenditure of Obese Children, *Am. J. Diseases of Children,* **60**:1082, 1940.

Bullen, B. A., R. B. Reed, and J. Mayer: Physical Activity of Obese and Nonobese Adolescent Girls Appraised by Motion Picture Sampling, *Amer. J. Clin. Nutr.,* **14**:211, 1964.

Cathcart, E. P., and W. A. Burnett: Influence of Muscle Work on Metabolism in Varying Conditions of Diet, *Proc. Roy. Soc. (Biol.),* **99**:405, 1926.

Chaveau, A.: Source et Nature du Potential Directement Utilisé dans le Travail Musculaire d'après les Exchanges Respiratoires, chez l'homme en Etat d'Abstinence, *C. R. A. Sci. (Paris),* **122**:1163, 1896.

Chirico, A. M., and A. J. Stunkard: Physical Activity and Human Obesity, *New Engl. J. Med.,* **263**:935, 1960.

Christensen, E. H., and O. Hansen: Arbeitsfähigkeit und Ehrnährung, *Skand. Arch. Physiol.,* **81**:160, 1939.

Christophe, J., and J. Mayer: Effect of Exercise on Glucose Uptake in Rats and Men, *J. Appl. Physiol.,* **13**:269, 1958.

Crittenden, R. H.: "Physiological Economy in Nutrition," Frederick A. Stokes Company, New York, 1904.

Darling, R. C., R. E. Johnson, G. C. Pitts, R. C. Consolazio, and P. F. Robinson: Effects of Variations in Dietary Protein on the Physical Well-being of Men Doing Manual Work, *J. Nutr.,* **28**:273, 1944.

Döbeln, W. von: Anthropometric Determination of Fat-free Body Weight, *Acta Med. Scand.,* **165**:37, 1959.

Dole, V. P.: Fat as an Energy Source, in K. Rodahl and B. Issekutz, Jr. (eds.), "Fat as a Tissue," McGraw-Hill Book Company, New York, 1964.

Durnin, J. V. G. A.: Activity Patterns in the Community, *Canad. Med. Ass. J.*, **96**:882, 1967.

Fordtran, J. S., and B. Saltin: Gastric Emptying and Intestinal Absorption during Prolonged Severe Exercise, *J. Appl. Physiol.*, **23**:331, 1967.

Fredholm, B. B.: Inhibition of Fatty Acid Release from Adipose Tissue by High Arterial Lactate Concentrations, *Acta Physiol. Scand.*, **77**(Suppl. 330):1969.

Fry, R. C.: A Comparative Study of "Obese" Children Selected on the Basis of Fat Pads, *Am. J. Clin. Nutr.*, **1**:453, 1953.

Garrow, J. S.: "Energy Balance and Obesity in Man," North-Holland Publ. Co., Amsterdam/American Elsevier Publ. Inc., New York, 1974.

Garza, C., N. S. Scrimshaw, and V. R. Young: Human Protein Requirements: the Effect of Variations in Energy Intake within the Maintenance Range, *Amer. J. clin. Nutr.*, **29**:280, 1976.

Gasnier, A., and A. Mayer: Recherches sur la Régulation de la Nutrition, I, Qualités et Côtes des Méchanismes Régulateurs Généraux; III, Méchanismes Régulateurs de la Nutrition et Intensité du Métabolisme, *Ann. Physiol.*, **15**:145, 1939.

Greene, J. A.: Clinical Study of the Etiology of Obesity, *Ann. Intern. Med.*, **12**:1797, 1939.

Grunt, J. A., J. F. Crigler, Jr., D. Slone and J. S. Soeldner: Changes in Serum Insulin, Blood Sugar and Free Fatty Acid Levels Four Hours after Administration of Human Growth Hormone to Fasting Children with Short Stature, *Yale J. Biol. and Med.*, **40**:68, 1967.

Hamilton, C. L.: Control of Food Intake, in W. S. Yamamoto and J. R. Brobeck (eds.), "Physiological Controls and Regulations," p. 274, W. B. Saunders Company, Philadelphia, 1965.

Hartley, L. H.: Growth Hormone and Catecholamine Response to Exercise in Relation to Physical Training, *Med. Sci. Sports*, **7**:34, 1975.

Hedman, R.: The Available Glycogen in Man and the Connection between Rate of Oxygen Intake and Carbohydrate Usage, *Acta Physiol. Scand.*, **40**:305, 1957.

Hermansen, L., E. Hultman, and B. Saltin: Muscle Glycogen during Prolonged Severe Exercise, *Acta Physiol. Scand.*, **71**:129, 1967.

Hultman, E.: Studies on Muscle Metabolism of Glycogen and Active Phosphate in Man with Special Reference to Exercise and Diet, *Scand. J. Clin. Lab. Invest.*, **19**(Suppl. 94):1967.

Hultman, E., and L. Nilsson: Liver Glycogen as Glucose-supplying Source during Exercise, in J. Keul (ed.): "Limiting Factors of Physical Performance," p. 179, Georg Thieme, Stuttgart, 1973.

Hutchinson, R. C.: Meal Habits and Their Effects on Performance, *Nutr. Abstr. Rev.*, **22**:283, 1952.

Ingle, D. J., and J. E. Nezamis: The Effect of Insulin on the Tolerance of Normal Male Rats to the Overfeeding of a High Carbohydrate Diet, *Endocrinology*, **40**:353, 1947.

Issekutz, B., Jr.: Effect of Exercise on the Metabolism of Plasma Free Fatty Acids, in K. Rodahl and B. Issekutz, Jr. (eds.), "Fat as a Tissue," chap. 11, McGraw-Hill Book Company, New York, 1964.

Issekutz, B., Jr., and H. Miller: Plasma Free Fatty Acids during Exercise and the Effect of Lactic Acid, *Proc. Soc. Exp. Biol. Med.*, **110**:237, 1962.

Issekutz, B., Jr., N. C. Birkhead, and K. Rodahl: Effect of Diet on Work Metabolism, *J. Nutr.*, **79**:109, 1963.

Issekutz, B., Jr., H. I. Miller, P. Paul, and K. Rodahl: Source of Fat Oxidation in Exercising Dogs, *Am. J. Phys.*, **207**:583, 1964.

Issekutz, B., Jr., H. I. Miller, P. Paul, and K. Rodahl: Aerobic Work Capacity and Plasma FFA Turnover, *J. Appl. Physiol.*, **20**:293, 1965.

Issekutz, B., Jr., and M. Allen: Effect of Metylprednisolone on Carbohydrate Metabolism of Exercising Dogs, *J. Appl. Physiol.*, **31**:813, 1971.

Issekutz, B., Jr., W. A. S. Shaw, and T. B. Issekutz: Effect of Lactate on FFA and Glycerol Turnover in Resting and Exercising Dogs, *J. Appl. Physiol.*, **39**(3):349, 1975.

Johnson, M. L., B. S. Burke, and J. Mayer: Relative Importance of Inactivity and Overeating in the Energy Balance of Obese High School Girls, *Am. J. Clin. Nutr.*, **4**:37, 1956.

Juel-Nielsen, N.: On Psychogenic Obesity in Children, *Acta Paediat. (Stockholm)*, **42**:130, 1953.

Keys, A.: Physical Performance in Relation to Diet, *Fed. Proc.*, **2**:164, 1943.

Krogh, A., and J. Lindhard: Relative Value of Fat and Carbohydrate as Source of Muscular Energy, *Biochem. J.*, **14**:290, 1920.

Ladell, W. S. S.: Effects of Water and Salt Intake upon Performance of Men Working in Hot and Humid Environments, *J. Physiol. (London)*, **127**:11, 1955.

Larsson, S.: Diet, Exercise, and Body Composition, in G. Blix (ed.), "Nutrition and Physical Activity," p. 132, Almqvist & Wiksell, Uppsala, 1967.

Mayer, J.: "Overweight Causes, Cost and Control," Prentice-Hall, Inc., Englewood Cliffs, N.Y., 1968.

Mayer, J., N. B. Marshall, J. J. Vitale, J. H. Christensen, M. B. Mashayekhi, and F. J. Stare: Exercise, Food Intake and Body Weight in Normal Rats and Genetically Obese Adult Mice, *Am. J. Physiol.*, **177**:544, 1954.

Mayer, J., P. Roy, and K. P. Mitra: Relation between Caloric Intake, Body Weight and Physical Work in an Industrial Male Population in West Bengal, *Am. J. Clin. Nutr.*, **4**:169, 1956.

Mayer, J., and B. Bullen: Nutrition and Athletic Performance, *Physiol. Rev.*, **40**:369, 1960.

Mayer, J., and D. W. Thomas: Regulation of Food Intake and Obesity, *Science*, **156**:328, 1967.

Metropolitan Life Insurance Company: "How to Control Your Diet," p. 4, 1963.

McGilvery, R. W.: The Use of Fuels for Muscular Work, in N. Howald and J. R. Poortmans (eds.), "Metabolic Adaptation to Prolonged Physical Exercise," p. 12, Birkhäuser Verlag, Basel, 1975.

Miller, H. I., B. Issekutz, Jr., P. Paul, and K. Rodahl: Effect of Lactic Acid on Plasma Free Fatty Acids in Pancreatectomized Dogs, *Am. J. Physiol.*, **207**:1226, 1964.

Nisbett, R. E.: Determinants of Food Intake in Obesity, *Science*, **159**:1254, 1968.

Owen, O. E., A. P. Morgan, H. G. Kemp, J. M. Sullivan, M. G. Herrera, and G. F. Cahill, Jr.: Brain Metabolism during Fasting, *J. Clin. Invest.*, **46**:1589, 1967.

Paul, P.: Effects of Long Lasting Physical Exercise and Training on Lipid Metabolism, in H. Howald and J. R. Poortmans (eds.), "Metabolic Adaptation to Prolonged Physical Exercise," p. 156, Birkhäuser Verlag, Basel, 1975.

Peckos, P. S.: Caloric Intake in Relation to Physique in Children, *Science*, **117**:631, 1953.

Pettenkofer, M. von, and C. Voit: Üntersuchungen über dem Stoffverbrauch des normalen Menschen, *Z. Biol.*, **2**:459, 1866.

Pollock, M. L., T. Hickman, Z. Kendrick, A. Jackson, A. C. Linnerrud, and G. Dawson: Prediction of Body Density in Young and Middle-aged Men, *J. Appl. Physiol.*, **40:**300, 1976.

Pruett, E. D. R.: "Fat and Carbohydrate Metabolism in Exercise and Recovery, and Its Dependence upon Work Load Severity," Institute of Work Physiology, Oslo, 1971.

Reichard, G. A., B. Issekutz, Jr., P. Kimbel, R. C. Putnam, N. J. Hochella, and S. Weinhouse: Blood Glucose Metabolism in Man during Muscular Work, *J. Appl. Physiol.*, **16:**1001, 1961.

Robinson, S., and A. H. Robinson: Chemical Composition of Sweat, *Physiol. Rev.*, **34:**202, 1954.

Rodahl, K.: "Nutritional Requirements under Arctic Conditions," Norsk Polarinstitutts Skrifter no. 118, Oslo University Press, 1960.

Rodahl, K., S. M. Horvath, N. C. Birkhead, and B. Issekutz, Jr.,: Effects of Dietary Protein on Physical Work Capacity during Severe Cold Stress, *J. Appl. Physiol.*, **17:**763, 1962.

Rodahl, K., H. I. Miller, and B. Issekutz, Jr.,: Plasma Free Fatty Acids in Exercise, *J. Appl. Physiol.*, **19:**489, 1964.

Rodahl, K., and B. Issekutz, Jr.,: Nutritional Effects on Human Performance in the Cold, in L. Vaughan (ed.), "Nutritional Requirements for Survival in the Cold and at Altitude," p. 7, Arctic Aeromedical Laboratory, Fort Wainwright, Alaska, 1965.

Rognum, T. O., and E. Kindt: Fettcellestörrelse hos 14 kvinnelige overvektige pasienter, *T. norske laegeforen.*, **93:**1737, 1973.

Rony, H. R.: "Obesity and Leanness," Lea & Febiger, Philadelphia, 1940.

Salans, L. B., S. W. Cushman, and R. E. Weismann: Studies of Human Adipose Tissue, Adipose Cell Size and Number in Non-obese and Obese Patients, *J. Clin. Invest.*, **52:**929, 1973.

Saltin, B.: Aerobic Work Capacity and Circulation at Exercise in Man: With Special Reference to the Effect of Prolonged Exercise and/or Heat Exposure, *Acta Physiol. Scand.*, **62**(Suppl. 230):1964.

Shreeve, W. W., N. Baker, M. Miller, R. A. Shipley, G. E. Ingefy, and J. W. Craig: C[14] Studies in Carbohydrate Metabolism: Oxidation of Glucose in Diabetic Human Subjects, *Metabolism*, **5:**22, 1956.

Simonson, E.: Influence of Nutrition on Work Performance, Nutrition Fronts in Public Health, Nutrition Symposium Series no. 3, p. 72, National Vitamin Foundation, New York, 1951.

Staff, P. H., and S. Nilsson: Fluid and Glucose Ingestion during Prolonged Severe Physical Activity, *Tidsskrift for Den Norske Laegeforening*, **16:**1235, 1971.

Stefanik, P. A., F. P. Heald, Jr., and J. Mayer: Caloric Intake in Relation to Energy Output of Obese and Non-obese Adolescent Boys, *Am. J. Clin. Nutr.*, **7:**55, 1959.

Wahren, J., P. Felig, L. Hagenfeldt, R. Hendler, and G. Ahlborg: Splanchnic and Leg Metabolism of Glucose, Free Fatty Acids and Amino Acids during Prolonged Exercise in Man, in H. Howald and J. R. Poortmans (eds.), "Metabolic Adaptation to Prolonged Physical Exercise," p. 144, Birkhäuser Verlag, Basel, 1975.

Weis-Fogh, T.: Metabolism and Weight Economy in Migrating Animals, Particularly Birds and Insects, in G. Blix (ed.), "Nutrition and Physical Activity," p. 84, Almqvist & Wiksell, Uppsala, 1967.

Wilmore, J. H., and A. Behnke: An Anthropometric Estimation of Body Density and Lean Body Weight in Young Men, *J. Appl. Physiol.*, **27:**25, 1969.

Temperature Regulation 15

CONTENTS

HEAT BALANCE

METHODS OF ASSESSING HEAT BALANCE

MAGNITUDE OF METABOLIC RATE

EFFECT OF CLIMATE
Cold   Heat

EFFECT OF WORK

TEMPERATURE REGULATION

ACCLIMATIZATION
Heat   Cold

LIMITS OF TOLERANCE
Normal climate   Failure to tolerate heat   Upper limit of temperature
tolerance   Age   Sex   State of training

COORDINATED MOVEMENTS

MENTAL WORK CAPACITY

WATER BALANCE
Normal water loss   Thirst   Water deficit

PRACTICAL APPLICATION
Physical work   Warm-up   Radiation   Air motion   Clothing   Microclimate

Chapter 15

Temperature Regulation

The protected human being may well tolerate variations in environmental temperature between $-50°C$ and $100°C$. But a person can tolerate a variation of only about $4°C$ in deep body temperature without impairment of optimal physical and mental work capacity. Changes in body temperature affect cellular structures, enzyme systems, and numerous temperature-dependent chemical reactions and physical processes that take place in the body. The maximal limits which the living cell can tolerate range from about $-1°C$ at one end of the scale, when the ice crystals formed during freezing break the cell apart, to thermal heat coagulation of vital proteins in the cell at about $45°C$ at the other end of the scale. Only for shorter periods of time can they tolerate an internal temperature exceeding $41°C$. In fact, many animals, including humans, live their entire lives only a few degrees removed from their thermal death point.

The hot end of the scale is more of a problem than the cold end, for people can protect themselves more easily against overcooling than against overheating. Consequently, the controlling mechanism for temperature regulation is particularly geared to protect the body tissues against overheating (Hardy, 1967). (Various aspects of the physiology of temperature regulation have been presented by Hardy et al., 1970; Bligh and Moore, 1972; Wyndham, 1973; Cabanac, 1975.)

HEAT BALANCE

If the heat content of the body is to remain constant, heat production and heat gain must equal heat loss, according to the equation: $M \pm R \pm C - E = 0$ where M = metabolic heat production, R = radiant heat exchange (positive if the environment is hotter than the skin temperature, but negative if the temperature of the environment is lower than that of the skin), C = convective heat exchange (positive if the air temperature is higher than that of the skin, negative if the reverse), E = evaporative heat loss. This equation is valid only for conditions when the body temperature is constant. If the body temperature varies, a correction has to be introduced, and the following equation is applicable: $M \pm S \pm R \pm C - E = 0$, where S = storage of body heat (Winslow, Gagge, and Herrington, 1939). S is positive if the body heat content is falling, negative if the heat content increases. The specific heat of most tissues is about 0.83. Conductive heat exchange (K) is in most conditions negligible but increases in importance during such activities as swimming, since water has a heat-removing capacity which is some 20 times that of air.

One very important function of the blood circulation is to transport heat: to cool or to heat various tissues as may be needed, and to carry excess body heat from the interior of the body to the body surface, or skin. In this function the blood is very effective, for it has a high heat capacity (0.9), which means that the blood may carry a great deal of heat with only a moderate increase in temperature. Conductance of the tissue ($kJ \cdot m^{-1} \cdot hr^{-1} \cdot °C^{-1}$) is the amount of heat given off per square meter body surface per hour and per degree temperature difference between the interior of the body and its surroundings. When the skin blood flow increases, there is a rise in skin temperature, and the conductance increases. When the skin blood flow is reduced, there is a drop in skin temperature, and the conductance is reduced, i.e., the insulating value of the skin is increased.

This control of body temperature, the balance between overcooling and overheating, is the role of temperature regulation. This regulation endeavors to keep the temperature of certain tissues, such as the brain, heart, and guts, relatively constant. Within the body, the temperature is by no means uniform. The greatest gradient is found between the "shell" (the skin) and the "core" (deep central areas including heart, lungs, abdominal organs, and brain). The temperature of the core may be as much as 20°C higher than that of the shell, but the ideal difference between shell and core is about 4°C at rest. Even within the core the temperature varies from one place to another. This complicates the calculation of the heat content of the body and makes it difficult to study temperature regulation. Evidently the term *body temperature* is a misnomer. The maintenance of a normal body temperature is actually quite compatible with considerable gains or losses of heat. The problem is, which temperatures are being regulated.

METHODS OF ASSESSING HEAT BALANCE

Measurements of the *deep body temperature* (core temperature) may be accomplished with the aid of mercury thermometers, thermocouples, or thermistors. The classical site of measurement is the rectum. Since the temperature in the rectum (T_r) varies with distance from the anus, it is customarily measured at a depth of 5 to 8 cm. This rectal temperature is, in a resting individual, slightly higher than the temperature of the arterial blood; it is about the same as liver temperature, but slightly lower (0.2 to 0.5°C) than the part of the brain where the thermal regulatory center is located. During physical exertion or exposure to heat, the temperature of this part of the brain increases more rapidly than does the rectal temperature, and the time interval until a new temperature equilibrium is established has been found to be about 30 min (Nielsen, 1938). The temperature increase or decline in the brain and in the rectum is of the same magnitude, however. The rectal temperature is therefore a representative indicator for the purpose of assessing changes in the deep body temperature, provided the measurement is made under steady-state conditions, i.e., after some 30 to 40 min. It has been found that the eardrum temperature is a fairly good indication of the actual brain temperature. This may be obtained by placing a thermocouple, introduced through the ear, against the eardrum (Benzinger and Taylor, 1963). However, the eardrum temperature is not quite identical with the temperature in the thermoregulatory center. Another alternative is to measure the temperature in the esophagus, which is relatively accessible for such measurements. Although the temperature of the esophagus is not identical with any of the above-mentioned core temperatures, it generally changes parallel with these temperatures (Saltin and Hermansen, 1966; Nielsen, 1969). In work and heat studies, measurement of the oral temperature has its limitations (Strydom et al., 1965).

The skin temperature is measured with the aid of a radiometer, or by placing thermocouples or thermistors on the skin at certain locations. The mean skin temperature ($\overline{T}_s$) is calculated by assigning certain factors to each of the measurements in proportion to the fraction of the body's total surface area represented by each specific area, as follows (Hardy and DuBois, 1938):

Head	0.07
Arms	0.14
Hands	0.05
Feet	0.07
Legs	0.13
Thighs	0.19
Trunk	0.35
	1.00

For the calculation of the heat content of the body, the following equation may be applied:

$$\text{Heat content} = 0.83 \ W \ (0.65 T_r + 0.35 \ \overline{T}_s)$$

where W represents body weight, 0.83 is the specific heat of the body, and 0.65 and 0.35 are the factors assigned to the rectal and the mean skin temperatures, respectively (Burton, 1935). (The specific heat of the body has been assumed to be 0.83 for nearly a century, but it varies with the individual's body composition and may range from 0.70 up to 0.85; see Hardy et al., 1970, p. 345. It is also clear that the heat content of the body during exercise in a hot environment cannot be determined from any fixed ratio of $\overline{T}_s$ and $\overline{T}_r$; see Wyndham, 1973.)

The *metabolic rate*, or the magnitude of heat production, is assessed by the measurement of oxygen uptake. The volume of 1 liter oxygen consumed corresponds to approximately 20 kJ (4.9 kcal). Human calorimeters large enough to measure the rate of heat production by direct calorimetry have been constructed and are in use in certain laboratories.

The *evaporative heat loss* (E) plays a major role in the cooling of the skin and the blood during exposure to heat. At normal skin temperature, the evaporation of 1 liter of sweat requires 2.4 MJ (580 kcal). The magnitude of the sweat loss may be estimated simply by weighing the subject nude or dressed in dry clothing, before and after the experiment, and by weighing food and fluid ingested and stools and urine voided during the period of observation. Furthermore, weight loss due to respiratory gas exchange should be included in the calculation, and may be accounted for as follows (Snellen, 1966):

$$C_{ge} = \dot{V}_{o_2} \ (1.977 \cdot R - 1.429)$$

where C_{ge} = weight loss in g $\cdot$ min^{-1} due to respiratory gas exchange
$\dot{V}_{o_2}$ = oxygen uptake in liters $\cdot$ min^{-1} STPD
R = respiratory quotient
1.977 = weight in g of 1 liter CO_2 STPD
1.429 = weight in g of 1 liter O_2 STPD

Such measurements are valid for the calculation of heat balance only as long as all the sweat produced during the experiment is actually evaporated. On the other hand, whether the produced sweat is evaporated or part of it has run off the body, the sweat rate is an indication of the magnitude of the heat stress.

The *air temperature* affects convective heat loss or gain (C) and is most conveniently measured with the aid of the usual mercury thermometer. If the thermometer is exposed to radiation, it should be shielded. A piece of tinfoil may be used, but care should be taken to allow free air passage around the thermometer.

The *humidity of the air* may be measured with the aid of a sling psychrometer or an electronic device for measuring humidity. The rate of evaporation of the produced sweat is greatly dependent on the humidity of the air.

The *air movement*, which may be measured by a hot-wire animometer, affects both convective heat exchange (C) and evaporative heat loss (E).

The *radiant heat exchange* (R) depends on the temperature difference between the individual and the surroundings. This may be assessed by the values obtained from a mercury thermometer placed in a hollow, spheric black copper container with a diameter of 15 cm (a globe thermometer). Because of rapidly changing radiation intensities typical for many industrial operations, it is often almost impossible to obtain a true picture of the intensity of the radiation.

Afferent impulses from thermoreceptors in the skin, which respond to very rapid temperature changes, may signal peripheral thermal disturbances long before the central core temperature has been affected. Such impulses, integrated with various sensory input from the rest of the body, result in feelings of thermal comfort or discomfort (Hardy, Stolwijk, and Gagge, 1971). More specifically, the sensation of thermal comfort appears to be the result of the interaction between signals evoking temperature sensation, input signals for temperature regulation, and sensations arising from thermoregulatory activities associated with skin blood flow, sweating, and shivering (Hardy et al., 1971). Generally speaking, the state of thermal neutrality or thermal comfort is characterized by core temperature of 36.6 to 37.1°C, and skin temperatures between 32° and 35.5°C (Precht et al., 1973). Apparently, comfort temperature may be subject to some degree of adaptation or accustomization. Thus, according to Burton and Edholm (1969), the preferred indoor temperature in Britain is about 18°C as against 24°C in the United States.

MAGNITUDE OF METABOLIC RATE

Human beings may be considered to be tropical animals inasmuch as they require an ambient temperature of 28°C if they, when nude, are to remain in thermal balance, at rest, i.e., if they are to maintain a resting metabolic rate, and at the same time be in the so-called comfort zone. The oxygen uptake under these conditions is about 0.20 to 0.30 liter $\cdot$ min^{-1}. It is slightly higher when the body size is larger. This corresponds to a production of 60 to 90 kcal $\cdot$ hr^{-1}, or 70 to 100 watts. This energy is the by-product of metabolic processes which are essential for the maintenance of life. This produced heat makes up for the heat lost through convection (C), radiation (R), and evaporation (E). Under these conditions, $C + R$ accounts for about 75 percent of the heat loss, and E accounts for only 25 percent. Heat loss through the lungs, through the saturation of the air with water vapor during respiration, accounts for about two-fifths of E. Not all the rest of E is due to the evaporation of sweat, for part of the water loss through the skin occurs without the involvement of the sweat glands, the so-called perspiratio insensibilis. The total water loss through the skin amounts to a minimum of 0.5 liter $\cdot$ day^{-1}.

Muscular work is associated with an increase in metabolic rate. Since the mechanical efficiency (the ratio of external work to the extra energy used) may vary from 0 to 25 percent depending on the kind of work, at least 75 percent of the energy used is converted into heat. Well-trained athletes may, during short work periods of 5- to 10-min duration, attain an oxygen uptake of up to about 6 liters · min^{-1} (2,000 watts), and during more prolonged work as much as 4 to 5 liters · min^{-1}. The amount of heat thus produced during 1 hr could theoretically increase the body temperature of a 70-kg individual from 37°C to about 60°C if the excess heat were not dissipated. With fever, or during shivering due to cold exposure, the heat production may be increased two- to fourfold.

EFFECT OF CLIMATE

Cold

For a nude resting individual, the ideal ambient temperature is about 28°C. Under such conditions the mean skin temperature is about 33°C, and the temperature of the core is about 37°C. The temperature gradient from core to skin is then adequate to facilitate the transfer of the excess heat from the metabolically active tissues to the surroundings. Of the total amount of blood (the circulating blood volume) pumped by the heart, about 5 percent flows through the blood vessels of the skin. If the ambient temperature drops, the temperature difference between the skin and the environment is increased; this causes an increased heat loss through convection and radiation. A reduced heat flow to the skin would result in a gradual lowering of the skin temperature. This would produce a reduced temperature gradient between the skin and the environment. Actually, a reduction in the "conductance of the tissue" occurs, partly because of *vasoconstriction of the skin's blood vessels* causing a reduction in blood flow, partly because the blood in the veins of the extremities is deviated from the superficial to the deep veins. Because of the proximity of the deep veins to the arteries, a heat exchange occurs (Fig. 15-1). Owing to this system of countercurrent, in a subject exposed to an ambient temperature of 9°C, the blood leaving the heart will have a temperature of about 37°C. As it is flowing through the arm, it will be gradually cooled so that by the time it reaches the hand, it may have dropped to about 21°C. The returning venous blood absorbs a considerable part of the heat as the blood flows through the arm. In other words, cooling of arterial blood flowing through the arteries of the limbs depends on the rewarming of cold blood returning in adjacent veins from more distal areas. Thus, a cooling of the body core is prevented (Bazett et al., 1948; Schmidt-Nielsen, 1963). In a hot environment, on the other hand, the blood from the limbs returns primarily through superficial veins, thus facilitating further cooling of the blood.

The major effect of the vasoconstriction of the skin is a sudden displacement of the blood volume from the skin to the central circulation, as evidenced by a sudden rise in central blood volume, and a redistribution of blood from superficial to deep veins (Rowell, 1974).

The heat exchange between arterial and venous blood is still more important for arctic animals. Sea birds swim in the icy sea, and the large surfaces of their bare feet are exposed to the cooling effect of the water. However, little body heat is lost because their feet cool down to the temperature of the water, thus reducing the loss of metabolic heat (Irving, 1967). Warm

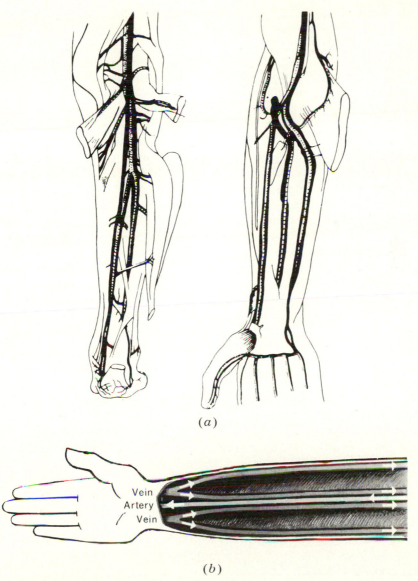

(a)

(b)

Figure 15-1 (a) The anatomic relationship between the arteries and the deep veins in the forearm which constitute a heat exchange system in the extremity. *(From Todt, 1919.)* (b) Schematic illustration of the two possible ways that venous blood flow from the hand, the superficial veins, or the deeper ones anatomically close to the arteries can make a heat exchange possible.

feet of a gull or a duck standing on snow or ice would cause it to melt, and soon the feet would be frozen solid to the ground where they stood! Hogs, naked as a man in a cold environment, and narwhal, walrus, or seals in arctic waters can prevent the loss of heat by having a very cold skin. These animals have a considerable layer of subcutaneous fatty tissue as an effective insulation when their blood vessels constrict. It is an interesting observation that fats in the peripheral regions have a lower melting point than those in the warmer internal tissues; if they did not, the peripheral tissues and legs would become too inflexible in cold weather (Irving, 1967).

Thickly furred animals can use their bare extremities to release excess heat from the body (heat can also be dissipated by evaporation from the mouth and tongue during panting).

Through peripheral vasoconstriction, a sixfold increase in the insulating capacity of the skin and subcutaneous tissues is possible (Burton, 1963). This vascular constriction is particularly active in the fingers and toes. It has been estimated that the blood flow through the fingers may vary a hundredfold or more (from 0.2 to 120 ml blood $\cdot$ min^{-1} $\cdot$ 100 g^{-1} of tissue) (Robinson, 1963). The disadvantage of this vasoconstriction is that the temperature in the peripheral tissues may approach that of the environment (Fig. 15-2). For this reason, one is apt to suffer cold fingers and toes. The blood vessels of the head are far less subject to active vasoconstriction.

Another protective mechanism to maintain heat balance is an *increase of the metabolic rate*, mediated through muscle activity in the form of shivering by a reflex mechanism. Shivering consists of a synchronous activation of practically all muscle groups; antagonists are made to contract against one another. Since the mechanical efficiency of shivering is 0 percent, the heat production is relatively high, and the metabolic rate may increase to 2 to 4 times that of

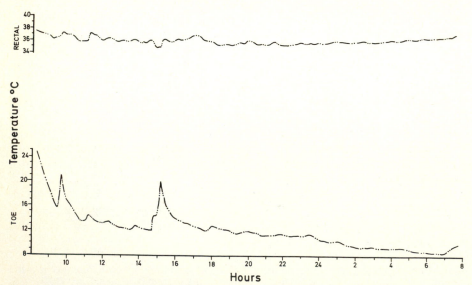

Figure 15-2 Rectal and toe temperatures in a nude subject exposed to 8°C continuously for 3 days. By the end of 24 hours, the skin temperature of the big toe had dropped to about 8°C, the temperature of the ambient air.

resting metabolic rates. Ordinary dynamic muscular work, on the other hand, may easily increase the metabolic rate tenfold or more.

Even though the resources for maintaining the core temperature may be quite effective, this maintenance does to some extent take place at the expense of the peripheral tissues, the shell. Local cold injury may be the result in extreme conditions. During prolonged severe cold exposure, even the core temperature may drop. Under certain circumstances, especially when exposed to cold water, obese individuals may be better off in the cold than lean individuals because of the insulating value of the adipose tissue (see Keatinge, 1969; Holmér and Bergh, 1974). An unclothed individual of average body build will be helpless from hypothermia after approximately 20 to 30 min in water at 5°C and after $1^1/_2$ to 2 hrs in water at 15°C. With thick conventional clothing, these times can be substantially prolonged (see Keatinge, 1969). Keatinge points out that one should not exercise in cold water in an attempt to keep warm; it will have the reverse effect. In one study on swimming in water at 18°C the oxygen uptake was elevated by approximately 0.5 liter $\cdot$ min^{-1} compared with swimming at the same speed in warmer water. For lean subjects there was, however, a significant drop in esophageal temperature. The maximal oxygen uptake and heart rate were markedly reduced. The effect of exposure to moderately cold water for 20 to 30 min will thus seriously impair physical performance and swimming may be hazardous (Holmér and Bergh, 1974; Davies et al., 1975).

Vangaard (1975) investigated the extremity temperature during general cold stress. The changes in local temperatures were found equal to those seen under circulatory arrest. He also explored the relationship between local temperature and nervous conduction velocity in a peripheral motor nerve in subjects exposed to a minor cold stress. The decrease in conduction velocity was 15 m $\cdot$ s^{-1} for each 10°C fall in temperature. At a local temperature of 8 to 10°C, a complete nervous block was established. This may explain the common finding that a local cooling in the extremities may be accompanied by a rapid onset of physical impairment. (For a more detailed discussion, see LeBlanc, 1975.)

Summary When a resting person is exposed to a cold environment, there are two main mechanisms by which a lowering of the body temperature can be prevented: (1) a reduction in the peripheral blood flow with a secondary drop in the skin temperature (reducing the heat loss by radiation and convection); and (2) an increase in the metabolic heat production by shivering. In a hypothermic person the combined effect of reduced efficiency and reduced maximal aerobic power will impair physical performance.

Heat

When the nude resting body is exposed to heat (when the ambient temperature exceeds 28°C), or during muscular work, the heat content of the body tends to increase. Under such conditions, the blood vessels of the skin dilate, venous return in the extremities takes place through superficial veins, and the conductance of the tissue increases. In the comfort zone, the skin blood flow, as mentioned, amounts to about 5 percent of the cardiac minute volume; in extreme heat, it may increase to 20 percent or more. The increased heat flow to the skin increases the skin temperature. If the temperature of the surroundings is lower than that of the skin, heat loss is facilitated through $C + R$. If the heat load is sufficiently large, the sweat glands are activated, and as the produced sweat is evaporated, the skin is cooled. It has been calculated that there are at least 2 million sweat glands in the skin. Recruitment of sweat glands from different areas of the body is not consistent between individuals (see Nadel et al., 1971). The activity of individual sweat glands follows a cyclic pattern. The sweat starts to drop off the skin when the sweat intensity has reached about one-third the maximal evaporative capacity (Kerslake, 1963).

The individual difference in the capacity for sweating is quite large; some people have no sweat glands at all. As a person becomes accustomed to heat, the amount of sweat produced in response to a standard heat stress increases. A person may produce several liters of sweat per hour. Workers exposed to intense heat may lose as much as 6 to 7 liters of sweat in the course of the working day. Sweat loss up to 10 to 12 kg in 24 hr has been reported (Leithead and Lind, 1964).

> During prolonged exposure to a hot environment, there is a gradual reduction in the sweat rate, even if the body water loss is replaced at the same rate. This decline in sweat rate is greater in humid than in dry heat, greater "when the men wore Army tropical uniforms than when they wore only broadcloth shorts" (Gerking and Robinson, 1946). The explanation of this "fatigue" of the sweat mechanism is presently not clear. Ahlman and Karvonen (1961) report that exercise could again induce sweating after the sweating had ceased during repeated thermal stimuli in a sauna bath. The suppression in sweating is related to the wetting of the skin. Drying the skin with a towel at regular intervals or superimposing an increase in air velocity around an exercising subject will enhance the sweating rate. When water is evaporated from the skin surface, the solutes are left behind. An increase in osmotic pressure on the skin surface seems to produce an increase in the sweat secretion rate (see Nadel and Stolwijk, 1973).

The sweat contains different salts, notably NaCl, in varying concentrations, and excessive sweating may therefore cause a considerable salt loss.

Exposure of resting subjects to direct whole-body heating causes a rise in skin temperature which is accompanied by an almost immediate rise in heart rate and cardiac output, and a drop in total peripheral resistance and splanchnic blood flow. Cardiac output may in some cases rise to levels 3 to 4 times the baseline values (Rowell, 1974). The excess cardiac output is diverted to the skin. This increased skin blood flow is further supplemented by a reduction in splanchnic, renal, and perhaps even muscle blood flow. As the skin temperature rises, cutaneous resistance vessels relax and cutaneous venous pressure and volume increase until a new level of wall tension is reached. This volume displacement is augmented by splanchnic vasoconstriction, which reduces distending pressure in the splanchnic veins, allowing them to empty passively. In this way, blood may be displaced from central to cutaneous venous beds

(Rowell, 1974). The skin blood flow will increase by local heating. This blood flow can be further elevated if the whole body skin temperature and/or core temperature rise. Keeping the local skin temperature high will not abolish skin vasoconstriction response to lower body negative pressure (simulating the effect of a mild hemorrhage by pooling blood in the leg veins). Thus, local factors and reflex influences to a skin area can interact so as to modify the degree but not the pattern of the skin vasomotor response (Johnson et al., 1976).

Exposure to dry, hot air in the form of the Finnish sauna bath has been studied by Eisalo (1956). Both in healthy and in hypertensive subjects, he found that the cardiac output increased by an average of 73 percent (65 percent in the hypertensive group), the mean circulation time decreased by almost 60 percent, and the pulse rate increased by more than 60 percent. There was a slight but significant decrease in systolic blood pressure in healthy subjects, but in the hypertensive subjects it decreased by 29 mm Hg on the average. Twenty minutes after the sauna, the mean decrease was 54 mm Hg. The diastolic blood pressure remained practically unchanged in the healthy subjects, while it decreased significantly in the hypertensive subjects. There was a statistically highly significant decrease in peripheral resistance in both groups of subjects.

Summary   The person who is exposed to a hot environment experiences (1) a vasodilation in the skin, making an increased heat transfer from "core" to "shell" possible; and perhaps (2) an activation of the sweat glands, with the evaporation of the sweat taking heat from the body and causing an evaporative heat loss.

EFFECT OF WORK

Since the mechanical efficiency of the human body is only about 25 percent, roughly 75 percent of the total energy utilized is converted into heat. The greater the work intensity, the greater the total amount of heat produced. This excess heat has to be removed and dissipated in order to prevent overheating and hyperthermia.

Figure 15-3 shows an example of how the thermal balance is maintained during muscular work of different intensity over a 1-hr period. The body temperature increases during work, and this temperature elevation may be interpreted as the result of an active regulation (Christensen, 1931; Nielsen, 1938; Berggren and Christensen, 1950). The difference between "energy output" and "heat production" in Fig 15-3 is an expression of the mechanical efficiency (about 23 percent), and the difference between "heat production" and "total heat loss" is a consequence of the elevated body temperature. Note that the convective and radiative heat losses are almost constant despite the large variations in heat production. Evaporation takes care of the extra heat loss as the work load increases.

Figure 15-4 shows how the rectal temperature, measured after about 45

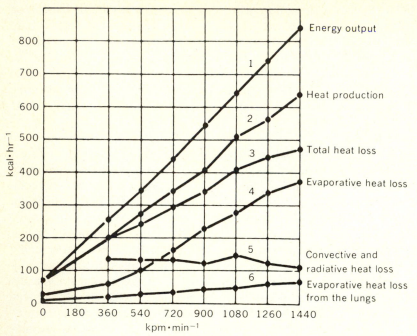

Figure 15-3 Heat exchange at rest and during increasing work intensities (expressed in kilopond meters per minute along the abscissa) in a nude subject at a room temperature of 21°C. Further explanation in text. *(From M. Nielsen, 1938.)*

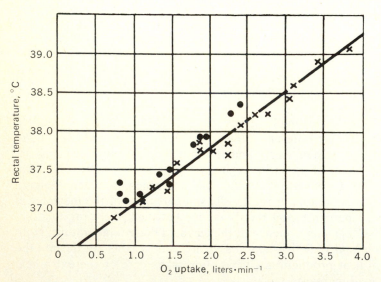

Figure 15-4 The relationship between oxygen uptake and body temperature in work with the legs (X) and work with the arms (●). *(From M. Nielsen, 1938.)*

min of work on the bicycle ergometer, increases linearly with the O_2 uptake, at least up to an energy demand of about 75 percent of the individual's maximal aerobic power. At higher metabolic rates the increase in core temperature seems to be curvilinearly related to the oxygen uptake (Davies et al., 1976). This end temperature does not depend on the absolute magnitude of the energy output but on the level of metabolism relative to the individual's maximal aerobic power. A subject with a maximal O_2 uptake of 2.0 liters $\cdot$ min^{-1} attains a body temperature of about 38°C during a work load which demands an O_2 uptake of 1.0 liter $\cdot$ min^{-1}, i.e., 50 percent of the person's maximal aerobic power. A subject with a maximum of 5.0 liters O_2 $\cdot$ min^{-1} may expend 2½ times more energy (O_2 uptake of 2.5 liters $\cdot$ min^{-1}) without the body temperature exceeding 38°C (I. Åstrand, 1960).

Saltin and Hermansen (1966) have further studied the relationship between body temperature (esophageal temperature) and oxygen uptake. Figure 15-5 illustrates their data. There is a large scatter of the individual curves when temperature is related to the oxygen uptake (left panel), but the curves come closer together when the oxygen uptake is expressed as a percentage of the individual's maximal oxygen uptake (right panel). If by training there is an increase in the individual's maximal oxygen uptake, the core temperature will become reduced at a given submaximal oxygen uptake. Therefore, the core temperature is approximately 38°C, when the oxygen uptake is 50 percent of the maximal in both untrained and trained subjects. In studies on the effect of an acute reduction in the maximal oxygen uptake by exposing the subjects to high altitude, the data are conflicting. In some subjects the core temperature was related to the absolute oxygen uptake; in others the relationship followed very closely the changes in the subject's highest oxygen uptake (i.e., it behaved as in Fig. 15-5, right panel). (See Saltin in Hardy et al., 1970, p. 316.) A reduction

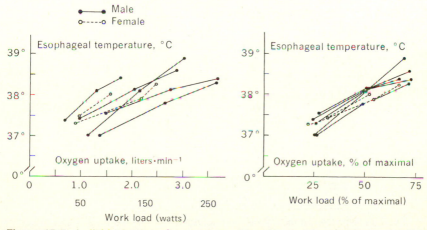

Figure 15-5 Individual values for esophageal temperature in relation to oxygen uptake or external work load (left panel) and to oxygen uptake in percent of the individual's maximal oxygen uptake (right panel). (From Saltin and Hermansen, 1966.)

in maximal oxygen uptake induced by a moderate CO poisoning increased the core temperature at a given submaximal oxygen uptake, but it still correlated with the relative work load (Nielsen, 1971).

Figure 15-6 shows the relationship between muscle, rectal, and esophageal temperatures, measured simultaneously, and the oxygen uptake as a percentage of the individual's maximal oxygen uptake. As expected, the highest temperatures are seen in the working muscles where most of the heat is produced.

Nielsen (1938) had a subject perform a certain amount of work, 150 watts, at ambient temperatures varying between 5°C and 36°C. After 30 to 40 min of work, the rectal temperature was the same, regardless of the room temperature. Since the mechanical efficiency was constant, the heat dissipation had to be the same in all these experiments. Figure 15-7 shows how radiation and convection (R + C) accounted for about 70 percent of the heat loss at the lower ambient temperatures. The skin temperature was then 21°C. In the experiment carried

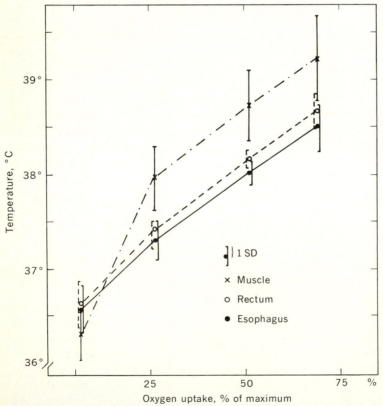

Figure 15-6 Average temperature measured simultaneously in the esophagus, the rectum, and the working muscle in relation to the oxygen uptake in percent of the individual's maximal oxygen uptake. Seven subjects were working for 60 min on a bicycle ergometer. To the left, data obtained at rest. (SD = standard deviation.) *(From Saltin and Hermansen, 1966.)*

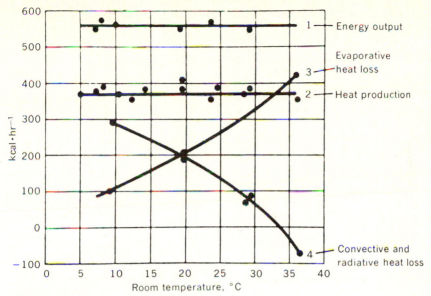

Figure 15-7 Heat exchange during work (150 watts) at different room temperatures in a nude subject. *(Modified from M. Nielsen, 1938.)*

out at the highest ambient temperature, the skin temperature was 35°C, and the body absorbed heat from the environment (i.e., 36°C air temperature). This was completely counteracted by an increasing evaporative heat loss. In the cold, the subject evaporated 150 g sweat; in the heat he evaporated 700 g. The greater variation in the rectal temperature at the end of the work period in all these experiments was only 0.11°C. This difference in body temperature may be brought about merely by evaporating about 11 g sweat. These findings have been confirmed by Nielsen (1969) and Stolwijk et al. (1968). One may add that the average skin temperature is a linear function of the ambient air temperature and is relatively independent of the level of exercise (Nielsen, 1969.)

During maximal exercise the rectal temperature may exceed 40°C and the muscle temperature 41°C without causing any discomfort for the working person.

External heat load is not a factor in cardiovascular response during brief exposures (4 to 6 min) to work in high ambient temperatures (Rowell, 1974; Drinkwater et al., 1976.) During prolonged light work in a hot environment, the heart rate rises markedly while cardiac output increases more gradually for 30 to 40 min in spite of a progressive fall in stroke volume. During prolonged moderate to heavy muscular work, the heart rate does not reflect changes in cardiac output (CO) or in skin blood flow (SBF). Actually, stroke volume (SV) decreases while cardiac output is maintained by increased heart rate. During graded exercise, submaximal cardiac output is maintained by increased heart rate in spite of reduced stroke volume (SVR) (Fig. 15-8) (Rowell, 1974).

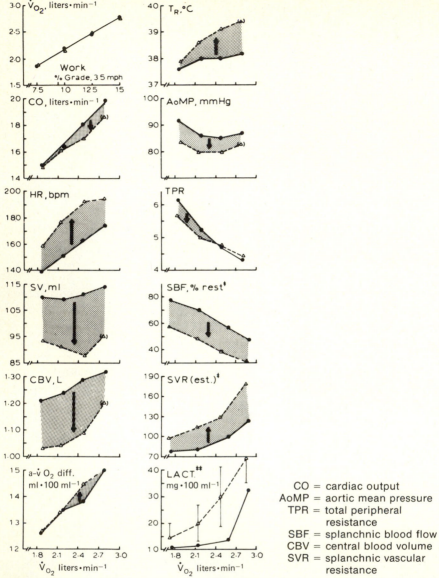

Figure 15-8 Cardiovascular responses to graded exercise in hot (43.3°C △--△) and neutral (25.6°C ● —— ●) environments. In upper left-hand graph, $\dot{V}_{O_2}$ is plotted against work load to show absence of temperature effects on $\dot{V}_{O_2}$. All other data are plotted against $\dot{V}_{O_2}$ during each of four levels of exercise. Arrows show direction of temperature-induced change in each variable. Largest reductions in CO and SV and increase in a-v̄ O_2 difference were seen at highest work load at 43.3°C. Two of six subjects were unable to reach this level in the heat; as a result, this effect is not seen in averaged responses. Data for SBF (*) were from 11 different subjects in an experiment of identical design. SVR (‡) was estimated from aortic MP (AoMP) values taken from the six subjects in another study. Average blood lactate (lact. ‡‡) concentrations were higher at 43.3°C but not significantly so, as indicated by large standard deviation. *(From Rowell, 1974.)*

Summary   Muscular work may increase the heat production from 10 to 20 times the heat production at rest. During work in a "neutral" environment, there is an increase in body temperature up to a maximum of about 40°C or slightly higher at maximal work loads. The body temperature is not related to the absolute heat production but to the relative work load, i.e., actual oxygen uptake in relation to the individual's maximal aerobic power; at a 50 percent load, the deep body temperature is about 38°C. The deep body temperature at rest and during work is, within a wide range, not affected by the environmental temperature; but the skin temperature is. In a given environment, the sweating rate is mainly dependent on the actual heat production and not primarily on the skin or rectal temperature.

TEMPERATURE REGULATION

In the hypothalamus and the adjacent preoptic region, as shown in animal experiments, there are nerve cells which by local heating and cooling may elicit the same reactions which occur during exposure to heat or cold (Hammel, 1965; Hardy, 1967). These cells belong to the temperature regulatory center which is connected via nervous pathways with receptors in the skin, the central nervous system, and possibly elsewhere in the body, such as in the deep leg veins, the muscles, the abdomen, and the spinal cord (Hensel, 1974). These receptors consist of a net of fine nerve endings which are specifically activated by heat or cold stimuli (Zotterman, 1959; Hansel, 1963). These temperature receptors are especially sensitive to rapid changes in temperature and are highly susceptible to adaptation. In the heat receptors, the maximal frequency of the impulses occurs in a steady-state condition between 38 to 43°C; in the cold receptors, the maximal impulse frequency occurs at 15 to 34°C (Zotterman, 1959). At temperatures above 45°C, the cold receptors may again be activated. This may explain the paradoxical cold sensation experienced at the first contact with very hot water. The receptors register not only temperature changes but also temperature levels, particularly if the skin temperature is below 32°C in the case of the cold receptors, and above 37°C in the case of the heat receptors (Kenshalo et al., 1961). The number of active receptors determines to some extent the sensation of temperature. The fact that temperature sensation is a relative matter is best illustrated by the simple experiment of putting one finger in warm water, another finger in cold water. When both fingers are then simultaneously put into lukewarm water, the finger which previously was exposed to cold water will sense the lukewarm water as warm; the other finger will interpret it to be cold.

For the regulation of body temperature, the hypothalamic temperature regulatory center and the temperature-sensitive receptors in the skin play a dominating role (Smiles et al., 1976). Precisely how the temperature regulatory mechanism works is unknown. In his recent review of the subject, Cabanac (1975) concludes that the general way of looking at the short-term temperature regulation has not changed fundamentally during the last 10 years. It appears that several temperature sensors are capable of triggering defense reactions independently in order to maintain thermal balance, and that the response is a function of several inputs combined at the same time. According to Cabanac (1975), the temperature sensors in the spinal cord may be at least one-quarter to one-half as sensitive as the hypothalamic temperature sensors. Warm stimulation of the spinal cord in animals is followed by all the warmth defense reactions (skin vasodilation, reduced heat production, increase in evaporative heat loss) proportional to the spinal cord temperature. Cooling of the spinal cord in unanesthetized dogs induces shivering, peripheral vasoconstriction, and piloerection proportional to the magnitude of the stimulus (for references, see Cabanac, 1975). Nevertheless, it appears that the hypothalamus remains the main center of temperature regulation, for studies of the effects of spinal cord lesions have shown that the spinal network does not possess complete thermoregulatory capability (Walther et al., 1971). On the basis of the available evidence, Cabanac (1975) concludes that it is probably more accurate to consider the control system for the regulation of body temperature as consisting of a number of networks operating independently of one another.

According to Benzinger et al. (1963), an increase of the brain temperature above 37°C, considered to be the normal "set point of the thermostat," will elicit sweating, and the vasoconstrictor impulses to the cutaneous blood vessels decline in frequency or are totally absent. He finds that the sweat intensity diminishes at a certain skin temperature below 33°C. This is interpreted as being the result of cold receptors inhibiting the heat loss center. If, on the other hand, the skin temperature is above 33°C, the sweat production is independent of whether the skin temperature is 33°C or 39°C. The heat receptors, according to Benzinger, do not affect the sweating. Other investigators do not support Benzinger on this point (Hardy, 1967).

The *anterior hypothalamus* and the preoptic region are sensitive to changes in the local temperature. Many neurons have been observed to increase their discharge rate when heated; only a few cells increase their discharge frequency when the local tissue temperature is lowered (Hardy, 1967). Preoptic heating in animals exposed to a neutral environment can induce vasodilation in the skin and, eventually, panting and sweating. With the animal in a cool environment, such a local heating can inhibit the normal response of shivering and vasoconstriction of peripheral blood vessels. The depression of the metabolic rate causes the body temperature to drop.

An intact *posterior hypothalamus* is required to induce the reactions to a cold environment, namely shivering and an increase of metabolic rate, and to restrict flow of heat to the skin by a vasoconstriction of skin blood vessels. This posterior center, however, is temperature blind: it is essentially insensitive to local temperature changes. The function of this area is largely coordinating and it receives, rather than generates, temperature signals. Afferent impulses from cold receptors in the skin seem to be the main drive for this center.

The two thermoregulatory centers are in a way connected so that the response to stimulation of the anterior center includes stimulation of sweating, but inhibition of shivering and vasoconstriction. Conversely, the action of a stimulation of the posterior center involves a stimulation of vasoconstriction and an increase of heat production, but a simultaneous inhibition of responses to heat. The final common pathways include not only the motor pathways of the synaptic and somatic systems, but also blood humoral transmissions. Cold may, via the thermoregulatory center in the hypothalamus, affect the pituitary gland and the release of hormones which in turn act on their target organs to release thyrotropic and adrenal hormones, increasing the heat production in the tissues. According to Hardy (1967), many large animals, however, show relatively little of the neuroendocrine response to cold. There are functional connections between the central control of body temperature and the areas regulating water and food intake (Andersson, 1967).

We may conclude that cold acts primarily on the periphery, stimulating cold receptors which signal to the central nervous system, but the internal temperature can influence these thermal signals. Heat has a direct central effect, but peripheral signals from thermosensitive receptors can modify the response. In other words, "chemical" thermoregulation (increased metabolic rate) originates mainly from cold reception in the skin, but "physical" thermoregulation (sweating, increased heat conductance of the skin) is to a high degree elicited by central warm reception. [In a warm environment of 33°C, only a small increase in oxygen uptake is observed in dogs when the hypothalamic region is cooled to 33.5°C. The same degree of cooling in a neutral environment of 23°C, however, caused a fourfold increase in the heat production (Hammel et al., 1963). This experiment illustrates that the peripheral, rather than the central, sensing receptors determine the response to a cold environment. Benzinger (1967) points out that shivering ceases immediately when one takes a warm shower, even if the rectal temperature is low. Nielsen (1976a) reports that with high mean skin temperatures, above 31 to 32°C during the warming of a hypothermic subject, very little metabolic increase was measured even with core temperatures below 35°C. She emphasizes that during rewarming, the response from the skin receptors may exert an inhibition on central cold receptors. When the shivering mechanism is switched on, the increased metabolic rate is superimposed on the energy demand of exercise (Holmér and Bergh, 1974; Nielsen, 1976a). On the other hand, it is difficult to suppress sweating completely even during light exercise by any manipulation of the ambient air or skin temperature and still keep the subject acceptably comfortable (Stolwijk et al., 1968).]

To simplify the understanding of the thermoregulation, we can assume that the center for thermoregulation has a "set point," and adjustments are made to minimize the deviation of the actual body temperature from this set point (Hammel, 1965; Hardy, 1965). The set point is not constant but may change with many physiological conditions. If the body temperature exceeds the set point, the thermoregulating center switches on the cooling actions; if the body temperature is below the set point, the metabolic rate increases and heat conservation mechanisms are switched on. For example, when the skin temperature falls in a cold environment, the afferent nerve impulses from the cold receptors elevate the set point so that the hypothalamic temperature will be below it, and therefore will start driving the heat-conserving mechanisms. In fact, the internal body temperature in humans may *rise* during mild exposure to cold and still cause shivering and cutaneous vasoconstriction (Hardy, 1965, 1967). Conversely, in a hot environment, the set point goes below the hypothalamic temperature so as to drive the mechanisms which promote heat loss.

The diurnal variations in body temperature may be due to variations in the set point; sleep decreases the set point and gradually lowers the temperature of the body, which upon awakening increases the set point again (see Cabanac et al., 1976). Werger et al. (1976) noticed that during exercise at night, thresholds for sweating and vasodilation were shifted toward lower core temperatures (the shifts averaged 0.6°C).

From this point of view, exercise should lower the set point. Within seconds after the onset of work, an increase in sweat secretion has been observed (Meyer et al., 1962; Beaumont and Bullard, 1963). On the other hand, it appears that the body regulates at a higher temperature during exercise than at rest. The optimal temperature may range from just above 37°C up to 40°C, depending on the severity of work. The knowledge of the thermoregulation during exercise is deficient (Cabanac, 1975; B. Nielsen, 1976b). The facts are: (1) Deep body temperature is mainly a function of the energy output and is, within a wide range, independent of the ambient air temperature (M. Nielsen, 1938; B. Nielsen, 1969; Stolwijk et al., 1968). Actually, the total aerobic energy production seems to be more decisive for the final body temperature than the heat production (e.g., when comparing positive and negative work, arm and leg work, Nielsen, 1969). (2) Skin temperature is principally related to the ambient air temperature but not to metabolic rate (and body temperature) (for references, see Saltin et al., 1968). (3) At a *constant work load* the sweat rate increases with the ambient temperature (and skin temperature) and is then unrelated to the body temperature (Fig. 15-7); in a *constant environment* the skin sweating is a linear function of the heat production but unrelated to the skin temperature [for references see (1) above]. Therefore skin temperature and hypothalamic temperature may independently modify the sweating response to exercise. This is illustrated in Fig. 15-9. However, the *local* skin temperature can modify the effect on the output from a central controller. Thus, local heating of the skin can decrease the T_{es} threshold for active sweating, and local skin cooling can have the opposite effect (Nadel et al., 1971).

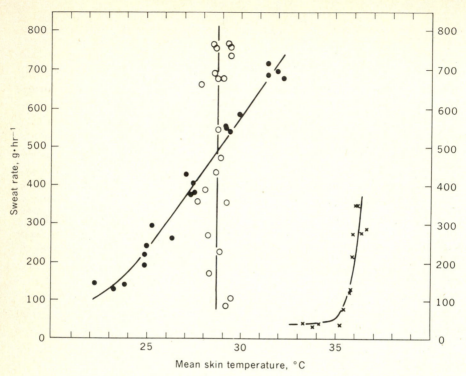

Figure 15-9 Steady-state values of sweat rate plotted against the corresponding values of mean skin temperature: (O) work intensity from 90 watts to 235 watts at constant environmental temperature of 20°C; (●) constant work intensity (150 watts) at environmental temperatures from 5 to 30°C; (X) experiments at rest at environmental temperatures from 25 to 44°C. *(From B. Nielsen, 1969.)*

On the basis of the available evidence, it appears that the change in the setting of the body thermostat during work may conceivably operate as expressed by the following equation:

$$R = aT_h + bT_s$$

where R = response, T_h = hypothalamic temperature, and T_s = skin temperature; a and b are constants.

Nielsen (1969) noticed a remarkable similarity in the thermoregulatory responses to passive heating by diathermia and to active heating by muscular exercise. Thus, at the same level of heat production, rectal and skin temperatures and estimated skin blood flow were increased to the same level in the two kinds of experiments. Therefore, she concludes, a work factor of nervous origin (from mechanoreceptors or cortical irradiation) may operate on the thermoregulatory centers at the start of exercise but hardly in the later phase of exercise. In intermittent work with the same heat production over a given period of time as in continuous work (with different intensity of work during the periods of activity), Nielsen (1969) found the same body temperature and sweat rate, which should also exclude nervous impulses related to the severity of work as the "work factor." She proposes that a chemical factor, liberated during work in proportion to the engagement of the aerobic processes, may be responsible for the temperature resetting. Since the work load in relation to the individual's maximal aerobic power dominates the adjustment of the body thermostat more than the absolute aerobic power (I. Åstrand, 1960), the attention should be focused at some stress factor, e.g., epinephrine and norepinephrine. For discussions on various candidates, see Gale, 1973; Schönbaum and Lomax, 1973; B. Nielsen, 1976b; Sobocinska and Greenleaf, 1976.)

Summary The temperature-regulating center is located mainly in the hypothalamus. It behaves like a thermostat, and its set point may change during different physiological conditions. Thermosensitive receptors, particularly in the skin, contribute to the regulation of the set point. In a cold environment, stimulation of the cold receptors may elevate the set point, and the heat-conserving mechanisms are switched on. In a hot environment, the set point becomes lowered, and the heat loss can increase by means of vasodilation in the skin and sweating. The factors involved in the regulation of the body temperature during exercise are largely unknown.

The sweating and the dilation of the skin blood vessels do not run parallel. The produced sweat may in itself cause a vasodilation; on the other hand, the local skin temperature may affect the diameter of the blood vessels in that area. At high evaporative sweat rates, the skin may cool, causing vasoconstriction and reduced blood flow.

It is probable that other impulses to the temperature regulatory center exist. Fever is caused by the "thermostats" being set at a higher level. The reason for the diurnal variations in the body temperature is unknown.

ACCLIMATIZATION

Continuous or repeated exposure to heat, and possibly also to cold, causes a gradual adjustment or acclimatization resulting in a better tolerance of the temperature stress in question. Many plants prepare for the winter by increasing their carbohydrate content, and certain types of apple trees may, during the winter, tolerate air temperatures below −40°C, but during the month of July, they fail to survive air temperatures below −3°C. Certain insects accumulate the "antifreeze" glycerol in the fall, which enables them to survive cold. (For literature, see Dill, 1964.)

Heat

After a few days' exposure to a hot environment, the individual is able to tolerate the heat much better than when first exposed. This improvement in heat tolerance is associated with increased sweat production, a lowered skin and body temperature, and a reduced heart rate (Robinson et al., 1943; Kuno, 1956; Bass, 1963; Leithead and Lind, 1964; Wyndham, 1967; 1973; Rowell, 1974). An example is illustrated in Fig. 15-10. Usually the skin blood flow is reduced; in one experiment it declined from 2.6 to 1.5 liters $\cdot$ m^{-2} $\cdot$ min^{-1}, i.e., to about 60 percent of the original value (Bass, 1963). The increased sweat rate provides the possibility for a more effective cooling of the skin through the evaporative heat loss, and the resultant lowered skin temperature provides for a better cooling of the blood flowing through the skin. Thus, the body can afford to cut down on the skin blood flow. In acute experiments, the sweat

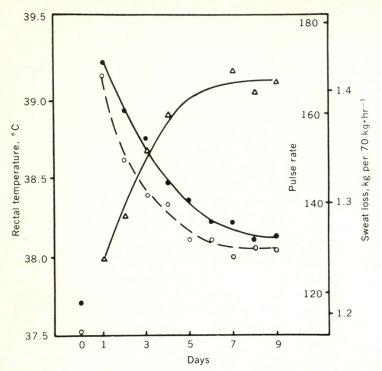

Figure 15-10 Mean rectal temperature (●), heart rates (O), and sweat losses (Δ) in a group of men during a 9-day acclimatization to heat. On day 0 they worked for 100 min at a rate of energy expenditure of 1.2 MJ (300 kcal) · hr^{-1} in a cool climate. On the following day they worked in a hot climate (48.9°C dry-bulb and 26.7°C wet-bulb temperature). *(Modified from Lind and Bass, 1963.)*

glands have the capacity to produce more sweat than they do under ordinary circumstances. The reason why this capacity is not fully utilized until after several days' exposure to a hot environment is not known. The increase in sweat production may go as high as 100 percent (Leithead and Lind, 1964).

It is possible, then, to demonstrate objectively physiological alterations in response to prolonged exposure to heat. It is also possible to explain why the heat tolerance gradually increases. It is found that the skin blood flow in acute experiments increases at the expense of the blood flow through other tissues (Rowell et al., 1965; Rowell, 1974). As the individual becomes acclimatized, normal distribution of the blood flow is once more established. Within 4 to 7 days' exposure to a hot environment, most of the changes have taken place, and at the end of 12 to 14 days, the acclimatization is complete. Even a relatively short daily heat exposure will have some effect.

During acute exposure to heat and the first phase of acclimatization, the heart rate during a standardized prolonged exercise is increased and the stroke volume reduced. Gradually the heart rate decreases and the stroke volume increases but seems to remain depressed throughout the acclimatization. The cardiac output remains relatively constant. Thus, the major cardiovascular adjustments during heat acclimatization are reciprocal changes in heart rate and stroke volume while cardiac output and also arterial blood pressure remain essentially unaltered (Rowell, 1974; Wyndham et al., 1976). Rowell (1974) concludes that the circulatory adjustments to chronic heat stress appear to depend primarily on the reduction in skin and core temperatures caused by the enhanced sweating. The mechanisms are, however, far from revealed. The reduced skin temperature could trigger a lowering of the heart rate. We have described situations in which a "manipulation" with the heart rate is followed by reciprocal variations in stroke volume (see page 190). Rowell has discussed other possible mechanisms. Measurements of total blood volume during acclimatization have given conflicting results. Some investigators have reported an increase in the blood volume, but apparently this is transitory. Senay et al. (1976) concluded that the most critical event in the first phase of heat acclimatization is an expansion of the plasma volume.

A well-trained individual adjusts better to heat than one who is in poor physical condition, but training cannot replace acclimatization (see below). If muscular work is involved in the exposure to the hot climate, physical work should be included in the acclimatization period. The effect is the same whether the climate is hot and dry or less hot but damp (high humidity). Similarly, the response seems to be the same whether the work in question is heavy and of short duration or less heavy but carried out for a longer period. During the period of acclimatization and during heat exposure, it is important that fluid and salt losses be replaced.

The effect of heat acclimatization persists several weeks following heat exposure, although some impairment in heat tolerance may be detected after a few days following cessation of exposure, such as after a long weekend, especially if the individual is fatigued and alcohol has been consumed (Bass, 1963; Williams et al., 1967).

Cold

Acclimatization to heat may be conveniently studied in subjects living in a hot climate or in laboratory experiments. Studies of adaptation to cold, on the other hand, require climatic chamber experiments inasmuch as individuals, when moving to a cold climate, normally bring their semitropical climate with them. They have the ability to protect themselves against the cold by clothing or adequately insulated dwellings, even in arctic regions. Animals habitually exposed to cold develop an effective protection in the form of fur and an effective heat exchange system in their peripheral blood vessels. In certain types of seals, the skin may be cooled to 0°C without an increase in oxygen uptake. In the human, a lowering of the skin temperature a few degrees may cause a doubling of the metabolic rate. The human, then, has, with respect to reaction to cold, taken a path somewhat different from that followed by arctic animals.

According to some investigators, the metabolic rate of "cold-acclimatized" individuals is often elevated when they are exposed nude to a standardized cold stress, even though shivering is said to be less pronounced. Nevertheless, the metabolic rate is unchanged at normal room temperature. It is possible that hormones, notably norepinephrine, play a role in the elevation of the metabolic rate, but the mechanism is unclear. The aborigines of Central Australia and the bushmen of the Kalahari Desert, wearing little or no clothing, may be exposed to night temperatures of about 0°C or below. They do not shiver even though their core and skin temperatures keep falling throughout the night, and they can sleep. Control subjects were "fighting" against the cold environment by shivering, and they could not sleep. There are similar observations on other primitive populations and also on women divers of the Korean peninsula who do not increase their heat production by shivering to the same extent as individuals who have not been exposed to cold environment for longer periods of time (for references, see LeBlanc, 1975, p. 116.) The essential feature of cold adaptation seems to be reduced shivering. The energy saved is not very important, but on other grounds this effect has advantages. Shivering is disturbing and uncomfortable. "Shivering is the first line of defense but nobody likes it" (LeBlanc, 1975).

While it is difficult to demonstrate definite evidence of general physiological acclimatization to cold in humans, such acclimatization can be produced in animals exposed to severe cold. If rats are placed in a refrigerator kept at 5°C for as long as 6 weeks, the metabolic rate of the animals will increase twofold. They will double their food intake, and their thyroid function will increase. The main feature of the cold-acclimatized rat is its ability to maintain a high rate of heat production. This ability is absent in the nonacclimatized rat, which is unable to survive in the cold. If these findings in the cold-acclimatized rats are compared with findings in human beings, such as Eskimo or Caucasians habitually exposed to cold, it is observed that the changes found in the rat do not necessarily occur in the human. A person's food intake is not materially increased in a cold environment (Rodahl, 1963), nor is the rate of heat production. It is true that the Eskimo's rate of heat production is higher than that of Caucasians, but much of this is due to the specific dynamic effect of diet, for Eskimo living on the white person's diet do not, as a rule, have any higher rate of heat production than the white person (Rodahl, 1952). When whites move to the Arctic, their metabolic rate is no higher than it was at home. Normal Eskimo do not show increased thyroid function compared with normal Caucasians (Rodahl and Bang, 1957). The most probable reason is that in clothed individuals habitually exposed to cold environments, the degree of cold exposure is not sufficiently severe to cause an appreciable increase in metabolism. Without increased metabolism, there is no need for increased food intake.

On the other hand, when the human is exposed to cold stress which is far more severe than that normally encountered by the clothed individual living in the Arctic, certain physiological changes do occur which may be interpreted as hormonally induced adjustments to cold (Rodahl et al., 1962). When young men, dressed in shorts and sneakers only, were confined continuously to cold-chamber temperatures of 8°C for 3 to 10 days, they responded by violent shivering which lasted more or less continuously night and day throughout the experiment, even at night when they slept under a blanket. They soon learned to continue to sleep in spite of the shivering. As a result, the increased heat production due to the shivering was maintained even during sleep, so that the body temperature remained normal and the subjects slept relatively comfortably. Occasionally, however, the same subjects, for some reason, failed to keep up the vigorous shivering during the night while they slept. Consequently their body temperature continued to drop and approached dangerously low levels. Under these conditions, the subjects occasionally objected to being disturbed, and wanted to be left in peace to continue to sleep. If this were allowed, the subject might continue to cool and, conceivably, might eventually die from hypothermia. It thus appears that an exposed person sleeping in the cold may actually freeze to death if the body's rate of cooling is slow and gradual, so that violent shivering is not produced which would wake the person up.

These subjects in the climatic chamber doubled their metabolic rates because of their constant shivering. Their resting heart rate was markedly elevated, and the nitrogen loss in the urine was greatly increased. These changes were interpreted as being the result of hormonal changes brought about by the cold stress (Issekutz et al., 1962). It should be borne in mind that the cold exposure of these subjects in the climatic chamber (8°C) was far greater than the cold exposure of the Eskimo or any clothed group of individuals living in the Arctic. Consequently, all the cold-room subjects suffered ischemic cold injury of their feet, although the temperature never approached the freezing temperature of the tissue.

When nutritional deficiency or nutritional stress was superimposed on the cold stress in these subjects, there was a marked impairment in physical work capacity. After 9 days' constant exposure to 8°C in the nude, maximal oxygen uptake showed no deterioration when the individuals were fed an adequate diet consisting of 12.5 MJ (3000 kcal) and 70 g protein. When the caloric intake was reduced to 6.3 MJ (1500 kcal) or the protein intake was reduced to 4 g · day^{-1}, or when both energy and protein intakes were reduced, a marked deterioration in physical performance capacity occurred within 3 to 5 days. The effect was most pronounced in the treadmill running time; maximal oxygen uptake in some of the subjects was less affected.

However, any physiological adaptation to cold in humans is of little practical value compared with the importance of know-how, experience, and state of physical fitness. There is definitely a limited capacity of the automatic thermostatic system. The success of the Eskimo in getting along in the cold depends on their ability to avoid the extreme cold. Since most arctic clothing (except fur clothing) is inadequate in terms of insulation to maintain thermal balance when an individual is exposed without shelter to extreme prolonged cold, the only way to survive is to remain active in order to increase the heat production. The fitter the Eskimo are, the longer they can do this. Thus, survival in the Arctic is a matter of survival of the fittest.

When a person, whether an Eskimo, a Caucasian, or a Negro, allows his or her hands to be repeatedly exposed to cold for about $^{1}/_{2}$ hr daily for a few weeks, this cold stress will cause an increased blood flow through the hands, so that they will remain warmer and are not so apt to become numb when exposed to cold. Similar findings have been made in fish filleters who habitually expose their hands to cold water (Hellström, 1965). This may be termed local acclimatization to cold. While it inevitably will cause a greater amount of heat to be lost from the hands, it will improve the ability of the hand and fingers to perform work of a precise nature in the cold. To this extent this local cold acclimatization is beneficial (LeBlanc, 1975; Nelms and Soper, 1962; Strömme et al., 1963).

LeBlanc et al. (1975) found that repeated exposure of the hands and face to severe cold activates some adaptive mechanisms characterized by a diminution of the sympathetic response and a concomitant enhancement of the vagal activation normally observed when the extremities and the face are exposed to cold.

LIMITS OF TOLERANCE

Normal Climate

Optimal function requires that the body temperature be maintained between 36.5 to 39.5°C. The ideal room temperature is about 20°C for clothed individuals who are sitting or standing still. The more active the individual, the lower the room temperature should be. Thus, when performing heavy physical work, a person may prefer a room temperature of 15°C or even lower. Not only may acclimatization play a role, but habits and established traditions may affect the so-called comfort temperature. This may explain the fact that the preferred room temperature is higher in the United States than in England and higher in the summer than in winter in both the United States and England. The preferred room temperature in Singapore is higher than in the United States (Pepler, 1963).

Failure to Tolerate Heat

The most serious consequence of exposure to intense heat is heat stroke, which may be fatal. It is caused by a sudden collapse of temperature regulation leading to a marked rise in body heat content. The rectal temperature may be 41°C or higher. The skin is hot and dry. There are tachycardia and hypotension, metabolic acidosis, and disseminated intravascular coagulation. The victim is confused or unconscious. This form of temperature-regulatory failure is rare. The risk is higher in nonacclimatized than in acclimatized individuals. Obese persons and older individuals are most susceptible. The treatment is rapid cooling until the rectal temperature has dropped below 39°C.

Another type of temperature-regulation failure is the so-called anhidrotic heat exhaustion. The victim may have a body temperature of 38 to 40°C, and may sweat very little or not at all. He or she feels very tired, may be out of breath, and has tachycardia. The main trouble is reduced sweat production. When the patient stops working and is removed to a cool place, this condition rapidly improves.

A third type of serious disturbance due to heat exposure is excessive loss of fluid and salt, usually because of failure to replace fluid and salts lost through sweating. After several weeks' exposure, the patient may eventually experience cramps, the so-called miner's cramps, which in rare cases may be fatal. Intravenous administration of NaCl will promptly relieve the cramps.

Heat syncope is a less serious affliction due to heat exposure. This is primarily caused by an unfavorable blood distribution. A large proportion of the blood volume is distributed to the peripheral vessels, especially in the lower extremities as the result of prolonged standing, or by a reduction in blood volume due to dehydration. The result is a fall in blood pressure and inadequate oxygen supply to the brain, which may lead to unconsciousness. If the victim is placed in the horizontal position, preferably with the legs elevated, he or she quickly regains consciousness. This type of heat collapse is a form of built-in safety mechanism of the body.

Certain individuals exhibit an untoward reaction to heat in the form of heat rash. This condition may make the individual unsuited for work in a hot environment (Minard and Copman, 1963).

Upper Limit of Temperature Tolerance

It is not feasible to quote exact permissible limitations for the working environment. These limitations depend on the combination of the different climatic factors, on the nature and type of work in question, on the severity of the work load, and on the duration of the work. Finally, there are wide individual variations in the tolerance of climatic stress, apart from the effect of acclimatization.

Leithead and Lind (1964) have suggested the following categories: (1) intolerable conditions; (2) just tolerable conditions, i.e., tolerable for intermittent exposure only; (3) easily tolerable conditions. The first two conditions may be encountered in cases of emergency, accidents, fire, and military operations. At air temperatures above 120°C, the heat pain may be the limiting factor. An individual who can tolerate 120°C for about 10 min may tolerate 200°C for about 2 min, if the air is dry. Under favorable conditions, temperatures of 50 to 60°C may be tolerated for hours. At higher air temperatures the sweating may not be able to prevent a continuous gradual increase of the body temperature. Under such conditions the body's capacity to store heat may be the limiting factor.

Wyndham et al. (1965) suggest the following criteria for the strain on the person at work in hot conditions, based on the measurements of rectal temperature: Conditions of work and heat should be judged to be "easy" when the T_r does not exceed 38°C; it should be considered to be "excessive" when the T_r exceeds 39.2°C; with the T_r between these two limits, the conditions should be graded as increasingly "difficult" as T_r approaches 39.2°C. It is evident that the person who is acclimatized to heat can maintain a higher energy output at a given body temperature than one who is unacclimatized.

For the assessment of the thermal heat load, e.g., in industrial situations, so-called heat stress indices have been introduced during the past 50 years. By such indices one tries to predict the heat load on the individual. They are usually based on data from physical measurements of air temperature (dry bulb), radiant temperature (globe), relative humidity (wet bulb), and air velocity. A measure of the individual's metabolic rate is also often included, i.e., the energy demand of the work is measured or estimated (Kerslake, 1972; I. Åstrand et al., 1975). The American Conference of Governmental Industrial Hygienists (ACGIH, 1976) has recommended limit values for industrial application. These are mainly based on wet bulb and globe temperatures. In Sweden, guidelines were introduced to modify the index for air velocity (above or below 0.5 m · s⁻¹). The Swedish Wet Bulb-Globe Temperature Index (SWBGT) is determined as follows (I. Åstrand et al., 1975):

1   At air velocities ≥ 0.5 m · s⁻¹
    SWBGT = 0.7 · pWB + 0.5 · GT
2   At air velocities < 0.5 m · s⁻¹
    SWBGT = 0.7 · pWB + 0.5 · GT + 2

where pWB = psychrometric wet-bulb temperature (°C)
      GT = globe temperature (°C).

Generally, formula (1) will be used because at most work places with high heat exposure the air velocity usually exceeds 0.5 m · s⁻¹.

Fig. 15-11 presents limit values which are supposed to represent conditions to which most workers can be exposed daily without harmful effects. The limit values are based on the assumption that almost all acclimatized and fully clothed workers with an appropriate fluid and salt intake should be able to work under the given work conditions without their body temperatures rising above 38°C. Due to large interindividual variations in heat tolerance, however, some people may be uncomfortable or exposed to potential injury even below the specified limit values. No worker should be allowed to continue working when his body temperature rises above 38.0°C. The limit values are specified for heat exposure with different degrees of work severity and with varying periods of work and rest. The values apply to reasonably acclimatized persons. Values 1 − 2°C less should be applied to partially acclimatized or unacclimatized persons.

Age

Although the experimental data are still limited, the available evidence suggests that heat tolerance is reduced in older individuals (Leithead and Lind, 1964; Robinson, 1963; Lind et. al., 1970). They start to sweat later than do young individuals. Following heat exposure, it takes longer for their body temperature to return to normal levels. Older people react with a higher peripheral blood flow, but their maximal capacity is probably lower. In one study, it was found that 70 percent of all individuals who suffered heat stroke were over sixty years of age (Minard and Copman, 1963).

Sex

Women have a lower tissue conductance in cold and a higher tissue conductance in heat than do men. This fact suggests a greater variation in the peripheral reaction to climatic stress in women. It appears that this fact is of no importance for the performance of work.

State of Training

As mentioned, it appears that a trained individual is better able to adjust to heat than one who is untrained. Physical training will enhance the sweating mechanism at a given level of central sweating drive. The increased metabolic rate during training raises a high thermoregulatory demand, and apparently this demand will induce an increased peripheral sensitivity of the sweat glands to the central sweating drive. It has been mentioned that for an individual who increases his maximal aerobic power by training, a given rate of exercise will require a lower percentage of his maximal oxygen uptake. Concomitantly the core temperature will decrease during a standardized exercise. The increased capability for heat dissipation behind this adaptation is due to an enhanced sweating response. In addition, heat acclimatization results in a further enhancement of the sweating response at a given level of central sweating drive by lowering of the zero point of the central nervous system drive for sweating. In other words, physical training seems to increase the slope of the sweating rate-core temperature curves, i.e., the activity of the sweat glands increases at a given core temperature. In contrast, acclimatization to heat seems to lower the threshold core temperature at which sweating starts (it moves the curves "to the left" without changing the slope. For further discussion, see Nadel et al., 1974). For an optimal heat acclimatization, simultaneous exposure to both heat and exercise is recommended. Physical training will also improve the circulatory potential. Thereby the trained individual may be able to maintain a cardiac output sufficient to meet metabolic requirements and the demand for peripheral bloodflow for a longer period of time than untrained people (Drinkwater et al., 1976a.) A convalescent patient during recovery from illness is particularly sensitive to heat stress. It may be wise in such cases to subject the patient to a period of acclimatization prior to the assumption of full duties during an 8-hr working day in a hot environment.

COORDINATED MOVEMENTS

The speed of nervous impulses and the sensitivity of the receptors are affected by the temperature of the tissues. At about 5°C the skin receptors for pressure and touch do not react on stimulation. The execution of coordinated motions depends upon the inflow from these receptors to the central nervous system. The numbness in the cold is the result of this lack of sensitivity of the skin receptors. Irving (1966) reports that the skin at a temperature of 20°C was only one-sixth as sensitive as at 35°C; i.e., an impact on the skin had to be six times greater to be felt at the lower skin temperature. The muscle spindles show an increased sensitivity at moderately lowered muscle temperatures, but at 27°C the activity in response to a standardized stimulus is reduced to 50 percent; at a temperature of 15 to 20°C, it is completely abolished (Stuart et al., 1963). This phenomenon also contributes to the difficulty of performing fine coordinated movements in the cold. It may be partially responsible for an increased accident rate in certain types of manual work operations in cold environments.

MENTAL WORK CAPACITY

An evaluation of the mental or intellectual work capacity during exposure to heat or cold is hampered by subjective variations and lack of suitable objective testing methods (Pepler, 1963). As a rule, a deterioration is observed when the room temperature exceeds 30 to 35°C if the individual is acclimatized to heat. For the unacclimatized, clothed individual, the upper limit for optimal function is about 25°C.

The observed deterioration in performance capacity refers to precise manipulation requiring dexterity and coordination, ability to observe irregular, faint optical signs, the ability to remain alert during prolonged, monotonous tasks, and the ability to make quick decisions. During a 3-hr drilling operation, the best results were achieved at 29°C, but at a room temperature of 33°C, the performance was reduced to 75 percent; at 35.5°C, to 50 percent; and at 37°C, to 25 percent. A high level of motivation may to some extent counteract the detrimental effect of the climate.

WATER BALANCE

Normal Water Loss

Reasonable figures for the daily water loss are as follows: from gastrointestinal tract, 200 ml; respiratory tract, 400 ml; skin, 500 ml; kidneys, 1,500 ml = 2,600 ml. This loss is balanced by an intake as follows: as fluid, 1,300 ml; as water in the food, 1,000 ml; water liberated during the oxidation in the cells, 300 ml = 2,600 ml. However, the water loss can increase considerably when the individual exercises or is exposed to a hot environment.

Water loss through the respiratory tract varies roughly with the pulmonary ventilation (dryness and temperature of the inspired air have some influence). The ventilation varies within a wide range directly with the production of CO_2, and this production is, in turn, proportional to the metabolic rate. Therefore, the water volume from oxidation, proportional to the metabolic rate, equals, by coincidence, roughly the water loss through the respiratory tract.

During very heavy exercise, glycogen is the preferred fuel. About 2.7 g of water is stored together with each gram of glycogen (Chap. 14), and this water becomes free as the glycogen is combusted. If during such heavy exercise, 5 MJ (1200 kcal) is totally consumed, 80 percent, or 4 MJ (960 kcal), may be derived from glycogen. The liberated volume of water (including the water of oxidation) will be close to 800 ml. Assuming a mechanical efficiency of about 25 percent, 3.8 MJ (900 kcal) of the 5 MJ (1200 kcal) should be dissipated as heat if the body temperature should be maintained unchanged. An exclusive evaporative heat loss demands the evaporation of about 1,500 ml of water to eliminate 3.8 MJ (900 kcal). Under these conditions, only approximately half the necessary water volume must be taken from body "stores," for the rest is apparently liberated in the processes producing the heat. It should be emphasized that more sweat may be secreted than is evaporated from the skin. On the other hand, radiative and convective heat exchange may reduce the demand on the evaporative heat loss. When the glycogen depots are again restored, extra water is certainly needed.

Thirst

In adults about 70 percent of the lean body weight is water, so there is a substantial buffer to cover water losses over limited periods of time. However, in the long run, water intake must balance water loss by the several routes mentioned. Hypothalamus and adjacent preoptic regions play the essential role in the thirst mechanism (Stevenson, 1965; Andersson, 1967). There are some sort of osmoreceptors reacting on an increase in the osmolarity of the intracellular fluid. Any change in the internal environment leading to cellular hypohydration (dehydration) will elicit thirst. A second effect of a rise in body fluid osmolarity is an increased secretion of antidiuretic hormone (ADH) from the neurohypophysis, an effect mediated from the same center (Verney, 1947). The kidneys must excrete a minimal amount of water as a vehicle for the elimination of solids. When water is in excess in the body, little or no ADH is brought to the kidneys and more water is excreted. A water deficit will, as stated, stimulate ADH secretion, causing an increased reabsorption of water by increasing the water permeability of the wall of the collecting ducts in the kidneys. It is a common observation that the volume of urine is reduced when sweating is profuse.

It has been shown that injection of minute amounts of hypertonic saline into, and electrical stimulation within, the anterior parts of the hypothalamus may elicit excessive drinking in the goat (Andersson, 1967). Similar stimulations will not only induce drinking but inhibit feeding. The intracellular fluid volume in the specific cells of the hypothalamus may be the crucial factor in thirst. The osmotic pressure across the cell membrane will of course influence this volume. There are volume receptors (stretch receptors) in the walls of the atria and great veins which can detect and reflexively adjust variations in the volume of intravascular fluid (see Gauer et al., 1970; Goetz et al., 1975).

Satiety is to a large extent a matter of behavior. Stevenson (1965) points out that a person usually waits until taking food to replace the last part of a water deficit.

Sensation of oral-pharyngeal dryness can elicit the urge to drink, but this reflex is not essential for the maintenance of a normal water intake. When one drinks, there is a temporary relief of thirst. This negative feedback operates partly from the oral-pharyngeal level but is also induced from stomach distension. The rapid relief of thirst after water intake is also explained by a normalization of the osmolarity of the blood. Not only does water move out of the gastrointestinal tract, but salts diffuse in the opposite direction along a concentration gradient.

The salt content of the sweat is less than that of the blood, and sweat loss therefore causes an increase in the salt concentration of the blood. As discussed earlier, increased salt concentration of the body fluids leads to the sensation of thirst and reduced urine volume. Certain studies have indicated that the salt content of sweat may be reduced as the result of acclimatization to heat, which should result in increased osmolarity and increased thirst at a given sweat loss (Robinson, 1963). The acclimatized individual is better able to maintain the fluid balance than one who is not acclimatized. Here again, experience may be an important factor, for a water deficit will negatively influence the physical condition.

It is a common observation that a voluntary water intake does not necessarily cover the water loss induced by excessive sweating (Pitts et al., 1944; Adolph, 1947; Leithead and Lind, 1964). The risk of a voluntary hypohydration is greatest in an individual unaccustomed to heat. The risk is also greater when a large portion of the food consists of dried or dehydrated rations, since a considerable volume of the daily water intake comes normally with the regular meals.

Summary   Osmometric, volumetric, and thermal excitations appear to be nature's signals which feed information into the control system for water intake, located mainly in the hypothalamus. Sweat loss increases the osmotic pressure of the body fluids and thereby elicits the urge to drink. However, the sensation of thirst does not always "force" the individual to cover the water loss, particularly not when this loss is pronounced because of profuse sweating or because the individual does not eat normal meals containing a large amount of water.

Water Deficit

We have concluded that high sweat rates with excessive loss of body fluids may cause a deficit of body water (hypohydration or dehydration). The regulation of body temperature has priority over the regulation of body water. Therefore, a hypohydration can be driven very far, and may in fact be a threat to life if the environment is very hot and water is not available.

Prolonged exposure to heat and/or prolonged exercise certainly causes a hypohydration. In both situations a decrease in plasma volume has been noticed. Costill and Fink (1974) report a 16 to 18 percent reduction in plasma volume at a hypohydration equivalent to a 4 percent decrease in body weight. There is a shrinkage of the red cells during hypohydration (therefore changes in hematocrit are not a reliable measure of changes in plasma volume; see Costill et al., 1974; Harrison et al., 1975). During exercise and/or exposure to heat stress there is a movement of protein from the interstitial spaces, particularly in skeletal muscles, to the vascular volume. The increase in plasma protein concentration will raise plasma oncotic (osmotic) pressure thereby helping to maintain blood volume by reducing water loss and enhancing water gain (Senay, 1972; Harrison et al., 1975). As suggested by Senay (1972), heat acclimatization may increase the ability to shift protein and fluid from the interstitial to the intravascular volume, which could improve the efficiency of the cardiovascular system. (The effect of exercise and heat stress on the protein translocation also creates methodological problems when plasma volume is evaluated from plasma protein concentrations.) Irrespective of the cause of sweating, hypohydration is associated with a decrease in stroke volume during exercise and a concomitant increase in heart rate during

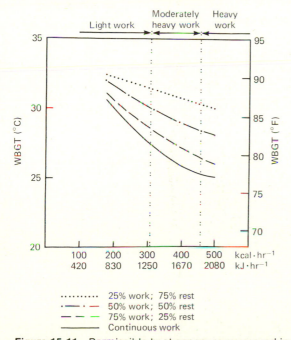

Figure 15-11 Permissible heat exposure expressed in SWBGT °C for different types of work. Modified from ACGIH (1976). Light work corresponds to an oxygen uptake of ≤ 1.0 liters $\cdot$ min^{-1} (20 kJ $\cdot$ min^{-1}; 5 kcal $\cdot$ min^{-1}), moderately heavy work $> 1.0 - 1.5$ liters $O_2 \cdot$ min^{-1} (20-30 kJ $\cdot$ min^{-1}; 5-7.5 kcal $\cdot$ min^{-1}), and heavy work > 1.5 liters $\cdot$ min^{-1} (30 kJ $\cdot$ min^{-1}; 7.5 kcal $\cdot$ min^{-1}). (From I. Åstrand et al., 1975.)

submaximal work (Fig. 15-12). It is remarkable, however, that during maximal exercise, oxygen uptake, cardiac output, and stroke volume are not modified by a sweat loss of up to 5 percent of body weight. However, endurance, i.e., the work time which can be tolerated on a standardized maximal work load, is definitely reduced after dehydration (Fig. 15-13). It has also been found that rectal temperatures are significantly higher in the dehydrated subject (Fig. 15-14), and the rise is related to the weight loss incurred (Gisolfi and Copping, 1974). The excessive rise in core temperature with hypohydration is probably due to inadequate sweating (Greenleaf and Castle, 1971).

Unpublished results from a study in a Norwegian cement factory, where the ambient temperature was fairly high even in the winter, showed that the heart rate of the workers at a standardized submaximal work load was significantly lower (more than 10 beats $\cdot$ min^{-1}) at the end of the work shift when 2 liters of fluid was taken in the course of the shift, compared to the days when no fluid was taken between meals.

The explanation for the gradual decrease in physical performance as a hypohydration develops is presently not available. It apparently cannot be primarily a modification of the aerobic energy yield, because the maximal aerobic power was not impaired in Saltin's experiments, despite a pronounced hypohydration. The explanation should be sought at the cellular level where changes may occur during a hypohydration. The maximal isometric strength after a water deficit is reported to be unaffected (Saltin, 1964) or slightly decreased in connection with progressive hypohydration (Bosco et al., 1968). In any case, a reduced water content within the muscle cell and a disturbed electrolyte balance can easily influence the muscle cell's ability to contract and its susceptibility to metabolites. The reduction in work performance is more

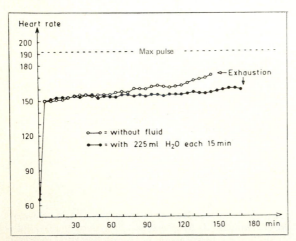

Figure 15-12 Mean heart rates in subjects running on the treadmill at 70 percent of the maximal O$_2$ uptake until exhaustion, with and without fluid. *(From Staff and Nilsson, 1971.)*

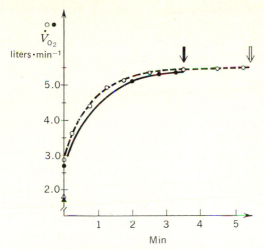

Figure 15-13 Oxygen uptake (liters·min⁻¹) at a work load which could be tolerated for 5¹/₂ min during normal conditions (unfilled symbols) but only for 3¹/₂ min after hypohydration (filled symbols). Arrows indicate maximal work time. *(From Saltin, 1964.)*

marked if a water deficit is caused by extended heavy work than after exposure to hot environment without exercise being involved.

Dehydration causes a reduced tilt-table tolerance. A person who normally could tolerate prolonged 45° head-up tilt with a heart rate of 90 beats · min⁻¹ fainted within 7.5 min after a fluid loss corresponding to 3 percent of his body

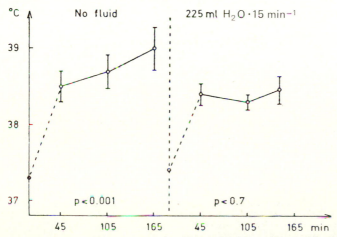

Figure 15-14 Mean rectal temperature in subjects running on the treadmill at 70 percent of their maximal oxygen uptake, with and without fluid. When no fluid was taken, rectal temperature rose significantly from the 45th to the 165th min. *(From Staff and Nilsson, 1971.)*

weight. Following a fluid loss corresponding to 6 percent of his body weight, he fainted within 1.5 min. His heart rate before fainting was 115 and 135 respectively. A low stroke volume was a characteristic finding (Adolph, 1947).

During prolonged and physically heavy training or during participation in certain competitive sports, the sweat rate may be very high. In some cases it may be as high as 2 liters $\cdot$ hr^{-1}. An adequate water balance plays an important role in maintaining optimal performance capacity. It is unfortunate that in certain athletic events such as marathon running or walking, the established rules may actually limit the available fluid supply to the athletes. Similarly, the current practice of weighing in wrestlers at least 2 hr before the start of the day's match may permit those who have purposely become dehydrated in order to qualify for a lower weight class to replace their lost body water before the match if they are to compete late; this may not be possible for those who are to compete first. In any case, such dehydration is not only harmful to the individual, but also poor sportmanship. (See the American College of Sports Medicine: Position Stand on Weight Loss in Wrestlers, *Med. Sci. Sports,* **8** (2): xi, 1976.)

Well-trained subjects are less affected in their performance by a hypohydration than untrained subjects (Buskirk et al., 1958; Saltin, 1964). Acclimatization to heat does not seem to protect from the deteriorating effect of a hypohydration.

The simplest method of determining whether the fluid intake has been adequate is by weighing the individual under standard conditions. Even a reduction in body weight of 1 to 2 percent may represent a deterioration in work capacity (Pitts et al., 1944; Adolph, 1947; Ladell, 1955; Saltin, 1964; Gisolfi and Copping, 1974).

It should be emphasized that the degree of body hypohydration is overestimated from measurements of body weight when large amounts of glycogen have been metabolized, which will deliver a "surplus" of water. The fluid loss during prolonged heat exposure should preferably be replaced by drinking 100 to 150 ml water several times per hour. The water temperature should be about 15°C. In the case of heat exposure lasting for several weeks, the ingestion of salt tablets is advisable, 5 to 15 g per day depending on diet, climate, and degree of physical activity.

Excessive fluid loss can also occur in cold environments. Lennquist (1972) has shown that the cold-induced diuresis may persist for several days and may lead to a considerable fluid deficit, accompanied by hemoconcentration and reduction in blood volume. The cold-induced rise in osmolar excretion could largely be accounted for by significant increases in the excretion of sodium, chloride, and calcium. The increased excretion of sodium was dependent on a reduced tubular sodium reabsorption. According to Lennquist (1972), the releasing mechanism of cold diuresis is to be sought in the renal tubules and not in an increased glomerular filtration rate. Wallenberg (1974) has published results indicating that tubular sodium reabsorption in a cold-exposed person is influenced by plasma osmotic pressure and changes in arterial blood pressure. Cold-induced suppression of distal tubular sodium reabsorption could be almost completely abolished by an albumin infusion of 0.4 to 0.5 g per kg.

Summary   An individual tolerates heavy physical work less well if subjected to a water deficit, even if the water loss is only about 1 percent of the body weight. At a submaximal work load, the heart rate is increased, the stroke volume is reduced, and the body temperature is higher than normal. Drinking water to satiety may not fully compensate for a water loss.

PRACTICAL APPLICATION

Physical Work

It should be borne in mind that heat exposure in itself represents an extra load on the blood circulation. Exhaustion occurs much sooner during heavy physical work in the heat because the blood, in addition to carrying oxygen to the working muscle, also has to carry heat from the interior of the body to the skin. This represents an extra burden on the heart, which has to pump that much harder. This is convincingly demonstrated in Table 15-1, which shows the difference in work pulse in a subject performing the same work in a hot environment and in a cool one. The stress of heat and the hydrostatic factors in prolonged standing work may be added to the stress of work itself.

Williams et al. (1967) observed no difference in maximal O_2 uptake in subjects working in the heat and at comfort temperature. At submaximal work loads, however, they found that the major change in hemodynamics in the heat was an increase in heart rate and a fall in stroke volume. Neither cardiac output nor arteriovenous difference was significantly altered compared with comfortable conditions (Rowell, 1974). Williams et al. (1967) also demonstrated a larger lactate production in the subject who worked in the heat as compared with one working in a neutral environment. This finding can be explained as a result of a reduced muscle blood flow.

In the armed forces and in certain industries, the problem of an efficient

Table 15-1   Effect of Environmental Temperature on Human Response to Standard Work on a Bicycle Ergometer for 45 Min

Environmental temp.	Heart rate, beats · min⁻¹	Rectal temp., °C T_r	Mean skin temp., °C T_s	O_2 uptake, liters · min⁻¹	Weight loss	
					kg	% of body weight
Cool	104	37.7	32.8	1.5	0.25	0.3
Hot steel mill, air temp. 40 to 50°C + radiation	166	38.8	37.6	1.5	1.15	1.6

At the same work load, the temperature difference between core and shell is 4.9°C in the cool environment, but only 0.9°C in the heat. This necessitates a much greater skin blood flow in the heat. Hence, the markedly elevated heart rate in the heat.

and rapid method of acclimatizing a large number of people is often of practical importance. Daily exposure to a hot environment for about an hour will, after a week, result in some acclimatization. However, studies by Wyndham and coworkers (1973) to establish the minimal number of days required for acclimatization showed that a person cannot be acclimatized adequately for a normal shift of 6 to 8 hours in less than 4 hours per day and in less than 8 to 9 days. They apply a step test with a gradual increase in rate of work up to an oxygen uptake of 1.4 liters $\cdot$ min^{-1} combined with heat stress conditions (air temperature 31.7°C, air saturated with water vapor). This program is applied to recruits for the gold mines in South Africa.

Warm-up

The benefit of the higher temperature during work lies in the fact that the metabolic processes in the cell can proceed at a higher rate, since these processes are temperature-dependent. For each degree of temperature increase, the metabolic rate of the cell increases by about 13 percent. At the higher temperature, the exchange of oxygen from the blood to the tissues is also much more rapid. Physical work capacity is increased following warm-up (Simonson et al., 1936; Asmussen and Böje, 1945). Furthermore, the nerve messages travel faster at higher temperatures. At the temperature of the human body, which is much higher than that of a frog, our nerve messages go up to 8 times as fast as those of the frog (Hill, 1927). Thus, there is a very good reason for a person to keep the body temperature up as he or she does, even at considerable expense, in order to be able to move more quickly. This is also the reason why athletes have discovered that it pays to warm up before an athletic event. This warming up may make a difference of 3 s in a 400-yard dash. The warming up may profitably consist of rather vigorous exercise, such as running at a rate of about 12 km/hr for 15 to 30 min just before the event (Högberg and Ljunggren, 1947). In the case of ordinary exercise, a 5-min warm-up consisting of light to moderate exercise is usually adequate.

Högberg and Ljunggren (1947) examined the effect of warm-up in the form of running at moderate speed combined with calisthenics on the speed of running 100, 400, or 800 m in well-trained athletes. They compared this effect with the effect of heating the body passively in a sauna bath for a period of 20 min prior to the race and found that the beneficial effect of passively elevating the body temperature by such a bath was much less than that of elevating the body temperature by a warm-up through physical exercise. In the 100-m dash, the improvement after a proper warm-up was in the order of 0.5 to 0.6 s, corresponding to 3 to 4 percent, compared with the results without any warm-up. In the 400-m race, the improvement amounted to 1.5 to 3.0 s, corresponding to 3 to 6 percent. In the 800-m race, the improvement was 4 to 6 s, or 2.5 to 5.0 percent. Thus the percentage improvement was roughly the same at all distances examined. Similar results have been obtained in swimming (Muido, 1946).

With regard to the duration of the warm-up, Högberg and Ljunggren (1947) observed better results after a 15-min warm-up than after a 5-min one, but no further significant improvement occurred in the 100-m race when the warm-up was extended from 15 to 30 min. The authors observed no deterioration in performance attributable to fatigue as a consequence of rather vigorous warm-up. They recommend a warm-up period of 15 to 30 min at a relatively high rate of energy expenditure (in their experiments about 3.0 to 3.4 liter O_2 uptake $\cdot$ min^{-1}, equivalent to running at a speed of 12 to 14 km $\cdot$ hr^{-1}). The duration and intensity of warm-up should be adjusted according to the environmental temperature and amount of clothing. The higher the environmental temperature and the greater the amount of clothing, the sooner the desired body temperature of about 38.5°C is attained (muscle temperature 39°C or higher). Ideally, the rest period between warm-up and the start of the race should be no more than a few minutes, in any case no more than 15 min. After 45 min rest, the beneficial effect of the warm-up is abolished, at which time the muscle temperature has also returned to pre-warm-up levels. The use of warm clothing is recommended during warm-up, and this clothing should be worn until the athlete is ready to start the race.

Bernard et al. (1973) report that strenuous exercise, without prior warm-up, induced abnormal ECG changes in 70 percent of their 44 normally asymptomatic subjects (aged from twenty-one to fifty-two years). Two minutes of jogging-in-place as a warm-up just prior to the exercise eliminated or reduced these abnormal ECG responses.

Figure 15-15 presents a summary of this discussion of the beneficial effect of warm-up for the physical performance. The improvement is particularly related to the increase in muscle temperature. The higher the muscle temperature (i.e., the heavier the preceding warm-up exercise), the better the performance.

Radiation

Figure 15-16 illustrates how the radiant heat may be reduced from 5,4 MJ (1300 kcal) $\cdot$ hr^{-1} to about 54 kJ (kcal) $\cdot$ hr^{-1} by placing an aluminum shield between the worker and the heat source, which in this particular case had a temperature of 188°C. An inexpensive protection against radiation may be provided by a sheet of masonite covered with tinfoil. It is important that the surface be kept clean. If the shield has to be transparent, substances which will reflect infrared light, such as glass, should be used.

It should be emphasized that behavior is an important aspect of the protection of the body in hot or cold environments. One can modify radiation to or from the body by changing the posture. The surface area of the human silhouette will increase by a factor of 3 when a person changes from crouching to an expanded body position. Just a change in orientation to the direction of the wind or sun can alter the heat exchange markedly.

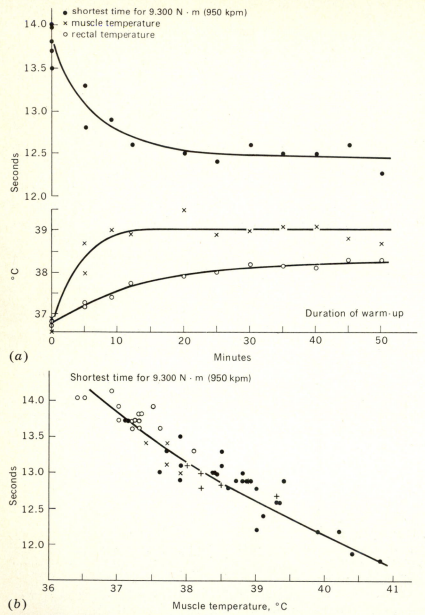

Figure 15-15 (*a*) Two lower curves show the temperature in the lateral vastus muscle (upper curve) and in the rectum (lower curve) after warm-up at a work load of 160 watts of different durations (abscissa). The top curve shows the shortest period of time required for the completion of an energy output corresponding to 9.300 N · m on a bicycle ergometer following warm-up of different durations, as shown on the abscissa.
(*b*) The relationship between the time required for a spurt on the bicycle ergometer (9.300 N · m) and the temperature measured in the lateral vastus muscle immediately prior to the spurt. O = no warm-up; ● = 30-min warm-up of different intensity; X = warm-up in the form of warm showers; + = warm-up with the aid of diathermy. *(From Asmussen and Böye, 1945.)*

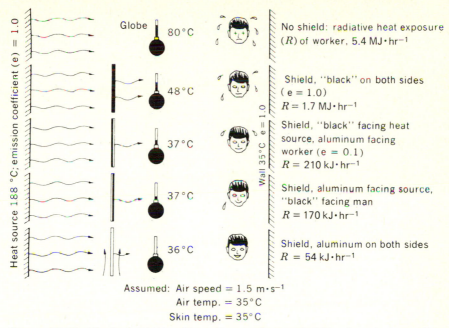

Figure 15-16 Effect of shields with various surface emissivities in reducing radiant heat load. *(Modified from Hertig and Belding, 1963.)*

Air Motion

Air motion increases the evaporation of sweat. However, if the air temperature is higher than that of the skin, the air motion may serve to increase the heat load in that it will cause the skin to pick up more heat through convection.

Clothing

In a moist, hot climate where the temperature of the environment is lower than that of the skin, it is advisable to wear as little clothing as possible. If the ambient temperature is higher than that of the skin, the clothing may protect the individual from the radiant heat of the environment. Loose-fitting clothing which permits free circulation of air between the skin and the clothing is preferable. Workers habitually exposed to intense heat in their work have learned to dress in heavy clothing for protection. This allows some of the radiant heat to be absorbed in the clothing a distance away from the skin. However, it also impairs the facility for evaporative heat loss.

The problem of clothing in the cold when heavy physical work is alternated with rest periods is even more complicated, since there is no single item of clothing capable both of protecting against cold at rest and of facilitating heat dissipation during heavy work. The conventional method is to unbutton the coat during work and to button it up during inactivity.

It should also be pointed out that the surface of the hands represents about

5 percent of the total surface area of the body. In a nude individual, about 10 percent of the heat produced may be eliminated through the hands. In a clothed individual, up to 20 percent of the heat produced may be eliminated through the hands (Day, 1949).

In order to remain in heat balance, a person sleeping outdoors at −40°C needs protective clothing with an insulation value of about 12 Clo units. (A "Clo" unit equals the amount of insulation provided by the clothing a person usually wears at room temperature.) However, when the same individual is physically active, moving about or walking along, only the equivalent of 4 Clo units will be needed because the person's body heat production is now at least 3 times greater than it was when sleeping because of the increased metabolic rate associated with the increased physical activity (Fig. 15-17). This requirement is adequately met by the original double-layer caribou clothing of the Eskimo. Two layers of caribou fur, amounting to a thickness of 3 in., have a total insulation value of about 12 Clo units (Fig. 15-18). This is adequate to maintain heat balance under practically any condition likely to be encountered by the Eskimo. Temperature measurements inside the Eskimo clothing, taken in the field, have confirmed that the Eskimo's body inside the clothing is indeed comfortably warm. There is, therefore, some truth to the old statement that the Eskimo, by virtue of their clothing, are really surrounded by a tropical climate.

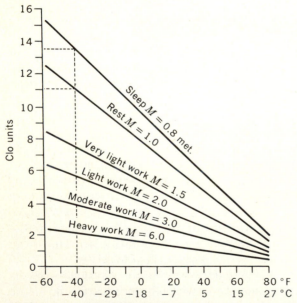

Figure 15-17 Insulating requirements in the cold when protected from the wind during different rates of heat production. A Clo unit is the thermal insulation which will maintain a resting man indefinitely comfortable in an environment of 21°C, relative humidity less than 50 percent, and air movement 6 m · min⁻¹. The unit referred to as "met" is the metabolic rate of a resting man. *(From Burton and Edholm, 1969.)*

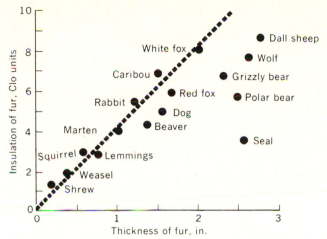

Figure 15-18   Insulating value of different furs. *(Redrawn from Scholander et al., 1950.)*

The ordinary uniform usually worn by airmen and soldiers in the Arctic, on the other hand, has an insulation value of only 4 Clo units, which is only one-third that of the Eskimo's clothing. The arctic uniform offers adequate protection for an active person at temperatures as low as −40°C, but the person would be in negative heat balance if inactive (Fig. 15-17). Temperatures below −40°C occur on an average of about 2 days per month in the winter in the interior of Alaska.

The insulation value of most materials is proportional to the amount of air which is trapped within the material itself, since air is such a superb insulator. In the case of fur, air is trapped in the space between each hair, but the superior insulation quality of caribou fur, over and above other fur, lies in the fact that the caribou hair is hollow and contains trapped air inside each hair as well as in the spaces between them.

A physiological assessment of the insulating value of clothing may be attained simply by measuring heat production (oxygen uptake) and heat loss as evidenced by changes in stored body heat (by applying the formula and procedures for obtaining rectal and mean skin temperatures described earlier in this chapter) in normal subjects under controlled climatic-chamber conditions. An example of such a study is presented in Fig. 15-19, in which similar garments made of nylon pile and wool pile were compared in paired experiments at rest for 1 hr and during 2 hr of fairly strenuous physical activity (treadmill walking at 100 m/min, 5° incline) followed by a 2-hr rest in a climatic chamber at −20°C (the values represent the means of five subjects). Evaporative weight loss was determined by weighing the subjects in the nude before and after the experiment. The accumulation of moisture in the experimental clothing was assessed by weighing the garments before and after the experiment on a scale with an accuracy of ±10 g. In this study, no significant

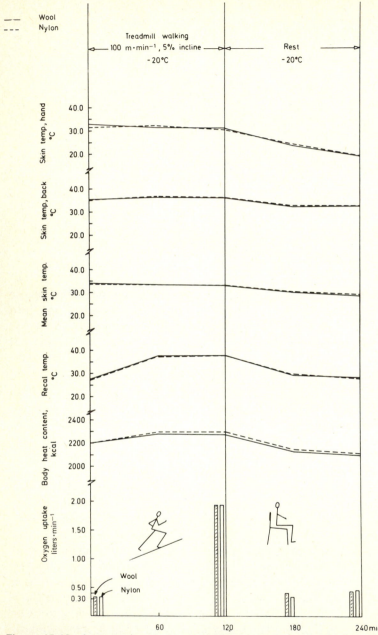

Figure 15-19 A comparison between similar garments made of nylon pile and of wool pile during 2-hr treadmill walking and 2-hr rest in a climatic chamber at −20°C. There was no statistically significant difference in O₂ uptake, skin or rectal temperature, or stored heat resulting from the two types of garments worn. Figures represent means of five subjects. *(From Rodahl et al., 1974.)*

difference could be detected between the two types of garments in terms of thermal insulation, nor in the ability of the two types of fabric to allow free escape of moisture produced by sweating during the physical activity (Rodahl et al., 1974).

Microclimate

The solution to the problem of providing optimal working environment may be to create local microclimates by cooling or heating the clothing, by providing environmental suits to be worn under special circumstances, or by enclosing the work area in a suitable artificially made environment which will facilitate an effective climatic control. Radiant heaters may be used, and exposure suits may be applicable. With any solution, there will be certain complications. In the ideal climate, the temperature of the skin is about 33°C, but not uniformly so. The feet are normally colder than the trunk, and a person may accept a greater lowering of the temperature of the feet without feeling cold. A bath, with a water temperature of 33°C, feels cold. In order for the bather to feel comfortable, the water temperature has to be about 35°C. However, such a water temperature does not produce temperature equilibrium; it will cause the body temperature to rise. As Burton (1963) puts it: man is not constructed to spend much time in water.

A local heating of the floor may represent an unphysiological manner of regulating room temperature (Burton, 1963) due to the fact that the receptors in the skin of the feet exert a relatively dominating influence upon the temperature-regulating center. An induced vasodilation of the feet is effective in increasing heat loss. In spite of warm feet, the subjects of one study eventually became cold. With this background, it might appear unphysiological to heat our dwellings, factories, etc., by keeping the floor hotter than the room air.

It is thus possible, by improper clothing or by local heating or cooling of limited skin areas, to upset the normal physiological temperature regulation. Local heating of hands and feet may, for example, bring about shivering and sweating at the same time. Even if the air temperature is high, heat loss by outgoing radiation to cold surfaces, like a cold window, may cause a most unpleasant sensation, commonly referred to as draft. Because of radiative heat loss to the night sky, a person exposed, unshielded, in the arctic environment may actually be exposed to a cold stress 10 to 20° colder than that which the air thermometer might indicate. In a small enclosed area, such as the cabin of an aircraft or a car, it is difficult to satisfy the requirement for adequate ventilation, hot or cold, without causing certain areas of the body to be too hot or too cold. It is evident that much money, as well as many heart beats and much sweating, could be saved by proper planning of lecture halls, office and factory buildings, and machinery, taking into account all the factors which constitute the optimal climate.

The secret of the success of an experienced arctic traveler or hunter lies in the ability to avoid the extreme cold. The Eskimo's dwelling, whether it be a skin tent, a peat-covered house, or a log cabin, is comfortably warm at all times. The temperature is kept around 21°C (70°F) in the day and may drop to about 10°C (50°F) during the night when the sleeping Eskimo is well covered by fur. Although, in the summer, the Eskimo may spend as much as 9 hr or more out of doors, the average amount of time spent outside in the winter is only 1 to 4 hr. Furthermore, experience has taught arctic dwellers to take advantage of the characteristic temperature distribution in their environment, produced by the so-called temperature inversion during the winter. The coldest spot is at the surface of the snow, and especially in a depression in the terrain, such as a riverbed where cold, heavy air is trapped. A few feet up the hillside, the temperature is usually many degrees warmer. While the temperature at the actual snow surface may be as cold as −50°C, the temperature under the snow cover, in the narrow air space between the ground and the snow crust, is usually maintained at −9 to −6°C throughout the winter. It is here that the field mice and other arctic rodents which do not hibernate run around perfectly comfortable all winter long. Actually, smaller animals, like weasels and mice, are not able to carry a fur thick enough for insulation and must therefore spend the winter mostly underneath the snow. By taking advantage of this characteristic temperature distribution, the experienced traveler may successfully escape the extreme degrees of cold stress.

REFERENCES

ACGIH (American Conference of Governmental Industrial Hygienists): "Threshold Limit Values for Chemical Substances and Physical Agents in the Workroom Environment with Intended Changes for 1976," Cincinnati, Ohio, 1976.

Adolph, E. F., and Members of the Rochester Desert Unit: "Physiology of Man in the Desert," Interscience Publishers, Inc., New York, 1947.

Ahlman, K., and M. J. Karvonen: Stimulating of Sweating by Exercise after Heat Induced "Fatigue" of the Sweating Mechanism, *Acta Physiol. Scand.*, **53**:381, 1961.

Andersson, B.: The Thirst Mechanism as a Link in the Regulation of the "Milieu Intérieur," in "Les Concepts de Claude Bernard sur le Milieu Intérieur," p. 13, Masson et Cie, Paris, 1967.

Asmussen, E., and O. Böje: Body Temperature and Capacity for Work, *Acta Physiol. Scand.*, **10**:1, 1945.

Åstrand, I.: Aerobic Work Capacity in Men and Women with Special Reference to Age, *Acta Physiol. Scand.*, **49**(Suppl. 169):67, 1960.

Åstrand, I., O. Axelson, U. Eriksson, and L. Olander: Heat Stress in Occupational Work, *AMBIO*, **3**:37, 1975.

Barnard, R. J., G. W. Gardner, N. V. Diaco, R. N. MacAlpin, and A. A. Kattus: Cardiovascular Responses to Sudden Strenuous Exercise—Heart Rate, Blood Pressure, and ECG, *J. Appl. Physiol.*, **34**:833, 1973.

Bass, E. E.: Thermoregulatory and Circulatory Adjustments during Acclimatization to Heat in Man, in J. D. Hardy (ed.), "Temperature: Its Measurement and Control in Science and Industry," vol. 3, part 3, p. 299, Reinhold Book Corporation, New York, 1963.

Bazett, H. C., L. Love, M. Newton, L. Eisenberg, R. Day, and R. Forster: Temperature Changes in Blood Flowing in Arteries and Veins in Man, *J. Appl. Physiol.*, **1**:3, 1948.

Beaumont, W. van, and R. W. Bullard: Sweating: Its Rapid Response to Muscular Work, *Science*, **141**:643, 1963.

Benzinger, T. H.: The Thermal Homeostasis of Man, in "Les Concepts de Claude Bernard sur le Milieu Intérieur," p. 325, Masson et Cie, Paris, 1967.

Benzinger, T. H., C Kitzinger, and A. W. Pratt: The Human Thermostat, in J. D. Hardy (ed.), "Temperature: Its Measurement and Control in Science and Industry," vol. 3, part 3, p. 637, Reinhold Book Corporation, New York, 1963.

Benzinger, T. H., and G. W. Taylor: Cranial Measurements of Internal Temperature in Man, in J. D. Hardy (ed.), "Temperature: Its Measurement and Control in Science and Industry," vol. 3, part 3, p. 111, Reinhold Book Corporation, New York, 1963.

Berggren, G., and E. H. Christensen: Heart Rate and Body Temperature as Indices of Metabolic Rate during Work, *Arbeitsphysiol.*, **14**:255, 1950.

Bligh, J., and R. E. Moore (eds.): "Essays on Temperature Regulation," North-Holland, Amsterdam, 1972.

Bosco, J. S., R. L. Terjung, and J. E. Greenleaf: Effects of Progressive Hypohydration on Maximal Isometric Muscular Strength, *J. Sports Med.*, **8**:81, 1968.

Burton, A. C.: Human Calorimetry, II—The Average Temperature of the Tissues of the Body, *J. Nutr.*, **9**:261, 1935.

Burton, A. C.: The Pattern of Response to Cold in Animals and the Evolution of Homeothermy, in J. D. Hardy (ed.), "Temperature: Its Measurement and Control in Science and Industry," vol. 3, part 3, p. 363, Reinhold Book Corporation, New York, 1963.

Burton, A. C., and O. G. Edholm: "Man in a Cold Environment," Hafner Publishing Company, Inc., New York–London, 1969.

Buskirk, E. R., P. F. Iampietro, and D. E. Bass: Work Performance after Dehydration: Effects of Physical Conditioning and Heat Acclimatization, *J. Appl. Physiol.*, **12**:189, 1958.

Cabanac, M.: Temperature Regulation, *Ann. Rev. of Physiol.*, **37**:415, 1975.

Cabanac, M., G. Hildebrandt, B. Massonnet, and H. Strempel: A Study of the Nycthemeral Cycle of Behavioural Temperature Regulation in Man, *J. Physiol.*, **257**:275, 1976.

Christensen, E. H.: Beiträge zur Physiologie schwerer Körperlicher Arbeit, Die Körpertemperatur während und unmittelbar nach schwerer körperlicher Arbeit, *Arbeitsphysiol.*, **4**:154, 1931.

Costill, D. L., L. Branam, D. Eddy, and W. Fink: Alterations in Red Cell Volume Following Exercise and Dehydration, *J. Appl. Physiol.*, **37**:912, 1974.

Costill, D. L., and W. J. Fink: Plasma Volume Changes Following Exercise and Thermal Dehydration, *J. Appl. Physiol.*, **37**:521, 1974.

Davies, C. T. M., J. R. Brotherhood, and E. Zeidifard: Temperature Regulation during Severe Exercise with Some Observations on Effects of Skin Wetting, *J. Appl. Physiol.*, **41**:772, 1976.

Davies, M., B. Ekblom, U. Bergh, and I-L. Kanstrup-Jensen: The Effect of Hypothermia on Submaximal and Maximal Work Performance, *Acta Physiol. Scand.*, **95**:201, 1975.

Davis, T. R. A.: Acclimatization to Cold in Man, in J. D. Hardy (ed.), "Temperature: Its

Measurement and Control in Science and Industry," vol. 3, part 3, p. 443, Reinhold Book Corporation, New York, 1963.

Day, R.: Regional Heat Loss, in L. W. Newburgh (ed.), "1, Physiology of Heat Regulation," p. 240, W. B. Saunders Company, Philadelphia, 1949.

Dill, D. B. (ed.): "Handbook of Physiology," Sec. 4, Adaptation to the Environment, American Physiological Society, Washington, D.C., 1964.

Drinkwater, B. L., J. E. Denton, I. C. Kupprat, T. S. Talag, and S. M. Horvath: Aerobic Power as a Factor in Women's Response to Work in Hot Environments, *J. Appl. Physiol.,* **41:**815, 1976a.

Drinkwater, B. L., J. E. Denton, P. B. Raven, and S. M. Horvath: Thermoregulatory Response of Women to Intermittent Work in the Heat, *J. Appl. Physiol.,* **41:**57, 1976b.

Eisalo, A.: Effects of the Finnish Sauna on Circulation, *Annals Medicinae Experimentalis et Biologicae Finniae,* Suppl. No. 4, vol. 34, Helsinki, 1956.

Gale, C. C.: Neuroendocrine Aspects of Thermoregulation, *Ann. Rev. Physiol.,* **35:**391, 1973.

Gauer, O. H., J. P. Henry, and C. Behn: The Regulation of Extracellular Fluid Volume, *Ann. Rev. Physiol.,* **32:**547, 1970.

Gerking, S. D., and S. Robinson: Decline in the Rates of Sweating of Men Working in Severe Heat, *Am. J. Physiol.,* **147:**370, 1946.

Gisolfi, C. V., and J. R. Copping: Thermal Effects of Prolonged Treadmill Exercise in the Heat, *Med. Sci. Sports,* **6:**108, 1974.

Goetz, K. L., G. C. Bond, and D. D. Bloxham: Atrial Receptors and Renal Function, *Physiol. Rev.,* **55:**157, 1975.

Greenleaf, G. E., and B. I. Castle: Exercise Temperature Regulation in Man during Hypohydration and Hyperhydration, *J. Appl. Physiol.,* **30:**847, 1971.

Hammel, H. T.: Neurons and Temperature Regulation, in W. S. Yamamoto and J. R. Brobeck (eds.), "Physiological Controls and Regulations," p. 71, W. B. Saunders Company, Philadelphia, 1965.

Hammel, H. T., S. Strömme, and R. W. Cornew: Proportionality Constant for Hypothalamic Proportional Control and of Metabolism in Unanesthetized Dogs, *Life Sciences,* **12:**933, 1963.

Hansel, H.: Electrophysiology of Thermosensitive Nerve Endings, in J. D. Hardy (ed.), "Temperature: Its Measurement and Control in Science and Industry," vol. 3, part 3, p. 191, Reinhold Book Corporation, New York, 1963.

Hardy, J. D.: The "Set-Point" Concept in Physiological Temperature Regulation, in W. S. Yamamoto and J. R. Brobeck (eds.), "Physiological Controls and Regulation," p. 98, W. B. Saunders Company, Philadelphia, 1965.

Hardy, J. D.: Central and Peripheral Factors in Physiological Temperature Regulation, in "Les Concepts de Claude Bernard sur le Milieu Intérieur," p. 247, Masson et Cie, Paris, 1967.

Hardy, J. D. and E. F. DuBois: "The Technique of Measuring Radiation and Convection," *J. Nutr.,* **15:**461, 1938.

Hardy, J. D., A. P. Gagge, J. A. J. Stolwijk (eds.): "Physiological and Behavioral Temperature Regulation," Charles C Thomas, Springfield, Ill., 1970.

Hardy, J. D., J. A. J. Stolwijk, and A. P. Gagge: in G. C. Whittow (ed.), "Comparative Physiology of Thermoregulation," vol. II, p. 327, Academic Press, Inc., New York, 1971.

Harrison, M. H., R. J. Edwards, and D. R. Leitch: Effect of Exercise and Thermal Stress on Plasma Volume, *J. Appl. Physiol.*, **39**:925, 1975.

Hellström, B.: "Local Effects of Acclimatization to Cold in Man," Universitetetsforlag, Oslo, 1965.

Hensel, H.: Thermoreceptors, *Ann. Rev. Physiol.*, **36**:233, 1974.

Hertig, B. A., and H. S. Belding: Evaluation and Control of Heat Hazards, in J. D. Hardy (ed.), "Temperature: Its Measurements and Control in Science and Industry," vol. 3, part 3, p. 347, Reinhold Book Corporation, New York, 1963.

Hill, A. V.: "Living Machinery," Harcourt, Brace & World, Inc., New York, 1927.

Högberg, P., and O. Ljunggren: Uppvärmningens inverkan på löpprestationerna, *Svensk Idrott*, **40**, 1947.

Holmér, I., and U. Bergh: Metabolic and Thermal Response to Swimming in Water at Varying Temperatures, *J. Appl. Physiol.*, **37**:702, 1974.

Irving, L.: Adaptations to Cold, *Sci. Am.*, **214**(1):94, 1966.

Irving, L.: Ecology and Thermoregulation, in "Les Concepts de Claude Bernard sur le Milieu Intérieur," p. 381, Masson et Cie, Paris, 1967.

Issekutz, B., Jr., K. Rodahl, and N. C. Birkhead: Effect of Severe Cold Stress on the Nitrogen Balance of Men under Different Dietary Conditions, *J. Nutr.*, **78**:189, 1962.

Johnson, J. M., G. L. Brengelmann, and L. B. Rowell: Interactions between Local and Reflex Influences on Human Forearm Skin Blood Flow, *J. Appl. Physiol.*, **41**:826, 1976.

Keatinge, W. R.: "Survival in Cold Water," Blackwell, Oxford, 1969.

Kenshalo, D. R., J. P. Nafe, and B. Brooks: Variations in Thermal Sensitivity, *Science*, **134**:104, 1961.

Kerslake, D. McK.: Errors Arising from the Use of Mean Heat Exchange Coefficients in Calculation of the Heat Exchanges of a Cylindrical Body in a Transverse Wind, in J. D. Hardy (ed.), "Temperature: Its Measurement and Control in Science and Industry," vol. 3, part 3, p. 183, Reinhold Book Corporation, New York, 1963.

Kerslake, D. McK.: "Monographs of the Physiological Society: The Stress of Hot Environment," Cambridge University Press, Cambridge, 1972.

Kuno, Y.: "Human Perspiration," Charles C Thomas, Publisher, Springfield, Ill., 1956.

Ladell, W. S. S.: The Effects of Water and Salt Intake upon the Performance of Men Working in Hot and Humid Environments, *J. Physiol.*, **127**:11, 1955.

LeBlanc, J.: "Man in the Cold," American Lecture Series, Charles C Thomas, Springfield, Ill., 1975.

LeBlanc, J., S. Dulac, J. Côté, and B. Girard: Autonomic Nervous System and Adaptation to Cold in Man, *J. Appl. Physiol.*, **39**:181, 1975.

Leithead, C. S., and A. R. Lind: "Heat Stress and Heat Disorders," Cassell & Co., Ltd., London, 1964.

Lennquist, S.: Cold Induced Diuresis, *Scand. J. Urology and Nephrology*, (Suppl. 9): 1972.

Lind, A. R., and D. E. Bass: Optimal Exposure Time for Development of Acclimatization to Heat, *Fed. Proc.*, **22**:704, 1963.

Lind, A. R., P. W. Humphreys, K. J. Collins, K. Foster, and K. F. Sweetland: Influence of Age and Daily Duration of Exposure on Responses of Men to Work in Heat, *J. Appl. Physiol.*, **28**:50, 1970.

Meyer, F. R., S. Robinson, J. L. Newton, C. H. Ts'ao, and L. O. Holgersen: The

Regulation of the Sweating Response to Work in Man, *The Physiologist*, **5**:182, 1962.

Minard, D., and L. Copman: Elevation of Body Temperature in Disease, in J. D. Hardy (ed.), "Temperature: Its Measurement and Control in Science and Industry," vol. 3, part 3, p. 253, Reinhold Book Corporation, New York, 1963.

Muido, L.: The Influence of Body Temperature on Performances in Swimming, *Acta Physiol. Scand.*, **12:102, 1946.**

Nadel, E. R., J. W. Mitchell, B. Saltin, and J. A. J. Stolwijk: Peripheral Modifications to the Central Drive for Sweating, *J. Appl. Physiol.*, **31**:828, 1971.

Nadel, E. R., K. B. Pandolf, M. F. Roberts, and J. A. J. Stolwijk: Mechanisms of Thermal Acclimation to Exercise and Heat, *J. Appl. Physiol.*, **37**:515, 1974.

Nadel, E. R., and J. A. J. Stolwijk: Effect of Skin Wettedness on Sweat Gland Response, *J. Appl. Physiol.*, **35**:689, 1973.

Nelms, J. D., and J. G. Soper: Cold Vasodilatation and Cold Acclimatization in the Hands of British Fish Filleters, *J. Appl. Physiol.*, **17**:444, 1962.

Nielsen, M.: Die Regulation der Körpertemperatur bei Muskelarbeit, *Skand. Arch. Physiol.*, **79**:193, 1938.

Nielsen, B.: Thermoregulation in Rest and Exercise, *Acta Physiol. Scand.*(Suppl. 323): 1969.

Nielsen, B.: Thermoregulation during Work in Carbon Monoxide Poisoning, *Acta Physiol. Scand.*, **82**:98, 1971.

Nielsen, B.: Metabolic Reactions to Changes in Core and Skin Temperature in Man, *Acta Physiol. Scand.*, **97**:129, 1976a.

Nielsen, B.: Physical Effort and Thermoregulation in Man, *Israel J. Med. Sci.*, **12**:974, 1976b.

Nielsen, B., and M. Nielsen: On the Regulation of Sweat Secretion in Exercise, *Acta Physiol. Scand.*, **64**:314, 1965.

Nishi, Y., and A. P. Gagge: Humid Operative Temperature, a Biophysical Index of Thermal Sensation and Discomfort, *J. Physiol. (Paris)*, **63**:365, 1971.

Pepler, R. D.: Performance and Well-being in Heat, in J. D. Hardy (ed.), "Temperature: Its Measurement and Control in Science and Industry," vol. 3, part 3, p. 319, Reinhold Book Corporation, New York, 1963.

Pitts, G. C., R. E. Johnson, and F. C. Consolazio: Work in the Heat as Affected by Intake of Water, Salt and Glucose, *Amer. J. Physiol.*, **142**:253, 1944.

Precht, H., J. Christophersen, H. Hensel, and W. Larcker: "Temperature and Life," Springer-Verlag, Berlin, 1973.

Robinson, S.: Circulatory Adjustments of Men in Hot Environments, in J. D. Hardy (ed.), "Temperature: Its Measurement and Control in Science and Industry," vol. 3, part 3, p. 287, Reinhold Book Corporation, New York, 1963.

Robinson, S., E. S. Turrell, H. S. Belding, and S. M. Horvath: Rapid Acclimatization to Work in Hot Climates, *Am. J. Physiol.*, **140**:168, 1943.

Rodahl, K.: Basal Metabolism of the Eskimo, *J. Nutr.*, **48**:359, 1952.

Rodahl, K.: Nutritional Requirements in the Polar Regions, U.N. Symposium on Health in the Polar Regions, Geneva, 1962, *WHO Public Health Paper*, **18**:97, 1963.

Rodahl, K., and G. Bang: "Thyroid Activity in Men Exposed to Cold," *Arctic Aeromed. Lab. Tech. Report* **57–36,** 1957.

Rodahl, K., S. M. Horvath, N. C. Birkhead, and B. Issekutz, Jr.: Effects of Dietary Protein on Physical Work Capacity during Severe Cold Stress, *J. Appl. Physiol.*, **17**:763, 1962.

Rodahl, K., F. A. Giere, P. H. Staff, and B. Wedin: A Physiological Comparison of the Protective Value of Nylon and Wool in a Cold Environment in A. Borg and J. H. Veghte (eds.), AGARD Report, no. 620, 1974.

Rowell, L. B.: Human Cardiovascular Adjustment to Exercise and Thermal Stress, *Physiol. Rev*, **54**:75, 1974.

Rowell, L. B., H. J. Marx, R. A. Bruce, R. D. Conn, and F. Kusumi: Reduction in Cardiac Output, Central Blood Volume and Stroke Volume with Thermal Skin in Normal Man during Exercise, *J. Clin. Invest.*, **45**:1801, 1965.

Saltin, B.: Aerobic Work Capacity and Circulation at Exercise in Man, *Acta Physiol. Scand.*, **62**(Suppl. 230): 1964.

Saltin, B., and L. Hermansen: Esophageal, Rectal and Muscle Temperature during Exercise, *J. Appl. Physiol.*, **21**:1757, 1966.

Saltin, B., A. P. Gagge, and J. A. J. Stolwijk: Muscle Temperature during Submaximal Exercise in Man, *J. Appl. Physiol.*, **25**:679, 1968.

Schmidt-Nielsen, K.: Heat Conservation in Counter-current Systems, in J. D. Hardy (ed.), "Temperature: Its Measurement and Control in Science and Industry," vol. 3, part 3, p. 143, Reinhold Book Corporation, New York, 1963.

Scholander, P. F., V. Walters, R. Hook, and L. Irving: Body Insulation of Some Arctic and Tropical Mammals and Birds, *Biol. Bull.*, **99**:225, 1950.

Schönbaum, E., and P. Lomax (eds.): "The Pharmacology of Thermoregulation," Karger, Basel, 1973.

Senay, L. C., Jr.: "Changes in Plasma Volume and Protein Content during Exposures of Working Men to Various Temperatures before and after Acclimatization to Heat: Separation of the Roles of Cutaneous and Skeletal Muscle Circulation, *J. Physiol.*, **224**:61, 1972.

Senay, L. C., D. Mitchell, and C. H. Wyndham: Acclimatization in a Hot, Humid Environment: Body Fluid Adjustments, *J. Appl. Physiol.*, **40**:786, 1976.

Simonson, E., N. Teslenko, and M. Gorkin: Einfluss von Vorübungen auf die Leistung beim 100 m. Lauf, *Arbeitsphysiol.*, **9**:152, 1936.

Smiles, K., R. S. Elizondo, and C. C. Barney: Sweating Responses during Changes of Hypothalamic Temperatures in the Rhesus Monkey, *J. Appl. Physiol.*, **40**:653, 1976.

Snellen, J. W.: Mean Body Temperature and the Control of Thermal Sweating, *Acta Physiol. Pharmacol. Neerl*, **14**:99, 1966.

Sobocinska, J., and J. E. Greenleaf: Cerebrospinal Fluid [Ca^{2+}] and Rectal Temperature Response during Exercise in Dogs, *Am. J. Physiol.*, **230**:1416, 1976.

Staff, P. H., and S. Nilsson: Vaeske og sukkertilförsel under langvarig intens fysisk aktivitet, *Tidsskrift for Den norske laegeforening,* **16**:1235, 1971.

Stevenson, J. A. F.: Control of Water Exchange, in W. S. Yamamoto and J. R. Brobeck (eds.), "Physiological Controls and Regulations," p. 253, W. B. Saunders Company, Philadelphia, 1965.

Stolwijk, J. A. J., B. Saltin and A. P. Gagge: Physiological Factors Associated with Sweating during Exercise, *J. Aerospace Med.*, **39**:1101, 1968.

Strömme, S., K. L. Andersen, and R. W. Elsner: Metabolic and Thermal Responses to Muscular Exertion in the Cold, *J. Appl. Physiol.*, **18**:756, 1963.

Strydom, N. B., C. H. Wyndham, C. G. Williams, J. F. Morrison, G. A. G. Bredell, A. J. S. Benade, and M. von Rahden: Acclimatization to Humid Heat and the Role of Physical Conditioning, *J. Appl. Physiol.*, **21**:636, 1966.

Stuart, D. G., E. Eldred, A. Hemingway, and Y. Kawamura: Neural Regulation of the Rhythm of Shivering, in J. D. Hardy (ed.), "Temperature: Its Measurement and

Control in Science and Industry," vol. 3, part 3, p. 545, Reinhold Book Corporation, New York, 1963.

Todt, C.: "An Atlas of Human Anatomy," The Macmillan Company, New York, 1919.

Vangaard, L.: Physiological Reactions to Wet-Cold, *Aviation Space and Environ. Med.*, **46**:33, 1975.

Verney, E. B.: The Antidiuretic Hormone and the Factors Which Determine Its Release, *Proc. Roy. Soc.*, ser. B., **135**:25, 1947.

Wallenberg, L. R.: Reduction in Cold-induced Natriuresis Following Hyperoncotic Albumin Infusion in Man Undergoing Water Diuresis, *Scand. J. Clin. Lab. Invest.*, **34**:233, 1974.

Walther, E. E., E. Simon, and C. Jessen: Thermoregulatory Adjustments of Skin Blood Flow in Chronically Spinalized Dogs. *Pflüger Arch.*, **332**:323, 1971.

Werger, B. C., M. F. Roberts, J. A. J. Stolwijk, and E. R. Nadel: Nocturnal Lowering of Thresholds for Sweating and Vasodilation, *J. Appl. Physiol.*, **41**:15, 1976.

Winslow, C.-E. A., A. P. Gagge, and L. P. Herrington: Influence of Air Movement upon Heat Losses from Clothed Human Body, *Amer. J. Physiol.*, **127**:505, 1939.

Wyndham, C. H.: Effect of Acclimatization on the Sweat Rate/Rectal Temperature Relationship, *J. Appl. Physiol.*, **22**:27, 1967.

Wyndham, C.: The Physiology of Exercise under Heat Stress, *Ann. Rev. Physiol.*, **35**:193, 1973.

Wyndham, C. H., G. G. Rogers, L. C. Senay, and D. Mitchell: Acclimatization in a Hot Humid Environment: Cardiovascular Adjustments, *J. Appl. Physiol.*, **40**:779, 1976.

Wyndham, C. H., N. B. Strydom, J. F. Morrison, C. G. Williams, G. A. G. Bredell, J. S. Maritz, and A. Munro: Criteria for Physiological Limits for Work in Heat, *J. Appl. Physiol.*, **20**:37, 1965.

Zotterman, Y.: Thermal Sensations, in J. Field (ed.), "1, Handbook of Physiology: Neurophysiology," vol. 1, p. 431, American Physiological Society, Washington, D.C., 1959.

Applied Sports Physiology

CONTENTS

Chapter 16

Applied Sports Physiology

INTRODUCTION

The development of lightweight electronic instruments and devices capable of recording and transmitting impulses by telemetry has made it possible to study a variety of physiological functions in the person exposed to different types of work stress, including athletic events. These studies in recent years have led to the accumulation of considerable data concerning physiological characteristics of the individual athlete, as well as physiological requirements of the specific athletic event. Such information may provide a foundation for the selection of athletes, for the analytical evaluation of technique and methods of training, and for the evaluation of training progress.

An attempt to present schematically the major components of physical or athletic performance in general is shown in Fig. 16-1.

Energy Yield

Work which engages large muscle groups continuously during 1 min or more may often tax the *aerobic power* to a maximal degree and thereby also impose a maximal load on the circulation (see Fig. 9-4). In many types of exercise (walking, running, swimming, rowing, cycling, cross-country skiing, skating, or calisthenics), the individual may set a pace which calls for an aerobic power which may vary from low to maximal (Fig. 9-3b). These types of activities are

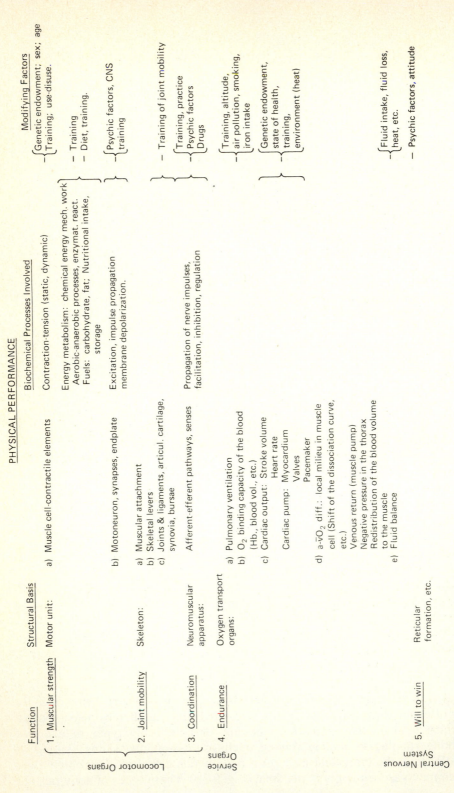

PHYSICAL PERFORMANCE

Function	Structural Basis	Biochemical Processes Involved	Modifying Factors

1. Muscular strength — Motor unit: — a) Muscle cell-contractile elements — Contraction-tension (static, dynamic) — { Genetic endowment; sex; age / Training; use-disuse.

Energy metabolism: chemical energy mech. work
Aerobic-anaerobic processes, enzymat. react.
Fuels: carbohydrate, fat; Nutritional intake, storage
— Training
— Diet, training.

b) Motoneuron, synapses, endplate — Excitation, impulse propagation membrane depolarization. — { Psychic factors, CNS / training

2. Joint mobility — Skeleton: — a) Muscular attachment
b) Skeletal levers
c) Joints & ligaments, articul. cartilage, synovia, bursae — — Training of joint mobility

3. Coordination — Neuromuscular apparatus: — Afferent-efferent pathways, senses — Propagation of nerve impulses, facilitation, inhibition, regulation — { Training, practice / Psychic factors / Drugs

4. Endurance — Oxygen transport organs: — a) Pulmonary ventilation
b) O_2 binding capacity of the blood (Hb., blood vol., etc.)
c) Cardiac output: Stroke volume / Heart rate / Cardiac pump: Myocardium / Valves / Pacemaker
d) $a-\bar{v}O_2$ diff.: local milieu in muscle cell (Shift of the dissociation curve, etc.) / Venous return (muscle pump) / Negative pressure in the thorax / Redistribution of the blood volume to the muscle
e) Fluid balance — — { Training, altitude, / air pollution, smoking, / iron intake
— { Genetic endowment, / state of health, / training, / environment (heat)

— { Fluid intake, fluid loss, / heat, etc.

5. Will to win — Reticular formation, etc. — — — Psychic factors, attitude

Locomotor Organs
Service Organs
Central Nervous System

Figure 16-1 Physical performance.

therefore excellent examples of exercises suitable for developing general physical fitness.

However, the same types of activities may also tax the *anaerobic processes* to a great extent when performed at maximal intensity. The feeling of exertion depends largely on the rate at which glycogen is broken down to lactic acid. Naturally, maximal work in which small muscle groups are engaged brings about the accumulation of lactic acid in the muscles involved and a feeling of exertion, although the total amount of energy applied may be quite small.

In work involving great intensity lasting for a few seconds, interrupted by periods of rest or light work, the peak load is moderate with respect to both aerobic processes and anaerobic energy yield, as evidenced by the formation of small amounts of lactic acid (Fig. 9-5; Tables 12-2, 12-3). Various kinds of ball games are typical examples of this type of activity.

In athletic events which require heavy work lasting more than 1 hr, the availability of muscle glycogen will eventually determine the level of achievement (Chap. 14). Special measures may therefore be necessary in order to attain improvement. Such events carried out in a hot climate place an additional demand on the heat-dissipating systems and on continuous replacement of lost water (Chap. 15).

Neuromuscular Function

Any evaluation of the engagement of individual muscle groups is impossible without the use of *electromyography*. Electromyography furnishes information pertaining to (1) which muscle or which parts of a muscle are activated; (2) the chronological order of the participation of the respective muscles in the activity; (3) the degree and duration of the contraction of the respective muscles in each movement. Such studies may also facilitate the development of an individual muscle training program. As a general rule, strength and technique are best trained through the utilization of the respective athletic event.

The *kinematic* analysis describes the geometrical form of a movement. In order to arrive at an idea of the forces which produce movement, i.e., a *kinetic* analysis, force-platforms with strain gauges as force-sensitive devices may be used. It is typical for an elite athlete, such as a champion golf player, to be able to repeat precisely, again and again, a certain motion or force, while the path of movement, the force developed, and the electromyogram are practically identical each time (Carlsöö, 1967). A discussion of the neuromuscular function as applied to sport activities falls primarily within the area of kinesiology and is therefore beyond the scope of this book.

ANALYSIS OF SPECIFIC ATHLETIC EVENTS

Walking

All active individuals have to walk, and occasionally even run, in order to move about in the course of their normal daily life and activity, whether they otherwise engage in any recreational physical activity or not. The energy cost of walking may vary within wide limits, not only among individuals but also in the same individual, depending on the circumstances. It certainly depends on total body weight, including clothing, speed of walking, type of surface, and gradient (Fig. 16-2), and on whether the person limps or not (Molbech, 1966).

There is a fair amount of data accumulated from different countries on the energy cost of walking on the level, and generally these data are in good agreement. Figure 16-3, based on data from Passmore and Durnin (1955), shows the combined effects of varying speed and varying body weight on the energy expenditure of walking.

A comprehensive treadmill–grade walking study was carried out by Margaria (1938), who found that going down a slope of 1 in 10 at varying speeds involved an energy expenditure of up to 25 percent less than walking on the level. However, on very steep declines, particularly at low speeds, energy expenditure may be considerably higher than when walking on the level.

The type of surface may affect the energy cost of walking from 23 kJ (5.5 kcal) $\cdot$ min^{-1} on an asphalt road to 31 kJ (7.5 kcal) $\cdot$ min^{-1} on a ploughed field for a 70-kg man walking at a speed of about 5.5 km $\cdot$ hr^{-1} (Granati and Busca, 1945). Walking up and down stairs may represent an energy expenditure of as much as 42 kJ (10 kcal) $\cdot$ min^{-1} for a 75-kg person (Passmore et al., 1952).

Going down stairs involves only about one-third the energy used in going up stairs. Going, and especially running, up stairs thus represents fairly heavy work, and it is therefore not surprising that so many housewives find this activity very tiring, especially when their care of sick children on an upper floor necessitates frequent ascents from the kitchen or elsewhere on the first floor. On the other hand, stair-climbing may be effective in improving physical fitness.

After the Olympic Games in London, 1948, the winner of the 10-km walking event, John Mikaelsson, was studied when walking on a treadmill.

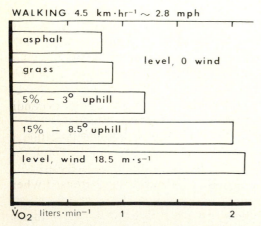

WALKING 4.5 km$\cdot$hr^{-1} ∼ 2.8 mph

asphalt

level, 0 wind

grass

5% − 3° uphill

15% − 8.5° uphill

level, wind 18.5 m$\cdot$s^{-1}

$\dot{V}_{O_2}$ liters$\cdot$min^{-1} 1   2

Figure 16-2 The energy expenditure of walking under different conditions. (Data from Pugh, 1971; 1976.)

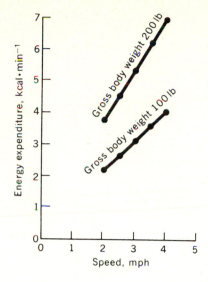

Figure 16-3 Effect of speed (mph) and gross body weight (lb) on energy expenditure (kcal · min⁻¹) of walking. *(Data from Passmore and Durnin, 1955.)*

When simulating the race by adjusting the speed of the treadmill so that the walking speed during the race was attained (13.3 km · hr⁻¹ = 10 km in 45 min 13.2s), his measured oxygen uptake was 4.0 liters · min⁻¹, or 58 ml · kg⁻¹ · min⁻¹ (unpublished results). The oxygen uptake during competitive walking seems to be approximately 75 percent of the maximum; uphill it even exceeds that level. In 1972 the maximal oxygen uptake in Swedish male competitive walkers was around 72 ml · kg⁻¹ · min⁻¹ (see Fig. 12-5).

Running

The energy expenditure of running varies tremendously, as is to be expected. In adults, individual variations are fairly small at submaximal speeds. Under these conditions, O_2 uptake per kilogram body weight is the same regardless of sex or athletic rating (P.-O. Åstrand, 1956). On the other hand, the oxygen uptake per kilogram body weight is higher for children than for adults when they both run at a certain speed (P.-O. Åstrand, 1952). It is not clear whether the reason for this inferior efficiency of running depends on inferior technique or is due to different dimensions.

There is very little difference in the measured values for energy expenditure per meter in elite runners (Margaria et al., 1975), indicating that any influence of running technique on energy expenditure must be small. Thus, in well-trained elite runners running at a speed of 20 km · hr⁻¹ on the level, the oxygen uptake varied only between 67 and 71 ml · kg⁻¹ · min⁻¹ (Karlsson et al., 1972). The energy expenditure is, on the other hand, greatly increased when running against the wind (Fig. 16-4).

As long as the speed is kept below a certain level, which varies with the individual, it is more economical to walk than to run (Fig. 16-5). In elite walkers, this level is higher than in less expert walkers. It may cost the same

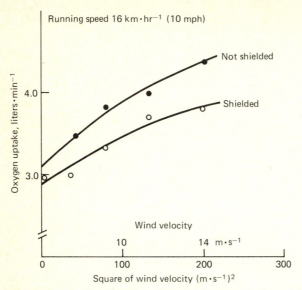

Figure 16-4 Showing the effect of increased air resistance due to wind on the energy expenditure of running. The curve "Not shielded" was obtained when the subject was running alone on the treadmill, and "Shielded" when he was about 1 m behind another runner. The wind was achieved by a big fan. An extrapolation of the data indicates that running 1500 m at about 4-min speed on a track in calm air close behind a pace-maker or a faster competitor may save up to 6 percent of the energy cost. Running behind and to one side of another runner gives a gain of about 1 s per lap. (From Pugh, 1971.)

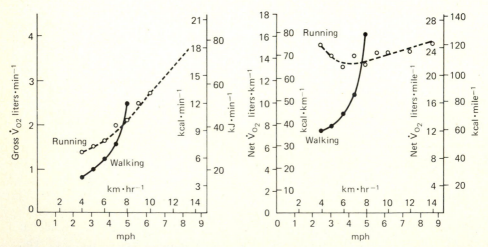

Figure 16-5 Energy cost of walking and running at different speeds. For calculation of the net energy cost (right panel) the oxygen uptake on the sitting subject was subtracted from the oxygen uptake measured during walking and running respectively. The body weight of the subject was 75 kg.

amount of energy to run at a rate of 14 km $\cdot$ hr^{-1} (8.7 mph) as it does to walk at a rate of only 10 km $\cdot$ hr^{-1} (6.2 mph).

The O$_2$ uptake depends upon the stride length, as is evident from Fig. 16-6. The subject ran at a given speed paced by a metronome, and the oxygen uptake was measured at steady state. In some experiments he was free to choose the stride frequency. In general, the stride length which is natural for the individual is also the most economical one. The energy cost of running is greatly increased with a further increase of the stride lengths. This fact has considerable practical significance. In long-distance running, the factor or economy and energy efficiency is important, so that it becomes essential to maintain the stride length most efficient for the individual. In short-distance running, such as the 100-m dash, on the other hand, where speed is more important than economy, the man who can run with rapid, long strides has an advantage, and he can afford to disregard energy cost for the limited period of time involved.

The well-known runner Nurmi had an unusually long stride length. It was thought that this was the key to his success, and efforts were made by others to copy his style. This, however, resulted in reduced efficiency and inferior results. It appears that the long stride was natural for Nurmi, and for him this represented the most economical style. This example clearly shows the danger of generalization on the basis of single cases and emphasizes the need for objective physiological studies.

Otherwise, an increase in the speed of running is brought about primarily by an increase in the stride length. An experienced 800-m runner ran on the treadmill at different speeds from 8 up to 30 km $\cdot$ hr^{-1}. The length of the stride increased more or less rectilinearly from about 80 to 220 cm, while the stride frequency increased from only about 170 to 230 steps $\cdot$ min^{-1} (Högberg, 1952). When increasing the running speed, the time spent in contact with the ground decreases markedly, whereas the time spent in the air first increases and then remains relatively constant. The increase in step frequency with the speed of

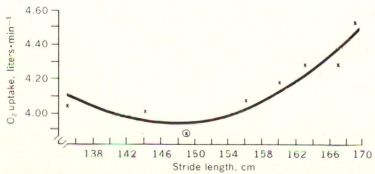

Figure 16-6 Oxygen uptake during running at a speed of 16 km $\cdot$ hr^{-1} with different lengths of stride. The encircled cross represents the freely chosen length of stride. *(From Högberg, 1952.)*

running is therefore mainly due to a decrease in the time spent in contact with the ground (Cavagna et al., 1976). During the time of contact the body "bounces" on the supporting limb.

It should be emphasized that when running within a wide range of speeds the energy demand per kilometer is practically the same (jogging at lower speeds, running at higher speeds; see Fig. 16-6). Walking at lower speeds is less costly but at higher speeds the energy cost approaches and even exceeds that of running (Zunts and Schumburg, 1901; Böje, 1944; Margaria et al., 1963). As a rule of thumb the energy cost of jogging or running is approximately $2kJ \cdot kg^{-1} \cdot km^{-1}$ (1 kcal) but for walking at 4 to 5 km $\cdot hr^{-1}$ the energy demand is only half the figure, or $1 kJ \cdot kg^{-1} \cdot min^{-1}$ (0.5 kcal). As pointed out, the surface, wind, and gradient will modify these figures. At about 4 km $\cdot hr^{-1}$ the work done at each step to lift the center of mass of the body equals the work done to increase its forward speed. The total mechanical energy involved (potential plus kinetic) is at this speed at a minimum, as is the energy cost (Cavagna et al., 1976).

In order to attain a level of achievement equivalent to the world elite in middle- and long-distance running, a maximal oxygen uptake close to, or preferably above, 80 ml $\cdot kg^{-1} \cdot min^{-1}$ is necessary in the case of men (Saltin and P.-O. Åstrand, 1967). In women runners of 400 and 800 m, a maximal aerobic power of 65 ml $\cdot kg^{-1} \cdot min^{-1}$ or higher is necessary in order to attain results qualifying for the elite class. An athlete with a high maximal oxygen uptake has the advantage of being better able to tolerate bouts of forced tempo than his competitor with a lower aerobic power in that he does not have to utilize the anaerobic energy yield to the same extent during the bouts of increased tempo. Ideally, the pace should be selected corresponding to the intensity when the oxygen uptake, following a linear increase with the increasing intensity, begins to level off (Fig. 9-3b).

On the basis of our present state of knowledge, it appears that a top runner who has had a proper warm-up attains his or her maximal oxygen uptake after about 45 s. Accordingly, it may be assumed that in a 400-m dash lasting 40 to 50 s, the runner taxes his or her anaerobic metabolic capacity maximally. In these cases, blood lactate concentrations of 20 to 25 mM have been measured. At shorter distances, i.e., in 100- to 200-m races, it is estimated that some 90 percent of the energy is derived from anaerobic metabolic processes. At distances exceeding 400 m, aerobic metabolic processes assume an increasingly important role. In a 800-m race, aerobic metabolic processes account for about 40 percent of the total energy utilized, and in a 1,500-m race, for 65 percent.

Swimming

Swimming engages practically all muscle groups of the body. It is therefore not surprising that very high oxygen uptakes have been obtained on swimmers (P.-O. Åstrand et al., 1963; Holmér, 1974). A maximal oxygen uptake of 3.75 liters $\cdot min^{-1}$ was attained by the female silver-medal winner in the 400-m freestyle in the Olympic Games in Rome, 1960. In male swimmers of good world standard, a maximum of around 6 liters $\cdot$ min^{-1} has been measured.

Since the specific gravity of the body is not much different from that of water, the weight of the body submerged in water is reduced to a few kilograms. The obese individual especially may keep afloat with very little energy expenditure. Swimming may therefore be an easy task when performed at a low level of intensity. For this reason, swimming and various exercises performed in water are a very common form of training for physically handicapped individuals.

The functional demands of competitive swimming were evaluated in 22 girl swimmers on the basis of the relationship between oxygen uptake during swimming at competitive speed and maximal oxygen uptake during work on a bicycle ergometer (P.-O. Åstrand et al., 1963). A very high correlation was observed, but the oxygen uptake during swimming averaged only 92.5 percent of the maximal oxygen uptake reached during cycling. However, five girls reached a higher value in the former case. The good correlation is also evident from the fact that the blood lactate concentration after swimming was of the same order of magnitude as after maximal cycling (10.3 and 10.5 mM respectively). Similar results were reported by P.-O. Åstrand and Saltin (1961) and Holmér (1974a); 12.8 mM after maximal swimming, and 12.9 mM after running for male swimmers. The quotient of pulmonary ventilation to oxygen uptake ($\dot{V}_E/\dot{V}_{O_2}$) was significantly lower during maximal swimming than during cycling (27.7 and 35.5 respectively). The reason for this relative hypoventilation may be the different mechanical conditions of breathing. The water pressure on the thorax makes the respiration more difficult. Furthermore, the breathing is not as free during swimming as in most other types of work, in that the respiration during competitive swimming is synchronized with the swimming strokes. Holmér et al. (1974a) also noted a similar difference in the ratio between maximal pulmonary ventilation and maximal oxygen uptake when comparing the data obtained during swimming and running (29.8 and 37.4 respectively for 5 subjects). However, despite the relative hypoventilation in the swimming subjects compared with running, the arterial oxygen pressure and content were the same in the two types of exercise.

In recent years, world swimming records have been attained by girls at an increasingly younger age. In Chap. 9 an analysis of their physiological possibilities was presented. It was concluded that the girls, by the age of thirteen to fourteen years, had almost reached the maximal power of their aerobic processes. During puberty, the organism may respond more strongly to the training. In the mentioned study of girl swimmers (P.-O. Åstrand et al., 1963), it was found that they had significantly greater functional dimensions than girls who had not taken part in competitive sport or undergone any special physical training. Vital capacity and heart volume (Fig. 6-24) were highly correlated to the maximal oxygen uptake.

Thus young girls may exhibit a very high motor power. It has been shown that women during breaststroke swimming at a certain speed have a lower oxygen uptake (greater mechanical efficiency) than men. This may be explained by the fact that the lower specific gravity in women, due to their greater fat content, reduces the effort required to keep the body floating. However, considerable individual variations in technique are typical for swimming (see Holmér, 1974b).

The energy requirement of different types of swimming is best mirrored in the record tables, which show that crawling is the most economical type of swimming as long as the swimmer masters the proper technique. It is possible to study the different swimming techniques in detail in a specially constructed swimming flume, such as that installed at the College of Physical Education in Stockholm. In this swimming flume, the water can be made to flow at different speeds through a channel, thus providing opportunity for studies similar to those on walking and running on a treadmill.

Holmér (1974a) studied the physiological responses to swimming in 87 subjects of varying states of skill and training in the Swedish swimming flume, compared with running and cycling. He found that oxygen uptake during swimming at a given submaximal speed depended on the degree of swimming training, body dimensions, swimming technique, and swimming style. Thus, oxygen uptake at a given submaximal speed was higher for untrained than for trained swimmers, and for tall subjects than for short subjects. It is the arm stroke that has the highest efficiency, not the leg kick. In fact, the maximal speed with arm strokes was almost the same in freestyle (1.31 m · s^{-1}) as for the whole stroke (1.34 m · s^{-1}) but at a significantly lower oxygen uptake. One might speculate that over longer distances the leg kicks should be deemphasized for it may waste oxygen and blood flow. In breaststroke the leg kicks are probably as important or more important than the arm strokes. The mechanical efficiency was 6 to 7 percent in freestyle and 4 to 6 percent in breaststroke in elite swimmers. Most costly from the standpoint of energy expenditure is the butterfly stroke. (For details, see Holmér, 1974b; Di Pramperio et al., 1974). The increase in oxygen uptake with increasing swimming speed was linear or slightly exponential. Maximal oxygen uptake during swimming was, for elite swimmers, 6 to 7 percent lower than during running and approximately the same as during cycling. For subjects untrained in swimming their maximal oxygen uptake during swimming was on the average 80 percent of the running maximum. When the body was submerged, vital capacity was reduced by 10 percent and the expiratory reserve volume was less than 1 liter as compared with 2.5 liters in air. The increase in tidal volume in water was exclusively achieved by the use of the inspiratory reserve volume. Heart rate, cardiac output, and stroke volume during submaximal swimming were of the same magnitude and increased with increasing speed in approximately the same way as during running. Heart rate was significantly lower in maximal swimming than in maximal running. The mean intra-arterial blood pressure at submaximal as well as maximal rates of work was higher in swimming than in running. (The circulatory data were observed on five subjects who were studied when swimming and running respectively, see Holmér et al., 1974). Breaststroke and butterfly swimming required 1 to 2 liters $\cdot$ min^{-1} higher oxygen uptake at a given submaximal speed than did freestyle and backstroke. The mechanical efficiency in swimming was 4 to 6 percent in breaststroke and 6 to 7 percent in freestyle in elite swimmers. Maximal oxygen uptake measured during swimming varied as a consequence of swimming training, whereas the maximal oxygen uptake measured during treadmill running remained relatively unchanged. Therefore, maximal oxygen uptake measured during running is not representative for performance in swimming, and training for the purpose of increasing a swimmer's maximal oxygen uptake should, to the greatest possible be done by swimming. (See Fig. 12-13.)

In view of the fact that A. Rodahl et al. (1976) have shown that the performance of competitive swimmers is significantly better in the evening than in the morning, presumably owing to the effects of circadian rhythms on body temperature, attempts to achieve new records should preferably take place in the evening.

When analyzing the world records, one finds that the average speed in swimming 1500 m is 83 percent of the best performance at 100 m (time 15 min and 50 s respectively). In running 5000 m the average speed is only 69 percent of the world record in 400 m (time 13 min 13 s and about 44 s respectively; for data on male athletes, see Jokl et al., 1976). The reason for the greater decline in speed with distance running compared to swimming is not known.

Speed Skating

Ekblom et al. (1967) have made a study of Swedish speed-skating athletes including the 1968 Olympic champion in 10,000 m, J. Höglin, the 1968 bronze medal winner, Ö. Sandler, and the 1964 World champion and Olympic champion in 10,000 m, J. Nilsson. They were studied when running on a treadmill as well as when skating at submaximal and maximal speeds.

Table 16-1 summarizes some of the maximal data attained in the two types

Table 16-1

Subject	Best time 10,000 m	Maximal values									
		Oxygen uptake				Pulmonary ventilation, liters $\cdot$ min^{-1}		Blood lactates, mM		Heart rate, beats $\cdot$ min^{-1}	
		liters $\cdot$ min^{-1}		ml $\cdot$ kg^{-1} $\cdot$ min^{-1}							
		R*	S*	R	S	R	S	R	S	R	S
Sandler, Ö.	15,20.6	5.77	5.48	79.0	75.1	184	183	20.3	19.9	186	186
Nilsson, J.	15,47.0	5.70	4.80	79.2	66.7	172	139	17.0	18.0	188	186
Höglin, J.	15,23.6	5.39	4.69	71.9	62.5	137	159	15.4	15.4	185	
Claesson	17,08.0	5.39	4.89	64.9	58.9	141	141	17.4	17.4		
Nilsson, I.	16,19.4	5.20	4.38	76.5	64.4	138	121	16.9			
Mean value		5.49	4.85	74.3	65.5	154	149	17			

*R = running; S = speed skating.

of activities. The average value for maximal oxygen uptake was 5.49 liters · min^{-1} during treadmill running, as against 4.85 liters · min^{-1} when skating, showing a difference of about 12 percent. Otherwise, the achieved maximal values were rather uniform. It should be pointed out that the extra burden imposed by the equipment which the skaters had to carry for the collection of the expired air might have caused the oxygen uptake values to be higher than they would have been when skating without equipment. Most of the determinations were made on Sandler who, relatively speaking, had the highest oxygen uptake when skating (95 percent of that attained during treadmill running). The values for blood lactate and heart rate which were attained in connection with skating competitions were similar to those obtained during the determination of maximal oxygen uptake.

When the ice condition is excellent, the maximal oxygen uptake expressed in liters per min is probably of greater practical importance than the maximal oxygen uptake expressed as ml O_2 per kg body weight per min. During speed skating the center of gravity of the body is moving relatively parallel to the ice surface without the marked vertical movements made during running. The potential to perfom work is from a dimensional point of view related to the body mass (L^3) but the air resistance is proportional to the body surface area (L^2) (Chap. 11). Therefore, a better performance can be expected from the larger subject, and more so at higher speeds (see Di Prampero et al., 1976). When the ice surface is soft, it is a drawback for the skater to be heavy, since weight increases the friction. The results from the mentioned study show that the elite long-distance skaters have a maximal aerobic power of about 5.5 liters · min^{-1}. The 500-m race and especially the 1,500-m race (skating time, a little over 2.0 min) impose particularly high demands on the anaerobic power (the peak blood lactate concentration for three of the skaters at the end of the race at the Swedish championship competition in 1965 averaged as follows: 500 m, 13.6 mM; 1,500 m, 17.3 mM; 5,000 m, 15.1 mM; 10,000 m, 13.3 mM). The technique is also different in the shorter distances compared with the longer ones. The skaters who come first in the 500-m race, and to some extent also in the 1,500-m race, infrequently win the 10,000-m race, thus illustrating that the physiological requirements are different in the two types of races. The skaters referred to in Table 16-1 are all typical long-distance skaters.

Figure 16-7 presents oxygen uptakes (Douglas bag method) during skating at different speeds at two different skating rinks. As is evident, the curve is not linear. An increase in the speed from 4 to 6 m · s^{-1} requires an additional 0.7 liter · min^{-1}, while the increase from 8 to 10 m · s^{-1} necessitates an increase in the aerobic power of 2.0 liters · min^{-1}. (A possible contribution from anaerobic processes was not taken into consideration in this study.) The explanation must be sought primarily in the increased air resistance at the higher speed. It increases with the speed raised to the second power, and accounts for a large part of the energy expenditure at high speeds. Thoman (personal communication) has calculated, from experiments with Sandler in a wind tunnel, that at a speed of 10 m · s^{-1}, 70 percent of the external work is devoted to overcoming

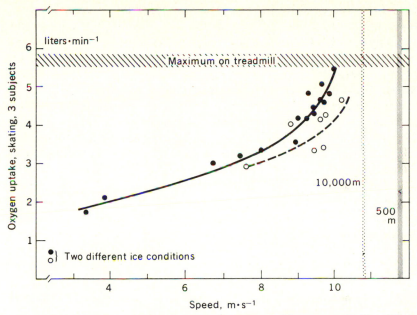

Figure 16-7 Oxygen uptake during speed skating at different speeds. The different symbols represent experiments performed at different speed-skating rinks. In excellent ice conditions as well as at high altitude (reduced atmospheric pressure = reduced air resistance), the curve is shifted to the right; i.e., the same expenditure of energy gives greater speed. Vertical pillars indicate speed during the skaters' best performances on 500 and 10,000 m respectively. Subjects J. Höglin, J. Nilsson, and Ö. Sandler. *(From Ekblom et al., 1967.)*

the air resistance, while the remaining 30 percent is required to overcome the friction of the ice. Thus, a more ideal aerodynamic profile would to a large extent reduce the air resistance. The speed-skating style, with the arms held on the back, is undoubtedly the result of experience. It is also known that the type of clothing is very important in regard to the air resistance. If the conventional speed-skating suit of a competitive speed skater were replaced with a suit made of a material similar to that used by many skin divers, the time on a 5,000-m speed-skating race might, theoretically speaking, be improved by 10 to 20 s.

The increase in speed of the skater is perhaps best illustrated by comparing the world records for the 10,000-m speed skating events from 1913 through 1976:

		Time Min: S	Average Speed
1913	Oscar Mathisen	17:22,6	34.53 km • hr⁻¹
1952	Hjalmar Andersen	16:32,6	36.27 km • hr⁻¹
1960	Knut Johannessen	15:46,6	38.03 km • hr⁻¹
1971	Ard Schenk	14:55,9	40.18 km • hr⁻¹
1976	Sten Stensen	14:38,8	41.00 km • hr⁻¹

If the 1913 record holder were to compete with the 1976 record holder, the former would still have four rounds to go (1,600 m) when the latest record holder had completed his final round.

The style of speed skating appears rather rigorous, with many muscle groups involved in static contraction. This fact may explain why the maximal oxygen uptake is lower during skating than during running. Figure 16-8 also shows that the blood lactate concentration at a given oxygen uptake is considerably higher during skating than when cycling. From a training point of view, it appears essential that the speed skater allows the muscle groups engaged in skating to become accustomed to tolerating a high lactic acid concentration.

The maximal isometric and dynamic muscle strength of skaters is not particularly great. It has been observed, however, that the leg muscles of the skaters who are the best sprinters and specialize in 500-m skating are much stronger than those of long-distance skaters.

Cross-country Skiing

The use of skis to transport the body across the snow has several advantages: It enables a person to move comparatively easily across loose, deep snow where walking on foot may be extremely difficult or almost impossible (Ramaswamy

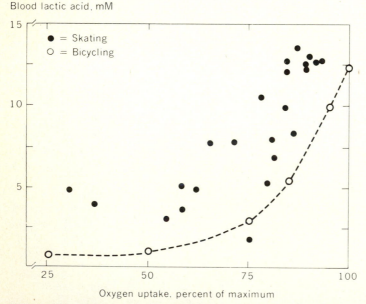

Figure 16-8 Blood lactate concentration at different work loads expressed in percentage of maximal oxygen uptake during speed skating and bicycling. The bicycling data are from well-trained noncyclists. The speed-skating data are obtained from well-trained speed skaters. At corresponding oxygen uptakes, more lactic acid is produced during speed skating than during bicycling. *(From Ekblom et al., 1967.)*

et al., 1966); it is possible to move much faster downhill on skis than on foot; finally, it allows the skier to move into the wilderness independently of roads.

Since more muscle groups generally are engaged in skiing than in walking (the use of the arms to pull and push on the ski-poles), the overall energy expenditure involved in transporting the body on skis from one place to another may be as high as, or higher than, the energy expenditure when moving the body the same distance on foot. In fact, the energy expenditure of skiing, especially uphill, may occasionally be surprisingly high. Meen et al. (1972) measured the oxygen uptake in the same subjects when skiing uphill at maximal speed and when running at maximal effort on the treadmill. They found slightly higher oxygen uptakes during skiing than during treadmill running. The mean difference was about 2.5 to 3.0 percent. However, Åstrand and Saltin (1961) did not find any significant difference in maximal oxygen uptake in the two activities.

Because skiing engages about all major muscle groups of the body, cross-country skiing is an excellent method of training for physical fitness and dynamic muscular endurance. For the same reason, top cross-country skiers generally have exceedingly high maximal oxygen uptakes. A maximal oxygen uptake of 7.4 liters $\cdot$ min^{-1} has been reported in the Finnish cross-country skier Mieto (Bergh, 1974). An additional advantage of cross-country skiing is that the work load varies greatly with the changing features of the terrain, with very high loads when climbing uphill and lighter loads when sliding downhill (see Fig. 16-9).

Rönningen (1976) compared the energy cost of walking on foot on a hard, snow-covered level road and of skiing on a level trail alongside the road. He used six young war college cadets, dressed in military arctic uniforms and carrying 22.5 kg on their backs. In both cases, the subjects moved at a speed of about 5 km $\cdot$ hr^{-1}. He obtained identical values for oxygen uptake (about 1.5 liters O_2 $\cdot$ min^{-1}) although the subjects stated that skiing felt easier than walking, probably because, in skiing, the work is accomplished by a larger muscle mass, since the arm muscles are being used to aid locomotion through the ski-poles. At higher speeds on hard snow the skiing is more economical. On loose snow the energy cost when skiing at 3.5 to 4.0 km $\cdot$ hr^{-1} was about 70

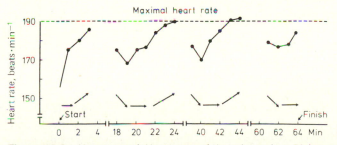

Figure 16-9 *Upper panel:* Heart rate of the winner in a 21-km cross-country ski race. *Lower panel:* Arrows indicate terrain profile: → = level ground, ↗ = uphill, ↓ = downhill. *(From Bergh, 1974.)*

percent of the demand during walking. For walking with snow shoes, the energy cost was somewhere between these extremes (Christensen and Högberg, 1950). Skiing uphill at an incline of 7 to 8 percent on a cross-country trail at a speed of about 4 km $\cdot$ hr^{-1} required an oxygen uptake of about 2.3 liters $\cdot$ min^{-1} on the average. One of the cadets was studied during a winter maneuver covering 120 km across the mountains in Norway in 5 days. Moving on skis with a load of 22.5 kg on the back, at a speed slightly less than 4 km $\cdot$ hr^{-1}, required an average oxygen uptake of 1.8 liters $\cdot$ min^{-1}, corresponding to roughly 40 percent of the individual's maximal oxygen uptake. When two men pulled a 72-kg sled while on skis, the oxygen uptake increased to 2.3 liters $\cdot$ min^{-1} on the average, corresponding to more than 50 percent of the individual's maximal oxygen uptake. Christensen and Högberg (1950) compared the energy cost of carrying an extra load of 30 kg in a rucksack and the transportation of the same load on a sled. The oxygen uptake in the rucksack experiment was 70 to 80 percent of the figures when the sled was pulled (at 4.0 km^{-1} $\cdot$ hr^{-1}).

The efficiency of skiing is illustrated by the following examples: A good skier covered 30 km on a snow-covered lake in 1 hr 20 min (speed 6.25 m $\cdot$ s^{-1}). The best time in the Swedish Vasa race, 86 km (with some 10,000 participants), is 4 hrs 10 min (average speed 5.75 m $\cdot$ s^{-1}). The terrain is relatively flat. In the world championship in 1974, on a hilly track, the 50 km winner finished after 2 hrs 22 min (average speed 5.89 m $\cdot$ s^{-1}). For comparison, the best time in track running 10,000 m gives a speed of 6.13 m $\cdot$ s^{-1}, and for marathon about 42.2 km, 5.47 m $\cdot$ s^{-1}.

It is thus clear that skiing, even at submaximal speeds, requires a high aerobic work capacity. Accordingly, most elite cross-country skiers have maximal oxygen uptakes of 5.5 liters $\cdot$ min^{-1} or more (in excess of 80 ml $\cdot$ kg^{-1} $\cdot$ min^{-1}), with a maximum of 94 ml obtained in an Olympic Champion (15 km race), see Fig. 12-15, an extremely high aerobic work capacity. The corresponding figures for the top Swedish female cross-country skiers are 3.5 to 4.4 liters $\cdot$ min^{-1} or 70 to 75 ml $\cdot$ kg^{-1} $\cdot$ min^{-1}. Recording of the heart rate during competitive skiing (Fig. 16-9) shows that it reaches maximal levels during uphill skiing, and drops, at the most, to only some 20 beats $\cdot$ min^{-1} below maximal values during downhill skiing and only slightly below maximal values during skiing on fairly long stretches of level ground. It appears that top competitive cross-country skiers may need to tax some 85 percent or more of their maximal oxygen uptake during a race.

From Fig. 16-10 it is evident that the longer the duration of the race, the lower the blood lactate concentration measured at the end of the race (Åstrand et al., 1963). It is thus evident that a high anaerobic capacity is important in the shorter-distance, 5- to 10-km races, and particularly in the relay races on skis where the tempo may be very uneven. This reduced ability to liberate energy by glycogenolysis with a lactate production after prolonged exercise may be related to reduced glycogen stores (Asmussen et al., 1974), inhibitory effects on key enzymes of other factors are unknown (Costill et al., 1971).

Measurements of the maximal oxygen uptake in Swedish elite skiers

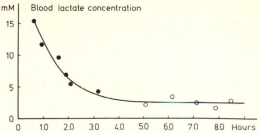

Figure 16-10 Blood lactate concentration at the end of races of distances from 10 to 85 km. Filled circles denote mean values, open circles represent individual values (1 mM is the equivalent of about 9 mg percent). *(From P.-O. Astrand et al., 1963.)*

throughout the year consistently show the highest values at the end of January, and the lowest values in May and June. The difference may be in the order of 5 to 25 percent (see Fig. 12-6) (Bergh, 1974).

In recent years, roller-skis have been widely used by competitive skiers as a training device during the seasons of the year when there is no snow on the ground. It appears, on the basis of a comparison between the oxygen uptake during maximal roller-skiing and maximal treadmill running, that roller-skiing may tax the oxygen-transporting capacity during prolonged roller-skiing as much as, or more than, during running. Thus roller-skiing and running may conceivably be of equal value as a method of endurance training (Fig. 16-11).

Alpine Skiing

Agnevik et al. (1969) have studied the Swedish elite athletes in this event, both in the laboratory and in connection with regular international competitions. One of the foremost skiers in the world in special slalom, Bengt-Erik Grahn, was included in this study. His maximal oxygen uptake measured on the bicycle ergometer was 3.9 liters $\cdot$ min^{-1} or 66 ml $\cdot$ kg^{-1} $\cdot$ min^{-1}; his maximal heart rate was 207 beats/min (see Fig. 16-12). With the aid of telemetry his heart

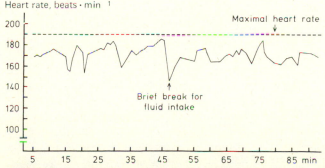

Figure 16-11 Heart rate during 90 min roller-skiing on a hilly road. In this subject a heart rate of 180 beats $\cdot$ min^{-1} corresponds to 90 to 93 percent of his maximal oxygen uptake. *(From Bergh, 1974.)*

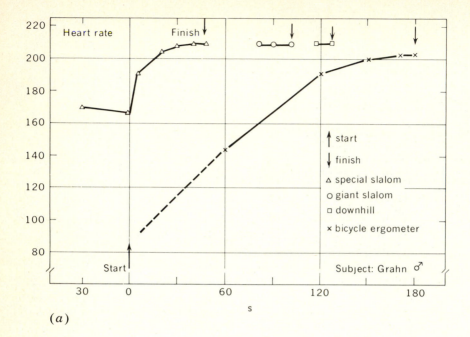

(a)

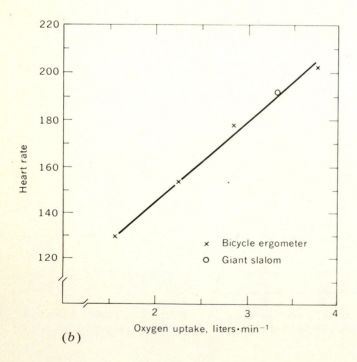

(b)

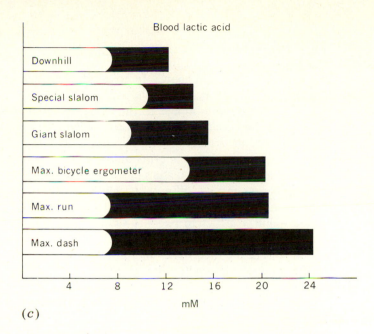

(c)

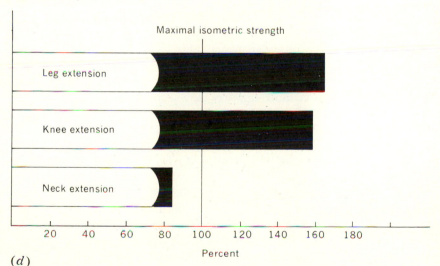

(d)

Figure 16-12 Data on male top athletes in Alpine skiing.
(a) Heart rate before and during competitions in Alpine skiing and during maximal work on a bicycle ergometer. Note the high heart rate before start and the rapid increase in heart rate after start.
(c) Peak blood lactate concentration after various activities; mean values of three to five determinations. "Max. run" = 3 × 1,000 m at maximum speed with a few minutes' rest in between; "Max. dash" = 5 × 50 s in a similar manner.
(b) Heart rate in relation to oxygen uptake during cycling and giant slalom the day after actual race on the same course.
(d) Maximal isometric muscle strength; 100 percent = data obtained on a group of recruits. *(From Agnevik et al., 1969.)*

rate was recorded before and during a competitive slalom race. Owing to technical difficulties, his heart rate could be followed only during the end of the race in both giant slalom and downhill skiing. Figure 16-12 shows a heart rate of more than 160 beats · min^{-1} at the start. This high heart rate no doubt was due to emotional factors and some degree of nervousness at the start of the race. The heart rate quickly rose to over 200 beats · min^{-1}, to the same maximal heart rate which was obtained at a maximal load on the bicycle ergometer. The same heart rate was recorded at the end of the other competitive events. The figure also shows how the heart rate increases at "supermaximal" work loads on the bicycle ergometer. In this case, the increase occurs more slowly than it does during competitions in Alpine skiing, a difference which must be attributed to psychic factors. (It should be pointed out that prior to the start of the race, the skier skis the track uphill in order to become familiar with the track. This physical effort does not explain the high heart rate, since an ample period of rest precedes the actual start of the race.) The day following the giant slalom race, some of the participants covered exactly the same track, but then the oxygen uptake was determined by the Douglas bag method as well. The subjects completed the run in almost exactly the same time as they had during the actual race. Figure 16-12 presents an example of the data collected. The oxygen uptake in this particular subject was 3.3 liters · min^{-1} or 87 percent of his maximal O_2 uptake measured in laboratory experiments. In another subject the oxygen uptake was 3.9 liters · min^{-1} or 78 percent of his maximal aerobic power. On this occasion the recorded heart rates were submaximal. Figure 16-12 shows that the heart rate measured lies on the regression line which was obtained for the heart rate–oxygen uptake relationship during tests on the bicycle ergometer. This observation supports the assumption that the competition itself represents an extra psychic stress leading to an elevated heart rate. Figure 16-12 also presents mean values for peak blood lactic acid in connection with competitions, maximal work on the bicycle ergometer, and running. The high lactic acid levels during skiing, in spite of a relatively short work time, may be explained by the assumption that certain muscle groups are engaged in intense static work. The best skier, Grahn, attained in his special event the highest lactic acid levels of the entire group. Finally, it is evident from Fig. 16-12 that the Alpine skiers are characterized by a great isometric strength in the stretch muscles of the legs. A group of military recruits is shown for comparison (values taken as 100 percent). The skiers had even greater muscle strength than a group of weight lifters. Recent studies (Eriksson et al., 1977) have confirmed these results. The winner of the World Cup in alpine skiing in 1976 and 1977 (Ingemar Stenmark) had a maximal oxygen uptake of 5.2 liters · min^{-1}, or 70 ml · kg^{-1} · min^{-1} when running on the treadmill. The static muscular strength of the legs' extensor muscles was very high in the skiers, on an average 2900 N (Stenmark leading with 3430 N). Athletes from other sports events studied had mean values at or below 2500 N. It can be calculated that the force the alpine skier must develop during actual skiing can reach several thousand Newtons. Muscle lactate concentrations up to 24 mmole · kg^{-1} wet muscle have been measured, and concentrations around 15 mM in the blood are common findings after competitions. A relatively wide scatter in muscle fiber composition of the quadriceps muscle has been noted, with a mean value of 43 percent slow twitch fibers (type I) (Eriksson et al., 1977).

It is thus clear that competitive Alpine skiing places heavy demands on both aerobic and anaerobic motor power. Great strength in the muscle groups involved is required. In addition to these basic requirements, the technique will obviously determine the level of achievement. From the training standpoint it is important to learn to master a good technique in spite of a high lactic acid concentration in the muscles.

Canoeing

Canoeing is rather unique in that it is about the only competitive sport in which mainly the arm and trunk muscles are engaged in endurance efforts. In addition, the same individual competes in 500-, 1,000-, and 10,000-m races, lasting from about 1.45 to 45 minutes.

For the success of the canoeist, maximal aerobic power is more critical than body weight, for the slightly greater resistance caused by the friction of the canoe in the water by a few kilos in extra body weight is insignificant. It is therefore more meaningful to express the canoeist's aerobic power in terms of liters $O_2 \cdot min^{-1}$ than in ml $O_2 \cdot kg^{-1} \cdot min.^{-1}$ According to Tesch et al. (1974, 1976) the mean maximal oxygen uptake of the present Swedish elite group of canoeists (including two world champions) is 5.40 liters $\cdot min^{-1}$ (range 4.7 to 6.1) or 68 ml $\cdot kg^{-1} \cdot min^{-1}$ (64 to 75); see Fig. 12-5. From Fig. 16-13 it appears that some of the canoeists utilize all their aerobic work capacity when paddling the canoe.

Although back, abdominal, chest, and shoulder muscles are also engaged in canoeing, the major part of the work is performed by the arms. A high maximal oxygen uptake in arm work must therefore be a major requirement for the canoeist. Tesch et al. (1974) compared the maximal oxygen uptake during

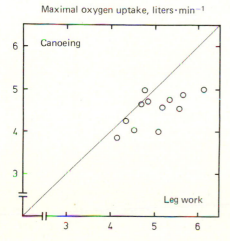

Figure 16-13 Relationship between maximal oxygen uptake during canoeing and during work with the legs in elite canoeists *(From Tesch et al., 1974.)*

leg work (bicycle ergometer work) with the maximal oxygen uptake during arm cranking in Swedish elite canoeists and found very high values for arm work, both in absolute terms (liters · min⁻¹) and in percentage of the maximal oxygen uptake measured during leg work. Swedish male junior and senior canoeists combined had a mean maximal oxygen uptake in work with the legs of 5.13 liters · min⁻¹ as against 4.45 liters · min⁻¹ in arm work (87 percent of that in leg work), and 4.48 liters · min⁻¹ during canoeing. In elite women canoeists, the maximal oxygen uptake during arm work was 90 percent of that obtained during work with the legs. The corresponding figures for a group of Swedish weight lifters was 78 percent, and 71 percent for physical education students. A comparison between maximal oxygen uptake in elite Swedish canoeists during leg exercise (bicycle ergometer) and arm cranking, and during 500-m, 1,000-m, and 10,000-m canoe racing, is presented in Fig. 16-14. One of the canoeists who had started systematic canoe training when very young showed, during repeated tests, consistently maximal oxygen uptakes in arm work that were as high as, or higher than, in leg work. The fact that he started canoe training at an early age might explain, at least in part, his exceptionally high aerobic work capacity with the arms.

For comparison, it may be pointed out that in a group of Greenland Eskimo kajak-hunters studied by Vokac and Rodahl (1977), the maximal oxygen uptake in arm cranking averaged 89 percent (81 to 104 percent) of that of bicycling on a bicycle ergometer. This is approximately the same rate between maximal oxygen uptakes in arm and leg exercise as found by Vrijens et al. (1975) in various Caucasian subjects, including highly trained athletes specialized in kajak paddling.

Studies of Swedish elite canoeists (Tesch et al., 1976) indicate that canoeists generally are unable to tax their maximal oxygen uptake fully during the 500-m race. In 1,000-m canoe races, almost maximal heart rates have been recorded (189 beats per min compared to a maximum of 195) during the last half of the race. In two Swedish canoeists studied during a 10,000-m race lasting 45 min, the mean heart rate during the entire race was kept at a level correspond-

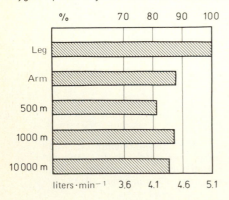

Oxygen uptake in junior and senior canoeists.

Figure 16-14 Oxygen uptake (liters · min⁻¹ and in percentage of the maximal oxygen uptake measured during leg work) in Swedish elite canoeists (junior and senior) during the final minutes of 500-m, 1,000-m and 10,000-m races. The maximal oxygen uptake in arm work is included for comparison. *(From Tesch et al., 1974.)*

ing to 96 and 98 percent of the maximal heart rate. In one of them, the heart rate was maximal during the last 2,000 m of the race.

According to Tesch et al. (1976), the higher the maintained speed during the race, the greater the lactate levels in the blood, the highest values being obtained at the end of the 500-m race (Fig. 16-15). Under comparable circumstances, blood lactate values are higher in senior than in junior elite canoeists, and higher in the final heat than in the same canoeist's semifinal or trial heat, perhaps indicating the positive relationship between degree of exertion and blood lactate levels.

While the maximal oxygen uptake measured during maximal leg work tends to remain more or less unchanged throughout the year in elite canoeists, the maximal oxygen uptake measured during arm cranking was, on the average, 8 percent higher during the active canoeing season than during the winter (Tesch et al., 1976).

Rowing

Secher (1973) has compiled the winning times in international rowing championships during the period from 1893 to 1971. The results for the eight are presented in Fig. 16-16. As is evident from the figure, there is a considerable scatter of the data owing to varying wind and water conditions. Because these external conditions may change very quickly (in a matter of minutes), it is quite meaningless to compare individual results from one race to another even when the time interval is as short as 1 hr. Mean results of races from several regattas, on the other hand, may be an indication of the general level of performance for any one year. It is observed that, on this basis, the mean improvement in rowing performance is 0.66 s per year, corresponding to a 1.6 percent increase in speed. Secher (1973) explains this improvement as the result of a combination of factors: the increasing body size and consequently an increase in the

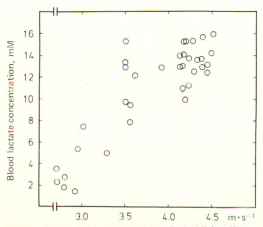

Figure 16-15 Blood lactate levels (mM) in elite canoeists, related to mean speed (m · sec⁻¹). *(From Tesch et al., 1974.)*

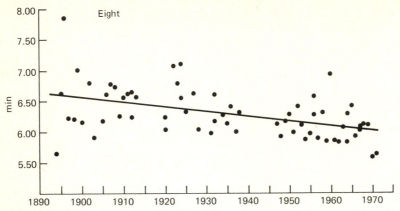

Figure 16-16 Improvements of results in international rowing championships. *(From Secher, 1973.)*

maximal aerobic power of the general population; better selection and training of rowers; better rigging of the boats, allowing better technique and greater mechanical efficiency; and better boats, causing less water resistance.

The water resistance against the boat is equal to the force (F) that moves the boat, and it increases proportionally to the velocity raised to the second power (F $\propto v^2$). Since work (W) is force times distance ((L), i.e. W = F $\cdot$ L) and power is force times distance divided by time (power = F $\cdot$ L $\cdot$ t^{-1} = F $\cdot$ v) that is force times velocity, the energy expenditure (or power) of the rower increases proportionally to velocity raised to the third power (power = F $\cdot$ v $\propto$ v^2 $\cdot$ v $\propto$ v^3).

By actually measuring the oxygen uptake of the rower while rowing at submaximal work intensity during near steady-state conditions, Secher (in unpublished data) observed that the mean results for single, double, and pair show an increase in energy expenditure proportional to velocity raised to the third power (Fig. 16-17), or to the 3.1 power, to be exact. Di Prampero et al. (1971) concluded from their data that "the mechanical power output necessary to maintain the boat progression as well as the energy expenditure appears to increase as the 3.2 power function of the average speed," which is very similar to Secher's figure.

By extrapolating the lines of best fit in Fig. 16-16 to 1971, we arrive at a mean value for all boats of 4.9 m $\cdot$ s^{-1}. At this speed the mean energy demand in the case of single, double, and pair amounts to 6.25 liters O$_2$ per min. One may therefore expect that the maximal aerobic power of the rower may be a limiting factor in rowing performance. This supposition is strengthened by the finding of a positive correlation between the average oxygen uptake of the crew and their placing in international championships (Fig. 16-18). The mean maximal oxygen uptake was 6.1 liters $\cdot$ min^{-1} for the crew taking first place (which is close to the figure 6.25 liters $\cdot$ min^{-1} referred to above), 5.7 liters $\cdot$ min^{-1} for the crew taking sixth place, and 5.1 liters $\cdot$ min^{-1} for the crew

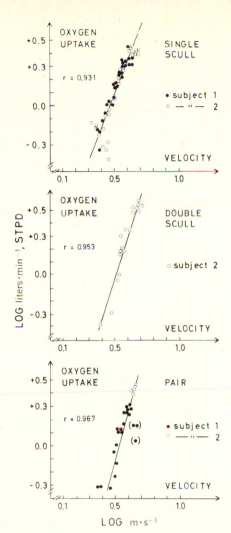

Figure 16-17 Relationship between energy expenditure and velocity, expressed logarithmically. *(By courtesy of N. H. Secher.)*

attaining the twelfth place. Furthermore, a comparison of the mean maximal oxygen uptakes of individual international champion oarsmen shows the following figures for the maximal oxygen uptakes in liter $\cdot$ min^{-1}: German: 5.9 ($n = 5$); Norwegian: 5.8 ($n = 6$); Danish: 5.7 ($n = 4$) (Vaage and Secher, personal communication). In a study by Secher et al. (1976a), the mean maximal oxygen uptake of a group of 20 top rowers (members of national teams) was 5.0 liters $\cdot$ min^{-1} as against 4.3 liters $\cdot$ min^{-1} in a group of 44 beginners. All these findings point to the maximal oxygen uptake as a limiting factor in rowing performance.

So far, we have considered the performance in rowing championships, expressed in terms of mean speed during the entire race. However, a closer

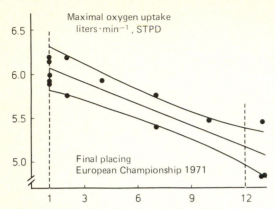

Figure 16-18 Showing the relationship between the maximal oxygen uptake of individual Scandinavian rowers and their final placing in the European championship in 1971; the higher the maximal oxygen uptake, expressed in liters · min⁻¹, the better the performance. The outer lines indicate 95 percent confidence limit. *(From Secher, Vaage, and Jackson, 1976.)*

analysis of the speed attained at different points during the race shows a general speed pattern starting with an initial very high speed followed by a decline to a steady level after the first quarter of the race, and then increasing again in a final spurt. Secher et al. (1976b) have measured oxygen uptake during simulated racing conditions on a bicycle ergometer leading to about the same state of exhaustion in all the subjects at the end of 6 min. Oxygen uptake increased more rapidly during the simulated racing experiment than during the control experiment when the speed was kept constant during the entire 6-min period. They conclude that the greater energy expenditure early in the simulated race was at least partially covered by the more rapid attainment of the maximal oxygen uptake. This appears to provide a physiological basis for the procedure generally practiced during racing of this kind. It may also supply the scientific basis for the development of an optimal speed profile for any type of race. In this connection it should be noted that the heart rate, generally speaking, attains its peak value more rapidly during a real race than during a training race (Fig. 16-19). It should also be kept in mind that the greater the degree of exertion in events lasting several minutes, the greater the anaerobic contribution to the overall energy metabolism, evidenced by a high blood lactate level (Table 16-2).

Ishiko (1971) measured the force acting on the oars of rowers of different international standing. He found that the force varied greatly, with peak values in the range of 800 N (Fig. 16-20). According to the force-velocity relationship, the greatest power is attained at 30 to 40 percent of the maximal voluntary isometric strength of the muscle (Monod and Scherrer, 1972; see Fig. 4-24, p. 104). One would therefore expect that the rowing strength of a rower ought to be in excess of 800 · 2.5 = about 2000 N. Secher (1975) finds an average rowing strength of 2000 N in seven international competitive rowers, 1800 N in national Danish oarsmen, and 1600 N in beginners. However, his results reveal

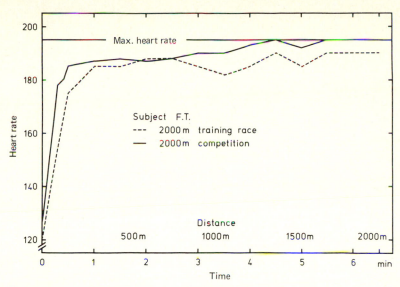

Figure 16-19   Heart rate of a rower during a real race and during a training race. *(By courtesy of E. D. R. Pruett.)*

that champion rowers do not show any greater muscle strength when their different muscle groups are tested by conventional tests of muscle strength (Secher, 1975).

The best rowers show a clear tendency to be heavy. This can be explained by the positive correlation between body weight and maximal oxygen uptake, discussed in Chap. 11, and the fact that there is no statistical significant

Table 16-2   Peak blood lactate levels in Norwegian elite rowers during the semifinal in the eight in the European championship regatta in 1969 and in a Norwegian national regatta the same year. (*By courtesy of O. Vaage.*)

	Peak blood lactate levels (mM)		
		Rowing	
Subjects	Maximum tread-mill running	Norwegian national regatta	European championship regatta, Klagenfurt
O. N.	10.2	13.4	15.9
T. W.	6.4	10.1	12.0
S. N.	11.6	13.7	17.2
A. H.	11.8	17.6	18.6
T. T.	14.7	20.4	18.6
K. S. J.	10.2	16.7	18.8
E. H.	13.1	11.8	17.4
T. A.	7.1	13.2	14.6
Mean ± SD	10.6 ± 2.8	14.6 ± 3.4	16.6 ± 2.4

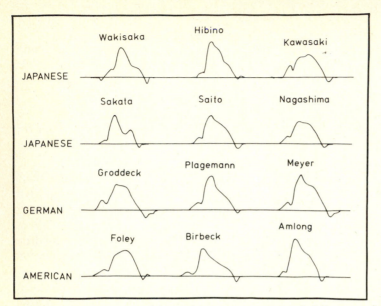

Figure 16-20 Examples of force-time curves, measured at the oar, of Japanese, German, and American oarsmen. The horizontal line gives time course, and the vertical deviations are proportional to the force developed during one stroke. *(From Ishiko, 1971.)*

correlation between maximal oxygen uptake in ml per kg body weight and rowing performance (Vaage, personal communication). In addition, Secher (1975) finds a positive correlation between rowing strength and body weight. This means that maximal oxygen uptake divided by body weight is not a good indicator of rowing performance (Secher et al., 1976a). It also appears that the larger body mass may be associated with a larger anaerobic capacity. It may therefore be concluded that the three attributes of greater maximal oxygen uptake, greater rowing strength, and greater anaerobic capacity together make a superior rower, and that these attributes more than compensate for the stronger water resistance caused by the greater body weight.

From their theoretical analysis of the oar movement and the forces involved Celentano et al. (1974) estimated that at a given mean speed the work required to cover a given distance decreased slightly when decreasing the speed oscillation at each stroke. Thus, it should be advantageous to increase the frequency of strokes, within the limits set by the efficiency of muscular contraction.

Ball Games

Football (Soccer) Generally speaking, most types of ball games represent more or less intermittent work with frequent interchanges of short bursts of physical effort interspaced with brief pauses. For this reason, various ball games, including football, have not, in the past, required the same level of physical endurance or aerobic power in the players as required in long-distance runners, cross-country skiers, or athletes engaged in similar events requiring continuous, long-lasting effort of near maximal intensity.

In the past, studies of superior individual football players or national teams have, with some exceptions, shown maximal oxygen uptakes significantly below the levels found in the endurance athletes just mentioned. In a recent report by Agnevik (1970a), maximal oxygen uptakes of Swedish top football players are presented. The mean value for the 11 members of the national team was 4.3 liters $O_2 \cdot min^{-1}$ (56.5 ml $\cdot$ kg^{-1} $\cdot$ min^{-1}). The corresponding values for a larger group of about 50 Swedish top football players were 4.2 liters $\cdot$ min^{-1} or 58.6 (50 to 69) ml $\cdot$ kg^{-1} $\cdot$ min^{-1}.

In Fig. 16-21, the heart rate in one of the top players on the Swedish team during a major football match is presented. It is observed that the heart rate, with the exception of a few brief occasions, is kept well below the maximal heart rate for this player, and reflects a fairly regular pattern, with periods of high effort interspaced with brief pauses, during which the heart rate may drop as much as 50 beats $\cdot$ min^{-1}. His mean heart rate during the entire match (90 min) was 175 beats $\cdot$ min^{-1}, while his maximal heart rate was 189 beats $\cdot$ min^{-1}. Similar studies have been made in other players, and, whereas the actual level of the mean heart rate may differ somewhat from one player to another during the match, the alternating pattern with changing heart rate is a fairly common feature found in most football players studied.

Blood lactate values up to 16 mM have been measured, but 11 mM may be fairly common among football players at the end of a match.

Table tennis   Lundin (1973) has reported the results of measurements of the maximal oxygen uptakes in seven male Swedish elite table tennis players, including three world champions. The mean maximal oxygen uptake was 4.42 liters $\cdot$ min^{-1} (3.6 to 5.1) corresponding to 65.0 ml $\cdot$ kg^{-1} $\cdot$ min^{-1} on the average.

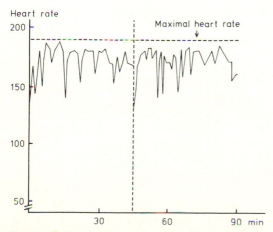

Figure 16-21   Heart rate in a top Swedish football (soccer) player on the national team during an important match. *(From Agnevik, 1970a.)*

It is interesting to note that, according to Lundin, the mean maximal oxygen uptake in Swedish elite table tennis players has been increasing in recent years, perhaps indicating more emphasis on physical fitness and endurance in this athletic event also. According to Lundin, there are small changes in maximal oxygen uptake in the same player from month to month, possibly because of the fact that the elite table tennis player is engaged in systematic training all year round.

Measurements of the actual oxygen uptake with the Douglas bag technique during a simulated table tennis match in seven elite players showed that, on the average, they taxed slightly more than 70 percent of their maximal aerobic power, corresponding to roughly 50 ml $O_2 \cdot kg^{-1} \cdot min^{-1}$.

The heart rate during important matches varied considerably. In some cases, the heart rate was maintained close to the maximal level (Fig. 16-22). In other cases, the heart rate dropped far below the maximal level, as in Fig. 16-23, when the losing player appeared to have lost interest in the game. In general, however, Lundin (1973) concludes that among top Swedish table tennis players, the heart rate during a match is, on the whole, 20 to 30 beats $\cdot min^{-1}$ below the maximal level.

During the actual playing, the blood lactate concentrations were around 2 to 3 mM with peak values up to 5 mM. Similar concentrations were observed in the muscles. In four players the proportion of slow and fast twitch fibers in the deltoid and quadriceps muscles varied markedly, with a mean value of approximately 45 percent fast twitch fibers (type II).

Badminton   A study of a group of Swedish top badminton players (Agnevik, 1970b) showed that the mean maximal oxygen uptake of the women players was 2.57 (2.3 to 2.7) liters $\cdot min^{-1}$ or 46.9 ml $\cdot kg^{-1} \cdot min^{-1}$. For the male players, the corresponding figures were 3.74 (3.0 to 4.4) liters $\cdot min^{-1}$, or 55.6 ml $\cdot kg^{-1} \cdot min^{-1}$. This indicates that the maximal oxygen uptake of these badminton players is quite low compared with that of other elite athletes in endurance events.

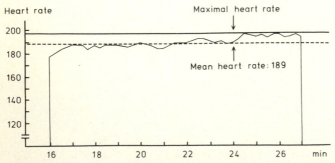

Figure 16-22   Heart rate in a Swedish elite table tennis player during a match. *(From Lundin, 1973.)*

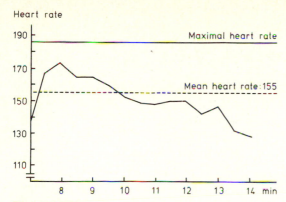

Figure 16-23 Heart rate in a Swedish elite table tennis player during a losing match. *(From Lundin, 1973.)*

Measurements of the oxygen uptake during a simulated badminton match revealed oxygen uptakes up to 3.9 liters · min^{-1} for the men (as against .39 liters · min^{-1} during maximal work on the bicycle ergometer) and 2.6 liters · min^{-1} for the women (as against 2.6 liters · min^{-1} during maximal bicycling) showing that during a hard game the players may actually tax all their aerobic power maximally.

Recordings of heart rates during important international matches showed that the players attained near maximal values in single matches, whereas doubles and mixed-doubles matches clearly showed lower heart rates (Fig. 16-24). Blood lactate values around 12 to 13 mM were obtained.

Ice Hockey According to Forsberg et al. (1974), the mean maximal oxygen uptakes of the Swedish elite ice hockey team during the 1973–1974 season was 4.9 liters · min^{-1} (4.5 to 5.5) or 65 ml · kg^{-1} · min^{-1} (61 to 69). The values have shown a tendency to increase since 1960, probably because of more intense endurance training. The values are somewhat higher than for Swedish elite football or bandy players.

The oxygen uptake during simulated matches, measured by the Douglas bag technique, showed that the players at maximal intensity utilized at the most 85 to 90 percent of their maximal oxygen uptakes measured during maximal running on the treadmill.

Heart rates recorded during important international ice hockey matches showed mean heart rates of about 180 beats · min^{-1}, with peaks reaching maximal heart rate values (204 beats · min^{-1}). Heart rates were recorded and blood lactate concentrations were measured in a Swedish forward player during all three periods in the international championship match against Soviet Russia in 1974. As is evident from Fig. 16-25, the more intense the play, the higher the blood lactate level. Again, the intermittent nature of this type of ball-game activity is reflected in the heart rate record shown in the upper panel

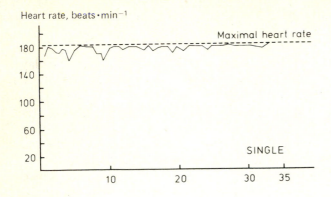

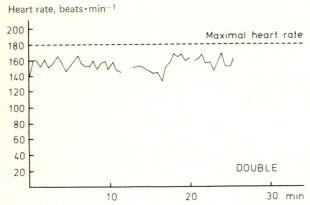

Figure 16-24 Heart rates recorded in an elite badminton player during a match. Upper panel: singles; lower panel: doubles. *(From Agnevik, 1970b.)*

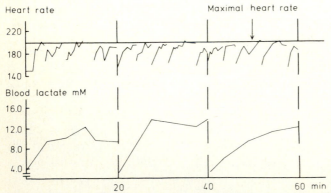

Figure 16-25 Relationship between heart rate and blood lactate concentration in a Swedish ice hockey player during a match against Russia, March 29, 1974. *(From Forsberg et al., 1974.)*

of Fig. 16-25. Similar findings have been reported by Green et al. (1976) in Canadian ice hockey players.

Summary It appears that most ball players, and especially those who play as team members, have lower maximal oxygen uptakes than endurance athletes (Fig. 12-5). Thus, the mean maximal oxygen uptakes of Swedish elite team players in football, bandy, ice hockey, and basketball is 60 ml · kg^{-1} · min^{-1}, as against some 75 to 94 ml · kg^{-1} · min^{-1} in the Swedish top athletes in endurance events such as skiing and orienteering. The explanation may be sought in the fact that team ball games usually involve varying degrees of intermittent work, with brief pauses between short bouts of strenous effort, while endurance events, such as long-distance running, represent a continuous effort of more or less high intensity. It thus appears that while ball games, and especially table tennis and badminton, may offer an excellent activity for the development of physical fitness for the ordinary person, they hardly represent a suitable training activity for the development of maximal aerobic power in top endurance athletes.

REFERENCES

Agnevik, G.: "Fotboll," *Idrottsfysiologi*, Rapport no. 7, Trygg-Hansa, Stockholm, 1970a.

Agnevik, G.: "Badminton," *Idrottsfysiologi*, Rapport no. 8, Trygg-Hansa, Stockholm, 1970b.

Agnevik, G., B. Wallström, and B. Saltin: A Physiological Analysis of Alpine Skiing, *J. Appl. Physiol.*, 1969.

Asmussen, E., K. Klausen, L. Egelund Nielsen, O. S. A. Techow, and P. J. Tönder: Lactate Production and Anaerobic Work Capacity after Prolonged Exercise, *Acta Physiol. Scand.*, **90**:731, 1974.

Åstrand, P.-O.: "Experimental Studies of Physical Working Capacity in Relation to Sex and Age," Munksgaard, Copenhagen, 1952.

Åstrand, P.-O.: Human Physical Fitness with Special Reference to Sex and Age, *Physiol. Rev.*, **36**:307, 1956.

Åstrand, P.-O., and B. Saltin: Maximal Oxygen Uptake and Heart Rate in Various Types of Muscular Activity, *J. Appl. Physiol.*, **16**:977, 1961.

Åstrand, P.-O., L. Engström, B. Eriksson, P. Karlberg, I. Nylander, B. Saltin, and C. Thorén: Girl Swimmers, *Acta Paediat.* (Suppl. 147):1963.

Åstrand, P.-O., I. Hallbäck, R. Hedman, and B. Saltin: Blood Lactates after Prolonged Severe Exercise, *J. Appl. Physiol.*, **18**:619, 1963.

Bergh, U.: "Längdlöpning," *Idrottsfysiologi*, Rapport no. 11, Trygg-Hansa, Stockholm, 1974.

Böje, O.: Energy Production, Pulmonary Ventilation, and Length of Steps in Well-trained Runners Working on a Treadmill, *Acta Physiol. Scand.*, **7**:362, 1944.

Carlsöö, S.: A Kinetic Analysis of the Golf Swing, *J. Sports Med.*, **7**:76, 1967.

Cavagna, G. A., H. Thys, and A. Zamboni: The Sources of External Work in Level Walking and Running, *J. Physiol.*, **262**:639, 1976.

Celentano, F., G. Cortili, P. E. Di Prampero, and P. Cerretelli: Mechanical Aspects of Rowing, *J. Appl. Physiol.,* **36**:642, 1974.

Christensen, E. H. and P. Högberg: Physiology of Skiing, *Arbeitsphysiologie,* **14**:292, 1950.

Costill, D. L., K. Sparks, R. Gregor, and C. Turner: Muscle Glycogen Utilization during Exhaustive Running, *J. Appl. Physiol.,* **31**:353, 1971.

Di Prampero, P. E., G. Cortili, F. Celentano, and P. Cerretelli: Physiological Aspects of Rowing, *J. Appl. Physiol.,* **31**:853, 1971.

Di Prampero, P. E., G. Cortili, P. Mognoni, and F. Saibene: Energy Cost of Speed Skating and Efficiency of Work against Air Resistance, *J. Appl. Physiol.,* **40**:584, 1976.

Di Prampero, P. E., D. R. Pendergast, D. W. Wilson, and D. W. Rennie: Energetics of Swimming in Man, *J. Appl. Physiol.,* **37**:1, 1974.

Ekblom, B., L. Hermansen, and B. Saltin: Hastighetsåkning på Skridsko, *Idrotts-fysiologi,* Rapport no. 5, Framtiden, Stockholm, 1967.

Eriksson, A., A. Forsberg, L. Källberg, P. Tesch, and J. Karlsson: Alpint, *Idrottsfysio-logi,* Rapport nr. 17, Framtiden, Stockholm, 1977.

Forsberg, A., B. Hultén, G. Wilson and J. Karlsson: "Ishockey," *Idrottsfysiologi,* Rapport no. 14, Trygg-Hansa, Stockholm, 1974.

Granati, A., and L. Busca: Il Lavoro della Tribiatura, *Boll. Soc. Ital. Biol. Sper.,* **20**:51, 1945.

Green, H., P. Bishop, M. Houston, R. McKillop, R. Norman, and P. Stothart: Time-motion and Physiological Assessments of Ice Hockey Performance, *J. Appl. Physiol.,* **40**:159, 1976.

Högberg, P.: How Do Stride Length and Stride Frequency Influence the Energy Output during Running? *Arbeitsphysiol.,* **14**:437, 1952.

Holmér, J.: Physiology of Swimming Man, *Acta Physiol. Scand.* (Suppl. 407):1974.

Holmér, I.: Energy Cost of Arm Stroke, Leg Kick, and the Whole Stroke in Competitive Swimming Styles, *Europ. J. Appl. Physiol.,* **33**:105, 1974b.

Holmér, I., E. M. Stein, B. Saltin, B. Ekblom, and P.-O. Åstrand: Hemodynamic and Respiratory Responses Compared in Swimming and Running, *J. Appl. Physiol.,* **37**:49, 1974.

Ishiko, T.: "Biomechanics of Rowing," in *Medicine and Sport,* vol. 6, Biomechanics II, p. 249, Karger, Basel, 1971.

Jokl, E., P. Jokl, R. Green, and B. Reinhardt: Running and Swimming World Records, *Am. Corr. Ther. J.,* **30** (no. 5), 1976.

Karlsson, J., L. Hermansen, G. Agnevik, and B. Saltin: "Löping," *Idrottsfysiologi,* Rapport no. 4, Trygg-Hansa, Stockholm, 1972.

Lundin, A.: "Bordtennis," *Idrottsfysiologi,* Rapport no. 12, Trygg-Hansa, Stockholm, 1973.

Margaria, R.: Sulla Fisiologia e Specialmente sul Consumo Energetico, della Marcia e della Corsa a Varie Velocita ed Inclinazioni del Terreno, *Atti dei Lincei,* **7**:299, 1938.

Margaria, R., P. Cerretelli, P. Aghemo, and G. Sassi: Energy Cost of Running, *J. Appl. Physiol.,* **18**:367, 1963.

Margaria, R., P. Aghemo, and F. Piñera Limas: A Simple Relation between Performance in Running and Maximal Aerobic Power, *J. Appl. Physiol.,* **38**(2):351, 1975.

Meen, H. D., R. Gullestad, and S. B. Strömme: En sammenligning av maksimalt oksygenopptak under skisprint i motbakke og löp på tredemölle, *Kroppsöving*, no. 7, 134, 1972.

Molbech, S.: "Energy Cost of Level Walking in Subjects with an Abnormal Gait," in K. Evang and K. L. Andersen (eds.), "Physical Activity in Health and Disease," Universitetsforlaget, Oslo, 1966.

Monod, H., and I. Scherrer: "How Muscles Are Used in the Body," in C. H. Bourne (ed.), "The Structure and Function of Muscle," vol. 1, Academic Press, Inc., New York, 1972.

Passmore, R., J. G. Thomson, and G. M. Warnock: Balance Sheet of the Estimation of Energy Intake and Energy Expenditure as Measured by Indirect Calorimetry, *Brit. J. Nutr.*, **6**:253, 1952.

Passmore, R., and J. V. G. A. Durnin: Human Energy Expenditure, *Physiol. Rev.*, **35**:801, 1955.

Pugh, L. G. C. E.: The Influence of Wind Resistance in Running and Walking and the Mechanical Efficiency of Work Against Horizontal or Vertical Forces, *J. Physiol.*, **213**:255, 1971.

Pugh, L. G. C. E.: Air Resistance in Sport, in E. Jokl, R. L. Anand, and H. Stoboy (eds.) "Advances in Exercise Physiology," S. Karger, Basel, 1976.

Ramaswamy, S. S., G. L. Dua, V. K. Raizada, G. P. Dimri, K. R. Viswanatan, J. Madhaviah, and T. N. Srivastava: Effect of Looseness of Snow on Energy Expenditure in Marching on Snow-covered Ground, *J. Appl. Physiol.*, **21**:1747, 1966.

Rodahl, A., M. O'Brien, and R. G. R. Firth: Diurnal Variation in Performance of Competitive Swimmers, *J. Sports Med. and Physical Fitness*, **16**:72, 1976.

Rönningen, H.: Unpublished results, 1976.

Saltin, B., and P.-O. Åstrand: Maximal Oxygen Uptake in Athletes, *J. Appl. Physiol.*, **23**:353, 1967.

Secher, N.: Development of Results in International Rowing Championships 1893–1971, *Med. Sci. Sports*, **5**:195, 1973.

Secher, N.: Isometric Rowing Strength of Experienced and Inexperienced Oarsmen, *Med. Sci. Sports*, **7**:280, 1975.

Secher, N. H., O. Vaage, and R. C. Jackson: Rowing Performance and Maximal Oxygen Uptake, unpublished results, 1976a.

Secher, N. H., R. A. Binkhorst, and W. Ruberg-Larsen: Oxygen Uptake during Simulated Racing Conditions, unpublished results, 1976b.

Tesch, P., K. Piehl, G. Wilson, and J. Karlsson: "Kanot," *Idrottsfysiologi*, Rapport no. 13, Trygg-Hansa, Stockholm, 1974.

Tesch, P., K. Piehl, G. Wilson, and J. Karlsson: Physiological Investigations of Swedish Elite Canoe Competitors, *Med. Sci. Sports*, **8**:214, 1976.

Vokac, Z., and K. Rodahl: Maximal Aerobic Power and Circulatory Strain in Eskimo Hunters in Greenland, *Nordic Council for Arctic Med. Res.*, Report No. 1, 1977, Oulu Univ. Press, Oulu, Finland.

Vrijens, J., J. Hoekstra, J. Bouckaert, and P. van Tyvanck: Effects of Training on Maximal Working Capacity and Haemodynamic Response during Arm and Leg Exercise in a Group of Paddlers, *Europ. J. Appl. Physiol.*, **34**:113, 1975.

Zuntz, L., and W. Schumburg: "Studien zu einer Physiologie des Marsches," Bibliothek v. Coler, Band 6., Verlag von A. Hirschwald, Berlin, 1901.

Factors Affecting Performance

17

CONTENTS

Chapter 17

Factors Affecting Performance

INTRODUCTION

Figure 9-1 is a schematic presentation of the components of the aerobic work capacity and the factors which affect this capacity. Most of these factors have been discussed earlier rather extensively from various viewpoints and in different connections. Thus the effect of the type of exercise was examined in Chap. 9, oxygen-transporting functions were discussed in Chaps. 6 and 9, metabolic aspects in Chap. 14, and training and deconditioning in Chap. 12. As far as the effect of the environment is concerned, the effect of temperature and water balance was explored in Chap. 15. The present chapter will be devoted to a combined discussion of the remaining factors mentioned in Fig. 9-1, including the effect of altitude.

HIGH ALTITUDE

The climbing of high mountains has always fascinated humans. The quest for new discoveries has taken men beyond the earth's atmosphere into outer space. Permanent human habitation is encountered up to an altitude of over 4,500 m. It is becoming increasingly popular to spend summer and winter vacations in high mountain areas undertaking the fairly heavy physical work involved in hiking, mountaineering, or skiing. In spite of pressure cabins, passenger flights at high altitudes involve a certain lack of oxygen. The air pressure in the airplane cabin usually corresponds to an altitude of 1,000 to 1,500 m. The decision to hold the 1968 Olympic Games in Mexico City at an altitude of 2,300 m created a special interest in the problems concerning the effects of altitude on physical performance. Several international symposia have been held to discuss various problems related to physical performance at high altitude. (For reports, see Weihe, 1964; Dill, 1964; Luft, 1964b; *Schw. Zschr. Sportmed.*, **14**:1–329, 1966; Margaria, 1967; Goddard, 1967; Jokl and Jokl, 1968; Roskamm et al., 1968). Luft (1964a) has reviewed some historic events in the exploration of altitude and the facilities available in various parts of the world for research in this field.

Physics

In the nineteenth century, Bert (1878) recognized that the detrimental effects of high altitude were due to the *diminished partial pressure of oxygen* at reduced barometric pressure. The barometric pressure at a given altitude depends on the weight of the air column over the point in question. The atmosphere is compressed under its weight. Its pressure and density are therefore highest at the surface of the earth and decrease exponentially with altitude. Because of temperature differences and turbulence, any sedimentation of the gas molecules of different molecular weight is avoided, and the chemical composition of the atmosphere is practically uniform up to an altitude of more than 20,000 m.

Table 17-1 presents barometric pressure and the oxygen pressure of the inspired air (tracheal air) at various altitudes. With a constant oxygen concentration of 20.94 percent of the dry air, the oxygen pressure of the inspired air in the trachea, saturated with water vapor, can easily be calculated from the formula $P_{O_2} = (P_{Bar} - 47) \cdot 20.94 \cdot 100^{-1}$. This means that at an altitude of about 19,000 m, where the barometric pressure is 47 mm Hg, there should be nothing but water molecules in the trachea!

The oxygen tension of the alveolar air, and thereby also the oxygen tension of the arterial blood, is determined by the magnitude of the pulmonary ventilation in addition to the composition and pressure of the inspired air. The more frequently the air is exchanged, the more closely the composition of the pulmonary air resembles the inspired air (water vapor subtracted). This will be discussed below.

The reduced density of the air at high altitudes affects the mechanics of breathing. Part of the work of breathing is expended in moving the air against the resistance of the airways. The resistance is relatively high when the flow is turbulent, as it is during exercise. Therefore, the influence of reduced density is more noticeable at high rates of air flow as in hyperpnea, during heavy exercise, or in flow-dependent pulmonary function tests. The maximal breathing capaci-

Table 17-1 Barometric pressure (standard atmosphere) at various altitudes and the pressure of oxygen after the inspired gas has been saturated with water vapor at 37°C (tracheal air).*

Altitude				Altitude			
m	ft	Pressure, mm Hg	P_{O_2} tracheal air, mm Hg	m	ft	Pressure, mm Hg	P_{O_2} tracheal air, mm Hg
0	0	760	149	5,500	18,050	379	69
500	1,640	716	140	6,000	19,690	354	64
1,000	3,280	674	131	6,500	21,330	330	59
1,500	4,920	634	123	7,000	22,970	308	55
2,000	6,560	596	115	7,500	24,610	287	50
2,500	8,200	560	107	8,000	26,250	267	46
3,000	9,840	526	100	8,500	27,890	248	42
3,500	11,840	493	93	9,000	29,530	230	38
4,000	13,120	462	87	9,500	31,170	214	35
4,500	14,650	433	81	10,000	32,800	198	32
5,000	16,400	405	75	19,215	63,000	47	0

*Based on dry conditions for average temperature at altitude when the temperature at sea level is 15°C and the barometric pressure is 760 mm Hg.

ty is considered higher at high altitude than at sea level (Cotes, 1954; Miles, 1957; Ulvedal et al., 1963). The respiratory muscles' ability to exert pressure seems to be reduced at low barometric pressure; however, the net effect of the reduced resistance to air flow at lowered barometric pressure is a *diminished respiratory work* to move a given volume of air in and out of the lungs (Fenn, 1954). On several occasions, pulmonary ventilations of 200 liters · min⁻¹ have been measured during maximal exercise at high altitude (see below). Another effect of the reduced density of the air at a low barometric pressure is a diminished external air resistance. The air resistance changes with the wind speed raised to the second power. The external work is therefore reduced at high altitude in sprint-type activities, speed skating, cycling, and alpine skiing with high velocities. The *air temperature* is on the whole lower, the higher the altitude. With a mean annual temperature of 15°C at sea level, the air temperature decreases linearly by 6.5°C · 1,000 m⁻¹ to about 11,000 m. The air also becomes increasingly *dry* with increasing altitude. Therefore the water loss via the respiratory tract is higher at high altitudes than at sea level. If much work is performed at high altitude, this loss may give rise to a hypohydration and a sensation of soreness and dryness in the throat.

The *solar radiation* is more intense at high altitude and the ultraviolet radiation may cause difficulties in the form of sunburn or snow blindness. Finally, *the force of gravity* is reduced with the distance from the earth's center. High altitudes may therefore have some favorable effect in the case of athletic events involving jumping or throwing.

Physical Performance

The reduced work capacity at high altitudes is well established. It is already evident at an altitude of about 1,200 m in the case of heavy exercise engaging large muscle groups for about 2 min or longer. As an example of the physical strain experienced during the conquest of high mountain peaks, Somervell (1925) may be quoted: "It may be of interest to record one or two personal observations which I made while climbing in the neighborhood of 27,000 to 28,000 feet. *Pulse:* The heart rate during the actual motion upwards was found to be beating 160–180 per minute, sometimes even more, regular in rhythm and of good volume. . . . *Respiration:* About 50 to 55 per minute while climbing. Approaching 28,000 feet I found that for every single step forward and upward, seven to ten complete respirations were required." Norton (1925) writes that at an altitude of 8,500 m he required 1 hr to climb a distance of 35 m even though the climb was not particularly difficult.

Athletic competitions at different altitudes provide an experimental condition where highly motivated, well-trained athletes subject themselves to all-out tests. Leary and Wyndham (1966) report observations from South Africa where important track meetings take place at altitudes of 1,500 m or above as well as at sea level. They conclude that the best performances recorded in middle- and long-distance events were on the coast, whereas sprinters recorded better times at medium altitude. South African championships held at medium altitude have been consistently dominated by athletes domiciled at such altitudes and those who acclimatized themselves by training for medium altitude for 3 to 4 weeks before the championship competitions. In competitions in the mile event, five international class athletes all performed better at sea level. The average time was 4 min and 0.5 s in the best coastal performance and 4 min and 15.2 s in Johannesburg (altitude 1,760 m, or 5,780 ft).

In competitions in Mexico City (altitude 2,300 m), the same or better performance in running distances up to 400 m has been reported. In 1500-m running, there is an impairment of about 3 percent, and in 5,000- and 10,000-m, an impairment of roughly 8 percent compared with sea level. In swimming, one finds an impairment of the time for 100 m of about 2 to 3 percent and an impairment of about 6 to 8 percent in the case of 400-m or longer distances when comparing Mexico City altitude with that of sea level (Goddard and Favour, 1967; Craig, 1969; Shephard, 1973). In jumping and throwing, some differences have been recorded among athletes competing at different altitudes. The world record in long jumping (8.90 m) by Bob Beaman in the Mexico City Olympic Games should be particularly noted. It has frequently been stated that recovery time in Mexico City is considerably longer than at low altitudes. Various types of collapses occurred with relatively high frequency during the 1968 Olympic Games in Mexico City (see Jokl and Jokl, 1977).

This brief summary shows that in types of work requiring an intense activity of no more than 1-min duration, and especially in events of this nature where the technique is of primary importance, there is no noticeable difference in performance between sea level and high altitude, at least up to an altitude of

2,500 m. The capacity to perform heavy work for 2 min or more is, on the other hand, definitely reduced at high altitude, with the exception of types of activities in which the air resistance plays a major role.

What, then, is the physiological explanation for this phenomenon?

Limiting Factors

We may base our analysis on the factors listed in the table of factors affecting physical work capacity at the beginning of Chap. 9. If one disregards altitudes above 3,000 m, which may cause a disturbance of psychological functions, it is evident that it is the aerobic power which is affected by a reduced oxygen pressure in the inspiratory air. Studies have shown that work during acute exposure to high altitude causes a rise in the blood lactate concentration at lighter work loads than is the case at sea level. At a fixed work load, the lactate level is higher, but the maximal concentration attained during exhaustive work is roughly the same as at sea level (Edwards, 1936; Asmussen et al., 1948; P.-O. Åstrand, 1954; Stenberg et al., 1966; Buskirk et al., 1967; Hermansen and Saltin, 1967). Figure 17-1 gives an example of how the blood lactate concentration is affected during work at sea level, 2,300 m, and 4,000 m simulated altitude. In this subject, however, the lactate concentration was the same when it was

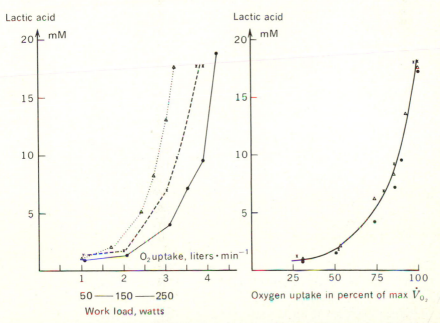

Figure 17-1 Blood lactate concentrations in one subject during exercise at sea level (filled dots) and acute exposure to 2,300 (crosses) and 4,000 m (triangles) (760, 580, and 462 mm Hg respectively). Work times for all submaximal work loads were 10 min, and all blood samples were taken between the ninth and tenth min. Work times for the maximal work loads were 4 to 5 min, and peak values for lactic acid are given.
In the left panel the absolute oxygen uptake is on the abscissa; on the right, oxygen uptake is in percentage of the maximum. *(From Hermansen and Saltin, 1967.)*

related to the relative, instead of the absolute, oxygen uptake. This must be interpreted as indicating that the anaerobic processes are brought into play at a relatively lower work load at high altitude. The maximal anaerobic power, at least the part of it which is determined by the glycogenolysis, on the other hand is not affected. The fact that the maximal oxygen debt is the same following maximal work at sea level and at altitude also points in the same direction (Saltin, 1966; Buskirk et al., 1967).

As an example, Saltin (1966) mentions that the sprinter Bodo Tümmler ran 1,500 m in Stockholm in 3.42 min and in Mexico City in 3.54 min. The O_2 uptake during the subsequent 60-min rest was 38 and 42 liters respectively; peak blood lactate concentration was 18.6 and 19.3 mM respectively. After a 3,000-m steeplechase, the values obtained for Bengt Persson in Stockholm were 28 liters O_2, blood lactates 18.2 mM, running time 8.34 min. Corresponding values in Mexico City were 33 liters O_2, 20.3 mM, and 9.32 min.

It is difficult to explain the higher lactic acid level in the blood at a standard submaximal work load at high altitude in view of the fact that the oxygen uptake is the same as at sea level (Christensen, 1937; Asmussen and Chiodi, 1941; P.-O. Åstrand, 1954; Pugh et al., 1964). Recent studies have indicated a slower increase in oxygen uptake at the beginning of exercise, i.e., the oxygen deficit is enlarged. Therefore, the demand of the anaerobic energy yield is correspondingly increased (see Chap. 9, p. 317; Raynaud et al., 1974).

With regard to the neuromuscular function, Christensen and Nielsen (1936) have shown that speed and strength are not affected by moderate hypoxia. In a test on a Hill's wheel, contraction time being less than 6 sec, their subjects attained the same maximum at sea level as at a barometric pressure of 440 and 390 mm Hg respectively.

From a psychological point of view, a stay at even moderate altitudes may represent a considerable stress, or in any case an unusual sensation. From experience one knows how a certain work intensity feels or affects the body under normal circumstances. The same physical effort at higher altitudes produces a higher pulmonary ventilation, a higher heart rate, and possibly other symptoms of fatigue which under normal conditions would not be customary at this work load. An adaptation to the new situation gradually takes place. For the athlete, the tactics may be different in training and competitions at high altitude compared with the situation at sea level. This will be further discussed at the end of this chapter.

Summary It may be concluded that it is the aerobic power which is directly affected during work under conditions of reduced oxygen pressure in the inspired air. We shall therefore summarize how the different steps in the oxygen transport from the air to the mitochondria of the exercising muscle cell are affected during exposure to high altitude.

Oxygen Transport

Figure 17-2 indicates the pressure levels at different distances along the transport chain from atmospheric air to the ultimate destination, the mitochon-

Oxygen tension, mm Hg

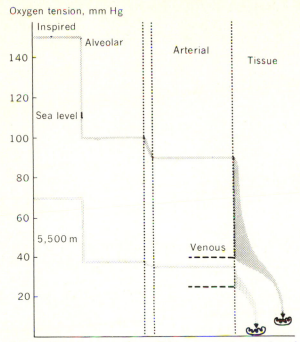

Figure 17-2 A comparison of the oxygen cascade from inspired air to tissue in the human at sea level and at 5,500 m (18,000 ft). *(Modified from Rahn, 1966.)*

dria. In the given example, the pressure gradient for oxygen between air and mixed venous blood at rest at sea level is about 110mm Hg (150 − 40 mm Hg). When one goes to an altitude of 5,500 m, P_{O_2} = 70 mm Hg, which is about the highest altitude to which people can become acclimatized and at which they can live and work for months and years (Rahn, 1966), the pressure-drop for the same oxygen delivered is reduced to about 50 mm Hg. A reduction of the oxygen pressure in the inspired air by 80 mm Hg is associated with a decrement of only 20 mm Hg in the mixed venous blood. Close to the mitochondria, the oxygen pressure may be 10 mm Hg in the sea-level situation and about 5 mm Hg at 5,400 m. This pressure is still adequate to provide optimal conditions for the oxidative enzyme reactions (Chap. 2). There are many factors explaining the increased "conductance" to keep the tissue oxygen pressure at this almost constant level. We shall first consider the *acute hypoxia*.

1   The *pulmonary ventilation* at a given oxygen uptake is markedly elevated (see Fig. 17-3). In this subject the ventilation at an oxygen uptake of 4.0 liters · min⁻¹ was 80 liters · min⁻¹ when pure oxygen was inhaled, 105 liters · min⁻¹ at sea level when the subject was breathing air, 140 liters · min⁻¹ at 2,000 m, and 160 liters · min⁻¹ at 3,000 m, or twice as high as when breathing oxygen. Even at sea-level conditions there exists a hypoxic drive which is evident when this subject's oxygen uptake exceeds 1.5 liters · min⁻¹. This hypoxic hyperpnea is elicited through the reflex pathway originating in the

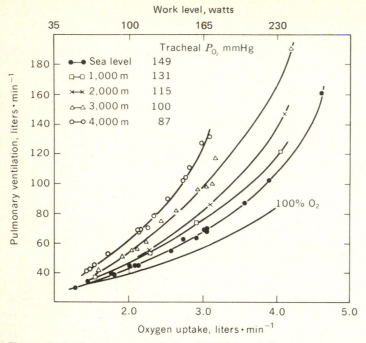

Figure 17-3 Pulmonary ventilation (BTPS) in relation to oxygen uptake in one subject at different work loads when breathing oxygen or air during exposure to various simulated altitudes. Note the high pulmonary ventilation of 190 liters · min⁻¹ reached during work at 3,000-m altitude. *(From P.-O. Åstrand, 1954.)*

chemoreceptors of the carotid body and aortic body (see Chap. 7). Since the production of carbon dioxide is roughly the same at a given oxygen uptake, this hyperventilation will inevitably wash out CO_2 from the blood into the inspired air, the dissolved CO_2 of the blood being more affected than the bicarbonate. The secondary effect of the hyperpnea will therefore be a rise in the pH of the blood, i.e., an uncompensated respiratory alkalosis. The reduced P_{CO_2} and elevated pH of the arterial blood exert an inhibitory influence on the respiratory center. On the other hand, the earlier accumulation of lactic acid in the blood must be considered to cause the pH to fall.

At sea level, the alveolar P_{CO_2} was about 38 mm Hg in the experiment, bringing the pulmonary ventilation to 80 liters · min⁻¹ (Fig. 17-3). The reduction in ventilation when breathing oxygen brought the alveolar P_{CO_2} to a somewhat higher level; at a simulated altitude of 4,000 m, the alveolar P_{CO_2} became reduced to 28 mm Hg during exercise, with a similar pulmonary ventilation of 80 liters · min⁻¹ (see also Dejours et al., 1963; Dempsey et al., 1972). The effect is that the hypoxic drive during work at high altitude must be stronger than reflected in the magnitude of the pulmonary ventilation. If the alveolar and arterial P_{CO_2} is maintained at about 40 mm Hg by the addition of a proper volume of CO_2 to the inspired air, the pulmonary ventilation during exercise in hypoxic conditions will be still higher than illustrated in Fig. 17-3. The ventilatory response must be viewed as a physiological compromise, with the call for an adequate oxygen supply matched against the need to maintain the acid-base balance as normal as possible.

Owing to the great increase in ventilation during work at high altitude and a given oxygen uptake, the alveolar P_{O_2} is higher than what it normally would be. This obviously facilitates the diffusion of oxygen to the blood in the pulmonary capillaries.

The maximal pulmonary ventilation during work at high altitude is the same as, or higher than, at sea level (Stenberg et al., 1966; Saltin, 1966; Grover and Reeves, 1967; Roskamm et al., 1968).

2   The diffusing capacity is reported to be unchanged in people who have lived at sea level after their new arrival at higher altitudes (West, 1962), or is slightly increased (Reeves et al., 1969; Gularia et al., 1971). The alveolar-arterial P_{O_2} gradient is greater for a given oxygen uptake compared with sea-level determinations (Cruz et al., 1973).

3   During submaximal work with reduced O_2 pressure in the inspired air, the lower O_2 saturation is compensated with an increased cardiac output (Asmussen and Nielsen, 1955; Stenberg, 1966; Hartley et al., 1973). This and other effects of high altitude on the oxygen transport in humans are illustrated by Fig. 17-4. The increase in the cardiac output is brought about by an increase in the heart rate; the stroke volume may even be reduced (see also McManus et al., 1974). The arterial blood pressure is largely unchanged. The lower arterial CO_2 pressure (hypocapnia) causes a venoconstriction which may preserve the cardiac output by increasing the central venous volume and cardiac filling pressure (see Cruz et al., 1976).

The previously mentioned hypocapnia during acute lack of oxygen produces a shift in the Hb dissociation curve (Fig. 6-16a), which means a net advantage for oxygen transport due to a higher arterial saturation. The arterial oxygen content is definitely reduced at high altitudes, however, and the arteriovenous O_2 difference drops.

It is rather interesting to note that the observed maximal values for heart rate, cardiac output, and stroke volume are the same at an altitude of 4,000 m (acute exposure) as at sea level. Apparently the lack of oxygen is not of such a magnitude that the pumping capacity of the heart muscle is reduced, in spite of the fact that the Pa_{O_2} is estimated to be lower than 50 mm Hg. In a comparable study, Blomqvist and Stenberg (1965) showed that there were no ECG signs of myocardial ischemia when their subjects were performing maximal work at the same simulated altitude of 4,000 m. The conclusion of this study by Stenberg et al. (1966) was that the reduced maximal oxygen uptake in moderate acute hypoxia compared with normoxia was closely related to the reduction of the arterial oxygen content. During maximal work at hypoxia, the oxygen uptake was on an average 72 percent, the arterial oxygen content was 74 percent, and the cardiac output was 100 percent of the values attained at sea level. In other words, the maximal oxygen uptake was highly correlated with the volume of oxygen offered to the tissue (arterial oxygen content times maximal cardiac output).

During maximal exercise at sea level, almost all the oxygen is extracted

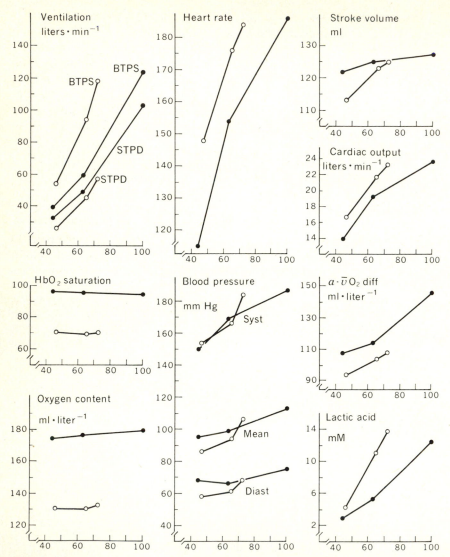

Figure 17-4 Mean values on six subjects studied at two submaximal work loads and one maximal work load (bicycle ergometer) at sea level (filled dots) and when acutely exposed to simulated altitude of 4,000 m in a decompression chamber (open dots). Note that the maximal oxygen uptake was reduced to 72 percent of the value at sea level. Abscissa = oxygen uptake in percentage of the maximum attained at sea level. *(From Stenberg et al., 1966.)*

from the blood passing the working muscles, so that there is nothing more to gain in this respect at acute exposure to high altitude.

The initial effect of high altitude often includes mountain sickness with various symptoms like headache, nausea, vomiting, physical and mental fatigue, interrupted sleep, digestive disorders. In rare cases previously healthy persons may develop pulmonary edema within a few hours after exposure to high altitude (see Reeves et al., 1969; Clarke et al., 1975).

Summary Acute exposure to a reduced oxygen pressure in the inspired air during exercise is associated with hyperpnea in excess of that at sea-level conditions for the same energy requirement. The cardiac output also rises out of proportion to the oxygen uptake. These factors, combined with the displacement of the physiological range of the O_2 dissociation curve to its steep part, enhance the oxygen transport. These compensatory responses cannot, however, fully compensate for the reduced oxygen pressure. The maximal oxygen uptake is reduced, and the importance of the anaerobic energy yield increases.

The quantitative effect on the oxygen transport during maximal work at different altitudes is illustrated in Fig. 17-5. At the altitude of Mexico City, 2,300 m, the reduction is in the order of 15 percent; at 4,000 m, it is about 30 percent (from 4.24 to 3.07 liters · min^{-1} in the study by Stenberg et al., 1966). The considerable scatter of the data observed when different studies are compared may be explained by several facts: (1) The effect of the reduced oxygen pressure on the physical work capacity is different in different

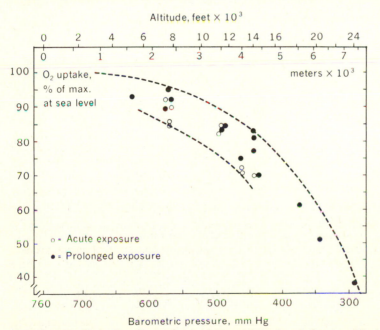

Figure 17-5 Reduction in maximal oxygen uptake in relation to altitude barometric pressure. Unfilled dots denote experiment in the acute hypoxic stage; filled dots, data obtained after various periods of acclimatization. In principle, the maximal aerobic power during acute exposure to reduced oxygen pressure falls at the lower part, within the dotted lines; during acclimatization, it is shifted toward the upper part of the field. *(Data from Balke, 1960; Pugh, 1964; Stenberg et al., 1966; Buskirk et al., 1967; Hansen et al., 1967b; Saltin, 1967; Roskamm et al., 1968.)*

individuals; (2) different techniques have been applied, especially concerning criteria for ascertaining that the maximal oxygen uptake has been reached; (3) conceivably, persons with a high maximal aerobic power are more affected than persons with a lower maximal aerobic power in that the diffusing capacity may be more critical for the former. The progressive fall in arterial oxygen saturation as the work level is raised at high altitude in spite of an increasing alveolar tension, and the resulting large alveolar-arterial oxygen differences, can be explained by diffusion limitations of the lung (West et al., 1962; Saltin, 1967; Grover and Reeves, 1967).

Adaptation to High Altitude

We shall now discuss the effect of a prolonged stay at high altitude, i.e., *acclimatization* to reduced oxygen pressure in the inspired air. It is customary to distinguish between short-term adaptation when it is a matter of days, weeks, or a few months at high altitude, and long-term adaptation when the stay consists of years at high altitude.

1 The first few days' exposure to reduced oxygen pressure entails a further increase in the pulmonary ventilation at a given work load. This hyperpnea will further raise the P_{O_2} and reduce the P_{CO_2} of the alveolar air. This ventilatory response is illustrated in Fig. 17-6. It shows data obtained at sea level, during acclimatization for 4 weeks to an altitude of 4,300 m (14,250 ft), and again at sea level during the reacclimatization. Four subjects were studied, but the figure presents the data on only one of the subjects. However, he represents the normal reaction reasonably well. Three comments should be made: (*a*) It is evident that within a week at high altitude, a new level for pulmonary ventilation is attained, exceeding the value noticed in acute exposure to the same degree of hypoxia. The prolonged exposure to this hypoxia caused a 40 to 100 percent increase in the pulmonary ventilation compared with the sea level controls, the increase being more pronounced the heavier the work loads. At the end of the stay, the ventilation at an oxygen intake of 2.7 liters $\cdot$ min^{-1} (200 watts) was about 120 liters $\cdot$ min^{-1} when breathing air, compared with 60 liters $\cdot$ min^{-1} at sea level. The alveolar P_{CO_2} was 24 mm Hg at 4,300 m, compared with 40 mm Hg at sea level. (*b*) Even when oxygen is inhaled during exercise at high altitude, blocking the peripheral chemoreceptor drive, there is a gradual increase in pulmonary ventilation (see Fig. 17-6). In the example chosen, the ventilation in the control experiment at sea level was 50 liters $\cdot$ min^{-1}, but at the end of the 4-week sojourn at high altitude, it was raised to 65 liters $\cdot$ min^{-1} during the standard exercise. If oxygen is used during work at high altitude, it is certainly oxygen-saving if the individual is unacclimatized. Since, on the other hand, oxygen can scarcely be supplied continuously for long periods of time, an acclimatization is desirable, especially at very high altitudes. Furthermore, the oxygen-providing equipment may fail. (*c*) When the subjects returned to sea level following exposure to altitude, it took several weeks before the control level was attained (see Buskirk et al., 1967; Forster et al., 1971). The return of the alveolar P_{CO_2} to control levels paralleled the shift in pulmonary ventilation.

A calculation of the pulmonary ventilation at STPD gives practically the same volumes at a given oxygen uptake in subjects acclimatized to various

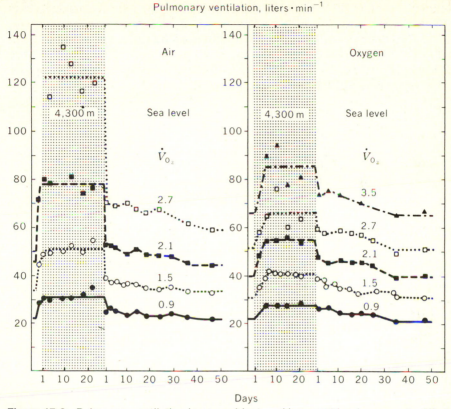

Figure 17-6 Pulmonary ventilation in one subject working on a bicycle ergometer (1) at sea level (base line to the left), (2) during a 4-week sojourn at an altitude of 4,300 m (shaded area), and (3) again after return to sea level including almost 50 days' observation time. Oxygen uptake is indicated on respective line. *Left panel:* Subject breathing room air; *Right panel:* Subject breathing oxygen during exercise. *(Unpublished data obtained by I. Åstrand and P.-O. Åstrand at White Mountain Research Station, California, 1957.)*

altitudes for about 4 weeks (Christensen, 1937; Dejours et al., 1963; Pugh et al., 1964). In other words, the number of oxygen molecules which are inhaled per unit of time is constant at the various altitudes. It should be remembered, however, that the alveolar oxygen pressure will still not be the same at high altitude as at sea level.

At high altitude, the energy cost for the respiratory muscles to move the greater volume of air may not be higher than at sea level because of the reduced density of the air. When sea level is again reached, the "abnormally" high ventilatory response at a given oxygen uptake must require extra respiratory work.

It has been suggested that the extra increase in ventilation in connection with work at high altitude must be attributed to a hypoxic drive via the peripheral chemoreceptors. This hypoxic drive is prevalent even during chronic hypoxia, at least for a considerable period of time (P.-O. Åstrand, 1954; Dejours et al., 1963). The induced alkalosis on sudden exposure to low P_{O_2}, reducing the central chemoreceptor drive, becomes gradually compensated by a proportionate decrease in the blood bicarbonate, and restoration of a normal pH occurs in the acclimatized person. Prolonged alteration in arterial P_{CO_2} in either direction tends to bring about alterations in the renal acid-base excretion that slowly tend to return the arterial H^+ toward normalcy. As the alkalosis is reduced (pH is lowered), there is a further increase in pulmonary ventilation, as illustrated in Fig. 17-6. However, a restoration of CSF [H^+] does not sufficiently explain the hyperventilation obtained upon sojourn to high altitude (Dempsey et al., 1972).

Evidence of a real change in the regulation of the body P_{PO_2} developing during the first week after ascent from sea level to high altitudes is presented in Fig. 17-7. The respiratory response to CO_2 was tested by breathing CO_2-O_2 mixtures during work with 100 watts (oxygen uptake, 1.5 liters · min^{-1}). There was a marked shift to the left so that, after acclimatization, a given pulmonary ventilation was attained at 15 to 20 mm Hg lower alveolar P_{CO_2} compared with controls at sea level. The difference in the response to CO_2 can be illustrated by the following example: At sea level, a pulmonary ventilation of 35 liters · min^{-1} was obtained when the end-expiratory P_{CO_2} was 45 mm Hg; after 1 week at 4,300 m, the same P_{CO_2} was recorded when ventilation was as high as 83 liters · min^{-1}. (In this subject, the high CO_2 mixture had an almost narcotic effect, with a reduced ventilatory response as a consequence; see days 15 and 19 at high altitude, and day 1 at sea level.)

After the return to sea level, there was a gradual return of the CO_2 response curve to the control level. This slow reacclimatization shows a pattern parallel with the one illustrated in Fig. 17-6. When the hypoxic drive was reduced by the increased oxygen pressure in the

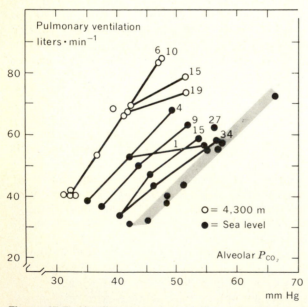

Figure 17-7 Ventilatory response during a standard work load (100 watts) to inhalation of various CO_2-O_2 mixtures at sea level (shaded area), during a prolonged sojourn at altitude of 4,300 m (unfilled dots), and at various intervals during the reacclimatization to sea-level conditions (filled dots). Figures at top of lines denote the day for the experiment after arrival at altitude and sea level respectively. (Same subject as in Fig. 17-6). *(Unpublished data by I. Åstrand and P.-O. Åstrand.)*

inspired air, the pulmonary ventilation was reduced. This produces a reduced washout of CO_2 and an uncompensated acidosis with increased central chemoreceptor drive as a consequence. (For a more detailed discussion of the regulation of respiration during hypoxia, see Kellogg, 1964.)

There is another effect of the *reduced alkaline reserve* in the blood of the person acclimatized to high altitude. The individual will have less ability to withstand acidosis from other acids which may arise in the course of metabolism (Roughton, 1964). During acute exposure of a few weeks' duration to high altitude, the person can apparently attain the same high blood lactate level as at sea level, but data by Edwards (1936) suggest a gradual decline in this maximum (Hansen et al., 1967a). The reduced alkaline reserve may be one factor behind this decrease in maximal anaerobic power.

Natives and long-term residents at high altitude have a lower pulmonary ventilation than the short-term acclimatized individual. Forster et al. (1971) found that the maximal ventilation increase in response to hypoxia usually occurred during the first 3 to 4 weeks at altitude, and then it began to decrease toward sea-level values. The highlanders were also less responsive to CO_2. With a lower exercise $\dot{V}_E$ the alveolar CO_2 tension is higher. The ventilatory response to a removal of hypoxemia is less in residents at high altitude than in individuals who have spent limited time at high altitude (see also Dempsey et al., 1972). There seems to be a gradual adaptation of the peripheral chemoreceptors (Severinghaus et al., 1966; Milledge and Lahiri, 1967).

2 With regard to the transport of O_2 from the alveolar air to the pulmonary capillary blood, the situation is somewhat controversial. West (1962) found no change in the diffusing capacity after a 6-month exposure to 5,800-m altitude. Natives and long-term residents at high altitude may have a greater diffusing capacity than comparable sea-level residents (Velasquez, 1959; DeGraff et al., 1965 Dempsey et al., 1971; Guleria et al., 1971; Cruz et al., 1975). Kreuzer et al. (1964) found that the values of the alveolar-arterial P_{O_2} gradient in the Andean natives was higher than in normal sea-level residents, which is in contrast to Hurtado's (1964) observation of a particularly small gradient in his native subjects in the same area. Similar findings of a relatively narrow alveolar-arterial O_2 pressure gradient in natives and long-term residents at high altitude are reported by Dempsey et al. (1971) and Cruz et al. (1975). Frisancho (1975) points out that the vital capacity and residual lung volume are larger in highland natives than in subjects from low altitudes. An exposure to hypoxia during the developmental period, i.e., during childhood, may be essential. A greater alveolar area and an increased capillary volume would facilitate gas diffusion in the lungs.

3 Pugh et al. (1964; Pugh, 1964) have done hemodynamic studies including maximal work at very high altitudes. They report that a prolonged sojourn at various altitudes brought the *cardiac output* at a given work load down to the level typical for the same work load performed at sea level. However, the maximal cardiac output was markedly reduced, and after several months' stay at 5,800 m, the values were 16 to 17 liters · min^{-1} compared with 22 to 25 liters · min^{-1} at sea level. This reduction of cardiac output was a combined effect of a lowered stroke volume and maximal heart rate (reduced from 192 down to 135 beats · min^{-1}). This study confirms the data by Christensen and Forbes (1937).

A number of measurements of cardiac output during work at altitudes between 3,000 and 4,300 m have been made with exposure up to a few weeks (Klausen, 1966; Alexander et al., 1967; Hartley et al., 1967; Vogel et al., 1967; Saltin et al., 1968; Vogel et al., 1974b). The results indicate that after a few days, the minute volume of the heart during submaximal work is already reduced, compared with the cardiac output during acute exposure to the hypoxic condition, and that it returns gradually to values typical for sea-level conditions or it may even become subnormal. During maximal work, the cardiac output is reduced. A reduced stroke volume appears to be the primary reason for the reduced cardiac output; the lowering of the heart rate, in any case during maximal work, is a more inconsistent finding. (Figure 17-4 shows that the stroke volume during light work is reduced at acute exposure to the hypoxia.) Grover et al. (1976) rule out myocardial hypoxia as a basis for the decrease in stroke volume.

Vogel et al. (1974a) studied eight subjects, who were native to an altitude of 4,350 m, at that altitude and also at sea level. At sea level, cardiac output was the same, heart rate was less, and stroke volume was greater at rest and during submaximal exercise than was observed at altitude. At sea level the maximal oxygen uptake increased from an average of 2.97 to 3.25 liters $\cdot$ min^{-1} (9 percent), which was due to a greater maximal cardiac output (an 8 percent increase). (It is an open question as to why the a-$\bar{v}$ O$_2$ difference was not greater during maximal exercise at sea level than at high altitude. The oxygen content of arterial blood was namely 23.1 ml $\cdot$ 100 ml^{-1} at sea level and 20.0 ml at altitude). In these subjects the arterial blood pressure and peripheral resistance were higher at altitude than at sea level. This is in contrast to other studies indicating that the systemic blood pressure in adult highland natives is lower than in lowland natives at sea level (see Frisanchno, 1975).

An example of the heart rate response to fixed work loads was presented by P.-O. Åstrand and I. Åstrand (1958); see Fig. 17-8. During acute exposure to a tracheal oxygen tension of about 85 mm Hg (4,300 m), the heart rate was 15 to 30 beats higher per minute than at sea level. When the hypoxia was prolonged, there was a gradual decrease in heart rate at a given oxygen uptake. In the later stage of acclimatization, the heart rate attained during lower levels of work fell in the same range as those recorded at sea level. At the heavier loads, however, the heart rate was even lower than in experiments with high tracheal P_{O_2}. In this subject, the normal maximal heart rate was about 190 beats $\cdot$ min^{-1}. At the high altitude, it gradually declined to 135 (a decline in maximal heart rate at prolonged exposure to altitudes exceeding 3000 m has also been reported by Christensen and Forbes, 1937; Cerretelli and Margaria, 1961; Pugh et al., 1964; Cerretelli, 1976). When the subject was allowed to breath 100 percent oxygen during almost maximal work, the heart rate increased within seconds by as much as 25 beats $\cdot$ min^{-1}. Pugh et al. (1964) noticed that a maximal heart rate of 130 to 150 was elevated to almost sea-level values when the subjects were breathing oxygen. Hartley et al. (1974) noted that intravenous atropine increased the maximal heart rate in subjects exposed to 4,600 m altitude (a mean decrease of 24 beats $\cdot$ min^{-1} was reduced to 13 beats $\cdot$ min $^{-1}$ by atropine). An increased parasympathetic activity during prolonged exposure to hypoxia may partly explain the reduction in maximal exercise heart rate.

Within the first few days at high altitude, the *hemoglobin concentration* in

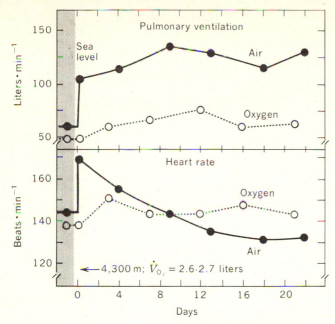

Figure 17-8 Pulmonary ventilation and heart rate during exercise on a bicycle ergometer, work load 200 watts at sea level, and during a 22-day sojourn at an altitude of 4,300 m. On some days, room air was inhaled during work; on other days, pure oxygen was taken. (Same subject as in Figs. 17-6 and 17-7.) *(Modified from P.-O. Åstrand and I. Åstrand, 1958.)*

the blood increases, but this increase is mainly due to a hemoconcentration secondary to a decrease in the plasma volume (Merino, 1950; Surks et al., 1966; Buskirk et al., 1967). Gradually the increased erythrocytopoiesis brings hemoglobin content to high levels so that the oxygen content per liter of arterial blood may be the same in the acclimatized person at 4,500 m as it is in a person at sea level (Christensen and Forbes, 1937; Hurtado et al., 1945; Reynafarje, 1967). At an altitude of 4,500 m at Morococha in Peru, the native residents had a hemoglobin concentration which averaged 20.8 g · 100 ml^{-1} blood (Hurtado et al., 1945).

As a consequence of an increased hemocentration during acclimatization to high altitude, the oxygen offered to the tissues per liter of arterial blood may be the same in the individual living at high altitude as in the resident at sea level. However, these values of oxygen are really not comparable. The oxygen pressure gradient between blood and tissue is most important for the final transfer of oxygen to the mitochondria, and the gradient is reduced during hypoxia. However, the gradual decline in cardiac output during prolonged exposure to a low oxygen pressure in the ambient air can be partially explained by the concomitant rise in the oxygen-combining capacity of the blood. The increased viscosity of the blood with the elevated hematocrit must necessitate an increased cardiac work at a given cardiac output, but the net effect of the hematologic response to prolonged hypoxia in terms of work of the heart cannot be evaluated at present. In any case, the increase in hemoglobin concentration and the mentioned shift in the operational range to the steeper slope of the oxygen-dissociation curve provide major contributions to the gradually increased oxygen conductance within the body at altitude. During chronic hypoxia there is also an increase in the 2, 3-DPG concentration in the blood which enhances the unloading of oxygen to the tissues (see p. 245). However, the opposite effect on the oxygen dissociation curve of reduced P_{CO_2} may cancel out the 2, 3-DPG effect (Morpurgo et al., 1972).

4   Barbashova (1964) has summarized various aspects of cellular adaptation to high altitude. It is concluded that the oxygen-utilization efficiency for an aerobic energy yield increases at a low O_2 tension; this increase must be interpreted as an adaptation at the enzyme level. The ability to tolerate lack of oxygen increases, furthermore, in connection with an acclimatization. Vannotti (1946) and Cassin et al. (1966) report that an increased capillarization takes place after a period of acclimatization to high altitudes. An increase in the number of capillaries reduces the distance between the capillary and the most distant cells within its tissue cylinder. A relatively low O_2 tension in the capillaries may therefore still provide the oxygen supply to these distant cells. Rahn (1966) emphasizes that an increase in the number of open capillaries plays a most important role not only in daily life at sea level but particularly during acclimatization to high altitude by changing the O_2 conductance.

Reynafarje (1962) reports that the myoglobin content in the skeletal muscles increases during altitude adaptation; this will have a favorable effect on the O_2 transport.

The critical alveolar P_{O_2}, at which an unacclimatized person loses consciousness within a few minutes in acute exposure to hypoxia, is 30 mm Hg, with minimal individual variations (Christensen and Krogh, 1936). This limit is set at an altitude of slightly more than 7,000 m. Down to this low P_{O_2}, the demand of the nerve cells for oxygen can apparently be maintained (Noell, 1944). The well-acclimatized individual can spend hours at an altitude above 8,000 m breathing the ambient air. This must be an example of adaptation on the cellular level. The comprehension of the last step in the oxygen transfer from air to tissue, i.e., from the capillary to the mitochondria, is, however, still very incomplete.

5   There is one more finding in high-altitude dwellers that should be mentioned. Residents at altitudes of around 3,500 m or above develop a pulmonary hypertension with increased pulmonary vascular resistance and hypertrophy of the right ventricle of the heart (Penaloza et al., 1963; Hultgren et al., 1965; Frisancho, 1975). This phenomenon does not change rapidly when returning to lower altitudes. Lockhart et al. (1976) noted that raising the arterial oxygen pressure to normal sea-level values had no effect on the pulmonary circulation at rest but prevented, to a large extent, the rise in pulmonary arterial pressure during exercise. This type of hypoxia modification of the cardiopulmonary system, as well as other changes in connection with adaptation to hypoxia, is believed to occur primarily in the fetal and infant period and is retained in adult life. The physiological consequences and the significance of these findings are not revealed. One effect might be a more even ventilation/perfusion ratio in the lungs which reduces the difference in oxygen pressure between alveolar air and arterial blood (Bisgard et al., 1974; Cruz et al., 1975; Chap. 7, p. 242).

Summary It may be stated that with prolonged exposure to reduced oxygen pressure in the inspired air, the compensatory devices are more slowly acquired, such as (1) a further increase in pulmonary ventilation (in long-term dwellers at high altitude it is followed by a significant reduction in ventilation); (2) an increased hemoglobin concentration in the blood; and (3) morphological and functional changes in the tissues (increased capillarization, myoglobin content, modified enzyme activity). The initially observed increase in cardiac output during exercise is replaced by a gradual decline to, or even below, the values observed at sea level. During submaximal, as well as maximal, work the stroke volume becomes reduced. If the sojourn has been at very high altitude, 4,000 m or higher, the maximal heart rate may become reduced compared with sea-level values. All these adaptive changes are reversible, but it may take several weeks before the values return to sea-level values in the case of sea-level dwellers who have stayed at high altitudes for a month or more.

The net effect of this acclimatization to high altitude is a gradual improvement in the physical performance in endurance events or prolonged work. An increased oxygen availability to the working muscles is important for this improvement (see Pugh, 1965; Saltin, 1966; Maher et al., 1974). The maximal anaerobic power after a prolonged period of acclimatization has not been carefully analyzed.

It is also true that the oxygen content per liter of blood eventually increases, but the maximal cardiac output apparently is reduced at about the same rate. It should be pointed out that the well-trained individual is not acclimatized to high altitude any sooner or any more effectively than the untrained individual. The fact that acclimatization of highland natives does not depend on hyperventilation is perhaps due in part to their enlarged lung volume which facilitates the receiving of an adequate oxygen supply at the alveolar level (Frisancho, 1975). The earlier the age when one becomes a sojourner at high altitudes, the greater the influence of the environment modifying the expression of inherited potential. The optimum seems to be during growth and development.

Saltin (1966) followed a top athlete during a 3-week period at the Mexico City altitude. His initial decrease in maximal oxygen uptake of 14 percent became less during the stay in Mexico City, but it was still 6 percent below the sea-level value after 19 days. The same trend was observed for the oxygen uptake measured during maximal efforts in a canoe. In a group of eight international top athletes, the reduction in maximal oxygen uptake at the altitude of 2,300 m averaged 16 percent (ranging from 9 to 22 percent); after 19 days at this altitude, the maximum was still 11 percent below the sea-level average, with a range of 6 to 16 percent (Saltin, 1967). This example illustrates the individual variations in response to hypoxic conditions.

Performance after Return to Sea Level

Opinions differ concerning the question of whether the performance capacity at sea level is improved following exposure to high altitude, or whether a certain amount of training at high altitude is more effective than the same amount of training at sea level (Goddard, 1967). The above-mentioned subject, studied by Saltin (1966), was the best adapted athlete in the Swedish test group sent to Mexico City in 1965. When he returned to Stockholm after the 3-week Mexican sojourn, his maximal oxygen uptake was no higher than before the trip. Buskirk et al. (1967), as well as Consolazio (1967), state that their subjects who stayed at altitudes up to about 4,000 m for 4 weeks or more did not attain any better results than usual when they returned to sea level. The measured maximal O_2 uptake was not improved. Buskirk et al. conclude that there is little evidence to indicate that performance on return from altitude is better than before going to high altitude, if training remains relatively constant. One should view with caution any statement about the physical superiority of the indigenous resident in any environment until that person can be compared with individuals of similar training and experience (and physical endowment).

Grover and Reeves (1967) also came to similar conclusions when studying the exercise performance at altitudes of 300 and 3,100 m in men native to those two altitudes. The maximal oxygen uptake was 26 percent less at the higher altitude for both groups; the sea-level athletes won all track competitions at both low and high altitudes; their performance after return to low altitude was not improved by their sojourn at medium altitude. (See also Hansen et al., 1967b; Kollias et al., 1968; Adams et al., 1975.)

Cerretelli (1976) reports studies on subjects acclimatized to high altitude (5,350 or higher) for about 4 months. When studied after their return to sea level (13 subjects), there was no significant increase in maximal oxygen uptake compared with the data obtained before departure. The hemoglobin concentration was still 12 percent higher than before the sojourn at high altitude. It is difficult to explain why an elevation of the hemoglobin content, after a prolonged sojourn at high altitude, does not effectively improve the maximal aerobic power. This is in contrast to the beneficial effect of an acute increase in the hematocrit by reinfusion of red cells (Chap. 6, p. 184; see Fig. 17-9).

It is true that both training and prolonged exposure to hypoxia produce similar changes with respect to increased vascularization in the skeletal

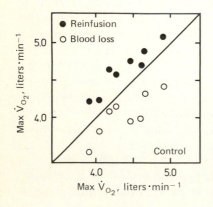

Figure 17-9 The effect of bloodletting (800 ml) and subsequent reinfusion one month later of the same red blood cells on maximal oxygen uptake in eight subjects (three in one study and five in a subsequent study). *(From Ekblom et al., 1976.)*

muscles, increased myoglobin content, and possibly similar changes in the oxidative transport system. On the other hand, training at sea level produces no increase in the Hb concentration. Exposure of more than a few days' duration to high altitudes evidently hinders the attainment of a maximal stroke volume, and the increased pulmonary ventilation following return to sea level represents no advantage. The same is true for the reduced alkaline reserve in the blood, with reduced buffer capacity for lactic acid as a consequence. The training intensity has to be reduced at high altitude. For these reasons the improvements attained at high altitude cannot be transferred directly to sea-level conditions.

It should be emphasized that few records were broken in swimming and middle and long distance running when the athletes returned to sea level after the pre-Olympic Games in Mexico. Apparently performance at sea level is not enhanced by prolonged exposure to hypoxia.

Practical Applications

In order to attain top achievement at altitudes of 2,000 m or higher in activities requiring the engagement of the maximal aerobic power, an acclimatization period of 2 to 3 weeks seems necessary. At lower altitudes, the time required is probably less. A longer exposure to high altitude would probably be beneficial from a physiological point of view, but this advantage must be considered against possible psychological, social, and economic factors. After the initial acclimatization, the improvement per week is so small that it may easily be concealed by day-to-day variations in the physical fitness.

Athletes competing in events where the technique is of prime importance or in events primarily involving anaerobic metabolic processes may arrive at more or less the time of the competition if the altitude is not so high that mountain sickness may be expected.

A theoretical consideration shows that the performance capacity in certain events ought to be better in competitions at high altitude (e.g., 100- to 400-m, bicycling). Dickinson et al. (1966) state, on the basis of ballistic calculations, that at an altitude corresponding to Mexico City, one might expect an improvement of 6 cm in shot-putting, 53 cm in throwing the hammer, 69 cm in javelin throwing, and 162 cm in discus throwing. In view of the fact that throwers are used to mastering different wind conditions, the reduced density of the air at high altitude certainly does not require any special period of preparation.

There is no evidence to suggest that it is necessary to take it easy during the initial period of exposure to high altitude. It is necessary, however, to become accustomed to the fact that the subjective feeling of fatigue is different, which must be reflected in the choice of tactics. Experience has shown (especially in the case of cross-country skiers) that if the effort is too intense, a considerably longer recovery time will be required than is needed at sea level.

One is forced to accept a slower tempo, and the intensity and duration of training activities must be reduced. Swimmers discover that they can remain under water for a shorter time after turning than they normally do, and they must adapt their swimming strokes to a different breathing rhythm. The ability to tolerate an intense tempo for long periods of time at high altitudes is different from one individual to the next. This complicates the selection of the athletes for a team. There are examples of outstanding athletes in long-distance events at sea level who consistently fail at high altitude. In competitions at high altitude, collapse from unknown causes occurs more frequently than it does at sea level. This fact, however, does not appear to represent an increased health hazard. Ingestion of additional fluid and possibly also carbohydrates may be required during exposure to high altitude when it is combined with heavy physical work. There is no evidence to suggest that the preparation for competitions which place heavy demands on the aerobic power should include training at an altitude which is higher than that of the actual place of competition. Pugh (1964) states that, after suitable acclimatization, people can, by their own efforts and without supplementary oxygen, ascend to about 8,600 m (28,200 feet). It is a pity, in a way, that Mount Everest has an altitude of 8,848 m; it is true that the peak has been reached by climbers breathing oxygen, but from a sporting point of view, Mount Everest is still unconquered!

HIGH GAS PRESSURES

Although humans can become acclimatized to low air pressures, there is no way to become biologically acclimatized to high air pressures, such as those encountered in deep sea diving and when a submarine crew tries to escape from the inside of the craft, where the pressure is normal, to the surface through the sea where the air pressure is higher. (For a more comprehensive review of the subject, see Lambertsen, 1967; Fagraeus, 1974.)

Pressure Effects

For every 10 m (33 ft) of sea water the diver descends, an additional pressure of 1 atm is acting upon his or her body. Small changes in sea depth thus bring about great pressure changes. The effect of changing pressures on the blood P_{O_2} and P_{CO_2} and its consequence for underwater swimming and breath holding were discussed in Chap. 7. The body may tolerate high pressures as long as the pressure is the same inside and outside the body. When diving with a snorkel connected to the mouth, one maintains the atmospheric pressure in the lungs, while the surface of the thorax, in addition, is exposed to the pressure of the water. At a depth of about 1 m, the pressure difference becomes so large that the inspiratory muscles no longer have the strength to overcome the external pressure, and normal breathing becomes impossible. For this reason, a snorkel system does not permit diving to depths exceeding about 1 m. At greater depths, breathing apparatus has to be used in which the pressure in the system corresponds to that prevailing at the depth in question. If there is an overpressure in the system, the lung tissues may be damaged, with hemorrhaging as a consequence.

As the pressure increases, more gases can be taken up by the diver's body and dissolved in the various tissues. At a depth of 10 m, twice as much gas will be dissolved in the blood and tissues as at sea surface. This is apt to give the diver trouble, mainly because of the nitrogen.

Nitrogen

The problem with nitrogen is that it diffuses into various tissues of the body very slowly, and once dissolved, it also leaves the body very slowly when the pressure is once more reduced to the normal atmospheric pressure. This is especially bad when the pressure is suddenly reduced from several atmospheres, as may be the case during submarine escape or deep sea diving. Then the nitrogen is released from the tissues in the form of insoluble gas bubbles. These bubbles congregate in the small blood vessels, where they obstruct the flow of blood. This, then, gives rise to symptoms such as pains in the muscles and joints, and even paralysis may develop if the bubbles become trapped in the brain. These symptoms are known as the *bends*. Obviously, the severity of the symptoms depends on the magnitude of the pressure, which relates to the depth to which the person has descended under water, the length of time spent at that depth, and the speed of ascent to the surface.

The bends can be avoided to a large extent by a slow return to normal pressure so as to allow time for the tissues to get rid of their excess nitrogen without the formation of bubbles. Another way to avoid the bends is to prevent the formation of nitrogen bubbles by replacing atmospheric nitrogen by helium, which is less easily dissolved in the body. This is done by having the diver breathe a helium-oxygen gas mixture. Another advantage of this method is that it is more apt to prevent the so-called nitrogen narcosis which occurs when air is breathed at 3 atm or more, when there is an onset of euphoria and impaired mental activity with lack of ability to concentrate. With increasing pressures, the individual is progressively handicapped and may be rendered helpless at 10 atm. Diving to depths exceeding 100 m while breathing ordinary atmospheric air may thus be fatal.

Pilots flying high-altitude aircraft may also suffer from bends if there is a sudden loss of pressure in the pressurized cabin, but the symptoms in these cases are usually not so severe as in the divers, and they usually do not occur at altitudes lower than 3,000 m in any case.

Oxygen

Prolonged breathing of 100 percent oxygen at 1 atm may also be quite harmful (Lambertsen, 1965); irritation of the respiratory tract may occur after 12 hr and frank bronchopneumonia after 24 hr, and the peripheral blood flow (the flow through the brain) may be reduced. In most individuals, no harmful effects result from breathing mixtures with less than 60 percent oxygen, but newborn infants are particularly susceptible to oxygen poisoning and may suffer harmful effects with oxygen concentrations over 40 percent. The remarkable thing is that oxygen poisoning is apparently no problem when breathing 100 percent oxygen at altitudes over 6,000 m, no matter for how long. Oxygen poisoning, therefore, is not much of a problem in aviation medicine, but it is indeed an important problem in deep sea diving where it may even affect the brain function when pure oxygen is breathed under increased pressure. This latter form of oxygen toxicity is apt to occur in divers at depths greater than 10 m, but there are great individual variations in sensitivity to 100 percent oxygen. The onset of symptoms may be hastened by vigorous physical activity at great depths; it starts with muscular twitchings and a jerking type of breathing and ends in unconsciousness and convulsions. The exact cause of this is unknown, but it is assumed that it is a matter of interference with certain enzyme systems in the tissues.

When a person breathes pure oxygen at a pressure of 3 atm or higher, the dissolved oxygen covers the oxygen need of the body at rest. No oxygen would be removed from HbO_2 during its passage through the capillary bed. Therefore the hemoglobin of the venous blood would still be saturated with oxygen, which would interfere with the amount of H^+ ions taken up by Hb, a weaker acid than HbO_2 (Chap. 5). Thus, CO_2 entering the blood from metabolizing cells would raise the blood P_{CO_2}, and the H^+ concentration would be higher than under normoxic conditions when the desaturation of HbO_2 simultaneously favors the removal of H^+ ions. The end result would be a CO_2 retention in the tissues and an acidosis.

When breathing air at a hyperbaric pressure during heavy exercise, the increased air density reduces the pulmonary ventilation. Secondarily the maximal oxygen uptake may be reduced. By substituting nitrogen with helium, with a lower density, the pulmonary ventilation goes up, and the oxygen uptake may be higher than the maximum measured during normal atmospheric conditions (Fagraeus, 1974).

Carbon Dioxide

If the pressure of the carbon dioxide in the respiratory air is increased, as it is when the absorption of CO_2 fails in a closed breathing system, CO_2 pressure may be reached (about 75 mm Hg or higher), which produces a narcotic effect.

One effect of *reduced* CO_2 pressure in the blood as a consequence of hyperventilation was discussed in Chap. 7 (p. 261): Breath holding during exercise can, after such a procedure, be prolonged until unconsciousness, which induces a potential danger when practiced during underwater swimming (see Craig, 1976).

Oxygen Inhalation in Sports

In athletic events, oxygen breathing is of very limited value. Whereas, with a lung volume of 5 liters, one normally has a supply of about 0.8 liter of oxygen, this oxygen volume may be increased to about 4.5 liters if 100 percent oxygen is inhaled. This volume may be consumed by the cells within less than a minute during heavy exercise. In most events the weight of the equipment for an adequate oxygen supply would be a definite handicap and, if used during competition, it would disqualify the athlete because the additional oxygen supply would be considered doping.

After a few breaths of atmospheric air, the oxygen is quickly diluted, since the nitrogen content of the air is as much as 79 percent. The elimination of anaerobic metabolites is not speeded up if pure oxygen is inhaled during recovery.

Oxygen inhalation may improve the performance capacity just prior to swimming under water and when there is a need to increase the supply of oxygen *during* heavy muscular work. (It increases the maximal oxygen uptake, reduces the anaerobic participation in the energy yield, and reduces the pulmonary ventilation, see Chap. 6, p. 184; Chap. 9, p. 317). The "sniffing" of oxygen practiced by football players during the rest periods may perhaps have a psychological effect. From a physiological point of view, however, the extra oxygen is definitely wasted.

TOBACCO SMOKING

Larson et al. (1961) have prepared a very extensive summary on this subject. Only a few aspects of tobacco smoking and physical performance will be discussed here.

Circulatory Effects

Tobacco smoke contains up to 4 percent by volume of carbon monoxide. By inhalation, some of this carbon monoxide is absorbed. The affinity of the hemoglobin to carbon monoxide is 200 to 300 times greater than to oxygen. The presence of even small amounts of carbon monoxide may, therefore, noticeably reduce the oxygen-transporting capacity of the blood. Carbon monoxide also interferes in a negative way with the unloading of oxygen in the tissues by shifting the oxyhemoglobin dissociation curve to the left (Chap. 6). A study has shown that subjects who smoked 10 to 12 cigarettes a day had 4.9 percent carbon monoxide hemoglobin, those who smoked 15 to 25 cigarettes a day had 6.3 percent, and those who smoked 30 to 40 cigarettes per day had 9.3 percent. Other studies have confirmed these findings; it may take 1 day or more for the carbon monoxide content of the blood to return to normal (Larson et al., 1961). This amount of carbon monoxide in the blood gives no subjective symptoms at rest, as long as the concentration in the blood is below 10 percent. The adverse effect is noticeable only during physical exertion. A blocking of 5 percent of the hemoglobin by CO will reduce the maximal oxygen uptake and performance (see Ekblom and Huot, 1972). There is no way to compensate for a reduced oxygen content in arterial blood during maximal exercise.

In a study, normal subjects worked on a bicycle ergometer without having smoked 12 hr prior to the experiment and immediately after smoking one or two cigarettes (Juurup and Muido, 1946). Oxygen uptake, pulmonary ventilation, and respiratory rates were unaffected by smoking. A slight rise in arterial blood pressure was noted, but the difference was not significant. At a fixed oxygen uptake, the heart rate was 10 to 20 beats $\cdot$ min^{-1} higher when the work test was preceded by smoking. The difference in heart rate between smokers and nonsmokers was greater the higher the work load. The effect of smoking could not be observed if the subject waited 10 to 45 min after smoking before the work test started (the exercise was at a submaximal level).

Respiratory Effects

In Chap. 7 (in the section "Airway Resistance"), it was pointed out that inhalation of smoke from a cigarette could, within seconds, cause a two- to threefold rise in the airway resistance. Da Silva and Hamosh (1973) also observed an increase in airway resistance (measured by body plethysmography) in subjects after smoking one cigarette. In addition to this acute effect, smoking also causes a more chronic swelling of the mucous membranes of the airways, leading to an increased airway resistance. At rest when the pulmonary ventilation is less than 10 liters $\cdot$ min^{-1}, the increased airway resistance is not noticeable, however. When the demand on respiration is elevated, the increased respiratory resistance caused by smoking may be noticeable. A reduced pulmonary ventilation capacity may cause a smaller volume of oxygen to reach the alveoli, resulting in an impaired gas exchange. The effect may be subjective symptoms of distress.

Smoking Habits among Athletes

A study of 285 top athletes in Great Britain (Report in *JAMA*, **170:**1106, 1959) showed that 16 percent of them smoked, but only 4 percent of them smoked more than 11 cigarettes per day. These findings are in agreement with the results of similar studies of top athletes elsewhere. Most of the smokers are among those engaged in athletic events requiring skill rather than endurance. None of the middle- and long-distance runners or swimmers were smokers. Among skiers, some of the ski jumpers and downhill skiers may smoke, but none of the cross-country skiers of the elite class are smokers. This may indicate that under conditions where the demand on the oxygen-transporting system is very great, smoking has been found to impair performance. If the current trends continue, even athletes participating in events primarily requiring skill will be required to undergo strenuous physical training which taxes the oxygen-transporting system to a considerable degree. This may make it desirable for the athlete to refrain from smoking while training, even though tobacco smoking at the time of the competitive event may not affect the outcome.

ALCOHOL AND EXERCISE

Neuromuscular Function

It is well established that alcohol may temporarily cause impaired coordination. The performance of rather simple movements is used to test whether or not an individual is under the influence of alcohol (walking on a straight line, touching the tip of the nose with the index finger with the eyes closed, etc.). A precise assessment in borderline cases is impossible, however, The tolerance of alcohol varies greatly from one individual to another. Ikai and Steinhaus (1961) noted that the maximal isometric muscle strength could actually be improved in some cases, especially with untrained subjects, after moderate alcohol consumption. They explain this on the basis of a depressing influence of alcohol on the central inhibition of the impulse traffic in the nerve fibers to the skeletal muscles during maximal effort. The result is an increased impulse activity and an increased strength (see Chap. 4). At times, individuals taking part in competitive events, such as target-shooting, maintain that they achieve better results following moderate alcohol consumption. They feel more relaxed. It is conceivable that the depression of the inhibiting effect of certain CNS centers may cause routine procedures to progress normally without the disturbing effect caused by the anxiety of the actual competition.

Aerobic and Anaerobic Power

Blomqvist et al. (1970) studied eight young male subjects during bicycle exercise at two submaximal and one maximal work load before and after peroral ethanol intake producing blood levels of 90 to 200 mg percent. Oxygen uptake and heart rate were determined. In four of the subjects, cardiac output, stroke volume, and intra-arterial pressure were also measured. Table 17-2 summarizes the most important changes. During maximal work with an O_2 uptake of about 3 liters $\cdot$ min^{-1} and a cardiac output of about 21 liters $\cdot$ min^{-1}, no difference was observed when the results before and after alcohol consumption were compared. Maximal heart rate, stroke volume, $a\text{-}\bar{v}O_2$ difference, and calculated peripheral resistance were also unaffected. During submaximal work, on the other hand, the heart rate was on an average 12 to 14 beats higher per minute in the alcohol experiments ($p < 0.01$). In the latter case, the cardiac output was greater while the stroke volume was unaffected. The O_2 uptake during submaximal work was slightly higher after alcohol, but the $a\text{-}\bar{v}O_2$ difference was nevertheless reduced (i.e., the cardiac output was more elevated than would be expected from the increased oxygen uptake). At rest and during submaximal work, the calculated total peripheral resistance was reduced.

Asmussen and Böje (1948) studied the effect of alcohol on the ability to perform one activity of 9.400 N $\cdot$ m and another of 96.700 N $\cdot$ m as fast as possible. The first type of work could be performed in 12 to 15 s, simulating a

Table 17-2   Hemodynamic Effects of Alcohol

	$\dot{V}_{02}$	Heart rate	Cardiac output	Stroke volume	$a\text{-}\bar{v}O_2$ difference	Total peripheral resistance
Rest, sitting	↑ *	↑	↑	O	↓	↓
Submaximal exercise	↑	↑	↑	O	↓	↓
Maximal exercise	O	O	O	O	O	O

*The arrows denote the observed changes, which are significant in all cases except the O_2 uptake.
Source: Blomqvist et al., 1970.

100-m sprint in its effect on the organism, whereas the latter, lasting about 5 min, simulated a 1,500-m run. A blood concentration of alcohol of up to 100 mg percent did not significantly affect the ability to perform this kind of maximal work.

Summary These studies have revealed an effect on the circulatory response to submaximal exercise in individuals with elevated blood alcohol level. Measurements during maximal work, however, showed no effect on oxygen uptake, cardiac output, heart rate, stroke volume, or total peripheral resistance. Nor was the maximal time the subjects could tolerate a standard load affected by the alcohol intake.

It should be pointed out that the question of dosage in the case of alcohol consumption in connection with athletic performance is a most difficult one. The tolerance varies greatly from individual to individual, and probably also from time to time in the same individual. Furthermore, the use of alcohol is definitely to be considered as a form of doping. It should also be kept in mind that alcohol ingestion may, in some cases, be followed by a subsequent spell of hypoglycemia.

DOPING

The use of stimulants of many kinds to improve physical performance has been practiced since ancient times. The Romans tried to increase the speed of their chariot racing by giving the horses a mixture of honey and water. The Indians of South America chewed coca leaves on their long, strenuous mountain journeys in order to enhance endurance and suppress the feeling of fatigue. With the increasing commercialism and professionalism in human competitive sports, the problem of doping no longer is limited to racing animals, but relates to athletes as well. This is a matter of considerable concern, for it represents an unnecessary risk for the health of the user, while any real advantage can be expected from it only in exceptional cases (Ariëns, 1965). This is true both in the use of drugs for the purpose of depressing the feeling of fatigue or of stimulating the central nervous system to attain greater "pep," and in the use of hormones or other products to augment muscle strength or physical capacity (amphetamine, cocaine, heroin, ephedrine, morphine and anabolic steroids).

The most commonly used drugs for doping are the psychostimulants, or "pep pills," such as amphetamine. In general, their main effect, at least as far as the athlete is concerned, is to suppress the feeling of fatigue and thus to permit individuals to exert themselves to complete exhaustion, and thereby to improve performance. Bucher and Smith (1965), in a double-blind study in highly trained swimmers, runners, and weight throwers, showed that amphetamine sulfate in a dose of 14 mg per 70 kg of body weight improved the measured performance significantly in about 75 percent of the cases. The weight throwers obtained the greatest improvement from amphetamine (3 to 4 percent); the runners obtained an improvement of approximately 1.5 percent; the swimmers showed varying degrees of improvement (0.6 to 1.2 percent). This does not mean, however, that amphetamine causes an improvement in performance which could not have been achieved without it had the individuals been properly motivated and able to mobilize all their capacities at their own will. It should be emphasized that maximal oxygen uptake (and cardiac output and arteriovenous oxygen difference) is attained at a submaximal effort. The maximal power of aerobic *and* anaerobic processes is difficult to measure in a reproducible manner. It is a general finding that the peak concentration of lactate in athletes' muscles and blood is lower in an "all out" experiment conducted in the laboratory as compared with a measurement in samples obtained after an important competition. Results from a laboratory test of the effect of various drugs cannot be directly applied to competitive situations.

Hanley (1972) reports observations from a world weight-lifting championship in 1970. In the first three classes eight of the nine who finished in the first three places turned in urine specimens positive for amphetamines. (They were disqualified, but those who finished fourth, fifth, and sixth replaced them as winners without having dope control tests.) In the next six classes there was only one specimen positive for amphetamines. It was pointed out that more records were actually broken in the classes where tests were negative than in the positive ones. The use of psychopharmaca is associated with the risk of addiction and of dangerous intoxications (Ariëns, 1965).

The use of anabolic steroids by athletes is based on the fact that these steroids, in controlled laboratory experiments, have been found to cause enhanced protein retention and increases in muscle mass and muscle strength. The reported results, however, are inconclusive. Thus several studies have indicated such positive relationships, but other investigators have failed to find such a relationship (for references, see Johnson et al., 1975; Lamb, 1975). One problem is the difficulty involved in measuring muscle strength, especially in standardizing the subject's state of motivation. An anabolic effect may be achieved by doses of 10 to 20 mg per day. The effect is of short duration, however, since the ingested steroids may cause the endogenous production to be reduced. According to Oseid (1976), athletes have been known to use doses up to 200 or 300 mg per day. The results of such overdosage include not only an increase in muscle mass, but also an increased appetite leading to vastly increased food intake, amounting to as much as 25 to 33.5 MJ a day. This causes weight gain, and the added weight is localized primarily to the trunk, especially the neck, shoulders, chest, and arm regions, a result that may appear favorable for wrestlers, weight throwers, etc. This has led to a considerable abuse of anabolic steroids by athletes. The medication may give a sense of well-being and may also give the athlete more "appetite" for training. The side effects are serious, however. The endogenous hormone production is suppressed. The production of sperm cells may be reduced, sterility may result, and the hormonal balance may be upset, thus producing a variety of complications. In children, these steroids may cause a premature closure of the epiphysis of the long bones, leading to cessation of growth. In women, they may cause various degrees of masculinity, with changes in hair growth, voice, clitoris, and the production of the female sex hormones. These steroids may also cause liver changes and undesirable changes in the fat and carbohydrate metabolism. They also cause a stronger tendency to injuries of the ligaments and tendons due to a too rapid increase in muscle strength without a corresponding development of ligaments and tendons. All these possibilities clearly show that the use of anabolic steroids is dangerous (see Johnson, 1975). As pointed out by Lamb (1975): "It is hard to believe that such powerful drugs will not have damaging side effects with prolonged use." To summarize, we may again quote Lamb: "Until more conclusive evidence is presented an evaluation of the scientific evidence and large mass of testimonial evidence leads one to believe that anabolic steroid administration for 3–6 weeks often will be accompanied by extra gains in strength, body weight, and lean body mass if the recipient also participates in a program of intensive strength training and ingests a protein-rich diet."

Another group of drugs which have been used for doping purposes lately are the anticholinesterases: pyridostigmine and neostigmine (Mestinon). The effect is rather complicated, and overdosage may cause dangerous side effects. These drugs affect the motor and plate, causing a delay in the action of the cholinesterase, which normally would cause an instantaneous breakdown of acetylcholine. The general effect is claimed to be diminished fatigue and enhanced endurance. The danger is that large doses may have a paralyzing effect on the central nervous system leading to collapse.

In a study of eight subjects, Ekblom et al. (1976) have shown that bloodletting (800 ml blood) followed by reinfusion of the same red blood cells a month later caused a significant increase in maximal oxygen uptake (Fig. 17-9). The effect may be explained on the basis of the achieved increase in the transporting capacity of the blood as the result of an ultimate increase in the number of red blood cells. (See discussion on p. 184.) (Williams et al., 1973, could not confirm that reinfusion of red cells improved physical performance. Unfortunately, they did not measure the oxygen content of arterial blood, and there is no proof that they were able to maintain the oxygen binding potential of the stored blood.)

In any case, a transfusion of blood or red cells to an athlete in connection with a competition is a violation of the rules: "Doping is defined as the administration or use of substances in any form alien to the body or of physiological substances in abnormal amounts and with abnormal methods by healthy persons with the exclusive aid of attaining an artificial and unfair increase in performance in competition. Furthermore, various psychological measures to increase performance in sports must be regarded as doping." (Statement by the International Olympic Committee.)

THE WILL TO WIN

Not everyone can win. Certainly the same person cannot win all the time. For every winner there is at least one loser, and to lose, at least at times, is part of life.

This is especially true in athletic competitions. Not all people are equally fit to win Olympic laurels or to rank among the best in any athletic competition. Some reach the top, but they are only a very small percentage of all those who try. The rest of them may still aspire to win for their own sake by striving to reach personal goals in performance. They can at least reach as high a level of performance as their own endowment will permit. Such an achievement is in itself a gratifying experience.

The attainment of athletic championship requires that the athlete meets a number of requirements. He or she must be endowed with the necessary talents to start with. But talent alone is not enough. Champion-quality athletes must master the proper techniques and have the suitable tools or equipment. And they must subject themselves to arduous training as well. But, above all, they must have the proper motivation, not only to apply all their resources in the final test during competition, but also to endure the hardship of their training. They need to prepare themselves, to gain the necessary competitive experience, and to master the art of competing. They have to nurse their health, and must be prepared to accept the day-to-day variations in their level of fitness. Finally, they need a certain amount of good luck. It is said that chance favors only the prepared mind. It is equally true to say that the athlete must be prepared to win when favored by chance.

The art of competing is the skillful application of all these factors with the single purpose of outperforming the competitors. This art includes the proper application of the stress hormones, aimed at preparing the organism for the supreme effort. The competitor must master the body's regulatory mechanisms, so as to allow the hormone-producing glands to be stimulated to the optimal level at the proper time. Some can, others cannot, do this; individuals differ also in this respect. It is in any case a matter of experience and adaptation to the stress of competing, which can be gained only through regular participation in competition.

So much depends on psychic factors. Some athletes are able to mobilize all their resources in an almost superhuman effort during an important competition. They can reveal an amazing ability to concentrate, to maintain a perfect control of the most intricate coordinated motions of the body and its various parts, to the almost complete exclusion of all extraneous disturbing factors, including the spectators. Others fail in this. They allow themselves to become tense; their muscles stiffen, their neuromuscular machinery fails at the crucial point. Such persons may win during training or at minor competitive events when they are naturally relaxed, but they fail when it matters, in the qualifying meet or during the major events. This possibility applies especially in events which require perfect coordination and timing, such as gymnastics. To some extent, it may also apply to runners, where the tense person tends to tighten the muscles, including the antagonists, so that both flexors and extensors of the same joint are contracted simultaneously, leading to clumsiness, waste of energy, and premature exhaustion. Examples of this are most clearly seen in events that require perfect coordination and exact timing, such as high jumping, diving, or ski jumping. If the jumper has only three trials and has failed twice, everything depends on that third and last chance. This crisis imposes a great strain on the performer. Some unique individuals will repeatedly succeed during the last trial, whereas others notoriously fail under such circumstances. Some may fail during the important qualifying test, and yet, only a week later at a less important competition, may attain a world record in the same event.

All these factors show the importance of participating regularly and frequently in competitions among top athletes where, by competing with those who are equal to or better than oneself, one learns by experience how to apply oneself to the full when it matters. The spectators' stimulation may have both positive and negative effects—negative in the sense that inexperienced competitors are encouraged to exceed their capacity to exhaust themselves prematurely, positive in the sense that it may inspire more mature competitors to mobilize all their resources to the utmost in the supreme effort to reach their goal.

REFERENCES

Adams, W. C., E. M. Bernauer, D. B. Dill, and J. B. Bomar, Jr.: Effects of Equivalent Sea-level and Altitude Training on $\dot{V}_{O_2}$ max and Running Performance, *J. Appl. Physiol.*, 39:262, 1975.

Alexander, J. K., L. H. Hartley, M. Modelski, and R. F. Grover: Reduction of Stroke Volume during Exercise in Man Following Ascent to 3,100 m Altitude, *J. Appl. Physiol.*, 23:849, 1967.

Ariëns, E. J.: General and Pharmacological Aspects of Doping, in A. De Schaepdryver and M. Hebbelinck (eds.), "Doping," Pergamon Press, Oxford and New York, 1965.

Asmussen, E., and O. Böje: The Effect of Alcohol and Some Drugs on the Capacity for Work, *Acta Physiol. Scand.*, 15:109, 1948.

Asmussen, E., and H. Chiodi: The Effect of Hypoxemia on Ventilation and Circulation in Man, *Am. J. Physiol.*, 132:426, 1941.

Asmussen, E., W. von Döbeln, and M. Nielsen: Blood Lactate and Oxygen Debt after Exhaustive Work at Different Oxygen Tensions, *Acta Physiol. Scand.*, 15:57, 1948.

Asmussen, E., and M. Nielsen: Cardiac Output during Muscular Work and Its Regulation, *Physiol. Rev.*, 35:778, 1955.

Åstrand, P.-O.: The Respiratory Activity in Man Exposed to Prolonged Hypoxia, *Acta Physiol. Scand.*, 30:343, 1954.

Åstrand, P.-O., and I. Åstrand: Heart Rate during Muscular Work in Man Exposed to Prolonged Hypoxia, *J. Appl. Physiol.*, 13:75, 1958.

Balke, B.: Work Capacity at Altitude, in W. R. Johnson (ed.), "Science and Medicine of Exercise and Sports," p. 339, Harper & Row, Publishers, Incorporated, New York, 1960.

Barbashova, Z. I.: Cellular Level of Adaptation, in D. B. Dill (ed.), "Handbook of Physiology," sec. 4, Adaptation to the Environment, p. 37, American Physiological Society, Washington, D.C., 1964.

Bert, P.: "La Pression Barométrique," Masson et Cie, Paris, 1878.

Bisgard, G. E., J. A. Will, I. B. Tyson, L. M. Dayton, R. R. Henderson, and R. F. Grover: Distribution of Regional Lung Function during Mild Exercise in Residents of 3100 m, *Resp. Physiol.*, 22:369, 1974.

Blomqvist, G., B. Saltin, and J. H. Mitchell: Acute Effects of Ethanol Ingestion on the Response to Submaximal and Maximal Exercise in Man, *Circulation*, **42**:463, 1970.

Blomqvist, G., and J. Stenberg: The ECG Response to Submaximal and Maximal Exercise during Acute Hypoxia, in G. Blomqvist, The Frank Lead Exercise Electrocardiogram, *Acta Med. Scand.*, **178**(Suppl. 440):82, 1965.

Bucher, H. K., and G. M. Smith: Drugs and Athletic Performance, in A. De Schaepdryver and M. Hebbelinck (eds.), "Doping," Pergamon Press, Oxford and New York, 1965.

Buskirk, E. R., J. Kollias, R. F. Akers, B. K. Prokop, and E. Picón-Reátegui: Maximal Performance at Altitude and on Return from Altitude in Conditioned Runners, *J. Appl. Physiol.*, **23**:259, 1967.

Cassin, S., R. D. Gilbert, and E. M. Johnson: Capillary Development during Exposure to Chronic Hypoxia, Report SAM-TR-66-16, USAF School of Aviation Medicine, Randolph Field, Tex., 1966.

Cerretelli, P.: Limiting Factors to Oxygen Transport on Mount Everest, *J. Appl. Physiol.*, **40**:658, 1976.

Cerretelli, P., and R. Margaria: Maximum Oxygen Consumption at Altitude, *Intern. Z. Angew. Physiol.*, **18**:460, 1961.

Christensen, E. H.: Sauerstoffaufnahme und Respiratorische Funktionen in Grossen Höhen, *Skand. Arch. Physiol.*, **76**:88, 1937.

Christensen, E. H., and A. Krogh: Fliegerundersuchungen; die Wirkung niedriger O_2-Spannung auf Höhenflieger, *Skand. Arch. Physiol.*, **73**:145, 1936.

Christensen, E. H., and H. E. Nielsen: Die Leistungsfähigkeit der menschlichen Skelettmuskeln bei niedrigen Sauerstoffdruck, *Skand. Arch. Physiol.*, **74**:272, 1936.

Christensen, E. H., and W. H. Forbes: Der Kreislauf in grossen Höhen, *Skand. Arch. Physiol.*, **76**:75, 1937.

Clarke, C., M. Ward, and E. Williams (eds.): "Mountain Medicine and Physiology," Alpine Club, London, 1975.

Consolazio, C. F.: Submaximal and Maximal Performance at High Altitude, in R. F. Goddard (ed.), "The International Symposium on the Effects of Altitude on Physical Performance," p. 91, The Athletic Institute, Chicago, 1967.

Cotes, J. E.: Ventilatory Capacity at Altitude and Its Relation to Mask Design, *Proc. Roy. Soc. (London)*, ser. B., **143**:32, 1954.

Craig, A. B.: Olympics, 1968: A Post-mortem, *Med. Sci. Sports*, **1**:177, 1969.

Craig, A. B., Jr.: Summary of 58 Cases of Loss of Consciousness during Underwater Swimming and Diving, *Med. Sci. Sports*, **8**:171, 1976.

Cruz, J., C. R. Grover, J. T. Reeves, J. T. Maher, A. Cymerman, and J. C. Denniston: Sustained Venoconstriction in Man Supplemented with CO_2 at High Altitude, *J. Appl. Physiol.*, **40**:96, 1976.

Cruz, J. C., H. Hartley, and J. A. Vogel: Effect of Altitude Relocations upon AaDO$_2$ at Rest and during Exercise, *J. Appl. Physiol.*, **39**:469, 1975.

Da Silva, A. M. T., and P. Hamosh: Effect of Smoking a Single Cigarette on the "Small Airways," *J. Appl. Physiol.*, **34**:361, 1973.

DeGraff, A. C., Jr., R. F. Grover, J. W. Hammond, Jr., J. M. Miller, and R. L. Johnson, Jr.: Pulmonary Diffusing Capacity in Persons Native to High Altitude, *Clin. Res.*, **13**:74, 1965.

Dejours, P., R. H. Kellogg, and N. Pace: Regulation of Respiration and Heart Rate Response in Exercise during Altitude Acclimatization, *J. Appl. Physiol.*, **18**:10, 1963.

Dempsey, J. A., W. G. Reddan, M. L. Birnbaum, H. V. Forster, J. S. Thoden, R. F. Grover, and J. Rankin: Effects of Acute through Life-long Hypoxic Exposure on Exercise Pulmonary Gas Exchange, *Resp. Physiol.*, **13:**62, 1971.

Dempsey, J. A., H. V. Forster, M. L. Birnbaum, W. G. Reddan, J. Thoden, R. F. Grover, and J. Rankin: Control of Exercise Hyperpnea under Varying Durations of Exposure to Moderate Hypoxia, *Resp. Physiol.*, **16:**213, 1972.

Dickinson, E. R., M. J. Piddington, and T. Brain: Project Olympics, *Schw. Zschr. Sportmed.*, **14:**305, 1966.

Dill, D. B. (ed.): "Handbook of Physiology," sec. 4, Adaptation to the Environment, American Physiological Society, Washington, D.C., 1964.

Edwards, H. T.: Lactic Acid in Rest and Work at High Altitude, *Am. J. Physiol.*, **116:**367, 1936.

Ekblom, B., and R. Huot: Response to Submaximal and Maximal Exercise at Different Levels of Carboxyhemoglobin, *Acta Physiol. Scand.*, **86:**474, 1972.

Ekblom, B., G. Wilson, and P.-O. Åstrand: Central Circulation during Exercise after Venesection and Reinfusion of Red Blood Cells, *J. Appl. Physiol.*, **40:**379, 1976.

Fagraeus, L.: Cardiorespiratory and Metabolic Functions during Exercise in Hyperbaric Environment, *Acta Physiol. Scand.*, Suppl. 414, 1974.

Fenn, W. O.: The Pressure Volume Diagram of the Breathing Mechanism, in W. M. Boothby (ed.), "Handbook of Respiratory Physiology," USAF School of Aviation Medicine, Randolph Field, Tex., 1954.

Forster, H. V., J. A. Dempsey, M. L. Birnbaum, W. G. Reddan, J. Thoden, R. F. Grover, and J. Rankin: Effects of Chronic Exposure to Hypoxia in Ventilatory Response to CO_2 and Hypoxia, *J. Appl. Physiol.*, **31:**586, 1971.

Frisancho, A. R.: Functional Adaptation to High Altitude Hypoxia, *Science*, **187:**313, 1975.

Goddard, R. F. (ed.): "The International Symposium on the Effects of Altitude on Physical Performance," The Athletic Institute, Chicago, 1967.

Goddard, R. F., and C. B. Favour: United States Olympic Committee Swimming Team Performance in International Sports Week, Mexico City, Oct., 1965, in R. F. Goddard (ed.), "The International Symposium on the Effects of Altitude on Physical Performance," p. 135, The Athletic Institute, Chicago, 1967.

Grover, R. F., R. Lufschanowski, and J. K. Alexander: Alterations in the Coronary Circulation of Man Following Ascent to 3,100 m Altitude, *J. Appl. Physiol.*, **41:**832, 1976.

Grover, R. F., and J. T. Reeves: Exercise Performance of Athletes at Sea Level and 3,100 Meters Altitude, in R. F. Goddard (ed.), "The International Symposium on the Effects of Altitude on Physical Performance," p. 80, The Athletic Institute, Chicago, 1967.

Guleria, J. S., J. N. Parde, P. K. Sethi, and S. B. Roy: Pulmonary Diffusing Capacity at High Altitude, *J. Appl. Physiol.*, **31:**536, 1971.

Hanley, D. F.: Pill Popping and Performance, *Modern Medicine*, **40:**81, 1972.

Hansen, J. E., G. P. Stelter, and J. A. Vogel: Arterial Pyruvate, Lactate, pH, and P_{CO_2} during Work at Sea Level and High Altitude, *J. Appl. Physiol.*, **23:**523, 1967a.

Hansen, J. E., J. A. Vogel, G. P. Stelter, and C. F. Consolazio: Oxygen Uptake in Man during Exhaustive Work at Sea Level and High Altitude, *J. Appl. Physiol.*, **23:**511, 1967b.

Hartley, L. H., J. K. Alexander, M. Modelski, and R. F. Grover: Subnormal Cardiac Output at Rest and during Exercise in Residents at 3,100 m Altitude, *J. Appl. Physiol.*, **23:**839, 1967.

Hartley, L. H., J. A. Vogel, and J. C. Cruz: Reduction of Maximal Heart Rate at Altitude and Its Reversal with Atropine, *J. Appl. Physiol.*, **36:**362, 1974.

Hartley, L. H., J. A. Vogel, and M. Landowne: Central, Femoral, and Brachial Circulation during Exercise in Hypoxia, *J. Appl. Physiol.*, **34:**87, 1973.

Hermansen, L., and B. Saltin: Blood Lactate Concentration during Exercise at Acute Exposure to Altitude, in R. Margaria (ed.), "Exercise at Altitude," p. 48, Excerpta Medica Foundation, Amsterdam, 1967.

Hultgren, H. N., J. Kelly, and H. Miller: Pulmonary Circulation in Acclimatized Man at High Altitude, *J. Appl. Physiol.*, **20:**233, 1965.

Hurtado, A.: Animals in High Altitudes: Resident Man, in D. B. Dill (ed.). "Handbook of Physiology," sec. 4, p. 843, Adaptation to the Environment, American Physiological Society, Washington, D.C., 1964.

Hurtado, A., C. Merino, and E. Delgado: Influence of Anoxemia on the Hemopoietic Activity, *Arch. Int. Med.*, **75:**284, 1945.

Ikai, M., and A. H. Steinhaus: Some Factors Modifying the Expression of Strength, *J. Appl. Physiol.*, **16:**157, 1961.

Johnson, L. F.: The Association of Oral Androgenic-anabolic Steroids and Life-threatening Disease, *Med. Sci. Sports*, **7:**284, 1975.

Johnson, L. C., E. S. Roundy, P. E. Allsen, A. G. Fisher, and L. J. Silvester: Effect of Anabolic Steroid Treatment on Endurance, *Med. Sci. Sports*, **7:**287, 1975.

Jokl, E., and P. Jokl (eds.): "Exercise and Altitude," S. Karger, New York, 1968.

Jokl, E., and P. Jokl: Heart and Sport, in D. Brumer and E. Jokl (eds.): "The Role of Exercise in Internal Medicine," Medicine and Sport, vol. 10, p. 36, S. Karger, Basel, 1977.

Jokl, Ernst, A. H. Frucht, M. J. Karvonen, D. C. Seaton, E. Simon, and Peter Jokl: Sports Medicine, *Ann. N. Y. Acad. Sci.*, **134:**908, 1966.

Juurup, A., and L. Muido: On Acute Effects of Cigarette Smoking on Oxygen Consumption, Pulse Rate, Breathing Rate and Blood Pressure in Working Organisms, *Acta Physiol. Scand.*, **11:**48, 1946.

Kellogg, R. H.: Central Chemical Regulation of Respiration, in W. O. Fenn and H. Rahn (eds.), "Handbook of Physiology," sec. 3, Respiration, vol. 1, p. 507, American Physiological Society, Washington, D.C., 1964.

Klausen, K.: Cardiac Output in Man in Rest and Work during and after Acclimatization to 3,800 m, *J. Appl. Physiol.*, **21:**609, 1966.

Kollias, J., E. R. Buskirk, R. F. Akers, E. K. Prokop, P. T. Baker, and E. Picón-Reátegui: Work Capacity of Long-time Residents and Newcomers to Altitude, *J. Appl. Physiol.*, **24:**792, 1968.

Kreuzer, F., S. M. Tenney, J. C. Mithoefer, and J. Remmers: Alveolar-arterial Oxygen Gradient in Andean Natives at High Altitude, *J. Appl. Physiol.*, **19:**13, 1964.

Lamb, D. R.: Androgens and Exercise, *Med. Sci. Sports*, **7:**1, 1975.

Lambertsen, C. J.: Effects of Oxygen at High Partial Pressure, in W. O. Fenn and H. Rahn (eds.), "Handbook of Physiology," sec. 3, Respiration, vol. 2, p. 1027, American Physiological Society, Washington, D.C., 1965.

Lambertsen, C. J. (ed.): "Underwater Physiology," The Williams & Wilkins Company, Baltimore, 1967.

Larson, P. S., H. B. Haag, and H. Silvette: "Tobacco," The Williams & Wilkins Company, Baltimore, 1961.

Leary, W. P., and C. H. Wyndham: The Possible Effect on Athletic Performance of Mexico City's Altitude, *S. Afr. Med. J.*, **40:**984, 1966.

Lockhart, A., M. Zelter, J. Mensch-Dechene, G. Antezana, M. Paz-Zamor, E. Vargas, and J. Coudert: Pressure-Flow-Volume Relationships in Pulmonary Circulation of Normal Highlanders, *J. Appl. Physiol.*, **41:**449, 1976.

Luft, U. C.: Laboratory Facilities for Adaptation Research: Low Pressures, in D. B. Dill (ed.), "Handbook of Physiology," sec. 4, p. 329, Adaptation to the Environment, American Physiological Society, Washington, D.C., 1964a.

Luft, U. C.: Aviation Physiology: The Effect of Altitude, in W. O. Fenn and H. Rahn (eds.), "Handbook of Physiology," sec. 3, Respiration, vol. 2, p. 1099, American Physiological Society, Washington, D.C., 1964b.

Maher, J. T., L. G. Jones, and L. H. Hartley: Effects of High-altitude Exposure on Submaximal Endurance Capacity of Men, *J. Appl. Physiol.*, **37:**895, 1974.

Margaria, R. (ed.): "Exercise at Altitude," Excerpta Med. Found., Amsterdam, 1967.

McManus, B. M., S. M. Horvath, N. Bolduan, and J. C. Miller: Metabolic and Cardio-Respiratory Responses to Long-term Work under Hypoxic Conditions, *J. Appl. Physiol.*, **36:**177, 1974.

Merino, C.: Studies on Blood Formation and Destruction in the Polycythemia of High Altitude, *Blood*, **5:**1, 1950.

Miles, S.: The Effect of Changes in Barometric Pressure on Maximum Breathing Capacity, *J. Physiol.*, **137:**85P, 1957.

Milledge, J. S., and S. Lahiri: Respiratory Control in Lowlanders and Sherpa Highlanders at Altitude, *Respir. Physiol.*, **2:**310, 1967.

Morpurgo, G., P. Battaglia, N. D. Carter, G. Modiano, and S. Passi: The Bohr Effect and the Red Cell 2, 3-DPG and HB Content in Sherpas and Europeans at Low and High Altitude, *Experientia*, **28:**1280, 1972.

Noell, W.: Über die Durchblutung und die Sauerstoffversorgung des Gehirns, VI, Einfluss der Hypoxämie und Anämie, *Arch. Ges. Physiol.*, **247:**553, 1944.

Norton, E. F.: "The Fight for Everest," Edward Arnold (Publishers) Ltd., London, 1925.

Oseid, S.: "Idrett og Stimulerende Midler, —Doping," Norges Idrettsforbund, Oslo, 1976.

Penaloza, D., F. Sime, N. Banchero, R. Gamboa, J. Cruz, and E. Marticorena: Pulmonary Hypertension in Healthy Men Born and Living at High Altitudes, *Am. J. Cardiol.*, **11:**150, 1963.

Pugh, L. G.: Animals in High Altitude: Man above 5,000 Meters-Mountain Exploration, in D. B. Dill (ed.), "Handbook of Physiology," sec. 4, p. 861, Adaptation to the Environment, American Physiological Society, Washington, D.C., 1964.

Pugh, L. G.: "Report of Medical Research Project into Effects of Altitude in Mexico City," report to the British Olympic Committee, 1965.

Pugh, L. G., M. B. Gill, S. Lahiri, J. S. Milledge, M. P. Ward, and J. B. West: Muscular Exercise at Great Altitudes, *J. Appl. Physiol.*, **19:**431, 1964.

Rahn, H.: Introduction to the Study of Man at High Altitudes: Conductance of O_2 from the Environment to the Tissues, in "Life at High Altitudes," Scientific Publ. 140, p. 2, Pan-American Health Organization, WHO, Washington, D.C., September 1966.

Raynaud, J., J. P. Martineaud, J. Bordachar, M. C. Tillous, and J. Durand: Oxygen Deficit and Debt in Submaximal Exercise at Sea Level and High Altitude, *J. Appl. Physiol.*, **37:**43, 1974.

Reeves, J. T., J. Halpin, J. E. Cohn, and F. Daoud: Increased Alveolar-arterial Oxygen Difference during Simulated High-altitude Exposure, *J. Appl. Physiol.*, **27:**658, 1969.

Reynafarje, B.: Myoglobin Content and Enzymatic Activity of Muscle and Altitude

Adaptation, *J. Appl. Physiol.*, **17**:301, 1962.

Reynafarje, C.: Humoral Control of Erythropoiesis at Altitude, in R. Margaria (ed.), "Exercise at Altitude," p. 165, Excerpta Medica Foundation, Amsterdam, 1967.

Roskamm, H., L. Samek, H. Weidemann, and H. Reindell: "Leistung und Höhe," Knoll AG, Ludwigshafen am Rhein, 1968.

Roughton, F. J. W.: Transport of Oxygen and Carbon Dioxide, in W. O. Fenn and H. Rahn (eds.), "Handbook of Physiology," sec. 3, Respiration, vol. 1, p. 767, American Physiological Society, Washington, D.C., 1964.

Saltin, B.: Aerobic and Anaerobic Work Capacity at 2,300 Meters, *Schw. Zschr. Sportmed.*, **14**:81, 1966.

Saltin, B.: Aerobic and Anaerobic Work Capacity at an Altitude of 2,250 Meters, in R. F. Goddard (ed.), "The International Symposium on the Effects of Altitude on Physical Performance," p. 97, The Athletic Institute, Chicago, 1967.

Saltin, B., R. F. Grover, C. G. Blomqvist, L. H. Hartley, and R. L. Johnson, Jr.: Maximal Oxygen Uptake and Cardiac Output after Two Weeks at 4,300 Meters, *J. Appl. Physiol.*, **25**:400, 1968.

Severinghaus, J. W., C. R. Bainton, and A. Carcelen: Respiratory Insensitivity to Hypoxia in Chronically Hypoxic Man, *Respir. Physiol.*, **1**:308, 1966.

Shephard, R. J.: Athletic Performance at Moderate Altitudes, *Medicina Dello Sport*, **26**:36, 1973.

Somervell, T. H.: Note on the Composition of Alveolar Air at Extreme Heights, *J. Physiol.*, **60**:282, 1925.

Stenberg, J., B. Ekblom, and R. Messin: Hemodynamic Response to Work at Simulated Altitude, *J. Appl. Physiol.*, **21**:1589, 1966.

Surks, M. I., K. S. Chinn, and L. O. Matoush: Alterations in Body Composition in Man after Acute Exposure to High Altitude, *J. Appl. Physiol.*, **21**:1741, 1966.

Ulvedal, F., T. E. Morgan, Jr., R. G. Cutler, and B. E. Welch: Ventilatory Capacity during Prolonged Exposure to Simulated Altitude without Hypoxia, *J. Appl. Physiol.*, **18**:904, 1963.

Vannotti, A.: The Adaptation of the Cell to Effort, Altitude and to Pathological Oxygen Deficiency, *Schweiz. Med. Wschr.*, **76**:899, 1946.

Velasquez, T.: Tolerance to Acute Anoxia in High Altitude Natives, *J. Appl. Physiol.*, **14**:357, 1959.

Vogel, J. A., J. E. Hansen, and C. W. Harris: Cardiovascular Responses in Man during Exhaustive Work at Sea Level and High Altitude, *J. Appl. Physiol.*, **23**:531, 1967.

Vogel, J. A., L. H. Hartley, and J. C. Cruz: Cardiac Output during Exercise in Altitude Natives at Sea Level and High Altitude, *J. Appl. Physiol.*, **36**:173, 1974a.

Vogel, J. A., L. H. Hartley, J. C. Cruz, and R. P. Hogan: Cardiac Output during Exercise in Sea-level Residents at Sea Level and High Altitude, *J. Appl. Physiol.*, **36**:169, 1974b.

Weihe, W. H. (ed.): "The Physiological Effects of High Altitude," The Macmillan Company, New York, 1964.

West, J. B.: Diffusing Capacity of the Lung for Carbon Monoxide at High Altitude, *J. Appl. Physiol.*, **17**:421, 1962.

West, J. B., S. Lahiri, M. B. Gill, J. S. Milledge, L. G. Pugh, and M. P. Ward: Arterial Oxygen Saturation during Exercise at High Altitude, *J. Appl. Physiol.*, **17**:617, 1962.

Williams, M. H., A. R. Goodwin, R. Perkins, and J. Bocrie: Effect of Blood Reinjection upon Endurance Capacity and Heart Rate, *Med. Sci. Sports*, **5**:181, 1973.

Appendix

CONTENTS

S. I. UNITS USED IN TEXT (S.I. = Système International d'Unités)

Basic units

Length: Meter (m)
Mass: Kilogram (kg)
Time: Second (s)

DEFINITIONS

Frequency	Hertz	Hz	(s^{-1})
Force	Newton	N	$(kg \cdot m \cdot s^{-2})$
Energy	Joule	J	$(kg \cdot m^2 \cdot s^{-2} = Nm)$
Power	Watt	W	$(kg \cdot m^2 \cdot s^{-3} = J \cdot s^{-1})$

UNIT ABBREVIATIONS, PREFIXES, PHYSICAL CONSTANTS

nm   nanometer (nano 10^{-9})
μm   micrometer (micro 10^{-6})
mm   millimeter (milli 10^{-3})
m   meter
mg   milligram
g   gram
kg   kilo
ml   milliliter
M   molar concentration
mM   millimolar concentration
s   second
min   minute
J   joule
kJ   kilojoule (= 10^3 joule; kilo 10^3)
MJ   megajoule (= 10^6 joule; mega 10^6)
mV   millivolt
V   volt
kp   kilopond: 1 kp is the force acting on the mass of 1 kg at normal acceleration of gravity

CONVERSION TABLES

Length and Weight

1 centimeter = 0.39370 in.
1 meter = 39.37 in.
1 kilometer = 0.62137 mile
1 inch = 2.54 cm
1 foot = 30.480 cm

1 milliliter = 0.03381 fl oz
1 liter = 1.0567 U.S. qt
1 kilogram = 2.2046 lb

Power

1 watt = 0.001 kilowatt
1 watt = 0.73756 ft-lb $\cdot$ s^{-1}
1 watt = 1 $\cdot$ 10^7 ergs $\cdot$ s^{-1}
1 watt = 0.056884 BTU 1 min = 3.41304 BTU-hr
1 watt = 0.01433 kilocalories $\cdot$ min^{-1}
1 watt = 1.341 $\cdot$ 10^{-3} hp (horsepower)
1 watt = 1 J $\cdot$ s^{-1}
1 watt = 6.12 kpm $\cdot$ min^{-1}

1 kilocalorie per minute = 69.767 watts
1 kilocalorie per minute = 51.457 ft-lb $\cdot$ s^{-1}
1 kilocalorie per minute = 6.9770 $\cdot$ 10^8 ergs $\cdot$ s^{-1}
1 kilocalorie per minute = 3.9685 BTU $\cdot$ min^{-1}
1 kilocalorie per minute = 0.093557 hp

1 horsepower = 745.7 watts
1 horsepower = 550 ft-lb $\cdot$ s^{-1}
1 horsepower = 7.457 $\cdot$ 10^9 ergs $\cdot$ sec^{-1}
1 horsepower = 42.4176 BTU $\cdot$ min^{-1}
1 horsepower = 10.688 kcal $\cdot$ min^{-1}
1 horsepower = 745.7 joules $\cdot$ s^{-1}
1 horsepower = 75 kpm $\cdot$ sec^{-1}

Work and Energy

1 kilocalorie = 4.186 $\cdot$ 10^{10} ergs
1 kilocalorie = 4,186 joules
1 kilocalorie = 3.9680 BTU
1 kilocalorie = 3087.4 ft-lb
1 kilocalorie = 426.85 kpm
1 kilocalorie = 1.5593 $\cdot$ 10^{-3} hp-hr

1 erg = 2.3889 $\cdot$ 10^{-11} kcal
1 erg = 1 $\cdot$ 10^{-7} joule
1 erg = 9.4805 $\cdot$ 10^{-14} BTU
1 erg = 7.3756 $\cdot$ 10^{-8} ft-lb
1 erg = 1.0197 $\cdot$ 10^{-8} kpm
1 erg = 3.7251 $\cdot$ 10^{-14} hp-hr

1 joule = 2.3889 $\cdot$ 10^{-4} kcal
1 joule = 1 $\cdot$ 10^7 ergs

1 joule $= 9.4805 \cdot 10^{-4}$ BTU
1 joule $= 0.73756$ ft-lb
1 joule $= 0.10197$ kpm
1 joule $= 3.7251 \cdot 10^{-7}$ hp-hr

1 BTU $= 0.25198$ kcal
1 BTU $= 1.0548 \cdot 10^{10}$ ergs
1 BTU $= 1054.8$ joules
1 BTU $= 777.98$ ft-lb
1 BTU $= 107.56$ kpm
1 BTU $= 3.9292 \cdot 10^{-4}$ hp-hr

1 foot-pound $= 3.2389 \cdot 10^{-4}$ kcal
1 foot-pound $= 1.35582 \cdot 10^{7}$ ergs
1 foot-pound $= 1.3558$ joules
1 foot-pound $= 1.2854 \cdot 10^{-3}$ BTU
1 foot-pound $= 0.13825$ kpm
1 foot-pound $= 5.0505 \cdot 10^{-7}$ hp-hr

1 kilogram-meter $= 2.3427 \cdot 10^{-3}$ kcal
1 kilogram-meter $= 9.8066 \cdot 10^{7}$ ergs
1 kilogram-meter $= 9.8066$ joules
1 kilogram-meter $= 9.2967 \cdot 10^{3}$ BTU
1 kilogram-meter $= 7.2330$ ft-lb
1 kilogram-meter $= 3.6529 \cdot 10^{-6}$ hp-hr

1 watt $= 6.12$ kpm $\cdot$ min^{-1} (approx. $= 6$ kpm $\cdot$ min^{-1})
50 watts $=$ approx. 300 kpm $\cdot$ min^{-1}
1 kpm $\cdot$ min^{-1} $= 0.1635$ watt
1 kp $= 9.80665$ newtons
1 kpm $= 9.80665$ joules

Speed

km · hr⁻¹	mph	m · s⁻¹	km · hr⁻¹	mph	m · s⁻¹
10	6.22	2.78	200	124	55.6
20	12.4	5.56	220	137	61.2
30	18.7	8.34	240	149	66.7
40	24.9	11.1	260	162	72.3
50	31.1	13.9	280	174	77.8
60	37.4	16.7	300	187	83.4
70	43.6	19.4	320	199	88.9
80	49.8	22.2	340	211	94.5
90	56.0	25.0	360	224	100
100	62.2	27.8	380	236	106
120	74.7	33.3	400	249	111
140	87.1	38.9	420	261	117
160	99.5	44.5	440	274	122
180	112	50.0	460	286	128

Temperature: Conversion of degrees centigrade into degrees Fahrenheit and vice versa

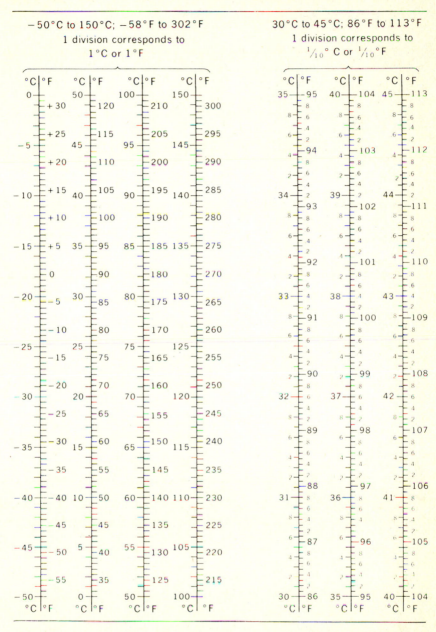

−50°C to 150°C; −58°F to 302°F
1 division corresponds to
1°C or 1°F

30°C to 45°C; 86°F to 113°F
1 division corresponds to
$\frac{1}{10}$° C or $\frac{1}{10}$°F

SOURCE: *Documenta Geigy*, "Scientific Tables," 5th ed., Geigy Pharmaceuticals, Ardsley, New York, 1956.

LIST OF SYMBOLS

$\bar{x}$	dash over any symbol indicates a mean value
$\dot{x}$	dot above any symbol indicates time derivate

Respiratory and Hemodynamic Notations

V	gas volume
$\dot{V}$	gas volume per unit time (usually liters $\cdot$ min^{-1})
R or RQ	respiratory exchange ratio (volume CO_2 $\cdot$ volume O_2^{-1})
I	inspired gas
E	expired gas
A	alveolar gas
F	fractional concentration in dry gas phase
f	respiratory frequency (breath $\cdot$ unit time^{-1})
TLC	total lung capacity
VC	vital capacity
FRC	functional residual capacity
RV	residual volume
T	tidal gas
D	dead space
FEV	forced expiratory volume
FEV$_{1.0}$	forced expiratory volume in 1 s
MVV	maximal voluntary ventilation
MVV$_{40}$	maximal voluntary ventilation at $f = 40$
D_L	diffusing capacity of the lungs (ml $\cdot$ min^{-1} $\cdot$ mm Hg^{-1})
P	gas pressure
B or Bar	barometric
STPD	0°C, 760 mm Hg, dry
BTPS	body temperature and pressure, saturated with water vapor
ATPD	ambient temperature and pressure, dry
ATPS	ambient temperature and pressure, saturated with water vapor
Q	blood flow or volume
$\dot{Q}$	blood flow $\cdot$ unit time^{-1} (without other notation, cardiac output; usually liters $\cdot$ min^{-1})
SV	stroke volume
HR	heart rate (usually beats $\cdot$ min^{-1})
BV	blood volume
THb or Hb$_T$	total amount of hemoglobin in body
Hb	hemoglobin concentration (g $\cdot$ 100 ml^{-1})
Hct	hematocrit
BP	blood pressure
R	resistance
C	concentration in blood phase
S	percent saturation of Hb
a	arterial

c	capillary
v	venous

Temperature Notations

T or t	temperature
r or re	rectal
s	skin
e or oe	esophageal (oesophageal)
m	muscle
ty	tympanic
M	metabolic energy yield
C	convective heat exchange
R	radiation heat exchange
E	evaporative heat loss
S	storage of body heat
°C	temperature in degrees centigrade
°F	temperature in degrees Fahrenheit

Dimensions

W	weight
H	height
L	length
LBM	lean body mass
BSA	body surface area

Statistical Notations

M	arithmetic mean
SD or S.D.	standard deviation
SE or S.E.	standard error of the mean
n	number of observations
r	correlation coefficient
range	smallest and largest observed value
Σ	summation
D or d	difference
P	probability
*	denotes a (probably) significant difference; $0.05 \geq P > 0.01$
**	denotes a significant difference; $0.01 \geq P > 0.001$
***	denotes a (highly) significant difference; $P \leq 0.001$

Examples

V_A	volume of alveolar gas
$\dot{V}_E$	expiratory gas volume $\cdot$ min^{-1}

$\dot{V}_{O_2}$	volume of oxygen $\cdot$ min^{-1} (oxygen uptake $\cdot$ min^{-1})
V_T	tidal volume
P_A	alveolar gas pressure
P_B	barometric pressure
$F_{I_{O_2}}$	fractional concentration of O_2 in inspired gas
$P_{A_{O_2}}$	alveolar oxygen pressure
pH_a	arterial pH
Ca_{O_2}	oxygen content in arterial blood
$Ca_{O_2} - C\bar{v}_{O_2}$	difference in oxygen content between arterial and mixed venous blood (often written a-$\bar{v}O_2$ diff.)
T_r	rectal temperature
$\overline{T}_s$	mean skin temperature

Index

Index